Primer
of Applied Regression
and Analysis of Variance

Primer
of Applied Regression
and Analysis of Variance

STANTON A. GLANTZ, Ph.D.

Professor of Medicine
Member, Cardiovascular Research Institute
Member, Institute for Health Policy Studies
University of California, San Francisco
San Francisco, California

BRYAN K. SLINKER, D.V.M., Ph.D.

Associate Professor of Veterinary and Comparative Anatomy, Pharmacology, and Physiology
Washington State University
Pullman, Washington

McGraw-Hill, Inc.
Health Professions Division

New York   St. Louis   San Francisco   Auckland   Bogotá   Caracas
Lisbon   London   Madrid   Mexico City   Milan   Montreal   New Delhi
Paris   San Juan   Singapore   Sydney   Tokyo   Toronto

PRIMER OF APPLIED REGRESSION
AND ANALYSIS OF VARIANCE

Copyright © 1990 by McGraw-Hill, Inc. All rights reserved. Printed in the
United States of America. Except as permitted under the United States
Copyright Act of 1976, no part of this publication may be reproduced or
distributed in any form or by any means, or stored in a data base or retrieval
system, without the prior written permission of the publisher.

34567890   DOCDOC   9876543

ISBN 0-07-023407-8

This book was set in Trump Medieval by Waldman Graphics Incorporated.
The editors were William Day and Muza Navrozov;
the production supervisor was Annette Mayeski;
the cover and text were designed by José Fonfrias.
R. R. Donnelley & Sons Company was printer and binder.

A disk containing all the data sets used to generate the examples and problems
as well as the control files used to generate the examples (given in Appendix B)
is available for $19.95 from Datadisk, SE 810 Meadowvale, Pullman, WA
99163. Specify disk size ($5\frac{1}{4}$ or $3\frac{1}{2}$ inch) and computer (IBM® PC or
Macintosh™). Discounts for bulk orders are available.

BMDP, Minitab, SAS, and SPSS-X are registered trademarks of BMDP
Statistical Software, Minitab Statistical Software, the SAS Institute, and SPSS
Statistical Software, respectively.

Library of Congress Cataloging-in-Publication Data
Glantz, Stanton A.
    Primer of applied regression and analysis of variance / Stanton A.
  Glantz, Bryan K. Slinker.
      p.   cm.
  Includes bibliographical references.
  ISBN 0-07-023407-8
    1. Biometry. 2. Regression analysis. 3. Analysis of variance.
  I. Slinker, Bryan K. II. Title.
    [DNLM: 1. Analysis of Variance. 2. Biometry. 3. Regression
Analysis. QH 323.5 G545p]
QH323.5.G56 1990
574'.01'5195 – dc20
DNLM/DLC
for Library of Congress   90-5413
                                                 CIP

To Marsha, Aaron, and Frieda Glantz for coming to Vermont
so that we could write this book

To Kathy and Kyle Slinker, who put up with this nonsense
with reasonably good humor

Contents

Preface .... **xix**

O N E   **Why Do Multivariate Analysis?** 1

    OUR FIRST VISIT TO MARS 2

    DUMMIES ON MARS 6

    SUMMARY 8

    PROBLEM 9

T W O   **The First Step: Understanding Simple Linear Regression** 10

    MORE ON MARS 10

    The Population Parameters 13

    HOW TO ESTIMATE THE LINE OF MEANS FROM A SAMPLE 14

    The Best Straight Line through the Data 16

    Variability about the Regression Line 22

    Standard Errors of the Regression Coefficients 22

    HOW CONVINCING IS THE TREND? 25

    Testing the Slope of the Regression Line 26

    Comparing Slopes and Intercepts of Two Regression Lines 27

    Testing the Regression as a Whole 28

    Dietary Fat and Breast Cancer 33

CORRELATION AND CORRELATION COEFFICIENTS 36

The Relationship between Regression and Correlation 37

DOING REGRESSION AND CORRELATION ANALYSIS
WITH A COMPUTER 39

HEAT EXCHANGE IN GRAY SEALS 42

SUMMARY 45

PROBLEMS 46

THREE   **Regression with Two or More Independent Variables** **50**

WHAT WE REALLY DID ON MARS 51

HOW TO FIT THE BEST PLANE THROUGH THE DATA 54

Computing the Regression Coefficients 55

Variability about the Regression Plane 57

Standard Errors of the Regression Coefficients 57

Muddying the Water: Multicollinearity 59

DOES THE REGRESSION EQUATION DESCRIBE THE DATA? 63

Incremental Sums of Squares and the Order of Entry 65

Relationship to t Tests of Individual Regression Coefficients 67

THE COEFFICIENT OF DETERMINATION AND THE MULTIPLE
CORRELATION COEFFICIENT 68

MORE DUMMIES ON MARS 69

MECHANISMS OF TOXIC SHOCK 72

PROTEIN SYNTHESIS IN NEWBORNS AND ADULTS 74

GENERAL MULTIPLE LINEAR REGRESSION 79

Multiple Regression in Matrix Notation 80

DIABETES, CHOLESTEROL, AND THE TREATMENT
OF HIGH BLOOD PRESSURE 83

BABY BIRDS BREATHING IN BURROWS 86

POLYNOMIAL (AND SOME OTHER NONLINEAR) REGRESSIONS .. 88

Heat Exchange in Gray Seals Revisited 89

Other Nonlinear Regressions 92

INTERACTIONS BETWEEN THE INDEPENDENT VARIABLES 94

How Bacteria Adjust to Living in Salty Environments 94

The Response of Smooth Muscle to Stretching 101

SUMMARY 103

PROBLEMS 104

FOUR   **Do the Data Fit the Assumptions?** **110**

ANOTHER TRIP TO MARS 112

LOOKING AT THE RESIDUALS 119

A QUANTITATIVE APPROACH TO RESIDUAL ANALYSIS 121

Standardized Residuals 123

Using Standardized Residuals to Test for Normality of the Residuals 125

Leverage 130

Studentized Residuals 134

Cook's Distance 136

WHAT DO YOU DO WITH AN INFLUENTIAL OBSERVATION
ONCE YOU HAVE FOUND IT? 142

Problems with the Data 142

Problems with the Model 145

Data Transformations 147

WATER MOVEMENT ACROSS THE PLACENTA 151

CHEAPER CHICKEN FEED 157

HOW THE BODY PROTECTS ITSELF FROM EXCESS ZINC
AND COPPER 164

Back to Square One 170

SUMMARY 176

PROBLEMS 177

FIVE   **Multicollinearity and What to Do about It** **181**

WHERE MULTICOLLINEARITY COMES FROM 182

BACK TO MARS 184

DETECTING AND EVALUATING MULTICOLLINEARITY 189

Qualitative Suggestions of Harmful Multicollinearity 190

Correlations among the Independent Variables 191

The Variance Inflation Factor 191

Auxiliary Regressions 193

The Correlations of the Regression Coefficients 194

The Consequences of Having Two Pumps in One Heart 194

FIXING THE REGRESSION MODEL 200

Centering the Independent Variables 200

Deleting Predictor Variables 207

More on Two Pumps in One Heart 208

FIXING THE DATA 210

Getting More Data on the Heart 211

USING PRINCIPAL COMPONENTS TO DIAGNOSE
AND TREAT MULTICOLLINEARITY 216

Standardized Variables, Standardized Regression,
and the Correlation Matrix 217

Principal Components of the Correlation Matrix 222

Principal Components to Diagnose Multicollinearity on Mars 225

Principal Components and the Heart 226

Principal Components Regression 228

More Principal Components on Mars 231

The Catch 231

Recapitulation 234

SUMMARY 235

PROBLEMS 236

SIX Selecting the "Best" Regression Model 239

SO WHAT DO YOU DO? 240

WHAT HAPPENS WHEN THE REGRESSION EQUATION
CONTAINS THE WRONG VARIABLES? 240

WHAT DOES "BEST" MEAN? 245

The Coefficient of Determination R^2 246

The Adjusted R^2 247

The Standard Error of the Estimate $s_{y|x}$ 247

Independent Validations of the Model with New Data 248

The Predicted Residual Error Sum of Squares, PRESS 249

Bias Due to Model Underspecification and C_p 250

But What Is "Best"? 254

SELECTING VARIABLES WITH ALL POSSIBLE
SUBSETS REGRESSION 255

What Determines an Athlete's Time in Triathlon? 256

SEQUENTIAL VARIABLE SELECTION TECHNIQUES 262

Forward Selection 262

Backward Elimination 264

Stepwise Regression 265

Interpreting the Results of Sequential Variable Selection 266

Another Look at the Triathlon 267

SUMMARY 269

PROBLEMS 270

SEVEN   **One-Way Analysis of Variance** 272

USING A t TEST TO COMPARE TWO GROUPS 273

Does Secondhand Tobacco Smoke Nauseate Martians? 274

Using Linear Regression to Compare Two Groups 277

THE BASICS OF ONE-WAY ANALYSIS OF VARIANCE 279

Expected Mean Squares 281

Using Linear Regression to Do Analysis of Variance with Two Groups 283

USING LINEAR REGRESSION TO DO ONE-WAY ANALYSIS
OF VARIANCE WITH ANY NUMBER OF TREATMENTS 284

Hormones and Depression 286

MULTIPLE COMPARISON TESTING 292

The Bonferroni *t* Test 294

More on Hormones and Depression 295

Dunnett's Test for Multiple Comparisons against a Single
Control Group 297

The Student-Newman-Keuls Test for All Pairwise Comparisons 300

What Is a Family? 302

Diet, Drugs, and Atherosclerosis 303

TESTING THE ASSUMPTIONS IN ANALYSIS OF VARIANCE 307

Formal Tests of Homogeneity of Variance 308

More on Diet, Drugs, and Atherosclerosis 309

SUMMARY 312

PROBLEMS 313

E I G H T **Two-Way Analysis of Variance** 316

PERSONALITY ASSESSMENT AND FAKING HIGH GENDER
IDENTIFICATION 319

AN ALTERNATIVE APPROACH FOR CODING
DUMMY VARIABLES 328

An Alternative Approach to Personality 329

Why Does It Matter How We Code the Dummy Variables? 332

The Kidney, Sodium, and High Blood Pressure 336

What Do Interactions Tell Us? 341

Multiple Comparisons in Two-Way Analysis of Variance 344

More on the Kidney, Sodium, and High Blood Pressure 345

UNBALANCED DATA 350

ALL CELLS FILLED, BUT SOME CELLS HAVE
MISSING OBSERVATIONS 352

The Case of the Missing Kidneys 353

Summary of the Procedure 357

What If You Use the Wrong Sum of Squares? 360

Multiple Comparisons with Missing Data 362

ONE OR MORE CELLS EMPTY 365

Multiple Comparisons with Empty Cells 368

More on the Missing Kidney 371

Multiple Comparisons for the Missing Kidney 375

Recapitulation 376

SUMMARY 377

PROBLEMS 378

N I N E   Repeated Measures **381**

ACCOUNTING FOR BETWEEN-SUBJECTS VARIABILITY
IN LINEAR REGRESSION 381

Measuring Heart Size with a Catheter 382

ACCOUNTING FOR BETWEEN-SUBJECTS VARIABILITY
IN ANALYSIS OF VARIANCE 390

ONE-WAY REPEATED-MEASURES ANALYSIS OF VARIANCE 391

Hormones and Food 392

Comparison with Simple Analysis of Variance 398

Multiple Comparisons in Repeated-Measures Analysis of Variance 399

Recapitulation 400

ASSUMPTIONS UNDERLYING REPEATED-MEASURES
ANALYSIS OF VARIANCE 400

TWO-FACTOR ANALYSIS OF VARIANCE WITH REPEATED
MEASURES ON ONE FACTOR 404

Partitioning the Variability 405

Testing the Nonrepeated-Measures Factor 407

Testing the Repeated-Measures Factor 408

Is Alcoholism Associated with a History of Childhood Aggression? 410

Multiple Comparisons 420

MISSING DATA IN TWO-FACTOR ANALYSIS OF VARIANCE
WITH REPEATED MEASURES ON ONE FACTOR 421

Expected Mean Squares and Random Factors 421

What Happens to the Expected Mean Squares When There Are Missing Data? 425

More on Drinking and Antisocial Personality 428

Multiple Comparisons 430

TWO-WAY ANALYSIS OF VARIANCE WITH REPEATED MEASURES ON BOTH FACTORS 431

Partitioning the Variability 432

Candy, Chewing Gum, and Tooth Decay 435

Multiple Comparisons 445

MISSING DATA IN REPEATED MEASURES ON BOTH OF TWO FACTORS 446

Expected Mean Squares 446

What Happens to the Expected Mean Squares When There Are Missing Data? 448

More on Chewing Gum 450

Multiple Comparisons 452

EQUALITY OF VARIANCES REVISITED 452

How Similar Are the Mechanical Properties of the Whole Heart to Heart Muscle? 453

SUMMARY 460

PROBLEMS 460

TEN **Nonlinear Regression** **464**

MARTIAN MOODS 468

Grid Searches 470

FINDING THE BOTTOM OF THE BOWL 471

The Method of Steepest Descent 473

The Gauss-Newton Method 478

Marquardt's Method 482

Where Do You Get a Good First Guess? 483

How Do You Tell You Are at the Bottom of the Bowl? 483

MATHEMATICAL DEVELOPMENT OF NONLINEAR
REGRESSION ALGORITHMS 485

The Method of Steepest Descent 487

The Gauss-Newton Method 488

Marquardt's Method 490

HYPOTHESIS TESTING IN NONLINEAR REGRESSION 491

REGRESSION DIAGNOSTICS IN NONLINEAR REGRESSION 493

EXPERIMENTING WITH DRUGS 494

KEEPING BLOOD PRESSURE UNDER CONTROL 500

Is the Model Parameterized in the Best Form? 505

SUMMARY 508

PROBLEMS 509

E L E V E N **Regression with a Qualitative Dependent Variable** **512**

OUR LAST VISIT TO MARS 513

Odds 517

The Multiple Logistic Equation 519

ESTIMATING THE COEFFICIENTS IN A LOGISTIC REGRESSION .. 519

Maximum Likelihood Estimation 520

HYPOTHESIS TESTING IN LOGISTIC REGRESSION 523

Testing the Logistic Equation 523

Testing the Individual Coefficients 525

Back to Mars 527

IS THE LOGISTIC REGRESSION EQUATION AN APPROPRIATE
DESCRIPTION OF THE DATA? 528

Regression Diagnostics for Logistic Regression 529

Goodness-of-Fit Testing 531

ARE BONE CANCER PATIENTS RESPONDING
TO CHEMOTHERAPY? 536

STEPWISE LOGISTIC REGRESSION 546

Nuking the Heart 548

SUMMARY 566

PROBLEMS 566

APPENDIX A   **A Brief Introduction to Matrices and Vectors** **569**

DEFINITIONS 569

ADDING AND SUBTRACTING MATRICES 570

MATRIX MULTIPLICATION 571

INVERSE OF A MATRIX 573

TRANSPOSE OF A MATRIX 573

EIGENVALUES AND EIGENVECTORS 574

APPENDIX B   **Statistical Package Cookbook** **575**

GENERAL COMMENTS ON SOFTWARE 576

A Note on User Interfaces 578

REGRESSION 579

Martian Height and Weight (Figs. 1-1 and 2-10; Data in Table 2-1) 579

Martian Weight, Height, and Water Consumption (Figs. 1-2, 3-1, 3-2, 3-6,
and 4-16; Data in Table 1-1) 581

Heat Exchange in Gray Seals (Figs. 2-11, 3-16, and 4-18;
Data in Table C-1, Appendix C) 581

Martian Secondhand Smoke Exposure (Figs. 1-3, 3-7, and 3-8;
Data in Table 3-1) 582

Mechanisms of Toxic Shock (Figs. 3-9 and 3-10; Data in Table C-2,
Appendix C) 582

Protein Synthesis in Newborns and Adults (Figs. 3-11 and 3-12;
Data in Table C-3, Appendix C) 583

Diabetes, Cholesterol, and the Treatment of High Blood Pressure
(Fig. 3-13; Data in Table C-4, Appendix C) 583

Baby Birds Breathing in Burrows (Figs. 3-14 and 3-15; Data in Table C-5,
Appendix C) 583

How Bacteria Adjust to Living in Salty Environments
(Figs. 3-17 through 3-20; Data in Table C-6, Appendix C) 584

The Response of Smooth Muscle to Stretching (Figs. 3-21 and 3-22; Data in Table C-7, Appendix C) 584

Martian Intelligence (Figs. 4-1 to 4-6, 4-10 to 4-15; Data in Table C-8, Appendix C) 584

Water Movement across the Placenta (Figs. 4-20 to 4-24; Data in Table C-9, Appendix C) 586

Cheaper Chicken Feed (Figs. 4-25 to 4-30; Data in Table C-10, Appendix C) 588

How the Body Protects Itself from Excess Copper and Zinc (Figs. 4-31 to 4-42; Data in Table C-11, Appendix C) 588

MULTICOLLINEARITY 590

Martian Weight, Height, and Water Consumption, Showing Severe Multicollinearity (Figs. 5-1, 5-2, and 5-16; Data in Table 5-2) 591

Interaction between Heart Ventricles (Figs. 5-6, 5-11 to 5-13; Data in Table C-12, Appendix C) 593

Effect of Centering on Multicollinearity (Figs. 5-7 to 5-10; Data in Table 5-4) 594

VARIABLE SELECTION METHODS 595

What Determines an Athlete's Time in a Triathlon (Figs. 6-3 to 6-7; Data in Table C-13, Appendix C) 595

ONE-WAY ANALYSIS OF VARIANCE 598

Does Secondhand Tobacco Smoke Nauseate Martians? (Figs. 7-1 and 7-2; Data in Table 7-1) 598

Hormones and Depression (Figs. 7-3 and 7-4; Data in Table 7-5) 599

Diet, Drugs, and Atherosclerosis (Figs. 7-6 to 7-9; Data in Table C-14, Appendix C) 599

TWO-WAY ANALYSIS OF VARIANCE 602

Personality Assessment and Faking High Gender Identification (Figs. 8-1 to 8-3; Data in Tables 8-2, 8-5, and 8-7) 602

The Kidney, Sodium, and High Blood Pressure (Figs. 8-4 to 8-8 and 8-10; Data in Table 8-12) 602

More on Missing Data 604

REPEATED MEASURES IN REGRESSION 607

Hypothetical Example Illustrating Repeated Measures in Regression (Figs. 9-2 to 9-4; Data in Table 9-1) 607

Measuring Heart Size with a Catheter (Figs. 9-1 and 9-5; Data in Table 9-2) 607

ONE-WAY REPEATED-MEASURES ANALYSIS OF VARIANCE 607

Hormones and Food (Figs. 9-7 and 9-8; Data in Table 9-4) 607

TWO-WAY REPEATED-MEASURES ANALYSIS OF VARIANCE
WITH REPEATED MEASURES ON ONE FACTOR 610

Is Alcoholism Associated with a History of Childhood Aggression?
(Figs. 9-10 to 9-15; Data in Table 9-8) 610

TWO-WAY REPEATED-MEASURES ANALYSIS OF VARIANCE
WITH REPEATED MEASURES ON BOTH FACTORS 615

Candy, Chewing Gum, and Tooth Decay (Figs. 9-18 to 9-21;
Data in Table 9-15) 615

How Similar Are the Mechanical Properties of the Whole Heart
to Heart Muscle? (Figs. 9-22 to 9-26; Data in Table C-17, Appendix C) 620

NONLINEAR REGRESSION 622

Martian Moods (Figs. 10-2 to 10-7; Data in Table C-18, Appendix C) 622

Experimenting with Drugs (Figs. 10-8 to 10-11; Data in Table C-19,
Appendix C) 623

Keeping Blood Pressure under Control (Figs. 10-12 to 10-15; Data
in Tables C-20 and C-21, Appendix C) 624

LOGISTIC REGRESSION 626

Martian Graduations (Figs. 11-1 to 11-6; Data in Table C-22,
Appendix C) 626

Are Bone Cancer Patients Responding to Chemotherapy?
(Figs. 11-7 to 11-13; Data in Table C-23, Appendix C) 627

Nuking the Heart (Figs. 11-14 to 11-22; Data in Table C-24,
Appendix C) 630

APPENDIX C   Data for Examples 633
APPENDIX D   Data for Problems 665
APPENDIX E   Statistical Tables 710
APPENDIX F   Solutions to Problems 721

Index 761

Preface

Although we arrived at our common interest in applications of statistics to biomedical data analysis from quite different backgrounds, we each first encountered multiple regression as graduate students. For one of us it was a collaboration with a political science graduate student and volunteer from Common Cause to analyze the effects of expenditures on the outcomes of political campaigns in California* that sparked the interest. For the other it was a need to quantify and unravel the multiple determinants of the way the heart responded to a sudden change in blood pressure.† These experiences convinced us that methods based on multiple regression could be very useful for describing and sorting out complex relationships, particularly when one cannot precisely control all the variables under study. Indeed, it is this capability that has made multiple regression and related statistical techniques so important to modern social science and economic research. As our interests in cardiovascular physiology developed, and ultimately converged, we found that these same statistical techniques offered powerful tools for understanding complex physiological problems, particularly when dealing with intact organisms. Since then, multiple regression has become an indispensable tool in our research.

The problem with our reliance on these methods is that, although they offer a great deal, they are rarely applied in the health and life sciences (beyond a few specialties, such as epidemiology and psychology). Indeed, many people who read our scientific papers exhibit a mixture of

---

*S. Glantz, A. Abramowitz, and M. Burkart, "Election Outcomes: Whose Money Matters?" *J Politics* 38: 1033–1038, 1976.

†B. K. Slinker, K. B. Campbell, J. A. Ringo, P. A. Klavano, J. D. Robinette, J. E. Alexander, "Mechanical Determinants of Transient Changes in Stroke Volume," *Am J Physiol* 246: H435–H447, 1984.

confusion and intimidation. In many ways, this book is a reaction to that reaction. In writing this book, we sought to present these methods in a context that will be both appealing and understandable to people working in the health and life sciences. Beyond the obvious frustration with getting people to understand our work, we believe that these methods are greatly underutilized in the health and life sciences.

Although this book is directed at people interested in the health and life sciences, nothing in this book limits its audience to these disciplines. The techniques we develop apply to a wide variety of disciplines in the social and behavioral sciences, engineering, and economics as well. Indeed, multiple regression methods are much more widely used in many disciplines other than the health and life sciences. We hope that people working in those other disciplines will find this book a useful applications-oriented text and reference.

This book also grows out of the very positive experience associated with writing the introductory text—*Primer of Biostatistics* (also published by McGraw-Hill). It has been very gratifying to see that text make introductory statistical methods comprehensible to a broad audience and, as a result, help improve the quality of statistical methods applied to biomedical problems. This book seeks to go the next step and make more advanced multivariable methods accessible to the same audience.

In writing this book, we have tried to maintain the spirit of *Primer of Biostatistics* by concentrating on intuitive and graphical explanations of the underlying statistical principles and using data from actual published research in a wide variety of disciplines to illustrate the points we seek to make. We also use real data in the problems at the end of each chapter. We have also concentrated on the methods that are of the broadest practical use to investigators in the health and life sciences.

This book should be viewed as a follow-on to *Primer of Biostatistics*, suitable for the individual reader who already has some familiarity with basic statistical terms and concepts, or as a text for a second course in applied statistics.

Multiple regression can be developed elegantly and presented using matrix notation, but few readers can be expected to be familiar with matrix algebra. We, therefore, present the central development of the book using only simple algebra and most of the book can be read without resorting to matrix notation. We do, however, include the equivalent relationships in matrix notation (in optional sections or footnotes) for the interested reader and to build the foundation for a few advanced methods—particularly principal components regression in Chapter 5— that require matrix notation. We also include an appendix that covers all necessary notation and concepts of matrix algebra necessary to read this book.

This focus on practical applications led us to present multiple regression as a way to deal with complicated experimental designs. Most investigators are familiar with simple (two-variable) linear regression, and this procedure can be easily generalized to multiple regression, which simply has more independent variables. We base our development of multiple regression on this principle.

Our focus on practical applications also led us to present analysis of variance as a special case of multiple regression. Most presentations of analysis of variance begin with classical developments and only mention regression implementations briefly at the end, if at all. We believe that the regression implementation of analysis of variance is much more appropriate for health and life sciences audiences for two reasons. First, the regression approach permits a development in which the assumptions of the analysis of variance model are much more evident than in traditional formulations. Second, and more important, regression implementations of analysis of variance can easily handle unbalanced experimental designs and missing data. While one can often design experiments and data collection procedures in other disciplines to ensure balanced experimental designs and no missing data, this situation often does not occur in the health and life sciences, particularly in clinical research.

Practical applications also dictated our selection of the specific topics in this book. We include a detailed discussion of repeated-measures designs — when repeated observations are made on the same experimental subjects — because these designs are very common in clinical research, yet they are rarely covered in texts at this level. We also include detailed treatment of nonlinear regression models because so many things in biology are nonlinear, and logistic regression because many things in clinical research are measured on dichotomous scales (e.g., dead or alive). We also devote considerable attention to how to be sure that the data fit the assumptions that underlie the methods we describe and how to handle missing data. Like any powerful tools, these methods can be abused, and it is important that people appreciate the limitations and restrictions on their use.

The universal availability of computer programs to do the arithmetic necessary for multiple regression and analysis of variance has been the key to making these methods accessible to anyone who wishes to use them. We use the four most widely available of these programs, BMDP, Minitab, SAS, and SPSS-X, to do the computations associated with all the examples in the book. We present outputs from these programs to illustrate how to read them, and we include the input control language files and data sets in the appendixes. We also include a comparison of the relevant capabilities of these programs and some hints on how to use them. This information should help the reader learn the mechanics of using these (and similar) statistical packages as well as reproduce the results in all the examples in the book.

As already mentioned, the examples and problems in this book are based on actual published research from the health and life sciences. We say "based on" because we selected the examples by reviewing journals for studies that would benefit from use of the techniques we present in this book. Many of the papers we selected did not use the methods we describe or analyze their data as we do. Indeed, one of our main objectives in writing this book is to popularize the methods we present. Few authors present their raw data, and we were forced to work backwards and simulate what the raw data may have looked like based on the summary statistics (i.e., means and standard deviations) presented in the papers. We have also taken some liberties with the data to simplify the presentation for didactic purposes, generally by altering the sample size or changing a few values to make the problems more manageable. We sought to maintain the integrity of the conclusions the authors reached, except in some cases where the authors would have reached different conclusions using the analytical techniques we present in this book. The bottom line of this discussion is that the examples should be viewed for their didactic value as *examples,* not as precise reproductions of the original work. *Readers interested in the original work should consult the papers we used to generate the examples and problems,* and compare the approaches of the original authors with those we illustrate. Keep in mind, however, that in most cases we are dealing with simulated or altered data sets.

Finally, we would like to thank all the people who helped us with this book. Robert Appleyard, Anne Cross, Julien Hoffman, Susan Sacks, and Ruth Mickey all provided helpful criticism of the manuscript. Trish Warshaw prepared the illustrations. We also thank those who made raw data available to us to develop some of the examples. William Day provided encouragement and good outside review of the manuscript. Finally, we thank our families for giving us the time to do this book. Like most worthwhile enterprises, it took longer than we thought it would.

We have high hopes for this book to become a significant force within the health and life sciences community to popularize multivariate statistical techniques, which have proved so powerful for people in the social and economic sciences. You, the reader, will be the judge on how well we have succeeded.

STANTON A. GLANTZ
BRYAN K. SLINKER

Why Do Multivariate Analysis?

Investigators in the health and biological sciences use a tremendous variety of experimental techniques to gather data. Molecular biologists examine fragments of DNA in test tubes and chromatographs. Physiologists measure the function of organ systems with transducers in whole animals. Clinical investigators observe the response of people to medical and surgical treatments. Epidemiologists collect information about the presence of diseases in whole populations. The types of data collected by each of these investigators are as different as the methods used to collect them. These differences are so great that the observations and methods of one discipline can be incomprehensible to people working in another. Once the data are collected, however, they are analyzed using common statistical methods.

Because few biomedical investigators have received much formal training in statistical analysis, they tend to use very basic statistical techniques, primarily the t test. The t test and related methods such as analysis of variance are known as *univariate* statistical methods because they involve analyzing the effects of the treatment or experimental intervention on a single variable, *one variable at a time*. Although there is nothing intrinsically wrong with such an analysis, it provides only a limited view of the data.

To understand the limitations of a univariate approach to analyzing a set of data, consider the following example: Someone approaches you, says that they are feeling short of breath, and asks you what you think the problem is. You first note that the person is male, then recall everything you know about the relationship between sex and shortness of breath. Second, ask the person's age; the person is 55. Knowing that the person is 55, recall everything you know about shortness of breath in people who are 55 *without taking into account the fact that you know the person is a man*. Third, ask if the person just exercised; the person walked up two flights of stairs. Knowing that the person walked up two flights of stairs, recall everything you know about exercise and shortness of breath *without taking into account the sex or age of the person.* Fourth, find out if the person smokes; the person does. Knowing that the person smokes, recall all you know about smoking and shortness of breath *without taking into account the sex or age of the person or the fact that the person just climbed two flights of stairs.* Fifth, ask the person his or her favorite food; it is deep-fat fried egg yolks. Knowing that the person loves deep-fat fried egg yolks, recall all you know about diet and shortness of breath *without taking into account the sex or age of the person, the fact that the person just climbed two flights of stairs, or the fact that the person smokes.*

So what can you say? Sex alone tells very little. The fact that the person is 55 could indicate that he or she is in the early stages of some disease, but most people are still healthy at age 55. The fact that the person just exercised by climbing stairs could mean that they are just out of shape. The fact that the person smoked might have something to do with the shortness of breath, but it is hard to draw any specific conclusions about why. The fact that the person likes deep-fat fried egg yolks may mean that he or she has indigestion. Given each fact in isolation, it is difficult to draw any meaningful conclusions about why the person is short of breath.

This process of ignoring everything else you know about the person seems a little silly. A much more reasonable approach—indeed one that you probably could not help but take in reading this example—would be to consider all the variables together and ask the question: What is probably wrong with a 55-year-old man who smokes, eats deep-fat fried egg yolks, just climbed two flights of stairs, and now feels short of breath? One highly likely possibility is that the person has heart disease.

In the first case, we did a univariate analysis—we examined each variable one at a time. While providing some information, it was not nearly as incisive as the second case in which we *considered the effects of several variables at once* in a so-called *multivariate analysis.* When you think about most problems, you implicitly do a multivariate analysis. As with the qualitative example above, statistical multivariate techniques allow you to consider the effects of several independent variables at once and so better describe (and hopefully understand) the underlying relationships in a set of data.

OUR FIRST VISIT TO MARS

Suppose we wish to study the determinants of the weight of Martians. After writing and rewriting our research grant several times, we finally obtain the money necessary to visit Mars and measure the heights and

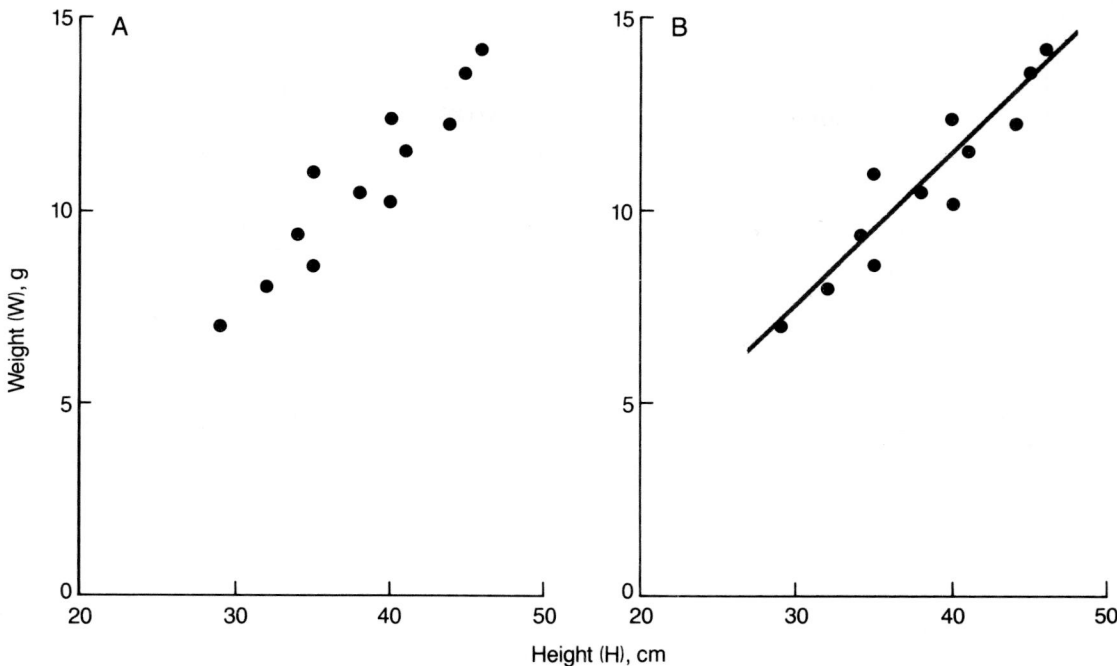

FIGURE 1-1 *A.* Data observed on the relationship between weight and height in a random sample of 12 Martians. While there is some random variation, these observations suggest that the average weight tends to increase linearly as height increases. *B.* The same data together with an estimate of the straight line relating mean weight to height, computed using a simple linear regression with height as the independent variable.

weights of a random sample of Martians. In addition to height and weight, we also count the number of cups of canal water they drink each day. Figure 1-1*A* shows a plot of weight (the dependent variable) versus height (the independent variable) for the 12 Martians in our sample.

Inspecting Fig. 1-1*B* reveals a linear (straight-line) relationship between these variables. One can describe this relationship by fitting the data with the straight line (computed using simple linear regression)

$$\hat{W} = -4.1 \text{ g} + .39 \text{ g/cm } H \tag{1.1}$$

in which $\hat{W}$ equals the predicted weight (in grams) and H equals the height (in centimeters). This line leads to the conclusion that, on the average, for every 1-cm increase in height, the average Martian's weight increases by .39 g.

This analysis, however, fails to use some of the information we have collected, namely, our knowledge of how much canal water C each Mar-

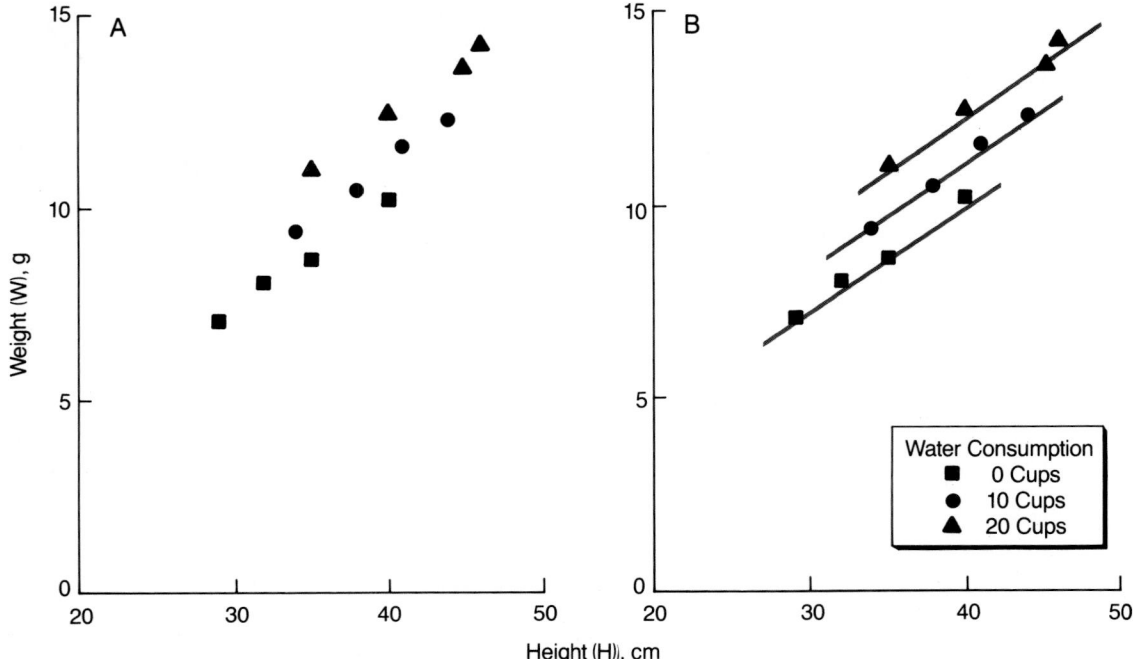

FIGURE 1-2 In addition to measuring the weight and height of the 12 Martians shown in Fig. 1-1, we also measured the amount of water they consumed each day. Panel *A* shows the same data as in Fig. 1-1*A*, with the different levels of water consumption identified by different symbols. This plot suggests that greater water consumption increases weight at any given height. Panel *B* shows the same data with separate lines fit through the different levels of water consumption or through the relationship between height and weight at different levels of water consumption, computed using a multiple linear regression with height and water consumption as the independent variables.

tian drinks every day. Because different Martians drink different amounts of water, it is possible that water consumption could have an effect on weight independent of height.

Fortunately, all the Martians drank 0, 10, or 20 cups of water each day, so we can indicate these differences graphically by replotting Fig. 1-1 with different symbols depending on the amount of water consumed per day. Figure 1-2 shows the results. Note that as the amount of water consumed increases among Martians of any given height, their weights increase. We can formally describe this effect by adding a term dependent on canal water C to the equation above, and then fit it to the data in Table 1-1 (and Fig. 1-2*A*) to obtain

$$\hat{W} = -1.2 \text{ g} + .28 \text{ g/cm } H + .11 \text{ g/cup } C \tag{1.2}$$

TABLE 1-1 Data of Martian Weight, Height, and Water Consumption

Weight W, g	Height H, cm	Water Consumption C, cups/day
7.0	29	0
8.0	32	0
8.6	35	0
9.4	34	10
10.2	40	0
10.5	38	10
11.0	35	20
11.6	41	10
12.3	44	10
12.4	40	20
13.6	45	20
14.2	46	20

This equation describes not only the effect of height on weight, but also the independent effect of drinking canal water.

Because there are three discrete levels of water consumption, we can use Eq. (1.2) to draw three lines through the data in Fig. 1-2A, depending on the water consumption. For example, for Martians who do not drink at all, on the average, the relationship between height and weight is

$$\hat{W} = -1.2 \text{ g} + .28 \text{ g/cm } H + .11 \text{ g/cup} \cdot 0 \text{ cups} = -1.2 \text{ g} + .28 \text{ g/cm } H$$

and for Martians who drink 10 cups of canal water per day, the relationship is

$$\hat{W} = -1.2 \text{ g} + .28 \text{ g/cm } H + .11 \text{ g/cup} \cdot 10 \text{ cups} = -.1 \text{ g} + .28 \text{ g/cm } H$$

and, finally, for Martians who drink 20 cups of canal water per day, the relationship is

$$\hat{W} = -1.2 \text{ g} + .28 \text{ g/cm } H + .11 \text{ g/cup} \cdot 20 \text{ cups} = 1.0 \text{ g} + .28 \text{ g/cm } H$$

Figure 1-2B shows the data with these three lines.

Comparing Figs. 1-1B and 1-2B reveals two interesting things. First, by taking into account not only height but also water consumption, we obtained a better prediction of the weight of any given Martian, as reflected by the fact that the data points cluster more closely around the lines in Fig. 1-2B than the line in Fig. 1-1B. Second, by accounting for the independent effect of drinking, we conclude that weight is less sensitive to height than before. When one accounts for water consumption, the independent effect of height is that of a .28-g increase in weight for each 1-cm increase in height in Eq. (1.2) compared with an estimate of .39 g/cm without accounting for water consumption in Eq. (1.1). Our estimate of the effect of height on weight in Fig. 1-1 and Eq. (1.1) was biased upward because we were erroneously attributing to height effects that were due to both height and drinking water.

In this simple example, all the Martians drank similar amounts of water from one of three discrete groups (0, 10, or 20 cups per day). There is nothing in the technique of multiple regression that requires that this occur. The method would have worked equally well had each Martian consumed a different amount of water. (It would, however, have made it harder to draw Fig. 1-2.) As we will see in Chap. 3, we can handle this eventuality by plotting Fig. 1-2 in three dimensions, with height and water consumption as separate axes. In fact, one of the benefits of multiple regression analysis of data such as this is that it does not require that each subject be observed under exactly the same conditions.

DUMMIES ON MARS

It is also common to examine the relationship between two variables in the presence or absence of some experimental condition. For example, while on Mars, we also investigated whether or not exposure to second-hand cigarette smoke affects the relationship between height and weight. (The tobacco companies had colonized Mars to expand their markets after everyone on Earth broke their addiction.) To keep things simple, we limited ourselves to Martians who do not drink water, so we located 16 Martians and determined whether or not they were exposed to secondhand smoke. Figure 1-3A shows the results.

Even though the effect of involuntary smoking is either present or absent, we can still use multiple regression analysis to test for this effect. To do this, we will use a *dummy variable* which indicates the presence or absence of exposure to secondhand smoke. Specifically, we will define the variable D to be 0 for Martians who are not exposed to secondhand smoke and 1 for Martians who are. Next we fit the data in Fig. 1-3A with the equation

$$\hat{W} = -.8 \text{ g} + .27 \text{ g/cm } H - 2.9 \text{ g } D \tag{1.3}$$

In this example the coefficient of D, -2.9 g, is negative, which indicates that exposure to secondhand smoke *decreases* the average weight of a Martian by 2.9 g at any given height. Specifically, for the unexposed group, $D = 0$ and Eq. (1.3) reduces to

$$\hat{W} = -.8 \text{ g} + .27 \text{ g/cm } H - 2.9 \text{ g}{\cdot}0 = -.8 \text{ g} + .27 \text{ g/cm } H$$

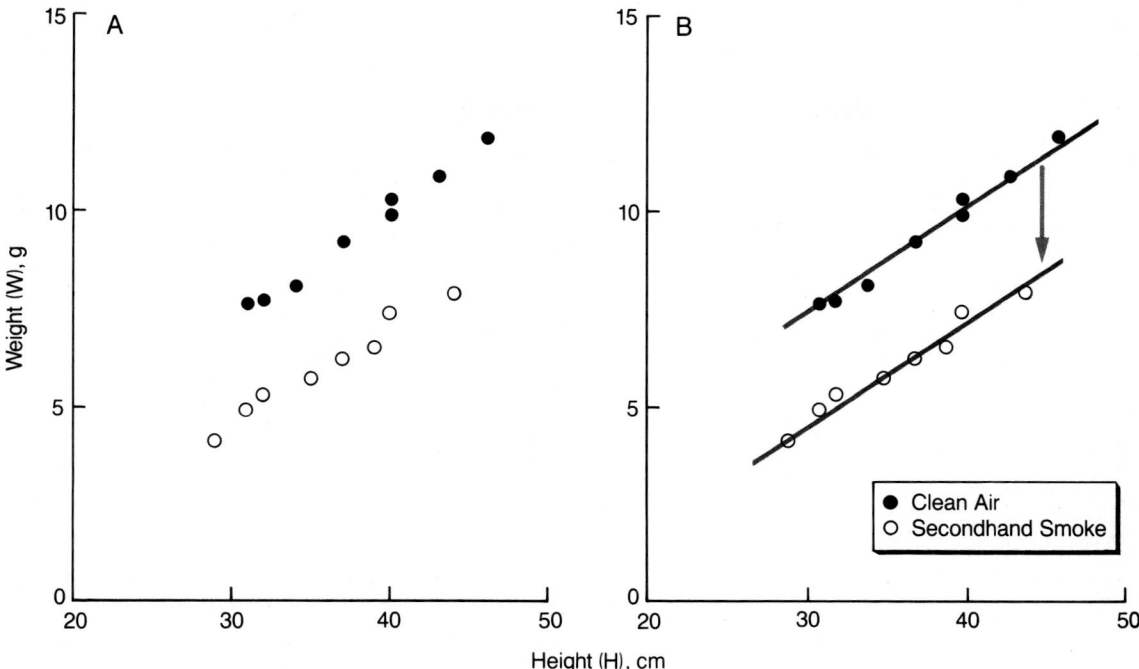

FIGURE 1-3 *A.* The relationship between weight and height in Martians, some of whom were exposed to secondhand tobacco smoke, suggests that exposure to secondhand tobacco smoke leads to lower weights at any given height. *B.* The data can be described by two parallel straight lines that have different intercepts. There is a reduction in average weight of 2.9 g at any given height for those Martians exposed to secondhand smoke compared to those who were not. These data can be described using a multiple regression in which one of the independent variables is a dummy variable, which takes on values 1 or 0, depending on whether or not the individual Martian was exposed to secondhand smoke.

and, for the exposed group, $D = 1$ and Eq. (1.3) becomes

$$\hat{W} = -.8 \text{ g} + .27 \text{ g/cm } H - 2.9 \text{ g·1} = -3.7 \text{ g} + .27 \text{ g/cm } H$$

Because the slopes are the same in these two equations, they represent two parallel lines. The coefficient of D quantifies the parallel shift of the line relating weight to height for the smoke-exposed and nonexposed Martians (Fig. 1-3*B*).

This is a very simple application of dummy variables. Dummy variables can be used to encode a wide variety of experimental conditions. In fact, as we will see, one can do complex analyses of variance as regression problems using appropriate dummy variables.

SUMMARY

When you conduct an experiment, clinical trial, or survey, you are seeking to identify the important determinants of a response and the structure of the relationship between the response and its (multiple) determinants, gauging the sensitivity of a response to its determinants, or predicting a response under different conditions. To achieve these goals, investigators often collect a large body of data, which they then analyze piecemeal using univariate statistical procedures such as *t* tests and one-way analysis of variance or two-variable simple linear regression. Although in theory there is nothing wrong with breaking up the analysis of data in this manner, doing so often results in a loss of information about the process being investigated.

While widely used in the social and behavioral sciences and economics, multivariate statistical methods are not yet widely used or appreciated in many biomedical disciplines. These multivariate methods are based on *multiple linear regression*, which is the multiple variable generalization of the simple linear regression technique used to draw the best straight line through a set of data in which there is a single independent variable. Multivariate methods offer several advantages over the more widely used univariate statistical methods, largely by bringing more information to bear on a specific problem. They provide a powerful tool for gaining additional quantitative insights about the question at hand because they allow one to assess the continuous functional (in a mathematical sense) relationship among several variables. This capability is especially valuable in (relatively) poorly controlled studies that are sometimes unavoidable in studies of organ systems physiology, clinical trials, or epidemiology.

In particular, compared to univariate methods, multivariate methods are attractive because

- *They are closer to how you think about the data.*
- *They allow easier visualization and interpretation of complex experimental designs.*
- *They allow you to analyze more data at once, so the tests are more sensitive (i.e., there is more statistical power).*
- *Multiple regression models can give more insight into underlying structural relationships.*
- *The analysis puts the focus on relationships among variables and treatments rather than individual points.*
- *They allow easier handling of unbalanced experimental designs and missing data than traditional analysis of variance computations.*

Multiple regression and related techniques provide powerful tools for analyzing data and, thanks to the ready availability of computers and statistical software, are accessible to anyone who cares to use them. Like all powerful tools, however, multiple regression and analysis of variance must be used with care to obtain meaningful results. Without careful thinking and planning, it is easy to end up buried in a confusing pile of meaningless computer printout. Look at the data, think, and plan before computing. Look at the data, think, and plan as preliminary results appear. Computers cannot think and plan; they simply make the computations trivial so that you have time to think and plan!

Once you have mastered the methods based on multiple regression, the univariate methods so common in biomedical research will often seem as simple-minded as one-symptom-at-a-time diagnosis.

PROBLEM

1.1 Find and discuss two articles from the last year which would benefit from a multivariate analysis of data that was originally analyzed in a univariate fashion.

The First Step: Understanding Simple Linear Regression

The purpose of this book is to teach how to formulate biological, clinical, and other problems in terms of equations that contain parameters that can be estimated from data and, in turn, can be used to gain insights into the problem under study. The power of multiple regression—whether applied directly or used to formulate an analysis of variance—is that it permits you to consider the simultaneous effects of several variables that act together to determine the value of some outcome variable. Although multiple linear regression models can become quite complicated, they are all, in principle, the same. In fact, one can present the key elements of any linear regression in the context of a simple two-variable linear regression.* We will estimate how much one variable increases (or decreases) on the average as another variable changes with a *regression line* and quantify the *strength of the association* between the two variables with a *correlation coefficient.* We will generalize these concepts to the multiple variable case in Chap. 3. The remainder of the book will explore how to apply these powerful tools intelligently.

MORE ON MARS

Simple *linear regression* is a parametric statistical technique used to analyze experiments in which the samples are drawn from populations characterized by a mean response that varies *continuously* with the mag-

---

*This chapter is revised from S. Glantz, *Primer of Biostatistics* (2nd ed.), New York, McGraw-Hill, 1987, Chap. 8, "How to Test for Trends." The notation has been revised to be consistent with that used in multiple regression, and a detailed discussion of sums of squares as they apply to regression analysis and the use of computer programs to do the computation in regression analysis has been added.

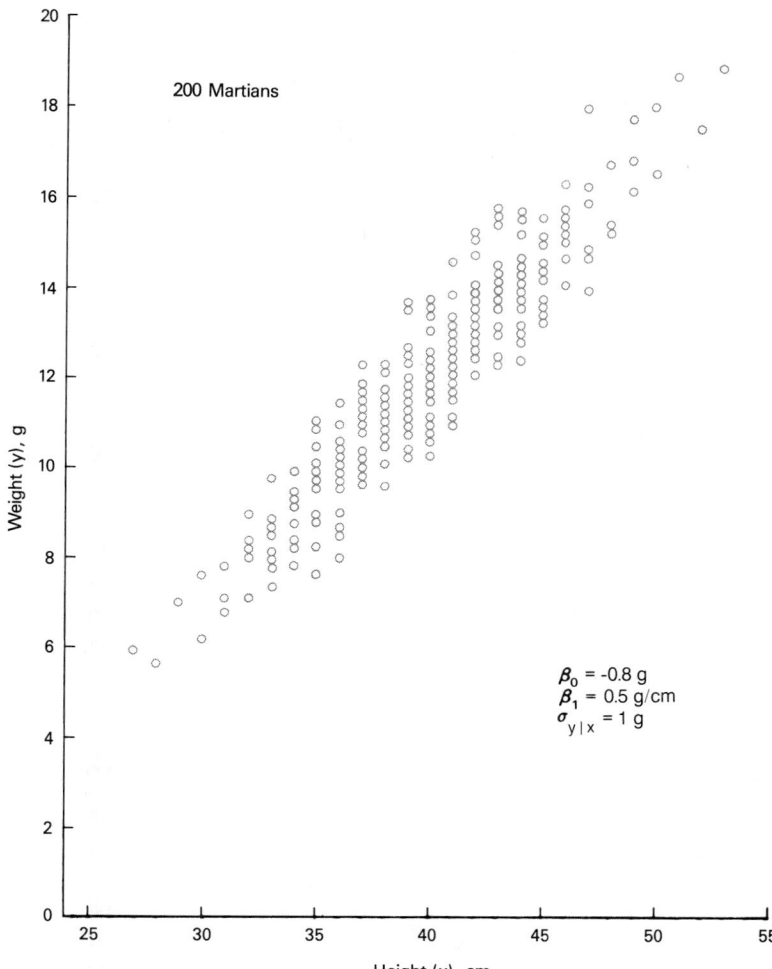

FIGURE 2-1 The relationship between weight and height in the entire population of 200 Martians with each Martian being represented by a circle. The mean weight of Martians at any given height increases linearly with height, and the variability in weight at any given height is the same and normally distributed about the mean at that height. A population must have these characteristics to be suitable for linear regression analysis.

nitude of the treatment. To understand the nature of this population and the associated random samples, we continue to explore Mars, where we now examine the entire *population* of Martians — all 200 of them.

Figure 2-1 shows a plot in which each point represents the height x and weight y of one Martian. Because we have observed the *entire population*, there is no question that tall Martians tend to be heavier than

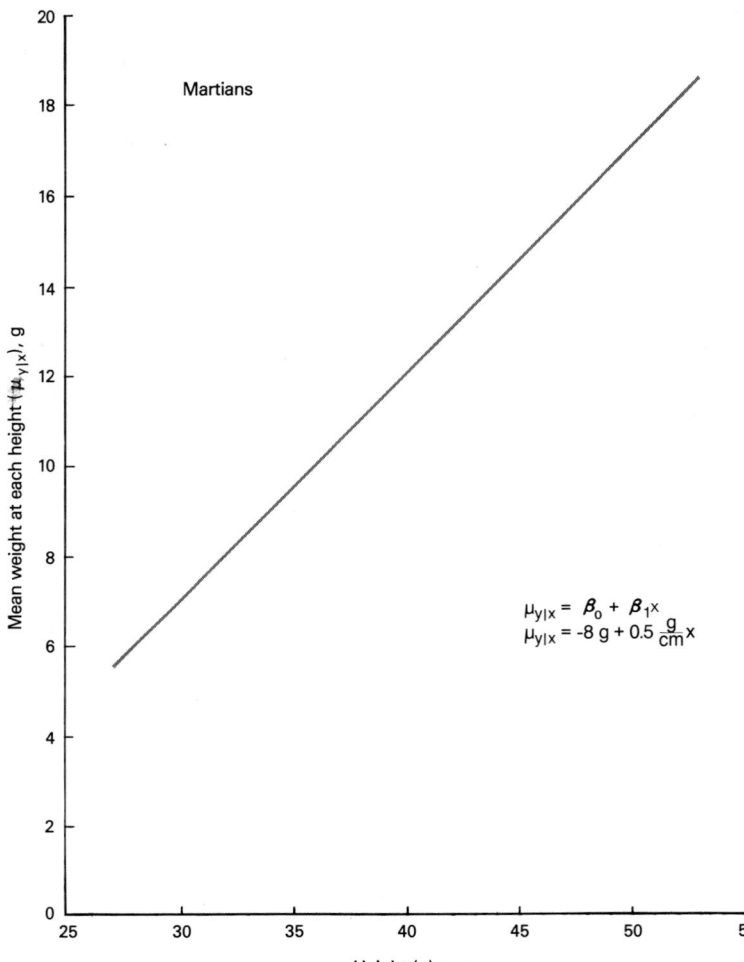

$$\mu_{y|x} = \beta_0 + \beta_1 x$$
$$\mu_{y|x} = -8 \text{ g} + 0.5 \frac{\text{g}}{\text{cm}} x$$

Figure 2-2 The line of means for the population of Martians in Fig. 2-1.

short Martians. (Because we are discussing simple linear regression, we will ignore the effects of differing water consumption.) For example, the Martians who are 32 cm tall weigh 7.1, 7.9, 8.3, and 8.8 g, so the mean weight of Martians who are 32 cm tall is 8 g. The 8 Martians who are 46 cm tall weigh 13.7, 14.5, 14.8, 15.0, 15.1, 15.2, 15.3, and 15.8 g, so the mean weight of Martians who are 46 cm tall is 14.9 g. Figure 2-2 shows that *the mean weight of Martians at each height increases linearly as height increases.*

This line does not make it possible, however, to predict the weight of *an individual* Martian if you know his height. Why not? Because there is variability in weights among Martians at each height. Figure 2-1 reveals that the standard deviation of weights of Martians with *any given height* is about 1 g.

The Population Parameters

Now, let us define some new terms and symbols so that we can generalize from Martians to other populations with similar characteristics. In regression analysis, we distinguish between the *independent, or predictor*, and *dependent, or response*, variables. The effects of one or more independent variables combine to determine the value of the dependent variable. For example, because we are considering how weight varies with height, height x is the independent variable and weight y is the dependent variable. In some instances, including the example at hand, we can only *observe* the independent variable and use it to *predict* the value of the dependent variable (with some uncertainty due to the variability in the dependent variable at each value of the independent variable). In other cases, including controlled experiments, it is possible to *manipulate* the independent variable to control, with some uncertainty, the value of the dependent variable. In the first case, it is only possible to identify an *association* between the two variables, whereas in the second case it is possible to conclude that there is a *causal* link.

For any given value of the independent variable x, it is possible to compute the value of the mean of all values of the dependent variable corresponding to that value of x. We denote this mean $\mu_{y|x}$ to indicate that it is the mean of all the values of y in the population at a given value of x. These means fall along a straight line given by

$$\mu_{y|x} = \beta_0 + \beta_1 x \tag{2.1}$$

in which β_0 is the intercept and β_1 is the slope of this so-called *line of means*. For example, Fig. 2-2 shows that, on the average, the weight of Martians increases by .5 g for every 1-cm increase in height, so the slope β_1 of the $\mu_{y|x}$ versus x line is .5 g/cm. The intercept β_0 of this line is -8 g. Hence,

$$\mu_{y|x} = -8 \text{ g} + .5 \text{ g/cm } x$$

There is variability about the line of means. For any given value of the independent variable x, the values of y for the population are normally distributed with mean $\mu_{y|x}$ and standard deviation $\sigma_{y|x}$. $\sigma_{y|x}$ is the standard deviation of weights y at any given value of height x, i.e., the standard deviation computed after allowing for the fact that mean weight varies with height. As noted above, the residual variation about the line of means for our Martians is 1 g; $\sigma_{y|x} = 1$ g. This variability is an important factor in determining how useful the line of means is for predicting the value of the dependent variable, e.g., weight, when you know the value of the independent variable, e.g., height. If the variability about the line of means is small, one can precisely predict the value of the dependent variable at any given value of the independent variable. The methods we develop below require that this standard deviation be *the*

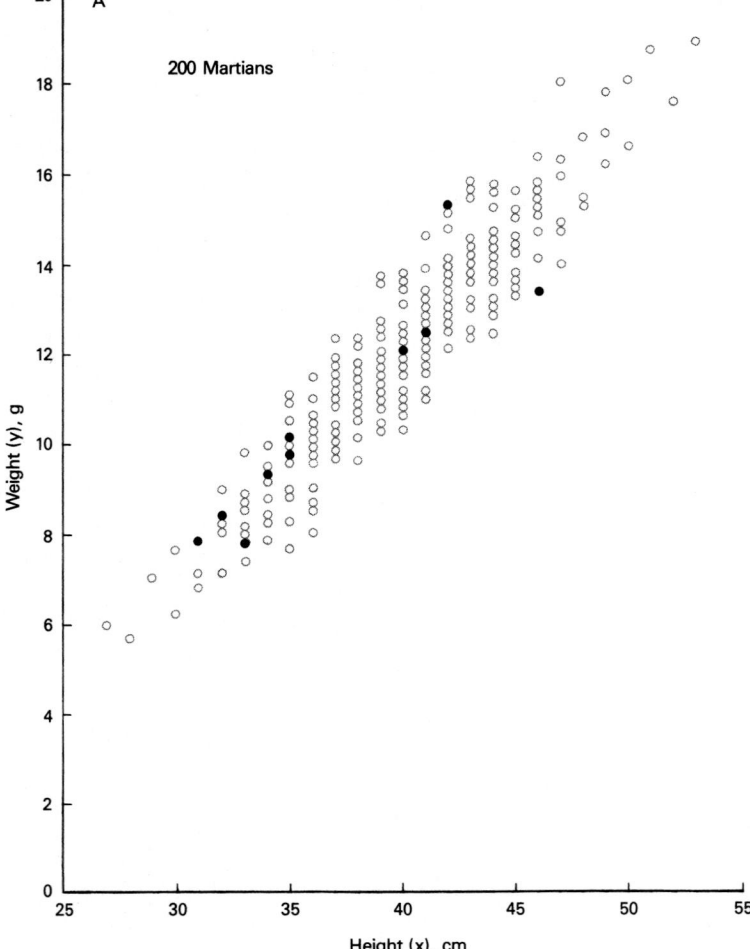

FIGURE 2-3 A random sample of 10 Martians, showing (*A*) the members of the population that were selected together with the entire population and (*B*) the sample as it appears to the investigator.

same for all values of *x*. In other words, *the variability of the dependent variable about the line of means is the same regardless of the value of the independent variable.*

We can also write an equation in which the random element associated with this variability explicitly appears:

$$y = \mu_{y|x} + \epsilon = \beta_0 + \beta_1 x + \epsilon \tag{2.2}$$

The value of the dependent variable *y* depends on the mean value of *y* at that value of *x*, $\mu_{y|x}$, plus a random term ϵ that describes the variability in the population. ϵ is called the *deviation*, or *error*, from the line of means. By definition, the mean of all possible values of ϵ is zero, and the

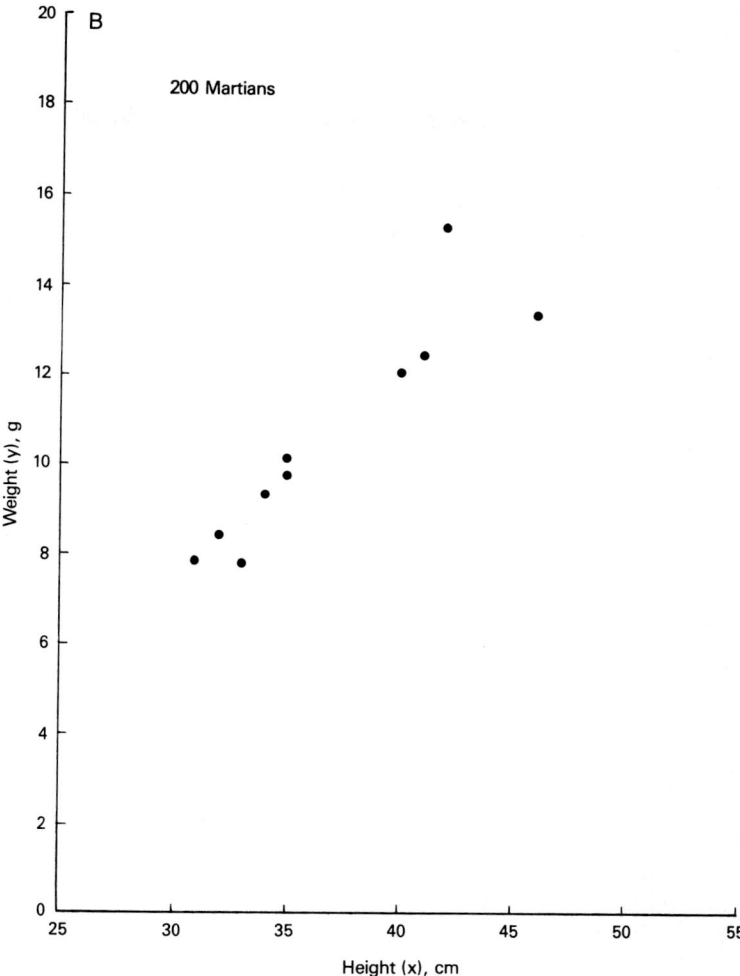

FIGURE 2-3 (*continued*)

standard deviation of all possible values of ϵ at a given value of x is $\sigma_{\epsilon} = \sigma_{y|x}$. This equation is a more convenient form of the linear regression model than Eq. (2.1), and we will use it for the rest of the book.

In sum, we will be analyzing the results of experiments in which the observations were drawn from populations with these characteristics:

- *The mean of the population of the dependent variable at a given value of the independent variable increases (or decreases) linearly as the independent variable increases.*

- *For any given value of the independent variable, the possible values of the dependent variable are distributed normally.*

- *The standard deviation of the population of the dependent variable about its mean at any given value of the independent variable is the same for all values of the independent variable.*
- *The deviations of all members of the population from the line of means are statistically independent; i.e., the deviation associated with one member of the population has no effect on the deviations associated with the other members.*

The parameters of this population are β_0 and β_1, which define the line of means, and $\sigma_{y|x}$, which defines the variability about the line of means. Now let us turn our attention to the problem of estimating these parameters from samples drawn at random from such populations.

HOW TO ESTIMATE THE LINE OF MEANS FROM A SAMPLE

Because we observed the entire population of Mars, there was no uncertainty about how weight varied with height. This situation contrasts with real problems in which we cannot observe all members of a population and must infer things about it from a limited sample that we assume is representative. To understand the information that such samples contain, let us consider a sample of 10 individuals selected at random from the population of 200 Martians. Figure 2-3A shows the members of the population who happened to be selected; Fig. 2-3B shows what an investigator or reader would see. What do the data in Fig. 2-3B allow you to say about the underlying population? How certain can you be about the resulting statements?

Simply looking at Fig. 2-3B reveals that weight increases as height increases among the 10 specific individuals in *this* sample. The real question of interest, however, is: Does weight vary with height in the population from which the sample was drawn? After all, there is always a chance that we could draw an unrepresentative sample. Before we can test the hypothesis that the apparent trend in the data is due to chance rather than a true trend in the population, we need to estimate the population trend from the sample. This task boils down to estimating the intercept β_0 and slope β_1 of the line of means, together with some measure of the precision of these estimates.

The Best Straight Line through the Data

We will estimate the two population parameters β_0 and β_1 with the intercept and slope, b_0 and b_1, of a straight line fit through the observed sample data points. Figure 2-4 shows the same sample as Fig. 2-3B with four proposed lines, labeled I, II, III, and IV. Line I is obviously not appropriate; it does not even pass through the data. Line II passes through the data but has a much steeper slope than the data suggest is really true.

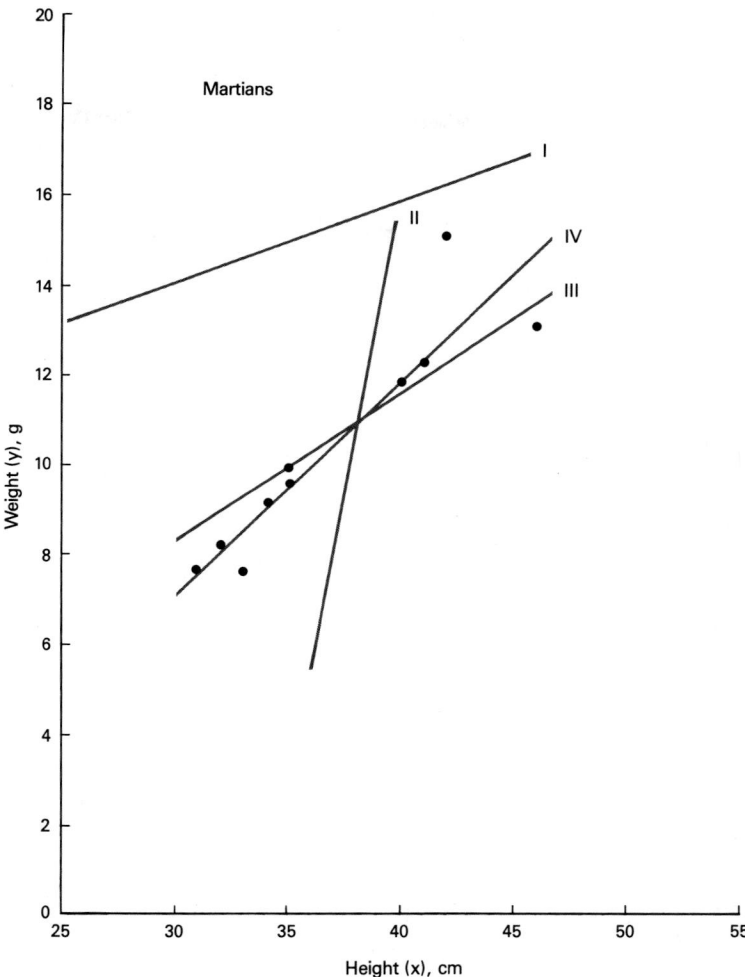

FIGURE 2-4 Four different possible lines to estimate the line of means from the sample in Fig. 2-3. Lines I and II are unlikely candidates because they fall so far from most of the observations. Lines III and IV are more promising.

Lines III and IV seem more reasonable; they both pass along the cloud defined by the data points. Which one is best?

To select the best line and so get our estimates b_0 and b_1 of β_0 and β_1, we need to define precisely what "best" means. To arrive at such a definition, first think about why line II seems better than line I and line III seems better than line II. The "better" a straight line is, the closer it comes to all the points taken as a group. In other words, we want to select the line that minimizes the total residual variation between the data and the line. The farther any one point is from the line, the more the line varies from the data, so let us select the line that leads to the smallest residual variation between the observed values and the values predicted from the straight line. In other words, the problem becomes

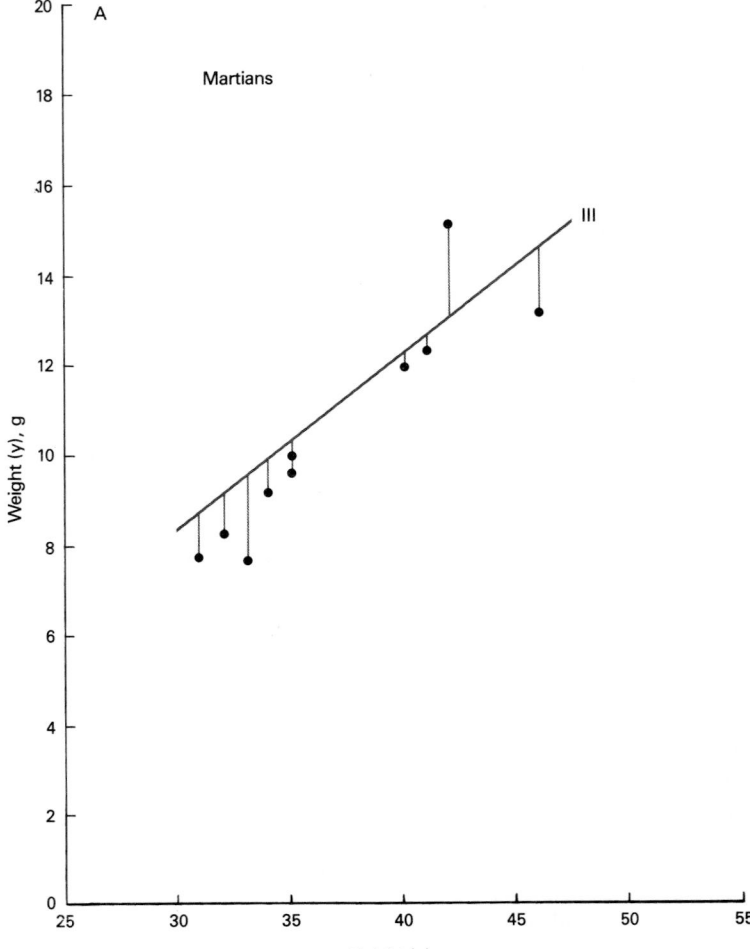

FIGURE 2-5 Lines III and IV from Fig. 2-4, together with the deviations between the lines and observations. Line IV is associated with the smallest sum of square deviations between the regression line and the observed values of the dependent variable, weight. The vertical lines indicate the deviations between the values predicted by the regression line and the observations. The light line is the line of means for the population of Martians in Fig. 2-1 and 2-2. The regression line approximates the line of means but does not coincide with it. Line III is associated with larger deviations from the observations than line IV.

one of defining a measure of this variation, then selecting values of b_0 and b_1 to minimize this quantity.

Variation in a population is quantified with the variance (or standard deviation) by computing the sum of the squared deviations from the mean and then dividing it by the sample size minus one. Now we will use the same idea and use the *residual sum of squares*, or SS_{res}, which is the *sum of the squared differences between each of the observed values of the dependent variable and the value on the line at the corresponding values of the independent variable*, as our measure of how much any given line varies from the data. We square the deviations so that positive and negative deviations contribute equally. Specifically, we seek the values of b_0 and b_1 that minimize the residual sum of squares:

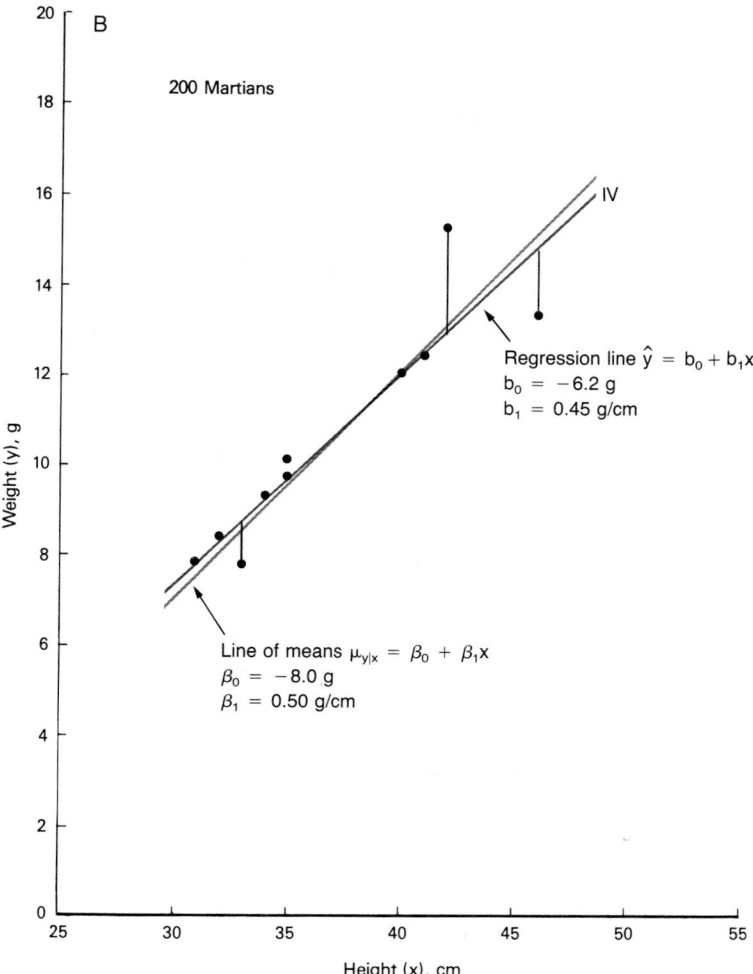

FIGURE 2-5 (*continued*)

$$SS_{res} = \Sigma \, [Y - \hat{y}(X)]^2 = \Sigma \, [Y - (b_0 + b_1 X)]^2 \qquad (2.3)$$

where the Greek sigma Σ indicates summation over all the observed data points (X, Y). $\hat{y}(X)$ is the value on the best-fit line corresponding to the observed value X of the independent variable $b_0 + b_1 X$, so $[Y - (b_0 + b_1 X)]$ is the amount (in terms of the dependent variable) that the observation deviates from the regression line.

Figure 2-5 shows the deviations associated with lines III and IV in Fig. 2-4. The sum of squared deviations is smaller for line IV than line III, so it is the best line. In fact, it is possible to prove mathematically that line IV is the one with the smallest sum of squared deviations between the observations and the line. For this reason, this procedure is often called the *method of least squares*, or least-squares regression.

TABLE 2-1 Computation of Regression of Martian Weight on Height in Fig. 2-5B

Observed Height X, cm	Observed Weight Y, g	X^2, cm^2	XY, cm·g	Estimated Weight $\hat{y}(X)$, g	Residual $Y - \hat{y}(X)$, g
31	7.8	961	241.8	7.76	0.04
32	8.3	1024	265.6	8.20	0.10
33	7.6	1089	250.8	8.65	−1.05
34	9.1	1156	309.4	9.09	0.01
35	9.6	1225	336.0	9.54	0.06
35	9.8	1225	343.0	9.54	0.26
40	11.8	1600	472.0	11.76	0.04
41	12.1	1681	496.1	12.20	−0.10
42	14.7	1764	617.4	12.65	2.05
46	13.0	2116	598.0	14.42	−1.42
369	103.8	13841	3930.1		
ΣX	ΣY	ΣX^2	ΣXY		

The resulting line is called the *regression line* of y on x (here the regression line of weight on height). It is our best estimate of the line of means. Its equation is

$$\hat{y} = b_0 + b_1 x \tag{2.4}$$

where $\hat{y}$ denotes the value of y on the regression line as a function of the independent variable x. $\hat{y}$ is the best estimate of the mean value of all values of the dependent variable, at any given value of the independent variable x. The slope b_1 is

$$b_1 = \frac{n\Sigma XY - \Sigma X \Sigma Y}{n\Sigma X^2 - (\Sigma X)^2} \tag{2.5}$$

and the intercept b_0 is

$$b_0 = \overline{Y} - b_1 \overline{X} \tag{2.6}$$

in which X and Y are the coordinates of the n points in the sample and $\overline{X}$ and $\overline{Y}$ are their respective means.*

Table 2-1 shows these computations for the sample of 10 points in Fig. 2-3B. From this table, $n = 10$, $\Sigma X = 369$ cm, $\Sigma Y = 103.8$ g, $\Sigma X^2 = 13,841$ cm^2, and $\Sigma XY = 3930.1$ g·cm. Substitute these values into Eq. (2.5) to find

$$b_1 = \frac{10(3930.1 \text{ g·cm}) - (369 \text{ cm})(103.8 \text{ g})}{10(13,841 \text{ cm}^2) - (369 \text{ cm})^2} = .44 \text{ g/cm}$$

*The values of b_0 and b_1 that minimize SS_{res} are those that satisfy the conditions $\partial SS_{\text{res}}/\partial b_0 = 0$ and $\partial SS_{\text{res}}/\partial b_1 = 0$. To find these values, differentiate Eq. (2.3)

TABLE 2-1 (*continued*) **Computation of Regression of Martian Weight on Height in Fig. 2-5B**

$[Y - \hat{y}(X)]^2$, g^2	$Y - \overline{Y}$, g	$(Y - \overline{Y})^2$, g^2	$\hat{y}(X) - \overline{Y}$, g	$[\hat{y}(X) - \overline{Y}]^2$, g^2
0.002	−2.6	6.7	−2.62	6.864
0.009	−2.1	4.3	−2.18	4.572
1.098	−2.8	7.7	−1.73	2.993
0.001	−1.3	1.6	−1.29	1.664
0.004	−0.8	0.6	−0.84	0.706
0.070	−0.6	0.4	−0.84	0.706
0.002	1.4	2.0	1.38	1.904
0.010	1.7	3.0	1.82	3.312
4.223	4.3	18.7	2.26	5.153
2.020	2.6	6.9	4.04	16.322
7.438		51.8		44.358
SS_{res}		SS_{tot}		SS_{reg}

Note: Some columns may not balance due to rounding errors.

$$\overline{X} = \Sigma X/n = 369/10 = 36.9 \text{ cm} \text{ and } \overline{Y} = \Sigma Y/n = 103.8/10 = 10.38 \text{ g},$$

so, from Eq. (2.6),

$$b_0 = (10.38 \text{ g}) - (.44 \text{ g/cm})(36.9 \text{ cm}) = -6.0 \text{ g}$$

(*footnote continued from page 20*)

with respect to b_0 and b_1:

$$\frac{\partial SS_{res}}{\partial b_0} = 2\Sigma [Y - (b_0 + b_1 X)](-1) = 0$$

$$\Sigma Y - b_0 \Sigma 1 - b_1 \Sigma X = 0$$

then note that $\Sigma 1 = n$, and rearrange to obtain

$$b_0 = \frac{\Sigma Y}{n} - b_1 \frac{\Sigma X}{n} = \overline{Y} - b_1 \overline{X}$$

which is Eq. (2.6).

$$\frac{\partial SS_{res}}{\partial b_1} = 2\Sigma [Y - (b_0 + b_1 X)] (-X) = 0$$

$$\Sigma XY - b_0 \Sigma X + b_1 \Sigma X^2 = 0$$

Substitute for b_0:

$$\Sigma XY - \left(\frac{\Sigma Y}{n} - b_1 \frac{\Sigma X}{n}\right) \Sigma X + b_1 \Sigma X^2 = 0$$

and solve for b_1 to obtain Eq. (2.5).

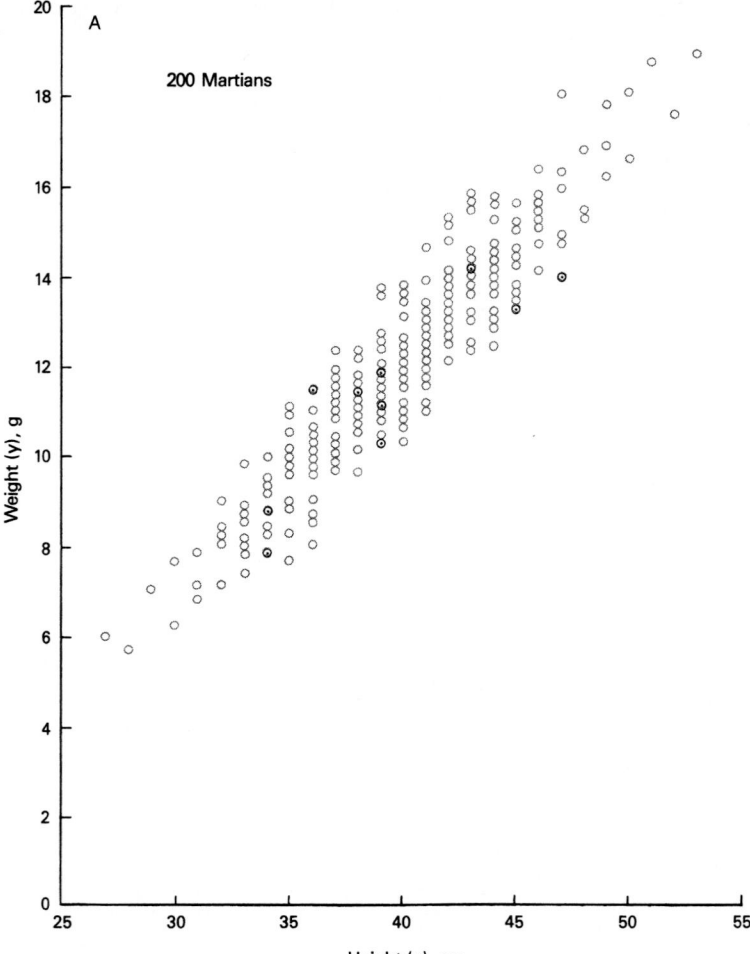

FIGURE 2-6 This figure illustrates a second random sample of 10 Martians drawn from the population of Fig. 2-1. This sample is associated with a different regression line than that computed from the first sample shown in Fig. 2-3.

Line IV in Figs. 2-4 and 2-5*B* is this regression line:

$$\hat{y} = -6.0 \text{ g} + (.44 \text{ g/cm})x$$

These two values are estimates of the population parameters $\beta_0 = -8$ g and $\beta_1 = .5$ g/cm, the actual intercept and slope of the line of means. The light line in Fig. 2-5*B* shows the true line of means.

Variability about the Regression Line

We have the regression line to estimate the line of means, but we still need to estimate the variability $\sigma_{y|x}$ of population members about the

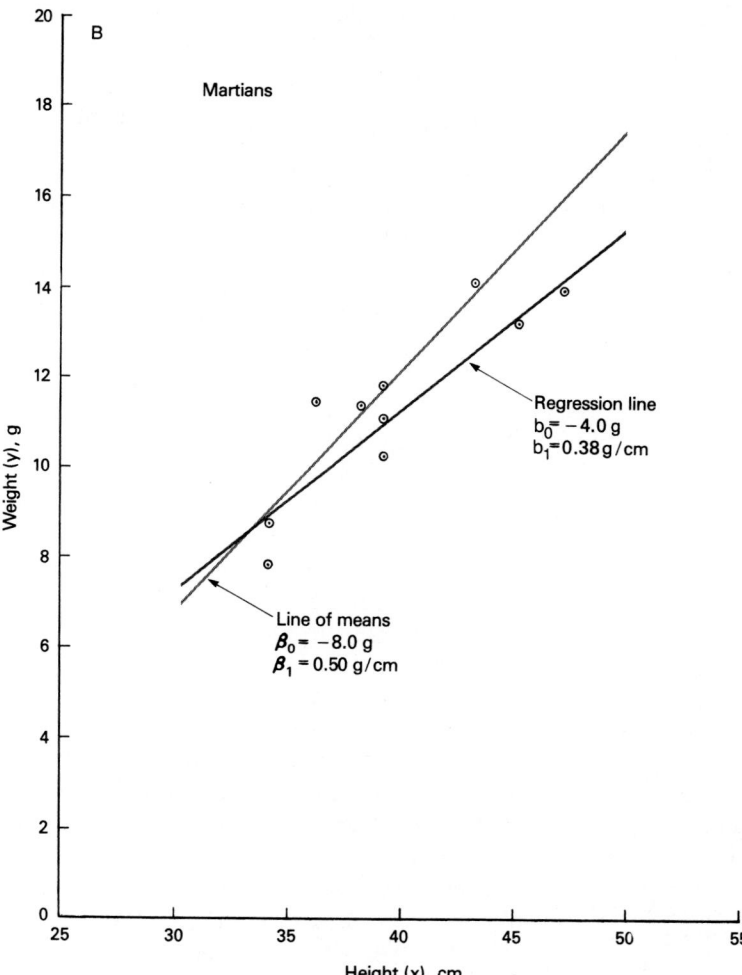

Figure 2-6 (continued)

line of means. We estimate this parameter by computing the square root of the "average" squared deviation of the data about the regression line:

$$s_{y|x} = \sqrt{\frac{\Sigma[Y - (b_0 + b_1 X)]^2}{n - 2}} = \sqrt{\frac{SS_{res}}{n - 2}} \tag{2.7}$$

$s_{y|x}$ is called the *standard error of the estimate*. For the sample shown in Fig. 2-3B (and Table 2-1),

$$s_{y|x} = \sqrt{\frac{7.44 \text{ g}^2}{10 - 2}} = .96 \text{ g}$$

This number is an estimate of the actual variability about the line of means in the underlying population, $\sigma_{y|x} = 1$ g.

Standard Errors of the Regression Coefficients

Just as the sample mean is only an estimate of the true population mean, the slope and intercept of the regression line are only estimates of the slope and intercept of the true population line of means. After all, there is nothing special about the sample in Fig. 2-3B. Figure 2-6A shows another sample of 10 individuals drawn at random from the population of all Martians. Figure 2-6B shows what you would see. Like the sample in Fig. 2-3B, the results of this sample also suggest that taller Martians tend to be heavier, but the relationship looks a little different from that associated with our first sample of 10 individuals. This sample yields $b_0 = -4.0$ g and $b_1 = .38$ g/cm as estimates of the intercept and slope of the true line of means. There are, in fact, over 10^{16} different ways to select 10 Martians at random from the population of 200 Martians, and each of these different samples can be used to compute a slope and intercept.

We can now consider the population of *all* slopes of regression lines computed based on samples of 10 Martians. Statisticians have demonstrated that the mean value of the distribution of all possible values of b_1 is β_1. Because the mean value of the distribution of all possible values of the regression line slope b_1 is equal to the slope of the line of means β_1, we say that the *expected value* of b_1 equals β_1 and, therefore, that the regression slope is an *unbiased estimator* of the slope of the line of means.

Likewise, the regression intercept b_0 is an unbiased estimator of the intercept of the line of means.

Because both b_0 and b_1 are computed from sums of the observations, the distributions of all possible values of b_0 and b_1 tend to be normally distributed about β_0 and β_1.* The distributions of all possible regression line slopes and intercepts have standard deviations σ_{b_1} and σ_{b_0}, called the *standard error of the slope* and *standard error of the intercept*, respec-

---

*This result is a consequence of the central-limit theorem, which states that a weighted sum of normally distributed random variables will also be normally distributed. It also states that a weighted sum of random variables drawn from any distribution will tend to be distributed normally as the number of elements added together to form the sum gets large. The proof of this theorem, which is one of the principal theorems of statistical theory, is beyond the scope of this book.

tively.* We will use these standard errors to test hypotheses about, and compute confidence intervals for, the regression coefficients and the regression equation itself, particularly when using multiple regression to assess the potential importance of several independent variables on a dependent variable.

The standard deviation of the population of all possible values of the regression line slope, the *standard error of the slope,* can be estimated from the sample with[†]

$$s_{b_1} = \sqrt{\frac{s_{y|x}^2}{(n-1)s_x^2}} = \sqrt{\frac{s_{y|x}^2}{\Sigma(X - \overline{X})^2}} \tag{2.8}$$

The *standard error of the intercept* of the regression line is the standard deviation of the population of all possible intercepts. Its estimate is

$$s_{b_0} = s_{y|x}\sqrt{\frac{1}{n} + \frac{\overline{X}^2}{(n-1)s_X^2}}$$

From the data in Fig. 2-3B and Table 2-1, it is possible to compute the standard errors for the slope and intercept as

$$s_{b_1} = \sqrt{\frac{(.96 \text{ g})^2}{(10 - 1)(5.0 \text{ cm})^2}} = .064 \text{ g/cm} \tag{2.9}$$

and

$$s_{b_0} = (.96 \text{ g})\sqrt{\frac{1}{10} + \frac{(36.9 \text{ cm})^2}{(10 - 1)(5.0 \text{ cm})^2}} = 2.4 \text{ g}$$

These standard errors can be used to compute confidence intervals and test hypotheses about the slope and intercept of the regression line using the *t* distribution.

HOW CONVINCING IS THE TREND?

There are many hypotheses we can test about regression lines, but the most common and important one is that the slope of the line of means is zero. This hypothesis is equivalent to estimating the chance that we

---

*These standard errors are exactly analogous to the standard error of the mean, which quantifies the variability in the distribution of all possible values of the sample mean. For a detailed discussion of the standard errors of the mean, see S. Glantz, *Primer of Biostatistics* (2nd ed.), New York, McGraw-Hill, 1987, pp. 21–27.

†For a derivation of these equations, see J. Neter, W. Wasserman, and M. H. Kutner, *Applied Linear Regression Models,* Homewood, Ill., Irwin, 1983, Chap. 3, "Inferences in Regression Analysis."

would observe a *linear* trend* as strong or stronger than the data show *when there is actually no relationship* between the dependent and independent variables. The resulting P value quantifies the certainty with which you can reject the hypothesis that there is no linear trend relating the two variables.

Testing the Slope of the Regression Line

Because the distribution of all possible values of the regression slope is approximately normal, we can use the t statistic:

$$t = \frac{b_1 - \beta_1}{s_{b_1}} \tag{2.10}$$

to test the hypothesis that the true slope of the line of means is β_1 in the population from which the sample was drawn.

To test the hypothesis of no linear relationship between the dependent and independent variables, set β_1 to zero in the equation above and compute

$$t = \frac{b_1}{s_{b_1}}$$

and then compare the resulting value of t with the two-tailed critical value t_α in Table E-1, Appendix E, defining the 100α percent most extreme values of t that would occur if the hypothesis of no trend in the population were true.† (Use the value corresponding to $v = n - 2$ degrees of freedom.)

For example, the data in Fig. 2-3B (and Table 2-1) yielded $b_1 = .44$ g/cm and $s_{b_1} = 0.064$ g/cm from a sample of 10 points. Hence, $t = .44/.064 = 6.905$, which exceeds 5.041, the value of t for $P < .001$ with $v = 10 - 2 = 8$ degrees of freedom (from Table E-1, Appendix E). Hence, it is unlikely that this sample was drawn from a population in which there was no relationship between the independent and dependent variables, height and weight. We can use these data to assert that as height increases, weight increases $(P < .001)$.

Of course, like all statistical tests of hypotheses, this small P value

---

*This restriction is important. As discussed in Chap. 3, it is possible for there to be a strong *nonlinear* relationship in the observations and for the procedures we discuss here to miss it.

†If you are using a table of critical values for the one-tailed t distribution, use the $t_{\alpha/2}$ critical value.

does not guarantee that there is really a trend in the population. It does indicate, however, that it is unlikely that such a strong linear relationship would be observed in the data if there were no such relationship between the dependent and independent variables in the underlying population from which the observed sample was drawn.

If we wish to test the hypothesis that there is no trend in the population using confidence intervals, we use the definition of t above to find the $100(1 - \alpha)$ percent confidence interval for the slope of the line of means:

$$b_1 - t_\alpha s_{b_1} < \beta_1 < b_1 + t_\alpha s_{b_1}$$

We can compute the 95 percent confidence interval for β_1 by substituting the value of $t_{.05}$ with $v = n - 2 = 10 - 2 = 8$ degrees of freedom, 2.306, into this equation together with the observed values of b and s_{b_1}:

$$.44 - 2.306(.064) < \beta_1 < .44 + 2.306(.064)$$

$$.29 \text{ g/cm} < \beta_1 < .59 \text{ g/cm}$$

Because this interval does not contain zero, we can conclude that there is a trend in the population ($P < .05$).* Note that the interval contains the true value of the slope of the line of means, $\beta_1 = .5$ g/cm.

It is likewise possible to test hypotheses about, or compute confidence intervals for, the intercept using the fact that

$$t = \frac{b_0 - \beta_0}{s_{b_0}}$$

is distributed according to the t distribution with $v = n - 2$ degrees of freedom.

These hypothesis testing techniques lie at the core of the procedures for interpreting the results of multiple regression.

Comparing Slopes and Intercepts of Two Regression Lines

It is also possible to use the t test to test hypotheses about two regression lines. For example, to test the hypothesis that two samples were drawn from populations with the same slope of the line of means, we compute

$$t = \frac{\text{difference of regression slopes}}{\text{standard error of difference of regression slopes}}$$

---

*The .1 percent confidence interval does not contain zero either, so we could obtain the same P value as with the first method using confidence intervals.

This test is exactly analogous to the definition of the t test to compare two sample means. The standard error of the difference of two regression slopes b_1 and b_2 is[*]

$$s_{b_1-b_2} = \sqrt{s_{b_1}^2 + s_{b_2}^2}$$ (2.11)

and so

$$t = \frac{b_1 - b_2}{s_{b_1-b_2}} \cdot$$

The resulting value of t is compared with the critical value of t corresponding to $v = (n_1 - 2) + (n_2 - 2) = n_1 + n_2 - 4$ degrees of freedom. Similar tests can be constructed to compare intercepts of different regression lines.[†]

Testing the Regression as a Whole

We have already stated that a linear regression analysis yields the "best" straight line through a set of data in the sense that it minimizes the sum of squared deviations (or errors) between the values $\hat{y}(X)$ of the independent variable predicted by the regression line at given values of the dependent variable X and the observed value of the dependent variable Y. In the preceding section, we used the standard error of the slope, which is an estimate of how precisely we can estimate the true slope of the line

---

[*]This equation provides the best estimate of the standard error of the differences of the slopes if both regression lines are computed from the same number of data points. If there are a different number of points in each regression, use the pooled estimate of the standard error of the differences of the slopes (which is analogous to the pooled variance estimate in a t test of the difference between two means):

$$s_{y|x_p}^2 = \frac{(n_1 - 2)s_{y|x_1}^2 + (n_2 - 2)s_{y|x_2}^2}{n_1 + n_2 - 4}$$

in

$$s_{b_1-b_2} = \sqrt{\frac{s_{y|x_p}^2}{(n_1 - 1)s_{x_1}^2} + \frac{s_{y|x_p}^2}{(n_2 - 1)s_{x_2}^2}}$$

where the subscripts 1 and 2 refer to data for the first and second regression lines.

[†]It is also possible to use dummy variables to conduct comparisons of regression lines. For example, we used dummy variables to test the hypothesis of different intercepts (with a common slope) for the data on involuntary smoking and weight in Fig. 1-3. We will make extensive use of dummy variables in this way throughout the rest of this book. For a detailed discussion of different strategies for specifically comparing regression lines, see D. G. Kleinbaum, L. L. Kupper, and K. E. Muller, *Applied Regression Analysis and Other Multivariable Methods* (2nd ed.), Boston, PWS-Kent, 1988, pp. 262–281.

relating y and x, to compute a t statistic to test whether the slope was significantly different from zero. We now present an alternative way, using the F test statistic, to conduct an overall test of whether the regression line describes the data.

As discussed above, when doing a linear regression, we select the slope and intercept that minimizes the sum of squared deviations between the regression line and observed data SS_{res} defined by Eq. (2.3). The more tightly the measured data points cluster about the estimated regression line, the smaller SS_{res}. To argue that the straight-line regression describes the relationship between the dependent and independent variables, we need to demonstrate that SS_{res} is "small." A small value of SS_{res} indicates that there is little residual variability about the regression line. Because "small" is a relative term, however, we need to compare SS_{res} with something. We will compare SS_{res} with the sum of squared differences between the regression line and the mean of the dependent variable (Fig. 2-7). This quantity is called the *sum of squares due to regression*, or SS_{reg}:

$$SS_{reg} = \Sigma[\hat{y}(X) - \overline{Y}]^2 = \Sigma[(b_0 + b_1 X) - \overline{Y}]^2 \tag{2.12}$$

where $\hat{y}(X)$ is the point on the regression line at data value X and $\overline{Y}$ is the mean of all observed values of the dependent variable.

To see the logic for comparing SS_{reg} with SS_{res}, note that the total deviation of a given observation from the mean of the dependent variable $Y - \overline{Y}$ can be divided into the difference between the observed value of the dependent variable and the corresponding point on the regression line, $Y - \hat{y}(X)$, and the difference between the value on the regression line and the mean of the dependent variable, $\hat{y}(X) - \overline{Y}$ (Fig. 2-7):

$$Y - \overline{Y} = [Y - \hat{y}(X)] + [\hat{y}(X) - \overline{Y}] \tag{2.13}$$

Next, square both sides of this equation and sum over all the data points to obtain*

$$\Sigma(Y - \overline{Y})^2 = \Sigma[Y - \hat{y}(X)]^2 + \Sigma[\hat{y}(X) - \overline{Y}]^2 \tag{2.14}$$

The expression on the left side of this equation is the sum of squared deviations of the observations about their mean, called the *total sum of*

---

*To obtain this result, square both sides of Eq. (2.13) and sum over all the observations:

$$\Sigma(Y - \overline{Y})^2 = \Sigma(Y - \hat{y})^2 + 2\Sigma(Y - \hat{y})(\hat{y} - \overline{Y}) + \Sigma(\hat{y} - \overline{Y})^2 \tag{1}$$

where $\hat{y} = \hat{y}(X)$. From Eq. (2.6)

$$\overline{Y} = b_0 + b_1 \overline{X} \tag{2}$$

(footnote continued on pp. 30 and 31)

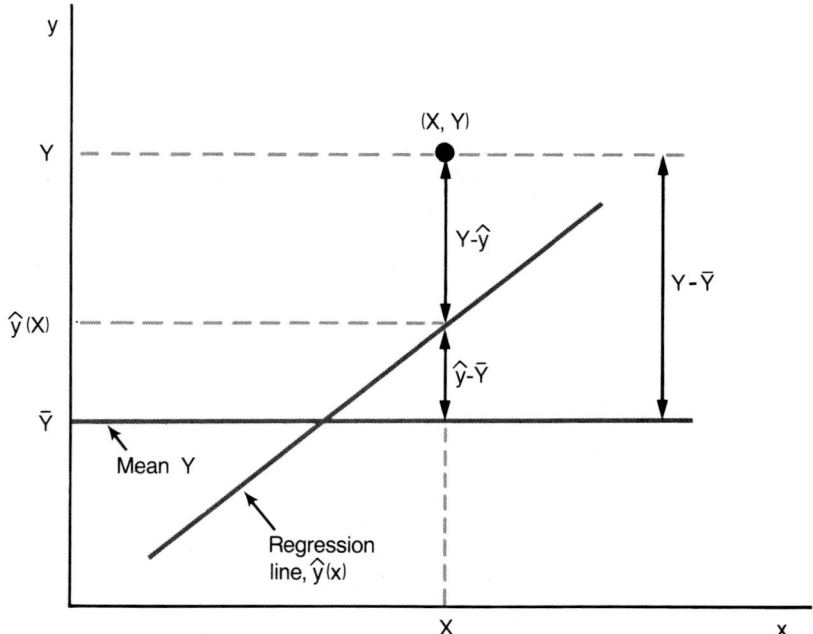

FIGURE 2-7 The total deviation of any observed value of the dependent variable Y from the mean of all observed values of the dependent variable $\hat{y}$, $(Y - \overline{Y})$ can be separated into two components: $(\hat{y} - \overline{Y})$, the deviation of the regression line from the mean of all observed values of the dependent variable, and $(Y - \hat{y})$, the deviation of the observed value of the dependent variable Y from the regression line. $(\hat{y} - \overline{Y})$ is a measure of the amount of variation associated with the regression line and $(Y - \hat{y})$ is a measure of the residual variation around the regression line. These two different deviations will be used to partition the variation in the observed values of Y into a component associated with the regression line and a residual component. If the variation of the regression line about the mean response is large compared with the residual variation around the line, we will conclude that the regression line provides a good description of the observations.

(footnote continued from page 29)

so

$$(\hat{y} - \overline{Y}) = (b_0 + b_1 X) - (b_0 + b_1 \overline{X}) = b_1(X - \overline{X})$$

Use Eqs. (2.4) and (2) to rewrite the middle term in Eq. (1) as

$$2b_1\Sigma(X - \overline{X})(Y - \hat{y}) = 2b_1\Sigma(X - \overline{X})[Y - (b_0 + b_1 X)] \tag{3}$$

$$= 2b_1\Sigma X[Y - (b_0 + b_1 X)] - 2b_1\overline{X}\Sigma[Y - (b_0 + b_1 X)] \tag{4}$$

The value of b_1 is selected to minimize

$$SS_{\text{res}} = \Sigma[Y - (b_0 + b_1 X)]^2$$

squares, SS_{tot}, which is a measure of the overall variability of the dependent variable[*]:

$$SS_{tot} = \Sigma(Y - \overline{Y})^2 \tag{2.15}$$

The first term on the right side of Eq. (2.15) is SS_{reg}, and the second term on the right side is SS_{res}, so Eq. (2.14) can be written

$$SS_{tot} = SS_{res} + SS_{reg}$$

In other words, we can divide the total variability in the dependent variable, as quantified with SS_{tot}, into two components: the variability of the regression line about the mean (quantified with SS_{reg}) and the residual variation of the observations about the regression line (quantified with SS_{res}). This so-called *partitioning of the total sum of squares* into these two sums of squares will form an important element in our interpretation of multiple regression and use of regression to do analysis of variance.

To be consistent with the mathematics used to derive the *F* distribution (which is based on ratios of variances rather than sums of squares), we do not compare the magnitudes of SS_{reg} and SS_{res} directly, but rather after we convert to "variances" by dividing by the associated degrees of freedom. In this case, there is 1 degree of freedom associated with SS_{reg} because there is 1 independent variable in the regression equation, so the so-called *mean square regression,* MS_{reg}, is

$$MS_{reg} = \frac{SS_{reg}}{1} \tag{2.16}$$

There are 2 coefficients in the regression equation (b_0 and b_1), so there are $n - 2$ degrees of freedom associated with the residual (error) term

---

[*]The total sum of squares is directly related to the variance (and standard deviation) of the dependent variable according to

$$s_Y^2 = \frac{\Sigma(Y - \overline{Y})^2}{n - 1} = \frac{SS_{tot}}{n - 1}$$

which occurs when

$$\frac{\partial SS_{res}}{\partial b_1} = 2\Sigma[Y - (b_0 + b_1 X)](-X) = 0 \tag{5}$$

and

$$\frac{\partial SS_{res}}{\partial b_0} = 2\Sigma[Y - (b_0 + b_1 X)](-1) = 0 \tag{6}$$

Substituting from Eqs. (5) and (6) demonstrates that Eq. (4), and hence Eqs. (3) and (1), equals zero, and yields Eq. (2.14).

and the so-called *residual mean square, MS_{res}, or mean square error,* is

$$MS_{res} = \frac{SS_{res}}{n - 2} \tag{2.17}$$

Note that the MS_{res} is a measure of the variance of the data points about the regression line; the square root of this value is the standard deviation of the residuals of the data points about the regression line, the standard error of the estimate $s_{y|x}$ defined by Eq. (2.7).

Finally, we compare the magnitudes of the variation of the regression line about the mean value of the dependent variable MS_{reg} with the residual variation of the data points about the regression line by computing the ratio:

$$F = \frac{MS_{reg}}{MS_{res}} \tag{2.18}$$

This F ratio will be large if the variation of the regression line from the mean of the dependent variable (the numerator) is large relative to the residual variation about the line (the denominator). One can then consult the table of critical values of the F test statistic in Table E-2, Appendix E, with the degrees of freedom associated with the numerator (1) and denominator $(n - 2)$ to see if the value is significantly greater than what one would expect by sampling variation alone. If this value of F is statistically significant, it means that fitting the regression line to the data significantly reduces the "unexplained variance," in the sense that the mean square residual from the regression line is significantly smaller than the mean square variation about the mean of the dependent variable.

Because in the special case of simple linear regression this F test is simply another approach to testing whether the slope of a regression line is significantly different from zero, it will yield exactly the same conclusion as we reached with the t test defined by Eq. (2.10).

To illustrate this point, let us test the overall goodness of fit of the linear regression equation, Eq. (2.4), to the data on heights and weights of Martians in Table 2-1. From Table 2-1, $SS_{reg} = 45.4$ g^2 and $SS_{res} = 7.4$ g^2. There are $n = 10$ data points, so, according to Eq. (2.16),

$$MS_{reg} = \frac{44.4 \text{ g}^2}{1} = 44.4 \text{ g}^2$$

and, according to Eq. (2.17),

$$MS_{res} = \frac{7.44 \text{ g}^2}{10 - 2} = .93 \text{ g}^2$$

So

$$F = \frac{MS_{reg}}{MS_{res}} = \frac{44.4 \text{ g}^2}{.93 \text{ g}^2} = 47.74$$

This value of F is associated with 1 numerator degree of freedom and 8 denominator degrees of freedom. According to Table E-2, Appendix E, the critical value of the F distribution is 11.26 for $\alpha = .01$. Because 47.74 exceeds 11.26, we conclude that there is a statistically significant relationship between the heights and weights of Martians. This procedure is simply another way of reaching the conclusion that the slope of the line relating the weight to the height of Martians is significantly different from zero ($P < .01$), as before. In fact, the value of F we just computed equals the square of the value of t we computed when testing the slope in Eq. (2.10).

Given the fact that the F and t tests provide identical information, why bother with the more complex test of overall fit with F? The reason is that, although these two tests provide equivalent information in simple linear regression with a single independent variable, they provide different information in multiple regression with several independent variables. In multiple regression, the t tests of the individual coefficients test whether *each* independent variable, considered one at a time, contributes to predicting the dependent variable. The F test for overall fit tests whether *all* the independent variables, taken together, contribute to predicting the dependent variable. In contrast to simple linear regression, this F test will provide important additional information beyond that associated with the t tests used to test whether or not the individual coefficients are significantly different from zero.[*]

Dietary Fat and Breast Cancer

Diet influences the development and growth of certain cancers, such as cancer of the breast, in experimental animals. To see whether or not people show the same tendency to develop breast cancer as laboratory mice and rats fed diets high in fat, Carroll[†] plotted the age-adjusted death rate for breast cancer against daily animal fat intake (Fig. 2-8A) and daily vegetable fat intake (Fig. 2-8B) for people in 39 different countries.

The regression line associated with the data in Fig. 2-8A is

$$\hat{y} = 2.5 + .16/(\text{g/day}) \, x_a$$

in which $\hat{y}$ represents the age-adjusted death rate (per 100,000 population) on the regression line and x_a represents the animal fat intake, in grams per day. The standard error of the slope s_{b_1} is .013/(g/day). To test the hypothesis that there is no relationship between animal fat intake and

[*]We will develop this distinction in more detail in Chap. 3.

[†]K. K. Carroll, "Experimental Evidence of Dietary Factors and Hormone-Dependent Cancers," *Cancer Res.* 35:3374–3383, 1975.

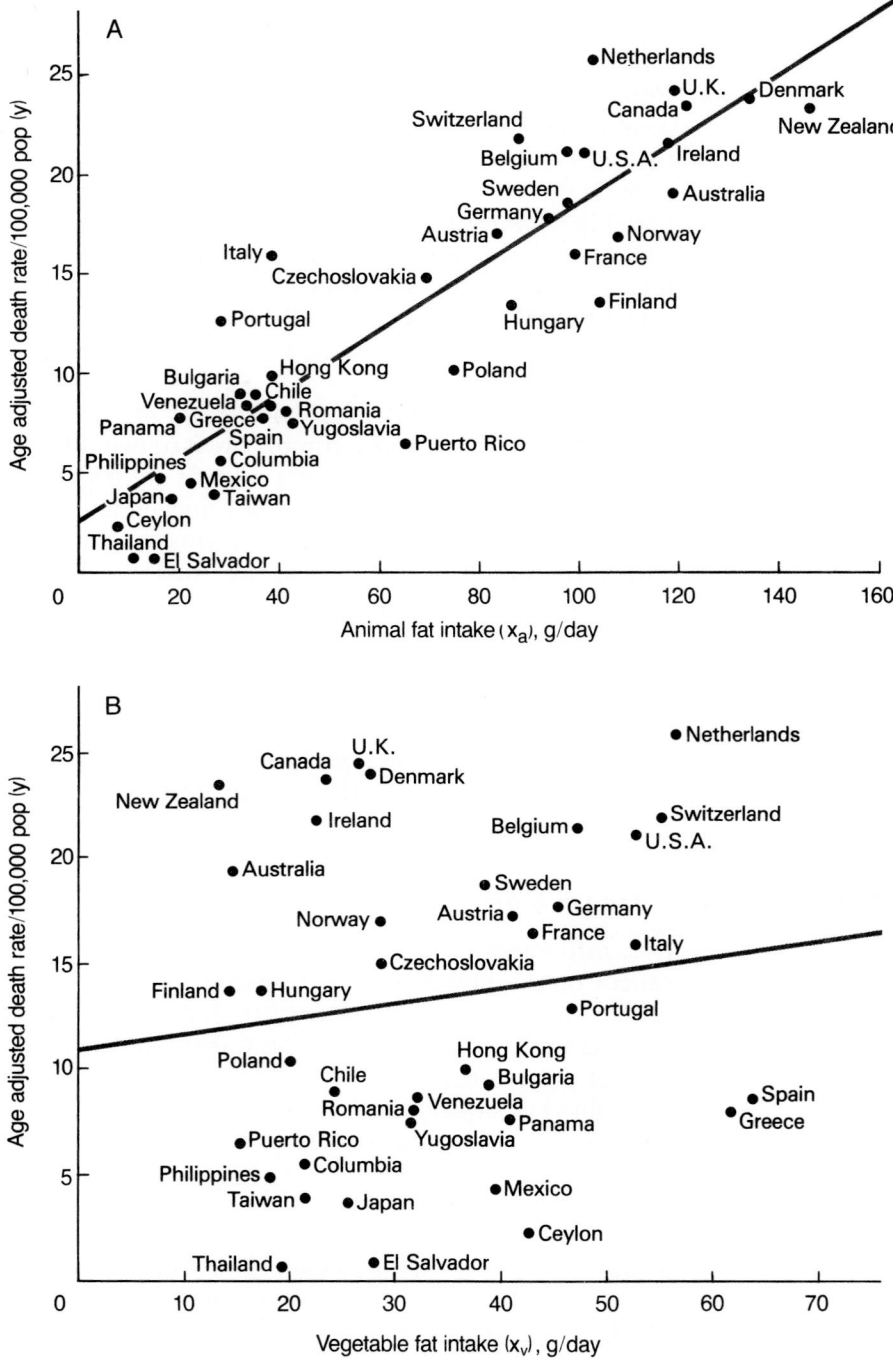

FIGURE 2-8 Relationship between per capita consumption of animal fat (*A*) and vegetable fat (*B*) and age-adjusted death rate from breast cancer in 39 countries. There appears to be a strong relationship for animal fat but not vegetable fat. (Adapted from Figure 4 of K. K. Carroll, "Experimental Evidence of Dietary Factors and Hormone-Dependent Cancers," *Cancer Res.* 35:3374–3383, 1975.)

death rate from breast cancer, compute

$$t = \frac{b_1}{s_{b_1}} = \frac{.16}{.013} = 12.31$$

This value exceeds 3.574, the value that defines the .1 percent most extreme values of the t distribution with $v = 39 - 2 = 37$ degrees of freedom. Thus, with less than 1 chance in 1000 of being wrong $(P < .001)$, we can report that the death rate from breast cancer increases as dietary animal fat consumption increases.

What about the relationship between vegetable fat and death rate from breast cancer? The observations in Fig. 2-8B are associated with the regression equation

$$\hat{y} = 10.4 + .084/(g/day) \, x_v$$

and the standard error of the slope is .056/(g/day). To test the hypothesis that there is no relationship between death rate and vegetable fat intake, compute

$$t = \frac{b_1}{s_{b_1}} = \frac{.084}{.056} = 1.5$$

which is not sufficient to reject the hypothesis of no linear trend. Notice that the regression line in Fig. 2-8B is essentially flat, indicating no trend between the variables.

These two studies suggest that animal fat, as opposed to vegetable fat, is an important environmental determinant of the likelihood that a woman will develop breast cancer. Does this analysis prove that eating animal fat causes cancer? Not at all. Carroll did not manipulate the diets of the women he observed in 39 countries, so these data are the results of observational rather than experimental studies. There may not be a direct causal link between animal fat consumption and the development of breast cancer; these two variables may both be related to some third, underlying variable that makes both of them change simultaneously.*

*For example, the incidence of breast cancer also increases with a wide variety of other variables, including income, automobiles, and television sets, although these factors are not as closely related to death rate as dietary-fat intake is. (For more details, see B. S. Drasar and D. Irving, "Environmental Factors and Cancer of the Colon and Breast," *Br. J. Cancer*, 27:167–172, 1973.) Does this mean that television causes cancer? Probably not, although there is always the chance that low-level x-ray or other radiation leaks may be a factor. Rather, all these observations, taken together, suggest that the constellation of factors related to a sedentary and affluent lifestyle is important in determining the risk of developing breast cancer. The multiple regression techniques we develop in Chap. 3 make it possible to consider several such factors as simultaneous determinants of the dependent variable, in this case breast cancer incidence.

The important thing that strengthens the link between fat intake and cancer in people is the fact that fat consumption affects cancer development in animal experiments where the investigator can actively manipulate the diet. Note, however, that this link in human beings rests on scientific, as opposed to purely statistical, reasoning.

When interpreting the results of regression analysis, it is important to keep the distinction between observational and experimental studies in mind. When investigators can actively manipulate the independent variable, while controlling other factors, and observe changes in the dependent variable, they can draw strong conclusions about how changes in the independent variable *cause* changes in the dependent variable. On the other hand, when investigators only observe the two variables changing together, they can only observe an *association* between them in which one changes as the other changes. It is impossible to rule out the possibility that both variables are independently responding to some third factor, called a *confounding variable,* and that the independent variable does not causally affect the dependent variable.

CORRELATION AND CORRELATION COEFFICIENTS

Linear regression analysis of a sample provides an estimate of how, on the average, a dependent variable changes when an independent variable changes and an estimate of the variability in the dependent variable about the line of means. These estimates, together with their standard errors, permit computing confidence intervals to show the certainty with which you can predict the value of the dependent variable for a given value of the independent variable. In addition, it is often desirable to quantify the *strength of the association* between two variables.

The *Pearson product-moment correlation coefficient,* a number between -1 and $+1$, is often used to quantify the strength of this association. The Pearson product-moment correlation coefficient r is defined as

$$r = \frac{\Sigma(\overline{X} - X)(\overline{Y} - Y)}{\sqrt{\Sigma(X - \overline{X})^2 \Sigma(Y - \overline{Y})^2}} \tag{2.19}$$

in which the sums are over all the observed (X, Y) points. The magnitude of r describes the strength of the association between the two variables, and the sign of r tells the direction of this association: $r = +1$ when the two variables increase together (Fig. 2-9A), and $r = -1$ when one decreases as the other increases (Fig. 2-9B). Figure 2-9C also shows the more common case of two variables that are correlated, though not perfectly. Figure 2-9D shows two variables that do not appear to relate to each other at all; $r = 0$. The tighter the linear relationship between the two variables, the closer the magnitude of r to 1; the weaker the linear relationship between the two variables, the closer r is to 0.

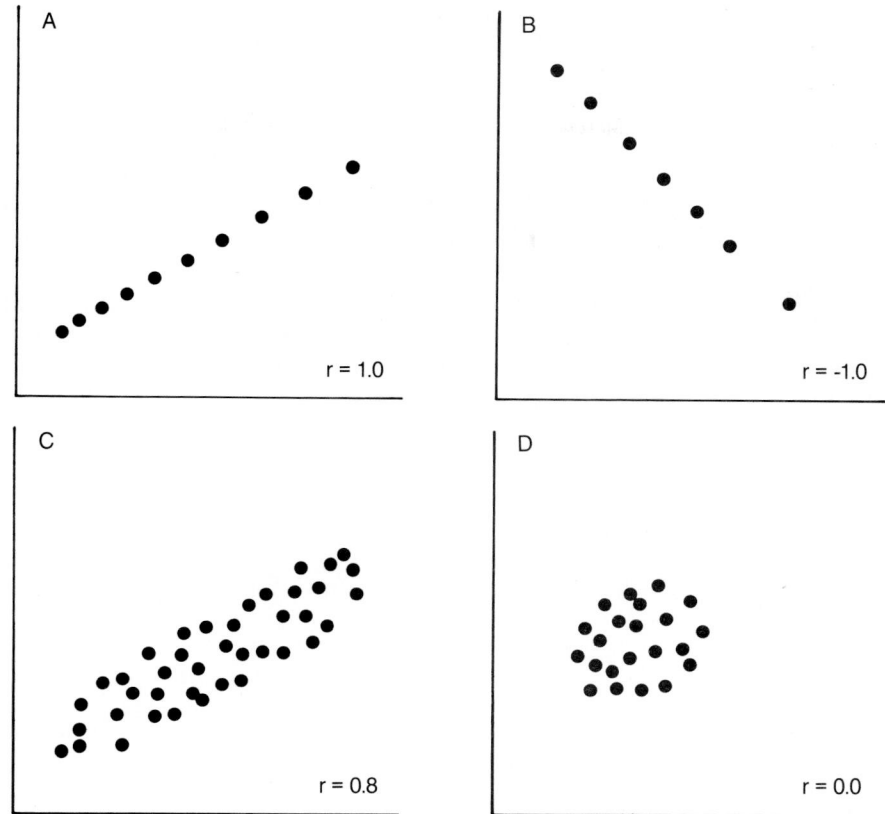

FIGURE 2-9 The closer the magnitude of the correlation coefficient is to 1, the less scatter there is in the relationship between the two variables. The closer the correlation coefficient is to 0, the weaker the relationship between the two variables.

The Relationship between Regression and Correlation

It is possible to compute a correlation coefficient for any data suitable for linear regression analysis.* In the context of regression analysis, it is possible to add to the meaning of the correlation coefficient. Recall that we selected the regression equation that minimized the sum of squared deviations between the points on the regression line and the value of the

*There are other ways to define a correlation coefficient that do not require as many assumptions as the Pearson product-moment correlation coefficient. For example, the *Spearman rank correlation coefficient* is used to quantify the strength of a trend between two variables that are measured on an *ordinal scale*. In an ordinal scale, responses can be graded, but there is no arithmetic relationship between the different possible responses. For a discussion of the Spearman rank correlation coefficient, see S. Glantz, *Primer of Biostatistics* (2nd ed.), New York, McGraw-Hill, 1987, pp. 230–236.

dependent variable at each observed value of the independent variable. It can be shown that the correlation coefficient also equals

$$r = \sqrt{1 - \frac{SS_{res}}{SS_{tot}}} \qquad (2.20)$$

When all the data points fall exactly on the regression line, there is no variation in the observations about the regression line, so $SS_{res} = 0$ and the correlation coefficient equals 1 (or -1), indicating the dependent variable can be predicted with certainty from the independent variable. On the other hand, when the residual variation about the regression line is the same as the variation about the mean value of the dependent variable, $SS_{res} = SS_{tot}$ and $r = 0$, indicating no trend in the data. When $r = 0$, the dependent variable cannot be predicted at all from the independent variable.

For example, we can compute the correlation coefficient using the sample of 10 points in Fig. 2-3B. (These are the same data used to illustrate the computation of the regression line in Table 2-1 and Fig. 2-5B.) Substitute from Table 2-1 into Eq. (2.20) to obtain

$$r = \sqrt{1 - \frac{7.44 \text{ g}^2}{51.8 \text{ g}^2}} = .925$$

The square of the correlation coefficient r^2 is known as the *coefficient of determination*. From Eq. (2.20)

$$r^2 = 1 - \frac{SS_{res}}{SS_{tot}} = \frac{SS_{reg}}{SS_{tot}}$$

Because SS_{tot} is a measure of the total variation in the dependent variable, people say that the coefficient of determination is the fraction of the total variance in the dependent variable "explained" by the regression equation. This is rather unfortunate terminology, because the regression line does not "explain" anything in the sense of providing a mechanistic understanding of the relationship between the dependent and independent variables. Nevertheless, the coefficient of determination is a good description of how well a straight line describes the relationship between the two variables.

The correlation coefficient is also related to the slope of the regression equation according to

$$r = b_1 \frac{s_X}{s_Y}$$

We can use the following intuitive argument to justify this relationship: When there is no relationship between the two variables under study, both the slope of the regression line and the correlation coefficient are zero. In fact, testing the hypothesis that the correlation coefficient is zero is exactly equivalent to testing the hypothesis that the slope of the regression line is zero.*

DOING REGRESSION AND CORRELATION ANALYSIS WITH A COMPUTER

By this point, you should be convinced that doing a regression or correlation analysis by hand is very tedious. This is even more so for multiple regression, which involves more than one independent variable. In fact, the widespread availability of computer programs to do multivariate analysis — not new theoretical breakthroughs — has been the key to the fact that these methods are gaining in popularity.

Because computer-based methods are essentially the only way to do multivariate analysis, we will concentrate on the meaning and interpretation of the results of such analysis (as opposed to the computational details) throughout the rest of this book. There are several widely available software packages used for multivariate analysis, including BMDP, Minitab, SAS, and SPSS. Because all these programs perform the same kinds of analysis, the inputs and outputs are, in theory, the same. There are, however, some differences in what they compute and how they present the results. To help you make the connection between these computer programs and the underlying concepts of multivariate analysis, we will be illustrating the examples in this book with the results the programs provide. Clearly understanding this connection is particularly important because, as the programs become more "user friendly," it has become easier for people who do not have a clue as to what the results mean to run the programs.

Figure 2-10 shows the output for all four of the programs listed above for the data in Table 2-1 relating Martian weight to Martian height, with all the statistics we have been discussing highlighted. (Appendix B contains the control statements used to generate these outputs and the outputs for all the remaining examples in this book.) All four programs present the slope and intercept of the regression equation and the analysis of variance table, although there are minor differences in the other information presented. For example, BMDP does not report the standard error of the intercept and associated t and P values, and Minitab reports the coefficient of determination r^2 but not the correlation coefficient r.

---

*For a detailed discussion of the relationship between hypothesis testing using the slope or correlation coefficient, including a proof that they are equivalent, see S. Glantz, *Primer of Biostatistics* (2nd ed.), New York, McGraw-Hill, 1987, pp. 228–230.

A **BMDP**

```
DEPENDENT VARIABLE. . . . . . . . . . . . . .      1 W
TOLERANCE . . . . . . . . . . . . . . . . . .  0.0100
ALL DATA CONSIDERED AS A SINGLE GROUP


MULTIPLE R           0.9254       STD. ERROR OF EST.        0.9643
MULTIPLE R-SQUARE    0.8564

ANALYSIS OF VARIANCE
                  SUM OF SQUARES    DF    MEAN SQUARE    F RATIO    P(TAIL)
    REGRESSION         44.3576       1      44.3576      47.706     0.0001
    RESIDUAL            7.4384       8       0.9298

                               STD.    STD. REG
    VARIABLE     COEFFICIENT    ERROR    COEFF      T    P(2 TAIL) TOLERANCE

 INTERCEPT          -6.00761
 H         2         0.4441    0.0643     0.93     6.91     0.00    1.0000
```

B **MINITAB**

```
The regression equation is
W = - 6.01 + 0.444 H

Predictor      Coef      Stdev    t-ratio        p
Constant     -6.008      2.392      -2.51    0.036
H           0.44411    0.06430       6.91    0.000

s = 0.9643       R-sq = 85.6%     R-sq(adj) = 83.8%

Analysis of Variance

SOURCE       DF          SS         MS         F        p
Regression    1      44.358     44.358     47.71    0.000
Error         8       7.438      0.930
Total         9      51.796
```

FIGURE 2-10 The results of a computer analysis of the data on Martian height and weight in Fig. 2-3 using the computer programs BMDP (*A*), Minitab (*B*), SAS (*C*), and SPSS (*D*). All four computer programs yield the same regression equation, sums of squares, and other statistics, although there are some differences in notation from program to program. The coefficients in the regression equation, associated standard errors, and tests of significance for the individual coefficients are highlighted in the output from each program.

C   **SAS**

DEP VARIABLE: W   Weight

ANALYSIS OF VARIANCE

SOURCE	DF	SUM OF SQUARES	MEAN SQUARE	F VALUE	PROB>F
MODEL	1	44.35755625	44.35755625	47.706	0.0001
ERROR	8	7.43844375	0.92980547		
C TOTAL	9	51.79600000			

ROOT MSE	0.9642642	R-SQUARE	0.8564
DEP MEAN	10.38	ADJ R-SQ	0.8384
C.V.	9.289636		

PARAMETER ESTIMATES

VARIABLE	DF	PARAMETER ESTIMATE	STANDARD ERROR	T FOR H0: PARAMETER=0	PROB > |T|
INTERCEP	1	-6.0076	2.39213153	-2.511	0.0363
H	1	0.44410849	0.06429857	6.907	0.0001

D   **SPSS**

Equation Number 1   Dependent Variable..   W   Weight

Beginning Block Number 1. Method: Enter   H

Variable(s) Entered on Step Number 1..   H   Height

		Analysis of Variance			
Multiple R	.92541		DF	Sum of Squares	Mean Square
R Square	.85639	Regression	1	44.35756	44.35756
Adjusted R Square	.83844	Residual	8	7.43844	.92981
Standard Error	.96426				

F =   47.70628   Signif F = .0001

------------------ Variables in the Equation ------------------

Variable	B	SE B	Beta	T	Sig T
H	.444108	.064299	.925413	6.907	.0001
(Constant)	-6.007603	2.392132		-2.511	.0363

FIGURE 2-10 (continued)

There are also differences in how things are labeled. For example,

- BMDP and SAS refer to the term b_0 in Eq. (2.4) as "intercept," whereas Minitab and SPSS refer to it as "constant."
- BMDP and Minitab label the coefficients in the regression equa-

tion "coefficient," SAS labels them "parameter estimate," and
SPSS labels them "B."

- BMDP and SAS label the standard errors of the regression coefficients "standard error," whereas Minitab labels them "standard
deviation" (presumably because the standard error is the standard
deviation of the distribution of all possible estimates of the
regression coefficient), and SPSS labels it "SE B."

- SAS refers to SS_{reg} and MS_{reg} as "model," whereas the other programs refer to these quantities as "regression."

- BMDP labels the standard error of the estimate $s_{y|x}$ "standard
error of the estimate," Minitab labels it "s," SAS labels it "root
MSE," and SPSS labels it "standard error."

Note that all these programs report the results to four to eight decimal places. The programs are written this way to permit analyzing problems for which the variables of interest take on a wide range of values,
from cell counts in the thousands to weights in small fractions of a gram.
The fact that the results are reported to so many decimal places does not
mean that the results have that level of precision. Thus, when reporting
the results of such computations, you should only report the number of
significant digits justified by the precision of the original measurements.

HEAT EXCHANGE IN GRAY SEALS

Linear regression and correlation analysis are based on the assumption
that if the two variables are related to each other, the relationship takes
the form of a consistently linear upward or downward trend. When the
relationship is not linear, both regression and correlation procedures can
provide misleading results or even fail to detect a relationship.

Warm-blooded animals that live in cold climates have developed special mechanisms to enable them to keep their body temperatures normal.
These mechanisms are especially well developed in marine mammals
that spend much of their life in cold water. Because water conducts heat
about 25 times better than air, life in cold water poses special problems
for those animals that need to maintain a constant body temperature.

Folkow and Blix[*] wanted to understand better the mechanisms that
aided the blood flow and body temperature regulation of the northern
gray seal. To get an overall look at body temperature regulation, they
calculated the seals' thermal conductivity (a measure of how easily the
body loses heat) at water temperatures between +20 and −40°C. Figure

[*]L. P. Folkow and A. S. Blix, "Nasal Heat and Water Exchange in Gray Seals," *Am. J.
Physiol.* 253:R883–R889, 1987.

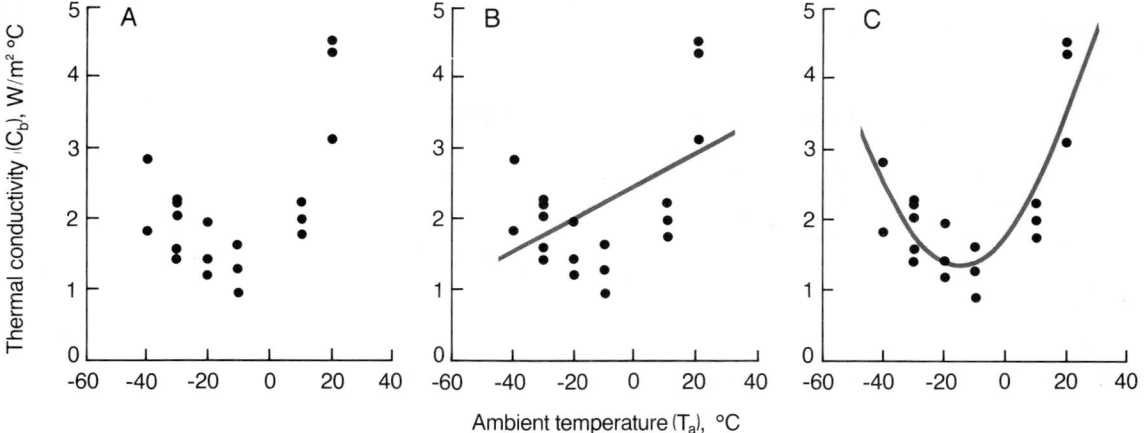

FIGURE 2-11 *A.* Data relating ambient air temperature (T_a) to total body conductance (C_b) in gray seals. Submitting these data to a simple linear regression analysis yields a linear relationship whose slope is significantly different from zero (see Fig. 2-12). Closer examination of these data, however, reveals that the relationship between total body conductance and temperature is not linear but rather initially falls with increasing temperature to a minimum around $-10°C$, then increases. *B.* The observed values of conductivity tend to be above the regression line at the low and high temperatures and below the regression line at the intermediate temperatures, violating the assumption of linear regression that the population members are distributed at random and with constant variance about the line of means. *C.* Allowing for the nonlinearity in the line of means leads to a much better pattern of residuals. We will return to this example in Chap. 3 to illustrate how to compute this nonlinear regression. (Data from Figure 3 of L. P. Folkow and A. S. Blix, "Nasal Heat and Water Exchange in Gray Seals," *Am. J. Physiol.* 253:R883–R889, 1987.)

2-11*A* shows the data relating body conductivity C_b to ambient temperature T_a. Figure 2-12 shows the results of using the simple linear regression

$$\hat{C}_b = C_0 + ST_a \tag{2.21}$$

to fit these data (the data are in Table C-1, Appendix C). The intercept C_0 is 2.41 W/(m^2 · °C). The slope of the line S—which gives the sensitivity of thermal conductivity to changes in ambient temperature—is .023 W/(m^2 · °C^2). Figure 2-11*B* shows this regression line.

The standard error of the slope is .009 W/(m^2 · °C^2). The associated value of the t test statistic to test the null hypothesis that the slope is zero is .023 [W/(m^2 · °C^2)]/.009 [W/(m^2 · °C^2)] = 2.535. This value exceeds the 5 percent critical value of the t distribution with 18 degrees of freedom, 2.101, so we conclude that thermal conductivity increases as temperature increases ($P < .05$). (The number of degrees of freedom is that associated with the error term in the analysis of variance table.) We

```
DEP VARIABLE: CB          Conductivity
                            ANALYSIS OF VARIANCE

                       SUM OF         MEAN
         SOURCE    DF   SQUARES       SQUARE      F VALUE     PROB>F

         MODEL      1   4.46921498   4.46921498    6.428     0.0207
         ERROR     18  12.51436002   0.69524222
         C TOTAL   19  16.98357500

             ROOT MSE     0.8338119     R-SQUARE    0.2631
             DEP MEAN     2.1025        ADJ R-SQ    0.2222
             C.V.        39.65811


                        PARAMETER ESTIMATES

                     PARAMETER       STANDARD      T FOR H0:
     VARIABLE  DF    ESTIMATE         ERROR       PARAMETER=0    PROB > |T|

     INTERCEP   1   2.40927210      0.22226552      10.840       0.0001
     TA         1   0.02272386      0.008962611      2.535       0.0207
```

FIGURE 2-12 Results of analyzing the data on heat exchange in gray seals in Fig. 2-11*A* using a simple linear regression. This analysis shows a statistically significant linear relationship between ambient temperature and body conductivity, although a closer examination of the quality of this fit reveals that there are systematic deviations from linearity (see Fig. 2-11*B* and *C*).

compute the F value for the overall goodness of fit of this regression equation using the values of MS_{reg} and MS_{res} reported in the analysis of variance table according to Eq. (2.18):

$$F = \frac{4.4692}{0.6952} = 6.43$$

This value exceeds the 5 percent critical value of F for 1 numerator and 18 denominator degrees of freedom ($F_{.05} = 4.41$), so we conclude that there is a statistically significant relationship between C_b and T_a, just as we concluded based on looking at the slope. As ambient temperature decreases, so does body conductivity; as it gets colder, the body makes adjustments so that it loses less heat.

This conclusion, however, is quite misleading. Closer examination of Fig. 2-11*B* reveals that there are systematic deviations between the regression line and the data. At low temperatures, the regression line systematically underestimates the data points, at intermediate temperatures the line systematically overestimates the data, and at high temperatures it again systematically underestimates the data. This situation violates one of the assumptions upon which linear regression is based—

that the members of the population are randomly distributed about the line of means.* A closer examination of the data in Fig. 2-11C reveals that as ambient temperature decreases, body conductivity does decrease, but only to a point. Below a certain temperature (about $-15°C$), body conductivity actually goes up again.† This result leads to a simple insight into the physiology. The seals can reduce heat loss from their bodies (i.e., lower their thermal conductivity) by decreasing blood flow to their skin and the underlying blubber. This tactic keeps heat in the blood deeper in the body and thus further from the cold water. However, at very low water temperatures (below about $-15°C$) the skin needs increased blood flow to keep it warm enough to protect it from freezing. This increase in blood flow to the skin is associated with an increase in heat loss. The simple linear regression in Fig. 2-11B would not have revealed this situation. Thus, although we found a statistically significant linear trend, this linear model does not appropriately quantify the relationship between body thermal conductivity and ambient temperature.

SUMMARY

Simple linear regression and correlation are techniques for describing the relationship between two variables—one dependent variable and one independent variable. The analysis rests on several assumptions, most notably that the relationship between these two variables is fundamentally linear, that the members of the population are distributed normally about the line of means, and that the variability of members of the population about the line of means does not depend on the value of the independent variable. When the data behave in accordance with these assumptions, the procedures we have developed yield a reasonable description of the data and make it possible to test whether or not there is, in fact, likely to be a relationship between the dependent and independent variables.

This material is important because it lays the foundation for the rest of this book. There is, in fact, no fundamental difference between the ideas and procedures we have just developed for the case of simple linear regression and those associated with multiple regression. The only difference is that in the latter there is more than one independent variable.

In addition to outlining the procedures used to describe the data and compute the regression line, we have also seen the importance of checking that the data are consistent with the assumptions made at the outset of the regression analysis. These assumptions must be reasonably satis-

*We formally discuss analysis of the residuals between the regression line and data in Chap. 4.

†Chapter 3 shows how to compute the nonlinear regression shown in Fig. 2-11C.

fied for the results to be meaningful. Unfortunately, this simple truth is often buried under pounds of computer printout, especially when dealing with complicated multiple regression problems. The example just done illustrates an important rule that should be scrupulously observed when using any form of regression or correlation analysis: Do not just look at the numbers. *Always look at a plot of the raw data* to make sure the data are consistent with the assumptions behind the method of analysis.

PROBLEMS

2.1 From the regression outputs shown in Fig. 2-10A to D relating Martian weight to height, find the following statistics to three significant figures: slope, standard error of the slope, intercept, standard error of the intercept, correlation coefficient, standard error of the estimate, and the F statistic for overall goodness of fit.

2.2 Benzodiazepine tranquilizers (such as Valium) exert their physiological effects by binding to specific receptors in the brain. This binding then interacts with a neurotransmitter, γ-amino butyric acid (GABA), to cause changes in nerve activity. Because most direct methods of studying the effects of receptor binding are not appropriate for living human subjects, Hommer and his coworkers[*] sought to study the effects of different doses of Valium on various readily measured physiological variables. They then looked at the correlations among these variables to attempt to identify those that were most strongly linked to the effect of the drug. Two of these variables were the sedation state S induced by the drug and the blood level of the hormone cortisol C (the data are in Table D-1, Appendix D). Is there a significant correlation between these two variables? (Use S as the dependent variable.)

2.3 The Surgeon General has concluded that breathing secondhand cigarette smoke causes lung cancer in nonsmokers, largely based on epidemiological studies comparing lung cancer rates in nonsmoking women married to nonsmoking men with those married to smoking men. An alternative approach to seeking risk factors for cancer is to see whether there are differences in the rates of a given cancer in different countries where the exposures to a potential risk factor differ. If the cancer rate varies with the risk factor across different countries, then the data suggest that the risk factor causes the cancer. This is the approach we used in Fig. 2-8 to associate animal, but not vegetable, fat with breast cancer. To investigate whether or not in-

[*]D. W. Hommer, V. Matsuo, O. Wolkowitz, G. Chrousos, D. J. Greenblatt, H. Weingartner, and S. M. Paul, "Benzodiazepine Sensitivity in Normal Human Subjects," *Arch. Gen. Psychiatry* 43:542–551, 1986.

voluntary smoking was associated with breast cancer, Horton[*] collected data on female breast cancer and male lung cancer rates in 36 countries (the data are in Table D-2, Appendix D). Because, on a population basis, virtually all lung cancer is due to smoking, Horton took the male lung cancer rate as a measure of exposure of women to secondhand tobacco smoke. Is female breast cancer associated with male lung cancer rates? In other words, is there evidence that breathing secondhand tobacco smoke causes breast cancer?

2.4 When antibiotics are given to fight infections, they must be administered in such a way that the blood level of the antibiotic is high enough to kill the bacteria causing the infection. Because antibiotics are usually given periodically, the blood levels change over time, rising after an injection, then falling back down until the next injection. The interval between injections of recently introduced antibiotics has been determined by extrapolating from studies of older antibiotics. To update the knowledge of dosing schedules, Vogelman and coworkers[†] studied the effect of different dosing intervals on the effectiveness of several newer antibiotics against a variety of bacteria in mice. One trial was the effectiveness of gentamicin against the bacterium *Escherichia coli*. As part of their assessment of the drug, they evaluated the effectiveness of gentamicin in killing *E. coli* as a function of the percentage of time the blood level of the drug remained above the effective level (the so-called mean inhibitory concentration M). Effectiveness was evaluated in terms of the number of bacterial colonies C that could be grown from the infected mice after treatment with a given dosing schedule (known as colony forming units CFU); the lower the value of C, the more efficacious the antibiotic. Table D-3, Appendix D, contains C and M values for two different dosing intervals, those with a code of 0 were obtained with dosing intervals of every 1 to 4 hours, whereas those with a code of 1 were obtained with dosing intervals of 6 to 12 hours. Analyze C as a function of M for the data obtained using 1- to 4-hour dosing intervals. Include the 95 percent confidence intervals for parameter estimates in your answer. Does this regression equation provide a significant fit to the data?

2.5 The kidney is important for eliminating substances from the body and for controlling the amount of fluid in the body, which, in turn, helps regulate blood pressure. One way the kidney senses blood volume is by the concentration of ions in the blood. When solutions

[*]A. W. Horton, "Indoor Tobacco Smoke Pollution. A Major Risk Factor for Both Breast and Lung Cancer?" *Cancer* 62:6–14, 1988.

[†]B. Vogelman, S. Gudmundsson, J. Leggett, J. Turnidge, S. Ebert, and W. A. Craig, "Correlation of Antimicrobial Pharmacokinetic Parameters with Therapeutic Efficacy in an Animal Model," *J. Infect. Dis.* 158:831–847, 1988.

with higher than normal concentrations of ions are infused into kidneys, a dilation of the renal arteries occurs (there is a decrease in renal vascular resistance). In addition, the kidney produces renin, a hormone that, among other actions, initiates a cascade of events to change vascular resistance by constricting or dilating arteries. Thus, the kidneys help control blood pressure both by regulating the amount of fluid in the vascular system and by direct action on the resistance of arteries. Because blood pressure is so highly regulated, it is reasonable to hypothesize that these two ways of regulating blood pressure interact. To study this potential interaction, Wilcox and Peart[*] wanted to see if the change in renal vascular resistance caused by infusing fluids with high concentrations of ions was related to the production of renin by the kidneys. To do this, they studied the effect of infusions of four solutions of different ion concentration on renal vascular resistance and made simultaneous measurements of renin production by the kidney. Using the raw data in Table D-4, Appendix D, the mean ($\pm$ SD) changes in renal vascular resistance and renin in four groups are

Fluid Infused	Group	Change in Renal Resistance ΔR, mmHg/(mL · min · kg)	Change in Renin Δr, ng/(mL · h)	n
NaCl plus arachiodonic acid	1	2.35 ± 2.25	−10.21 ± 9.13	8
NaCl	2	1.60 ± 1.48	−1.08 ± 4.21	9
Dextrose	3	−1.70 ± 1.14	6.53 ± 5.89	8
Na acetate	4	−3.38 ± 1.38	11.28 ± 11.35	5

A. Do these mean values indicate a significant relationship between the changes in renal resistance and the change in renin production? Include a plot in your analysis. *B.* Using the raw data in Table D-4, Appendix D, evaluate the relationship between the changes in renal resistance and renin, including a plot of the data. *C.* What features of the regression analysis changed when you used all the data compared to when you used the mean values? Did your overall conclusion change? Why or why not? *D.* Which one of these two analyses is incorrect, and why?

2.6 When patients are unable to eat for long periods, they must be given intravenous nutrients, a process called parenteral nutrition. Unfortunately, patients on parenteral nutrition show increased calcium loss via their urine, sometimes losing more calcium than they are

[*]C. S. Wilcox and W. S. Peart, "Release of Renin and Angiotensin II into Plasma and Lymph during Hyperchloremia," *Am. J. Physiol.* 253:F734–F741, 1987.

given in their intravenous fluids. Such a calcium loss might contribute to bone loss as the body pulls calcium out of bones to try to keep the calcium level in the blood within the normal range. In order to better understand the mechanisms of the calcium loss in the urine, Lipkin and his coworkers[*] measured urinary calcium U_{Ca} and related it to dietary calcium D_{Ca}, dietary protein level D_p, urinary sodium U_{Na}, and glomerular filtration rate G_{fr}, which is a measure of kidney function. These data are in Table D-5, Appendix D. Using simple (univariate) regression analyses, is there any evidence that U_{Ca} relates to any of these four other variables? What can you tell about the relative importance of these four variables in determining U_{Ca}?

2.7 There are two basic types of skeletal muscle in the body, fast-twitch and slow-twitch. Fast-twitch muscles are more suited to providing rapid, but short, bursts of activity, whereas slow-twitch muscles are more suited for slower, sustained activity. Different people have different mixes of these two types of muscle; for example, sprinters typically have about 75 percent fast-twitch fibers in their thigh muscles, and long-distance runners typically have only about 30 percent fast-twitch fibers. In order to predict the aptitude of athletes for certain types of activity, a small piece of muscle is often analyzed biochemically to determine the proportion of the two fibers. A new method of studying molecules in the body without the need to take a piece of tissue is magnetic resonance (MR) spectroscopy. MR spectroscopy works because a pulse of strong electromagnetic energy causes the nuclei in molecules to line up in the same direction. However, the nuclei quickly "relax" by losing energy to return to their normal state of random alignments. Different molecules have different behavior, and by monitoring characteristics of the relaxation, much can be learned about biochemical processes in the body. Spectroscopists typically measure two different relaxation times, one in the direction of the magnetic field of the MR spectrometer—the longitudinal relaxation time, or T_1—and one perpendicular to the direction of the magnetic field—the transverse relaxation time, or T_2. Different types or assemblages of molecules are characterized by their relaxation times. Kuno and coworkers[†] postulated that muscle fiber type F, expressed as a percentage of fast-twitch fibers, could be predicted using MR spectroscopy. *A.* How well is F predicted by T_1 (the data are in Table D-6, Appendix D)? *B.* Does the regression equation make physiological sense?

[*]E. W. Lipkin, S. M. Ott, C. H. Chesnut III, and A. Chait, "Mineral Loss in the Parenteral Nutrition Patient," *Am. J. Clin. Nutr.* 47:515–523, 1988.

[†]S. Kuno, S. Katsuta, T. Inouye, I. Anno, K. Matsumoto, and M. Akisada, "Relationship between MR Relaxation Time and Muscle Fiber Composition," *Radiology* 169:567–568, 1988.

Regression with Two or More Independent Variables

Chapter 2 laid the foundation for our study of multiple linear regression by showing how to fit a straight line through a set of data points to describe the relationship between a dependent variable and a single independent variable. All this effort may have seemed somewhat anticlimactic after all the arguments in Chap. 1 about how, in many analyses, it was important to consider the simultaneous effects of several independent variables on a dependent variable. We now extend the ideas of simple (one independent variable) linear regression to multiple linear regression, when there are several independent variables. The ability of a multiple regression analysis to quantify the relative importance of several (sometimes competing) possible independent variables makes multiple regression analysis a powerful tool for understanding complicated problems, such as commonly arise in biology, medicine, and the health sciences.

We will begin by moving from the case of simple linear regression—when there is a single independent variable—to multiple regression with two independent variables because the situation with two independent variables is easy to visualize. Next, we will generalize the results to any number of independent variables. We will also show how you can use multiple linear regression to describe some kinds of nonlinear (i.e., not proportional to the independent variable) effects as well as the effects of so-called interaction between the variables, when the effect of one independent variable depends on the value of another. We will present the statistical procedures that are direct extensions of those presented for simple linear regression in Chap. 2 to test the overall goodness of fit between the multiple regression equation and the data as well as the contributions of the individual independent variables to determining the value of the dependent variable. These tools will enable you to approach a wide variety of practical problems in data analysis.

WHAT WE REALLY DID ON MARS

When we first visited Mars in Chap. 1, we found that we could relate how much Martians weighed W to their height H and the number of cups of canal water C they consumed daily. Figure 3-1A shows the data we collected (reproduced from Fig. 1-2B). As we discussed in Chap. 1, because there are three discrete levels of water consumption ($C = 0$, 10, or 20 cups/day), we could use Eq. (1.2)

$$\hat{W} = -1.2 \text{ g} + .28 \text{ g/cm } H + .11 \text{ g/cup } C \tag{3.1}$$

to draw three lines through the data in Fig. 3-1A, depending on the level of water consumption. In particular, we noted that for Martians who did not drink at all, on the average, the relationship between height and weight was

$$\hat{W} = -1.2 \text{ g} + .28 \text{ g/cm } H + .11 \text{ g/cup} \cdot 0 \text{ cups} = -1.2 \text{ g} + .28 \text{ g/cm } H$$

and for Martians who drank 10 cups of canal water per day, the relationship was

$$\hat{W} = -1.2 \text{ g} + .28 \text{ g/cm } H + .11 \text{ g/cup} \cdot 10 \text{ cups} = -.1 \text{ g} + .28 \text{ g/cm } H$$

and, finally, for Martians who drank 20 cups of canal water per day, the relationship was

$$\hat{W} = -1.2 \text{ g} + .28 \text{ g/cm } H + .11 \text{ g/cup} \cdot 20 \text{ cups} = 1.0 \text{ g} + .28 \text{ g/cm } H$$

Figure 3-1B shows an alternative presentation of these data as a three-dimensional graph. The vertical and horizontal axes represent weight and height, as before, but we have added a third axis representing the number of cups of water consumed. The data collected at each level of water consumption cluster along a line in this three-dimensional space.

Closer examination of Fig. 3-1B reveals that these data not only cluster along the three lines, but also that the three lines all lie in a single *plane*, shown in Fig. 3-1C. Equation (3.1) describes this plane. In fact, the three lines in Fig. 3-1A can be thought of as contour lines which describe the projection of the three-dimensional plane in Fig. 3-1C into the H-W plane, just as contour lines are used on a topographical map to indicate the elevation of geographical features. Moreover, there is nothing about Eq. (3.1) which restricts the values of C to be 0, 10, and 20 cups/day, so, within the range of our data, we can reasonably expect this equation to describe the relationship between weight, height, and water consumption of, say, 3, 5, or 12.6 cups/day.

In fact, just as one does a simple linear regression by selecting the two coefficients in the equation of a straight line (which relates a dependent variable to a single independent variable) that minimizes the sum of squared deviations between the points on the line and the ob-

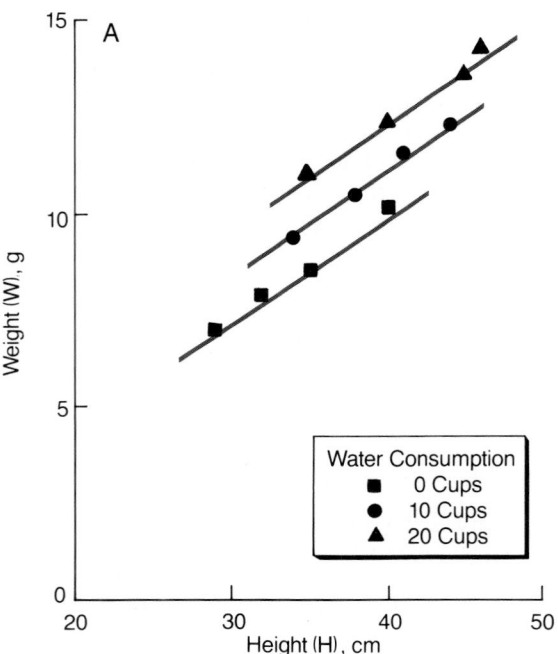

FIGURE 3-1 Three different ways of looking at the data on Martian height, water consumption, and weight in Fig. 1-2. Panel *A* reproduces the simple plot of height against weight with different water consumptions indicated with different symbols from Fig. 1-2*B*. Panel *B* shows an alternative way to visualize these data, with height and water consumption plotted along two axes. Viewed this way, the data points fall along three lines in a three-dimensional space. The problem of computing a multiple linear regression equation to describe these data is that of finding the "best" plane through these data which minimizes the sum of square deviations between the predicted value of weight at any given height and water consumption and the observed value (panel *C*). The three lines drawn through the data in panel *A* simply represent contour lines on the plane in panel *C* at constant values of water consumption of 0, 10, and 20 cups per day.

served data points, one does a multiple linear regression by selecting the coefficients in the equation of a plane (which relates a dependent variable to several independent variables) that minimizes the sum of squared deviations between the points on the plane and the observed data points. When there are two independent variables, as in the example we have been discussing, it is easy to visualize this plane in three dimensions (one dimension for the dependent variable and one for each of the two independent variables). When there are more than two independent variables, the ideas remain the same and the equations we will develop for two variables can be generalized directly, but the number of dimensions increases so it becomes difficult to draw pictures.

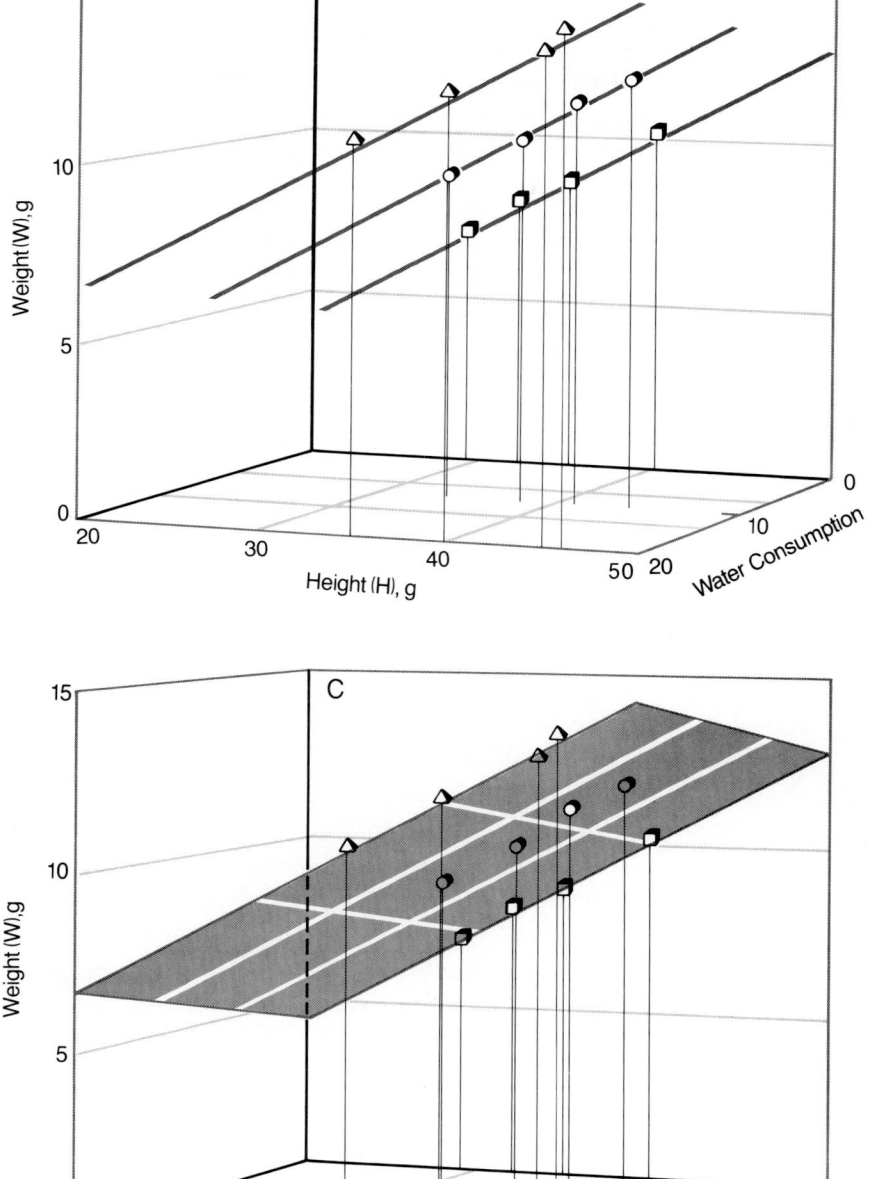

FIGURE 3-1 (*continued*)

HOW TO FIT THE BEST PLANE THROUGH A SET OF DATA

As before, we begin by specifying the population from which the observed sample was drawn. When we did a simple linear regression, we specified that the population was one in which the values associated with individual members of the population varied normally about a line of means given by Eq. (2.2):

$$y = \beta_0 + \beta_1 x + \epsilon$$

in which the values of ϵ vary randomly with mean equal to zero and standard deviation $\sigma_\epsilon = \sigma_{y|x}$. We now make the analogous specification for a population in which the values of the dependent variable y associated with individual members of the population are distributed about a *plane of means* which depends on the two independent variables x_1 and x_2 according to

$$y = \beta_0 + \beta_1 x_1 + \beta_2 x_2 + \epsilon \qquad (3.2)$$

As before, at any given values of the independent variables x_1 and x_2, the random component ϵ is normally distributed with mean zero and standard deviation $\sigma_\epsilon = \sigma_{y|x}$, which does not vary with the values of the independent variables x_1 and x_2.

As with simple linear regression, the population characteristics are

- *The mean of the population of the dependent variable at a given set of values of the independent variables varies linearly as the values of the independent variables vary.*
- *For any given values of the independent variables, the possible values of the dependent variable are distributed normally.*
- *The standard deviation of the dependent variable about its mean at any given values of the independent variables is the same for all values of the independent variables.*
- *The deviations of all members of the population from the plane of means are statistically independent, i.e., the deviation associated with one member of the population has no effect on the deviations associated with other members.*

The parameters of this population are β_0, β_1, and β_2, which define the plane of means, the dependent-variable population mean at each set of values of the independent variables, and $\sigma_{y|x}$, which describes the variability about the plane of means. Now we turn our attention to the problem of estimating these parameters from samples drawn from such populations and testing hypotheses about these parameters.

Computing the Regression Coefficients

We will estimate the plane of means with the *regression equation*

$$\hat{y} = b_0 + b_1 x_1 + b_2 x_2 \tag{3.3}$$

in which the values of b_0, b_1, and b_2 are estimates of the parameters β_0, β_1, and β_2, respectively. As in simple linear regression, we will compute the values of b_0, b_1, and b_2 that minimize the *residual sum of squares* SS_{res}, the sum of squared differences between the observed value of the dependent variable Y, and the value predicted by Eq. (3.3) at the corresponding observed values of the independent variables X_1 and X_2, $\hat{y}(X_1, X_2)$:

$$SS_{res} = \Sigma[Y - \hat{y}(X_1, X_2)]^2 = [Y - (b_0 + b_1 X_1 + b_2 X_2)]^2 \tag{3.4}$$

The estimates of the regression parameters which minimize this residual sum of squares are similar in form to the equations for the slope and intercept of the simple linear regression line given by Eqs. (2.5) and (2.6).[*] From this point forward, we will leave the arithmetic involved in solving these equations to a digital computer.

We are now ready to do the computations to fit the best regression plane through the data in Fig. 3-1 (and Table 1-1) that we collected on Mars. We will use the regression equation

$$\hat{W} = b_0 + b_H H + b_C C$$

Here we denote the dependent variable, weight, with W rather than y and the independent variables, height and cups of water, with H and C rather than x_1 and x_2. Figure 3-2 shows the results of fitting this equation to the data in Table 1-1. In this output, b_0 is labeled "Constant" and equals -1.2202. Likewise, the parameters b_H and b_C, labeled H and C, are 0.2834 and 0.1112, respectively. Rounding the results to an appropriate number of significant digits and adding the units of measurement yields $b_0 = -1.2$ g, $b_H = .28$ g/cm, and $b_C = .11$ g/cup, which is where we got the numbers in Eqs. (1.3) and (3.1).

Let us pause and consider the physical interpretation of these regression coefficients. The fact that the three lines in Fig. 3-1A corresponding to the three levels of water consumption had the same slope (i.e., were parallel) illustrated the point that the effect of changes in height on weight were the same at each level of water consumption. When we

---

[*]To derive these equations, take the partial derivatives of Eq. (3.4) with respect to b_0, b_1, and b_2, set these derivatives equal to zero, and solve the resulting simultaneous equations. The actual equations are derived, using matrix notation, later in this chapter.

```
The regression equation is
W = - 1.22 + 0.283 H + 0.111 C

Predictor         Coef        Stdev     t-ratio         p
Constant       -1.2202       0.3210       -3.80     0.004
H             0.283436      0.009142       31.00     0.000
C             0.111212      0.005748       19.35     0.000

s = 0.1305       R-sq = 99.7%      R-sq(adj) = 99.7%

Analysis of Variance

SOURCE           DF          SS          MS          F          p
Regression        2      54.213      27.107    1591.79      0.000
Error             9       0.153       0.017
Total            11      54.367

SOURCE           DF      SEQ SS
H                 1      47.839
C                 1       6.375
```

FIGURE 3-2 The results of fitting the data in Fig. 3-1 with the multiple linear regression model given by Eq. (3.1). The fact that the coefficients associated both with Martian height H and water consumption C are significantly different from zero means that the regression plane in Fig. 3-1C is tilted with respect to both the H and C axes.

replotted the data in Fig. 3-1A in Fig. 3-1B, it was to move toward the idea of the regression plane describing the relationship between several independent variables and a single dependent variable. We can now reexamine this figure from a slightly different perspective to illustrate further the meaning of the regression coefficients.

The points collected at each level of water consumption can be thought of falling along a line that represents the intersection of the regression plane and a plane parallel to the H-W plane *at a constant value of C*. This line represents the dependence of W on changes in H while *holding C constant*. It happens that we have illustrated these intersections at three specific values of C (0, 10, and 20 cups/day), but we could have put this "cutting plane" at any value of C. Regardless of where we put this plane, the slope of the line with respect to changes in H would have been the same. Thus, b_H, whose value is .28 g/cm, is the increase in weight for each unit of increase in height *holding water consumption constant*. Likewise, b_C, whose value is .11 g/cup, is the increase in weight for each unit of increase in water consumption *holding the height constant*. This discussion illustrates the fact that, in multiple linear regression, the regression coefficients indicate the sensitivity of the dependent variable to change in each independent variable, while holding the other variables constant.*

Variability about the Regression Plane

We now have the regression equation to estimate the location of the plane of means but still need an estimate of the variability of members of the population about this plane $\sigma_{y|x}$. As with simple linear regression, we estimate this parameter as the square root of the "average" squared deviation of the data from the regression plane:

$$s_{y|x} = \sqrt{\frac{\Sigma(Y - \hat{y})^2}{n - 3}} = \sqrt{\frac{[Y - (b_0 + b_1X_1 + b_2X_2)]^2}{n - 3}} = \sqrt{\frac{SS_{res}}{n - 3}} = \sqrt{MS_{res}} \qquad (3.5)$$

As before, this statistic is known as the *standard error of the estimate*.[†] The residual mean square is simply the square of the standard error of the estimate, which is the variance of the data points around the regression plane.

$$s_{y|x}^2 = \frac{.153\ g^2}{12 - 3} = .017\ g^2$$

and the standard error of the estimate is $s_{y|x} = .131$ g. All these values appear in the computer output in Fig. 3-2.

Standard Errors of the Regression Coefficients

The coefficients b_i that we compute when fitting a regression equation to a set of data are estimates of the parameters β_i. Like all statistics, these estimates depend on the specific members of the population that happen to have been selected in our random sample; different samples will yield different values of the b_i. Like all statistics, the possible values of b_i will have some distribution. We assumed that the underlying population is

---

*One can formally see that the regression coefficients equal the change in the predicted value of the dependent variable per change in the independent variable, holding all others constant by taking the partial derivative of Eq. (3.3) with respect to the independent variables: $\partial y/\partial x_i = b_i$.

[†]We divide by $n - 3$ because there are three parameter estimates in the regression equation $(b_0, b_1,$ and $b_2)$, so there are $n - 3$ degrees of freedom associated with the residual sum of squares. This situation compares with simple linear regression where we divided by $n - 2$ because there were two parameter estimates in the regression equation $(b_0$ and $b_1)$ and by $n - 1$ when computing the standard deviation of a sample, where there is one parameter estimate (the mean). These adjustments in the degrees of freedom are necessary to make $s_{y|x}$ an unbiased estimator of $\sigma_{y|x}$.

normally distributed about the plane of means; therefore, of all possible values of each of the estimated regression coefficients, b_i will be distributed normally with a mean value equal to β_i and a standard deviation s_{b_i}. s_{b_i} is the *standard error of* b_i and is equal to

$$s_{b_i} = \sqrt{\frac{s_{y|x}^2}{\Sigma(X_i - \overline{X}_i)^2 \, (1 - r_{x_1 x_2}^2)}} = \sqrt{\frac{MS_{res}}{\Sigma(X_i - \overline{X}_i)^2 \, (1 - r_{x_1 x_2}^2)}} \tag{3.6}$$

where $r_{x_1 x_2}$ is the correlation between the two independent variables x_1 and x_2.*

We can use the standard errors to test whether or not each regression coefficient is significantly different from zero using the t test statistic:

$$t = \frac{b_i}{s_{b_i}}$$

with $n - 3$ degrees of freedom. Likewise, we can use the standard errors to compute the $100(1 - \alpha)$ percent confidence interval[†] for the parameter β_i with

$$b_i - t_\alpha s_{b_i} < \beta_i < b_i + t_\alpha s_{b_i}$$

In addition, if this confidence interval does not include zero, we can conclude that b_i is significantly different from zero with $P < \alpha$.

The computer output in Fig. 3-2 contains the standard errors of the coefficients associating Martian weight with height and water in Eq. (3.1): $s_{b_H} = .00914$ g/cm and $s_{b_c} = .00575$ g/cup. The critical value of t for $\alpha = .05$ with $\nu = 12 - 3 = 9$ degrees of freedom is 2.262, so the 95 percent confidence interval for β_H is

$$b_H - t_{.05} s_{b_H} < \beta_H < b_H + t_{.05} s_{b_H}$$

$$.2834 \text{ g/cm} - 2.262 \cdot .00914 \text{ g/cm} < \beta_H < .2834 \text{ g/cm} + 2.262 \cdot .00914 \text{ g/cm}$$

$$.263 \text{ g/cm} < \beta_H < .304 \text{ g/cm}$$

and for β_C is

$$b_C - t_{.05} s_{b_c} < \beta_C < b_C + t_{.05} s_{b_c}$$

$$.1112 \text{ g/cup} - 2.262 \cdot .00575 \text{ g/cup} < \beta_C < .1112 \text{ g/cup} + 2.262 \cdot .00575 \text{ g/cup}$$

$$.098 \text{ g/cup} < \beta_C < .124 \text{ g/cup}$$

*The equation for the standard error of the slope of a simple linear regression [given by Eq. (2.8)] is just a special case of Eq. (3.6) with $r_{x_1 x_2} = 0$.

[†]This equation yields the confidence interval for each parameter, one at a time. It is also possible to define a *joint confidence region* for several parameters simultaneously. For a discussion of joint confidence regions, see J. Neter, W. Wasserman, and M. H. Kutner, *Applied Linear Regression Models*, Homewood, Ill., Irwin, 1983, pp. 147–154.

Neither of these 95 percent confidence intervals includes zero, so we conclude that both height and water consumption affect Martian weight $(P < .05)$.

We can also test whether these coefficients are significantly different from zero by computing

$$t_H = \frac{b_H}{s_{b_H}} = \frac{.2834 \text{ g/cm}}{.00914 \text{ g/cm}} = 31.00$$

and

$$t_C = \frac{b_C}{s_{b_C}} = \frac{.1112 \text{ g/cup}}{.00575 \text{ g/cup}} = 19.35$$

(In fact, the computer already did this arithmetic too.) Both t_H and t_C exceed 4.781, the critical value of t for $P < .001$ with 9 degrees of freedom, so we conclude that both height and water consumption exert independent effects on the weight of Martians. One of the great values of multiple regression analysis is to be able to evaluate simultaneously the effects of several factors and use the tests of significance of the individual coefficients to identify the important determinants of the dependent variable.

Muddying the Water: Multicollinearity

Note that the standard error of a regression coefficient depends on the correlation between the *independent* variables $r_{x_1 x_2}$. If the independent variables are uncorrelated (i.e., statistically independent of each other), the standard errors of the regression coefficients will be at their smallest. In other words, you will obtain the most precise possible estimate of the true regression equation. On the other hand, if the independent variables depend on each other (i.e., are correlated), the standard errors can become very large, reflecting the fact that the estimates of the regression coefficients are quite imprecise. This situation, called *multicollinearity*, is one of the potential pitfalls in conducting a multiple regression analysis.*

The qualitative reason for this result is as follows: When the independent variables are correlated, it means that knowing the value of one variable gives you information about the other. This implicit information about the second independent variable in turn provides some information about the dependent variable. This redundant information makes it difficult to assign the effects of changes in the dependent variable to one or the other of the independent variables and so reduces the precision with

*Techniques for detecting and dealing with multicollinearity will be treated in detail in Chap. 5.

which we can estimate the parameters associated with each independent variable.

To understand how multicollinearity introduces uncertainty into the estimates of the regression parameters, consider the extreme case in which the two independent variables are *perfectly* correlated. In this case, knowing the value of one of the independent variables allows you to compute the precise value of the other, so the two variables contain totally redundant information. Figure 3-3A shows this case. Because the independent variables x_1 and x_2 are perfectly correlated, the projections of the data points fall along a straight line in the x_1-x_2 plane. The regression problem is now reduced to finding the *line* in the plane defined by the data points and the projected line in the x_1-x_2 plane (the light line in Fig. 3-3A) which minimizes the sum of squared deviations between the line and the data. Note that the problem has been reduced from the three-dimensional one of fitting a (three-parameter) plane through the data to the two-dimensional problem of fitting a (two-parameter) line. *Any regression plane containing the line of points will yield the same* SS_{res} (Fig. 3-3B), and there are an infinite number of such planes. Because each plane is associated with a unique set of values of b_0, b_1, and b_2, there are an infinite number of combinations of these coefficients that will describe these data equally well. Moreover, because the independent variables are perfectly correlated, $r_{x_1 x_2} = 1$, the denominator in Eq. (3.6) equals zero, and the standard errors of the regression coefficients become infinite. One cannot ask for less precise estimates.

The extreme case shown in Fig. 3-3 never arises in practice. However, it is not uncommon for two (or more) predictor variables to vary together (and so contain some redundant information), giving the situation in Fig.

FIGURE 3-3 When one collects data on a dependent variable y that is determined by the values of two independent variables x_1 and x_2, but the independent variables change together in a precise linear manner, it becomes impossible to attribute the effects of changes in the dependent variable y uniquely to either of the independent variables (A). This situation, called *multicollinearity*, occurs when the two independent variables contain the same information; knowing one of the independent variables tells you the value of the other one. The independent variables are perfectly correlated and so the values of the independent variables corresponding to the observed values of the dependent variables fall along the line in the x_1-x_2 plane defined by the independent variables. The regression plane must go through the points that lie in the plane determined by the data and the line in the $x_1 - x_2$ plane. There are an infinite number of such planes, each with different values of the regression coefficients, that describe the relationship between the dependent and independent variable equally well (B). There is not a broad "base" of values of the independent variables x_1 and x_2 to "support" the regression plane. Thus, when perfect multicollinearity is present, the uncertainty associated with the estimates of the parameters in the regression equation is infinite, and regression coefficients cannot be reliably computed.

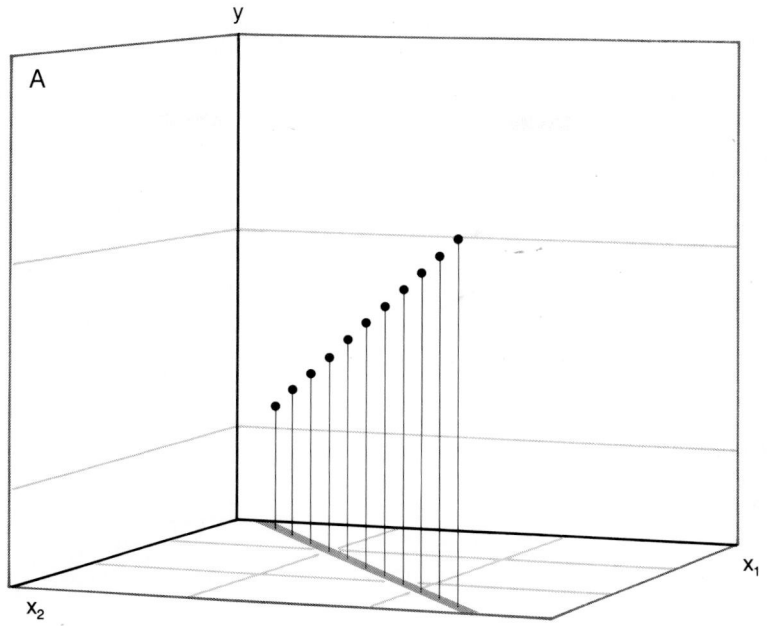

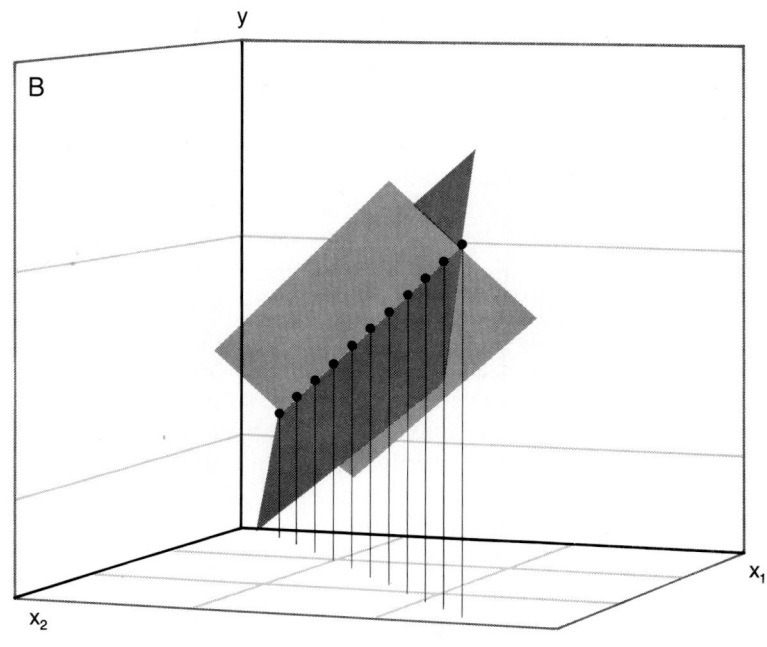

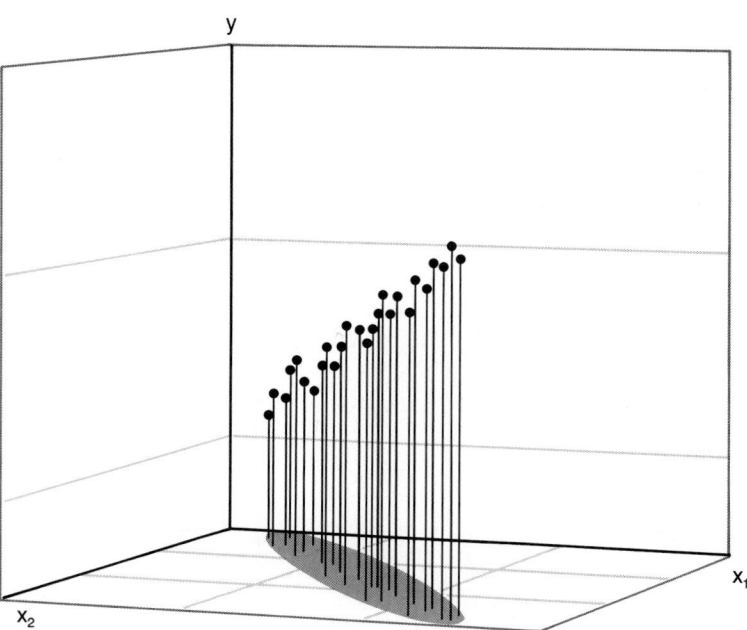

FIGURE 3-4 The extreme multicollinearity shown in Fig. 3-3 rarely occurs. The more common situation is when at least two of the independent variables are highly, but not perfectly, correlated, so the data can be thought of falling in a cigar-shaped cloud rather than in a line (compare with Fig. 3-3*A*). In such a situation it is possible to compute regression coefficients. However, there is still great uncertainty in these estimates, reflected by large standard errors of these coefficients, because much of the information contained in each of the independent variables is in common with the other.

3-4. We have more information about the location of the regression plane, but we still have such a narrow "base" on which to estimate the plane that it becomes very sensitive to the differences in the observations between the different possible samples. This uncertainty in the precise location of the plane is reflected in the large values of the standard errors of the regression parameters.

The usual source of multicollinearity is the inability to manipulate independently all the independent variables. This situation arises in observational studies (such as are common in epidemiology and economics) or in experimental situations when the independent variables are, in fact, coupled. For example, when studying in vivo physiological systems, it is common to perturb the system, measure the resulting changes in several variables, and then relate one of these variables to several of the others. For example, in studies of cardiac function, a common experiment is to constrict a large blood vessel near the heart and then relate the resulting change in the output of the heart (as the dependent variable) to changes in other hemodynamic variables, including heart rate, heart size and pressure at the beginning of the contraction, and the pressure in the aorta that the heart must pump against (as independent variables). In such studies the intervention causes the independent variables to change together, so they may be highly correlated and the separate effects of each variable on left ventricular output may be difficult to determine. The solution to this problem is to reduce the number of independent variables or change the experimental protocol by adding other interventions which break up (or at least produce different) patterns among the independent variables.

DOES THE REGRESSION EQUATION DESCRIBE THE DATA?

For simple linear regression with a single independent variable, testing the significance of the regression coefficient (slope) was equivalent to testing whether or not the regression equation described the data. Because of multicollinearity, this equivalence does not necessarily hold for multiple regression. In particular, testing whether or not the individual coefficients are significantly different from zero will only provide the same results as an overall test of fit for the regression plane *if the independent variables are uncorrelated (i.e., statistically independent)*. If the independent variables are correlated at all—which almost always occurs in practice—the results of the individual *t* tests on each of the regression coefficients will depend on the effects of all the other independent variables. If the independent variables are sufficiently correlated, it is possible for the tests of the individual coefficients to fail to reject the null hypothesis that they are zero (because of large standard errors associated with each coefficient) even though the regression plane provides a good fit to the data. To avoid this problem, we need an overall test for the significance of fit of the regression equation itself to the data rather than tests based on the values of the individual coefficients.

We will conduct this overall goodness-of-fit test by comparing the variation accounted for by the regression plane with the residual variation about the regression plane. As with simple linear regression, we can partition the total sum of squared deviations from the mean of the dependent variable $SS_{tot} = \Sigma(Y - \overline{Y})^2$ into components associated with the variability of the regression plane about the mean value of the dependent variable (SS_{reg}) and the residual between the regression plane and the observed value Y associated with each set of observed values of the independent variables X_1 and X_2 (SS_{res}) (Fig. 3-5), so that

$$SS_{tot} = SS_{reg} + SS_{res}$$

and so

$$SS_{reg} = SS_{tot} - SS_{res}$$

We can easily compute SS_{tot} from its definition [or the standard deviation of the dependent variable with $SS_{tot} = (n - 1)s_y^2$], and SS_{res} is just the residual sum of squares [Eq. (3.4)] that we minimized when estimating the regression coefficients.

Next, we convert these sums of squared deviations into mean squares (variance estimates) by dividing by the appropriate degrees of freedom.

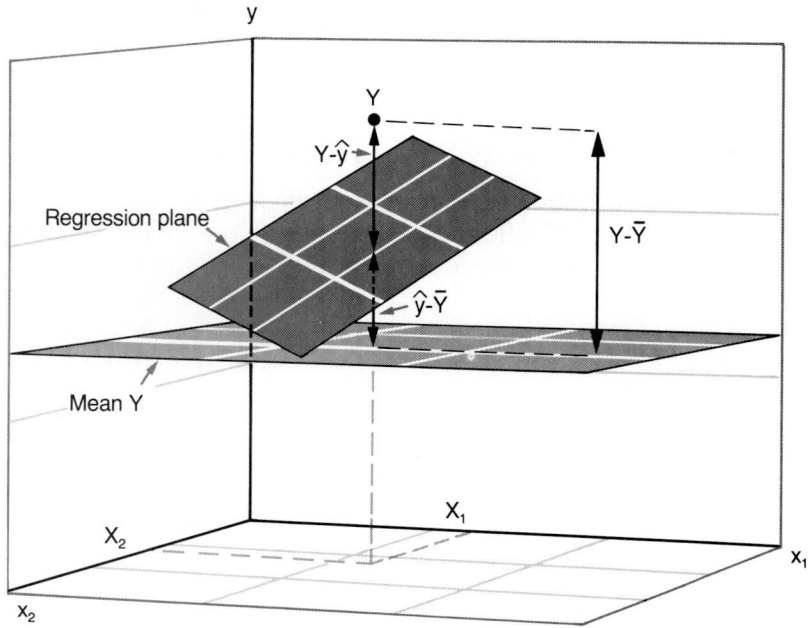

FIGURE 3-5 The deviation of the observed value of Y from the mean of all values of Y, $(Y - \bar{Y})$, can be separated into two components: the deviation of the observed value of Y from the value on the regression plane $(Y - \hat{Y})$ at the associated values of the independent variables X_1 and X_2, and the deviation of the regression plane the observed mean value of Y $(\hat{Y} - \bar{Y})$ (compare with Fig. 2-7).

In this case, there are 2 degrees of freedom associated with the regression sum of squares (because there are two independent variables) and $n - 3$ degrees of freedom associated with the residuals (because there are three parameters, b_0, b_1, and b_2, in the regression equation), so

$$MS_{reg} = \frac{SS_{reg}}{2}$$

and

$$MS_{res} = \frac{SS_{res}}{n - 3}$$

Finally, we test whether the variance "explained" by the regression plane is significantly larger than the residual variance by computing

$$F = \frac{MS_{reg}}{MS_{res}}$$

and comparing this value with the critical values of the F distribution with 2 numerator degrees of freedom and $n - 3$ denominator degrees of freedom.

For example, to test the overall goodness of fit of the regression plane given by Eq. (3.1) to the data we collected about the weights of Martians, use the data in Table 1-1 (or the computer output in Fig. 3-2) to compute $SS_{tot} = 54.367$ g^2 and $SS_{res} = .153$ g^2, so

$$SS_{reg} = 54.367 \text{ g}^2 - .153 \text{ g}^2 = 54.213$$

and so

$$MS_{reg} = \frac{SS_{reg}}{2} = \frac{54.213}{2} = 27.107$$

$$MS_{res} = \frac{SS_{res}}{n - 3} = \frac{.153}{9} = .017$$

and

$$F = \frac{MS_{reg}}{MS_{res}} = \frac{27.107}{.017} = 1594.5$$

All these results appear in the analysis-of-variance table of the computer output in Fig. 3-2. (The difference in the reported value of F and the one we just computed is due to round-off error.) The critical value of the F distribution for $P < .01$ with 2 numerator degrees of freedom and 9 denominator degrees of freedom is 8.02 (Table E-2, Appendix E), so we conclude that the regression plane significantly reduces the "unexplained" variance in the dependent variable. (An exact P value is reported by most computer programs, as shown in Fig. 3-2.)

Incremental Sums of Squares and the Order of Entry

In addition to the sums of squares we have already discussed—the sums of squares associated with the regression and residuals—many computer programs also print a sum of squares associated with each independent variable in the regression equation. These sums of squares, known as the *incremental sum of squares*, represent the increase in the regression sum of squares (and concomitant reduction in the residual sum of squares) that occurs when an independent variable is added to the regression equation *after the variables above it in the list are already taken into account in the regression equation*. This sum of squares can be used to construct an *F* statistic to test whether adding another potential independent variable to the regression equation significantly improves the quality of the fit.*

You can use this information to test whether or not adding additional independent variables adds significantly to your ability to predict (or "explain") the dependent variable by computing the *F* statistic:

$$F = \frac{\text{mean square associated with adding } x_2 \text{ given that } x_1 \text{ is in equation}}{\text{mean square residual for equation containing both } x_1 \text{ and } x_2} \tag{3.7}$$

$$F = \frac{MS_{x_2|x_1}}{MS_{\text{res}}}$$

This *F* test statistic is compared with the critical value for 1 numerator degree of freedom and $n - 3$ denominator degrees of freedom.[†]

For example, when we wrote the regression model for analyzing Martian weight, we specified that the two independent variables were height *H* and cups of water consumed *C*, *with H specified before C*. Thus, the analysis of variance tables in Fig. 3-2 show that the incremental sum of squares associated with height *H* is 47.839 g^2. Because each independent variable is associated with 1 degree of freedom, the mean square associated with each variable equals the sum of squares (because $MS = SS/1$). This value is the reduction in the residual sum of squares that one obtains by fitting the data with a regression line containing *only* the one independent variable *H*. The sum of squares associated with *C*, 6.375 g^2, is the increase in the regression sum of squares that occurs by *adding the variable C to the regression equation* already containing *H*. Again, there is 1 degree of freedom associated with adding each variable, so $MS_{C|H} = SS_{C|H}/1 = 6.375$ g^2. To test whether water consumption *C* adds information after taking into account the effects of height *H*, we compute

$$F = \frac{MS_{C|H}}{MS_{\text{res}}} = \frac{6.375}{.017} = 375$$

*These sums of squares will be particularly important for model-building techniques such as stepwise regression, which will be discussed in Chap. 6, and regression implementation of analysis of variance, which will be discussed in Chaps. 7, 8, and 9.

[†]Because there are only two independent variables, the mean square term with both independent variables in the equation is derived from the residual sum of squares. When there are more than two independent variables and one wishes to test the incremental value of including additional independent variables in the regression equation, the denominator mean square is based on the regression equation containing only the independent variables included up to that point. This mean square will generally be larger than MS_{res} and will be associated with a greater number of degrees of freedom. We discuss this general case in Chap. 6.

which exceeds 10.56, the critical value of the F distribution for $P < .01$ with 1 numerator and 9 denominator degrees of freedom. Therefore, we conclude that water consumption C affects Martian weight, even after taking the effects of height into account.

Note that the regression sum of squares equals the total of the incremental sums of squares associated with each of the independent variables, regardless of the order in which the variables are entered.

Because we have been analyzing the value of adding an additional variable, given that one variable is already in the regression equation, the results of these computations depend on the order in which the variables are entered into the equation. This result is simply another reflection of the fact that when the independent variables are correlated, they contain some redundant information about the dependent variable. The incremental sum of squares quantifies the amount of *new* information about the dependent variable contained in each additional independent variable. The only time when the order of entry does not affect the incremental sums of squares is when the independent variables are uncorrelated (i.e., statistically independent), so that each independent variable contains information about itself and no other variables.

For example, suppose that we had treated cups of water as the first variable and height as the second variable in our analysis of Martian heights. Figure 3-6 shows the results of this analysis. The increase in the regression sum of squares that occurs by adding height after accounting for water consumption is $SS_{H|C} = 16.368$ g^2, so $MS_{H|C} = 16.368/1 = 16.368$ g^2 and

$$F = \frac{MS_{H|C}}{MS_{\text{res}}} = \frac{16.368}{.017} = 962.82$$

which also exceeds the critical value of 10.56 of the F distribution for $P < .01$, with 1 numerator and 9 denominator degrees of freedom. Hence, we conclude that both water consumption and height contain independent information about the weight of Martians.

Relationship to t Tests of Individual Regression Coefficients

There is a direct connection between the t tests we did on the individual regression coefficients and the tests based on the incremental sum of squares we just presented; in fact, these two tests are equivalent. The t test is equivalent to conducting an F test based on the incremental sum of squares *with the variable of interest put into the regression equation last.*[*] As already discussed, the standard errors of the regression coeffi-

[*]For a proof of this fact, see J. Neter, W. Wasserman, and M. H. Kutner, *Applied Linear Regression Models,* Homewood, Ill., Irwin, 1983, pp. 278–281.

```
The regression equation is
W = - 1.22 + 0.111 C + 0.283 H

Predictor        Coef       Stdev      t-ratio        p
Constant      -1.2202      0.3210       -3.80      0.004
C             0.111212     0.005748     19.35      0.000
H             0.283436     0.009142     31.00      0.000

s = 0.1305       R-sq = 99.7%      R-sq(adj) = 99.7%


Analysis of Variance

SOURCE           DF          SS          MS         F         p
Regression        2       54.213      27.107    1591.79    0.000
Error             9        0.153       0.017
Total            11       54.367

SOURCE           DF       SEQ SS
C                 1       37.845
H                 1       16.368
```

FIGURE 3-6 Regression of Martian weight on height and water consumption to obtain the plane in Fig. 3-1*C*.

cients take into account the redundant information (due to multicollinearity) in the two independent variables used to predict the dependent variable. Hence, the individual *t* tests for the individual regression coefficients test the *marginal* value of including the information contained in each independent variable. Specifically, the *t* test for significance of each regression coefficient tests whether that independent variable contains significant predictive information about the dependent variable *after taking into account all the information in the other independent variables.*

For example, both *t* tests on the individual coefficients and *F* tests based on the incremental sum of squares led to the conclusion that both height and water consumption had statistically significant independent effects on the weight of Martians, even after taking into account any indirect effects that, say, water consumption had on weight through the effects of water consumption on height.

THE COEFFICIENT OF DETERMINATION AND THE MULTIPLE CORRELATION COEFFICIENT

To complete our extension of simple linear regression to multiple regression, we define the *multiple correlation coefficient* as a unitless measure of how tightly the data points cluster around the regression plane:

$$R = \sqrt{1 - \frac{SS_{res}}{SS_{tot}}} \qquad (3.8)$$

This definition is exactly the same as we used with simple linear regression. The multiple correlation coefficient is also equal to the Pearson product-moment correlation coefficient of the observed values of the dependent variable Y with the values predicted by the regression equation $\hat{y}$.

Likewise, the *coefficient of determination R^2* for multiple regression is simply the square of the multiple correlation coefficient. Because $SS_{tot} = SS_{reg} + SS_{res}$, the coefficient of determination is

$$R^2 = 1 - \frac{SS_{res}}{SS_{tot}} = \frac{SS_{reg}}{SS_{tot}} \qquad (3.9)$$

which we interpret as the fraction of the variance "explained" by the regression plane. The coefficient of determination, rather than the multiple correlation coefficient, is generally used when discussing multiple regression results.

It is also possible to construct hypothesis tests for overall goodness of fit based on the value of R^2, which, because they are equivalent to those just presented, will not be done here. We will use R^2 as a measure of goodness of fit and, later when we discuss stepwise regression and model building techniques in Chap. 6, as a measure of the additional information we gain by including more independent variables in the regression model.

MORE DUMMIES ON MARS

We now have the tools to analyze formally the data we reported in Chap. 1 on the effects of secondhand tobacco smoke on the height of Martians. Figure 3-7A presents these data. The question we have is: Does exposure to secondhand tobacco smoke affect the relationship between height and weight of Martians? To answer this question, we first define the *dummy* (or *indicator*) *variable D* according to

$$D = \begin{cases} 1 \text{ if exposed to secondhand tobacco smoke} \\ 0 \text{ if not exposed to secondhand tobacco smoke} \end{cases}$$

and fit the data in Fig. 3-7A with the multiple regression equation

$$\hat{W} = b_0 + b_H H + b_D D$$

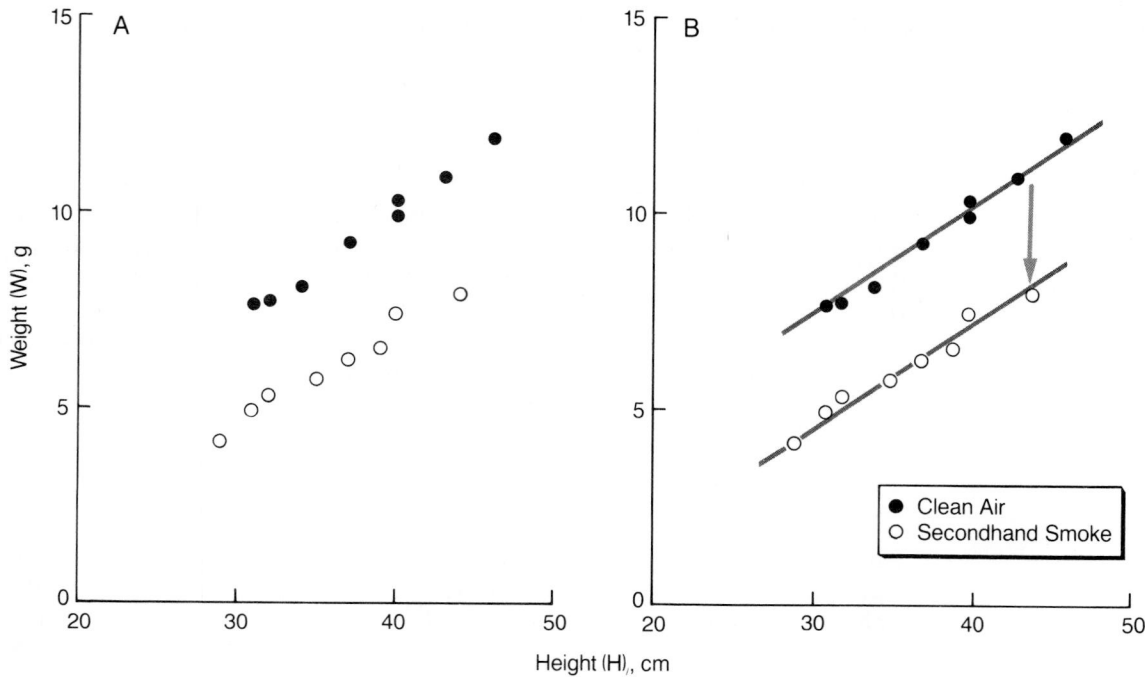

FIGURE 3-7 There appears to be a linear relationship between height and weight of Martians which is changed by exposure to secondhand smoke (A). This shift in the linear relationship between height and weight with exposure to secondhand smoke can be quantified using multiple regression and a dummy variable D defined to be 0 for the control cases and 1 for the data collected in the presence of secondhand smoke (B). The estimated value of the coefficient associated with the dummy variable, $b_D = -2.9$ g, is an estimate of the magnitude of the shift in the line associated with exposure to secondhand smoke.

Table 3-1 shows the data, together with the appropriately coded dummy variable. In this equation we assume that the dependence of weight on height is independent of the effect of secondhand smoke; in other words we assume that the effect of involuntary smoking, if present, would be to produce a parallel shift in the line relating weight to height.

Figure 3-8 shows the results of these computations; the regression equation is

$$\hat{W} = -.8 \text{ g} + .27 \text{ g/cm } H - 2.9 \text{ g } D$$

Figure 3-7B shows this regression equation superimposed on the data.

Because our primary interest is in the effect of breathing secondhand smoke, we focus on the coefficient b_D associated with the dummy variable D. Its value is -2.9 g, indicating that, on the average, at any given height, exposure to secondhand tobacco smoke is associated with being

TABLE 3-1 Data of Effect of Secondhand Smoke on Martians

D	Height H, cm	Weight W, g	
0	31	7.6	
0	32	7.7	
0	34	8.1	
0	37	9.2	Not exposed to
0	40	10.3	secondhand smoke
0	40	9.9	
0	43	10.9	
0	46	11.9	
1	29	4.1	
1	31	4.9	
1	32	5.3	
1	35	5.7	Exposed to
1	37	6.2	secondhand smoke
1	39	6.5	
1	40	7.4	
1	44	7.9	

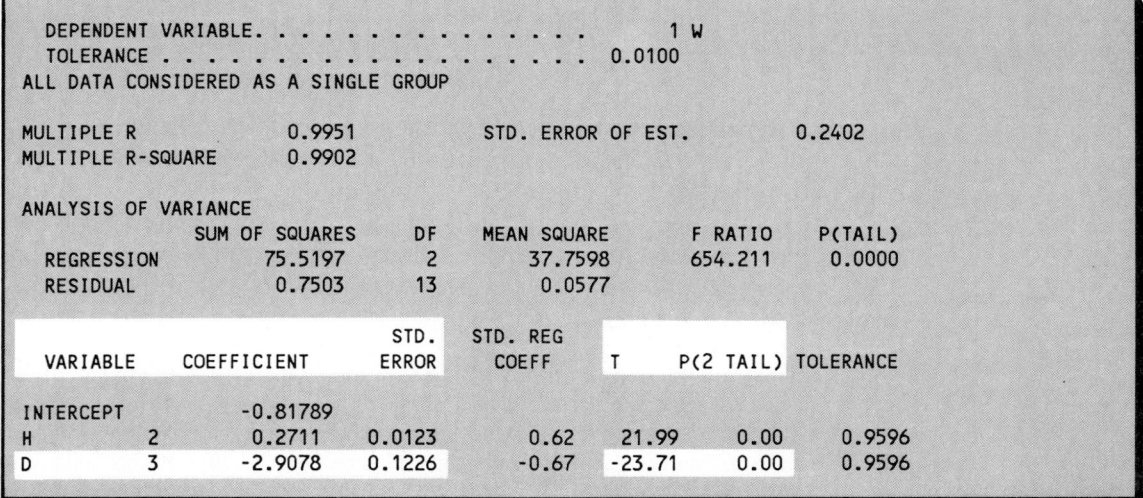

```
DEPENDENT VARIABLE. . . . . . . . . . . . . .        1 W
TOLERANCE . . . . . . . . . . . . . . . . . .  0.0100
ALL DATA CONSIDERED AS A SINGLE GROUP

MULTIPLE R           0.9951      STD. ERROR OF EST.        0.2402
MULTIPLE R-SQUARE    0.9902

ANALYSIS OF VARIANCE
             SUM OF SQUARES    DF    MEAN SQUARE    F RATIO    P(TAIL)
  REGRESSION     75.5197        2      37.7598      654.211    0.0000
  RESIDUAL        0.7503       13       0.0577

                            STD.    STD. REG
  VARIABLE     COEFFICIENT   ERROR    COEFF      T    P(2 TAIL) TOLERANCE

INTERCEPT        -0.81789
H        2        0.2711    0.0123    0.62    21.99    0.00     0.9596
D        3       -2.9078    0.1226   -0.67   -23.71    0.00     0.9596
```

FIGURE 3-8 Analysis of the data in Fig. 3-7. Exposure to secondhand smoke causes a 2.9-g downward shift in the line relating height and weight of Martians. The standard error associated with this estimate is .12 g, which means that the coefficient is significantly different from 0 ($P < .005$). Hence, this analysis provides us with an estimate of the magnitude of the shift in the curve with secondhand smoke exposure together with the precision of this estimate, as well as a P value to test the null hypothesis that secondhand smoke does not affect the relationship between height and weight in Martians.

2.9 g lighter. The standard error associated with this regression coefficient s_{b_D} is 0.1226 g. We use this standard error to test whether or not this shift in the regression line describing nonsmokers and the line describing involuntary smokers is statistically significant by computing

$$t = \frac{b_D}{s_{b_D}} = \frac{-2.9078}{0.1226} = -23.71$$

This value exceeds 4.221, the critical value of the t distribution with 13 degrees of freedom for $P < .001$, so we conclude that exposure to secondhand smoke stunts Martians' growth.[*]

MECHANISMS OF TOXIC SHOCK

Shock is the condition in which blood pressure drops to dangerously low levels, and can cause death. Certain kinds of bacteria have components in their cell walls, known as endotoxins, that lead to toxic (or septic) shock when the bacteria infect blood. Suttorp and his colleagues[†] hypothesized that one way these bacteria bring about shock is by causing the endothelial cells that line blood vessels to produce a chemical called prostacyclin which acts on the blood vessels to make them dilate, and so dropping blood pressure. Prostacyclin is made from arachidonic acid. To test this hypothesis, they grew endothelial cells in cell culture, exposed them to endotoxin, and measured prostacyclin production. They used four different levels of endotoxin exposure: 0, 10, 50, and 100 ng/mL. After these cells had been exposed to endotoxin for several hours, their ability to produce prostacyclin was evaluated by stimulating the cells with three levels of the prostacyclin precursor, arachidonic acid: 10, 25, and 50 μM. Because arachidonic acid is a precursor to prostacyclin production, prostacyclin production should depend on the amount of arachidonic acid present. The question is: Did prostacyclin production also depend on the level of endotoxin?

[*]An alternative interpretation for the analysis we just completed is to consider it a test of the effect of exposure to secondhand smoke on weight *controlling for* the effect of height. From this perspective, we are not interested in the effect of height on weight, but only need to "correct" for this effect in order to test the effect of secondhand smoke, "adjusted for" the effect of height. Viewed this way, the procedure we have just described is called an *analysis of covariance*. The model formulation and computations are not different from what we have presented. For more discussion of analysis of covariance, see D. G. Kleinbaum, L. L. Kupper, and K. E. Muller, *Applied Regression Analysis and Other Multivariable Methods* (2nd ed.), Boston, PWS-Kent Publishing, 1988, Chapter 15, "Analysis of Covariance and Other Methods for Adjusting Continuous Data."

[†]N. Suttorp, C. Galanos, and H. Neuhof, "Endotoxin Alters Arachidonate Metabolism in Pulmonary Endothelial Cells," *Am. J. Physiol.* 253:C384–C390, 1987.

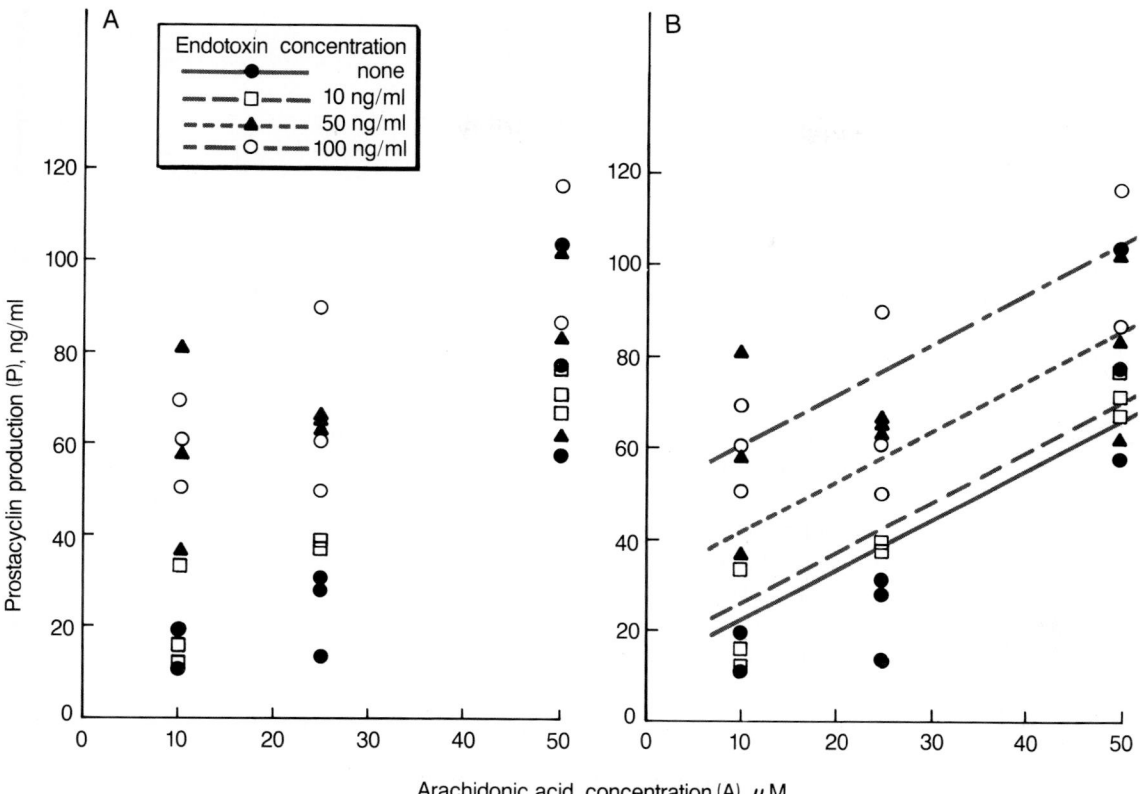

FIGURE 3-9 *A.* Data on prostacyclin production as a function of the arachidonic acid concentration at several different levels of endotoxin exposure. The different symbols indicate different levels of endotoxin exposure. It appears that at any given arachidonic acid concentration, prostacyclin production increases with the level of endotoxin exposure. *B.* These data can be well described with a multiple linear regression, with arachidonic acid *A* and endotoxin level *E* as independent variables. The lines represent the values predicted by this regression equation.

Figure 3-9*A* shows data from this study (which are in Table C-2, Appendix C). To use multiple linear regression to analyze the dependence of prostacyclin production on substrate (arachidonic acid) and the level of endotoxin exposure, we fit these data with the regression equation

$$\hat{P} = b_0 + b_A A + b_E E$$

where *P* is the prostacyclin production (dependent variable), *A* is the arachidonic acid concentration, and *E* is the endotoxin concentration (independent variables). If endotoxin exerts an independent effect on prostacyclin production, the coefficient b_E will be significantly different from zero.

```
The regression equation is
P = 10.9 + 1.11 A + 0.372 E

Predictor        Coef       Stdev     t-ratio         p
Constant       10.854       5.708        1.90     0.066
A               1.1140      0.1550        7.19     0.000
E               0.37159     0.06496       5.72     0.000

s = 15.35       R-sq = 71.9%      R-sq(adj) = 70.2%
```

FIGURE 3-10 Results of the multiple regression analysis of the data in Fig. 3-9.

Figure 3-10 shows the computer printout for this problem. As expected, prostacyclin production increases with the concentration of arachidonic aid, with $b_A = 1.1$ (ng · mL)/μM. More importantly, this analysis reveals that prostacyclin production depends on the concentration of endotoxin, independently of the effect of arachidonic acid, with $b_E = 0.37$. Because the standard error of b_E is only 0.065, the t statistic for testing the hypothesis that $\beta_E = 0$ (i.e., that endotoxin has no effect on prostacyclin production) is $t = 0.37159/0.06496 = 5.72$ ($P < .001$).

Hence, the regression equation to describe these data is

$$\hat{P} = 10.9 \text{ ng/mL} + 1.1 \text{ (ng · mL)/μM } A + 0.37E$$

Because we have data at discrete levels of endotoxin exposure, we can draw this regression as a set of lines of Fig. 3-9B, just as we represented contours on the three-dimensional regression plane surface in two dimensions in our study of Martian heights (Fig. 3-1). These data indicate that endotoxin increases the production of prostacyclin. Because Suttorp and his colleagues observed no cell damage and because increased prostacyclin production probably required an increase in a key enzyme that converts arachidonic acid to prostacyclin, they concluded that endotoxin exerts its effect by modifying the endothelial cells to increase the activity of this enzyme.

PROTEIN SYNTHESIS IN NEWBORNS AND ADULTS

Because infants are growing faster than adults, and so manufacture more skeletal muscle, Denne and Kalhan[*] hypothesized that a rapidly growing newborn would have different protein metabolism than an adult. To test

[*]S. C. Denne and S. C. Kalhan, "Leucine Metabolism in Human Newborns," *Am. J. Physiol.* 253:E608–E615, 1987.

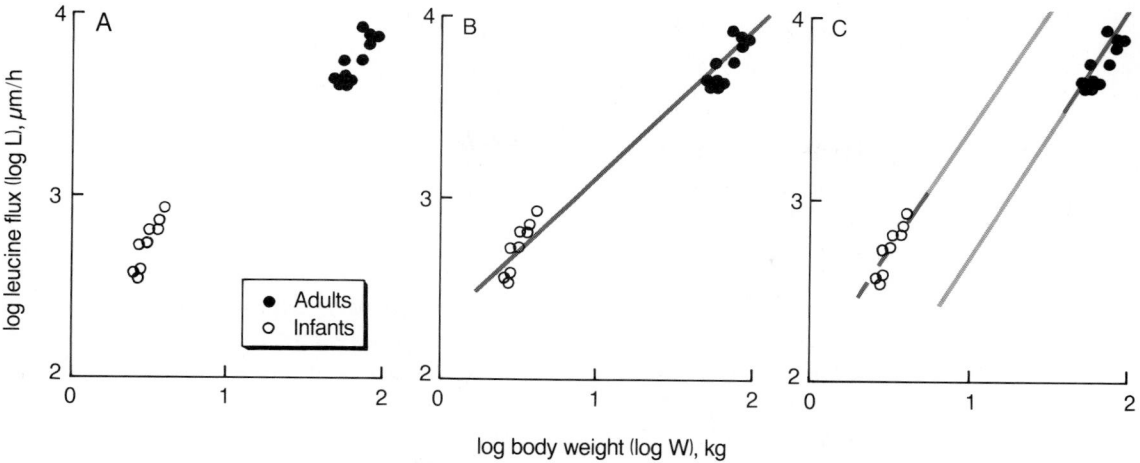

FIGURE 3-11 *A.* Data on leucine flux in infants and adults as a function of body weight. These data suggest that the log of leucine flux increases with the log of body weight. *B.* Fitting a linear regression to these data yields a highly significant slope with a high correlation coefficient of r = .99. *C.* Closer examination of the data, however, suggests that there may be two parallel lines, one for infants and one for adults. We test the hypothesis that there is a shift in the relationship between the log of leucine flux and the log of body weight by introducing a dummy variable A, defined to be 1 if the subject is an adult and 0 if the subject is an infant. Fitting the data to a multiple linear regression equation with log leucine flux as the dependent variable and log body weight and A as the independent variables reveals that one obtains a better description of these data by including the dummy variable. Hence, rather than there being a single relationship between body weight and leucine flux, there are actually two parallel relationships, which are different between infants and adults. (Data from Figure 5 of S. C. Denne and S. C. Kalhan, "Leucine Metabolism in Human Newborns," *Am. J. Physiol.* 253:E608–E615, 1987.)

this hypothesis, they measured leucine turnover in newborn infants and adults as a marker for protein metabolism. Leucine is an important amino acid that helps regulate body protein metabolism. In particular, leucine has been shown to stimulate protein synthesis and retard protein degradation in isolated skeletal muscle. Because metabolism depends on body size, they analyzed leucine metabolism as a function of body weight to see if newborns and adults had the same intrinsic metabolic activity for leucine.

Metabolic activity usually varies in proportion to the logarithm of body weight, so Fig. 3-11A shows these data plotted as the logarithm of leucine flux as a function of the logarithm of body weight. One approach to the data in Fig. 3-11A (Table C-3, Appendix C) would be to fit it with the simple linear regression

$$\hat{y} = b_0 + b_W X$$

where y equals the logarithm of the leucine turnover L and X equals the logarithm of body weight W. The resulting regression equation is

$$\log \hat{L} = 2.30 + 0.80 \log W \tag{3.10}$$

with a high correlation coefficient of .989. This result appears to indicate that differences in leucine metabolism between newborns and adults are due to differences in body size.

However, a close inspection of Fig. 3-11B reveals a serious problem with this conclusion. The data really do not fall on a line, but group in two small clusters of points that are very widely separated. Because there is so broad a range of body weights for which there are no data, one must assume that the relationship between $\log L$ and $\log W$ is linear over the region of missing data. Moreover, close examination of the relationship between the observations and the regression line in Fig. 3-11B reveals that the data are not randomly distributed about the regression line. The values of leucine flux ($\log L$) consistently fall below the regression line at lower weights ($\log W$) and the values of leucine flux fall above the line at higher weights *within both the newborn and adult groups*. This pattern in the residuals suggests that the population from which these samples were drawn does not meet the assumptions of linear regression analysis. These data actually suggest that there are two different lines, one for newborns and one for adults, rather than data scattered randomly at the extremes of a single line.

Inspection of the data in Fig. 3-11A suggests that the relationships between $\log L$ and $\log W$ may be parallel lines for the newborns and adults, so we will fit these data with the regression equation

$$\log \hat{L} = c_0 + c_W \log W + c_A A$$

where A is an indicator (or dummy) variable to encode age:

$$A = \begin{cases} 0 \text{ if newborn} \\ 1 \text{ if adult} \end{cases}$$

The results of fitting this equation to these data are shown in Fig. 3-12. The regression equation, shown in Fig. 3-11C, is

$$\log \hat{L} = 2.05 + 1.35 \log W - 0.76A \tag{3.11}$$

The dependence of leucine metabolism on body weight, quantified with c_W, is much greater than that estimated with the simple linear regression in Eq. (3.10): 1.35 compared to .80.

We now have two competing descriptions of the relationship between $\log L$ and $\log W$, Eqs. (3.10) and (3.11). Is there an objective test we can use to see if it is worth including the dummy variable A? Yes. We will test whether adding A significantly reduces the residual sum of squares

```
Equation Number 1     Dependent Variable..   LOGL   log Leucine

Beginning Block Number  1.  Method:  Enter      LOGW     A

Variable(s) Entered on Step Number
   1..    A          dummy variable for adult (=1)
   2..    LOGW       log Body Weight

Multiple R              .99210
R Square                .98426
Adjusted R Square       .98241
Standard Error          .07070

Analysis of Variance
                       DF      Sum of Squares      Mean Square
Regression              2          5.31452           2.65726
Residual               17           .08497            .00500

F =      531.62368      Signif F =  .0000

------------------ Variables in the Equation ------------------
Variable              B          SE B         Beta         T    Sig T

A              -.760525      .274278     -.728181     -2.773   .0130
LOGW           1.349520      .207036     1.711784      6.518   .0000
(Constant)     2.045855      .107738                  18.989   .0000
```

FIGURE 3-12 Multiple linear regression analysis of the data in Fig. 3-11. Note that the coefficient of A is significantly different from zero $(P = .013)$, which indicates that there is a significant downward shift in the curve of .76 log leucine flux units between adults and infants.

using the t test for the significance of c_A, which is equivalent to testing whether adding A to the equation that already contains log W significantly reduces the residual sum of squares.

From Fig. 3-12, we see that the computer reported this t statistic to be -2.773, which is associated with a P value of .013 and we conclude that there is a significant difference in the relationships between leucine flux and weight in adults and newborns.

These two regression models lead to fundamentally different conclusions. The simple linear regression given by Eq. (3.10) leads to the conclusion that the differences in leucine flux between adults and newborns are simply a reflection of the fact that production of the protein depends on body size and that adults are larger than newborns. Moreover, since the sum of logarithms is the logarithm of a product, Eq. (3.10) is equivalent to

$$\hat{L} = 200W^{.80} \tag{3.12}$$

because log 200 = 2.30.* It turns out that many biological processes scale between species of broadly different sizes according to weight to the 0.75 power, so that this result seems particularly appealing. The qualitative conclusion that one draws from this analysis is that the differences in leucine flux are simply a result of biological scaling. In particular, there is no fundamental difference between newborns and adults beyond size.

The problem with this tidy conclusion is that it ignores the fact that there is a significant shift between the lines relating leucine flux and body size in the adults and newborns. Accounting for this term will lead us to very different conclusions about the biology. To see why, rewrite Eq. (3.11) and take the antilogs of both sides to obtain

$$\hat{L} = 10^{(2.05 - .76A)} W^{1.35} = 112 \cdot 10^{-.76A} W^{1.35}$$

For newborns, $A = 0$, so $10^{-.76A} = 10^0 = 1$, and the relationship between leucine flux and weight becomes

$$\hat{L} = 112W^{1.35}$$

For the adults, $A = 1$, so $10^{-.76A} = 10^{-.76} = .17$, and the relationship between leucine flux and weight becomes

$$\hat{L} = .17 \cdot 112W^{1.35} = 19.5W^{1.35}$$

Comparing these two equations reveals that, as before, leucine production increases in both adults and newborns as a power of the weight, but the exponent of 1.35 is no longer near .75, so the argument that biological scaling is the primary reason for differences in leucine production no longer is as compelling. More importantly, these equations show that, for a comparable weight, the adults produce only 19.5/112 = 17 percent as much leucine as a comparable newborn would (if the newborn could be as large as an adult). Thus, this analysis leads to the conclusion that there is a qualitative difference between newborns and adults in terms of leucine production, with newborns producing much more leucine, even allowing for the differences in weights. This result is consistent with the fact that newborns are growing faster than adults.

There is one word of caution in drawing this conclusion: It requires implicit extrapolations outside the range of the observations for both the newborns and adults. There is always the chance that there is some nonlinearity that could account for the observations, although this situation appears to be unlikely.

*Recall that if $y = \log x$, $x = 10^y$.

GENERAL MULTIPLE LINEAR REGRESSION

The extension of what we have done from the regression model containing two independent variables given by Eq. (3.2) to the general case of k independent variables requires no new concepts. We estimate the regression parameters in the equation

$$y = \beta_0 + \beta_1 x_1 + \beta_2 x_2 + \beta_3 x_3 + \cdots + \beta_k x_k + \epsilon \tag{3.13}$$

This equation represents a plane (actually a hyperplane) in a $k + 1$ dimensional space in which k of the dimensions represent the k independent variables $x_1, x_2, x_3, \ldots, x_k$, and one dimension represents the dependent variable y. We estimate this plane from the data with the regression equation

$$\hat{y} = b_0 + b_1 x_1 + b_2 x_2 + b_3 x_3 + \cdots + b_k x_k \tag{3.14}$$

in which the regression coefficients $b_0, b_1, b_2, \ldots, b_k$ are computed to minimize the sum of squared residuals between the regression plane and the observed values of the dependent variable.

The *standard error of the estimate* remains an estimate of the variability of members of the population about the plane of means and is given by

$$s_{y|x} = \sqrt{\frac{\Sigma(Y - \hat{y})^2}{n - k - 1}} = \sqrt{\frac{SS_{\text{res}}}{n - k - 1}} = \sqrt{MS_{\text{res}}} \tag{3.15}$$

The *standard errors of the regression coefficients* s_{b_i} are

$$s_{b_i} = \sqrt{\frac{MS_{\text{res}}}{\Sigma(X_i - \overline{X}_i)^2 (1 - R_i^2)}} \tag{3.16}$$

where R_i^2 is the squared multiple correlation of x_i with the remaining independent variables $(x_1, x_2, \ldots, x_{i-1}, x_{i+1}, \ldots, x_k)$. We test whether an individual coefficient is significantly different from zero using

$$t = \frac{b_i}{s_{b_i}}$$

and compare the resulting value with a table of critical values of the t distribution with $n - k - 1$ degrees of freedom.

Overall goodness of fit of the regression equation is tested as outlined earlier in this chapter, except that the variance estimates MS_{reg} and MS_{res} are computed with

$$MS_{\text{reg}} = \frac{SS_{\text{reg}}}{k}$$

and

$$MS_{res} = \frac{SS_{res}}{n - k - 1}$$

The value of

$$F = \frac{MS_{reg}}{MS_{res}}$$

is compared with the critical value of the F distribution with k numerator degrees of freedom and $n - k - 1$ denominator degrees of freedom. The multiple correlation coefficient and coefficient of determination remain as defined by Eqs. (3.8) and (3.9), respectively.

The treatment of incremental sums of squares as additional variables are added to the regression equation, and associated hypothesis tests, are direct generalizations of Eq. (3.7):

$$F = \frac{\text{mean square associated with adding } x_j \text{ given that } x_1, \ldots, x_{j-1} \text{ are in the equation}}{\text{mean square residual for the equation containing } x_1, \ldots, x_j} \quad (3.17)$$

$$F = \frac{MS_{x_j|x_1,\ldots,x_{j-1}}}{MS_{res|x_1,\ldots,x_j}}$$

This F test statistic is compared with the critical value for 1 numerator degree of freedom and $n - j - 1$ denominator degrees of freedom.

Multiple Regression in Matrix Notation*

We can write the multiple linear regression equation, Eq. (3.13), in matrix notation by letting

$$y = \mathbf{x}\boldsymbol{\beta} + \boldsymbol{\epsilon}$$

where $\boldsymbol{\beta}$ is the *vector* of parameters

$$\boldsymbol{\beta} = \begin{bmatrix} \beta_0 \\ \beta_1 \\ \beta_2 \\ \vdots \\ \beta_k \end{bmatrix}$$

*This section is not necessary for understanding most of the material in this book and may be skipped with no loss of continuity. It is included for completeness and because matrix notation greatly simplifies derivation of results necessary in multiple regression analysis. Matrix notation is also crucial for a few methods, such as principal components regression (discussed in Chap. 5). Appendix A explains the relevant matrix notation and manipulations.

and $\mathbf{x}$ is the row matrix of independent variables

$$\mathbf{x} = [1 \quad x_1 \quad x_2 \cdots x_k]$$

where we have added a vector of ones in the first column to allow for the constant term b_0 in the regression equation. The problem is estimating the parameter vector $\boldsymbol{\beta}$ with the coefficient vector $\mathbf{b}$:

$$\mathbf{b} = \begin{bmatrix} b_0 \\ b_1 \\ b_2 \\ \vdots \\ b_k \end{bmatrix}$$

to obtain the regression equation

$$\hat{y} = \mathbf{xb} \tag{3.18}$$

which is equivalent to Eq. (3.14).

Suppose we have n observations:

$$
\begin{array}{ccccc}
x_1 & x_2 & \cdots & x_k & y \\
X_{11} & X_{12} & \cdots & X_{1k} & Y_1 \\
X_{21} & X_{22} & \cdots & X_{2k} & Y_2 \\
\vdots & \vdots & \vdots & \vdots & \vdots \\
X_{n1} & X_{n2} & \cdots & X_{nk} & Y_n
\end{array}
$$

Substituting each row into Eq. (3.18) yields

$$\hat{\mathbf{y}} = \mathbf{Xb}$$

where $\hat{\mathbf{y}}$ is the vector of predicted values of the dependent variable for each observation:

$$\hat{\mathbf{y}} = \begin{bmatrix} \hat{y}_1 \\ \hat{y}_2 \\ \vdots \\ \hat{y}_n \end{bmatrix}$$

and $\mathbf{X}$ is the *design matrix*

$$\mathbf{X} = \begin{bmatrix} 1 & X_{11} & X_{12} & \cdots & X_{1k} \\ 1 & X_{21} & X_{22} & \cdots & X_{2k} \\ \vdots & \vdots & \vdots & \vdots & \vdots \\ 1 & X_{n1} & X_{n2} & \cdots & X_{nk} \end{bmatrix}$$

which contains all the observed values of the independent variable.

We define the vector of observed values of the dependent variable as

$$\mathbf{Y} = \begin{bmatrix} Y_1 \\ Y_2 \\ \vdots \\ Y_n \end{bmatrix}$$

so the vector of residuals is

$$\mathbf{e} = \begin{bmatrix} e_1 \\ e_2 \\ \vdots \\ e_n \end{bmatrix} = \begin{bmatrix} Y_1 - \hat{y}_1 \\ Y_2 - \hat{y}_2 \\ \vdots \\ Y_n - \hat{y}_n \end{bmatrix} = \mathbf{Y} - \hat{\mathbf{y}}$$

The sum of squared residuals is

$$SS_{\text{res}} = \Sigma e_i^2 = \mathbf{e}^T \mathbf{e} = (\mathbf{Y} - \hat{\mathbf{y}})^T(\mathbf{Y} - \hat{\mathbf{y}}) = \Sigma(Y_i - \hat{y}_i)^2$$

The regression problem boils down to finding the value of $\mathbf{b}$ that minimizes SS_{res}. It is*

$$\mathbf{b} = (\mathbf{X}^T\mathbf{X})^{-1}\mathbf{X}^T\mathbf{Y} \tag{3.19}$$

The properties of $(\mathbf{X}^T\mathbf{X})^{-1}$ will be particularly important when discussing multicollinearity and how to deal with it in Chap. 5. $s_{\mathbf{b}}^2$, which is the *variance-covariance matrix of the parameter estimates*, is given as

$$\mathbf{s}_{\mathbf{b}}^2 = s_{y|x}^2(\mathbf{X}^T\mathbf{X})^{-1} = [s_{b_i b_j}^2]$$

The standard errors of the regression coefficients are the square roots of the corresponding diagonal elements of this matrix.

The correlation between the estimates of β_i and β_j is

$$r_{b_i b_j} = \frac{s_{b_i b_j}}{\sqrt{s_{b_i} s_{b_j}}} \tag{3.20}$$

*Here is the derivation:

$$\frac{\partial SS_{\text{res}}}{\partial \mathbf{b}} = \frac{\partial}{\partial \mathbf{b}}[(\mathbf{Y} - \hat{\mathbf{y}})^T(\mathbf{Y} - \hat{\mathbf{y}})] = \frac{\partial}{\partial \mathbf{b}}[(\mathbf{Y} - \mathbf{Xb})^T(\mathbf{Y} - \mathbf{Xb})] = -2\mathbf{X}^T\mathbf{Y} + 2(\mathbf{X}^T\mathbf{X})\mathbf{b}$$

For minimum SS_{res}, $\partial SS_{\text{res}}/\partial \mathbf{b} = 0$, so

$$-2\mathbf{X}^T\mathbf{Y} + 2(\mathbf{X}^T\mathbf{X})\mathbf{b} = 0$$

and, thus

$$(\mathbf{X}^T\mathbf{X})\mathbf{b} = \mathbf{X}^T\mathbf{Y}$$

and

$$\mathbf{b} = (\mathbf{X}^T\mathbf{X})^{-1}\mathbf{X}^T\mathbf{Y}$$

where $s_{b_i b_j}$ is the square root of the ijth element of the variance-covariance matrix of the parameter estimates. If the independent variables are uncorrelated, $r_{b_i b_j}$ will equal 0 if $i \neq j$. On the other hand, if the independent variables are correlated, they contain some redundant information about the regression coefficients and $r_{b_i b_j}$ will not be 0. If there is a great deal of redundant information in the independent variables, the regression coefficients will be poorly defined and $r_{b_i b_j}$ will be near 1 (or -1). Large values of this correlation between parameter estimates suggest that multicollinearity is a problem.

DIABETES, CHOLESTEROL, AND THE TREATMENT OF HIGH BLOOD PRESSURE

People with diabetes cannot properly regulate their blood glucose levels. In addition, diabetes is often associated with high blood pressure (hypertension) and raised cholesterol and triglyceride levels, all of which predispose diabetics to a high risk of heart disease and stroke. Hypertension is often treated with β-blocking drugs to lower blood pressure. These drugs act by blocking the β-receptor to which adrenaline binds to make the arteries contract, thereby increasing blood pressure. Although β-blockers are often the treatment of choice for hypertension, it is also known that β-blockers adversely affect the cholesterol profile in people with normal blood pressure. Cholesterol and triglycerides are assembled into a number of related compounds for transport in the blood, known as lipoproteins. In general, a high triglyceride level is associated with a high risk of heart disease, whereas the presence of high-density lipoprotein (HDL) is associated with a reduced risk of heart disease. Although there are other types of lipoproteins, for the purposes of this example a bad cholesterol profile consists of a high triglyceride level and a low HDL level. Thus, treating hypertension with β-blockers in diabetics, who already have a bad cholesterol profile, may actually increase the risk of heart disease by making the cholesterol profile even worse. To explore this possibility, Feher and his coworkers* studied the effect of β-blockers on a HDL subfraction, HDL-2, in 71 male hypertensive diabetics, while taking into account other potential determinants of HDL-2 (H). These other determinants were smoking, drinking, age, weight, and other metabolically related compounds, triglycerides, C-peptide, and blood glucose. This example illustrates how multiple regression can be used to account for many simultaneously acting factors in a clinical study.

They used the regression equation

$$\hat{H} = c_0 + c_B B + c_D D + c_S S + c_A A + c_W W + c_T T + c_C C + c_G G$$

*M. D. Feher, S. G. H. Rains, W. Richmond, D. Torrens, G. Wilson, J. Wadsworth, P. S. Sever, and R. S. Elkeles. "β-Blockers, Lipoproteins and Noninsulin-Dependent Diabetes," *Postgrad. Med. J.* 64:926–930, 1988.

to analyze their data. The independent variables include the continuous variables age A, weight W, triglycerides T, C-peptide C, and blood glucose G, as well as the dummy variables

$$B = \begin{cases} 1 \text{ if receiving } \beta\text{-blocker} \\ 0 \text{ if no } \beta\text{-blocker} \end{cases}$$

$$D = \begin{cases} 1 \text{ if drinks alcohol} \\ 0 \text{ if does not drink} \end{cases}$$

$$S = \begin{cases} 1 \text{ if smokes} \\ 0 \text{ if does not smoke} \end{cases}$$

Because a high HDL is good in that it is associated with a reduced risk of heart disease, a positive coefficient for any independent variable indicates that increasing its value has a beneficial effect on the cholesterol profile, whereas a negative value indicates an adverse effect. In terms of the principal question of interest—do β-blockers adversely affect cholesterol?—the coefficient c_B is of principal importance. The other variables are of lesser importance and are in the regression model primarily to account for potential confounding variables that might obscure an effect of the β-blockers.

The computer output for fitting these data (in Table C-4, Appendix C) to this regression equation is in Fig. 3-13. The R^2 is .595, and the F value for the test of overall goodness of fit is 11.4 with 8 numerator and 62 denominator degrees of freedom, which is associated with $P < .001$. Although statistically significant, this low R^2 indicates that the effects of the independent variables are not so striking as to make the regression equation suitable for making predictions in individual patients. Rather, the results describe average changes in the sample as a whole.

The coefficients c_D, c_W, c_C, and c_G are nowhere near significantly different from zero, indicating that drinking, weight, C-peptide concentration, and blood glucose level do not affect the level of HDL-2. The other regression coefficients are statistically significant. c_A is estimated to be $-.005$ mmol/(L · year), which means that, on the average, HDL-2 concentration decreases by .005 mmol/L with each year increase in age $(P < .001)$. c_S is estimated to be $-.040$ mmol/L, which indicates that, on the average, smoking decreases HDL-2 concentration, everything else being constant $(P = .059)$. c_T is estimated to be $-.044$ $(P < .001)$, which is consistent with other observations that triglyceride and HDL tend to vary in opposite directions.

Most important to the principal question asked by this study, c_B is $-.082$ mmol/L and is highly significant $(P < .001)$. The standard error associated with this coefficient is $s_{c_B} = .023$ mmol/L, so the 95 percent

```
The regression equation is
H = 0.711 - 0.0824 B - 0.0173 D - 0.0399 S - 0.00455 A - 0.00214 W - 0.0444 T
         + 0.00463 C - 0.00391 G

Predictor      Coef      Stdev    t-ratio         p
Constant      0.7110     0.1102      6.45      0.000
B            -0.08244    0.02293     -3.59      0.001
D            -0.01726    0.02121     -0.81      0.419
S            -0.03995    0.02078     -1.92      0.059
A            -0.004549   0.001179    -3.86      0.000
W            -0.002140   0.002722    -0.79      0.435
T            -0.044372   0.009411    -4.71      0.000
C             0.004633   0.007811     0.59      0.555
G            -0.003907   0.003239    -1.21      0.232

s = 0.07745      R-sq = 59.5%     R-sq(adj) = 54.3%

Analysis of Variance

SOURCE        DF         SS          MS          F          p
Regression     8     0.546959'   0.068370     11.40      0.000
Error         62     0.371915    0.005999
Total         70     0.918873
```

FIGURE 3-13 Multiple regression analysis of the determinants of the HDL-2 lipoprotein subfraction against whether or not the patient is receiving a β-blocking drug, whether or not the patient drinks alcohol or smokes, age, weight, triglyceride levels, C-peptide levels, and blood glucose. Controlling for all of the other variables, the presence of a β-blocking drug significantly reduces the level of HDL-2 ($P = .001$).

confidence interval for the change in HDL concentration with β-blockers is

$$c_B - t_{.05}s_{c_B} < \Delta H_{\beta\text{-blocker}} < c_B + t_{.05}s_{c_B}$$

There are 62 degrees of freedom associated with the residual (error) term, so $t_{.05} = 1.999$ and the 95 percent confidence interval is

$$-.082 \text{ mmol/L} - 1.999 \cdot .023 \text{ mmol/L} < \Delta H_{\beta\text{-blocker}}$$
$$< -.082 \text{ mmol/L} + 1.999 \cdot .023 \text{ mmol/L}$$

or

$$-.13 \text{ mmol/L} < \Delta H_{\beta\text{-blocker}} < -.04 \text{ mmol/L}$$

When compared with the average HDL-2 concentration observed in all patients, .20 mmol/L, this is a large change.

Thus, taking into account everything else in the regression model, β-blocking drugs lower HDL in hypertensive diabetics. So, although the β-blockers have the desired effect of lowering blood pressure, they have the undesired effect of lowering HDL. Thus, these results suggest that some other means besides β-blockers should be used to control hypertension in diabetics.

BABY BIRDS BREATHING IN BURROWS

Many birds and mammals live underground in burrows, where the ventilation can be quite poor. As a result, they are routinely subjected to environments with lower oxygen concentrations and higher carbon dioxide concentrations then they would encounter if they spent all their time living above ground. Most animals adjust their breathing patterns—how often and deeply they breathe—in response to changes in both oxygen and carbon dioxide concentrations. To investigate whether similar controls were active in birds who lived underground, Colby and his colleagues[*] studied breathing patterns in nestling bank swallows. When first born, these birds remain in burrows for about 2 weeks, where they are exposed to increasing levels of carbon dioxide and falling levels of oxygen. Colby and his colleagues exposed the nestlings to simulated burrow gas of varying composition and measured the so-called minute ventilation (the total volume of air breathed per minute).

To investigate the potential roles of oxygen and carbon dioxide as determinants of minute ventilation, we fit the data (in Table C-5, Appendix C) with the regression equation

$$\hat{V} = b_0 + b_O \cdot O + b_C C$$

where V is the percentage increase in minute ventilation, O is the percentage of oxygen, and C is the percentage of carbon dioxide in air baby birds breathe. Figure 3-14 shows the data with the resulting regression plane, and Fig. 3-15 shows the results of the computations.

The F statistic for the test of the overall regression equation is 21.44 with 2 and 117 degrees of freedom, which is associated with $P < .001$, so we conclude that this regression equation provides a better description of the data than simply reporting the mean minute ventilation. The coefficient $b_O = -5.33$ with a standard error of 6.43; we test whether this coefficient is significantly different from zero by computing

$$t = \frac{-5.33}{6.43} = -.83$$

[*]C. Colby, D. L. Kilgore, Jr., and S. Howe, "Effects of Hypoxia and Hypercapnia on V_T, f, and $\dot{V}_I$ of Nestling and Adult Bank Swallows," *Am. J. Physiol.* 253:R854–R860, 1987.

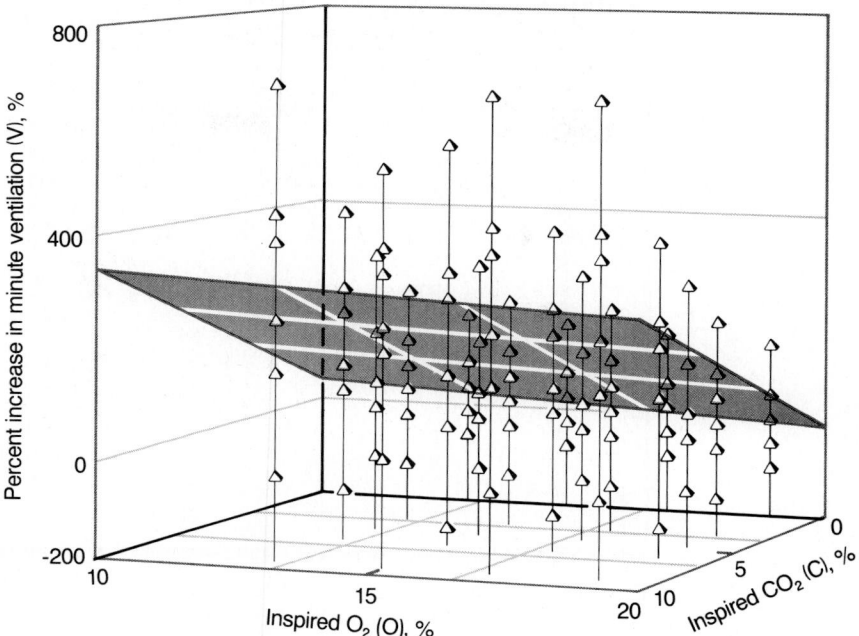

FIGURE 3-14 The percent change in minute ventilation for nestling bank swallows breathing various mixtures of oxygen O and carbon dioxide C. Ventilation increases as inspired carbon dioxide increases, but no trend is apparent as a function of inspired oxygen.

which does not come anywhere near -1.981, the critical value of t for $P < .05$ with $n - 3 = 120 - 3 = 117$ degrees of freedom (the exact P value of .408 is shown in the computer output in Fig. 3-15). Therefore, we conclude that there is no evidence that ventilation depends on the oxygen concentration. In contrast, the coefficient associated with the carbon dioxide concentration is $b_C = 31.1$ with a standard error of 4.79; we test whether this coefficient is significantly different from zero by computing

$$t = \frac{31.1}{4.79} = 6.5$$

which greatly exceeds 3.376, the critical value of t for $P < .001$ with 117 degrees of freedom, so we conclude that ventilation is sensitive to the carbon dioxide concentration.

The overall conclusion from these data is that ventilation in the nestling bank swallows is controlled by carbon dioxide but not oxygen concentration in the air they breathe. Examining the regression plane in Fig. 3-14 reveals the geometric interpretation of this result. The regres-

```
The regression equation is
V = 86 - 5.33 O + 31.1 C

Predictor        Coef        Stdev      t-ratio          p
Constant         85.9        106.0         0.81      0.419
O               -5.330       6.425        -0.83      0.408
C               31.103       4.789         6.50      0.000

s = 157.4        R-sq = 26.8%      R-sq(adj) = 25.6%

Analysis of Variance

SOURCE          DF           SS           MS          F          p
Regression       2        1061819       530909      21.44     0.000
Error          117        2897567        24766
Total          119        3959385

SOURCE          DF         SEQ SS
O                1          17045
C                1        1044773
```

FIGURE 3-15 Analysis of the minute ventilation in the nestling bank swallows data in Fig. 3-14. Minute ventilation depends on the level of carbon dioxide they are exposed to, C, but not the level of oxygen O. Geometrically this means that the regression plane is significantly tilted with respect to the C axis, but not the O axis.

sion plane shows only a small tilt with respect to the oxygen (O) axis, indicating that the value of O has little effect on ventilation V but has a much steeper tilt with respect to the carbon dioxide (C) axis, indicating that the value of C has a large effect on ventilation. The characteristics of breathing control are given by the regression coefficients. b_C is the only significant effect and indicates a 31.1 percent increase in ventilation with each 1 percent increase in the concentration of carbon dioxide in the air these newborn swallows breathed. This gives us a quantitative basis for comparison: if these babies have adapted to their higher carbon dioxide environment, one would expect the value of $b_C = 31.1$ to be lower than in adults of the same species, who are not confined to the burrow. Colby and his coworkers repeated this study in adults and found this to be the case; the value of b_C in newborns was significantly less than the value of b_C in the adults.

POLYNOMIAL (AND SOME OTHER NONLINEAR) REGRESSIONS

So far we have limited ourselves to regression models in which the changes in the dependent variable were proportional to changes in each of the independent variables. The surface described by the regression equation is a plane. The fact that this simple model describes so many situations is what makes regression analysis such a useful technique. As we saw in our study of heat exchange in seals at the end of Chap. 2, however, straight lines and flat planes do not always describe the relationships between dependent and independent variables. In that example, we concluded that a curved line better described the data than a straight one (Fig. 2-11). We now extend what we have done so far to describe such *nonlinear* relationships.

Consider, for example, the quadratic polynomial

$$\hat{y} = b_0 + b_1 x + b_2 x^2$$

This equation describes a curve, not a straight line—it is clearly a nonlinear relationship. (Depending on the values of the coefficients, the curve could be concave upward or concave downward.) We can, however, use the methods of this chapter to estimate the parameters in this equation by converting it into a multiple linear regression problem.

To do so, simply let $x_2 = x^2$ and rewrite this equation as

$$\hat{y} = b_0 + b_1 x + b_2 x_2$$

which is now a straightforward multiple linear regression problem with the two independent variables x and x_2. Had we started with a cubic polynomial, we would simply have defined a third independent variable $x_3 = x^3$. The process of estimating the parameters and testing hypotheses proceeds just as before.*

Heat Exchange in Gray Seals Revisited

In Chap. 2 we began an analysis of the relationship between body thermal conductivity in northern gray seals and the water temperature in which these marine mammals lived. We used simple linear regression to conclude that there was a statistically significant relationship between thermal conductivity and ambient water temperature. However, this linear relationship did not appropriately describe the data points, which seemed to fall on a curved line (Fig. 2-11). We will now use multiple linear regression to estimate that nonlinear relationship.

The data in Fig. 2-11 look as if they fall on a quadratic function (a

---

*There are some special nuances to polynomial regression because of the logical connections between the different terms. There are often reasons to believe that the lower order (i.e., lower power) terms should be more important, so the higher order terms are often entered stepwise until they cease to add significantly to the ability of the equation to describe the data. In addition, because the independent variables are powers of each other, there is almost always some multicollinearity, which will be discussed in Chap. 5.

second-order polynomial), so we will attempt to describe it with the equation

$$\hat{C}_b = C_0 + ST_a + S_2T_a^2$$

in which C_b is body conductivity and T_a is ambient temperature. [Compare this equation with Eq. (2.21).] We convert this nonlinear regression problem (with one independent variable T_a) to a multiple linear regression problem (with two independent variables) by defining the new variable $T_2 = T_a^2$. Table 3-2 shows the data set we will submit to the multiple regression program, including the new variable T_2 to estimate the parameters in the equation

$$\hat{C}_b = C_0 + ST_a + S_2T_2$$

Figure 3-16 shows the results of these computations; the regression equation is

$$\hat{C}_b = 1.72 \text{ W/(m}^2 \cdot {}^\circ\text{C)} + .058 \text{ W/(m}^2 \cdot {}^\circ\text{C}^2) \, T_a + .0019 \text{ W/(m}^2 \cdot {}^\circ\text{C}^3) \, T_2$$

TABLE 3-2 Data for Quadratic Fit to Seal Thermal Conductivity

Ambient Temperature T_a, °C	$T_a^2 = T_2$, °C²	Thermal Conductivity C_b, W/(m² · °C)
−40	1600	2.81
−40	1600	1.82
−40	1600	1.80
−30	900	2.25
−30	900	2.19
−30	900	2.02
−30	900	1.57
−30	900	1.40
−20	400	1.94
−20	400	1.40
−20	400	1.18
−10	100	1.63
−10	100	1.26
−10	100	0.93
10	100	2.22
10	100	1.97
10	100	1.74
20	400	4.49
20	400	4.33
20	400	3.10

```
DEP VARIABLE: CB        Conductivity

                        ANALYSIS OF VARIANCE

                        SUM OF          MEAN
           SOURCE   DF   SQUARES        SQUARE    F VALUE     PROB>F

           MODEL     2  12.22142562   6.11071281   21.814     0.0001
           ERROR    17   4.76214938   0.28012643
           C TOTAL  19  16.98357500

              ROOT MSE    0.5292697    R-SQUARE     0.7196
              DEP MEAN      2.1025     ADJ R-SQ     0.6866
              C.V.        25.17335

                        PARAMETER ESTIMATES

                        PARAMETER     STANDARD    T FOR H0:
         VARIABLE  DF    ESTIMATE       ERROR    PARAMETER=0    PROB > |T|

         INTERCEP   1   1.72490040   0.19190980     8.988       0.0001
         TA         1   0.05769199   0.008749326    6.594       0.0001
         T2         1   0.001880393  0.000357448    5.261       0.0001
```

FIGURE 3-16 Analysis of thermal conductivity of gray seals as a function of ambient temperature, including a quadratic term, $T_2 = T_a^2$, to allow for a nonlinear relationship. This computer run was used to fit the quadratic curve in Fig. 2-11C. Both the linear and quadratic terms are significantly different from 0, indicating significant curvature in the relationship. Note that the R^2 for this quadratic model is .72, compared with only .26 for the simple linear regression (in Figs. 2-11B and 2-12).

or, equivalently,

$$\hat{C}_b = 1.72 \text{ W}/(\text{m}^2 \cdot {}°\text{C}) + .058 \text{ W}/(\text{m}^2 \cdot {}°\text{C}^2)\, T_a + .0019 \text{ W}/(\text{m}^2 \cdot {}°\text{C}^3)\, T_a^2$$

The F statistic for overall goodness of fit is 21.814 with 2 numerator and 17 denominator degrees of freedom, which leads us to conclude that there is a statistically significant quadratic relationship between body thermal conductivity and ambient temperature ($P < .0001$). The t statistics associated with both the linear (T_a) and quadratic (T_a^2) terms are associated with P values below .0001, so we conclude that both terms contribute significantly to the regression fit.

There are several other ways to see that the quadratic equation provides a better description of the data than the linear one. From the simple linear regression, we concluded that there was a statistically significant relationship with an R^2 of .26 (see Fig. 2-12). Although statistically significant, this low value indicates only a weak association between thermal conductivity and ambient temperature. In contrast, the R^2 from the quadratic fit is .72, indicating a much stronger relationship between these two variables. The closer agreement between the quadratic regression line and the data than the linear regression line is also reflected by the smaller value of the standard error of the estimate associated with the quadratic equation, .53 W/(m^2 · °C) (from Fig. 3-16) compared with .83 W/(m^2 · °C) (from Fig. 2-12).

The final measure of the improvement in the fit comes from examining the residuals between the regression and data points. As already noted in Chap. 2, the data points are not randomly distributed around the linear regression line in Fig. 2-11B; the points tend to be above the line at low and high temperatures and below the line at intermediate temperatures. In contrast, the data points tend to be randomly distributed above and below the quadratic regression line along its whole length in Fig. 2-11C.

As we already noted in Chap. 2, the quadratic regression is not only a better statistical description of the data, but it also leads to a different interpretation than does a linear regression. The linear model led to the conclusion that body thermal conductivity decreases as ambient temperature decreases. The more appropriate quadratic model leads to the conclusion that this is true only to a point, below which body thermal conductivity begins to rise again as ambient temperature falls even lower. This result has a simple physiological explanation. The seals are able to reduce heat loss from their bodies (i.e., lower their thermal conductivity) by decreasing blood flow to their skin and the underlying layer of blubber, which keeps heat in the blood deeper in the body and further from the cold water. However, at really cold water temperatures—below about −15°C—the skin needs to be kept warmer so that it will not freeze. Thus, the seals must start to increase blood flow to the skin, which causes them to lose more body heat.

Other Nonlinear Regressions

Polynomials may be the most common nonlinear regression models used, but they are by no means the only ones. One can convert any nonlinear equation in which *the nonlinearities only involve the independent variables* into a corresponding multiple linear regression problem by suitably defining new variables.

For example, consider the nonlinear regression equation with dependent variable z and independent variables x and y:

$$\hat{z} = b_0 + b_1 x + b_2 \sin x + b_3 (x^3 + \ln x) + b_4 e^y$$

Simply let $x_1 = x$, $x_2 = \sin x$, $x_3 = (x^3 + \ln x)$, and $x_4 = e^y$, to obtain the straightforward multiple linear regression equation

$$\hat{z} = b_0 + b_1 x_1 + b_2 x_2 + b_3 x_3 + b_4 x_4$$

and estimate the parameters as before.

It is also sometimes possible to *transform* the independent variable using some mathematical function to convert the nonlinear problem to one that is linear in the regression parameters. For example, we can transform the nonlinear regression problem of fitting

$$\hat{y} = a e^{bx}$$

into a linear regression problem by taking the natural logarithm of both sides of this equation to obtain

$$\ln \hat{y} = \ln a + bx$$

Let $\hat{y}' = \ln \hat{y}$, the transformed dependent variable, and let $a' = \ln a$, to obtain

$$\hat{y}' = a' + bx$$

which is just a simple linear regression of y' on x with intercept a' and slope b. After doing the fit, one converts the parameter a' back to a using $a = e^{a'}$. Indeed, we already used this approach in the study of leucine metabolism earlier in this chapter, when we related the logarithm of leucine production to the logarithm of body weight. The resulting linear regression equation, Eq. (3.10), was a transformed form of the nonlinear Eq. (3.12).

Unfortunately, it turns out that the number of such simple transformations is quite limited. Transforming the independent variable can also introduce implicit assumptions about the nature of the underlying population which violate the assumptions upon which multiple linear regression is based.* For example, the logarithmic transformation weights deviations from small values of the dependent variable more heavily than the same deviation from large values of the dependent variable.

In addition, although such tricks will often simplify a problem, they will not always work. If the regression parameters appear nonlinearly in the equation, then methods of nonlinear regression described in Chap. 10 generally must be used. For example,

$$\hat{y} = a(e^{bx} - 1)$$

in which a and b are the regression parameters, cannot be converted to a multiple linear regression problem.

---

*Interestingly, such transformations can also be used to help data meet the assumptions of multiple linear regression. Chapter 4 discusses such use of transformations.

INTERACTIONS BETWEEN THE INDEPENDENT VARIABLES

So far we have assumed that each independent variable acted independently of the others, i.e., that the effect of a given change in one independent variable was always the same no matter what the values of the other independent variables. In this case, the regression surface is a flat plane (or hyperplane). While this is often the case, it is not always true. Sometimes the independent variables *interact* so that the effect of one is different depending on the value of one or more of the other independent variables. In this case, the response surface is not flat, but bent.

Interactions are commonly represented in multiple regression models by including a term (or terms) consisting of a product of two (or more) of the independent variables. For example, to include an interaction term in Eq. (3.3), we add the term $x_1 x_2$ to obtain

$$\hat{y} = b_0 + b_1 x_1 + b_2 x_2 + b_{12} x_1 x_2$$

This equation is still linear in the regression coefficients b_i, so we let $x_{12} = x_1 x_2$ and solve the resulting multiple regression problem with the three independent variables x_1, x_2, and x_{12}. If the coefficient associated with the interaction term b_{12} is significantly different from zero, it indicates that there is a significant interaction between the independent variables, i.e., the effects of one depend on the value of the other. We can make this interdependence explicit by rearranging the previous equation as

$$\hat{y} = b_0 + b_1 x_1 + (b_2 + b_{12} x_1) x_2$$

In this case, the sensitivity of $\hat{y}$ to changes in x_2 depends on the value of x_1.

As a general rule, one should always include the first-order terms for all variables included in interaction terms in a regression equation. In other words, most regression equations including the interaction term $x_1 x_2$ should also contain the first-order terms x_1 and x_2.

How Bacteria Adjust to Living in Salty Environments

Sodium chloride—salt—is a key element to the functioning of all cells; every cell has mechanisms to adjust the salt concentration inside the cell. Most cells do this by moving water in or out of the cell to adjust the difference in salt concentration across the cell membrane. When the concentration of salt outside the cell is high relative to the concentration inside, water moves out of the cell to reestablish the balance. If the salt concentration outside the cell is too high, enough water will leave the cell to cause cell damage or death.

Certain kinds of bacteria, called halophilic bacteria, can grow in environments with extremely high salt concentrations that would kill other bacteria. Halophilic bacteria have specialized systems for surviving in these high salt concentrations. Some of these bacteria have enzymes that can withstand the high salt, and the salt concentration inside these bacteria is kept high, similar to that in the environment outside the bacteria, so that normal cell function is maintained. Alternatively, some of these bacteria actively transport salt out of themselves to keep their intracellular salt concentration at a level similar to normal bacteria, where most ordinary enzymes can function. These bacteria usually manufacture substances that prevent water from moving out of the cell despite the differences in salt concentrations inside and outside the cell.

Rengpipat and colleagues[*] were interested in determining which of these strategies was used by a newly described species of halophilic bacteria, *Halobacteroides acetoethylicus.* They measured the activity of an enzyme called a hydrogenase (that gets rid of hydrogen gas, which is a by-product of other metabolic activity in this bacterium) at different levels of salt concentration, ranging from 0.5 to 3.4 *M* (Fig. 3-17A). Because Rengpipat and colleagues collected data at discrete values of salt concentration, we can think of the points in Fig. 3-17A as falling along the contours of the three-dimensional surface in Fig. 3-17B, with salt concentration being the third dimension. However, the surface is not flat. In contrast to the roughly parallel contours we encountered before, these contours are not parallel. Specifically, at lower salt concentrations (the upper curves), the slopes of the contour lines are steeper than at higher salt concentrations (lower curves). What this means is that the activity of the hydrogenase (the slope of the line relating enzyme catalyzed product formation to time) appears to decrease with increasing salt concentration. The question at hand is: Does this enzyme activity change systematically with salt concentration?

To answer this question, we formulate the regression model

$$\hat{T} = b_0 + b_S S + (b_t + b_{st}S)t \qquad (3.21)$$

in which the estimated amount of tritiated water produced $\hat{T}$ depends on both salt concentration S and time t, and *the sensitivity of tritiated water production to time (i.e., the rate of hydrogenase activity) depends on the salt concentration.* For purposes of computing the multiple linear regression, we define the new variable I (for interaction), where $I = S \cdot t$ and rewrite Eq. (3.21) as

$$\hat{T} = b_0 + b_S S + b_t t + b_{st}I$$

[*]S. Rengpipat, S. E. Lowe, and J. G. Zeikus, "Effect of Extreme Salt Concentrations on the Physiology and Biochemistry of *Halobacteroides acetoethylicus,*" *J. Bacteriol.* 170:3065–3071, 1988.

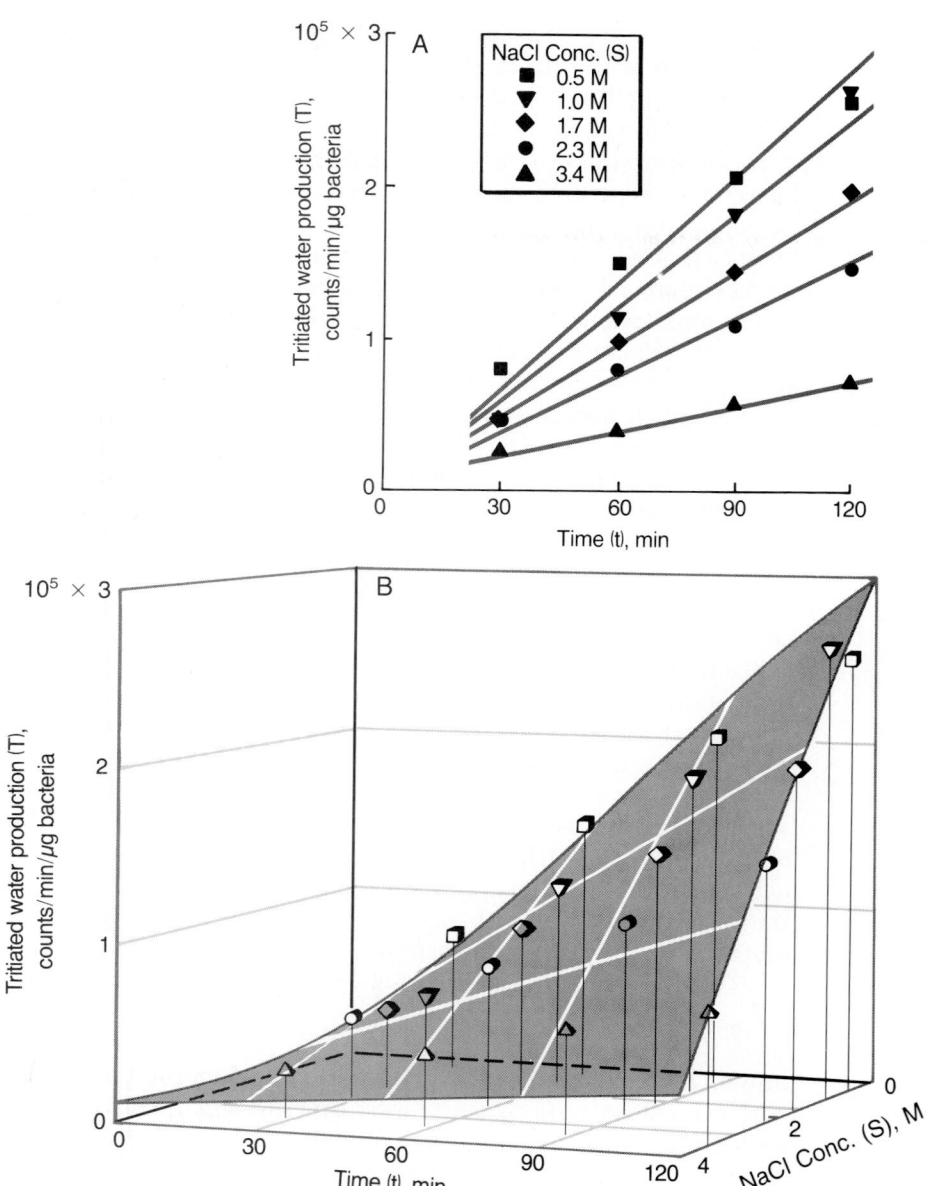

FIGURE 3-17 Panel *A* shows data on the production of tritiated water by the bacterium *H. acetoethylicus* over time in media with different concentrations of salt (NaCl). The rate of tritiated water production appears to increase more slowly as the salt concentration in the medium increases. In contrast to earlier examples, as the NaCl concentration changes, the *slope* of the relationship between tritiated water production and time changes. There is no shift in the intercept. These changes in slope are manifestations of an *interaction* between the two independent variables, time and NaCl concentration. Panel *B* shows the same data as panel *A* plotted in

DEP VARIABLE: W   Tritiated H2O production

ANALYSIS OF VARIANCE

SOURCE	DF	SUM OF SQUARES	MEAN SQUARE	F VALUE	PROB>F
MODEL	3	9.69601670	3.23200557	331.353	0.0001
ERROR	16	0.15606330	0.009753956		
C TOTAL	19	9.85208000			

ROOT MSE	0.09876212	R-SQUARE	0.9842	
DEP MEAN	1.184	ADJ R-SQ	0.9812	
C.V.	8.341395			

PARAMETER ESTIMATES

VARIABLE	DF	PARAMETER ESTIMATE	STANDARD ERROR	T FOR HO: PARAMETER=0	PROB > |T|
INTERCEP	1	0.003811189	0.10922900	0.035	0.9726
S	1	0.02482517	0.05331099	0.466	0.6477
T	1	0.02580154	0.001329493	19.407	0.0001
ST	1	-0.00598588	0.0006488807	-9.225	0.0001

FIGURE 3-18 Multiple regression analysis of the data in Fig. 3-17. Note that the interaction term is significantly different from zero, which means that different salt concentrations have different effects at different times so that the effects of salt concentration and time are not simply additive.

Figure 3-18 shows the results of these computations. (The data are in Table C-6, Appendix C.) The high value of R^2 = .984 indicates that this model provides an excellent fit to the data, as does inspection of the regression lines drawn in Fig. 3-17A. The high value of F associated with the overall goodness of fit, 331.4, confirms that impression.

The interesting conclusions that follow from these data come from examining the individual regression coefficients. First, the intercept term b_0 is not significantly different from zero (t = .04 with 16 degrees of freedom; P = .97), indicating that, on the average, no tritiated water has been produced at time zero, as should be the case. Likewise, the coeffi-

FIGURE 3-17 (*continued*) three dimensions with the two independent variables, time and NaCl concentration, plotted in the horizontal plane and the amount of tritiated water produced (the dependent variable) plotted on the vertical axis. Because of the interaction between time and NaCl concentration, the surface is no longer flat but rather twisted. (Data from Figure 5 of S. Rengpipat, S. E. Lowe, and J. G. Zeikus, "Effect of Extreme Salt Concentrations on the Physiology and Biochemistry of *Halobacteroides acetoethylicus*," *J. Bacteriol.* 170:3065–3071, 1988.)

cient b_S, which tests for an independent effect of salt concentration on tritiated water production, is also not significantly different from zero ($t = .47$; $P = .65$). This situation contrasts with that of the coefficients associated with time and the salt by time interaction, b_t and b_{st}, both of which are highly significant ($P < .0001$), indicating that the amount of tritiated water produced increases with time (because b_t is positive) at a rate which decreases as salt concentration increases (because b_{st} is negative). There is, however, physiologically significant hydrogenase activity at salt concentrations far above those usually found inside cells. This result indicates that *Halobacteroides acetoethylicus* has adapted to high salt environments by evolving enzymes that can withstand high intracellular salt concentrations, rather than by pumping salt out to keep salt concentrations inside the cell more like that of normal bacteria.

The presence of the interaction term in Eq. (3.21) means that the regression surface is no longer a plane in the three-dimensional space defined by the original two independent variables. (If we viewed this equation in a four-dimensional space in which the three independent variables S, T, and I are three of the dimensions, the surface would again be a plane.) Figure 3-17B shows the surface defined by the regression equation.

In order to demonstrate more clearly the importance of including interaction effects, suppose we had barged forward without thinking and fit the data in Fig. 3-17 to a multiple regression model that does not contain the interaction term:

$$\hat{T} = c_0 + c_S S + c_t t \tag{3.22}$$

The computer output for the regression using this equation appears in Fig. 3-19. The F statistic is 76.42 ($P < .0001$) and the R^2 is .90, both of which seem to indicate a very good model fit to the data. The regression parameter estimates indicate that the tritiated water is produced over time ($b_t = 0.015 \times 10^5$ counts per microgram of bacteria) and that increasing salt concentration is associated with less tritiated water production ($b_S = -0.42 \times 10^5$ counts per minute per microgram of bacteria per mole).

However, these numbers do not really tell the whole story. Figure 3-20 shows the same data as Fig. 3-17 with the regression line defined by Eq. (3.22) superimposed. By failing to include the interaction term, the "contours" are again parallel, so that the residuals are not randomly distributed about the regression plane, indicating that, despite the high values of R^2 and F associated with this equation, it does not accurately describe the data. Furthermore, there is a more basic problem: The original question we asked was whether the *rate* at which tritiated water was produced (i.e., the hydrogenase enzyme activity) depended on salt concentration. The term corresponding to this effect does not even appear

```
DEP VARIABLE: W          Tritiated H20 production

                      ANALYSIS OF VARIANCE

                      SUM OF          MEAN
          SOURCE   DF  SQUARES        SQUARE      F VALUE    PROB>F

          MODEL    2   8.86596009   4.43298004    76.421    0.0001
          ERROR   17   0.98611991   0.05800705
          C TOTAL 19   9.85208000

            ROOT MSE    0.2408465   R-SQUARE     0.8999
            DEP MEAN        1.184   ADJ R-SQ     0.8881
            C.V.        20.34177

                      PARAMETER ESTIMATES

                      PARAMETER     STANDARD    T FOR H0:
          VARIABLE DF  ESTIMATE      ERROR    PARAMETER=0   PROB > |T|

          INTERCEP  1  0.80292677   0.16225722     4.948      0.0001
          S         1  -0.424116    0.05307514    -7.991      0.0001
          T         1  0.01514667   0.001605644    9.433      0.0001
```

FIGURE 3-19 Analysis of the data in Fig. 3-17 without including the interaction between salt concentration and time.

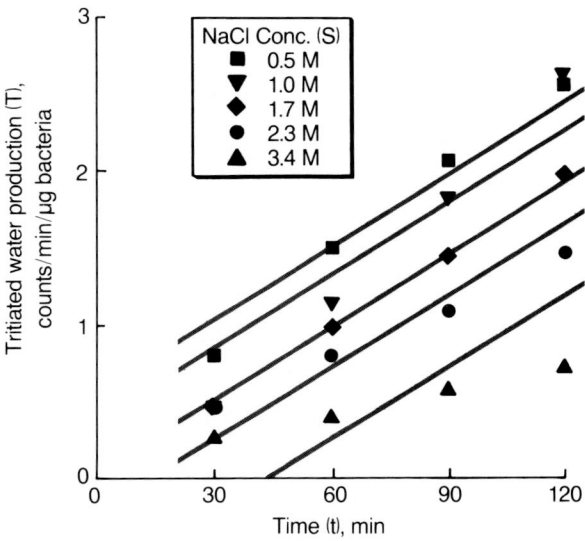

FIGURE 3-20 Plot of the data on tritiated water production against salt concentration and time from Fig. 3-17 together with the regression results from Fig. 3-19. By leaving the interaction term out of the model, we again have the result in which the differences in tritiated water production at any given time are represented by a parallel shift in proportion to the salt concentration. This model ignores the "fanning out" in the curves and is a misspecified model for these data (compare with Fig. 3-17*A*).

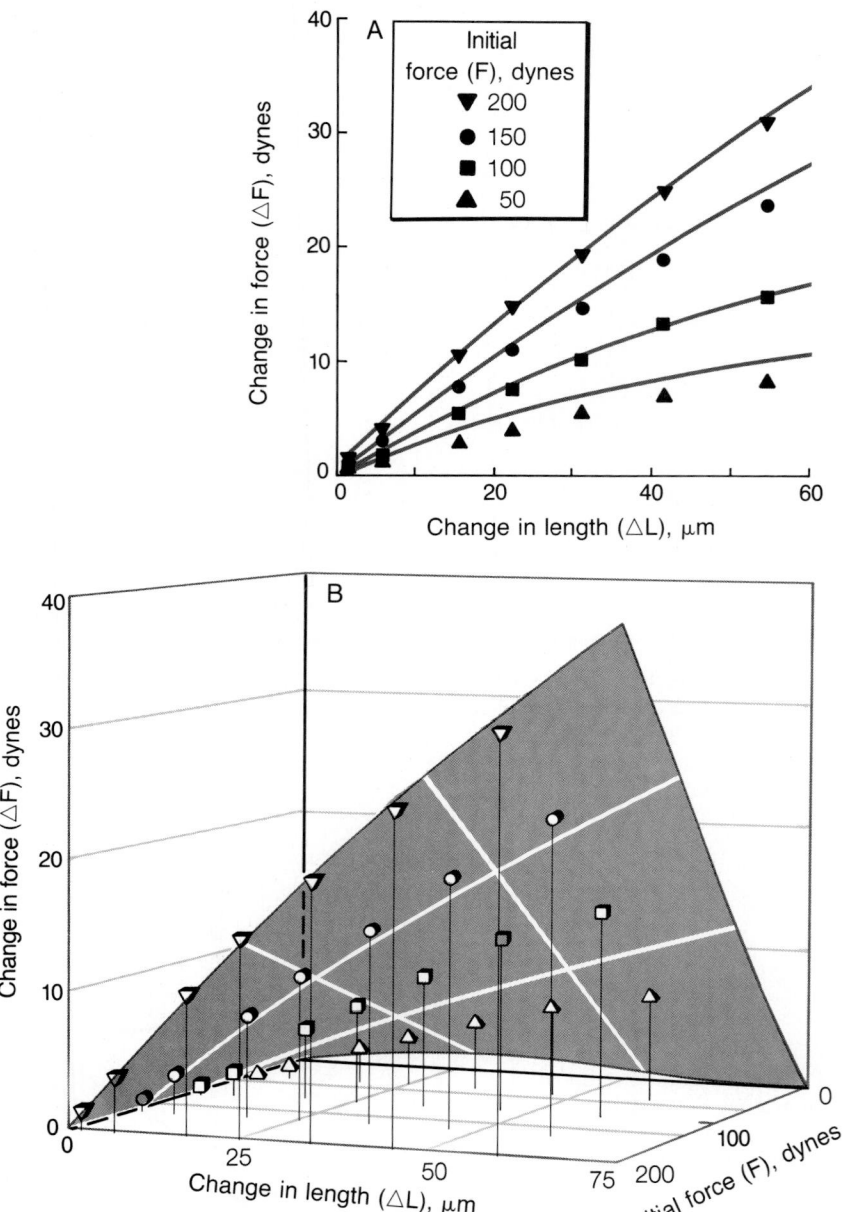

FIGURE 3-21 Panel *A* shows the change in force as a function of change in length in smooth muscles at several different initial force levels. Notice that these data appeared to exhibit both a nonlinearity in the relationship and also an interaction between the effects of changing length and changing initial force on the force increment accompanying a stretch. Panel *B* shows the same data plotted in three dimen-

in Eq. (3.22). This example again underscores the importance of looking at the raw data and how the regression equation fits it.

In the general discussion of interaction above, we noted that most of the time, one should include the main effects in a model if an interaction is significant. The lack of significance for the coefficients b_0 and b_S in Eq. (3.21) implies the regression model

$$\hat{T} = d_t t + d_{st} St$$

which contains the interaction St and the main effect t, but not the main effect S. In this case, it is appropriate to include the interaction without both of the main effects because there is no constant term.

The Response of Smooth Muscle to Stretching

Muscles serve two functions in the body: they produce positive motion by actively contracting and produce resistance to motion by passively stretching (like a spring). This resistance to extension is particularly important in smooth muscle, which is found in many of the body's internal organs, such as blood vessels and the digestive tract. Like most biological tissues, smooth muscle becomes stiffer as it is stretched, i.e., as the resistive forces increase, a greater increase in stiffness occurs. Meiss[*] studied this phenomenon by collecting data on the influence of developed force F on the relation between the change in force ΔF in response to small length changes and the magnitude of the length changes ΔL of contracting smooth muscle. The change in force per change in length is the stiffness of the muscle. He reported the family of slightly nonlinear relationships shown in Fig. 3-21A. The question is: Does the stiffness depend on the level of developed force F after allowing for the nonlinear relationship between ΔF and ΔL?

We can address this question by using a multiple regression model allowing for both some nonlinearity in and interaction between the variables (notice that for theoretical reasons, we expect no constant term

[*]R. A. Meiss, "Nonlinear Force Response of Active Smooth Muscle Subjected to Small Stretches," *Am. J. Physiol.* 246:C114–C124, 1984.

FIGURE 3-21 (*continued*)   sions, with initial force and change in length as the independent variables in the horizontal plane and the change in force that accompanies the change in length as the dependent variable on the vertical axis. Because of the nonlinear and interaction terms included in the model, the surface is both curved and twisted. The lines in panel *A* represent contour lines of this surface at constant levels of initial force. (Data from Figure 6B of R. A. Meiss, "Nonlinear Force Response of Active Smooth Muscle Subjected to Small Stretches," *Am. J. Physiol.* 246:C114–C124, 1984.)

and have omitted it from the model; i.e., ΔF will be zero when ΔL is zero).

$$\Delta \hat{F} = C_{\Delta L} \Delta L + C_{\Delta L^2}(\Delta L)^2 + C_{\Delta LF} \Delta L \cdot F$$
$$= (C_{\Delta L} + C_{\Delta LF} F) \Delta L + C_{\Delta L^2}(\Delta L)^2$$

The parameter estimates $C_{\Delta L}$ and $C_{\Delta L^2}$ describe the quadratic relationship between $\Delta \hat{F}$ and ΔL, whereas $C_{\Delta LF}$ quantifies how the relationship between $\Delta \hat{F}$ and ΔL changes in response to changes in developed force F.

Figure 3-22 shows the results of conducting this regression analysis (the data are in Table C-7, Appendix C). The R^2 of .9995 indicates that the regression equation accounts for 99.95 percent of the variation in ΔF associated with variation in ΔL and F, and the overall test of the regression equation confirms that the fit is statistically significant ($F = 15{,}262.3$ with 3 numerator and 25 denominator degrees of freedom; $P < .0001$). The parameter estimates $C_{\Delta L} = .092$ dyn/μm ($P < .0001$) and

```
DEP VARIABLE: DELTAF    change in Force

                         ANALYSIS OF VARIANCE

                       SUM OF          MEAN
        SOURCE    DF    SQUARES         SQUARE      F VALUE      PROB>F

        MODEL      3    4365.77625    1455.25875   15262.284    0.0001
        ERROR     25    2.38374988      0.09535
        U TOTAL   28    4368.16000

            ROOT MSE    0.3087879      R-SQUARE      0.9995
            DEP MEAN    9.585714       ADJ R-SQ      0.9994
            C.V.        3.221335
NOTE: NO INTERCEPT TERM IS USED. R-SQUARE IS REDEFINED.


                        PARAMETER ESTIMATES

                    PARAMETER      STANDARD     T FOR H0:
        VARIABLE  DF  ESTIMATE       ERROR     PARAMETER=0    PROB > |T|

        DELTAL    1   0.09239141   0.008184809    11.288       0.0001
        DELTAL2   1  -0.00169353   0.0001542442  -10.980       0.0001
        DELTALXF  1   0.002860597  .00003422563   83.581       0.0001
```

FIGURE 3-22 Multiple regression analysis of the data in Fig. 3-21. Note that all of the terms in the regression model are significantly different from zero, indicating that there is both curvature in the relationship between change in force and change in length and also a change in the relationship between change in force and change in length as the initial force changes.

$C_{\Delta L^2} = -.002$ dyn^2/μm $(P < .0001)$ confirm that the muscle becomes stiffer as the length perturbation increases (but in a nonlinear fashion such that the rate of change of ΔF as ΔL increases becomes smaller at larger values of L). Moreover, the fact that $C_{\Delta LF} = .003$ μm^{-1} is significantly different from zero $(P < .0001)$ and positive indicates that, independent of the effects of length change, the muscle is stiffer as the force it must develop increases.

These results indicate that the stiffness is smallest when the muscle is unloaded (when $F = 0$), and increases as the active force increases. The active stiffness of the contracting muscle is related to the number of cross-links between proteins in the muscle cells, so it appears that these cross-links not only play a role in the active development of forces, but also the passive stiffness to resist stretching.

SUMMARY

Data like those in the examples presented in this chapter are typical of well-controlled, isolated (in vitro) physiological experiments. Experimental designs of this kind allow one to make optimum use of modeling and multiple linear regression methods. However, as noted previously, not all interesting questions in the health sciences are easily studied in such isolated systems, and multiple linear regression provides a means to extract information about complex physiological systems when an experimenter cannot independently manipulate each predictor variable.

Multiple linear regression is a direct generalization of the simple linear regression developed in Chap. 2. The primary difference between multiple and simple linear regression is the fact that in multiple regression the value of the dependent variable depends on the values of several independent variables, whereas in simple linear regression there is only a single independent variable. The coefficients in a multiple linear regression equation are computed to minimize the sum of squared deviations between the value predicted by the regression equation and the observed data. In addition to simply describing the relationship between the independent and dependent variables, multiple regression can be used to estimate the relative importance of different potential independent variables. It provides estimates of the parameters associated with each of the independent variables, together with standard errors that quantify the precision with which we can make these estimates. Moreover, by incorporating independent variables that are nonlinear functions of the natural independent variables (e.g., squares or logarithms), it is possible to describe many nonlinear relationships using linear regression techniques.

Although we have assumed that the dependent variable varies normally about the regression plane, we have not had to make any such

assumptions about the distribution of values of the independent variables. Because of this freedom, we could define dummy variables to identify different experimental conditions and so formulate multiple regression models that could be used to test for shifts in lines relating two or
more variables in the presence or absence of some experimental condition. This ability to combine effects due to continuously changing variables and discrete (dummy) variables is one of the great attractions of
multiple regression analysis for analyzing biomedical data. In fact, in
Chaps. 7, 8, and 9, we will use multiple regression equations in which
all the independent variables are dummy variables to show how to formulate analysis of variance problems as equivalent multiple regression
problems. Such formulations are often easier to handle and interpret than
traditional analysis of variance formulations.

As with all statistical procedures, multiple regression rests on a set
of assumptions about the population from which the data were drawn.
The results of multiple regression analysis are only reliable when the
data we are analyzing reasonably satisfy these assumptions. We assume
that the mean value of the dependent variable for any given set of values
of the independent variables is described by the regression equation and
that the population members are normally distributed about the plane
of means. We assume that the variance of population members about the
plane of means for any combination of values of the independent variables is the same regardless of the values of the independent variables.
We also assume that the independent variables are statistically independent, i.e., that knowing the value of one or more of the independent
variables gives no information about the others. Although this latter
condition is rarely strictly true in practice, it is often reasonably satisfied.
When it is not, we have *multicollinearity*, which leads to imprecise estimates of the regression parameters and often very misleading results.
We have started to show how to identify and remedy these problems in
this chapter and will next treat them in more depth.

PROBLEMS

3.1   In the study of the effect of secondhand smoke on the relationship
between Martian weight and height in Fig. 3-7*A*, exposure to secondhand smoke shifted the line down in a parallel fashion as described by Eq. (1.3). If secondhand smoke affected the slope as well
as the intercept, the lines would not be parallel. What regression
equation describes this situation?

3.2   Draw a three-dimensional representation of the data in Fig. 3-7*A*
with the dummy variable that defines exposure to secondhand
smoke (or lack thereof) as the third axis. What is the meaning of
the slope of the regression plane in the dummy variable dimension?

3.3 For the study of the effects of Martian height and water consumption on weight in Fig. 3-1, the amounts of water consumed fell into 3 distinct groups, 0, 10, and 20 cups/day. This fortuitous circumstance made it easy to draw Fig. 3-1 but rarely happens in practice. Substitute the following values for the column of water consumption data in Table 1-1, and repeat the regression analysis used to obtain Eq. (3.1): 2, 4, 6, 10, 8, 12, 18, 14, 16, 20, 22, 24.

3.4 What is $s_{y|x}$ in regression analysis of the smooth muscle data in Fig. 3-22?

3.5 Tobacco is the leading preventable cause of death in the world today but attracts relatively little coverage in the print media compared with other, less serious, health problems. Based on anecdotal evidence of retaliation by tobacco companies against publications that carry stories on smoking, many observers have asserted that the presence of tobacco advertising discourages coverage of tobacco-related issues. To address this question quantitatively, Warner and Goldenhar[*] took advantage of the "natural experiment" that occurred when broadcast advertising of cigarettes was banned in 1971, which resulted in cigarette advertisers greatly increasing their reliance on print media, with a concomitant increase in advertising expenditures there. For the years from 1959 to 1983, they collected data on the number of smoking-related articles in 50 periodicals, 39 of which accepted cigarette advertising and 11 of which did not. They used these data to compute the average number of articles related to smoking published each year per magazine, with separate computations for those that did and did not accept cigarette advertising. They also collected data on the amount of cigarette advertising these magazines carried and their total advertising revenues, and computed the percentage of revenues from cigarette ads. To control for changes in the "newsworthiness" of the smoking issue, they observed the number of smoking-related stories in the *Christian Science Monitor*, a respected national newspaper that does not accept cigarette advertising (the data are in Table D-7, Appendix D). Do these data support the assertion that the presence of cigarette advertising discourages reporting on the health effects of smoking?

3.6 In the baby birds data shown in Fig. 3-14 and analyzed in Fig. 3-15, what was the R^2 and what was the F statistic for testing overall goodness of fit? What do these two statistics tell you about the overall quality of fit to the data? Does a highly significant F statistic mean that the fit is "good"?

[*]K. E. Warner and L. M. Goldenhar, "The Cigarette Advertising Broadcast Ban and Magazine Coverage of Smoking and Health," *J. Pub. Health Policy* 10:32–42, 1989.

3.7   Drugs that block the β-receptors for adrenaline and related compounds, generally known as catecholamines, in the heart are called β-blockers. These drugs are often used to treat heart disease. Sometimes β-blockers have a dual effect whereby they actually stimulate β-receptors if the levels of catecholamines in the body are low, but exert their blocking effect if catecholamine levels are elevated, as they are often in patients with heart disease. Xamoterol is a β-blocker that has such a dual stimulation-inhibition effect. Before xamoterol can be most effectively used to treat heart disease, the level of body catecholamines at which the switch from a stimulating to a blocking effect occurs needs to be determined. Accordingly, Sato and his coworkers[*] used exercise to stimulate catecholamine release in patients with heart disease. They measured a variety of variables related to the function of the cardiovascular system and related these variables to the level of the catecholamine norepinephrine in the blood with patients who did not exercise (group 1) and in three groups who engaged in increasing levels of exercise (groups 2 to 4), which increases the level of norepinephrine. Patients were randomly assigned to either no xamoterol (6 patients per group) or xamoterol (10 patients per group). Sato and his coworkers then related systolic blood pressure to the level of norepinephrine under the two conditions, drug and no drug. The mean values ($\pm$ SD) of systolic blood pressure and blood norepinephrine under all treatment conditions are

Xamoterol	Exercise Group	Systolic blood Pressure P, mmHg	Blood Norepinephrine N, pg/mL	n
No	1	125 ± 22	234 ± 87	6
No	2	138 ± 25	369 ± 165	6
No	3	150 ± 30	543 ± 205	6
No	4	162 ± 31	754 ± 279	6
Yes	1	135 ± 30	211 ± 106	10
Yes	2	146 ± 27	427 ± 257	10
Yes	3	158 ± 30	766 ± 355	10
Yes	4	161 ± 36	1068 ± 551	10

[*]H. Sato, M. Inoue, T. Matsuyama, H. Ozaki, T. Shimazu, H. Takeda, Y. Ishida, and T. Kamada, "Hemodynamic Effects of the β_1-Adrenoceptor Partial Agonist Xamoterol in Relation to Plasma Norepinephrine levels during Exercise in Patients with Left Ventricular Dysfunction," *Circulation* 75:213–220, 1987.

A. Is there any evidence that xamoterol had an effect, compared to control? At what level of plasma norepinephrine does xamoterol appear to switch its action from a mild β-stimulant to a β-blocker? Include a plot of the data in your analysis. *B.* Repeat your investigation of this problem using all of the data points listed in Table D-8, Appendix D. Do you reach the same conclusions? Why or why not? *C.* Which analysis is correct?

3.8 In Prob. 2.6, you used four simple regression analyses to try to draw conclusions about the factors related to urinary calcium loss U_{Ca} in patients receiving parenteral nutrition. (The data are in Table D-5, Appendix D.) *A.* Repeat that analysis using multiple regression to consider the simultaneous effects of all four variables, D_{Ca}, D_p, U_{Na}, and G_{fr}. Which variables seem to be important determinants of U_{Ca}? *B.* Compare this multivariate result to the result of the four univariate correlation analyses done in Prob. 2.6. *C.* Why are there differences, and which approach is preferable?

3.9 Cell membranes are made of two "sheets" of fatty molecules known as a lipid bilayer, which contains many proteins on the surface of the bilayer and inserted into it. These proteins are mobile in the bilayer, and some function as carriers to move other molecules into or out of the cell. The reason the proteins can move has to do with the "fluidity" of the bilayer, which arises because the lipids are strongly attracted to one another but are not covalently bound. Different kinds of lipids attract each other differently, so the fluidity of the membranes can change as the lipid composition changes. Zheng and coworkers[*] hypothesized that changes in membrane fluidity would change carrier-mediated transport of molecules so that as the membrane became less fluid, transport rate would slow. To investigate this hypothesis, they took advantage of the fact that increasing cholesterol content in a membrane decreases its fluidity, and they measured the accumulation of the amino acid, leucine, over time in cells with either no cholesterol or one of three amounts of cholesterol (the data are in Table D-9, Appendix D). Is there evidence that increasing the cholesterol content of the cell membrane slows the rate of leucine transport?

3.10 Inhibin is a hormone thought to be involved in the regulation of sex hormones during the estrus cycle. It has been recently isolated and characterized. Most assays have been bioassays based on cultured cells, which are complicated. Less cumbersome radioimmu-

[*]T. Zheng, A. J. M. Driessen, and W. N. Konings, "Effect of Cholesterol on the Branched-Chain Amino Acid Transport System of *Streptococcus cremoris*," *J. Bacteriol.* 170:3194–3198, 1988.

nological assays (RIAs) have been developed, but they are less sensitive than bioassays. Robertson and coworkers[*] developed a new RIA for inhibin and, as part of its validation, compared it to standard bioassay. They made this comparison in two different phases of the ovulation cycle, the early follicular phase and the midluteal phase (the data are in Table D-10, Appendix D). If the two assays are measuring the same thing, there should be no difference in the relationship between RIA R and bioassay B in the two different phases. A. Is there a difference? B. If the two assays are measuring the same thing, the slope should be 1.0. Is it?

3.11 In Prob. 2.4, you used linear regression to describe the relationship between gentamicin efficacy (in terms of the number of bacteria present after treatment C) and the amount of time the level of gentamicin in the blood was above its effective concentration M for dosing intervals of 1 to 4 hours. These investigators also used dosing intervals of 6 to 12 hours. (The data are in Table D-3, Appendix D.) Does the relationship between C and M depend on dosing schedule? Use multiple regression, and explain your answer.

3.12 In Prob. 2.7 you examined the ability of the MR spectrum relaxation time T_1 to predict muscle fiber composition in the thighs of athletes. Does adding the other MR spectrum relaxation time T_2 to the predictive equation significantly improve the ability of MR spectroscopy to precisely predict muscle fiber type (the data are in Table D-6, Appendix D)?

3.13 Ventilation of the lungs is controlled by the brain, which processes information sent from stretch receptors in the lungs and sends back information to tell the muscles in the chest and diaphragm when to contract and relax so that the lungs are inflated just the right amount. To explore the quantitative nature of this control in detail, Zuperku and Hopp[†] measured the output of stretch receptor neurons from the lungs during a respiratory cycle. To see how control signals from the brain interacted with this output, they stimulated nerves leading to the lungs at four different frequencies F, 20, 40, 80, and 120 Hz, and measured the change in output in the stretch receptor nerves ΔS. As part of the analysis, they wanted to know how the normal stretch receptor nerve discharge S affected the

[*]D. M. Robertson, C. G. Tsonis, R. I. McLachlan, D. J. Handelsman, R. Leask, D. T. Baird, A. S. McNeilly, S. Hayward, D. L. Healy, J. K. Findlay, H. G. Burger, and D. M. de Kretser, "Comparison of Inhibin Immunological and *in Vitro* Biological Activities in Human Serum," *J. Clin. Endocrinol. Metab.* 67:438–443, 1988.

[†]E. J. Zuperku and F. A. Hopp, "Control of Discharge Patterns of Medullary Respiratory Neurons by Pulmonary Vagal Afferent Inputs," *Am. J. Physiol.* 253:R809–R820, 1987.

change ΔS and how this relationship between ΔS and S was affected by the frequency of stimulation F (the data are in Table D-11, Appendix D). *A.* Find the best-fitting regression plane relating ΔS to S and F. *B.* Interpret the results.

3.14 In the baby birds example, we analyzed the dependence of ventilation on the oxygen and carbon dioxide content of the air the nestling bank swallows breathed. Table D-12, Appendix D, contains similar data for adult bank swallows. Find the best-fitting regression plane.

3.15 Combine the data sets for baby birds (Table C-5, Appendix C) and adult birds (Table D-12, Appendix D). Create a dummy variable to distinguish the baby bird data from the adult data (let the dummy variable equal 1 for baby birds). Using multiple linear regression, is there any evidence that the adult birds differ from the baby birds in terms of their ventilatory control as a function of oxygen and carbon dioxide?

3.16 Calculate the 95 percent confidence interval for the sensitivity of adult bird ventilation to oxygen in Prob. 3.14. What does this tell you about the statistical significance of this effect?

3.17 In Prob. 2.3 we concluded that female lung cancer rates were related to male lung cancer rates in different countries and that breathing secondhand smoke increased the risk of getting breast cancer. In Fig. 2-8, we showed that breast cancer rate also varied with the amount of animal fat in the diet of different countries, an older and more accepted result than the conclusion about involuntary smoking. Until recently, when the tobacco industry began invading the Third World to compensate for declining sales in the developed countries, both high animal fat diets and smoking were more prevalent in richer countries than poorer ones. This fact could mean that the dependence of female breast cancer rates on male lung cancer rates we discovered in Prob. 2.3 is really an artifact of the fact that diet and smoking are changing together, so that the effects are *confounded*. *A.* Taking the effect of dietary animal fat into account, does the breast cancer rate remain dependent on male lung cancer rate? *B.* Is there an interaction between the effects of diet and exposure to secondhand smoke on the breast cancer rate? The data are in Table D-2, Appendix D.

CHAPTER

FOUR

Do the Data Fit
the Assumptions?

Up to this point we have formulated models for multiple linear regression and used these models to describe how a dependent variable depends on one or more independent variables. By defining new variables as nonlinear functions of the original variables, such as logarithms or powers, and including interaction terms in the regression equation, we have been able to account for some nonlinear relationships between the variables. By defining appropriate dummy variables, we have been able to account for shifts in the relationship between the dependent and independent variables in the presence or absence of some condition. In each case, we estimated the coefficients in the regression equation, which, in turn, could be interpreted as the sensitivity of the dependent variable to changes in the independent variable. We also could test a variety of statistical hypotheses to obtain information on whether or not different treatments affected the dependent variable. All these powerful techniques rest on the assumptions that we made at the outset concerning the population from which the observations were drawn.

We now turn our attention to procedures to ensure that the data reasonably match these underlying assumptions. When the data do not match the assumptions of the analysis, it indicates that either there are erroneous data or that the regression equation does not accurately describe the underlying processes. The size and nature of the violations of the assumptions can provide a framework in which to correct erroneous data or to revise the regression equation.

Before continuing, let us recapitulate the assumptions that underlie what we have accomplished so far:

- *The mean of the population of the dependent variable at a given set of values of the independent variables varies linearly as the independent variables vary.*

- *For any given values of the independent variables, the possible values of the dependent variable are distributed normally.*
- *The standard deviation of the dependent variable about its mean at any given values of the independent variables is the same for all values of the independent variables.*
- *The deviations of all members of the population from the plane of means are statistically independent, i.e., the deviation associated with one member of the population has no effect on the deviations associated with other members.*

The first assumption boils down to the statement that *the regression model (i.e., the regression equation) is correctly specified* or, in other words, that the regression equation reasonably describes the relationship between the mean value of the dependent variable and the independent variables. We have already encountered this problem—for example, our analysis of the relationship between heat loss and temperature in gray seals in Chap. 2 and the study of protein synthesis in newborns and adults in Chap. 3. In both cases, we fit a straight line through the data, then realized that we were leaving out an important factor—in the first case a nonlinearity and in the second case another important independent variable. Once we modified the regression equations to remedy these oversights, they provided better descriptions of the data. *If the model is not correctly specified, the entire process of estimating parameters and testing hypotheses about these parameters rests on a false premise* and may lead to a misunderstanding of the system or process being studied. In other words, an incorrect model is worse than no model.

The remaining assumptions deal with the variation of the population about the regression plane and are known as the *normality, equal variance,* and *independence* assumptions. While it is not possible to observe all members of the population to test these assumptions directly, it is possible to gain some insight into how well you are meeting them by examining the distribution of the *residuals* between the observed values of the dependent variables and the values predicted by the regression equation. If the data meet these assumptions, the residuals will be normally distributed about zero, with the same variance, independent of the value of the independent variables, and independent of each other.*

We will now develop formal techniques based on the residuals to examine whether the regression model adequately describes the data and whether the data meet the assumptions of the underlying regression anal-

---

*Strictly speaking, the residuals cannot be independent of each other because the average residual must be zero because of the way the regression parameters are estimated. When there are a large number of data points, however, this restriction is very mild and the residuals are nearly independent.

ysis. These techniques, collectively called *regression diagnostics*, can be used in combination with careful examination of plots of the raw data to help you ensure that the results you obtain with multiple regression are not only computationally correct, but sensible.

ANOTHER TRIP TO MARS

Flush with several publications from our previous trips to Mars, we renew our grant and return again, this time to study the relationship between foot size and intelligence among the Martians. We collect data on 11 Martians and submit these data to a linear regression analysis using the equation

$$I = a_0 + a_F F \qquad (4.1)$$

in which I = intelligence (in zorp) and F = foot size (in centimeters). We submit the data (in Table 4-1) to linear regression analysis and obtain the computer output in Fig. 4-1, together with the values of the regression diagnostics we will discuss in this chapter. From this printout, the regression equation is

$$\hat{I} = 3.0 \text{ zorp} + .5 \text{ (zorp/cm)} F$$

The R^2 is .667, yielding a correlation coefficient r of .82, and the standard error of the estimate is $s_{I|F} = 1.2$ zorp. Because there is only one independent variable, we can use the methods described in Chap. 2 to test whether the overall regression equation ($F = 17.99$; $P = .002$) or the slope is significantly different from zero ($t = 4.24$; $P = .002$) to conclude that the slope of the regression line is significantly different from zero.

TABLE 4-1 Martian Intelligence vs. Foot Size Data Set Shown in Fig. 4-2A

Foot Size F, cm	Intelligence I, zorp
10	8.04
8	6.95
13	7.58
9	8.81
11	8.33
14	9.96
6	7.24
4	4.26
12	10.84
7	4.82
5	5.68

```
The regression equation is
I = 3.00 + 0.500 F


Predictor        Coef       Stdev     t-ratio         p
Constant        3.000       1.125       2.67      0.026
F               0.5001      0.1179      4.24      0.002


s = 1.237        R-sq = 66.7%     R-sq(adj) = 62.9%

Analysis of Variance

SOURCE          DF          SS          MS          F         p
Regression       1       27.510      27.510      17.99     0.002
Error            9       13.763       1.529
Total           10       41.273


 ROW    RawRes     StanRes     StudRes    StDelRes    Leverage    CookDist

   1    0.03900    0.03324     0.03324     0.03134    0.100000    0.000061
   2   -0.05082   -0.04332    -0.04332    -0.04084    0.100000    0.000104
   3   -1.92127   -1.63771    -1.77793    -2.08110    0.236364    0.489209
   4    1.30909    1.11588     1.11029     1.12680    0.090909    0.061637
   5   -0.17109   -0.14584    -0.14810    -0.13980    0.127273    0.001599
   6   -0.04136   -0.03526    -0.04051    -0.03820    0.318182    0.000383
   7    1.23936    1.05645     1.10190     1.11696    0.172727    0.126756
   8   -0.74045   -0.63117    -0.72516    -0.70458    0.318182    0.122700
   9    1.83882    1.56743     1.63487     1.83833    0.172727    0.279030
  10   -1.68073   -1.43267    -1.45488    -1.56846    0.127273    0.154341
  11    0.17945    0.15297     0.16607     0.15681    0.236364    0.004268
```

FIGURE 4-1 Results of regression analysis of Martian intelligence against foot size.

These numbers indicate that for every 1-cm increase in foot size, the average Martian increases in intelligence by .5 zorp. The correlation coefficient of .82 indicates that the data are clustered reasonably tightly around the linear regression line. The standard error of the estimate of 1.2 zorp indicates that most of the points fall in a band about 2.4 zorp $(2s_{I|F})$ above and below the regression line (Fig. 4-2A).

This pattern of data is, however, not the only one that can give rise to these results.* For example, Fig. 4-2B, C, and D shows three other widely different patterns of data *which give exactly the same regression equation, correlation coefficient, and standard error* as the results in Figs. 4-1 and 4-2A (the data are in Table C-8, Appendix C). Figures 4-3, 4-4, and 4-5 give the corresponding results of the regression analysis.

*The following discussion is based on a F. J. Anscombe, "Graphs in Statistical Analysis," *Am. Statist.* 27:17–21, 1973.

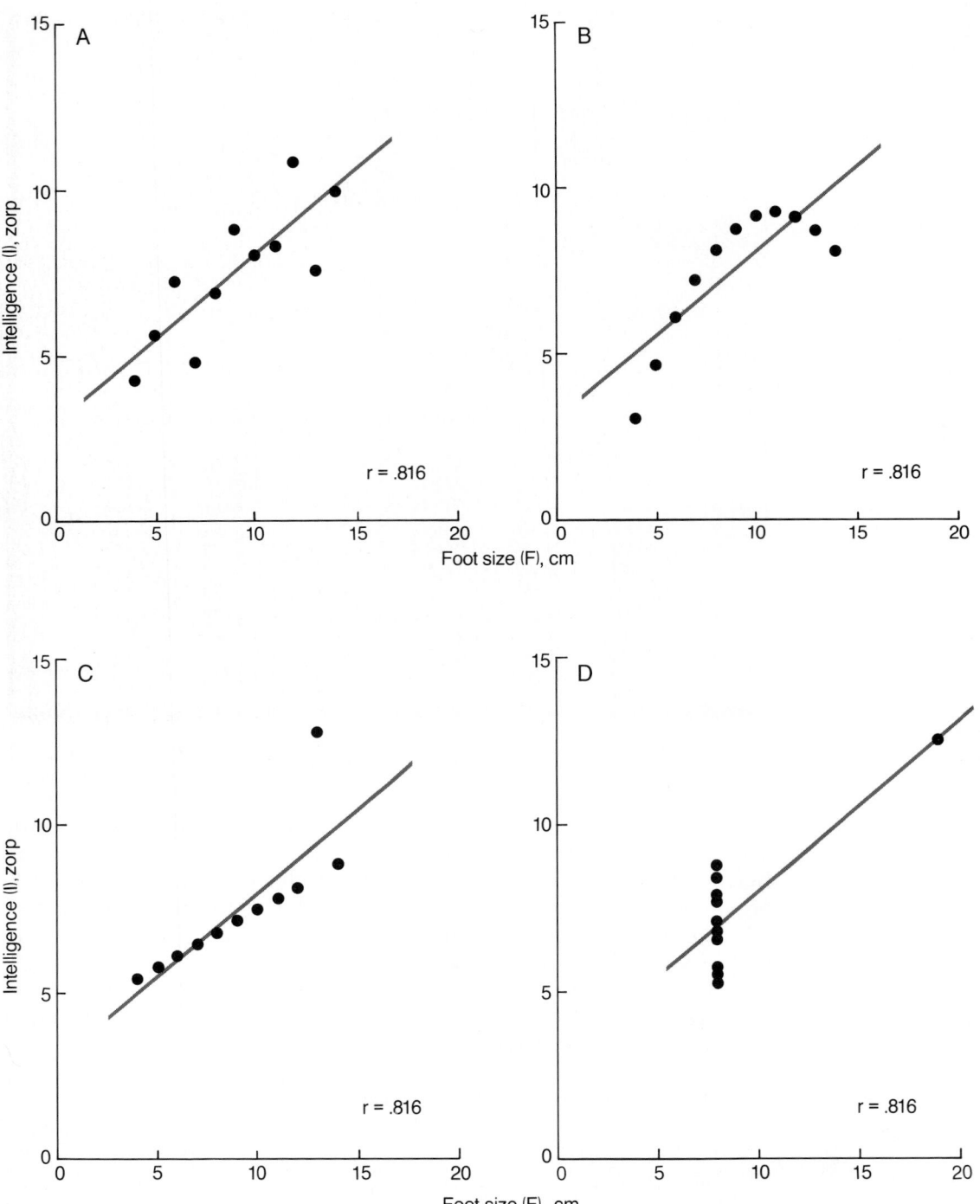

```
The regression equation is
I = 3.00 + 0.500 F

Predictor        Coef      Stdev     t-ratio        p
Constant        3.001      1.125        2.67     0.026
F               0.5000     0.1180       4.24     0.002

s = 1.237        R-sq = 66.6%     R-sq(adj) = 62.9%

Analysis of Variance

SOURCE          DF          SS          MS         F        p
Regression       1       27.500      27.500     17.97    0.002
Error            9       13.776       1.531
Total           10       41.276


ROW     RawRes    StanRes   StudRes   StDelRes   Leverage   CookDist

  1    1.13909    0.97049   0.97049    0.96699   0.100000   0.052325
  2    1.13909    0.97049   0.97049    0.96698   0.100000   0.052325
  3   -0.76091   -0.64829  -0.70379   -0.68259   0.236364   0.076657
  4    1.26909    1.08125   1.07583    1.08657   0.090909   0.057871
  5    0.75909    0.64674   0.65676    0.63460   0.127273   0.031452
  6   -1.90091   -1.61955  -1.86073   -2.23647   0.318182   0.807869
  7    0.12909    0.10998   0.11472    0.10823   0.172727   0.001374
  8   -1.90091   -1.61955  -1.86073   -2.23647   0.318182   0.807869
  9    0.12909    0.10998   0.11472    0.10824   0.172727   0.001374
 10    0.75909    0.64674   0.65676    0.63460   0.127273   0.031452
 11   -0.76091   -0.64829  -0.70379   -0.68259   0.236364   0.076657
```

FIGURE 4-3 Regression analysis for the data in Fig. 4-2*B*. The regression equation and analysis of variance are identical to that in Fig. 4-1. The pattern of the residuals presented at the bottom of the output, however, is different.

◀ **FIGURE 4-2** *A*. Raw data relating Martian intelligence to foot size, together with the regression line computed in Fig. 4-1. *B*. Another possible pattern of observations of the relationship between Martian intelligence and foot size that reflects a nonlinear relationship. These data yield exactly the same regression equation as that obtained from panel *A*. *C*. Another possible pattern of observations of the relationship between Martian intelligence and foot size that yields the same regression equation and correlation coefficient as that obtained from panel *A*. *D*. Still another pattern of observations that yields the same regression equation and correlation coefficient as the data in panel *A*. In this case, the regression equation is essentially defined by a single point at a foot size of 19 cm.

```
The regression equation is
I = 3.00 + 0.500 F

Predictor        Coef       Stdev     t-ratio         p
Constant        3.002       1.124        2.67     0.026
F               0.4997      0.1179       4.24     0.002

s = 1.236          R-sq = 66.6%    R-sq(adj) = 62.9%

Analysis of Variance

SOURCE          DF          SS          MS          F          p
Regression       1       27.470      27.470      17.97     0.002
Error            9       13.756       1.528
Total           10       41.226
```

ROW	RawRes	StanRes	StudRes	StDelRes	Leverage·	CookDist
1	-0.53973	-0.46018	-0.46018	-0.44	0.100000	0.01176
2	-0.23027	-0.19633	-0.19633	-0.19	0.100000	0.00214
3	3.24109	2.76339	2.99999	1216.69	0.236364	1.39285
4	-0.39000	-0.33252	-0.33085	-0.31	0.090909	0.00547
5	-0.68945	-0.58784	-0.59695	-0.57	0.127273	0.02598
6	-1.15864	-0.98787	-1.13497	-1.16	0.318182	0.30057
7	0.07918	0.06751	0.07042	0.07	0.172727	0.00052
8	0.38864	0.33136	0.38070	0.36	0.318182	0.03382
9	-0.84918	-0.72402	-0.75518	-0.74	0.172727	0.05954
10	-0.08055	-0.06867	-0.06974	-0.07	0.127273	0.00035
11	0.22891	0.19517	0.21188	0.20	0.236364	0.00695

FIGURE 4-4 Regression analysis for the data in Fig. 4-2C. The results of the regression computation and associated analysis of variance are identical to those in Fig. 4-2A. The pattern of residuals, however, is quite different, with all of the raw residuals quite small except for that associated with point 3, which corresponds to the one *outlier* in Fig. 4-2C.

Although the numerical regression results are identical, not all of these data sets are fit well by Eq. (4.1), as can be seen by looking at the raw data (Fig. 4-2). These graphical differences are also quantitatively reflected in the values of the regression diagnostics associated with the individual data points (Figs. 4-1 and 4-3 to 4-5). These differences are the key to identifying problems with the regression model or errors in the data. *The fact that we can fit a linear regression equation to a set of data—even if it yields a reasonably high and statistically significant correlation coefficient and small standard error of the estimate—does not ensure that the fit is appropriate or reasonable.*

Figure 4-2B shows a situation we have already encountered (in the study of heat transfer in gray seals in Chap. 2) when the data exhibit a

```
The regression equation is
I = 3.00 + 0.500 F

Predictor        Coef       Stdev     t-ratio          p
Constant        3.002       1.124        2.67      0.026
F               0.4999      0.1178       4.24      0.002

s = 1.236        R-sq = 66.7%    R-sq(adj) = 63.0%

Analysis of Variance

SOURCE           DF          SS           MS          F          p
Regression        1       27.490       27.490      18.00      0.002
Error             9       13.742        1.527
Total            10       41.232


 ROW  RawRes    StanRes    StudRes   StDelRes   Leverage   CookDist

    1  -0.421   -0.35913   -0.35913  -0.34104        0.1   0.007165
    2  -1.241   -1.05862   -1.05862  -1.06669        0.1   0.062259
    3   0.709    0.60480    0.60480   0.58217        0.1   0.020321
    4   1.839    1.56873    1.56873   1.73515        0.1   0.136718
    5   1.469    1.25311    1.25311   1.30031        0.1   0.087238
    6   0.039    0.03327    0.03327   0.03137        0.1   0.000061
    7  -1.751   -1.49367   -1.49367  -1.62382        0.1   0.123947
    8  -1.441   -1.22922   -1.22922  -1.27047        0.1   0.083944
    9   0.909    0.77541    0.77541   0.75678        0.1   0.033403
   10  -0.111   -0.09469   -0.09469  -0.08932        0.1   0.000498
   11   0.000    0.00000        *          *         1.0        *
```

FIGURE 4-5 Regression analysis for the data in Fig. 4-2*D*. The results of the regression computations and associated analysis of variance are identical to those in Fig. 4-2*A*, but the pattern of residuals is different. Point 11 is a leverage point.

nonlinear relationship. In this example, despite the reasonably high correlation coefficient, the straight line is not a good description of the data. This situation is an example of *model misspecification*, in which the regression equation fails to take into account important characteristics of the relationship under study (in this instance, a nonlinearity).

Figure 4-2*C* shows an example of an *outlier*, in which most of the points fall along a straight line, with one point falling well off the line. Outliers can have strong effects on the result of the regression computations. We estimate the regression line by minimizing the sum of squared deviations between the regression line and the data points, and so large deviations from the regression line contribute very heavily to determining the location of the line (i.e., the values of the regression coefficients). When the data points are near the "center" of the data, their effect is reduced because of the need for the regression line to pass through the other points at extreme values of the independent variable. In contrast, when the outlier is near one end or the other of the range of the independent variable, it is free to exert its effect, "pulling" the line up or down, with the other points having less influence. Such points are called *leverage points*. The outlier in Fig. 4-2C could be evidence of a nonlinear relationship with the true line of means curving up at higher foot sizes, it could be an erroneous data point, or it could be from a Martian that is somehow different from the others, making it unreasonable to include all the points in a single regression equation. Although the statistical techniques we will develop in this chapter will help detect outlier and leverage points, only knowledge of the substantive scientific question and quality of the data can help decide how to treat such points.

Figure 4-2D shows an even more extreme case of an outlier and leverage point. Of the 11 Martians, 10 had the same foot size (8 cm) but exhibited a wide range of intelligences (from 5.25 to 8.84 zorp). If we only had these data, we would have not been able to say anything about the relationship between foot size and intelligence (because there is no variability in foot size). The presence of the eleventh point at the 19-cm foot size produced a significant linear regression, with the line passing through the mean of the first 10 points and exactly through the eleventh point. This being the case, it should be obvious that, given the 10 points at a foot size of 8 cm, the eleventh point exerts a disproportionate effect on the coefficients in the regression equation. In fact, it is the only point that leads us to conclude there is a relationship. As in the last example, one would need to carefully examine the eleventh point to make sure that it was valid. Even if the point is valid, you should be extremely cautious when using the information in this figure to draw conclusions about the relationship between Martian foot size and intelligence. Such conclusions are essentially based on the value of a single point. It is essential to collect more data at larger foot sizes before drawing any conclusions.

The difficulties with the data in Fig. 4-2B, C, and D may seem contrived, but they represent common problems. In the case of simple (two-variable) linear regression, these problems become reasonably evident by simply looking at a plot of the data with the regression line. However, because we cannot graphically represent all of the data in one plot when there are many independent variables, we need additional methods to help identify problems with the data.*

---

*You should still look at plots of the dependent variable graphed against each of the independent variables. Nonlinearities can be detected, and influential points will sometimes be obvious. The general problem, however, is that a point might be highly influential, without being graphically evident in any of the plots of single variables.

There are two general approaches to dealing with the problem of testing the validity of the assumptions underlying a given regression model. Both are based on the residuals between the observed and predicted values of the dependent variable. The first approach is to examine plots of the residuals as functions of the independent or dependent variables. One can then see if the assumptions of constant variance appear satisfied or whether there appear to be systematic trends in the residuals that would indicate incorrect model specification. The second, and related approach, is to compute various statistics designed to normalize the residuals in one way or another to identify outliers or points which exercise undue influence on the values of the regression coefficients. We will develop these two approaches in the context of our Martian data, and then apply them to some real examples.

LOOKING AT THE RESIDUALS

According to the assumptions that underlie regression analysis, the members of the population under study should be normally distributed with constant variance around the plane of means. In other words, the deviations ϵ in

$$y = \beta_0 + \beta_1 x_1 + \beta_2 x_2 + \cdots + \beta_k x_k + \epsilon$$

should be normally distributed with a mean of zero and a constant standard deviation of $\sigma_{y|x}$. We cannot observe all members of the population to test this assumption, but we can use the *residuals* from the regression equation

$$e_i = Y_i - \hat{y}_i$$

as estimates of the deviations ϵ_i where Y_i is the observed value of the dependent variable, and

$$\hat{y}_i = b_0 + b_1 X_{1_i} + b_2 X_{2_i} + \cdots + b_k X_{k_i}$$

is the predicted value of the dependent variable for the ith observation.

If the assumption of constant variance holds, the residuals plotted against any independent variable should fall in a band around zero with constant width, regardless of the value of the independent variable. Figure 4-6A shows a plot of the residuals between observed and predicted Martian intelligence e_i vs. observed foot size F_i for the data in Fig. 4-2A. These points appear to fall in a constant-width band centered on zero, so the data appear consistent with the regression model. In contrast, if the model was misspecified or the data were not consistent with the assumptions of regression analysis, the residuals would exhibit some other pattern, such as a trend up or down, a curve, or a "megaphone" shape with one end of the band narrower than another.

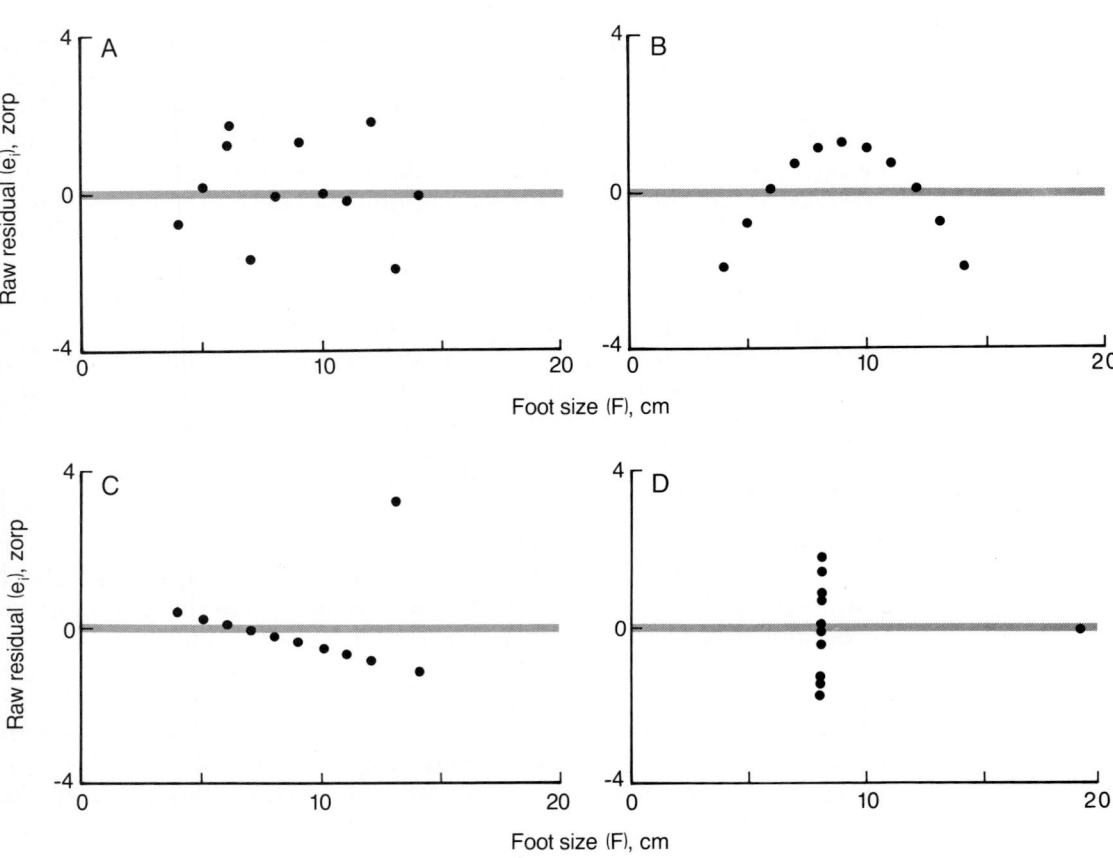

FIGURE 4-6 *A*. Plot of residuals as a function of foot size for the data in Fig. 4-2*A* (regression analysis in Fig. 4-1). The residuals are reasonably randomly scattered about zero with constant variance regardless of the value of foot size *F*. None of the residuals is large. *B*. Plot of the residuals as a function of foot size for the data in Fig. 4-2*B* (regression analysis in Fig. 4-3). In contrast to Panel *A*, the residuals are not randomly distributed about zero but rather show systematic variations, with negative residuals at high and low values of foot size and positive residuals at intermediate foot sizes. This pattern of residuals suggests a nonlinearity in the relationship between intelligence and foot size. *C*. Plot of the residuals as a function of foot size for the data in Fig. 4-2*C* (regression analysis in Fig. 4-4). The residuals are not randomly scattered about zero but rather all tend to be negative and fall on a trend line except for one large positive residual at a foot size of 13 cm. This pattern of residuals suggests the presence of an *outlier* which pulls the regression line away from the rest of the points. *D*. Plot of the residuals as a function of foot size for the data in Fig. 4-2*D* (regression analysis in Fig. 4-5). All the residuals, except one, are at a foot size of 8 cm and are randomly scattered about zero. However, there is a single point at a foot size of 19 cm which has a residual of 0. This pattern of residuals suggests a so-called *leverage point*, in which a single point located far from the rest of the data can pull the regression line through that point.

The residual pattern in Fig. 4-6A contrasts with those in Fig. 4-6B, C, and D, which show the residual plots for the data in Fig. 4-2B, C, and D, respectively. The band of residuals is bent in Fig. 4-6B, with negative residuals concentrated at low and high values and positive residuals concentrated at intermediate values of F. Such a pattern, when the residuals show trends or curves, indicates that there is a nonlinearity in the data.

The presence of an outlier is highlighted in Fig. 4-6C. Note that the outlier twists the regression line, as indicated by the fact that the residuals are not randomly scattered around 0, but are systematically positive at low values of foot size and systematically negative at high foot sizes (except for the outlier). This pattern in the residuals suggests that this outlier is also a leverage point, which is exerting a substantial effect on the regression equation. Figure 4-6D also highlights the presence of a leverage point, which, in this case, is manifested by a violation of the equal variance assumption. The plots in Fig. 4-2B, C, and D all suggest violations of the assumptions upon which the regression model is based and point to the need for remedial action.

To see how to use residual plots when there is more than one independent variable, let us return to our study of how Martian weight depends on height and water consumption in Chap. 3. Recall that when we fit these data (in Fig. 3-1) to a multiple regression model given by Eq. (3.1) we obtained an R^2 of .997, and the coefficients associated with both height and water consumption were significantly different from zero. However, now that we know that even a seemingly good regression model fit can be obtained even when one or more of the underlying assumptions is false, let us return to those data and examine the residuals. Figure 4-7 shows plots of the residuals vs. each independent variable (height H and cups of water C), the observed values of the dependent variable W, and the predicted values of the dependent variable $\hat{W}$. The residuals are uniformly distributed in a symmetric band centered on a residual value of zero. This lack of systematic trends and uniformity of pattern around zero strongly suggests that the model has been correctly specified and that the equal variance assumption is satisfied. By adding this to our previous information about the goodness of fit, we have no reason to doubt the appropriateness of our regression results, or the conclusions we drew about the relationship between Martian weight, height, and water consumption.

A QUANTITATIVE APPROACH TO RESIDUAL ANALYSIS

So far we have concentrated on a purely graphical approach to the problem of residual analysis. We saw that, although the data sets graphed in Fig. 4-2 yield identical regression results (Figs. 4-1, 4-3, 4-4, and 4-5), there were differences in the residuals. We will now develop several quanti-

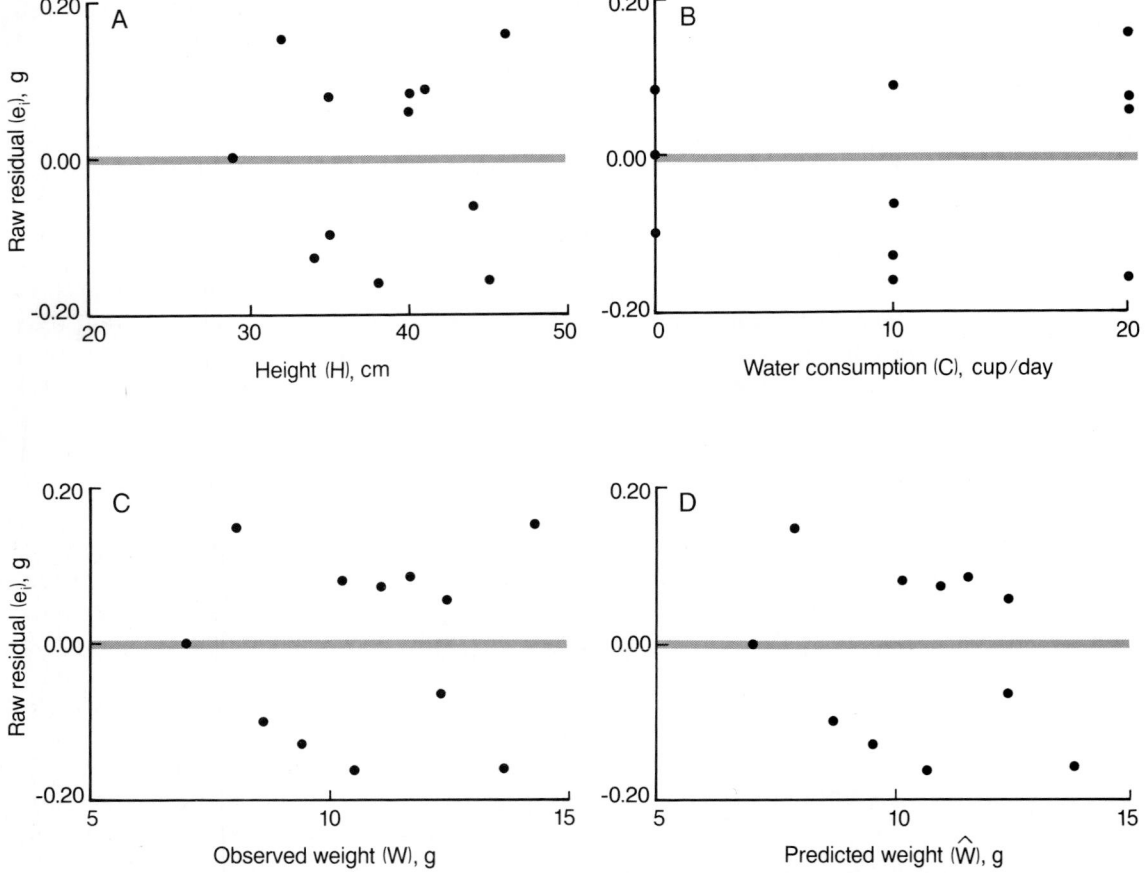

Figure 4-7 Plots of the residuals between predicted and observed Martian weight as a function of height *H*, water consumption *C*, observed weight *W*, and predicted weight *Ŵ* for the data and regression lines in Fig. 4-2. Such residual plots can be generated by plotting the residuals against any of the independent variables, the observed value of the dependent variable, or the predicted value of the dependent variable. In this case, none of the residual plots suggests any particular problem with the data or the quality of the regression equation. In each case the residuals are distributed evenly about zero with reasonably constant variance regardless of the values plotted along the horizontal axis.

tative statistics that can be computed to assess the pattern in the residuals, to locate outliers, and to identify points that may be exerting excessive influence on the parameters in the regression equation. Reducing these effects to numbers makes it easier to identify potentially important points for further examination.* We have already examined the *residual e*; we will now consider the *standardized residual, Studentized residual, leverage,* and *Cook's distance.*

Standardized Residuals

As we have already seen, the raw residuals can be used to locate outliers. The problem with using the raw residuals is that their values depend on the scale and units used in the specific problem. Because the residuals are in the units of the dependent variable, there are no a priori cutoff values to define a "large" residual. The most direct way to deal with this problem is to normalize the residuals by the standard error of the estimate $s_{y|x}$ to obtain *standardized residuals*

$$e_{si} = \frac{e_i}{s_{y|x}}$$

Recall that the standard error of the estimate $s_{y|x} = \sqrt{MS_{res}}$ is an estimate of $\sigma_{y|x}$, the standard deviation of the population about the line (or plane) of means, and the residuals themselves are estimates of the deviations of population members about the plane of means. $s_{y|x}$ is also the standard deviation of the residuals [see Eq. (2.7)]. Therefore, because the standard deviation has the same units as the underlying variable, e_{si} is unitless and does not depend on the scale of measurement. Because we have normalized the residual by its standard deviation, a standardized residual of 0 means that the point falls on the regression line or plane. Likewise, a standardized residual of 1 means that the residual is 1 standard deviation off the regression plane, a standardized residual of 2 is 2 standard deviations off the regression plane, and so forth.

Moreover, one of the assumptions of regression analysis is that the population members are normally distributed about the plane of means. One consequence of this assumption is that the residuals are normally distributed about the regression line or plane. In any normally distributed population, about two-thirds of the members fall within 1 standard deviation either side of the mean, and about 95 percent of the members fall within 2 standard deviations either side of the mean. Conversely, if the

---

*It is possible that no one data point would be considered influential by any of the statistics we will discuss in this section, and yet some group of data points acting together could exert great influence. Identification of such groups utilizes many of the same concepts we discuss, but the computations are more involved. For a further discussion of such methods, see D. A. Belsley, E. Kuh, and R. E. Welsch, *Regression Diagnostics: Identifying Influential Data and Sources of Collinearity*, New York, Wiley, 1980, pp. 31–39.

population is normally distributed about the plane of means, we would expect about two-thirds of the standardized residuals to have absolute values below 1 and almost all of the standardized residuals to have absolute values below 2. Thus, points with standardized residuals exceeding 2 warrant consideration as potential outliers.

Does this mean that any standardized residual whose absolute value exceeds 2 *is* an outlier? No. After all, about 5 percent of the standardized residuals should exceed 2 simply because of the nature of the normal distribution. Of course, very large values of the standardized residual, say above 3, warrant consideration in any event because very few points of a normally distributed population are more than 3 standard deviations above or below the mean.

The standardized residuals from the analysis of the data graphed in Fig. 4-2A are shown along with the regression output in Fig. 4-1. For all data points, the absolute value of the corresponding standardized residual is less than 1.64. Thus, in terms of the rule of thumb we just proposed — that absolute values exceeding 2 warrant a look and absolute values exceeding 3 warrant serious evaluation — none of these points seem extreme. This information, coupled with our previous graphical evaluation, further convinces us that the residuals from fitting this set of data to Eq. (4.1) indicate no violation of linear regression's assumptions. This situation contrasts with the data in Fig. 4-2C, in which one point is well off the trend defined by the other data points. The standardized residual corresponding to this outlier is 2.76 (point 3 in Fig. 4-4) indicating that this residual is 2.76 standard deviations off the regression line. In contrast, the next largest (in magnitude) standardized residual in this data set is only $-.99$. The large standardized residual identifies an outlier and serves as a warning that we should go back to the data set and experimental notes and closely examine this data point to see if there is some identifiable reason for the discrepancy, either an error in the data point or a problem with the model. Remember, this large standardized residual does not mean that this outlying point should be thrown out.

Although the standardized residuals are helpful for identifying outliers, they do not reveal all possible problems with the data or regression model. For example, the standardized residuals computed from the regression fit to the data in Fig. 4-2B show no extreme values (Fig. 4-3). However, our graphical evaluation of the pattern in the residuals from this data set led us to conclude that the linear model to which we fit the data was incorrect and that some nonlinear model would better represent the process that generated the data. Likewise, the standardized residuals computed from the regression fit to the data graphed in Fig. 4-2D show no extreme values (Fig. 4-5) when, in fact, a graphical analysis of these residuals (and other statistics discussed below) indicates a severe problem — there would be no relationship between intelligence and foot size *except* for the one point at the 19-cm foot size. This is a very important lesson: the residuals can all be small enough so as to not cause us to worry about any one point being an outlier when, in fact, there is a serious problem with the model fit to the data. For the data graphed in Fig. 4-2*B* and *C*, we would be seriously mistaken if we used the values of the standardized residuals alone to conclude that there was no problem with our analysis.

Using Standardized Residuals to Test for Normality of the Residuals

Another of the assumptions of regression analysis is that the members of the underlying population are normally distributed about the regression plane. We can use the residuals to test this assumption by examining the distribution of the residuals using two related methods. The simplest procedure is to plot the frequency distribution of the residuals to see if the distribution looks normal. If the distribution of residuals seriously deviates from normality, it often will be evident from this plot. Unfortunately, many bell-shaped curves are not normal, so simply looking at this plot may not be sufficient. We obtain a better graphical test or normality by constructing a so-called *normal probability plot* of the residuals. A normal probability plot is a plot of the cumulative frequency of the distribution of the residuals vs. the residuals themselves on a special scale (Fig. 4-8) which produces a straight line if the distribution is normal, just as an exponential function plots as a straight line on semilogarithmic graph paper.

To illustrate the evaluation of normality, let us return to the four data sets collected in our Martian intelligence study (Fig. 4-2). We first take the residuals tabulated in the regression diagnostics in Figs. 4-1, 4-3, 4-4, and 4-5 and plot frequency histograms of the residuals. Two of the histograms in Fig. 4-9 look bell-shaped (Fig. 4-9*A* and *D*, corresponding to the data in Fig. 4-2*A* and *D*, respectively), and so do not suggest any obvious deviation of the distribution of residuals from normality. In contrast, the pattern of residuals in Fig. 4-9*B*, corresponding to the data in Fig. 4-2*B*, shows a flat (uniform) distribution rather than a bell-shaped (normal) one. This pattern in the residuals is another indicator that there is something wrong with the regression model. Finally, the residuals in Fig. 4-9*C*, corresponding to the data in Fig. 4-2*D*, are skewed toward positive values because of the outlier we have already discussed. This distribution also suggests a problem with the regression model. In sum, examining a histogram of the residuals can help identify misspecified models (e.g., Fig. 4-9*B*) or outliers (e.g., Fig. 4-9*C*) when these difficulties lead to obvious deviations of the distributions from normality. It is, however, difficult to assess more subtle deviations from normality by simply looking at the histogram.

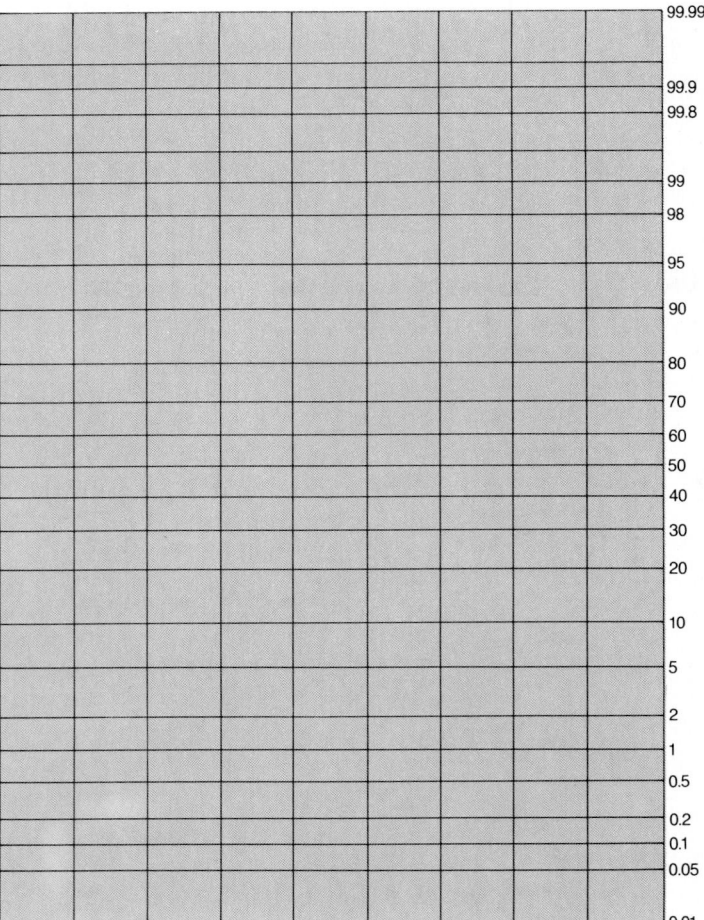

FIGURE 4-8 Normal probability paper can be used to construct a plot of the cumulative frequency of observations against the magnitude of the observations. Just as semilogarithmic graph paper causes exponential functions to plot as straight lines, normal probability graph paper causes data drawn from a normal distribution to appear as a straight line. Normal probability plots are commonly used to provide a qualitative check for normality in the residuals from a regression analysis. The numbers down the right edge are cumulative frequencies. Note the distorted scale.

A normal probability plot of the residuals is a better way to examine whether or not the residuals are normally distributed. We will now use the same residuals as before to illustrate how to construct and interpret normal probability plots. (Many computer programs that do multiple regression have options to automatically produce such a plot.)

1. Tabulate the residuals (from Fig. 4-1) in order, from the most negative to the most positive. The first column in Table 4-2 shows the ordered residuals. (You can use either the raw residuals or the standardized residuals; it does not have any effect on the result.)

2. Number the residuals from 1 through n, where 1 corresponds to the most negative residual and n corresponds to the most positive residual. The second column in Table 4-2 shows these values.

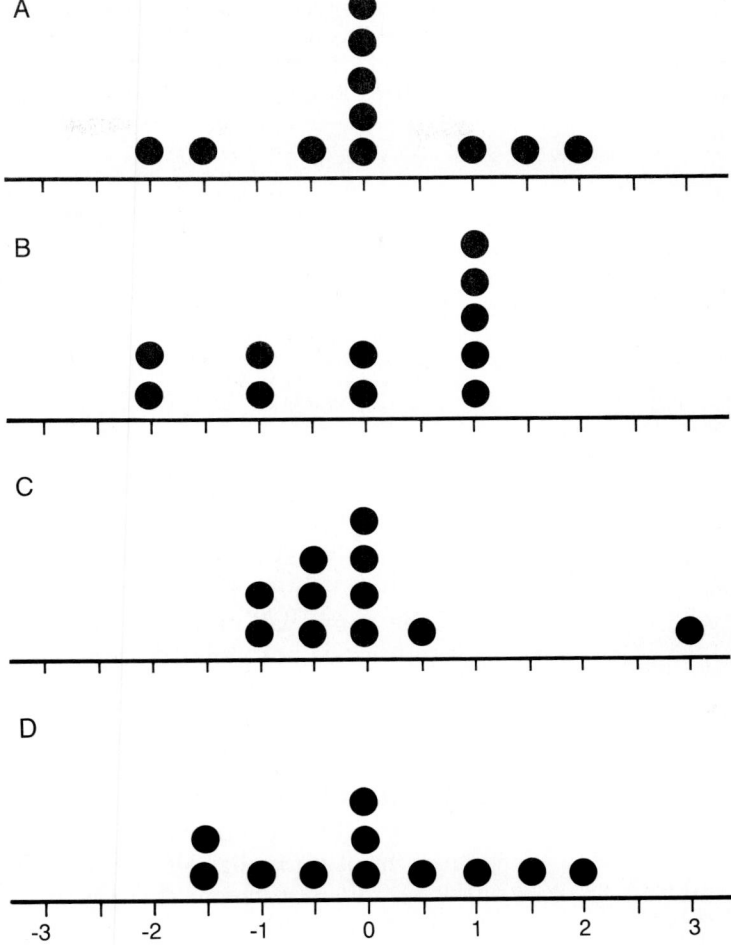

FIGURE 4-9 One way to assess whether or not the residuals from a regression analysis reasonably approximate the normal distribution is to prepare a histogram of the residuals, as in these histograms of the residuals from the regression analyses of the data in Fig. 4-2. If the distribution of residuals appears reasonably normal (*A*), there is no evidence that the normality assumption is violated. Flat distribution of residuals often suggests a nonlinearity (*B*) and a heavy-tailed distribution suggests an outlier (*C*). The presence of the leverage point is difficult to detect in the histogram of the residuals (*D*).

3. Compute the cumulative frequency of the residuals up to each point,

$$F_i = \frac{i - .5}{n}$$

where F_i is the cumulative frequency up to point i.* The third column in Table 4-2 shows these values.

4. Plot the value of the ith residual on the horizontal axis of the normal probability paper (from the first column of Table 4-2) vs.

*The cumulative frequency is the fraction of the residuals at or below point i.

TABLE 4-2 Ranked Residuals and Their Cumulative Frequency for Normal Probability Plot

Standardized Residual	Rank of Residual	Cumulative Frequency
−1.92	1	.045
−1.68	2	.136
− .74	3	.227
− .17	4	.318
− .05	5	.409
− .04	6	.500
.04	7	.591
.18	8	.682
1.24	9	.773
1.31	10	.864
1.84	11	.955

the cumulative frequency at that point on the vertical axis (from the third column of Table 4-2).

5. If the points fall on a line, this suggests that the normality assumption is not violated.

Figure 4-10A shows the results of such a plot for the Martian intelligence data of Figs. 4-1 and 4-2A. Because this line is reasonably straight (we do not require that it be perfectly straight), we conclude that the data are consistent with the normality assumption.

What if the line on the normal probability plot is not straight? Deviations from normality show up as curved lines. A simple curve with only one inflection means that the distribution of residuals is skewed in one direction or the other. A curve with two inflection points (e.g., an S shape) indicates a symmetric distribution whose shape is too skinny and long-tailed or too fat and short-tailed to follow a normal distribution. The S-shaped pattern often occurs when the assumption of equal variances is violated. Outliers will be highlighted as single points that do not match the trend of the rest of the residuals. As with the other diagnostic procedures we have discussed, substantial deviation from normality (i.e., from a straight line in a normal probability plot) indicates a need to examine the regression equation and data.

The normal probability plots of the residuals corresponding to the fits of Eq. (4.1) to the Martian intelligence data in Fig. 4-2B, C, and D are shown in panels B, C, and D of Fig. 4-10, respectively. The normal probability plots in panels A and D of Fig. 4-10 are reasonably linear, so we have no particular reason to question the normality assumption with regard to the residuals from these data sets. The residuals plotted in Fig.

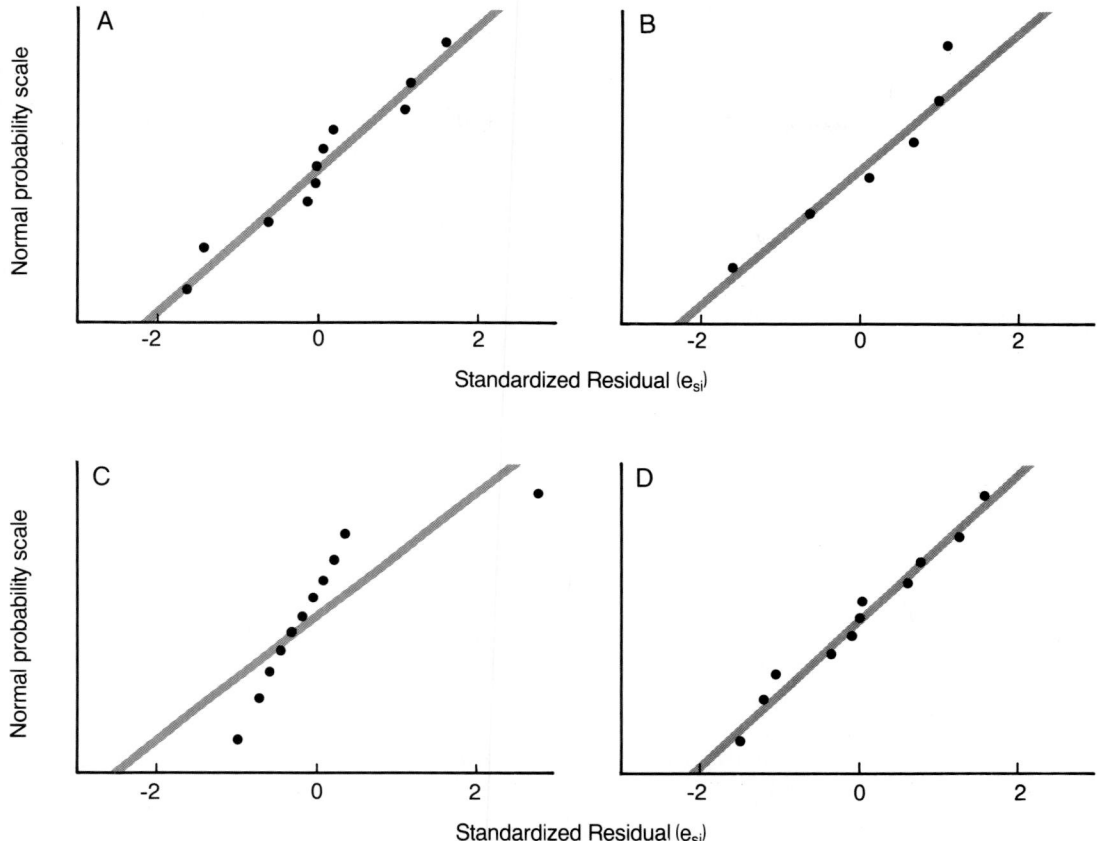

FIGURE 4-10 Normal probability plots of the residuals are a more sensitive qualitative indicator of deviations from normality than simple histograms such as those included in Fig. 4-9. The four panels of this figure correspond to the frequency histogram of the residuals given in Fig. 4-9 to the raw data in Fig. 4-2. Deviations from a straight diagonal line suggest nonnormally distributed residuals, particularly as illustrated in panels *B* and *C*.

4-10*B* are from the obviously nonlinear relationship shown by the data in Fig. 4-2*B*; in this instance, the points in the normal probability plot tend to fall along a curve rather than a straight line, which reflects the fact that the distribution of residuals tends to be uniform (compare with Fig. 4-9*B*). The curvature is not striking, however, so the plot suggests only a mild problem. In contrast, the normal probability plot in Fig. 4-10*C* of the residuals from the data in Fig. 4-2*C* clearly identifies the outlier. (In this example there is one extreme point, which corresponds to the value with the standardized residual of 2.76 in Fig. 4-4.) In fact, Fig. 4-10*C* exemplifies the classic pattern for how outliers appear in a normal probability plot: a good linear relationship with extreme points that are far removed from the line defined by the rest of the residuals.

Like the other procedures for examining residuals, we have used these results to highlight potentially erroneous data or point to possible problems in the formulation of the regression equation itself, rather than to formally test hypotheses about the residuals. It is, nonetheless, possible to formally test the hypothesis of normality of the residuals using the fact that normally distributed residuals will follow a straight line when plotted on a normal probability plot. To formalize this test, simply compute the correlation coefficient between the residuals and the corresponding cumulative frequency values (i.e., the first and third columns of Table 4-2) and test whether this correlation is significantly different from zero by computing

$$t = \frac{r}{\sqrt{(1 - r^2)/(n - 2)}}$$

and comparing this value with the one-tailed critical value of the t distribution with $n - 2$ degrees of freedom.[*] If the correlation is significantly different from zero, you can conclude that the residuals were drawn from a normal distribution. As a practical matter, in many biological and clinical studies, there are only a few data points, and these formal tests are not very powerful. When there is adequate power, the results of this formal test will generally agree with the results of simply examining the curve most of the time, and looking at the plot is probably more informative.[†]

Leverage

To understand the concept of *leverage,* we need to pause and consider the relationship between the observed and predicted values of the dependent variable, the independent variables, and the regression equation. In Chaps. 2 and 3 we used the observed values of the dependent and independent variables, Y_i and X_i (where X_i represents all the independent variables, if there are more than one), to compute the best estimate of the parameters in a regression equation. We then used the regression equation, together with the values of the independent variables, to compute an estimate of the dependent variable $\hat{y}_i$. It is possible to show that

[*]This situation is one of the rare instances when a one-tail t test would be appropriate because the slope of the normal probability plot is always positive.

[†]There are other formal statistical tests of normality, such as the Kolmogorov-Smirnov test and Shapiro-Wilk test, but they also lack power when the sample size is small. For further discussion of formal tests of normality, see J. H. Zar, *Biostatistical Analysis* (2nd ed.), Englewood Cliffs, N.J., Prentice-Hall, 1984, pp. 88–96.

this process can be replaced with one in which the estimates of the dependent variable $\hat{y}_i$ are computed directly from the n observed values of the dependent variable Y_i with the equation[*]

$$\hat{y}_i = h_{i1}Y_1 + h_{i2}Y_2 + h_{i3}Y_3 + \cdots + h_{in}Y_n = \Sigma h_{ij}Y_j$$

where h_{ij} quantifies the dependence of the predicted value of the dependent variable associated with the ith data point on the observed value of the dependent variable for the jth data point. The values of h_{ij} depend only on the values of the independent variables. Ideally, the value associated with each dependent variable should have about the same influence (through the computed values of the regression coefficients) on the values of other dependent variables and the h_{ij} should all be about equal. Furthermore, it can be shown that[†]

$$\sum_{j=1}^{n} h_{ij} = 1$$

and

$$h_{ii} = \sum_{j=1}^{n} h_{ij}^2$$

In practice, therefore, people have concentrated on the term h_{ii}, which is called the *leverage* and quantifies how much the observed value of a

---

[*]This relationship is most evident when we present the regression equation in matrix notation. The basic linear regression model is

$$\hat{\mathbf{y}} = \mathbf{X}\mathbf{b}$$

where $\hat{\mathbf{y}}$ is the vector of predicted values of the dependent variable, $\mathbf{X}$ is the data matrix, and $\mathbf{b}$ is the vector of regression coefficients. Let $\mathbf{y}$ be the vector of observed dependent variables corresponding to the rows of $\mathbf{X}$. According to Eq. (3.19)

$$\mathbf{b} = (\mathbf{X}^T\mathbf{X})^{-1}\mathbf{X}^T\mathbf{y}$$

contains the coefficients which minimize the sum of squared deviations between the predicted and observed value of the dependent variable. Substitute the second equation into the first to obtain

$$\hat{\mathbf{y}} = \mathbf{X}(\mathbf{X}^T\mathbf{X})^{-1}\mathbf{X}^T\mathbf{y} = \mathbf{H}\mathbf{y}$$

where $\mathbf{H}$ is the $n \times n$ *hat matrix* defined by

$$\mathbf{H} = \mathbf{X}(\mathbf{X}^T\mathbf{X})^{-1}\mathbf{X}^T = [h_{ij}]$$

which transforms the observed values of the dependent variable $\mathbf{y}$ into the predicted values of the dependent variable $\hat{\mathbf{y}}$. ($\mathbf{H}$ is called the hat matrix because it transforms Y into y-hat.) Note that $\mathbf{H}$ depends only on the values of the independent variables $\mathbf{X}$.

[†]For further discussion of this result, see J. Fox, *Linear Statistical Models and Related Methods: With Applications to Social Research*, New York, Wiley, 1984, p. 162.

given dependent variable affects the estimated value. These two equations require that the leverage is a number between 0 and 1. In the case of a simple linear regression with one independent variable, the leverage is

$$h_{ii} = \frac{1}{n} + \frac{(X_i - \overline{X})^2}{\Sigma(X_j - \overline{X})^2}$$

Note that as the value of the dependent variable associated with point i, X_i, moves farther from the mean of the observed values of X, $\overline{X}$, the leverage increases.

Ideally, all the points should have the same influence on the prediction, so one would like all the leverage values to be equal and small. If h_{ii} is near 1, $\hat{y}_i$ approaches Y_i, and we say that point i has high leverage. However, to use h_{ii} as a diagnostic, we need a more specific definition of a large h_{ii}. It is possible to show that the expected (average) value of the leverage is $(k + 1)/n$, where k is the number of independent variables in the regression equation. Hence, our assessment of whether a given point has high leverage will depend on the sample size and the number of independent variables,[*] and it has been suggested that a cutoff for identifying a large leverage is twice the expected value of h_{ii}. So, when $h_{ii} > 2(k + 1)/n$, we will say a point has high leverage and warrants further investigation. It is important to remember, however, that leverage merely represents the *potential* for the point exerting a strong influence on the results of the regression because the actual influence also depends on the observed value Y_i, whereas the leverage only depends on the values of the independent variables.

Leverage is best illustrated by the data in Fig. 4-2D, in which there are 10 intelligence values recorded at a foot size of 8 cm and an eleventh intelligence value recorded at a foot size of 19 cm. As we discussed earlier, given the 10 values at a foot size of 8 cm, only the eleventh point determines the regression equation—without this point there would have been no variation in foot size and therefore we could not compute a regression equation at all. Because the regression line goes through the mean of the 10 intelligence observations at a foot size of 8 cm (7 zorp) and the eleventh point, moving the eleventh point will have a much greater effect on the regression equation than moving any other point (Fig. 4-11). For example, suppose the intelligence observed in this eleventh Martian had been 7 zorp rather than 12.5 zorp. In this case, the slope of the regression line and the correlation would both be zero (Fig. 4-11, line *B*). Had the intelligence corresponding to a foot size of 19 cm been

[*]For a derivation of this result, see D. A. Belsley, E. Kuh, and R. E. Welsch, *Regression Diagnostics: Identifying Influential Data and Sources of Collinearity*, New York, Wiley, 1980, p. 17.

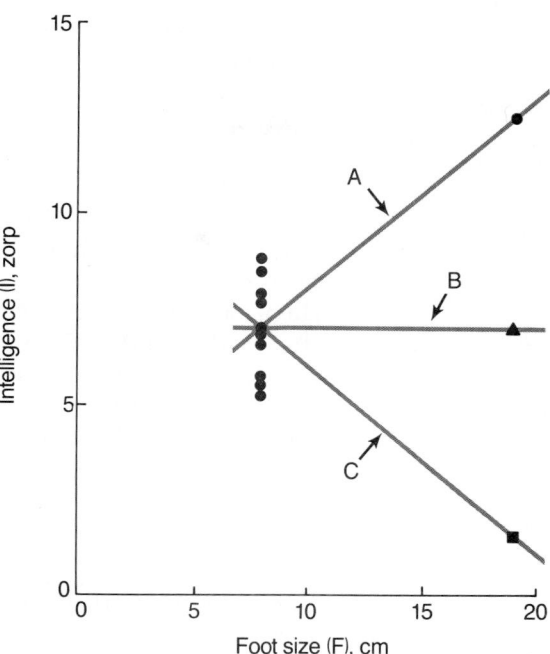

FIGURE 4-11 A single point far removed from the other points along the *X* axis has a tremendous potential influence, or *leverage*, on the regression equation. The data plotted at a foot size of 8 cm and the uppermost point at 19 cm correspond to the data and regression equation given in Fig. 4-2*D*. If we leave all of the observations at a foot size of 8 cm the same and simply move the point located at a foot size of 19 cm from an intelligence value of 12.5 zorp down to 7.0 zorp, the slope of the corresponding regression line will change from 0.5 zorp/cm to 0 zorp/cm. If we change the value of intelligence at this leverage point to 1.5 zorp, the slope of the regression line then falls to -0.5 zorp/cm. Because of its distance from the other data along the horizontal axis (i.e., in terms of the independent variable), this point can exert a tremendous influence on the estimate of the regression equation even though none of the other values is changing.

1.5 zorp, the slope would have been $-.5$ zorp/cm (instead of $+.5$ zorp/cm) and the correlation is again .82 (Fig. 4-11, line *C*). Although changing any of the other 10 points will change the regression equation, the effect will be much smaller because the other points are not too distant (here, not at all distant) from the others in terms of the independent variables (i.e., along the foot size axis). Because the eleventh point has such a great potential effect on the regression equation, due to its distance from the other points (in terms of the independent variable), we say it has high leverage.

In this example we have $k = 1$ independent variables and $n = 11$ data points, so the expected value of the leverage is $(k + 1)/n = (1 + 1)/11 = .18$. Using our rule of thumb that leverage is "large" if it exceeds twice the expected value, we will take our threshold for further investigation at .36. The leverage associated with the eleventh point is 1, which exceeds this threshold, while all other h_{ii} values are only .1. Thus, as we expect based on the foregoing discussion, the potential influence of this point, measured by leverage, is high. (In fact, 1 is the maximum value for h_{ii}.)

Points with high leverage warrant special attention because they have the potential for exerting disproportionate influence on the results of the regression analysis. This capacity does not, in itself, mean that there is necessarily a problem, but it does indicate that we need to evaluate the point to see that it is accurate and representative of the same population as the other members of the sample used to compute the regression equation.

Studentized Residuals

When we computed standardized residuals, we normalized them by dividing by $s_{y|x}$, which is constant for all values of X. However, in regression one can be more confident about the location of the regression line near the middle of the data than at extreme values, because of the constraint that the line must pass "through" all the data. We can account for this effect, and thus refine our normalization of the residuals, by dividing the residual e_i by its specific standard error, which is a function of X, or, more precisely, how far X is from $\overline{X}$. The Studentized residual is defined as

$$r_i = \frac{e_i}{s_{y|x}\sqrt{1 - h_{ii}}}$$

Hence, points farther from the "center" of the independent variables, which have higher leverages h_{ii}, will also be associated with larger Studentized residuals r_i than points near the center of the data that are associated with the same raw residual e_i.

This definition is more precisely known as the *internally Studentized residual* because the estimate of $s_{y|x}$ is computed using *all the data.* An alternative definition, known as the *externally Studentized residual* or *Studentized deleted residual*, is defined similarly, but $s_{y|x}$ is computed *after deleting the point associated with the residual:*

$$r_{-i} = \frac{e_i}{s_{y|x(-i)}\sqrt{1 - h_{ii}}}$$

where $s_{y|x(-i)}$ is the square root of MS_{res} computed *without* point i. The logic for the Studentized deleted residual is that if the point in question is an outlier, it would introduce an upward bias in $s_{y|x}$ and so understate the magnitude of the outlier effect for the point in question. On the other hand, if the point is not an outlier, deleting it from the variance computation would not have much effect on the resulting value.* As a result,

---

* The two ways of defining the Studentized residuals are related to each other according to

$$r_{-i} = r_i\sqrt{\frac{n - k - 2}{n - k - 1 - r_i^2}}$$

For a derivation of this result, see S. Weisberg, *Applied Linear Regression* (2nd ed.), New York, Wiley, 1985, pp. 113–116.

the Studentized deleted residual is a more sensitive detector of outliers. (Although most regression programs report Studentized residuals, they often do not clearly state which definition is used; to be sure, you should check the program's documentation to see which equation is used to compute the Studentized residual.)

To illustrate Studentized residuals, let us reconsider the four data sets from our study of the relationship between intelligence and foot size of Martians (data are graphed in Fig. 4-2), the results of which are in Figs. 4-1, 4-3, 4-4, and 4-5. For the data in Fig. 4-2A, we saw before that the absolute values of the standardized residuals ranged from .03 to 1.64 (Fig. 4-1). None of these values was particularly alarming because all are below the cutoff value of about 2. The absolute values of the r_i for this set of data range from .03 to 1.78, which again indicate no cause for concern. The r_i listed in Fig. 4-3 (corresponding to the data graphed in Fig. 4-2B) range in absolute value from .11 to 1.86 and give us no reason to consider any one data point as a potential outlier. For the data set graphed in Fig. 4-2C, the obviously outlying point at a foot size of 13 cm has an r_i of 3.0 (Fig. 4-4). As we saw for the standardized residuals, this value far exceeds the other standardized residuals for this data set—the next largest value is −1.13. In fact, in all four sets of data from our study of Martian intelligences, the values of r_i are similar to those of standardized residuals.

Likewise, for most of these sets of data the values of r_i and r_{-i} are similar. However, for the data set which has the obvious outlier at the foot size of 13 cm, the Studentized deleted residual r_{-i} of 1216.69 is much larger than the Studentized residual r_i of 3.0 (Fig. 4-4). The reason that r_{-i} is so much larger than r_i in this instance is because this point also has a high leverage value, h_{ii}.

The question now arises: Should you use r_i or r_{-i}? Most of the time it does not matter which one is used. If r_i is large, r_{-i} will also be large. However, as we just saw, the value of r_{-i} can greatly exceed r_i in some instances. Thus, r_{-i} is slightly preferred. If the computer program you use does not report r_{-i}, chances are it will report h_{ii} (or Cook's distance, discussed below), which can be used in combination with r_i to identify influential points.

The values of r_i tend to follow the t distribution with the number of degrees of freedom associated with MS_{res} in the regression equation, so one can test whether the Studentized residual is significantly different from zero using a t test. There are two problems with this approach. First, there are many data points and hence many residuals to test with t tests. This situation gives rise to a multiple comparisons problem and a resulting compounding of errors in hypothesis testing that we will discuss in Chap. 7. One can account for the multiple comparisons by doing Bonferroni corrected t tests, although, with moderately large samples (of, say, 100 points), this approach will become very conservative and perhaps not have adequate power to discern the significant differences. Second, the goal of residual examination is not so much hypothesis testing as a diagnostic procedure, designed to locate points that warrant further investigation. From this perspective, the value of the Studentized residual itself becomes the variable of interest, rather than some associated P value.

Cook's Distance

The techniques discussed so far have dealt with methods to locate points that do not seem consistent with the normality assumption or points that can *potentially* exert disproportionate influence on the results of the regression analysis. *Cook's distance* is a statistic which directly assesses the *actual* influence of a data point on the regression equation by computing how much the regression coefficients change when the point is deleted.

To motivate the Cook's distance statistic, let us return to the data we collected on Martian intelligence in Table 4-1. We sought to describe these data with Eq. (4.1) containing the two parameters a_0 and a_F. Fitting this equation to *all* the data yielded estimates for a_0 and a_F of 3.0 zorp and .5 zorp/cm, respectively. Suppose that we did not have the first data point in Table 4-1. We could still fit the remaining 10 data points with Eq. (4.1); doing so would yield parameter estimates of $a_{0(-1)} = 3.0$ zorp and $a_{F(-1)} = .5$ zorp/cm, respectively. The "-1" in parentheses in the subscripts indicates that the estimates were computed with point 1 de-

TABLE 4-3 Regression Coefficients and Cook's Distance Obtained by Deleting Each Point from Martian Intelligence vs. Foot Size Data Shown in Fig. 4-2A

Point Deleted	$a_{0(-i)}$	$a_{F(-i)}$	Cook's Distance D_i	R^2
None (all in)	3.00	.50	—	.67
1	3.00	.50	.0000	.66
2	3.01	.50	.0001	.66
3	2.41	.59	.4892	.78
4	2.89	.50	.0616	.70
5	2.99	.50	.0016	.66
6	2.98	.50	.0004	.60
7	2.50	.54	.1268	.71
8	3.54	.45	.1227	.56
9	3.34	.44	.2790	.67
10	3.49	.47	.1543	.69
11	2.90	.51	.0043	.64

leted from the data set. These values are not different from those obtained from all the data, which is what one would expect if the results of the regression analysis did not depend heavily on this one point. Now, repeat this exercise, computing the regression parameters when deleting the second data point, the third data point, and so forth. Table 4-3 shows the results of these computations. Note that, as expected, although there is some variability in the resulting estimates of $a_{0(-i)}$ and $a_{F(-i)}$, the values tend to fall near those computed when all the data were used.

Figure 4-12 shows an alternative presentation of these results, generated by plotting each value of $a_{0(-i)}$ vs. the corresponding value of $a_{F(-i)}$. Each point in this graph represents a set of coefficients, $a_{0(-i)}$ and $a_{F(-i)}$, estimated with one of the data points deleted. If none of the data points exerts undue influence, the points representing the sets of $a_{0(-i)}$ and $a_{F(-i)}$ will cluster around the point representing a_0 and a_F, the coefficients estimated using all 11 data points. This expected pattern is evident in Fig. 4-12, and so we conclude that none of the data points is exerting a disproportionate influence on the estimates of the regression coefficients.

Now, let us do this same analysis for the data in Fig. 4-2C, which contained an obvious outlier. Table 4-4 shows the results of computing the regression coefficients 11 times, each time with data point i omitted; Fig. 4-13 is a plot of the resulting $a_{0(-i)}$ vs. $a_{F(-i)}$. In contrast to the previous example where no point was influential, the values of the regression parameters obtained when deleting the third data point, $a_{0(-3)}$ and $a_{F(-3)}$, are far from the cluster of parameter estimates obtained when

TABLE 4-4 Regression Coefficients and Cook's Distance Obtained by Deleting Each Point from Martian Intelligence vs. Foot Size Data Shown in Fig. 4-2C

Point Deleted	$a_{0(-i)}$	$a_{F(-i)}$	Cook's Distance D_i	R^2
None (all in)	3.00	.50	—	.67
1	3.01	.51	.0118	.67
2	3.05	.50	.0021	.66
3	4.01	.35	1.3929	1.00
4	3.04	.50	.0055	.67
5	2.95	.51	.0260	.68
6	2.46	.58	.3006	.70
7	2.97	.50	.0005	.65
8	2.72	.53	.0338	.63
9	2.84	.53	.0595	.68
10	3.03	.50	.0004	.66
11	2.88	.51	.0070	.64

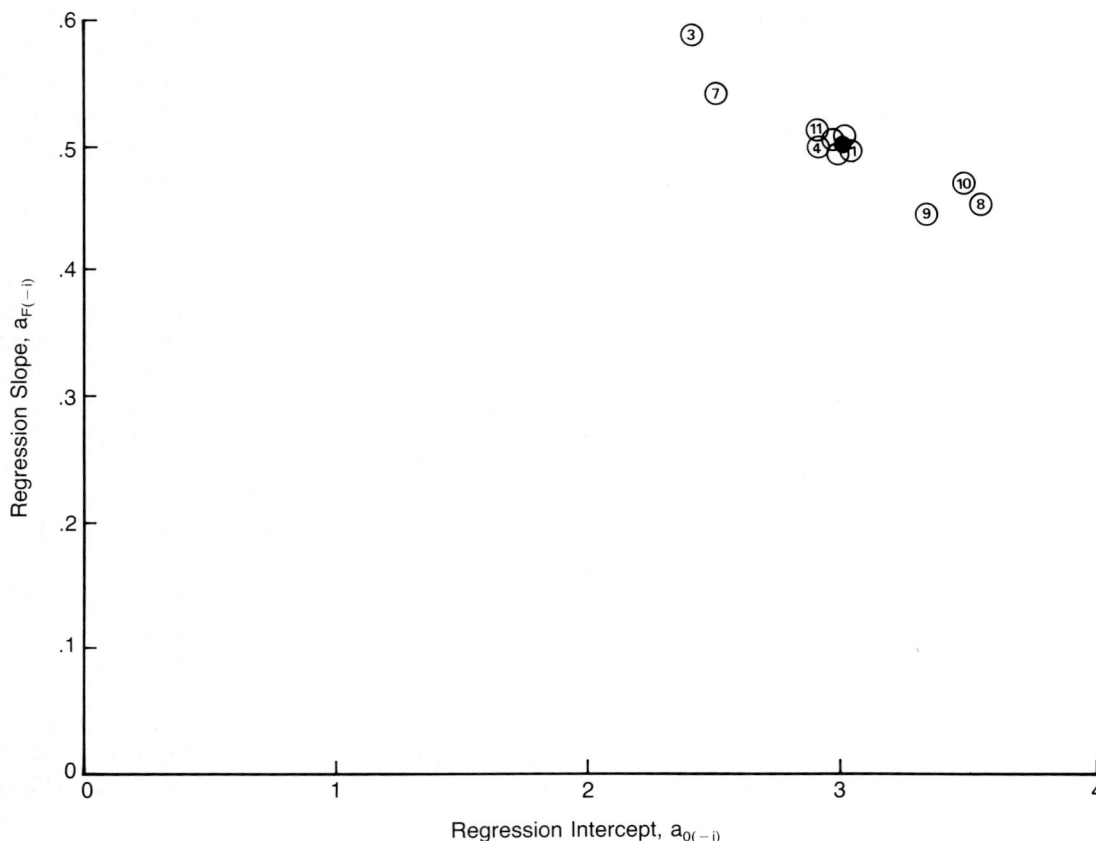

FIGURE 4-12 Cook's distance can be thought of as a normalized measure of the distance between the regression parameters estimated with all the data to the parameters estimated when deleting each point in turn. In this figure, the estimated value of the intercept is plotted against the estimated value of the slope for the line relating Martian intelligence and foot size using the data in Fig. 4-2*A*. The numbers indicate which data point was deleted in estimating the parameters. Because no one point is particularly influential, deleting any of the points results in small changes in the estimated slope and intercept, and all values tend to cluster around the slope ($a_F = .5$ zorp/cm) and intercept ($a_0 = 3$ zorp) estimated from all the data.

deleting any one of the other data points. This result leads to the conclusion that data point 3 is exerting a disproportionate influence on the regression coefficients.

We could simply rely on visual inspection of plots such as Figs. 4-12 and 4-13 to locate influential points when there is only one independent variable, but such graphical techniques become difficult if there are several independent variables. To deal with this problem, we need to define a statistic that measures how "far" the values of the regression coeffi-

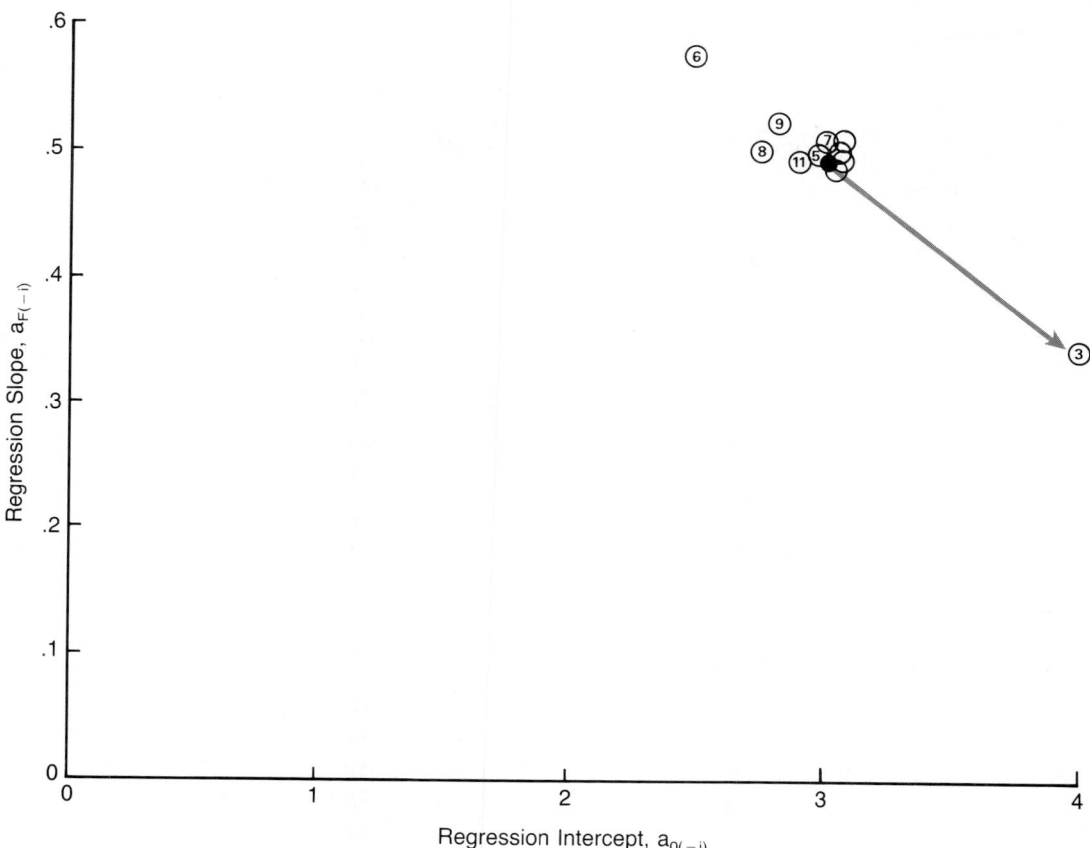

FIGURE 4-13 Similar to Fig. 4-12, this is a plot of the slopes and intercepts estimated from the data in Fig. 4-2C when deleting, in turn, each of the data points.
Deleting any point, except point 3, has only a small effect on the parameter estimates, whereas deleting point 3 has a large effect on the estimates of the slope and
intercept of the regression line relating Martian intelligence to foot size. The large
distance between the set of parameter estimates obtained with all data and the set
of estimates obtained when deleting point 3 normalized as the Cook's distance statistic, indicates that data point 3 is a very influential point. The Cook's distance
associated with point 3 (in Fig. 4-4) is much larger than any of the other points;
its value is 1.39, which is above our threshold of 1.0 for identifying a point as
influential.

cients move with deleting a point. One obvious statistic would be the
simple euclidean distance between the point representing the coefficients
computed with all the data (a_0, a_F) and the point representing the coefficients computed with data point i deleted $(a_{0(-i)}, a_{F(-i)})$:

$$d_i = \sqrt{(a_{0(-i)} - a_0)^2 + (a_{F(-i)} - a_F)^2}$$ (4.2)

There are two problems with this definition. First, the units do not work out. For example, here we are trying to add zorp^2 to (zorp/cm)^2. Second, the resulting statistic has units (of some sort) and so depends on the scale of measurement. It would be better to have a dimensionless measure of distance, so that the absolute value of the resulting statistic could be used for screening purposes.

The *Cook's distance* is such a statistic. It is defined, in matrix notation, as

$$D_i = \frac{(\mathbf{b}_{(-i)} - \mathbf{b})^T (\mathbf{X}^T\mathbf{X}) (\mathbf{b}_{(-i)} - \mathbf{b})}{(k + 1) s_{y|x}^2}$$

in which $\mathbf{b}$ is a vector containing the regression coefficients, $\mathbf{X}$ is the matrix of values of the independent variable, k is the number of independent variables, and $s_{y|x}$ is the standard error of the estimate. $\mathbf{b}$ can be thought of as the coordinate of the point in the $k + 1$ dimensional space spanned by the $k + 1$ coefficients in a regression equation with k independent variables (k variables plus one constant). For example, in the Martian intelligence example we have been discussing, there are $k = 1$ independent variables and $k + 1 = 2$ regression parameters; $\mathbf{b} = [a_0 \; a_F]^T$. In fact, the product $(\mathbf{b}_{(-i)} - \mathbf{b})^T(\mathbf{b}_{(-i)} - \mathbf{b})$ simply represents the square of the euclidean distance we presented in Eq. (4.2). The term $(\mathbf{X}^T\mathbf{X})$ represents the sums and products of the independent variables, similar to a variance estimate, and acts to convert the units of the numerator into the units of the dependent variable squared, which we then normalize out by dividing by $s_{y|x}^2$.[*] The resulting statistic D_i can be thought of as the dimensionless distance of a set of regression coefficients obtained by deleting point i from the data set from the set of regression coefficients obtained from the whole data set. When D_i is a big number, it indicates that point i is having a major effect on the regression coefficients.

It can be shown that[†]

$$D_i = \frac{r_i^2}{k + 1} \frac{h_{ii}}{1 - h_{ii}}$$

[*]An alternative (and equivalent) definition of Cook's distance is

$$D_i = \frac{(\hat{\mathbf{y}}_{(-i)} - \hat{\mathbf{y}})^T(\hat{\mathbf{y}}_{(-i)} - \hat{\mathbf{y}})}{(k + 1)s_{y|x}^2}$$

where $\hat{\mathbf{y}} = \mathbf{Xb}$ is the vector of predicted values of the dependent variable obtained from all the data and $\hat{\mathbf{y}}_{(-i)} = \mathbf{Xb}_{(-i)}$ is the vector of predicted values computed when the regression parameters are estimated without point i. This definition, which can be obtained by simple substitution into the definition of Cook's distance, illustrates that D_i is also the square of the distance that the predicted values of the dependent variable move when deleting point i, normalized by the residual variation.

[†]For a derivation and more detailed discussion of this result, see S. Weisberg, *Applied Regression Analysis* (2nd ed.), New York, Wiley, 1985, pp. 118–124, 293.

Thus, Cook's distance D_i depends on the internally Studentized residual for point i, r_i, which is a measure of the lack of fit of the regression model at point i, and the leverage (potential influence) of point i, h_{ii}. A large impact of a point on the regression coefficients, indicated by a large value of D_i, may be due to a large r_i, a large leverage, or both.

The Cook's distance statistic D_i tends to follow an F distribution with $k + 1$ numerator and $n - k - 1$ denominator degrees of freedom.[*] Because about 50 percent of the F distribution falls below 1 and virtually all of it falls below 4 or 5 (when the denominator number of degrees of freedom, which depends on the sample size, exceeds about 10), people consider points with D_i exceeding 1 to be worth further investigation and points with values exceeding 4 to be potentially serious outliers. Although it would be possible to formalize hypothesis tests on D_i, it, like the other diagnostics we have been discussing, should be used as a guide for finding points worth further consideration rather than for formal hypothesis tests.

None of the points in our original study of Martian intelligence has a D_i greater than .49 (Fig. 4-1 and Table 4-3). Thus, there is no indication of a problem with any of the points—an assessment which corresponds to our visual impression of the plot in Fig. 4-2A. In contrast, the data in Fig. 4-2C, which contains an outlier, have one data point with a D_i value of 1.39 (Fig. 4-4 and Table 4-4). By our rule of thumb, values of D_i greater than 1 warrant further investigation, and so we look at the data set and see that this observation corresponds to the visually obvious outlier in the scatter plot shown in Fig. 4-13.

The diagnostics we have discussed so far—residuals, standardized residuals, leverage, Studentized residuals, and Cook's distance—are all designed to provide different approaches to identifying points that do not fit with the assumed regression model.[†] No single one of these diagnostics tells the whole story. They should be used together to identify possibly erroneous data points or problems with the regression model (equation) itself.

---

[*]We say "tends to follow" because values of D_i should not be literally interpreted as F statistics. See J. Fox, *Linear Statistical Models and Related Methods: With Applications to Social Research*, New York, Wiley, 1984, p. 168.

[†]Three other influence diagnostics are the PRESS residual e_{-i}, DFFITS$_i$, and DFBETAS$_{j,i}$. The PRESS residual is a measure of how well the regression model will predict new data and will be discussed further in Chap. 6. DFFITS$_i$ is the number of estimated standard errors that a predicted value $\hat{y}_i$ changes if the ith point is removed from the data set. DFBETAS$_{j,i}$ is the number of standard errors that the coefficient b_j changes if the ith point is removed from the data set. For a derivation and discussion of DFFITS$_i$ and DFBETAS$_{j,i}$, see R. H. Myers, *Classical and Modern Regression with Applications*, Boston, Duxbury Press, 1986, pp. 284–288.

WHAT DO YOU DO WITH AN INFLUENTIAL OBSERVATION ONCE YOU HAVE FOUND IT?

The diagnostic techniques and statistics we have presented can all be used to identify outliers or potentially influential points. Having located such points, the question remains: What should you do with them?

There are two possible explanations for an influential observation:

1. Something is wrong with the data point (or points), and this problem is making it (or them) fall off the pattern evident from the rest of the data.
2. The model is not properly specified, so it cannot adequately describe the data.

Problems with the Data

The first step is to carefully check for data entry errors. Perhaps the single most common reason for an outlier is a data entry error in which two of the numbers associated with an observation are transposed. Such transposition errors can often exert major effects on the results of the regression analysis. Sometimes there is a transcription error in which one (or more) of the values is simply copied from the original data source incorrectly. Suppose that there had been a transposition error such that the dependent and independent variable values were reversed for the third data point in Table 4-1. If we fit these data (Table 4-5 and Fig. 4-14) using simple linear regression, compute the residuals, and plot them vs. foot size, we get the plot shown in Fig. 4-15. Ten of the residuals are in a

TABLE 4-5 Martian Intelligence vs. Foot Size Data Set Containing a Transposition Error

Foot Size F, cm	Intelligence I, zorp
10	8.04
8	6.95
7.58	13
9	8.81
11	8.33
14	9.96
6	7.24
4	4.26
12	10.84
7	4.82
5	5.68

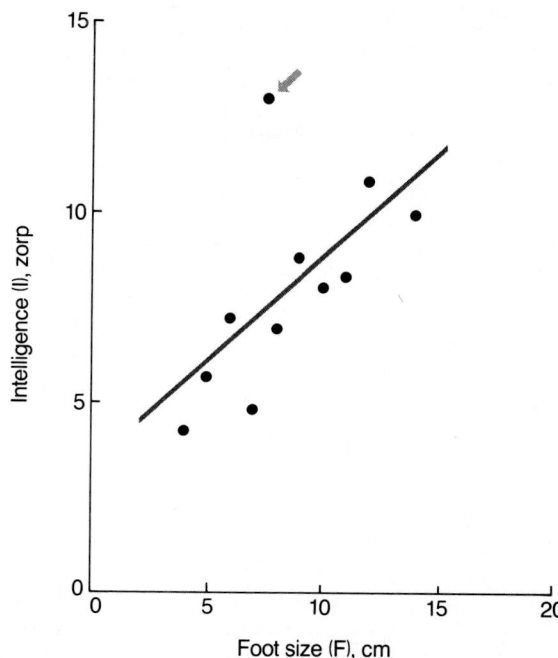

FIGURE 4-14 Plot of data on Martian intelligence versus foot size from Table 4-5, which contains a transposition error.

nice, symmetric band around zero, as we expect. However, one point is well above the rest. This residual has a raw magnitude of 5.5, compared with the next largest of 2.4 (Table 4-6). The outlying point has an r_{-i} of 5.49, which is also large enough to warrant attention. Cook's distance is only .4, and so is not large enough to indicate a problem on its own. It is, however, larger than most of the other values by, in most cases, 1 to 2 orders of magnitude, and so merits some attention. Here, most of the diagnostics flagged the third data point for further investigation; how-

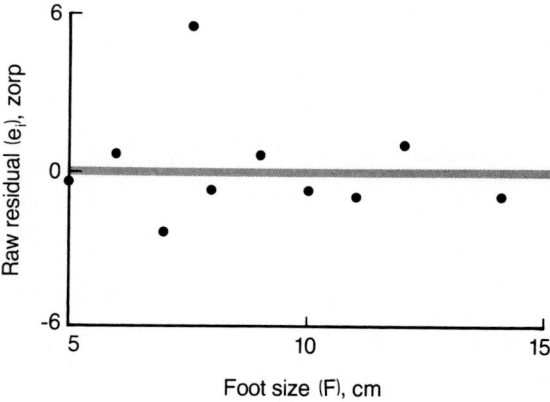

FIGURE 4-15 Residual plot for linear regression fit to the data in Fig. 4-14. This residual plot clearly identifies an outlier at a foot size of 7.58 cm. Close examination of the data reveals that there was a transposition error, which when corrected yields a better pattern in the residuals.

TABLE 4-6 Raw and Studentized Residuals and Cook's Distance
for Transposition Error in Martian Intelligence vs. Foot Size Data

Raw Residual	Studentized Residual r_{-i}	Cook's Distance
− .75	− .347	.0086
− .77	− .356	.0072
5.50	5.487	.3956
.55	.254	.0037
− .99	− .473	.0229
− .95	− .548	.1151
.58	.274	.0079
− 1.34	− .722	.1228
.99	.495	.0381
− 2.37	− 1.187	.0878
− .45	− .222	.0080

ever, in many instances only one or two diagnostics will identify such a point. The solution here is simple and uncontroversial: simply correct the data entry error and recompute the regression equation.

If there are no data entry errors, go back and double-check the laboratory or clinical notes to see if there was a technical error in the process of collecting the data. Perhaps there was a problem in the laboratory that invalidates the measurement. If so, the point should simply be deleted, unless it is possible to go back to original observations and correct the error. Although in theory there is nothing wrong with this approach, you must be very careful not to simply delete points because they do not fit your preconceived (regression) model.

If data are deleted, there must be a good, objectively applied reason for such deletions. This reason should be spelled out in any reports of the work, together with a description of the deleted data, unless the reasons for excluding the data are absolutely uncontroversial. Likewise, if it is possible to go back to original materials and recover the data, someone who is unaware of the nature of the problem should do the computations, to avoid introducing biases.

In the process of going back to the original sources, some factor may become evident that could explain why the experimental subject in question is unusual, such as being much older or younger than the other subjects. In such situations you may want to delete that subject as being not representative of the population under study or modify the regression model to take into account the things that make that subject different. Again, it is important that such post hoc changes be carefully justified and described in reports of the work. This situation shows the first way in which analysis of a problem in the data can lead to a reformulation of the regression model, typically by pointing to the need for additional independent variables or addition of an interaction term or nonlinear effect.

Problems with the Model

Once it has been determined that the data contain no errors, one has to turn to the regression model because, if the results violate the assumptions underlying the regression equation, it means that there is something wrong with the model itself. There are two common problems with regression models: leaving out important variables altogether or not taking into account nonlinearities or interactions among variables in the equation. Analysis of residuals usually can provide some leads concerning missing variables and can provide good guidance about the problem of nonlinearities.

When there are relatively large residuals after doing the initial analysis, it may be possible to identify potentially important additional independent variables by plotting the residuals against the other variables that may be influencing the dependent variable. If the suspected variable has a large influence, the residuals will show a systematic trend. To illustrate this procedure, let us reexamine our trip to Mars where we studied the relationship between Martian weight, height, and water consumption (Chap. 1). After we determine the relationship between Martian weight and height using Eq. (1.1), we suspect that the relationship between weight and height is affected by water consumption. To explore this hunch, we plot residuals of the regression of weight on height against water consumption (Fig. 4-16). Because there is a systematic positive linear trend, we have good evidence that we should add water consumption as an independent variable in the multiple linear regression equation we finally used to describe these data [Eq. (1.2)].

Identifying missing variables depends more on our knowledge of the system under study than it does on the pattern of residuals. In contrast, it is much easier to use the pattern of the residuals to identify situations in which the form of the variables included in the model may be inappropriate. The most common problem here is presence of a nonlinear effect when the regression equation is linear. For example, the data in both Fig. 4-2B (Martian intelligence vs. foot size) and Fig. 2-11 (thermal conductivity in seals) show obvious nonlinear trends in the residuals when they are plotted against the independent variable (Figs. 4-17A and 4-18A) or the predicted values of the dependent variable (Figs. 4-17B and 4-18B). Such trends in the residual plots are characteristic of a model misspecification where a nonlinear relationship is modeled with a linear

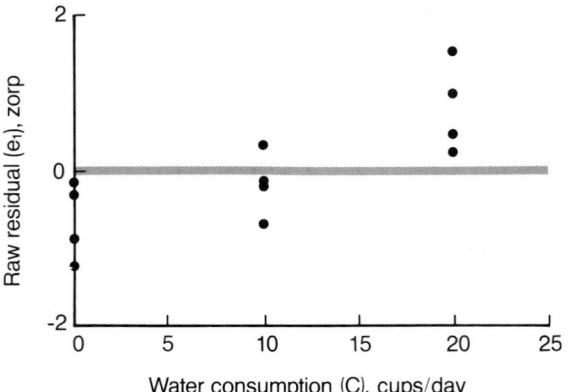

FIGURE 4-16 Examining the plot of the residuals between the predicted and observed values of the dependent variable and other independent variables not already in the regression equation can help identify variables that need to be added. For example, this figure shows a plot of the residual difference between predicted and observed Martian weight as a function of height leaving water consumption out of the regression equation (Fig. 1-1B). Because water consumption C is also an important determinant of Martian weight, a plot of the residuals from the regression just described against water consumption shows a linear trend. Including water consumption C in the regression equation will yield residual plots which show no trend and are uniformly distributed about zero.

equation. As we have already seen, it is often possible to transform the dependent variable, the independent variable, or both, to convert a non-linear relationship into a linear regression problem.

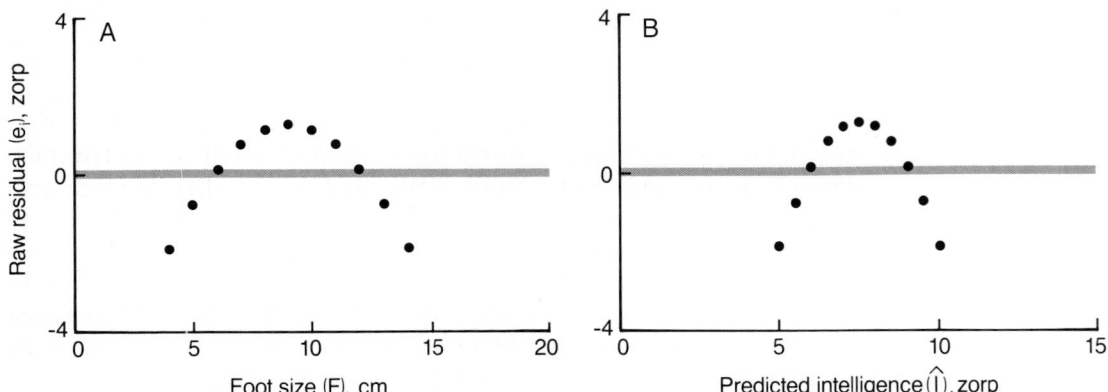

FIGURE 4-17 Plots of the residuals against the observed independent variable (A) or predicted value of the dependent variable (B) can be used to a identify a nonlinear relationship between the independent and dependent variable. These residuals are from the regression of intelligence on foot size using the data in Fig. 4-2B.

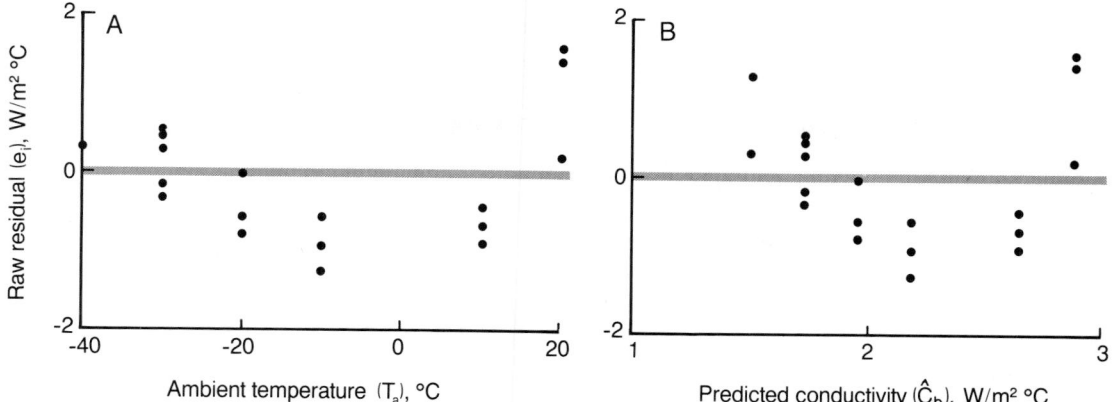

FIGURE 4-18 These plots of the residuals between predicted and observed thermal conductivity of gray seals as a function of the observed ambient temperature (the independent variable) and the predicted dependent variable, thermal conductivity, illustrate how nonlinearity in the relationship between thermal conductivity and ambient temperature appears as a systematic trend, rather than random scatter, in the residuals from a (misspecified) linear fit to the data.

Data Transformations

There are basically two ways in which the need for *introducing a nonlinearity* into the model presents itself. The first, and most obvious, is the situation we just discussed in which there is a nonlinear pattern in the residuals. Here, one uses the transformation to "flatten out" the residual plot. These transformations may be made on the dependent or independent variables. Table 4-7 shows some common linearizing transformations and Fig. 4-19 shows graphs of the associated functional forms for the case of one independent variable.

The second way in which a nonlinearity in the model is manifest is when the variance of residuals is not constant. Perhaps the most common example of this situation is when the random component of the measurement (say, random noise in the measurement instrument) is not simply additive, but rather a constant *percentage* error. If it is, the residual variance will increase with the dependent variable, giving the band of residuals a typical "megaphone" shape. Logarithm transformations of the dependent variable often resolve this problem. When the dependent variable is transformed to make the residuals more closely meet the assumption of constant variance, the procedure is called a *variance stabilizing transformation*. (Table 4-7 presents some commonly used transformations for stabilizing the variance.) Sometimes variance stabilizing transformations introduce nonlinearities in the residual pattern which require further modification of the regression model. On the other hand,

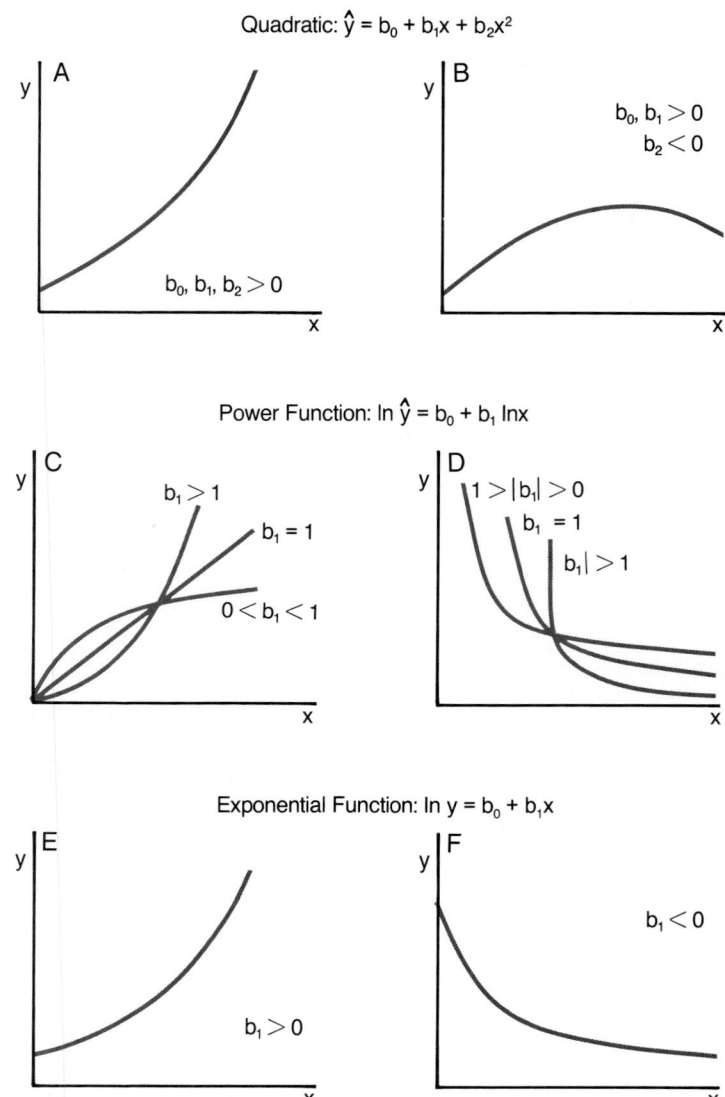

FIGURE 4-19 Commonly used transformations for changing nonlinear relationships between independent and dependent variables into corresponding linear regression problems. These transformations often not only linearize the problem but also stabilize the variance. In some cases, however, these transformations will linearize the relationship between the dependent and independent variables at the expense of introducing unequal variance because the transformation changes the error structure of the regression model. When such a transformation is used, it is important to carefully examine the residual plots to make sure that the error variance remains constant, regardless of the values of the independent and dependent variables. (Adapted from Figures 6.1 through 6.10 of R. H. Myers, *Classical and Modern Regression with Applications*, Boston, PWS-KENT Publishing Co., 1986.)

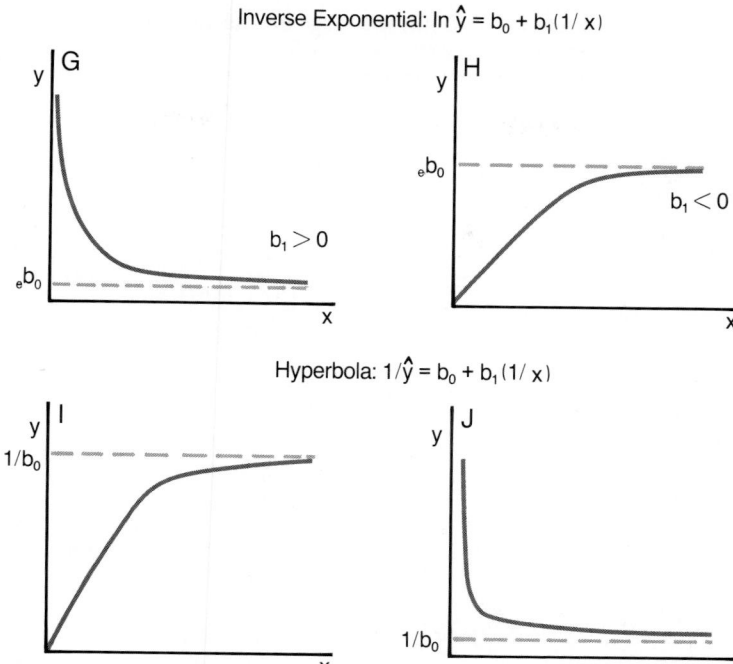

FIGURE 4-19 *(continued)*

often a single transformation will both stabilize the residual variance and remove nonlinearities from the residual plot.

Although introducing transformations of variables into the regression analysis greatly increases its flexibility and power, there are potential pitfalls. The interpretation of the basic linear regression model, given by Eq. (3.13), is quite simple. The effects of each of the independent variables are proportional to the regression coefficients and additive. The effect of random variation about the regression plane, whether due to biological variability or measurement error, simply adds to the effects of the independent variables. The problem is that by transforming the variables we change this relationship.

For example, if we transform the dependent variable by taking its logarithm, the equation relating the dependent and independent variables becomes

$$\ln y = b_0 + b_1 x + \epsilon$$

We can rewrite this equation in terms of y directly by

$$e^{\ln y} = e^{(b_0 + b_1 x + \epsilon)}$$

$$y = e^{b_0} e^{b_1 x} e^{\epsilon}$$

$$y = b_0' e^{b_1 X} \epsilon'$$

TABLE 4-7 Common Transformations for Linearization and/or Variance Stabilization

Transformation		Regression Equation[*]	Comments
Y	X^j	$\hat{y} = b_0 + \sum\limits_{j=1}^{n} b_{jX}^{j}$	nth-order polynomial; e.g., quadratic (Fig. 4-19A, B), cubic
$\ln Y$	$\ln X$	$\hat{y} = b_0 X^{b_1}$	Power function (Fig. 4-19C, D)
$\ln Y$	X	$\hat{y} = b_0 e^{b_1 X}$	Exponential (Fig. 4-19E, F); often used to stabilize variance; all Y must be positive
$\ln Y$	$\dfrac{1}{X}$	$\hat{y} = b_0 e^{b_1/X}$	Inverse exponential (Fig. 4-19G, H)
$\dfrac{1}{Y}$	$\dfrac{1}{X}$	$\hat{y} = \dfrac{X}{b_0 + b_1 X}$	Hyperbola (Fig. 4-19I, J)
$\dfrac{1}{Y}$	X	$\hat{y} = \dfrac{1}{b_0 + b_1 X}$	Often used to stabilize variances
Y	$\dfrac{1}{X}$	$\hat{y} = b_0 + \dfrac{b_1}{X}$	
$\sqrt{Y}$	X	$\hat{y} = (b_0 + b_1 X)^2$	Variance stabilization; particularly if data are counts

[*]The linearized forms of these regression equations appear in Fig. 4-19.

where $b_0' = e^{b_0}$ and $\epsilon' = e^{\epsilon}$. Note that, in terms of the dependent variable y, the random error term ϵ' is *multiplicative* rather than additive. Thus, the logarithmic transformation of the dependent variable has changed the assumed structure of the random component of the regression model. In fact, one often directly seeks such an effect. It is important, however, to consider exactly how the transformations affect the presumed structure of the error term and ensure that the result makes sense in terms of the underlying biological, clinical, or other situation that gave rise to the data in the first place. When conflicts between the need for transformation and the nature of the residuals in the model arise, you can seek other transformations or formulate a model which can be fit directly using the techniques of nonlinear regression discussed in Chap. 10.

Transformations of the dependent variable also introduce implicit weighting of different points into the process of estimating the parameters. Recall that the regression procedure minimizes the sum of squared deviations between the observed and predicted values of the dependent variable. If you transform the dependent variable, the regression will minimize the sum of squared deviations between the transformed dependent variable and the predicted values of the transformed dependent variable. For example, if we take logarithms of the dependent variable, the same absolute deviation at a small value of the dependent variable will contribute more to the sum of squared residuals than the same deviation at a large value of the dependent variable, simply because the value of the logarithm changes faster at small values than large ones (i.e., ln 10 − ln 9 = 2.30 − 2.20 = .10, whereas ln 100 − ln 99 = 4.61 − 4.60 = .01). This situation is appropriate for a constant percentage error in the observations, and so may be no problem at all in terms of the analysis or its interpretation. One simply needs to be aware of the impact of the transformation and make sure that the implicit weighting makes sense.[*] If a linearizing transformation does not make sense in this regard, it is still possible to analyze the data using the methods of *nonlinear regression* described in Chap. 10.

In sum, the pattern in the residuals taken as a whole, together with specific outlier points, can be used to identify things that need to be changed in the formulation of the regression equation itself. These procedures not only lead to an equation that better fits the assumptions underlying regression analysis, and so makes the attendant hypothesis tests valid, but it also can lead to equations that provide a better structural insight into the biological (or other) process that gave rise to the data in the first place.

WATER MOVEMENT ACROSS THE PLACENTA

The placenta transfers nutrients and fluids from the maternal circulation to the fetal circulation. Maternal blood reaches the placenta via the uterine artery, and fetal blood reaches the placenta through the umbilical artery in the umbilical cord. By measuring the blood flows in these arteries and their relationship to diffusion (usually measured by water transfer) across the placenta, it is possible to draw conclusions about the characteristics of the exchange between maternal and fetal circulations; specifically, to tell if the blood flows in the two sides of the placenta are in the same direction (concurrent exchange), the opposite direction (countercurrent exchange), or some intermediate arrangement (flowing at right angles to each other—so-called cross-current exchange).

Different animals have different kinds of placental exchange mechanisms. To identify the functional organization of placental exchange in the cow, Reynolds and Ferrell[†] measured the rates of uterine and umbilical artery blood flow and the transport of water across the placenta

[*]For a discussion of weighted least-squares regression, see R. H. Myers, *Classical and Modern Regression with Applications*, Boston, Duxbury Press, 1986, pp. 168–177.

[†]L. P. Reynolds and C. L. Ferrell, "Transplacental Clearance and Blood Flows of Bovine Gravid Uterus at Several Stages of Gestation," *Am. J. Physiol.* 253:R735–R739, 1987.

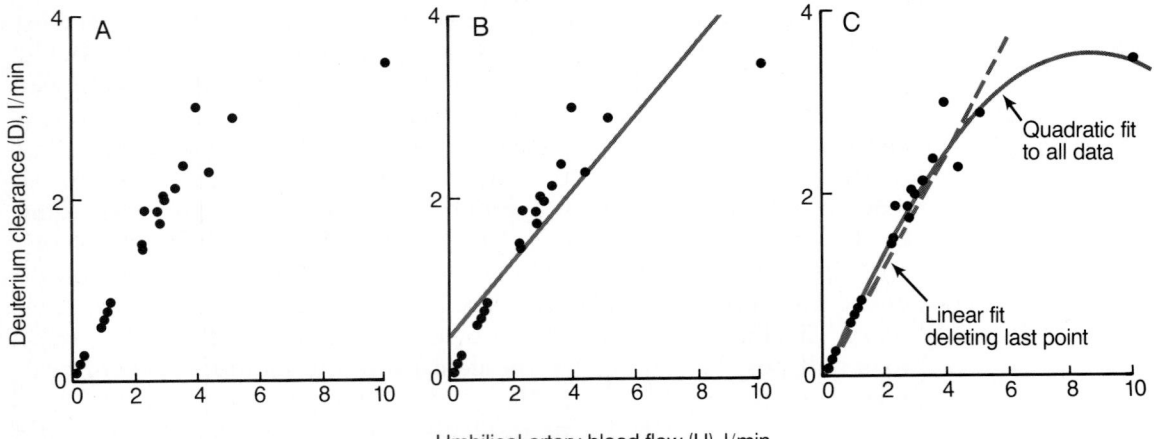

FIGURE 4-20 *A.* A plot of observations relating clearance of D_2O, a radioactive form of water, across the placenta in pregnant cows as a function of blood flow in the umbilical cord. *B.* The fit of the simple linear regression given by Eq. (4.3) to these data. *C.* The effect of fitting a line after deleting the observation at an umbilical flow of 10 L/min, as well as the fit of the quadratic function given by Eq. (4.5) to all the data. Note that the coefficient of the quadratic term (and so the nature of the curvature in the regression line) is *essentially determined by the single point* at an umbilical flow of 10 L/min. To be more confident in the shape of the curve, it would be desirable to collect more data at umbilical flows between 6 and 10 L/min.

at several times during the last half of pregnancy in cows. Transplacental transport of water was evaluated as the clearance of deuterium oxide (D_2O), a radioactive form of water. These data (in Table C-9, Appendix C) are plotted in Fig. 4-20*A*.

We begin the analysis of these data by fitting them with the simple linear regression equation

$$\hat{D} = b_0 + b_U U \tag{4.3}$$

where $\hat{D}$ is the estimated clearance of D_2O and U is umbilical artery blood flow. The computer output is in Fig. 4-21. The regression equation (Fig. 4-20*B*)

$$\hat{D} = .47 \text{ L/min} + .41U \tag{4.4}$$

has an R^2 of .818 ($F = 99.2$; $P < .01$), which seems to indicate a good overall model fit. We would interpret this equation as indicating that .47 L/min of D_2O passes through the placenta even when there was no umbilical blood flow ($U = 0$) and the D_2O clearance increased by .41 L/min for every 1 L/min increase in umbilical blood flow.

```
MULTIPLE R              0.9047
MULTIPLE R-SQUARE       0.8184
ADJUSTED R-SQUARE       0.8102

STD. ERROR OF EST.      0.4141

ANALYSIS OF VARIANCE
                      SUM OF SQUARES    DF    MEAN SQUARE      F RATIO
          REGRESSION    17.001844        1     17.00184        99.15
          RESIDUAL       3.7723517      22      0.1714705
```

	VARIABLES IN EQUATION FOR D							VARIABLES NOT IN EQUATION				
VARIABLE	COEFFICIENT	STD. ERROR OF COEFF	STD REG COEFF	TOLERANCE	F TO REMOVE	LEVEL		VARIABLE	PARTIAL CORR.	TOLERANCE	F TO ENTER	LEVEL
(Y-INTERCEPT	0.47188)											
U 2	0.40719	0.0409	0.905	1.00000	99.15	1						

CASE NO. LABEL	PREDICTED	RESIDUAL	WEIGHT	1 D	8 DSTRESID	12 HATDIAG	16 COOK
1	0.5818	-0.3918	1.000	0.1900	-0.9896	0.0866	0.0464
2	0.5492	-0.4492*	1.000	0.1000	-1.1453	0.0900	0.0639
3	0.6266	-0.3566	1.000	0.2700	-0.8948	0.0821	0.0361
4	0.8709	-0.2109	1.000	0.6600	-0.5170	0.0618	0.0091
5	0.9605	-0.1505	1.000	0.8100	-0.3667	0.0561	0.0042
6	0.8709	-0.2209	1.000	0.6500	-0.5419	0.0618	0.0100
7	0.9116	-0.1516	1.000	0.7600	-0.3701	0.0591	0.0045
8	0.8872	-0.1972	1.000	0.6900	-0.4828	0.0607	0.0078
9	0.9524	-0.2024	1.000	0.7500	-0.4944	0.0566	0.0076
10	0.9768	-0.1168	1.000	0.8600	-0.2840	0.0551	0.0025
11	0.8383	-0.2483	1.000	0.5900	-0.6110	0.0641	0.0132
12	1.6772	0.3128	1.000	1.9900	0.7656	0.0446	0.0139
13	1.8034	0.3266	1.000	2.1300	0.8021	0.0488	0.0168
14	1.3962	0.1038	1.000	1.5000	0.2506	0.0419	0.0014
15	1.5794	0.2706	1.000	1.8500	0.6591	0.0426	0.0099
16	1.6527	0.3773	1.000	2.0300	0.9289	0.0440	0.0200
17	1.6161	0.1139	1.000	1.7300	0.2753	0.0432	0.0018
18	1.3921	0.0779	1.000	1.4700	0.1879	0.0419	0.0008
19	1.9174	0.4626*	1.000	2.3800	1.1575	0.0542	0.0378
20	1.4369	0.4231*	1.000	1.8600	1.0459	0.0417	0.0237
21	2.5486	0.3414	1.000	2.8900	0.8701	0.1120	0.0483
22	2.0681	0.9319**	1.000	3.0000	2.6168	0.0637	0.1842
23	4.5601	-1.0801**	1.000	3.4800	-8.8915***	0.6086	13.5148
24	2.2554	0.0346	1.000	2.2900	0.0851	0.0793	0.0003

FIGURE 4-21 Results of regression analysis using a simple linear regression model to obtain the fit of the data on water transport across the placenta shown in Fig. 4-20*B*. Note that the 23rd data point appears to be a significant outlier: it has a very large Studentized residual r_{-i}, leverage h_{ii}, and Cook's distance D_i. (These diagnostics are labeled DSTRESID, HATDIAG, and COOK, respectively.)

The regression diagnostics, however, reveal a severe problem. Plots of the raw residuals against both U and $\hat{D}$ (Fig. 4-22) reveal that the residuals are not randomly scattered in a uniform band around zero, but rather systematically increase, except for one point which is far from the others. This is the classic pattern one finds if an outlier is distorting the regression line (compare with Fig. 4-6*C*).

This discrepant data point also stands out when we examine the regression diagnostics in Fig. 4-21. The 23rd data point has a very large

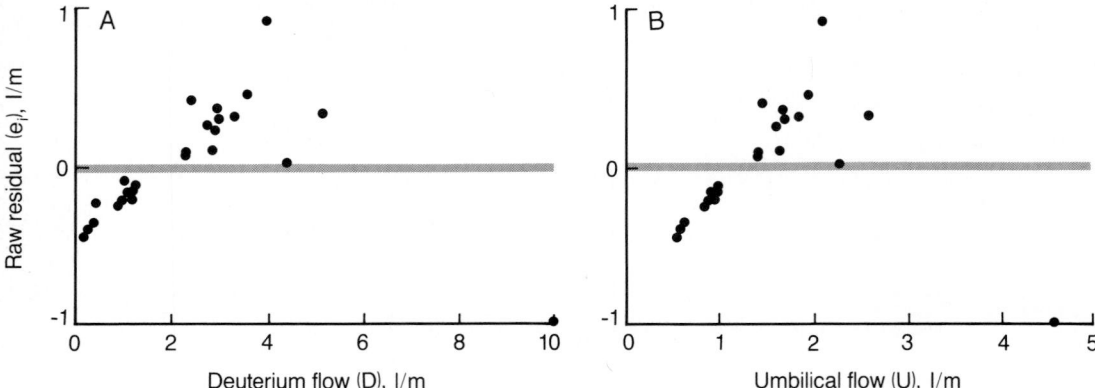

FIGURE 4-22 Plot of residuals against the independent variable (A) and dependent variable (B) for the linear fit to the data in Fig. 4-20. This residual plot highlights the outlier associated with the 23rd data point.

Studentized deleted residual r_{-i} of -8.89 (compared to the rule-of-thumb threshold value of 2), a large leverage of .61 [compared to the rule-of-thumb threshold value of $2(k + 1)/n = 2(1 + 1)/24 = .17$], and a large Cook's distance of 13.51 (compared to the rule-of-thumb threshold value of 4 for points meriting close attention). None of the other data points have values of the diagnostics that cause concern. Thus, all of the regression diagnostics indicate that the 23rd data point (the point with $D_2O = 3.48$ L/min and umbilical flow $= 10.04$ L/min) is potentially exerting undue influence on the results of this regression analysis. This problem is evident when we plot this regression line with the data (Fig. 4-20B); the point in question is clearly an outlier and leverage point which is greatly influencing the location of the regression line. Having identified this point as an influential observation, what do we do about it?

There are two possibilities: The datum could be in error, or the relationship between water clearance across the placenta and umbilical blood flow could be nonlinear. If the point in question was bad because of an irretrievable laboratory error, we simply delete this point and recompute the regression equation. The computer output for this recomputed regression is shown in Fig. 4-23 and the regression line appears in Fig. 4-20C. The new regression equation is

$$\hat{D} = .09 \text{ L/min} + .61U$$

Inspection of Fig. 4-20C also reveals that this line describes the data well. The R^2 of .952 indicates a large improvement in fit, compared to .818 when all points were included. None of the points has large leverage [although the 21st data point has a leverage of .261, which exceeds the threshold of $2(k + 1)/n = 4/23 = .17$], and two points (associated with

```
MULTIPLE R          0.9757
MULTIPLE R-SQUARE   0.9520
ADJUSTED R-SQUARE   0.9497

STD. ERROR OF EST.  0.1942

ANALYSIS OF VARIANCE
                    SUM OF SQUARES    DF    MEAN SQUARE      F RATIO
        REGRESSION  15.705316          1    15.70532         416.57
        RESIDUAL    0.79172754        21    0.3770131E-01
```

		VARIABLES IN EQUATION FOR D				.	VARIABLES NOT IN EQUATION			
		STD. ERROR	STD REG		F	.		PARTIAL		F
VARIABLE	COEFFICIENT	OF COEFF	COEFF	TOLERANCE	TO REMOVE	LEVEL.	VARIABLE	CORR. TOLERANCE	TO ENTER	LEVEL
(Y-INTERCEPT	0.09123)					.				
U 2	0.61239	0.0300	0.976	1.00000	416.57	1 .				

CASE NO. LABEL	PREDICTED	RESIDUAL	WEIGHT	1 D	8 DSTRESID	12 HATDIAG	16 COOK
1	0.2566	-0.0666	1.000	0.1900	-0.3582	0.1220	0.0093
2	0.2076	-0.1076	1.000	0.1000	-0.5843	0.1291	0.0261
3	0.3239	-0.0539	1.000	0.2700	-0.2884	0.1128	0.0055
4	0.6914	-0.0314	1.000	0.6600	-0.1638	0.0726	0.0011
5	0.8261	-0.0161	1.000	0.8100	-0.0835	0.0621	0.0002
6	0.6914	-0.0414	1.000	0.6500	-0.2162	0.0726	0.0019
7	0.7526	0.0074	1.000	0.7600	0.0385	0.0675	0.0001
8	0.7159	-0.0259	1.000	0.6900	-0.1349	0.0705	0.0007
9	0.8138	-0.0638	1.000	0.7500	-0.3324	0.0630	0.0039
10	0.8506	0.0094	1.000	0.8600	0.0488	0.0605	0.0001
11	0.6424	-0.0524	1.000	0.5900	-0.2745	0.0769	0.0033
12	1.9039	0.0861	1.000	1.9900	0.4490	0.0618	0.0069
13	2.0937	0.0363	1.000	2.1300	0.1899	0.0771	0.0016
14	1.4813	0.0187	1.000	1.5000	0.0959	0.0443	0.0002
15	1.7569	0.0931	1.000	1.8500	0.4836	0.0531	0.0068
16	1.8672	0.1628	1.000	2.0300	0.8594	0.0594	0.0236
17	1.8120	-0.0820	1.000	1.7300	-0.4263	0.0561	0.0056
18	1.4752	-0.0052	1.000	1.4700	-0.0269	0.0442	0.0000
19	2.2652	0.1148	1.000	2.3800	0.6121	0.0948	0.0202
20	1.5426	0.3174*	1.000	1.8600	1.7539	0.0454	0.0666
21	3.2144	-0.3244*	1.000	2.8900	-2.0938	0.2607	0.6657
22	2.4918	0.5082**	1.000	3.0000	3.4448	0.1240	0.5534
24	2.7735	-0.4835**	1.000	2.2900	-3.3211	0.1694	0.7610

FIGURE 4-23 Results of linear regression analysis to fit the data relating umbilical water transport to umbilical blood flow to a simple linear model after deleting the outlier identified in Figs. 4-21 and 4-22. Although the linear fit to the remaining data is better, there is still a cluster of points — particularly points 21, 22, and 24 — that could be considered outliers.

the original 22nd and 24th points) have Studentized deleted residuals r_{-i} greater than 3. Thus, there is a cluster of two or three points at the highest levels of umbilical flow that show unusually high scatter. On the whole, however, there is little evidence that these points are exerting serious influence on the regression result. (The Cook's distances for these three points are larger than for the others but are still less than 1.) In addition to these effects, the omission of the original 23rd data point has increased the estimated slope from .41 to .61, and the intercept is not significantly different from zero ($P > .50$). Thus, the data point we identified does, indeed, have great influence: deleting it increased the magnitude of the slope by about 50 percent and eliminated the intercept, indicating that there is no water clearance across the placenta when there is no umbilical blood flow (which is what we would expect).

What if we could not explain this outlying data point by a laboratory error? We would have to conclude that the regression model given by Eq. (4.3) is inadequate. Specifically, it needs to be modified to include a non-linearity in the relationship between D_2O clearance and umbilical flow. Let us try the quadratic relationship

$$\hat{D} = b_0 + b_U U + b_{U_2} U^2 \tag{4.5}$$

The results of this regression are in Fig. 4-24, and the line is plotted in Fig. 4-20C; the regression equation is

$$\hat{D} = -.099 \text{ L/min} + .85U - .05 \text{ min/L } U^2 \tag{4.6}$$

The data point at an umbilical flow = 10.04 L/min is no longer as great an outlier. However, it still has a very large leverage of .98, which greatly exceeds the threshold value of $2(k + 1)/n = 6/24 = .25$. Cook's distance for this point is greatly reduced, though still greater than 1. This reduction in Cook's distance reflects the reduced action of this point as an outlier. R^2 for this equation is .973, compared to .818 for the linear model. Thus, in terms of goodness of fit and residual distribution, the quadratic relationship of Eq. (4.5) is much better than the simple linear fit of Eq. (4.3).

The intercept in Eq. (4.6) is not significantly different from zero, and the other two coefficients are. The interpretation of this equation is that there is no water clearance across the placenta when there is no umbilical blood flow, that the rate of water clearance is .85 L/min for each 1 L/min increase in umbilical blood flow for small (near 0) blood flows, and that this rate falls off as umbilical blood flow increases. In contrast to those obtained with the original linear analysis, these results suggest that there is some sort of saturation or rate-limiting process in water clearance. Reynolds and Ferrell further conclude that this is evidence for an exchange system arranged in concurrent fashion, with umbilical and uterine blood flows in the same direction.

Although this result is plausible, and the fit of the equation to the data quite good, the fact remains that the only evidence we have for any nonlinear relationship is the *single* data point at 10.04 L/min umbilical blood flow. In such situations, where the data suggest a nonlinear relationship based on one (or even a few) extreme points with large gaps in data, one should collect additional data to fill in the gaps to provide convincing evidence that the relationship is, indeed, nonlinear.

```
MULTIPLE R              0.9864
MULTIPLE R-SQUARE       0.9731
ADJUSTED R-SQUARE       0.9705

STD. ERROR OF EST.      0.1632

ANALYSIS OF VARIANCE
                     SUM OF SQUARES    DF    MEAN SQUARE      F RATIO
        REGRESSION   20.214943          2    10.10747         379.53
        RESIDUAL      0.55925530       21    0.2663120E-01
```

		STD. ERROR	STD REG		F			PARTIAL		F	
VARIABLE	COEFFICIENT	OF COEFF	COEFF	TOLERANCE	TO REMOVE	LEVEL	VARIABLE	CORR.	TOLERANCE	TO ENTER	LEVEL

VARIABLES IN EQUATION FOR D   ·   VARIABLES NOT IN EQUATION

```
(Y-INTERCEPT    -0.09946 )
U     2     0.84950     0.0434     1.887     0.13806     383.61     1 .
U2    3    -0.04918     0.0045    -1.058     0.13806     120.65     1 .
```

CASE				1	9	13	17
NO. LABEL	PREDICTED	RESIDUAL	WEIGHT	D	DSTRESID	HATDIAG	COOK
1	0.1263	0.0637	1.000	0.1900	0.4151	0.1511	0.0106
2	0.0602	0.0398	1.000	0.1000	0.2610	0.1644	0.0047
3	0.2163	0.0537	1.000	0.2700	0.3465	0.1345	0.0065
4	0.6858	-0.0258	1.000	0.6600	-0.1605	0.0724	0.0007
5	0.8491	-0.0391	1.000	0.8100	-0.2417	0.0599	0.0013
6	0.6858	-0.0358	1.000	0.6500	-0.2227	0.0724	0.0014
7	0.7606	-0.0006	1.000	0.7600	-0.0040	0.0662	0.0000
8	0.7159	-0.0259	1.000	0.6900	-0.1605	0.0698	0.0007
9	0.8345	-0.0845	1.000	0.7500	-0.5249	0.0609	0.0062
10	0.8783	-0.0183	1.000	0.8600	-0.1129	0.0582	0.0003
11	0.6253	-0.0353	1.000	0.5900	-0.2199	0.0782	0.0014
12	1.9842	0.0058	1.000	1.9900	0.0361	0.0739	0.0000
13	2.1526	-0.0226	1.000	2.1300	-0.1413	0.0867	0.0007
14	1.5755	-0.0755	1.000	1.5000	-0.4662	0.0519	0.0041
15	1.8474	0.0026	1.000	1.8500	0.0164	0.0649	0.0000
16	1.9505	0.0795	1.000	2.0300	0.4963	0.0716	0.0066
17	1.8993	-0.1693*	1.000	1.7300	-1.0791	0.0682	0.0282
18	1.5692	-0.0992	1.000	1.4700	-0.6151	0.0517	0.0071
19	2.2965	0.0835	1.000	2.3800	0.5296	0.0989	0.0106
20	1.6376	0.2224*	1.000	1.8600	1.4361	0.0542	0.0375
21	2.9539	-0.0639	1.000	2.8900	-0.4195	0.1631	0.0119
22	2.4749	0.5251***	1.000	3.0000	5.0171**	0.1153	0.5081
23	3.4723	0.0077	1.000	3.4800	0.3020	0.9769	1.3415
24	2.6779	-0.3879**	1.000	2.2900	-3.0047	0.1349	0.3394

FIGURE 4-24 Results of linear regression analysis to fit the data relating umbilical water transport to umbilical blood flow using the quadratic function given by Eq. (4.5) and plotted in Fig. 4-21C. Allowing for the nonlinearity yields a better pattern in the residual diagnostics than with the simple linear model. Although the 23rd point is no longer an outlier, it still has a high leverage because it is so far removed from the other data along the umbilical flow axis.

CHEAPER CHICKEN FEED

Methionine is an essential amino acid, which means that it must be provided in the diet because the body cannot manufacture it. Some animal feed grains, for example, soybeans and corn, are deficient in methionine. Therefore, methionine is added to commercial farm animal feeds, especially poultry feed. Methionine is fairly expensive and, therefore, a similar but cheaper compound, hydroxy-methyl-thiobutyric acid—also known as methionine hydroxy analog, or MHA—is usually added to provide the methionine.

MHA does not substitute directly for methionine. After MHA is absorbed, it is converted to methionine. However, MHA can only replace methionine if it is absorbed with about the same efficiency. If the body cannot absorb MHA from the intestines as well as it can absorb methionine, the amount of MHA will have to be increased to make up for the relatively inefficient absorption. Because it is not known whether methionine and MHA are absorbed by the same mechanisms, Brachet and Puigserver* wanted to compare the absorption pathways of methionine and MHA.

They studied the rates of MHA uptake into isolated pieces of small intestines at different concentrations of MHA. These are classic transport kinetics studies designed to discover the speed and structure of the transport mechanisms. In the first part of their study, they found that there were two components to the uptake, a simple diffusional component and a component due to active transport where some other protein molecule in a cell membrane is necessary to bind the MHA, then "carry" it into the cell. This type of absorption was found for both methionine and MHA. They then repeated the study in the presence of 20-mM lactate, which they reasoned might act to inhibit the active component of the absorption. Under these conditions, methionine absorption is unaffected, but the active component of MHA absorption is inhibited and only the diffusional component remains, and the relationship between transport rate and MHA concentration is linear (Fig. 4-25). To obtain the diffusion constant—the speed with which MHA diffuses into the intestinal cells—they used linear regression fit of these data to the equation

$$\hat{R} = b_0 + b_M M \tag{4.7}$$

where R is the rate of uptake of MHA and M is the concentration of MHA. The regression results are summarized in the computer output shown in Fig. 4-26 (the data are in Table C-10, Appendix C). The intercept b_0 is estimated to be 36 nmol/g/min) with a standard error of 72 nmol/g/min) and is not significantly different from zero ($P > .50$). The slope b_M of 50 μL/g/min with a standard error of 1.7 μL/g/min, is highly significant, and R^2 is .939. Thus, we seem to have a good fit of these data to the equation

$$\hat{R} = 50 \ \mu L/g/min \ M \tag{4.8}$$

*P. Brachet and A. Puigserver, "Transport of Methionine Hydroxy Analog across the Brush Border Membrane of Rat Jejunum," *J. Nutr.* 117:1241–1246, 1987.

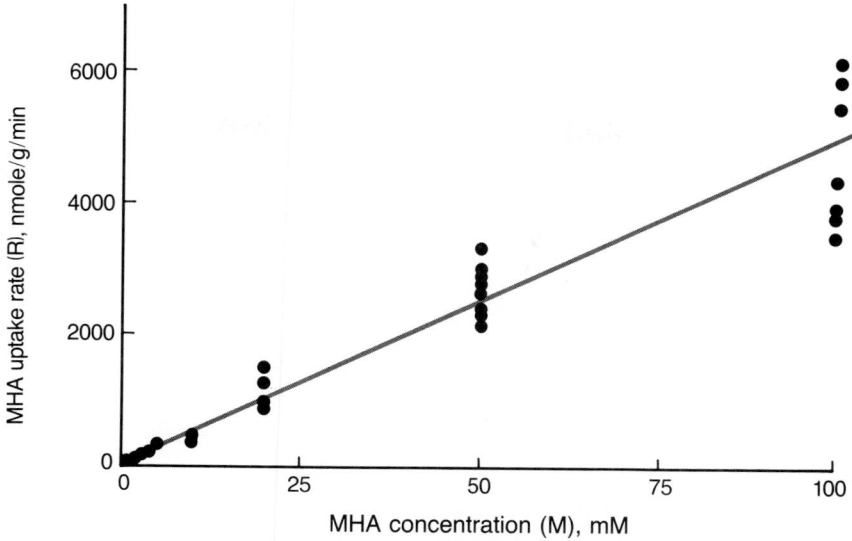

FIGURE 4-25 Relationship between transport of methionine hydroxy analog (MHA) and the concentration of MHA observed in rats. Although there appears to be a linear relationship, the variation of the observed MHA uptake increases as the MHA concentration increases. (Data adapted from Figure 6 of P. Brachet and A. Puigservier, "Transport of Methionine Hydroxy Analog across the Brush Border Membrane of Rat Jejunum," *J. Nutr.* 117:1241–1246, 1987.)

```
Variable(s) Entered on Step Number
  1..    M          Methionine concentration

Multiple R          .96879
R Square            .93855
Adjusted R Square   .93745
Standard Error   441.42064

Analysis of Variance
                DF      Sum of Squares        Mean Square
Regression       1     166661137.96686      166661137.96686
Residual        56      10911722.37797        194852.18532

F =     855.32086      Signif F =  .0000

----------------- Variables in the Equation -----------------
Variable           B          SE B        Beta        T  Sig T

M           50.223251     1.717277      .968788    29.246  .0000
(Constant)  35.548648    72.483558                   .490  .6257
```

FIGURE 4-26 Linear regression analysis of the data in Fig. 4-25 relating MHA uptake to MHA concentration in the rat jejunum.

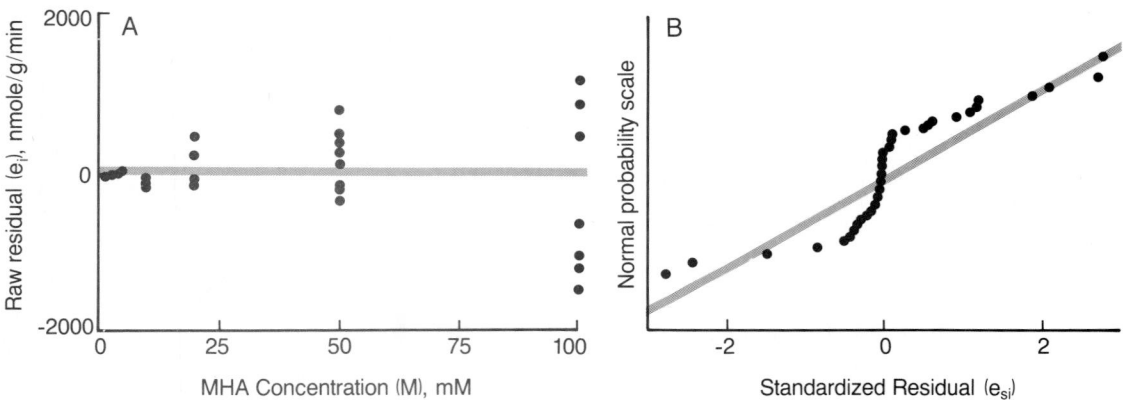

FIGURE 4-27 *A.* Analysis of the residuals from a linear regression model fit to the data in Fig. 4-25*A*. The residuals, while scattered symmetrically about 0, exhibit a "megaphone" shape where the magnitude of the residual increases as the value of the independent variable increases, suggesting that the constant variance assumption of linear regression is violated in these data. *B.* The normal probability plot of these residuals shows an S shape rather than a straight line, indicating that the residuals are not normally distributed.

This result is consistent with what we would expect from a simple diffusion process, in which there is no absorption when there is no MHA (because the intercept is 0) and the diffusion rate is directly proportional to the concentration of MHA, with the diffusion constant being 50 $\mu L/g/min$.

Despite the good fit, however, these data are not consistent with the equal variance assumption of linear regression. Examining Fig. 4-25 reveals that the scatter of data points about the regression line increases as the MHA concentration increases. This changing variance is even more evident from the "megaphone" pattern of the residual plot in Fig. 4-27*A*. Thus, the assumption of homogeneity of variances is violated (as it often will be for such kinetics studies). Moreover, a normal probability plot (Fig. 4-27*B*) reveals that the residuals are not normally distributed: they fall along an S-shaped curve characteristic of this violation of homogeneity of variance. So, what do we do?

The obvious answer is to transform the dependent variable using one of the transformations in Table 4-7 to stabilize the variance. The transformations recommended to correct the error structure, such as $\sqrt{R}$, $1/R$, and ln R, all tend to stabilize the variance (Fig. 4-28*A*, *B*, and *C*). Unfortunately, each of these transformations makes the relationship between R and M nonlinear. Thus, to maintain a linear relationship between the dependent and independent variables and meet the constant variance assumption, we also have to transform the independent variable M. The

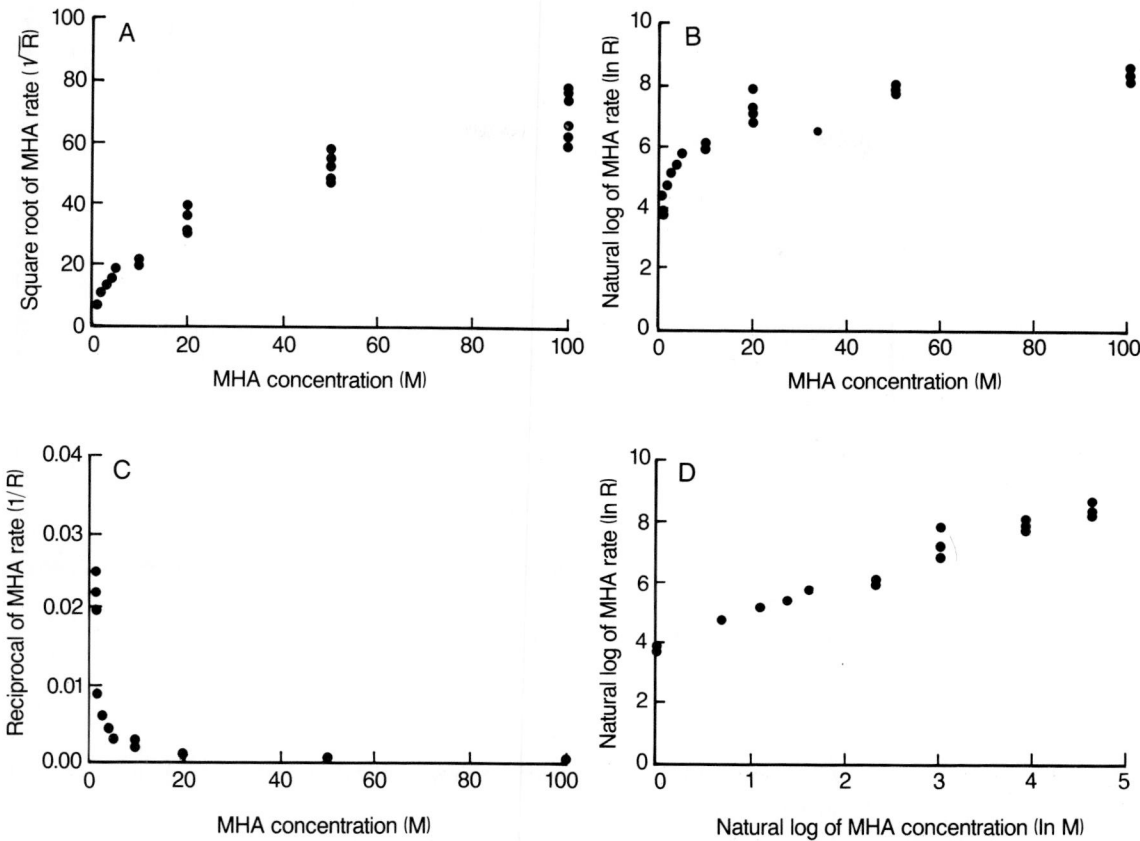

FIGURE 4-28 This figure shows four transformations that one could use in an attempt to equalize the variance in the data on MHA uptake as a function of MHA concentration shown in Fig. 4-25. *A*. Square root of MHA uptake versus MHA concentration. *B*. The log of MHA uptake rate versus MHA concentration. *C*. The inverse of MHA uptake rate versus MHA concentration. *D*. The log of the MHA uptake versus the log of the MHA concentration. Examination of these plots suggests that, although all transformations tend to stabilize the variance, only the log–log plot in panel *D* maintains a linear relationship between these two variables.

best transformation is $\ln R$ vs. $\ln M$ (Fig. 4-28*D*), to obtain the linear equation

$$\ln \hat{R} = \ln a_0 + a_M \ln M = a_0' + a_M \ln M \qquad (4.9)$$

which is equivalent to the power function model

$$\hat{R} = a_0 M^{a_M}$$

The computer output for the regression using Eq. (4.9) is in Fig. 4-29. Following this transformation, the residuals show little evidence of un-

```
Variable(s) Entered on Step Number
   1..    LOGM       natural log of Methionine

Multiple R            .99323
R Square              .98651
Adjusted R Square     .98627
Standard Error        .17716

Analysis of Variance
                    DF      Sum of Squares      Mean Square
Regression           1          128.56087        128.56087
Residual            56            1.75769           .03139

F =    4095.95632      Signif F =  .0000

----------------- Variables in the Equation -----------------
Variable             B        SE B        Beta        T   Sig T

LOGM           .996070     .015564     .993233    64.000  .0000
(Constant)    3.959733     .041629               95.120  .0000
```

FIGURE 4-29 Regression analysis of MHA uptake rate versus MHA concentration obtained after transforming the raw data in Fig. 4-25 by taking the natural logarithms of both the independent and dependent variables. The correlation is excellent; $r = .993$.

equal variance (Fig. 4-30A) and are normally distributed (Fig. 4-30B). The regression equation is

$$\ln \hat{R} = 3.95973 + .99607 \ln M$$

Transforming this equation back to the original power function form, we have

$$e^{\ln \hat{R}} = e^{3.95973 + .99607 \ln M} = e^{3.95973} e^{.99607 \ln M}$$

which yields

$$\hat{R} = 52M^{.99607}$$

Because the parameter estimate .99607 is not significantly different from 1 ($t = .2526$; $P > .50$), we have basically used a power function to estimate a straight line with a zero intercept and we can write

$$\hat{R} = 52 \ \mu L/g/min \ M$$

which is similar to the result obtained with the untransformed data [compare with Eq. (4.8)]. Because we have not violated the equality of variance or normality assumptions for the residuals, we now have an unbiased

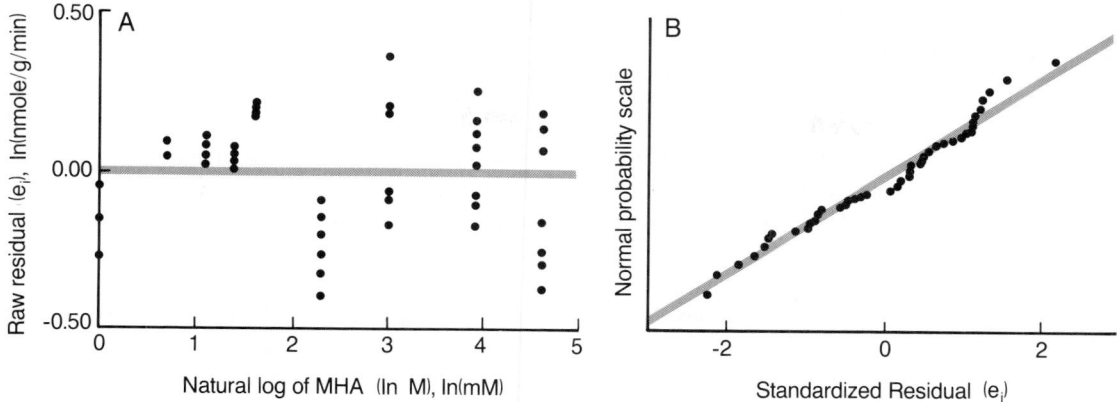

FIGURE 4-30 Analysis of the residuals from a linear regression model fit to the logarithmic transformed data of MHA uptake versus MHA concentration. The raw residuals plotted against the independent variable (now log M) (A) and the normal probability plot of the residuals (B) are both diagnostic plots which show that the residuals are now well behaved, with none of the suggestions of unequal variances or nonnormal distributions seen in Fig. 4-27. Hence, working with the transformed data provides a much better fit to the assumptions of the *linear regression model* than dealing with the raw data in Fig. 4-25.

estimate of the diffusion constant [52 compared to 50 µL/g/min], and the hypothesis tests we have conducted yield more meaningful P values than before because the data meet the assumptions that underlie the regression model. Because we were dealing with a transport process in which the rate is constrained by the underlying physics to be proportional to the concentration of MHA, there is no absorption when there is no MHA and the intercept is zero. Thus, the equations obtained with the untransformed [Eq. (4.7)] and transformed [Eq. (4.9)] data were very similar.* In general, however, there will not be a zero intercept and the transformed data can be expected to lead to a different equation than the untransformed data.

*Often kinetic analyses produce reaction or transport rates that are nonlinear functions of substrate concentration (when the transport is energy- or carrier-dependent, or the reaction is enzymatically catalyzed). Such processes are said to follow Michaelis-Menten kinetics, and there are standard transformations for these types of reaction analyses that go by the names of Lineweaver-Burk (1/rate vs. 1/substrate) and Eadie-Hofstee (rate/substrate vs. rate) plots. Although it is beyond the scope of this book to pursue these linearizing transformations in detail, they often will also have the effect of correcting the problem of unequal variance if it exists (e.g., the Lineweaver-Burk transformation corresponds to the $1/x$ vs. $1/y$ transformation suggested in Table 4-7).

HOW THE BODY PROTECTS ITSELF FROM EXCESS ZINC AND COPPER

The body needs trace amounts of chemical elements to properly carry out all of its metabolic activities, because they are necessary for the proper function of enzymes. The only way to obtain these elements is in the diet. However, too much of these elements can be toxic. Therefore, the body has mechanisms to keep the amounts of these elements at the correct level.

Two of these essential trace elements are copper, Cu, and zinc, Zn. The levels of these elements in the body are thought to be controlled (at least in part) by the enzyme metallothionein. As often happens, the amount of a regulatory enzyme such as metallothionein is, in turn, regulated by the amount of the substances that the enzyme acts on. If there is more Cu or Zn in the diet, the body reacts by producing more metallothionein so that correct levels are more quickly restored. Because the regulation of metallothionein by Cu and Zn is not well understood, Blalock and her coworkers[*] wanted to see how dietary copper and zinc levels affected the production of this protein.

A typical approach to such problems is to measure how the amount of genetic message for manufacture of the enzyme (rather than the enzyme itself) depends on the level of substrate the enzyme regulates. Thus, Blalock and her coworkers fed rats nine different diets containing one of three levels of copper (1, 6, or 36 mg/kg) and one of three levels of zinc (5, 30, 180 mg/kg) and measured the amount of the genetic message, messenger RNA (mRNA), for metallothionein production that appeared in the kidney for each combination of Cu and Zn. To gain a quantitative understanding of the sensitivity of mRNA to Cu and Zn, they used multiple linear regression analysis.

In any modeling activity, the simpler (or more parsimonious) the model, the better, so let us begin with the simplest possible model:

$$\hat{M} = a_0 + a_C\, \text{Cu} + a_Z\, \text{Zn} \tag{4.10}$$

where M is the amount of genetic message, mRNA, for metallothionein. The computer output for the fit of the data (Table C-11, Appendix C) to Eq. (4.10) is in Fig. 4-31. The regression equation is

$$\hat{M} = 9.4 \text{ molecules/cell} + .13 \text{ molecules/cell/mg/kg Cu}$$
$$+ .056 \text{ molecules/cell/mg/kg Zn}$$

[*]T. L. Blalock, M. A. Dunn, and R. J. Cousins, "Metallothionein Gene Expression in Rats: Tissue-Specific Regulation by Dietary Copper and Zinc," *J. Nutr.* 118:222–228, 1988.

```
DEP VARIABLE: M        Methallothionein mRNA

                              ANALYSIS OF VARIANCE

                        SUM OF           MEAN
            SOURCE   DF  SQUARES        SQUARE      F VALUE      PROB>F

            MODEL     2  818.75230297  409.37615149   6.967
            ERROR    33  1939.02659     58.75838139                0.0030
            C TOTAL  35  2757.77889

              ROOT MSE      7.665402    R-SQUARE      0.2969
              DEP MEAN     15.30556     ADJ R-SQ      0.2543
              C.V.         50.08248

                              PARAMETER ESTIMATES

                        PARAMETER      STANDARD     T FOR H0:
            VARIABLE  DF  ESTIMATE       ERROR     PARAMETER=0    PROB > |T|

            INTERCEP   1  9.44861111    2.10701736    4.484        0.0001
            CU         1  0.12777132    0.08265815    1.546        0.1317
            ZN         1  0.05617054    0.01653163    3.398        0.0018


               OBS    STUDRES    STDELRES   LEVERAGE    COOK

                1    -0.5486    -0.5427    0.069121   0.007449
                2    -0.9001    -0.8975    0.069121   0.020055
                3    -0.8055    -0.8011    0.069121   0.016059
                4    -0.8190    -0.8148    0.069121   0.016603
                5     0.1798     0.1771    0.056525   0.000645
                6     0.0723     0.0712    0.056525   0.000104
                7     1.0259     1.0267    0.056525   0.021018
                8     1.4288     1.4527    0.056525   0.040770
                9    -0.9624    -0.9613    0.103036   0.035468
               10    -0.6181    -0.6122    0.103036   0.014628
               11    -0.3288    -0.3243    0.103036   0.004140
               12     1.0349     1.0360    0.103036   0.041008
               13    -0.8590    -0.8555    0.056525   0.014737
               14    -0.6979    -0.6923    0.056525   0.009726
               15    -0.4830    -0.4773    0.056525   0.004659
               16    -0.3890    -0.3839    0.056525   0.003021
               17     0.4002     0.3951    0.043928   0.002453
               18     0.3602     0.3554    0.043928   0.001987
               19     1.3341     1.3507    0.043928   0.027261
               20    -0.1068    -0.1052    0.043928   0.000175
               21     0.7625     0.7575    0.090439   0.019268
               22     0.0922     0.0908    0.090439   0.000282
               23     1.3780     1.3978    0.090439   0.062938
               24    -0.3729    -0.3680    0.090439   0.004608
               25    -1.2162    -1.2254    0.103036   0.056636
               26    -1.2300    -1.2399    0.103036   0.057926
               27    -1.0233    -1.0241    0.103036   0.040099
               28    -1.1473    -1.1530    0.103036   0.050403
               29     2.1430     2.2745    0.090439   0.152211
               30     2.5534     2.8069    0.090439   0.216087
               31     0.6794     0.6737    0.090439   0.015297
               32     1.6916     1.7431    0.090439   0.094841
               33    -0.5700    -0.5641    0.136951   0.017185
               34    -0.5279    -0.5220    0.136951   0.014739
               35    -0.9070    -0.9045    0.136951   0.043515
               36    -0.7526    -0.7475    0.136951   0.029956
```

FIGURE 4-31 Analysis of metallothionein mRNA production as a function of dietary copper and zinc using a simple additive linear model. Although the model produces a significant reduction in the residual variance ($P < .003$), the quality of the fit is not very good, with $R^2 = .30$. Examining the residuals does not show any outliers or obviously influential points that can be used as guides for improving the model fit.

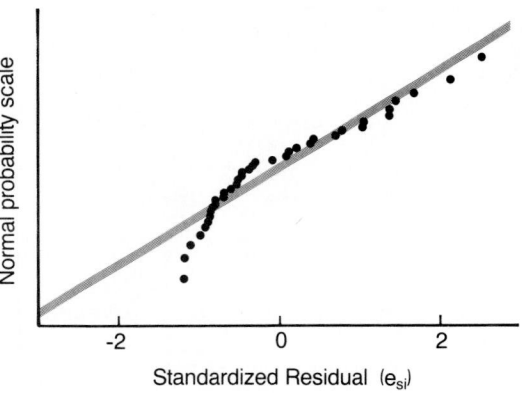

FIGURE 4-32 This normal probability plot for the residuals from the regression analysis shown in Fig. 4-31 shows that the residuals are not normally distributed because the pattern exhibits substantial curvature, particularly for negative residuals.

Messenger RNA production increases with both Cu and Zn levels, as expected. This model also indicates that messenger RNA production is 9.4 molecules/cell even when no copper or zinc is present.

This result is problematic because we would expect that there should be no metallothionein production if there is no Cu and Zn (assuming that there are no other substances present that might turn on metallo-thionein production). Such a physiologically implausible result points to an incorrectly or incompletely specified regression model. In addition, the R^2 is only 0.30 (although $F = 6.97$; $P = 0.003$), indicating that, although the equation provides some description of the data, it is not a very good fit. This high residual variability may be due to large popula-tion variability, large measurement errors, model misspecification, or some combination of these three factors. The normal probability plot of the residuals in Fig. 4-32 shows that the residuals are not normally dis-tributed—further evidence that this equation does not fit the data well. Two data points have Studentized residuals r_i or r_{-i} with magnitudes exceeding 2, but they are not huge. The other regression diagnostics are unremarkable. Thus, there are no obvious problems with the data. The question remains: Is the regression model correctly specified?

Plots of the residuals will help us investigate this issue.* Figure 4-33 shows plots of the residuals versus each of the independent variables (Cu and Zn). The residuals look randomly scattered about zero when plotted against Cu. There is slightly more variability at the highest level of copper, but this is not so marked as to be of concern. In contrast, the residuals plotted against Zn clearly show a systematic nonlinear trend,

---

*One should not ignore plots of the dependent variable against each of the independent variables for the raw data. In fact, as we have noted previously, looking at the raw data before starting the analysis is essential. The fact that the focus of this chapter is on examining residuals during an analysis does not change that advice.

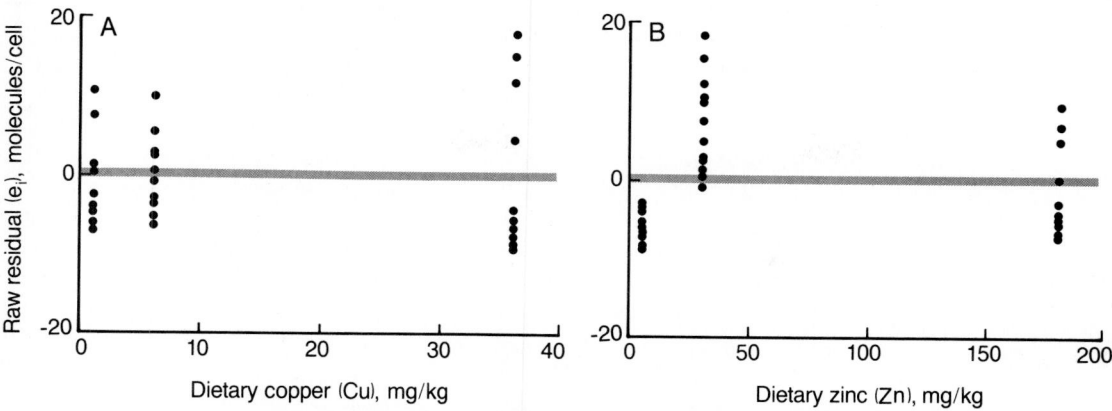

FIGURE 4-33 To further search for problems with the regression model used in Fig. 4-31, we plot the raw residuals against the two independent variables, dietary copper level and dietary zinc level. *A.* The residuals against dietary copper are reasonably distributed about zero with only a modest suggestion of unequal variance. *B.* The residuals plotted against dietary zinc level, however, suggest a nonlinearity in the effect of dietary zinc on metallothionein mRNA production.

with more positive residuals at the intermediate levels of zinc. This pattern, which is similar to that we encountered for our studies of Martian intelligence vs. foot size (Fig. 4-2*B*) and thermal conductivity in seals (Fig. 2-11), suggests a nonlinear effect of Zn. The simplest nonlinearity to introduce is a quadratic for the dependence of M on Zn, so we modify equation (4.10) to obtain

$$\hat{M} = a_0 + a_C \, \text{Cu} + a_Z \, \text{Zn} + a_{Z^2} \, \text{Zn}^2 \tag{4.11}$$

The computer output for this regression analysis is in Fig. 4-34. The regression equation is

$$\hat{M} = .06 \text{ molecules/cell} + .13 \text{ molecules/cell/mg/kg Cu}$$
$$+ .71 \text{ molecules/cell/mg/kg Zn}$$
$$- .0034 \text{ molecules/cell/(mg/kg)}^2 \, \text{Zn}^2$$

The intercept is not significantly different from zero, resolving the difficulty we had with the original model. Messenger RNA production still increases with Zn, but at a decreasing rate as Zn increases. Note that a_Z increased from .056 molecules/cell/mg/kg to .71 molecules/cell/mg/kg —an order of magnitude—when we allowed for the nonlinear effect of Zn. The reason for this change is that, by ignoring the nonlinear effect, we were forcing a straight line through a curve (as in our earlier examples in Figs. 2-11 and 4-2*B*). Including the nonlinearity also led to a better fit: R^2 increased from .30 to .70 and the standard error of the estimate fell

```
DEP VARIABLE: M              Methallothionein mRNA

                          ANALYSIS OF VARIANCE

                        SUM OF        MEAN
        SOURCE    DF    SQUARES      SQUARE     F VALUE    PROB>F

        MODEL      3  1918.63327  639.54442399   24.388    0.0001
        ERROR     32   839.14561693  26.22330053
        C TOTAL   35  2757.77889

            ROOT MSE     5.120869    R-SQUARE    0.6957
            DEP MEAN    15.30556     ADJ R-SQ    0.6672
            C.V.        33.45758

                        PARAMETER ESTIMATES

                   PARAMETER       STANDARD     T FOR H0:
VARIABLE  DF        ESTIMATE          ERROR    PARAMETER=0   PROB > |T|

INTERCEP   1      0.06294444     2.02028971        0.031       0.9753
CU         1      0.12777132     0.05521975        2.314       0.0273
ZN         1      0.71137778     0.10177049        6.990       0.0001
ZN2        1     -0.00338222     0.0005222434     -6.476       0.0001

        OBS    STUDRES    STDELRES   LEVERAGE     COOK

          1     0.4409     0.4352    0.104005    0.005640
          2    -0.0955    -0.0940    0.104005    0.000265
          3     0.0489     0.0481    0.104005    0.000069
          4     0.0283     0.0278    0.104005    0.000023
          5    -1.2147    -1.2241    0.104005    0.042819
          6    -1.3798    -1.4003    0.104005    0.055245
          7     0.0850     0.0837    0.104005    0.000210
          8     0.7039     0.6982    0.104005    0.014378
          9    -1.2285    -1.2387    0.104005    0.043794
         10    -0.7127    -0.7071    0.104005    0.014741
         11    -0.2795    -0.2754    0.104005    0.002267
         12     1.7629     1.8261    0.104005    0.090188
         13    -0.0414    -0.0407    0.091408    0.000043
         14     0.2045     0.2014    0.091408    0.001052
         15     0.5323     0.5262    0.091408    0.007125
         16     0.6757     0.6698    0.091408    0.011482
         17    -0.8660    -0.8625    0.091408    0.018860
         18    -0.9274    -0.9253    0.091408    0.021632
         19     0.5681     0.5620    0.091408    0.008118
         20    -1.6444    -1.6916    0.091408    0.068014
         21     1.3534     1.3720    0.091408    0.046072
         22     0.3496     0.3447    0.091408    0.003074
         23     2.2753     2.4461    0.091408    0.130212
         24    -0.3470    -0.3421    0.091408    0.003028
         25    -0.5542    -0.5481    0.137920    0.012285
         26    -0.5752    -0.5691    0.137920    0.013235
         27    -0.2598    -0.2559    0.137920    0.002699
         28    -0.4490    -0.4434    0.137920    0.008065
         29     1.7751     1.8401    0.137920    0.126027
         30     2.4061     2.6167    0.137920    0.231544
         31    -0.4753    -0.4695    0.137920    0.009037
         32     1.0810     1.0840    0.137920    0.046742
         33    -0.6366    -0.6306    0.137920    0.016208
         34    -0.5735    -0.5674    0.137920    0.013154
         35    -1.1414    -1.1470    0.137920    0.052102
         36    -0.9100    -0.9075    0.137920    0.033121
```

FIGURE 4-34 Based on the results of the residual plots in Fig. 4-33, we added a quadratic function in zinc to the regression equation to account for the nonlinearity in the effect of zinc. The effect of copper remains linear. The results of the regression analysis showed marked improvement in the quality of the fit, with R^2 increasing to .70. The residual diagnostics are unremarkable.

from 7.7 to 5.1 molecules/cell. These changes are due to the fact that ignoring the nonlinearity in the original model [Eq. (4.10)] constituted a gross misspecification of the model.

Now we want to see if the residuals have anything more to tell us. Including the nonlinearity reduced the residual variance. The Studentized residuals r_i or r_{-i} are somewhat smaller than before, with only one exceeding 2. Hence, in terms of our efforts to find the best model structure, the quadratic function seems to have taken care of the nonlinearity and we have no further evidence for systematic trends, although there is still a suggestion of unequal variances (Figs. 4-35 and 4-36). Although it is not perfect, there is no serious deviation from linearity in the normal probability plot of the residuals (Fig. 4-35). All residuals also have acceptable levels of leverage and Cook's distance, which indicates that no single point is having particularly major effects on the results.

This analysis provides us with what we wanted to know: How is metallothionein production regulated by these trace elements? Both Cu and Zn affect metallothionein production. The form of Eq. (4.11) indicates that the sensitivity of M to Cu is independent of Zn level, and vice versa. Hence, the effects of Cu and Zn are independent. However, Cu and Zn have qualitatively different affects: Cu increases metallothionein linearly at all levels studied, while Zn has a nonlinear effect such that metallothionein production increases with increasing Zn at lower levels, peaks at intermediate levels of Zn, and begins to decrease at the highest levels of Zn (Fig. 4-37A). Although it is possible to explain the decreased metallothionein production at the highest Zn levels by suggesting that Zn begins to be toxic at these high levels, this result is somewhat surprising.

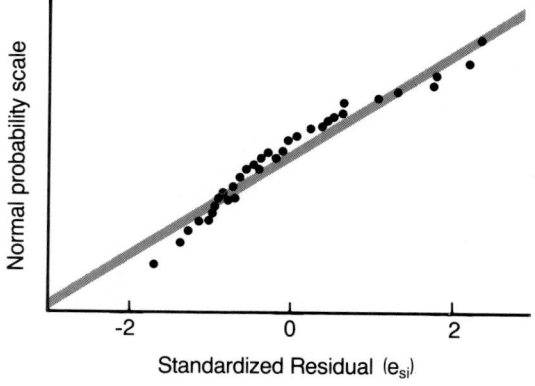

FIGURE 4-35 This normal probability plot of the residuals for the regression model in Fig. 4-34 reveals a pattern that appears as a straight line, indicating that the observed distribution of residuals is consistent with the assumption of a normal distribution (compare with Fig. 4-32).

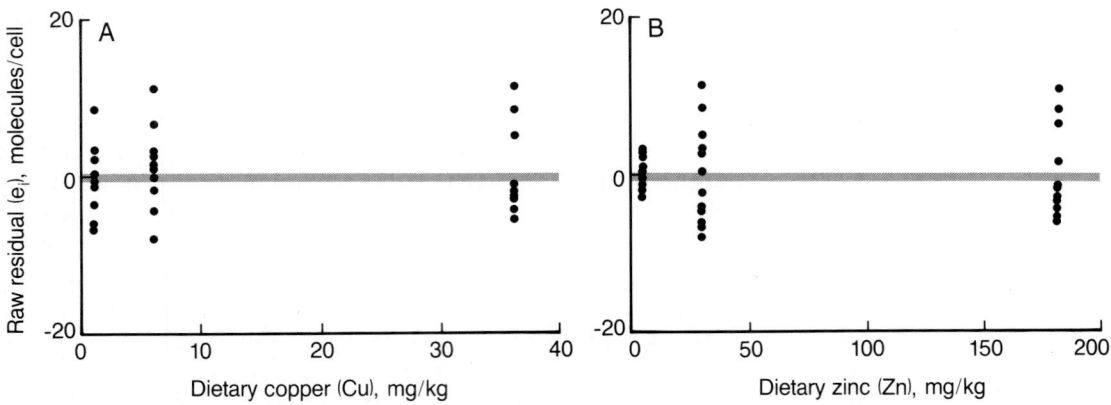

FIGURE 4-36 Plots of the residuals against both (*A*) copper and (*B*) zinc from the analysis in Fig. 4-34 reveal that including the nonlinear effect of zinc in the model removed the nonlinearity in the residual plot against dietary zinc level. The only remaining problem is unequal variance in the residuals, which is particularly evident when considering the plot of residuals against dietary zinc level.

Back to Square One

One of the messages we have been trying to send throughout the book is that *you should always look at the raw data*. So, let us do just that and plot metallothionein vs. both copper (Fig. 4-38*A*) and zinc (Fig. 4-38*B*). There is a small positive linear trend in the relation between metallothionein and copper, and a positive nonlinear trend in the relation between metallothionein and zinc, which we have already discovered through the residual plots. We took care of the nonlinearity by adding the quadratic term Zn^2 in Eq. (4.11).

However, a closer examination of the relationship between metallothionein and zinc shows that, although it is nonlinear, it seems to saturate — which actually makes more sense in terms of most biochemical reactions. A quadratic is really an inappropriate normalization for a saturable process because a quadratic can increase and then decrease, whereas a saturable function monotonically increases toward a maximum. Referring to the transformations described in Fig. 4-19, we find

FIGURE 4-37 *A*. A plot of the data and regression surface computed using Eq. (4.11) from Fig. 4-34. The quadratic term in zinc causes the surface to bend over as zinc level increases. Although this surface seems to fit the data well, we still have the problem of unequal variance in the residuals (in Fig. 4-36) and the potential difficulties associated with explaining the biochemical process, which reaches a maximum and then falls off. *B*. It is more common for biochemical processes to saturate, as shown in the regression surface for the model of Eq. (4.12) from Fig. 4-40.

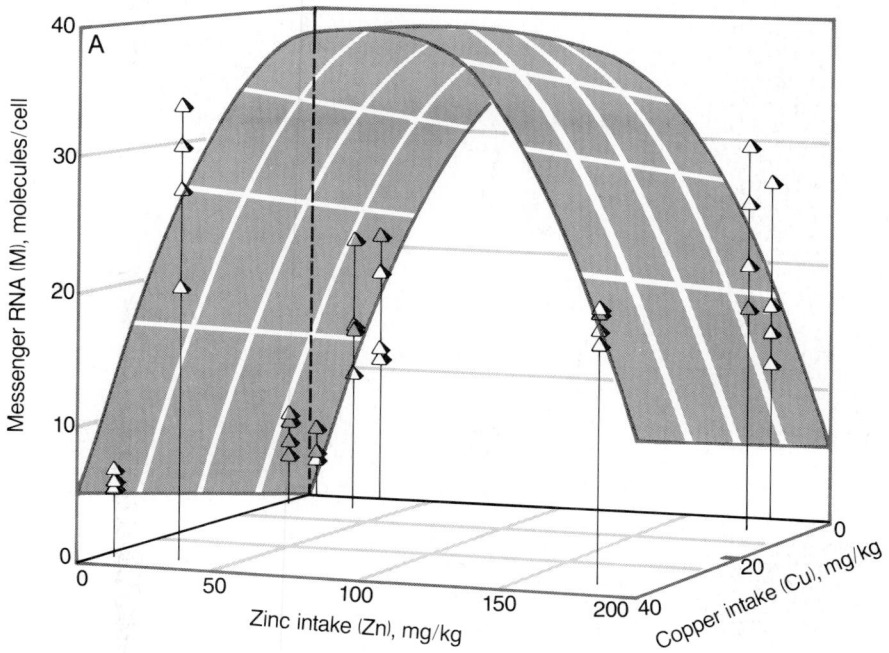

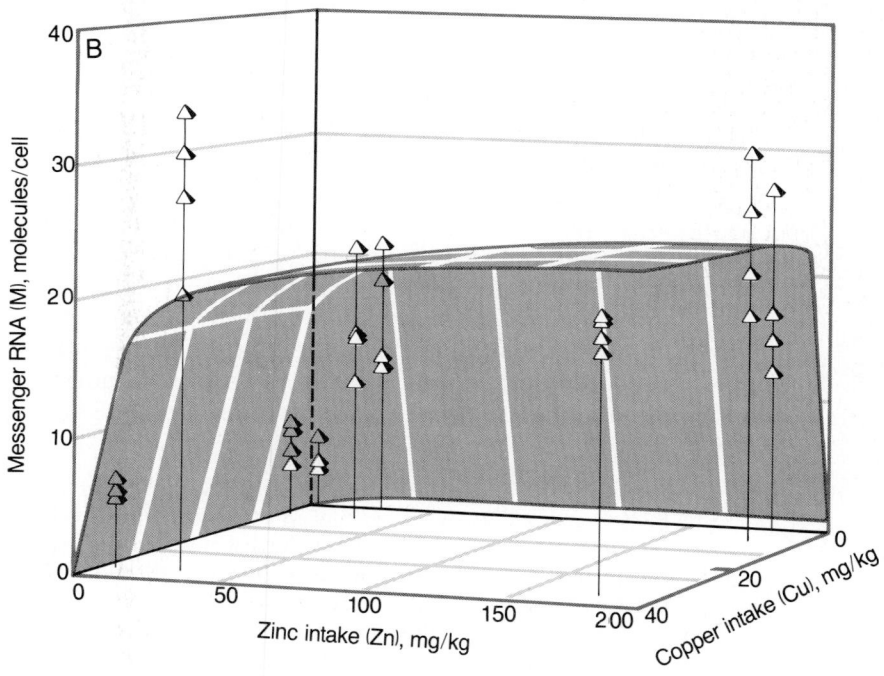

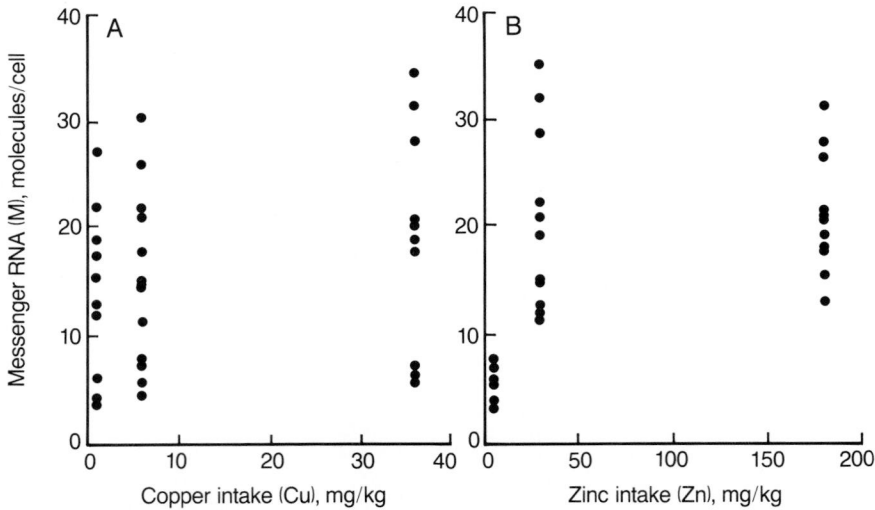

FIGURE 4-38 Plot of the original raw data of mRNA against (*A*) copper and (*B*) zinc levels used in the computations reported in Figs. 4-31 and 4-34. A closer examination of the plot of mRNA against zinc suggests a nonlinear relationship between mRNA and zinc, which increases from 0 and then saturates. The suggestion of a quadratic relationship is, perhaps, incorrect and is due to the relatively large variance at a zinc level of 36 mg/kg. Thus, a model that saturates may be useful in straightening out this relationship and also in equalizing the variances.

that there are three possible transformations for saturable functions, $1/y$ vs. $1/x$, $\ln y$ vs. $\ln x$, and $\ln y$ vs. $1/x$. To help keep a long story from getting too much longer, we will simply say that, of these three, we decided that the log metallothionein vs. $1/Zn$ transformation is the best for these data.

Now, how about copper? Figure 4-38*A* provides no evidence for a nonlinearity as a function of copper, but because we expect most reactions to saturate, it seems reasonable that this might also. So, we plotted metallothionein vs. copper at each of the three levels of zinc (Fig. 4-39). Nonlinear relationships are evident for both the low and high values of zinc. The middle level of zinc shows no such nonlinearity, and, thus, the plot of all these data lumped together (Fig. 4-38*A*) obscures what is most

FIGURE 4-39 To get a better look at the raw data as an aid in selecting a transformation, if any, for copper, we plot the observed mRNA values as a function of dietary copper levels for each level of dietary zinc: (*A*) high, (*B*) medium, and (*C*) low. Although the plot of all the data together suggests only a weak linear trend as a function of dietary copper (Fig. 4-38*A*), the plots of mRNA versus dietary copper at each level of zinc clearly show nonlinear trends that appear to saturate at the low and high levels of zinc.

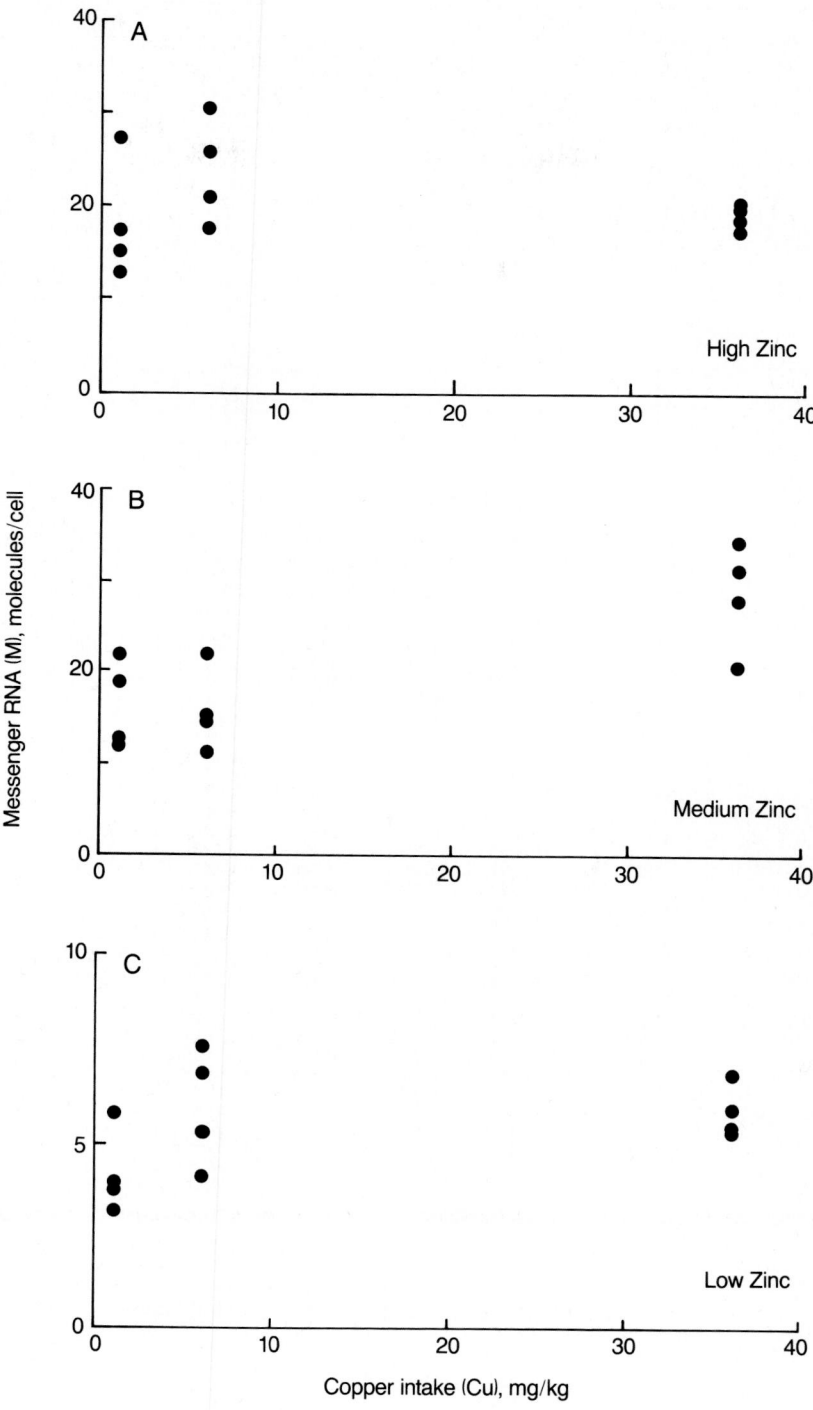

```
DEP VARIABLE: LOGM     natural log of Metallothionein mRNA

                           ANALYSIS OF VARIANCE

                      SUM OF        MEAN
         SOURCE    DF  SQUARES      SQUARE     F VALUE    PROB>F

         MODEL      2  14.36808252  7.18404126   92.362   0.0001
         ERROR     33   2.56677873  0.07778117
         C TOTAL   35  16.93486124

            ROOT MSE     0.2788928    R-SQUARE     0.8484
            DEP MEAN     2.523484     ADJ R-SQ     0.8392
            C.V.        11.0519

                           PARAMETER ESTIMATES

                      PARAMETER      STANDARD    T FOR H0:
         VARIABLE  DF   ESTIMATE       ERROR    PARAMETER=0    PROB > |T|

         INTERCEP   1   3.22300297   0.07666029     42.043       0.0001
         INVCU      1  -0.320868     0.10826561     -2.964       0.0056
         INVZN      1  -7.18032      0.54132804    -13.264       0.0001


            OBS   STUDRES   STDELRES   LEVERAGE    COOK

              1    1.1262    1.1309   0.136951   0.067086
              2   -1.1692   -1.1759   0.136951   0.072303
              3   -0.4056   -0.4004   0.136951   0.008703
              4   -0.5059   -0.5001   0.136951   0.013536
              5   -0.4853   -0.4797   0.090439   0.007807
              6   -0.7320   -0.7267   0.090439   0.017758
              7    1.0391    1.0404   0.090439   0.035784
              8    1.5929    1.6327   0.090439   0.084102
              9   -1.2139   -1.2230   0.103036   0.056427
             10   -0.5336   -0.5278   0.103036   0.010903
             11   -0.0437   -0.0430   0.103036   0.000073
             12    1.6695    1.7182   0.103036   0.106725
             13   -1.2124   -1.2214   0.090439   0.048718
             14   -0.2472   -0.2437   0.090439   0.002026
             15    0.7446    0.7395   0.090439   0.018378
             16    1.1079    1.1119   0.090439   0.040684
             17   -0.8391   -0.8352   0.043928   0.010783
             18   -0.9137   -0.9113   0.043928   0.012785
             19    0.5732    0.5673   0.043928   0.005032
             20   -1.9187   -2.0046   0.043928   0.056384
             21    0.4600    0.4544   0.056525   0.004225
             22   -0.3142   -0.3099   0.056525   0.001971
             23    1.0514    1.0531   0.056525   0.022074
             24   -0.9662   -0.9652   0.056525   0.018643
             25   -0.2774   -0.2735   0.103036   0.002947
             26   -0.3469   -0.3422   0.103036   0.004608
             27    0.5811    0.5752   0.103036   0.012931
             28    0.0520    0.0512   0.103036   0.000104
             29    1.7426    1.8008   0.056525   0.060643
             30    2.0794    2.1967   0.056525   0.086353
             31    0.2045    0.2015   0.056525   0.000835
             32    1.3327    1.3492   0.056525   0.035469
             33   -0.6447   -0.6389   0.069121   0.010288
             34   -0.5897   -0.5837   0.069121   0.008606
             35   -1.1173   -1.1216   0.069121   0.030896
             36   -0.8932   -0.8904   0.069121   0.019747
```

FIGURE 4-40 Analysis of the metallothionein mRNA production as a function of dietary copper and zinc using the transformation given by Eq. (4.12). Using these transformed data, R^2 improves to .85. The regression diagnostics are unremarkable.

likely a nonlinear, and saturable, relationship between metallothionein and copper. To keep things simple, we will take care of this nonlinearity using the same inverse transformation we used for Zn: 1/Cu.

Thus, we fit these data to the equation

$$\ln \hat{M} = d_0 + d_C \frac{1}{Cu} + d_Z \frac{1}{Zn} \tag{4.12}$$

the results of which are shown in Fig. 4-40. The R^2 is .85, up from .70 for the quadratic model in Eq. (4.11). Both regression coefficients are highly significant. Examination of the influence diagnostics shows small improvements, compared with the quadratic model, particularly in the normal probability plot of the residuals (Fig. 4-41). Although none of these diagnostics was terrible in the fit to Eq. (4.11), these diagnostics represent definite improvements. The plots of the residuals also show considerable improvement. The hint of unequal variances in the plot of residuals against zinc in Fig. 4-36 is less pronounced (Fig. 4-42), and the plots of residuals against measured and predicted metallothionein are more like the desired symmetric band around zero than they were before.

We can take the antilog of both sides of Eq. (4.12) to do the inverse transformation and rewrite this equation as

$$\hat{M} = e^{d_0} e^{d_C/Cu} e^{d_Z/Zn}$$

and, substituting for the regression coefficients, we have

$$\hat{M} = 25.1 \, e^{-.32/Cu} \, e^{-7.18/Zn} \tag{4.13}$$

The interpretation of this model is quite different from the one expressed by Eq. (4.11). The negative exponents for e indicate a logarithmic increase in metallothionein mRNA as both Cu and Zn increase (Fig. 4-37B). There is an interaction between Cu and Zn, expressed as the product of terms in Eq. (4.13), and so we end up with a multiplicative model, in contrast

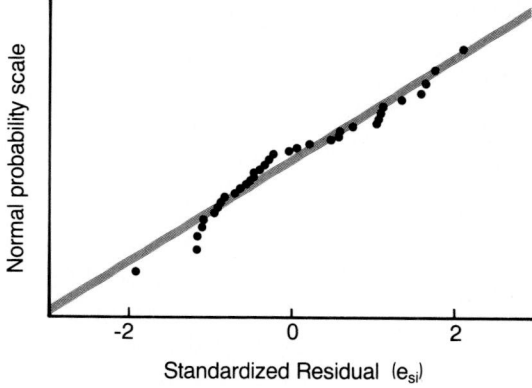

FIGURE 4-41 A normal probability plot of the residuals using the model given by Eq. (4.12) and the results of Fig. 4-40 follows a straight line, indicating that the residuals are consistent with the assumption of normality.

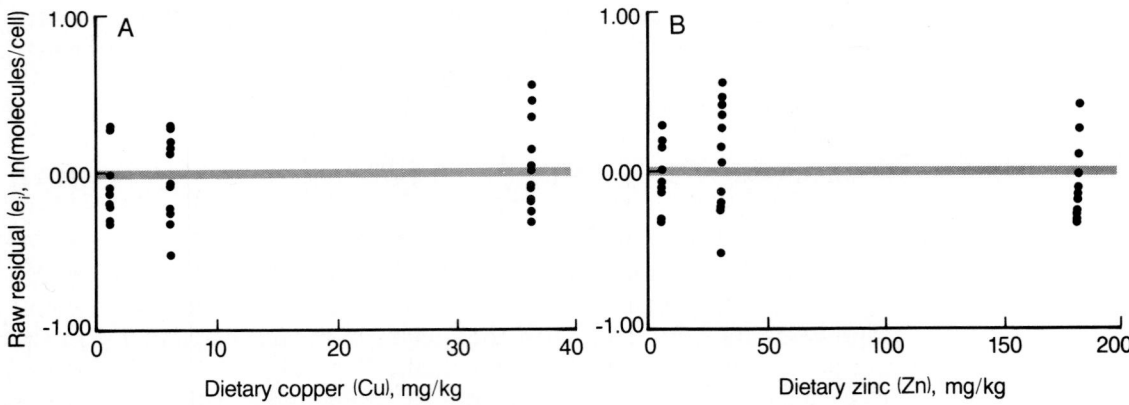

Figure 4-42 Examining the residual plots using the results in Fig. 4-41 reveals no evidence of nonlinearity or unequal variances (compare Fig. 4-33*B* to Fig. 4-36*B*). Taken together, the quality of the fit, the quality of the residuals, and the more reasonable physiological interpretation indicate that the model given by Eq. (4.12) is the best description of these data.

to Eq. (4.11), which had no interaction. More importantly, the dependence of the exponents on the reciprocal of the independent variables indicates a saturable process; the exponent gets smaller as Cu and Zn increase. This saturability makes more sense in terms of most kinetic models, and we no longer have to invoke some mysterious toxic effect, for example, to explain why metallothionein production actually went down at the higher levels of zinc when we used the quadratic model.

Thus, in terms of the statistics, we have one less parameter, higher R^2, smaller residuals, more randomly scattered residuals, and fewer possibly influential points than in the quadratic model. In terms of model structure and interpretation of the underlying physiology, we are able to make more reasonable inferences than with the quadratic model. This example underscores the fact that the choice of linearizing transformation can have great impact on a regression analysis. It also makes it abundantly clear how important it is to carefully dissect the raw data before and during an analysis.

Summary

Linear regression and analysis of variance are powerful statistical tools, which are derived from mathematical models of reality. For the results of these techniques to provide an accurate picture of the information contained in a set of experimental data, it is important that the experimental reality be in reasonable concordance with the assumptions used to derive the mathematical model upon which the analysis is based. A complete analysis of a set of data requires that, after completing the analysis, you go back to make sure that the results are consistent with the model assumptions that underlie regression analysis.

By examining the residuals and the various statistics derived from them, it is possible to identify influential points, which can then be double-checked for errors and corrected, if necessary. When there are no such errors, the patterns in the residuals can guide you in reformulating the regression model itself to more accurately describe the data. Of course, there are many cases in which these diagnostic procedures do not reveal any problems, and the analysis can continue along the lines of the initial formulation. In such cases, these diagnostic procedures simply increase the confidence you can have in your results.

PROBLEMS

4.1 Evaluate the reasonableness of the linear fit for the data on renal vascular resistance and renin production in Table D-4, Appendix D, that you analyzed in Prob. 2.5. Are there any influential points? Is some other regression model more appropriate?

4.2 Growth hormone G_h is severly elevated in subjects with acromegaly, a condition that causes enlarged and deformed bone growth, particularly in the face and jaw of adults. Although G_h is thought to exert its effects via another hormone, insulinlike growth factor–somatomedin-C (I_{gf}), a close correlation between G_h and I_{gf} has not been demonstrated. Because the body secretes G_h in cycles that have well-defined daily peaks, whereas I_{gf} is secreted continuously, Barkan and coworkers* postulated that the lack of correlation may be due to the differences in timing of secretion. They measured blood G_h every 10 or 20 min for 24 h in 21 subjects with acromegaly. As part of their analysis, they looked at the relationship between I_{gf} and G_h at a single time point, 8:00 A.M. Analyze these data (in Table D-13, Appendix D). *A.* Is there a good correlation between the blood levels of these two hormones? If not, how can one reconcile a poor correlation with the known physiological relationship between I_{gf} and G_h? *B.* Does the presence of influential points alter the basic biological conclusion? *C.* How can a point have a large leverage, yet a low Cook's distance?

*A. L. Barkan, I. Z. Beitins, and R. P. Kelch, "Plasma Insulin-Like Growth Factor-I/ Somatomedin-C in Acromegaly: Correlation with the Degree of Growth Hormone Hypersecretion," *J. Clin. Endocrinol. Metab.* 67:69–73, 1988.

4.3   The anion gap is the difference between the number of univalent positive ions (potassium and sodium) and negative ions (chloride) in plasma or urine. Because it was suggested that the urinary anion gap U_{ag} might be a useful indicator of the amount of ammonium ion secreted in the urine, Battle and his coworkers[*] measured urinary ammonium A and U_{ag} in normal subjects, patients with metabolic acidosis due to diarrhea, and patients with relatively common kidney defects. (The data are in Table D-14, Appendix D.) *A.* Is there any evidence that U_{ag} reflects ammonium in the urine? *B.* Is it possible to predict ammonium if you know U_{ag}? How precisely? *C.* Include an evaluation of influential points. *D.* Does a quadratic model provide a better description of these data than a linear model? Justify your conclusion.

4.4   Children with growth hormone H deficiency do not grow at normal rates. If they are given extra growth hormone, their growth rate can be improved. Hindmarsh and coworkers[†] made observations in growth hormone–deficient children that led them to hypothesize that children who have lower H at the outset of treatment respond more favorably and that this relationship should be nonlinear; that is, the treatment is more effective if treatment starts from a lower pretreatment H level. They measured a normalized growth rate G based on the change in height over time and the H level at the onset of treatment with additional growth hormone. (These data are in Table D-15, Appendix D.) Is there any evidence to support the investigator's hypothesis that the relationship between G and H is nonlinear? Are there other problems with these data? Can the nonlinearity be handled with a transformation so that linear regression can be used?

4.5   Certain kinds of bacteria, known as thermophilic bacteria, have adapted to be able to live in hot springs. Because of the very hot temperatures they live in (55°C, compared to a normal mammalian body temperature of 38°C), thermophilic bacteria must have made adaptations in their proteins, particularly the enzymes that catalyze the biochemical reactions the bacteria need to obtain energy. Thermophilic bacteria share with other closely related bacteria (and most other cells) an important enzyme called malate dehydrogenase (MDH). There is some evidence that the thermophilic bacterium

[*]D. C. Battle, M. Hizon, E. Cohen, C. Gutterman, and R. Gupta, "The Use of the Urinary Anion Gap in the Diagnosis of Hyperchloremic Metabolic Acidosis," *N. Engl. J. Med.* 318:594–599, 1988.

[†]P. C. Hindmarsh, P. J. Smith, P. J. Pringle, and C. G. D. Brook, "The Relationship Between the Response to Growth Hormone Therapy and Pre-treatment Growth Hormone Secretory Status," *Clin. Endocrinol.* 38:559–563, 1988.

Chloroflexus aurantiacus diverged early in its evolution from many other thermophilic bacteria. To explore this further, Rolstad and her coworkers[*] wanted to characterize the MDH of this bacterium. One of the features they wanted to determine was the optimum pH for the function of this enzyme. They measured enzyme activity A in terms of the reduction of the substrate oxalacetate expressed as a percentage of rate under standard conditions at pH = 7.5. Using buffers, the pH was varied over the range of 4.3 to 8.3 (the data are in Table D-16, Appendix D). Is there evidence of an optimum pH in this range? If so, discuss the strength of the evidence.

4.6 In Prob. 3.10 you analyzed the relationship between two methods of measuring the hormone inhibin. The new RIA method was compared against a standard bioassay method in two phases of the ovarian cycle. Suppose that the last two bioassay determinations of inhibin in the midluteal phase (the last two entries in Table D-10, Appendix D) were transposed. *A.* Does this single transposition affect the results? Why or why not? *B.* Would any of the regression diagnostics identify this transposition error?

4.7 Do a simple regression analysis of the early phase data from Table D-10, Appendix D. Compute the Studentized residuals r_i or r_{-i}, leverage values, and Cook's distances for the residuals from this regression fit. Are there any values with high leverage? If so, are these points actually influential? Explain why or why not.

4.8 Driving while intoxicated is a pressing problem. Attempts to deal with this problem through prevention programs where social drinkers learn to recognize levels of alcohol consumption beyond which they should not drive require that people be able to accurately judge their own alcohol consumption immediately prior to deciding whether or not to drive. To see if drinkers could actually judge their alcohol consumption, Meier and coworkers[†] administered a Breathalyzer test to measure blood alcohol and administered a short questionnaire to record recall of alcohol consumption, which was then used to estimate blood alcohol level based on the respondent's age, sex, and reported consumption (the data are in Table D-17, Appendix D). How well does drinker recall of alcohol consumption reflect actual blood alcohol level as determined by the Breathalyzer?

[*]A. K. Rolstad, E. Howland, and R. Sirevag, "Malate Dehydrogenase from the Thermophilic Green Bacterium *Chloroflexus aurantiacus*: Purification, Molecular Weight, Amino Acid Composition, and Partial Amino Acid Sequence," *J. Bacteriol.* 170:2947–2953, 1988.

[†]S. E. Meier, T. A. Brigham, and G. Handel, "Accuracy of Drinkers' Recall of Alcohol Consumption in a Field Setting," *J. Stud. Alcohol* 48:325–328, 1987.

4.9   Prepare a normal probability plot of the residuals from Prob. 4.8.

4.10   In Prob. 4.8, the data seem to reflect a threshold of about .10 blood alcohol where a relatively good correspondence between Breathalyzer-determined blood alcohol and self-reporting of consumption breaks down. Reanalyze these data after omitting all data points with Breathalyzer-determined blood alcohol levels greater than or equal to .10. Now, how well does self-reporting reflect actual blood alcohol?

4.11   Because the relationship between Breathalyzer- and self-reported blood alcohol levels in Prob. 4.8 appeared to change above a certain level, we reanalyzed the data after throwing out the points at high levels (Prob. 4.10). Another approach to this problem is to consider that the data define two different linear segments: one that describes the region of self-reported alcohol over a range of levels of 0 to .11 (determined by examining a plot of the data) and the other over the range of values greater than .11. Using only one regression equation, fit these two line segments. (*Hint:* you will need to use a dummy variable in a new way.) What are the slopes and intercepts of these two segments?

4.12   Repeat the regression analysis of Prob. 2.4 (data in Table D-3, Appendix D) and compute Studentized residuals, leverage values, and Cook's distances for the residuals and analyze the relative influence of the data points.

4.13   Evaluate the goodness of the fit of the linear model used to relate antibiotic effectiveness to blood level of the antibiotic in Prob. 2.4 (data in Table D-3, Appendix D). If the linear model is not appropriate, propose a better model.

4.14   *A.* Compare the regression diagnostics (Studentized residual, leverage, and Cook's distance) computed in Prob. 4.12 above with those obtained using a quadratic fit to the same data (see Prob. 4.13). *B.* Explain any differences.

4.15   Evaluate the appropriateness of the linear model used in Prob. 2.2. If indicated, propose and evaluate an alternative model. The data are in Table D-1, Appendix D.

Multicollinearity and What to Do about It

Multiple regression allows us to study how several independent variables act together to determine the value of a dependent variable. The coefficients in the regression equation quantify the nature of these dependencies. Moreover, we can compute the standard errors associated with each of these regression coefficients to quantify the precision with which we estimate how the different independent variables affect the dependent variable. These standard errors also permit us to conduct hypothesis tests about whether the different proposed independent variables affect the dependent variable at all. The conclusions we draw from regression analyses will be unambiguous when the independent variables in the regression equation are *statistically independent* of each other, i.e., when the value of one of the independent variables does not depend on the values of any of the other independent variables. Unfortunately, as we have already seen in Chap. 3, the independent variables often contain at least some redundant information and so tend to vary together, a situation called *multicollinearity*. Severe multicollinearity indicates that a substantial part of the information in one or more of the independent variables is redundant, which makes it difficult to separate the effects of the different independent variables on the dependent variable.

The resulting ambiguity is reflected quantitatively as reduced precision of the estimates of the parameters in the regression equation. As a result, the standard errors associated with the regression coefficients will be inflated and the values of the coefficients themselves may be unreasonable, and can even have a sign different from that one would expect based on other information about the system under study. Indeed, the presence of large standard errors or unreasonable parameter estimates are qualitative suggestions of multicollinearity.

Although multicollinearity causes problems in interpreting the regression coefficients, it does not affect the usefulness of a regression equation for purely empirical description of data or prediction of new observations if we make no interpretations based on the values of the individual coefficients. It is only necessary that the data used for prediction have the same multicollinearities and range as the data originally used to estimate the regression coefficients. In contrast, if the goal is meaningful estimation of parameters in the regression equation or identification of a model structure, one must deal with multicollinearity.

As we did with outliers in Chap. 4, we will develop *diagnostic* techniques to identify multicollinearity and assess its importance as well as techniques to reduce or eliminate the effects of multicollinearity on the results of the regression analysis.

WHERE MULTICOLLINEARITY COMES FROM

Multicollinearity among the independent variables arises for two reasons:

- *Structural multicollinearity* is a mathematical artifact due to creating new independent variables from other independent variables, such as by introducing powers of an independent variable or by introducing interaction terms as the product of two other independent variables.
- *Sample-based multicollinearity* is specific to a particular data set because of a poorly designed experiment, reliance on purely observational data, or the inability to manipulate the system adequately to separate the effects of different independent variables.

These different sources of multicollinearity require different strategies to deal with them. Structural multicollinearities can often be eliminated by rewriting the regression model in a different, but mathematically equivalent, form. Sample-based multicollinearities can often be resolved by collecting additional data under different experimental or observational conditions. When it is not possible to use additional data to eliminate sample-based multicollinearities, it is necessary to modify the regression model by eliminating one or more of the independent variables to reduce the multicollinearity or use a restricted regression technique, such as principal components regression, to analyze the data.

Structural multicollinearities can often be eliminated by changing the way that you write the regression model. For example, when the effect of an independent variable, say x, is nonlinear, we often include both x and x^2 in a quadratic regression equation to account for the nonlinear effect, then estimate coefficients for each of these terms. Unfortunately, the observed values of X and X^2 will be highly linearly correlated over a limited range of X values, which creates a multicollinearity problem when interpreting the coefficients associated with the linear and quadratic terms in the regression model. This problem can almost always be resolved by *centering the independent variables*, by using $(x - \overline{X})$ and $(x - \overline{X})^2$ in place of x and x^2 in the regression model. Centering does not alter the mathematical form of the regression model, yet it leads to new variables which are not so highly correlated. Alternatively, sometimes it is possible to replace the polynomial representation of the nonlinearity in x with a single *transformed* value of x, such as $\ln x$ or $1/x$, instead of including the two variables x and x^2, thus avoiding the introduction of multicollinearity. Centering also often eliminates or reduces structural multicollinearities associated with modeling the interactions between independent variables using a third independent variable that is the product of the first two. Because structural multicollinearities arise from the way you write the regression model, they are relatively easy to address.

Sample-based multicollinearities present more difficult problems. Sample-based multicollinearities arise because of the way the data were collected. The usual source of sample-based multicollinearities is the inability to manipulate independently all predictor variables.

When studying in vivo physiological systems, it generally is not easy to manipulate independently all potential independent variables. It is common to perturb the physiological system, measure the resulting changes in several variables, then relate one dependent variable to several independent variables, even though the independent variables cannot be controlled individually. For example, in studies of the function of the heart, a common method is to constrict a great thoracic blood vessel and then relate the resulting change in the volume of blood the heart pumps (the dependent variable) to changes in other hemodynamic variables, including heart rate, ventricular pressure or size at the beginning of a beat, and ventricular pressure or size at the end of a beat (the independent variables). In such studies the uncontrolled independent variables may vary together and thus be highly correlated, leading to difficulty in untangling their different effects on the dependent variable. Hence, we effectively have obtained redundant information.

Thus, sample-based multicollinearity can arise from a combination of factors relating to how the system is structured and perturbed. Often it is possible to design new experiments using different perturbations to *obtain new observations under different combinations of the independent variables*. Although the independent variables may still change together, the relationships between these variables during the response to the perturbation may be different than in the original experiment. When data from these two (or more) different experimental perturbations are combined in a single regression computation, the multicollinearity among the independent variables will often be reduced to an acceptable level.

Sample-based multicollinearities are also common in observational studies where there is no experimental control, when an investigator simply measures certain characteristics of subjects or systems as they are presented. Under these circumstances, some truly independent predictor variables may not be separable because, due to the lack of experimental control, both are driven by some other unknown process so that they vary in parallel. In socioeconomic systems, sample-based multicollinearity might arise because laws or regulations dictate certain responses in people or certain patterns of use or consumption. Sometimes it may be possible to redefine the population to incorporate additional sample information (for example, from other countries) to help break up the multicollinearity.

Getting more data under different experimental conditions is, by far, the best strategy for dealing with sample-based multicollinearity. Unfortunately, sometimes you are simply stuck with the sample-based multicollinearity. As a general rule, the less experimental control that can be exerted on a system, the more difficult it is to avoid sample-based multicollinearities. In such cases, you have simply included more related independent variables in the regression model than can be discerned with the available data. We say that *the model is overspecified.* When the model is overspecified, we can delete one or more of the variables from the regression model to remove the redundant information or impose restrictions on the possible solutions to the regression problem to compensate for the effects of the multicollinearity.*

BACK TO MARS

Before discussing how to detect and correct multicollinearity, let us return to Mars to examine the effects that multicollinearity has on the results of a regression analysis. In our original study of how the weight of Martians depended on their height and water consumption (in Chaps. 1 and 3), we fit the data shown in Fig. 3-1 and Table 1-1 with the regression equation

$$\hat{W} = b_0 + b_H H + b_C C \qquad (5.1)$$

Table 5-1 summarizes the result of this analysis; we found that the intercept b_0 was $-1.2 \pm .32$ g, the dependence of weight on height (inde-

---

*There are two widely used restricted regression techniques for dealing with multicollinearity: *principal components regression* and *ridge regression*. We will discuss principal components regression later in this chapter. For discussions of ridge regression, see R. H. Myers, *Modern and Classical Regression with Applications*, Boston, Duxbury, 1986, pp. 243–262, and D. C. Montgomery and E. A. Peck, *Introduction to Linear Regression Analysis*, New York, Wiley, 1982, pp. 310–334.

TABLE 5-1 Comparison of First (Chap. 3) and Second (Chap. 5) Studies
of the Dependence of Martian Weight on Height and Water Consumption

	First Study (Chap. 3)	Second Study (Chap. 5)
Summary statistics:		
Mean $W \pm$ SD,[*] g	10.7 ± 2.2	10.7 ± 2.2
Mean $H \pm$ SD, cm	38.3 ± 5.3	38.2 ± 3.0
Mean $C \pm$ SD, cup/day	10.0 ± 8.5	10.0 ± 8.5
Correlations of variables:		
W vs. H	.938	.820
W vs. C	.834	.834
H vs. C	.596	.992
Summary of regression analyses:		
$b_0 \pm s_{b_0}$, g	$-1.22 \pm .32$	20.35 ± 36.18
$b_H \pm s_{b_H}$, g/cm	$.28 \pm .009$	$-.34 \pm 1.04$
$b_C \pm s_{b_C}$, g/cup/day	$.11 \pm .006$	$.34 \pm .37$
R^2	.997	.700
$s_{W|H,C}$	.131	1.347

[*]SD = standard deviation.

pendent of water consumption) b_H was .28 ± .009 g/cm, and the dependence of weight on water consumption (independent of height) b_C was .11 ± .006 g/cup. In addition, we have previously judged this equation to be a good model fit because it had a high R^2 of .997 and because there were no apparent problems meeting the assumptions of regression analysis (Chap. 4). Figure 5-1A shows the data and associated regression plane.

However, because our results met with some skepticism back on earth, we decided to replicate our study to strengthen the conclusions. So, we recruited 12 more Martians and measured their weight, height, and water consumption. Table 5-2 presents the resulting data, and Fig. 5-1B shows a plot of these data superimposed on the plot of the data from the original study. Comparing Table 5-2 with Table 1-1 and examining Fig. 5-1B reveals that, as expected, the new sample is different from the original one, but overall these data appear to exhibit similar patterns.

We proceed with our analysis by fitting the new data with Eq. (5.1); Fig. 5-2 shows the results of this analysis, which are tabulated in the second column of Table 5-1. As before, the test for fit of the regression equation is highly significant ($F = 10.48$ with 2 numerator and 9 denominator degrees of freedom; $P = .004$). Nevertheless, the new results are

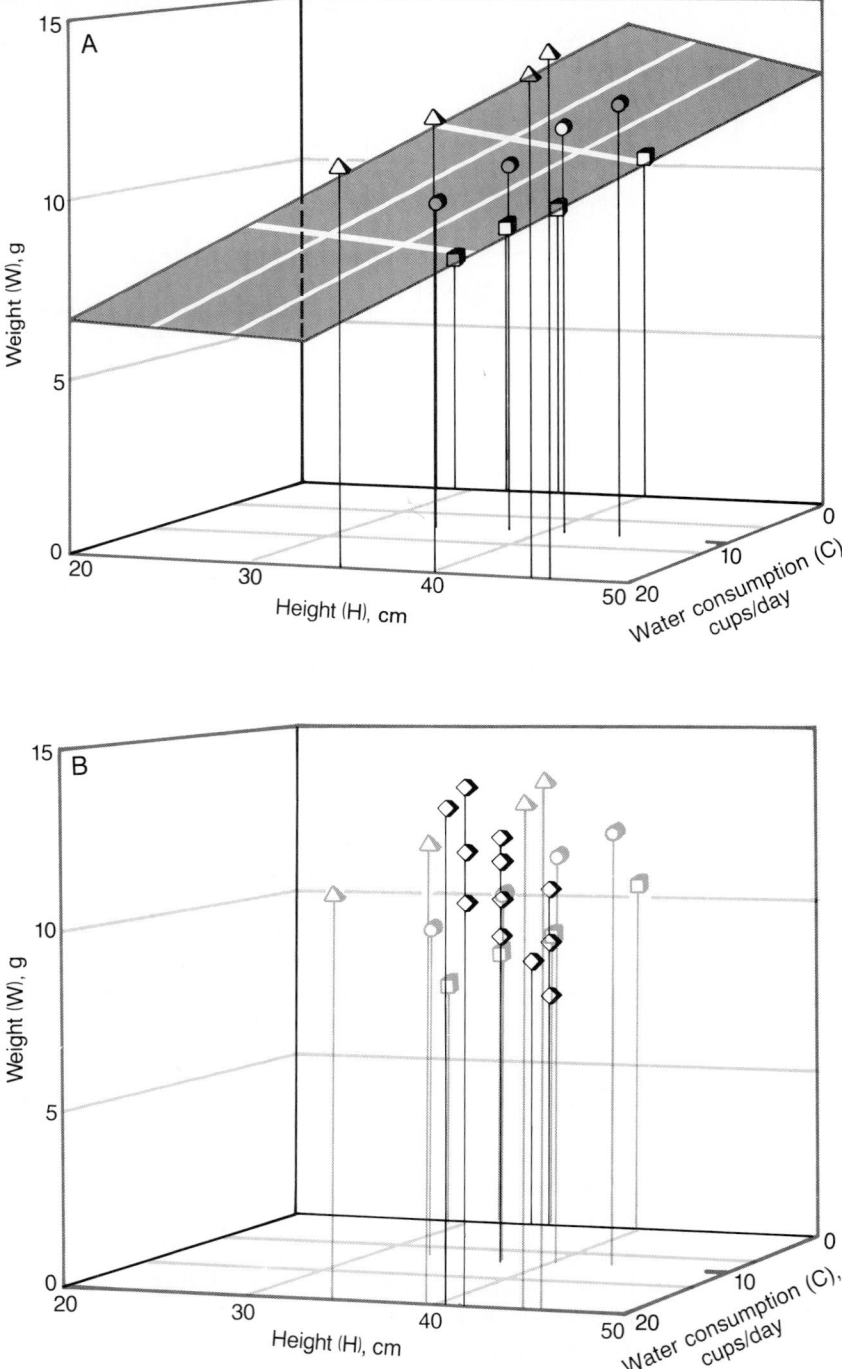

Figure 5-1 *A.* Plot of Martian weight as a function of height and water consumption from our original study discussed in Chaps. 1 and 3, together with the

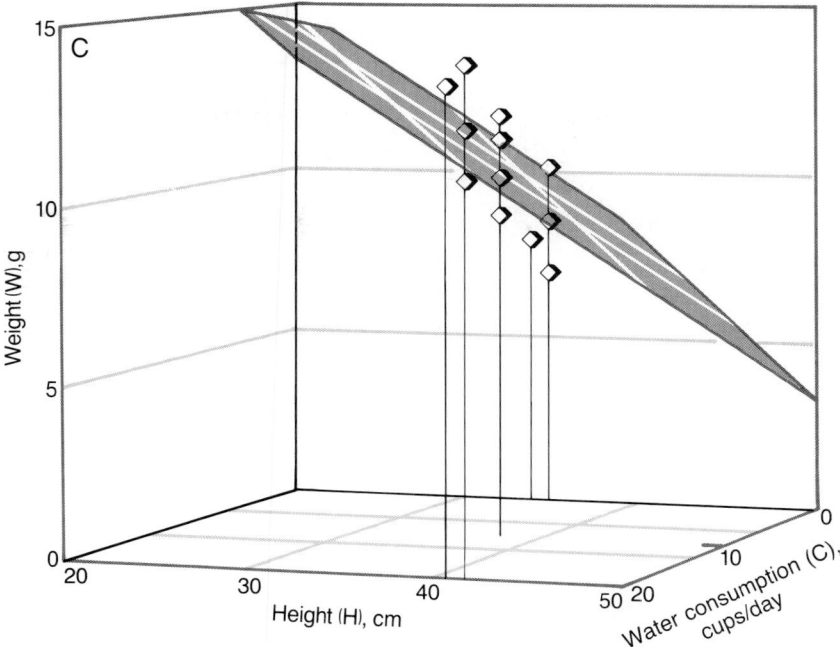

FIGURE 5-1 (*continued*)   associated regression plane (compare with Fig. 3-1*C*). *B.* Data points from another study of Martian weight, height, and water consumption conducted at a later time. The points from the original study are shown lightly for comparison. The new data do not appear to be very different from the data from our first study. *C.* Regression plane fit to the new data using the same equation as before, Eq. (5.1). Even though the data appear to have a pattern similar to the original study and the regression equation being used is the same, the regression plane fit to the data is quite different; it tilts in the opposite direction with respect to the height, *H*, axis. The reason for this difference is that, in contrast to the original set of data, the pattern of new observations of heights and water consumptions is such that height and water consumptions are highly correlated so the "base" of data points to "support" the regression plane is narrow and it becomes difficult to estimate accurately the location of the regression plane. This uncertainty means that, even though we can do the regression computations, it is not possible to obtain accurate estimates of regression parameters and the location of the regression plane. This uncertainty is reflected in large standard errors of the regression coefficients.

very different from those we originally found. The coefficients are far different from what we expected based on our earlier estimates: b_H is now *negative* ($-.34$ g/cm, compared to $+.28$ g/cm) and b_C is much larger ($.34$ g/cup/day, compared to $.11$ g/cup/day). These differences in the regression coefficients lead to a regression plane (Fig. 5-1*C*) that is far

TABLE 5-2 Data for Second Study of Martian Weight vs. Height and Water Consumption

Weight W, g	Height H, cm	Water Consumption C, cup/day
7.0	35	0
8.0	34	0
8.6	35	0
10.2	35	0
9.4	38	10
10.5	38	10
12.3	38	10
11.6	38	10
11.0	42	20
12.4	42	20
13.6	41	20
14.2	42	20

different from what we obtained before (Fig. 5-1*A*); it tilts in the opposite direction from the plane we originally estimated. Moreover, the *standard errors of the regression coefficients are extremely large*—2 to 3 orders of magnitude larger than those obtained in our original analysis. As a result,

```
The regression equation is
W = 20.3 - 0.34 H + 0.337 C

Predictor      Coef      Stdev    t-ratio       p       VIF
Constant      20.35      36.18       0.56    0.588
H            -0.340       1.043      -0.33    0.752      59.8
C             0.3365      0.3683      0.91    0.385      59.8

s = 1.347      R-sq = 70.0%     R-sq(adj) = 63.3%

Analysis of Variance

SOURCE      DF         SS         MS         F        p
Regression   2     38.038     19.019     10.48    0.004
Error        9     16.329      1.814
Total       11     54.367
```

FIGURE 5-2 Results of fitting the data in Fig. 5-1*B* to obtain the regression plane in Fig. 5-1*C*. The multicollinearity in the independent variable is reflected in the large variance inflation factors (VIF) and high standard errors of the estimates of the regression coefficients (labeled "Stdev" in the output). Compare these results with the results of analyzing the original data in Fig. 3-2.

in spite of this highly significant test for model goodness of fit, *none of the parameter estimates is significantly different from zero.*

How can this be? A significant F test for the overall goodness of fit means that the independent variables in the regression equation, taken together, "explain" a significant portion of the variance of the dependent variable.

The key to answering this question is to focus on the fact that the F test for goodness of fit uses the information contained in the independent variables *taken together*, whereas the t tests on the individual coefficients test whether the associated independent variable adds any information *after considering the effects of the other independent variables.*[*] In this case, the variables height H and water consumption C contain so much redundant information that neither one adds significantly to our ability to predict weight W when we have already taken into account the information in the other. The fact that the correlation between H and C is now .992 compared with .596 in the original sample (Table 5-1) confirms this conclusion. This high correlation between the two independent variables results in a narrow base for the regression plane (compare Fig. 5-1*C* with Fig. 5-1*A* and Fig. 3-3), so it is difficult to get a reliable estimate of the true plane of means or, in other words, it is difficult to estimate precisely how each affects the dependent variable independent of the other. The large standard errors associated with the individual parameter estimates in the regression equation reflect this fact. The characteristics we have just described—nonsignificant t tests for parameter estimates when the regression fit is highly significant, huge standard errors of the parameter estimates, strange sign or magnitude of the regression coefficients, and high correlations between the independent variables—are classic signs of multicollinearity.

DETECTING AND EVALUATING MULTICOLLINEARITY

As we have seen, multicollinearity exists when one (or more) of the independent variables can be predicted with high precision from one or more of the remaining independent variables. Quantifying this redundancy of information will form the basis of several numerical indices of the structure and severity of multicollinearity among several independent variables, in particular the so-called *variance inflation factor*, which can be used to detect multicollinearity. Before discussing these indices, however, we will consider two simpler clues to the presence of serious multicollinearity: strange results for the regression coefficients and the pairwise correlations between the independent variables. We will

[*]We discussed this distinction in detail in Chap. 3 in the section "Does the Regression Equation Describe the Data?"

then discuss the general case of a multicollinearity involving several variables. Checking for multicollinearity should be a routine part of every regression analysis when the purpose of the analysis is to use the regression coefficients to gain insight into the underlying nature of the system under study.

Qualitative Suggestions of Harmful Multicollinearity

We have already stressed the importance of carefully thinking about the results of a multiple regression analysis before accepting it, and multicollinearity is another reason to examine the results of any regression analysis closely. There are several questions you can ask to look for possible multicollinearity: Are the coefficients of reasonable magnitude? Do they have the expected sign? Are the values very sensitive to small changes in model structure or the presence or absence of a single or few data points? Is the overall regression significant even though the coefficients associated with the individual independent variables are not? Answering yes to any of these questions can suggest multicollinearity as a possible problem in interpreting your results.

Because we often use multiple linear regression as a tool to estimate the unknown magnitude of a response, there is often no prior information about what to expect as reasonable magnitudes of the parameter estimates. However, sometimes one does have prior experimental or theoretical knowledge that a parameter should have a certain sign or a magnitude within well-defined limits.

For example, based on what we know about the inhabitants of earth and on what we found in our original study of the dependence of Martian weight on height and water consumption, we would predict that b_H in Eq. (5.1) should be positive. When we estimated this coefficient using our second set of Martian weight, height, and water consumption data, we were surprised that the estimate for b_H was $-.34$ g/cm. This result alone does not mean that there is something wrong with the second set of data or regression model, but it serves as a strong warning that something is potentially wrong. (It is of course possible, though not probable, that neither of the two samples we obtained had any problem, but simply that one of them was unrepresentative of the true relationship between weight, height, and water consumption.) In this instance, there is a severe multicollinearity between height and water consumption that also led to imprecise estimates of all regression coefficients (quantified with the large standard errors). The fact that we expected a positive sign for b_H but found it to be negative was a good initial indication that we need to thoroughly evaluate the extent to which multicollinearity may be a problem.

Correlations among the Independent Variables

When there are only two independent variables, it is easy to diagnose multicollinearity: simply look at the correlation coefficient between the two independent variables. If the correlation is high, above approximately .8, there is a possible multicollinearity and if the correlation is very high, above approximately .95, there is a serious problem. Even when there are more than two independent variables, you should always examine the correlations between all the independent variables. High pairwise correlations indicate possible multicollinearity.

When there are more than two independent variables, it is, unfortunately, possible to have multicollinearities in the data even though the pairwise correlations between pairs of independent variables are relatively low. The reason for this is that the important issue is *whether or not there is redundant information in the independent variables taken as a whole*, not merely whether *two* variables happen to contain the same information. Thus, a set of three or more independent variables can exhibit multicollinearity because one of the variables can be predicted from some weighted sum of two or more of the others even though none of the offending variables is highly correlated with any one of the others. We will, therefore, turn our attention to developing a more general way to identify and assess the severity of multicollinearities. As with the diagnostics for influential data points, these statistics should be used as guides for the intelligent development of regression models, not as formal techniques for statistical hypothesis testing.

The Variance Inflation Factor

The effects of multicollinearity on the parameter estimates in multiple regression follow principally from the fact that the redundant information in the independent variables reduces the precision with which the regression coefficients associated with these variables can be estimated, which is reflected in large values of the standard errors of the coefficients. The principal diagnostic for multicollinearity, the so-called *variance inflation factor*, follows naturally from the expression for these standard errors.

From Eq. (3.16), the standard error of the coefficient b_i in the regression equation

$$\hat{y} = b_0 + b_1 x_1 + b_2 x_2 + \cdots + b_k x_k$$

is

$$s_{b_i} = \sqrt{\frac{MS_{\text{res}}}{\Sigma(X_i - \overline{X})^2} \frac{1}{(1 - R_i^2)}}$$

where R_i^2 is the multiple correlation of x_i with the remaining $k - 1$ independent variables $(x_1, x_2, \ldots, x_{i-1}, x_{i+1}, \ldots, x_k)$. If x_i is independent of the other independent variables, $R_i^2 = 0$ and this equation reduces to

$$s_{b_{i_{\min}}} = \sqrt{\frac{MS_{\text{res}}}{\Sigma(X_i - \overline{X})^2}}$$

and the standard error of b_i is at its smallest possible value. Now, square the last two equations and use them to obtain an expression relating the actual variance in the estimate of b_i, $s_{b_i}^2$, to the minimum possible variance if the independent variables contained no redundant information, $s_{b_{i_{\min}}}^2$:

$$s_{b_i}^2 = \frac{1}{1 - R_i^2} s_{b_{i_{\min}}}^2$$

The quantity

$$\text{VIF}_i = \frac{1}{1 - R_i^2}$$

is known as the *variance inflation factor*. It is a measure of how much the variance in the estimate of the regression parameter b_i is "inflated" by the fact that the other independent variables contain information redundant with x_i.

If there is *no redundant information* in the other independent variables, $R_i^2 = 0$ and $\text{VIF}_i = 1$. In contrast, if x_i shares information with the other independent variables, $R_i^2 \neq 0$, so $\text{VIF}_i > 1$ and $s_{b_i}^2$ is larger than the theoretical minimum value by a factor of VIF_i, indicating greater uncertainty in the estimate of the true regression parameter. This uncertainty makes it possible to obtain estimates b_i that are far from the true value of the parameter β_i in the equation that defines the plane of means. In addition, the large value of s_{b_i} undercuts hypothesis testing about the regression coefficient.*

---

*In addition to leading to difficult-to-interpret results, very severe multicollinearities can lead to large numerical errors in the computed values of the regression coefficients because of round-off and truncation errors during the actual computations. This situation arises because the programs do the actual regression computations in matrix form and estimate the regression coefficients using the equation

$$\mathbf{b} = (\mathbf{X}^T\mathbf{X})^{-1}\mathbf{X}^T\mathbf{y}$$

If there is an exact multicollinearity, $(\mathbf{X}^T\mathbf{X})$ cannot be inverted; if there is very serious multicollinearity, it cannot be inverted accurately. In this case, the regression coefficients $\mathbf{b}$ cannot be computed without serious numerical errors.

To protect against this situation, most statistical software routines for multiple linear regression use a related concept called *tolerance*, which is defined as $1 - R_i^2$

The variance inflation factor is the single most useful diagnostic for multicollinearity. Values of the variance inflation factor exceeding 10 (which corresponds to $R_i^2 = .9$ or $R_i = .97$) are clearly signs of serious multicollinearity. Values of the variance inflation factor exceeding 4 (which corresponds to $R_i^2 = .75$ or $R_i = .87$) also warrant investigation. For example, the variance inflation factors associated with our new study of determinants of weight on Mars are 59.8, compared with 1.6 for the original study. These values indicate that there is a problem with multicollinearity in the new study which will seriously undercut any effort to draw conclusions based on the values of the regression coefficients in Eq. (5.1).

The variance inflation factors identify independent variables with substantial multicollinearity with the other independent variables. They do not, however, answer the question of which variables contain redundant information. To answer this question, we need to investigate the relationships between the independent variables.

Auxiliary Regressions

Multicollinearity refers to the condition in which a *set* of independent variables contain redundant information. In other words, the value of one (or more) of the independent variables can be predicted with high precision from a linear combination (weighted sum) of one or more of the other independent variables. We can explore this relationship by examining the so-called *auxiliary regression* of the suspect variable x_i on the remaining independent variables. In other words, when there is multicollinearity, at least one of the coefficients in the regression equation

$$\hat{x}_i = f_0 + f_1 x_1 + f_2 x_2 + \cdots + f_{i-1} x_{i-1} + f_{i+1} x_{i+1} + \cdots + f_k x_k$$

will be significantly different from zero. The auxiliary regression equation will reveal the important redundancies in the data; the significant coefficients are associated with the set of variables that, taken together, contain much of the information in x_i. The coefficient of determination associated with this equation is R_i^2, which is used to compute the variance inflation factor.

All the variables with large variance inflation factors need not be part of the same multicollinear set. There may be two or more sets of mul-

---

(*Footnote continued from page 192*)
(the reciprocal of VIF_i) to warn of or guard against very serious multicollinearity. Although you can override the default values of tolerance set in these programs, doing so does not increase the computational accuracy of the results and is probably not a good idea. When the multicollinearities are so severe as to lead the programs to delete variables because of tolerance problems, it is time to examine the regression model and data, not force the computer program to ignore the problem.

ticollinear variables. Investigating all the possible auxiliary regression equations will help define what the sets of multicollinear variables are and provide some guidance in dealing with them, either by obtaining more information, deleting variables, or modifying the regression equation.* For example, when the multicollinearity is due to the fact that several variables tend to vary together during a physiological experiment because of the nature of the intervention being studied, it may be possible to include data collected during different interventions where the patterns of parallel changes in these variables is different with the new intervention. When the data are all analyzed together, the multicollinearity will be reduced. The patterns in the auxiliary regression equations can provide some guidance in designing these new experiments. Alternatively, when gathering new data under different conditions is impossible (as it is in much survey research), these equations can provide guidance on which variables should be dropped from the regression equation.

The Correlations of the Regression Coefficients

When there is multicollinearity among the independent variables, the redundant information reduces the precision with which we can estimate the regression coefficients associated with each of the independent variables. It also has the effect of introducing correlations between the regression coefficients because at least two of the coefficients will depend, at least in part, on the same information in several independent variables. We can estimate these correlations using Eq. (3.20). They can then be used as another, albeit nonspecific, diagnostic for multicollinearity. Although there are no well-developed guidelines for when a correlation between regression coefficients indicates serious multicollinearity, we take values above .9 as cause for concern.

The Consequences of Having Two Pumps in One Heart

The heart consists of two pumps: the right side pumps blood from the body into the lungs and into the left side of the heart, which pumps the blood into the aorta and the body (Fig. 5-3). These two sides of the heart interact mechanically in two ways. Because the output of the right side is the input to the left side, changes in right-side pump function affect the filling of the left side (after the blood passes through the lungs). In addition, the two sides of the heart share a common muscle wall, the septum, so that changes in the size of one side of the heart directly affect the other side of the heart by shifting the septum. This effect of one side

---

*Principal components, discussed later in this chapter, provide a more direct and elegant approach to this problem.

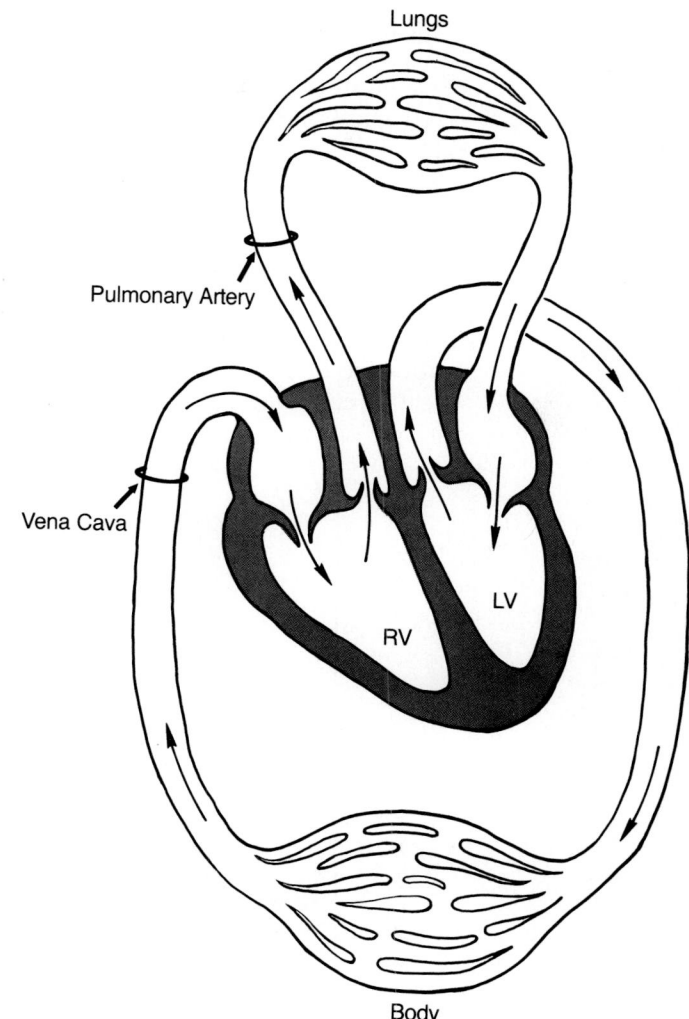

FIGURE 5-3 Schematic representation of the cardiovascular system. Blood flowing out of the right ventricle (RV) of the heart flows through the lungs into the left ventricle (LV) of the heart. Because of the close anatomical connection of the right and left ventricles, changes in right ventricular pressure and size affect the left ventricle both by changes in the amount of blood traveling through the lungs into the left ventricle and also via changes in the position of the wall, known as the intraventricular septum, which separates the two cardiac ventricles.

of the heart on the other is known as ventricular interaction. We wanted to study this interaction to find out how important the size of the right ventricle was in modifying the function of the left ventricle at end diastole when filling is completed, just before active contraction begins.[*]

Because the heart is a distensible (elastic) chamber, its properties during filling are often described by the relationship between the volume of blood in the ventricle and the pressure that results from that volume

[*]B. K. Slinker and S. A. Glantz, "End-Systolic and End-Diastolic Ventricular Interaction," *Am. J. Physiol.* 251:H1062–H1075, 1986.

of blood inside the distensible ventricle. Left ventricular volume is hard to measure, so we measured two dimensions of the ventricle and took their product (end-diastolic area, A_{ed}) as a substitute for volume, then related end-diastolic area to the pressure in the left ventricle at end diastole (P_{ed}). To generate a wide variety of left ventricular volumes, we occluded the pulmonary artery, the blood vessel leading away from the right side of the heart. Occluding the pulmonary artery increases the pressure the right ventricle must pump against, and so reduces right ventricular output and, a few beats later, left ventricular filling volume and pressure. The occlusion also increases right ventricular diastolic size. The question was: How large is the direct effect on left ventricular diastolic size of changing right ventricular size as opposed to the indirect effect of changing left ventricular filling?

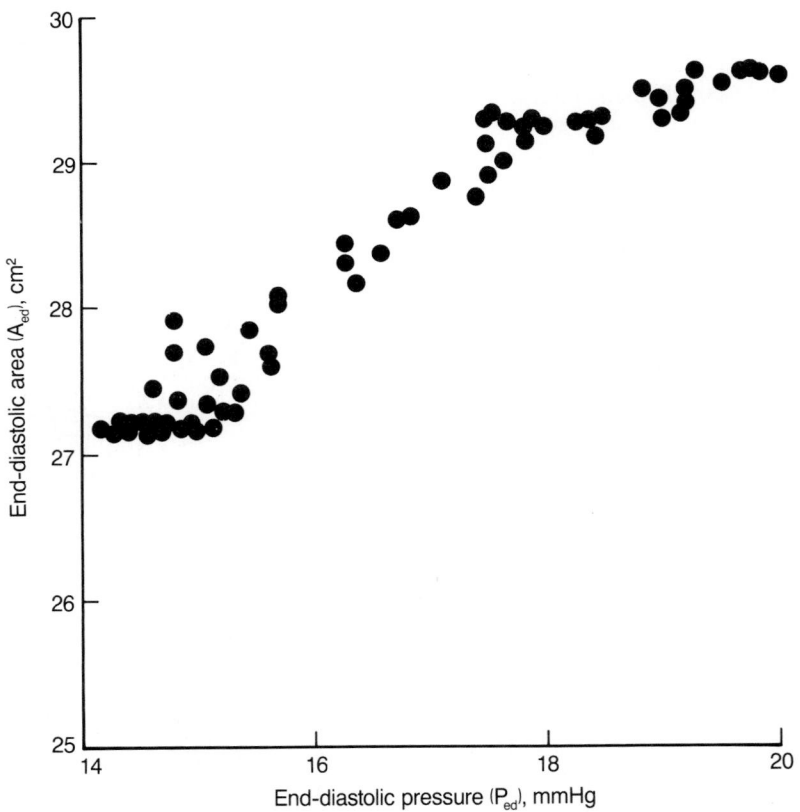

FIGURE 5-4 The relationship between left ventricular size during filling and left ventricular pressure during filling is nonlinear. (Adapted from Figure 4 of B. K. Slinker and S. A. Glantz, "Multiple Regression for Physiological Data Analysis: The Problem of Multicollinearity," *Am. J. Physiol.* 249:R1–R12, 1985.)

The relationship between left ventricular filling pressure and size is nonlinear (Fig. 5-4), so we used a quadratic function to describe the relation between left ventricular end-diastolic area and end-diastolic pressure. Thus, the regression mode that describes the diastolic properties of the left ventricle is

$$\hat{A}_{ed} = a_0 + a_P P_{ed} + a_{P^2} P_{ed}^2 \tag{5.2}$$

The physiological problem of interest is how the size of the right ventricle affects the diastolic function of the left ventricle, which is described by Eq. (5.2). Thus, the final regression model must incorporate the influence of right ventricular size (as judged by measuring one dimension of the right ventricle, D_R) on the parameters of Eq. (5.2). To model this effect, we assumed that, in addition to shifting the intercept of the relationship between A_{ed} to P_{ed}, changes in D_R would also change the slope of the relationship between P_{ed} and A_{ed} (Fig. 5-5). Specifically, we assumed that

$$a_0 = b_0 + b_R D_R$$

and

$$a_P = b_P + b_{PR} D_R$$

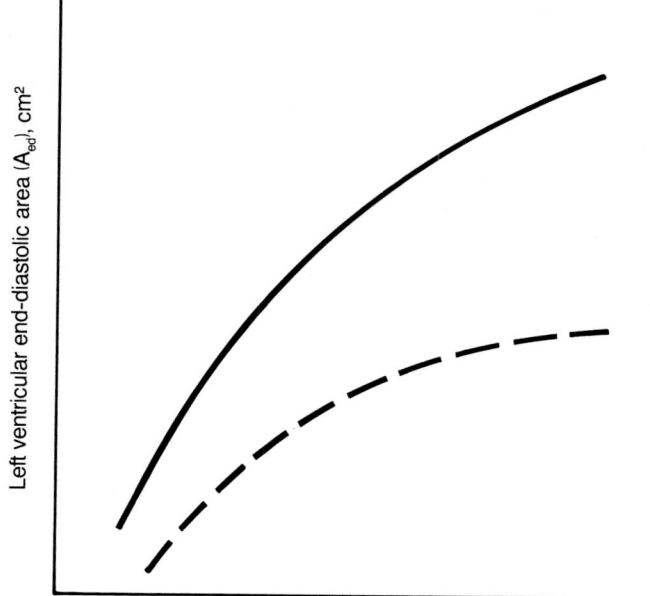

FIGURE 5-5 Schematic representation of the regression model given by Eq. (5.3). If ventricular interaction is present, increases in right ventricular pressure and size will shift the curve down and rotate it from the solid line to the dashed line. If such effects occur, the coefficients b_R and b_{PR} [Eq. (5.3)] will be significantly different from 0 and will quantify the magnitude of the shift and rotation of this curve as a function of right ventricular size.

Left ventricular end-diastolic area (A_{ed}), cm^2

Left ventricular end-diastolic pressure (P_{ed}), mmHg

Substitute from these equations into Eq. (5.2) to obtain our final regression model:

$$\hat{A}_{ed} = b_0 + b_P P_{ed} + b_{P^2} P_{ed}^2 + b_R D_R + b_{PR} P_{ed} D_R \tag{5.3}$$

The regression results are in Fig. 5-6 (the data are in Table C-12A, Appendix C). From the data plotted in Fig. 5-4, as well as knowledge gained about cardiac physiology over the last 100 years, we expect that the left ventricular size should increase with filling pressure. In other words, the estimate of b_P in this equation would be expected to be positive, yet it was estimated to be -2.25 cm^2/mmHg. This result is the first suggestion of a serious multicollinearity problem.

We can use the quantitative methods described above to evaluate the severity of the suspected multicollinearity. First, the simple correlations among the independent variables are very high (Table 5-3). The two

```
DEP VARIABLE: AED      Left Ventricle end-diastolic area

                              ANALYSIS OF VARIANCE

                                 SUM OF        MEAN
                  SOURCE    DF    SQUARES      SQUARE      F VALUE     PROB>F

                  MODEL      4   63.44972822  15.86243206  4584.661   0.0001
                  ERROR     64    0.22143308   0.003459892
                  C TOTAL   68   63.67116130

                     ROOT MSE    0.05882085    R-SQUARE     0.9965
                     DEP MEAN   28.37826       ADJ R-SQ     0.9963
                     C.V.        0.2072743

                              PARAMETER ESTIMATES

                 PARAMETER     STANDARD     T FOR H0:                VARIANCE   VARIABLE
VARIABLE   DF    ESTIMATE      ERROR        PARAMETER=0   PROB > |T| INFLATION  LABEL

INTERCEP   1    68.90126332   6.77178669      10.175       0.0001          0    INTERCEPT
PED        1    -2.24722      0.51930236      -4.327       0.0001   18761.67659 Left Ventricle end-diastolic pressure
PED2       1     0.02924297   0.007475779      3.912       0.0002    4491.13370 Squared end-diastolic pressure
DR         1    -1.1619       0.15851667      -7.330       0.0001    1782.02792 Right Ventricle end-diastolic dimension
PEDDR      1     0.04723354   0.009383851      5.033       0.0001    1339.96195 Pressure by dimension interaction
```

FIGURE 5-6 Analysis of left ventricular end-diastolic area as a function of left ventricular end-diastolic pressure and right ventricular dimension obtained during an occlusion of the pulmonary artery (data are in Table C-12A, Appendix C), which decreases right ventricular outflow and, thus, increases right ventricular size. Although providing an excellent fit to the observations ($R^2 = .9965$), there is severe multicollinearity in the model, indicated by the very large variance inflation factors. Thus, while this model provides a good description of the data, multicollinearity makes it impossible to uniquely attribute the changes in the dependent variable A_{ed} to any of the coefficients of the model and thus prevents us from meeting our goal of attributing changes in left ventricular end-diastolic area-pressure curve to changes in right ventricle size and output.

TABLE 5-3 Correlations among Independent Variables for Ventricular Interaction Data

UNCENTERED	P_{ed}	P^2_{ed}	D_R
P^2_{ed}	.999		
D_R	−.936	−.925	
$P_{ed}D_R$	.932	.940	−.747

CENTERED	$(P_{ed} - \overline{P}_{ed})$	$(P_{ed} - \overline{P}_{ed})^2$	$(D_R - \overline{D}_R)$
$(P_{ed} - \overline{P}_{ed})^2$	.345		
$(D_R - \overline{D}_R)$	−.936	−.105	
$(P_{ed} - \overline{P}_{ed})(D_R - \overline{D}_R)$	.150	.912	−.065

measured variables are highly correlated with $r = -.936$; thus there is potentially a multicollinearity in the data themselves. In addition, there are high correlations between all of the square and cross-product terms (as high as $r = .999$ for P_{ed} vs. P^2_{ed}). These correlations all far exceed our rule-of-thumb cut-off for concern of about .80.

The variance inflation factors in Fig. 5-6 also support our concern about multicollinearity. These values range from 1340 for $P_{ed}D_R$ to 18,762 for P_{ed}, and all exceed the recommended cutoff value of 10. For example, the VIF of 18,762 for P_{ed} means that the R^2 value for the auxiliary regression of P_{ed} on the remaining independent variables

$$\hat{P}_{ed} = 13.01 + .014P^2_{ed} - .3D_R + .018P_{ed}D_R$$

is .99995. It is hard to get closer to 1.0 than that. This auxiliary regression for P_{ed} demonstrates that there is a multicollinearity whereby the independent variables P^2_{ed}, D_R, and $P_{ed}D_R$ supply virtually the same information as P_{ed}. Similar auxiliary regressions could be developed for all the other independent variables, but all would tell essentially the same story: Each independent variable is highly correlated with a linear combination of all of the other independent variables.

Because this regression model contains quadratic and interaction terms, we should first look for problems of structural multicollinearity. If, after controlling the structural multicollinearity, problems still persist, we will need to look at sample-based multicollinearity that may arise from the nature of the experiments used to collect the data.

FIXING THE REGRESSION MODEL

There are two strategies for modifying the regression model as a result of serious multicollinearity. First, there are changes one can make, particularly centering the independent variables on their mean values before computing power and interaction terms, to greatly reduce or eliminate the attendant structural multicollinearities. Second, when it is impossible to collect more data in a way that will eliminate a sample-based multicollinearity, it may be necessary to delete one or more of the redundant independent variables in the regression equation. We now turn our attention to how to implement these strategies.

Centering the Independent Variables

If the multicollinearity is structural, it can often be dealt with by centering the measured independent variables on their mean values before computing the power (e.g., squared) and interaction (cross-product) terms specified by the regression equation. To center the independent variables, simply compute the mean values of each native independent variable, then replace each occurrence of the independent variable in the regression equation with the deviation from its mean. By "native" independent variables, we mean the independent variables that were actually measured. For example, in Eq. (5.3), left ventricular pressure P_{ed} and right ventricular size D_R are the native forms of the pressure and dimension variables. P_{ed}^2 and $P_{ed}D_R$ are not native variables. Hence the centered variables would be $(P_{ed} - \overline{P}_{ed})$, $(D_R - \overline{D}_R)$, $(P_{ed} - \overline{P}_{ed})^2$, and $(P_{ed} - \overline{P}_{ed}) \cdot (D_R - \overline{D}_R)$.

Although centering results in different values for the regression coefficients because they are written in terms of the deviations from the means of the associated independent variables, the resulting equation represents the same mathematical function as we would obtain by fitting the original equation using the uncentered variables. The new regression equation, written in terms of the centered independent variables will, however, usually exhibit less multicollinearity and provide more precise estimates of the regression coefficients.

It is easiest to illustrate how centering reduces the effects of structural multicollinearity by considering its effect on a quadratic fit to the data in Fig. 5-7 (and Table 5-4). Because these data obviously follow a nonlinear function, we will fit these data with the quadratic function

$$\hat{y} = b_0 + b_1 x + b_2 x^2 \tag{5.4}$$

The computer output for the fit of these data in Fig. 5-8A reveals that the regression equation is

$$\hat{y} = 4.88 - 1.43x + .23x^2 \tag{5.5}$$

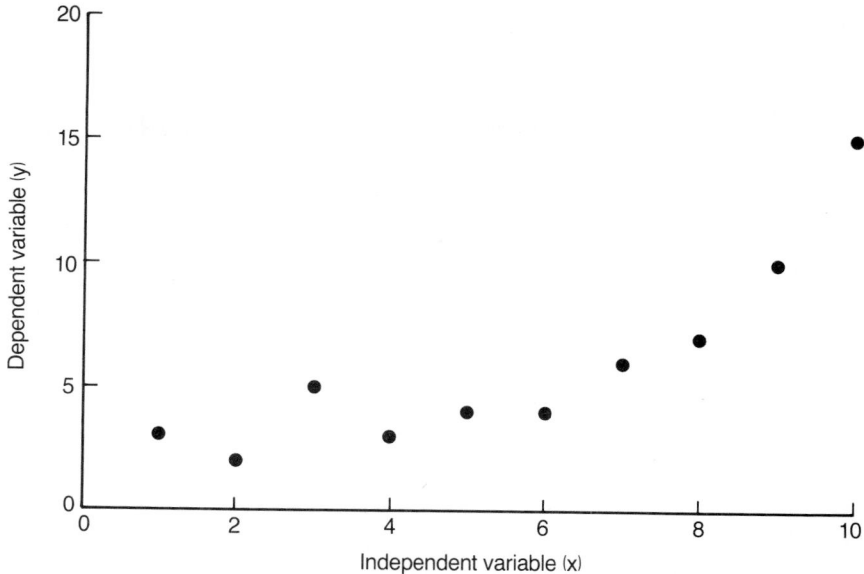

FIGURE 5-7 Illustration of data in which a nonlinear pattern can be described well by the quadratic function $\hat{y} = b_0 + b_1x + b_2x^2$.

The fit is reasonable, with an R^2 of .92 and a standard error of the estimate of 1.267, but both independent variables have large variance inflation factors of 19.9, indicating that multicollinearity is reducing the precision of the parameter estimates. The reason for this multicollinearity is directly evident when the two independent variables are plotted against each other (x^2 vs. x) as in Fig. 5-9A. The correlation between these two variables is .975, well above our threshold for concern.

TABLE 5-4 Data for Example of Effect of Centering

Y	X	X^2	X_c	X_c^2
3	1	1	-4.5	20.25
2	2	4	-3.5	12.25
5	3	9	-2.5	6.25
3	4	16	-1.5	2.25
4	5	25	$-.5$	.25
4	6	36	.5	.25
6	7	49	1.5	2.25
7	8	64	2.5	6.25
10	9	81	3.5	12.25
16	10	100	4.5	20.25

A

Equation Number 1   Dependent Variable..   Y   = 'Y value'

Beginning Block Number 1. Method: Enter   X   X2

Variable(s) Entered on Step Number 1..   X2   = 'squared X'
                                     2..   X   = 'X variable; raw'

		Analysis of Variance			
Multiple R	.95933				
R Square	.92031		DF	Sum of Squares	Mean Square
Adjusted R Square	.89754	Regression	2	129.67121	64.83561
Standard Error	1.26654	Residual	7	11.22879	1.60411

F =   40.41836   Signif F = .0001

------------------------------ Variables in the Equation ------------------------------

Variable	B	SE B	Beta	Tolerance	VIF	T	Sig T
X2	.231061	.055119	1.995640	.050235	19.906	4.192	.0041
X	-1.432576	.622136	-1.096198	.050235	19.906	-2.303	.0548
(Constant)	4.883333	1.489638				3.278	.0135

B

Equation Number 1   Dependent Variable..   Y   = 'Y value'

Beginning Block Number 1. Method: Enter   XC   XC2

Variable(s) Entered on Step Number 1..   XC2   ='squared centered X'
                                     2..   XC   ='X variable; centered'

		Analysis of Variance			
Multiple R	.95933				
R Square	.92031		DF	Sum of Squares	Mean Square
Adjusted R Square	.89754	Regression	2	129.67121	64.83561
Standard Error	1.26654	Residual	7	11.22879	1.60411

F =   40.41836   Signif F = .0001

------------------------------ Variables in the Equation ------------------------------

Variable	B	SE B	Beta	Tolerance	VIF	T	Sig T
XC2	.231061	.055119	.447288	1.000000	1.000	4.192	.0041
XC	1.109091	.139441	.848670	1.000000	1.000	7.954	.0001
(Constant)	3.993750	.605963				6.591	.0003

FIGURE 5-8 *A.* Result of fitting the data in Fig. 5-7 to Eq. (5.4). This equation provides an excellent description of the data, with $R^2 = .92$, but there is significant multicollinearity between the observed values of the two independent variables, x and x^2, reflected by the large variance inflation factors of 19.6. Thus, although this equation provides a good description of the data, it is impossible to interpret the individual coefficients. *B.* Result of analysis of the same data after centering the independent variable x on its mean and writing the regression equation in terms of $(x - \overline{X})$ and $(x - \overline{X})^2$ rather than x and x^2. The quality of the fit has not changed, but the variance inflation factors have been reduced to 1, indicating no multicollinearity and giving us the most precise possible estimates of the coefficients in the regression equation.

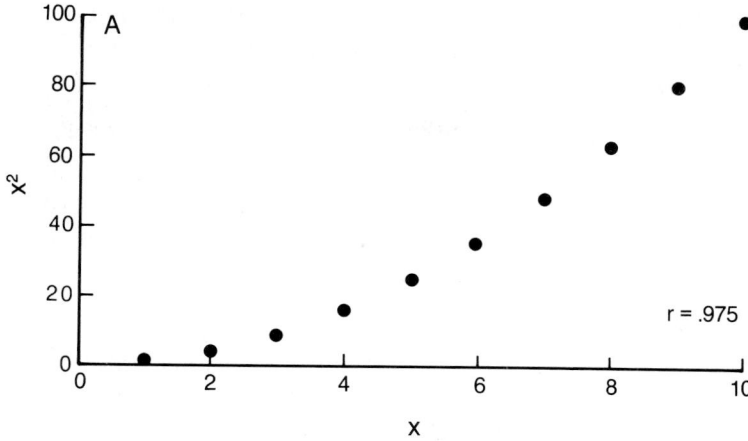

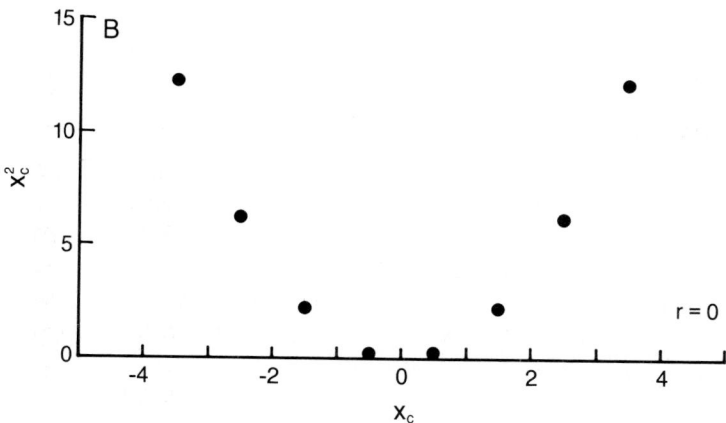

FIGURE 5-9 Illustration of how centering the data breaks up structural multicollinearities associated with quadratic (and other even powered) terms in a regression model. *A.* Over the range of observations in Fig. 5-7, the two independent variables in Eq. (5.4), x and x^2, are highly correlated, with $r = .975$. This high correlation between the independent variables gives rise to the multicollinearity evident in Fig. 5-8*A*. *B.* This situation contrasts with the situation we obtained by dealing with the centered variable $(x - \overline{X})$. Now the correlation between the two independent variables $(x - \overline{X})$ and $(x - \overline{X})^2$ in the regression equation is $r = 0$, indicating no multicollinearity. Centering does not change the functional relationship between the independent variables; it just makes the relationship more symmetric around zero and so breaks up any linear correlations that exist. Hence, centering will always reduce, generally to values near 0, the correlation between a centered independent variable and the even powers of this variable in the regression equation. Centering may also reduce structural multicollinearities that arise from inclusion of interaction terms in a regression equation.

To reduce this correlation, let us *center* the native x values by subtracting the observed mean value $\overline{X} = 5.5$ from each observed value of the independent variable to obtain the centered variable

$$x_c = x - \overline{X} = x - 5.5 \tag{5.6}$$

(Table 5-4 shows the centered values of the independent variable.) We now replace the original independent variable x in Eq. (5.4) with the centered variable x_c to obtain the equation

$$\hat{y} = a_0 + a_1 x_c + a_2 x_c^2 \tag{5.7}$$

Figure 5-8B shows the results of fitting this rewritten regression equation to the data in Table 5-4. Some things are the same as before; some things are different. The correlation coefficient and standard error of the estimate are the same as before, but the coefficients in the regression equation are different and the associated standard errors are smaller. The regression equation is

$$\hat{y} = 3.99 + 1.11 x_c + 0.23 x_c^2 \tag{5.8}$$

Note that the variance inflation factors associated with x_c and x_c^2 are now 1.0, the smallest they can be.

To see why, let us examine the plot of x_c^2 vs. x_c in Fig. 5-9B. Because we centered the independent variable, the quadratic term in the regression equation now represents a parabola rather than an increasing function (as in Fig. 5-9A), so the correlation between x_c and x_c^2 is reduced to 0, and the multicollinearity is eliminated. Although the magnitude of the effect will depend on how far the data are from the origin, *centering the data will always greatly reduce or eliminate the structural multicollinearity associated with quadratic (or other even-powered terms) in a regression equation.*

What of the differences between the two regression equations, Eqs. (5.5) and (5.8)? We have already noted that the values of the coefficients are different. These differences are due to the fact that the independent variables are defined differently and not to any differences in the mathematical function the regression equation represents. To demonstrate this fact, simply substitute from Eq. (5.6) into Eq. (5.8) to eliminate the variable x and obtain

$$\hat{y} = 3.99 + 1.11(x - 5.5) + .23(x - 5.5)^2$$
$$= 3.99 + 1.11 - 1.11 \cdot 5.5 + .23 x^2 - .23 \cdot 2 \cdot 5.5 x + .23 \cdot 5.5^2$$
$$= 4.88 - 1.43 x + .23 x^2$$

which is identical to Eq. (5.5). This result is just a consequence of the fact that, whereas multicollinearities affect the regression coefficients, they do not affect the mathematical function the regression equation represents.

What, then, accounts for the differences in the values of the regression coefficients and the fact that the formulation based on the centered data has standard errors so much smaller than the formulation using the raw variables? To answer these questions, consider what the coefficients in Eqs. (5.4) and (5.7) mean.

In the original formulation, Eq. (5.4), b_0 represents the value of the regression equation at a value of $x = 0$ and b_1 represents the slope of the tangent to the regression line at $x = 0$. (b_2 is an estimate of the second derivative, which is a constant, independent of x.) These two quantities are indicated graphically on Fig. 5-10A. Note that they are both *outside the range of the data* on the extrapolated portion of the regression curve. Because the interpretation of these coefficients is based on extrapolation outside the range of the data, there is relatively large uncertainty in their values, indicated by the large (compared with the results in Fig. 5-8B) standard errors associated with these statistics.

In the centered formulation, Eq. (5.7), a_0 represents the value of the regression line at $x_c = 0$ (or $x = \overline{X} = 5.5$) and a_1 represents the slope of the regression line at $x_c = 0$ (Fig. 5-10B). (a_2 is still an estimate of the second derivative, which is a constant, independent of x_c. The fact that it does not depend on x or x_c is why the value of the associated regression coefficient and standard error are the same in both formulations of the regression equation.) In contrast to the situation before, these quantities are estimated *in the middle of the data* where we can be most confident of how the regression function is behaving. This greater level of confidence is reflected by the smaller standard errors associated with these regression coefficients. The farther the data are located away from the origin of their measurement scale, the more dramatic the effect of centering will be.

This situation highlights another benefit of centering the data: *the coefficients will be estimated based on the "center" (mean) of the independent variables, where they can be estimated most precisely.* In addition, the coefficients often are most easily interpretable in this region as mean values, slopes, and other parameters at the middle of the data.* The disadvantage of centering is that the variables are no longer the natural variables of the problem. So when should you center the data? First, if there are serious structural multicollinearities, center the inde-

---

*In fact, some authors believe the independent variables should *always* be centered (and rescaled by dividing by their standard deviations). For a complete discussion from this perspective, see D. W. Marquardt, "Comment: A Critique of Some Ridge Regression Methods," *J. Am. Stat. Assoc.* 75:87–91, 1980.

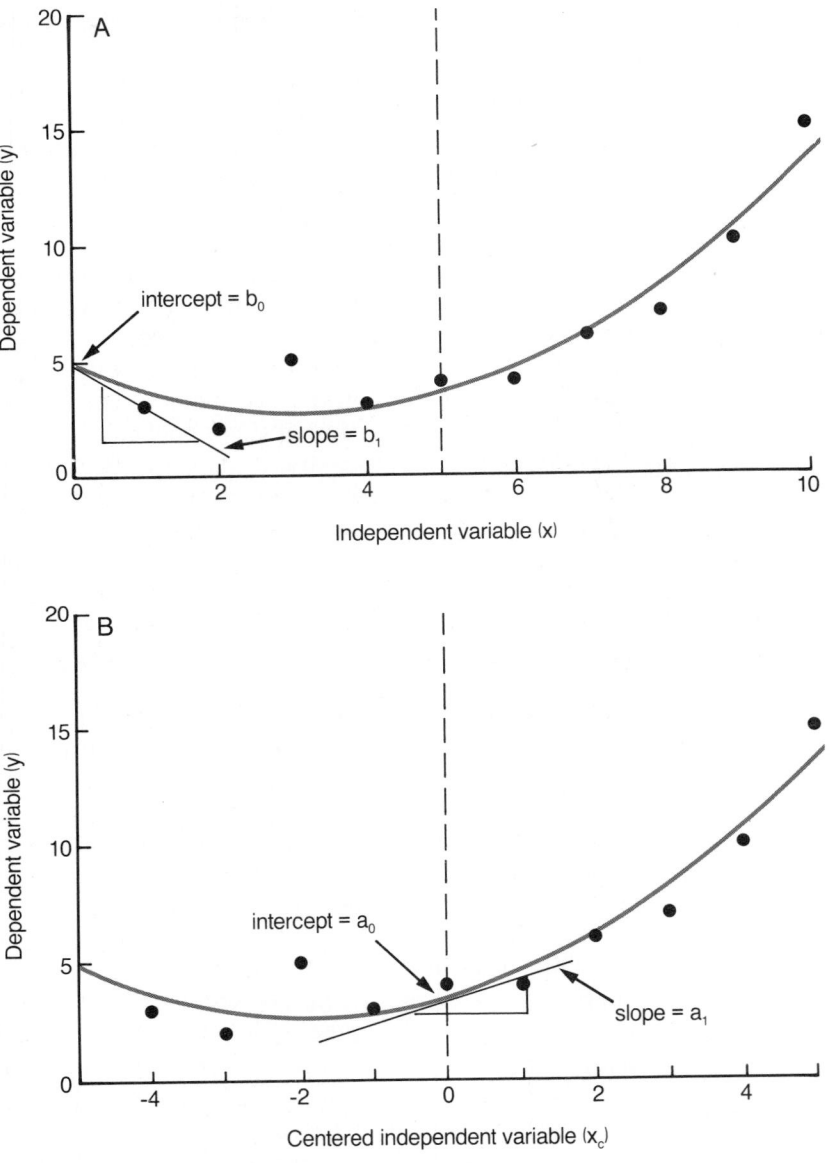

FIGURE 5-10 Another way to think about how centering in the presence of a multicollinearity improves the quality of the parameter estimates, together with an understanding of what those parameter estimates mean geometrically, can be obtained by examining the results of the two regression computations given in Fig. 5-8 for the data in Fig. 5-7. Panel *A* corresponds to the regression analysis of the raw data in Fig. 5-8*A*, whereas panel *B* corresponds to the analysis of the centered data in Fig. 5-8*B*. In both cases the mathematical function represented by the regression

pendent variables to reduce or eliminate the multicollinearity. Second, if centering will make the results easier to interpret, do it. Centering will always improve the precision of the regression parameter estimates.*

Although this discussion has concentrated on a simple structural multicollinearity associated with a quadratic term in a single independent variable, centering can be used with any number of independent variables as well as interaction terms. The basic rule is to center each of the native variables, then compute the powers, transformations, or interaction terms using the centered variables. The resulting regression equation will exhibit less multicollinearity and provide more precise parameter estimates than those based on the raw (uncentered) data.

Deleting Predictor Variables

Because multicollinearity implies redundancy of information in the independent variables, one can often eliminate serious multicollinearity by deleting one or more independent variables from the regression equation. In some instances of sample-based multicollinearity, this strategy may be the only practical option. Sometimes a structural multicollinearity is due to an extraneous term in the regression equation that can simply be dropped without impairing your ability to describe the data or interpret the results. Often this procedure will not substantially impair the ability of the regression equation to predict the dependent variable while improving the precision of the estimates and interpretability of the

*Note that the differences between Eqs. (5.4) and (5.7) get absorbed into the intercept. Therefore, centering will only work if an intercept term is included in the model. Most commercial statistical software packages provide options to fit a regression model without an intercept term. Thus, if you really want to fit a regression model that forces the regression through zero using a no-intercept option in the software, you should not center the data.

FIGURE 5-10 (*continued*)   equation is the same. The interpretation of the coefficients, however, is different. In each case the constant in the regression equation represents the value of y when the independent variable equals 0 and the coefficient of the linear term equals the slope of the line tangent to the curve when the independent variable equals 0. When using the uncentered data (panel *A*), the intercept and slope at a value of the independent variable of 0 are estimated beyond the range of the data. As with simple linear regression, the further one gets from the center of the data the more uncertainty there is in the location of the regression line and hence the more uncertainty there is in the slope and intercept of the regression line. This situation contrasts with that obtained using the centered data (panel *B*), where these two coefficients are estimated in the *middle* of the data where they are most precisely determined. The coefficient associated with the quadratic term is the same in both representations of the regression model.

remaining coefficients in the regression model. The trick is in determining which independent variables to drop.

Sometimes the multicollinearity diagnostics outlined above, particularly the results of the auxiliary regressions, can be combined with some theoretical understanding of the system under study to decide what terms to drop. Alternatively, a class of statistical decision procedures known as variable selection methods (discussed in Chap. 6) can be used to determine models by sequentially searching through different sets of possible independent variables and retaining only those that significantly contribute to explaining the variance of the response variable. Because of the redundancy of information provided by collinear independent variables, these variable selection techniques tend to avoid multicollinearity because they will not include variables that do not supply significant new (as opposed to redundant) information that is useful in explaining variation in the dependent variable.

The implicit assumption one makes in dropping variables is that the original regression model was overspecified, that it contained too many independent variables. While this may be a reasonable assumption—or even a required conclusion in the face of certain sample-based multicollinearities—dropping variables from a regression equation that was based on some prior hypothesis about the nature of the underlying process should be done with caution and thought as to the implications of these changes for the interpretations of the results.

More on Two Pumps in One Heart

When we fit Eq. (5.3) to data obtained from constricting the pulmonary artery to increase the size of the right ventricle, we encountered a severe multicollinearity. Because of the way we wrote Eq. (5.3), it is reasonable to think that much of our problem was due to the quadratic term P_{ed}^2 and the interaction term $P_{ed}D_R$. Thus, we need to do what we can to get rid of this structural multicollinearity using the techniques of centering and variable deletion we just discussed.

First, let us examine the impact of centering on the regression analysis using Eq. (5.3). We first center P_{ed} and D_R and then replace P_{ed}^2 and $P_{ed}D_R$ with the corresponding terms computed from these centered values to obtain

$$\hat{A}_{ed} = c_0 + c_P(P_{ed} - \overline{P}_{ed}) + c_{P2}(P_{ed} - \overline{P}_{ed})^2 + c_R(D_R - \overline{D}_R) + c_{PR}(P_{ed} - \overline{P}_{ed})(D_R - \overline{D}_R)$$

$$(5.9)$$

The correlation matrix for these centered data (the native data are in Table C-12A, Appendix C) appears in the lower half of Table 5-3. In terms of the pairwise correlations among independent variables, the centering

```
DEP VARIABLE: AED      Left Ventricle end-diastolic area

                              ANALYSIS OF VARIANCE

                                SUM OF        MEAN
                     SOURCE  DF  SQUARES      SQUARE      F VALUE    PROB>F

                     MODEL    4  63.44972822  15.86243206  4584.661  0.0001
                     ERROR   64   0.22143308   0.003459892
                     C TOTAL 68  63.67116130

                     ROOT MSE    0.05882085   R-SQUARE   0.9965
                     DEP MEAN   28.37826      ADJ R-SQ   0.9963
                     C.V.        0.2072743

                              PARAMETER ESTIMATES

           PARAMETER  STANDARD   T FOR H0:              VARIANCE   VARIABLE
VARIABLE DF ESTIMATE   ERROR     PARAMETER=0  PROB > |T|  INFLATION  LABEL

INTERCEP 1  28.43200587  0.01604937  1771.534   0.0001           0  INTERCEPT
CPED     1   0.13499458  0.01554421     8.685   0.0001  16.81002661  Centered end-diastolic pressure
CPED2    1   0.02924297  0.007475779    3.912   0.0002   9.39785950  Squared centered end-diastolic pressure
CDR      1  -0.372645    0.0142945    -26.069   0.0001  14.49113186  Centered end-diastolic dimension
CPEDDR   1   0.04723354  0.009383851    5.033   0.0001   7.33903175  Centered press. by dimension interaction
```

FIGURE 5-11 Result of analyzing the ventricular interaction data obtained during pulmonary artery occlusion after centering the independent variables. Centering significantly reduces the variance inflation factors from the order of thousands to around ten. Thus, most of the multicollinearity in the original computations (Fig. 5-6) was structural and was associated with the quadratic and interaction terms in the regression model itself. Unfortunately, even after we fix this structural multicollinearity, variance inflation factors still remain large enough to raise concern about multicollinearity, which means that there is a sample-based multicollinearity in the data as well.

produced a large and desirable reduction in the correlations between the independent variables. For example, the correlation between P_{ed} and P_{ed}^2 has been reduced from .999 in the raw data set to .345 between $(P_{ed} - \overline{P}_{ed})$ and $(P_{ed} - \overline{P}_{ed})^2$ in the centered data set. Similar dramatic reductions in pairwise correlations were realized for the other independent variables as well.

This reduction in multicollinearity is readily appreciated from the computer output from the regression results using this centered data to fit Eq. (5.9) (Fig. 5-11). Notice that, compared to the result in Fig. 5-6, the variance inflation factors and standard errors have been dramatically reduced. The variance inflation factor associated with end-diastolic pressure is reduced from 18,762 to 16.8, and the variance inflation factor associated with right ventricular dimension is reduced from 1782 to 14.5. The variance inflation factors associated with the other centered independent variables, $(P_{ed} - \overline{P}_{ed})^2$ and $(P_{ed} - \overline{P}_{ed})(D_R - \overline{D}_R)$, have been reduced to 9.4 and 7.3, respectively. These dramatic reductions in the value of the variance inflation factors indicate the severity of problems that structural multicollinearity causes. The reduction in multicollinearity is reflected in the standard error for the coefficient for P_{ed}, which decreased from .52 to .016 cm^2/mmHg. The standard error for the coefficient for D_R also decreased from .159 to .014 cm^2/mm. c_p is now positive, compared to the physiologically unrealistic negative value for b_p. In summary, the parameter estimates obtained after centering are more precise and better reflect what the data are trying to tell us.

In spite of the great strides we took toward reducing the multicollinearity by centering the data, we have not completely solved our problem. The variance inflation factors associated with $(P_{ed} - \overline{P}_{ed})$ and $(D_R - \overline{D}_R)$ both exceed 10. The associated auxiliary regressions

$$(P_{ed} - \overline{P}_{ed}) = -.44 - .88(D_R - \overline{D}_R) + .237(P_{ed} - \overline{P}_{ed})^2$$

and

$$(D_R - \overline{D}_R) = -.52 - 1.04(P_{ed} - \overline{P}_{ed}) + .195(P_{ed} - \overline{P}_{ed})^2$$

reveal an interrelationship among the independent variables P_{ed}, D_R, and P_{ed}^2 even after centering to reduce the structural multicollinearity. (The coefficient for $(P_{ed} - \overline{P}_{ed})(D_R - \overline{D}_R)$ is not significant in either auxiliary regression.) Thus, we still have a sample-based multicollinearity that requires attention.

What about deleting predictor variables as a way to reduce the multicollinearity in this example? The dramatic effect that centering had on this analysis tells us that much of our problem with multicollinearity was structural; it was introduced when we added the quadratic and cross-product terms to Eq. (5.2) to obtain Eq. (5.3). Thus, deleting one of these terms will reduce, and deleting both will eliminate, the structural multicollinearity. Unfortunately, in this case we have very sound physiological reasons for formulating Eq. (5.3) the way we did. The interaction term, and to a lesser extent the quadratic term, are in the model for a reason. In fact, the interaction is central to answering our original question about ventricular interaction. We cannot drop these terms from the equation without abandoning the original purpose of the study. We clearly need to look for other means to reduce the remaining multicollinearity.

FIXING THE DATA

As we just saw, the multicollinearity one often detects arises from structural problems that can be resolved by centering or deleting an obviously redundant independent variable. Unfortunately, this situation does not always exist. Perhaps there is no structural multicollinearity and one is faced with purely sample-based multicollinearity. Perhaps there is no single obviously redundant independent variable or you are confident, based on theoretical reasoning, that all the independent variables are potentially important and are reluctant to simply throw one or more of them out. Because sample-based multicollinearity arose from the way the data were collected, the most sensible way to fix the problem is to redesign the experiment in order to *collect more data under different circumstances* so that there are different patterns in the relationship between the independent variables in the new data than in the original data set. It would be even better to design the experiments to collect the data in a way to avoid sample-based multicollinearity in the first place.

You can accomplish this goal in an experimental setting by using different perturbations of the system under study and in an observational study by seeking out different situations for data collection. Precisely how to design the experiments or collect the observations depends more on knowledge of the system under study than statistical issues. Examining the structure of the multicollinearities in the existing data set can, however, provide some guidance in what changes you want to see in the new experiments or observations. Obviously, designing these new studies can require considerable cleverness, but when you can design and conduct such additional data-gathering procedures, you will often be rewarded with better estimates of the regression coefficients and stronger conclusions about their meaning.

Getting More Data on the Heart

In the ventricular interaction study there is enough sample-based multicollinearity in the data so that the multicollinearity is still unacceptably high, even after centering. Centering the data obtained from only the pulmonary artery constriction reduced the VIF associated with end-diastolic pressure to 16.8 — much, much better than 18,762, but still above our threshold of 10 — indicating cause for concern.

The source of this sample-based multicollinearity is obvious in Fig. 5-12A, which shows the high correlation between the two measured independent variables P_{ed} and D_R in the data obtained by constricting the pulmonary artery ($r = -.936$). Right ventricular size D_R increases because the right ventricle cannot empty as well and blood accumulates, and left ventricular filling pressure P_{ed} decreases because the reduction in flow out of the right ventricle causes less blood to reach the left ventricle through the lungs. If you think of yourself as being above the regression plane looking down on the axes defining the data space for these two independent variables, you can see that we have the situation schematically diagrammed in Fig. 3-3 in which there is too narrow a base for good estimation of the position of the regression plane. Recall that we indicated that one way to get around multicollinearity is to ensure that there is a good base of data. Extending our thinking about the data shown in Fig. 5-12A further, we can see that if we could generate more data in

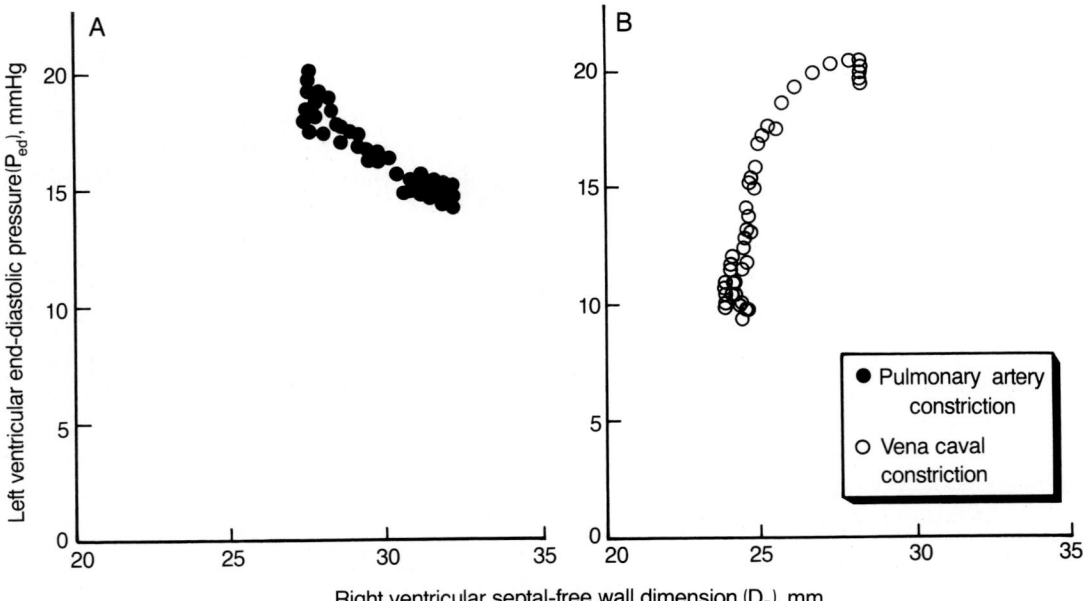

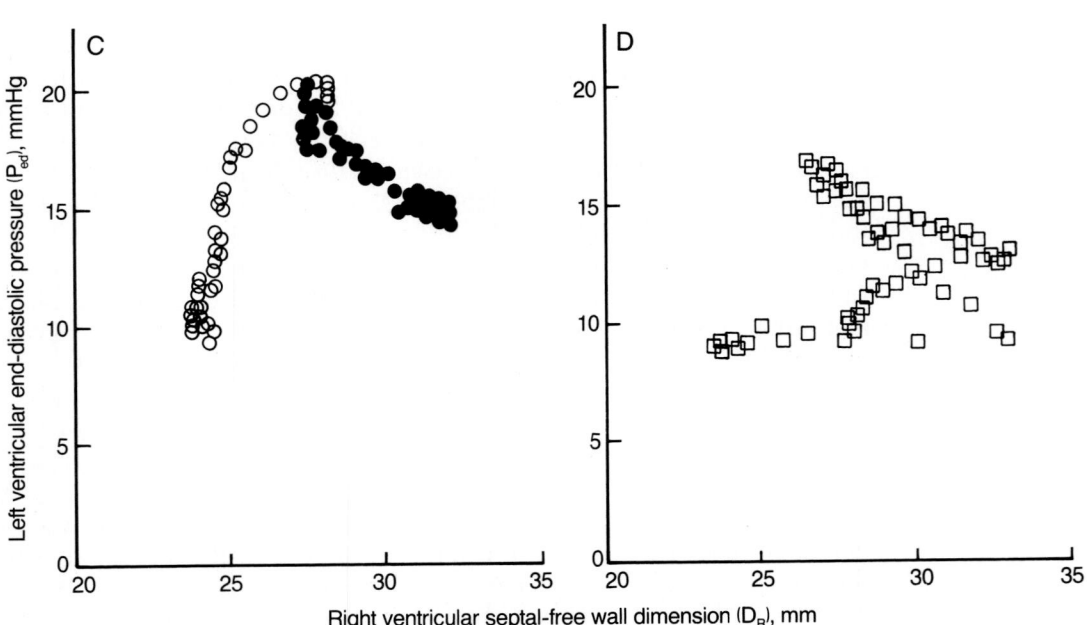

such a way that these two variables varied in some other direction—for example, were positively correlated—we would help break up the multicollinearity.

What we desire is a way to make both D_R and P_{ed} decrease simultaneously. Fortunately, in this experiment we can do that. If we constrict the big veins in the chest that return blood to the right heart so that less blood enters the heart, the size of the right ventricle will decrease, and so D_R will decrease. Because less blood reaches the right ventricle, it pumps less blood through the lungs to the left ventricle and P_{ed} will also decrease (after a delay of a few beats). The relationship between P_{ed} and D_R in this kind of experiment is shown in Fig. 5-12B. You can see that these data are, themselves, highly multicollinear (the linear correlation coefficient is .926), but the relationship is different from what we obtained with the pulmonary artery constriction (compare Fig. 5-12A and B). However, when these data are combined with the data in Fig. 5-12A,

---

◀ **FIGURE 5-12** *A*. The relationship between left ventricular end-diastolic pressure (P_{ed}) and right ventricular dimension (D_R) obtained during pulmonary artery occlusion illustrates the origins of the sample-based multicollinearity identified in Fig. 5-11. Because of the nature of the plumbing in the cardiovascular system, occluding the pulmonary artery reduces the outflow from the right ventricle and, thus, increases the size of the right ventricle. The fall in flow out of the right ventricle reduces the filling of the left ventricle and so reduces left ventricular size and pressure. Thus, because of the nature of the intervention, right ventricular diameter increases as left ventricular pressure decreases. *B*. An alternative intervention for studying ventricular interaction in the intact cardiovascular system would be to occlude the venae cavae which bring blood from the body to the right ventricle. By occluding the venae cavae, we decrease the pressure and size of the right ventricle because we reduce blood comingf to it from the body. The reduced right ventricular size causes a decrease in right ventricular output, which causes decreased filling of the left ventricle. Thus, during a vena caval constriction, right ventricular dimension and left ventricular pressure fall in parallel. Thus, although there is also a strong correlation between these two independent variables during vena caval constriction, the relationship between the two variables is different than that observed during pulmonary artery constriction in panel *A*, with one correlation being negative and the other being positive. *C*. By pooling the data obtained during the pulmonary artery constriction and vena caval occlusion, we are able to break up the multicollinearity and reduce the correlation between these two variables from .936 in panel *B* and .926 in panel *B* to .473. *D*. An even better way to break up the sample-based multicollinearity would be to randomly occlude and release both the pulmonary artery and venae cavae at the same time, thus obtaining data in which there is virtually no correlation between the independent variables in the model. Although this is a different procedure than commonly used in biomedical experiments, these random inputs provide the broadest possible base of information from which to estimate parameters in a regression model, such as Eq. (5.3), and hence the most reliable estimates of the coefficients in the model. The data in panel *D* were used for the subsequent regression analysis shown in Fig. 5-13.

the correlation between these two variables reduces to .473 (Fig. 5-12C). Both interventions cause P_{ed} to decrease, but the pulmonary artery constriction causes D_R to increase, whereas the constriction of the veins returning blood to the heart causes D_R to decrease. An even better way to design the experiment would be to constrict and release both vessels in one experimental run so that the two types of constrictions intermix and thus fill in even more of the data space for P_{ed} and D_R. The result of such an experiment is shown in Fig. 5-12D. The coverage of the data space is, indeed, better than shown in Fig. 5-12C; the correlation coefficient falls to .190. The sample-based multicollinearity has been greatly reduced because we have, in effect, provided a broader base of data from which to estimate the position of the regression plane.

The computer output from the regression analysis from fitting the data obtained using the combination of centering and better data collection strategies to Eq. (5.9) is in Fig. 5-13. (The native data are in Table C-12B, Appendix C.) The R^2 of .965 is highly significant ($F = 514$ with 4 and 74 degrees of freedom; $P < .0001$), as are all the regression coefficients (all $P < .0001$). The standard errors associated with these coefficients are greatly reduced (compare Figs. 5-6 and 5-13). This reduction follows directly from the reduced variance inflation factors, which range from 1.5 to 4.7, all well below 10. The final regression equation is

$$\hat{A}_{ed} = 27.5 + .48(P_{ed} - \overline{P}_{ed}) - .07(P_{ed} - \overline{P}_{ed})^2 - .30(D_R - \overline{D}_R) - .03(P_{ed} - \overline{P}_{ed})(D_R - \overline{D}_R)$$

The physiological interpretation is that there is a nonlinear relationship between left ventricular size and volume such that volume goes up as pressure goes up [by a factor of $.48(P_{ed} - \overline{P}_{ed})$ cm^2/mmHg], but that this increase in volume is smaller at higher pressures [by $-.07(P_{ed} - \overline{P}_{ed})^2$ cm^2/mmHg^2]. These terms describe the basic property of the left ventricle, which governs filling. This relationship is shifted downward as right ventricular size increases [by $.30(D_R - \overline{D}_R)$ cm^2/mm], which means that at any given pressure, the left ventricle will be smaller. The nature of this effect is a shift in the intercept, the left ventricular size at which pressure is zero. This intercept can be thought of as the "rest" volume from which the ventricle begins to fill. The effect of the interaction term also means that the left ventricle is harder to fill as right ventricular size increases; it also means that the slope of the relationship between left ventricular pressure and size decreases as right ventricular size increases [by $.03(P_{ed} - \overline{P}_{ed})(D_R - \overline{D}_R)$ cm^2/(mm·mmHg)]. In physiological terms, this effect is the same as if the muscle in the wall of the ventricle becomes stiffer as right ventricular size increases. Thus, the ventricular interaction effect we wanted to quantify is expressed as both a shift in the rest volume of the left ventricle and a stiffening of the walls of the chamber, both of which make it harder to fill between contractions.

```
DEP VARIABLE: AED      Left Ventricle end-diastolic area

                                    ANALYSIS OF VARIANCE

                                   SUM OF          MEAN
                   SOURCE    DF    SQUARES        SQUARE      F VALUE      PROB>F

                   MODEL      4  130.51089167   32.62772292   514.051     0.0001
                   ERROR     74    4.69690902    0.06347174
                   C TOTAL   78  135.20780068

                     ROOT MSE       0.251936     R-SQUARE     0.9653
                     DEP MEAN      27.01994      ADJ R-SQ     0.9634
                     C.V.           0.9324078

                                    PARAMETER ESTIMATES

             PARAMETER    STANDARD    T FOR H0:                  VARIANCE    VARIABLE
VARIABLE  DF  ESTIMATE      ERROR   PARAMETER=0   PROB > |T|    INFLATION    LABEL

INTERCEP   1  27.51396697  0.07447745   369.427     0.0001            0      INTERCEPT
CPED       1   0.47751073  0.01372093    34.802     0.0001     1.53534790    Centered end-diastolic pressure
CPED2      1  -0.069506    0.009734139   -7.140     0.0001     2.99692568    Squared centered end-diastolic pressure
CDR        1  -0.295939    0.02424428   -12.207     0.0001     4.68813201    Centered end-diastolic dimension
CPEDDR     1  -0.0313587   0.007304018   -4.293     0.0001     3.00511002    Centered press. by dimension interaction
```

FIGURE 5-13 Results of analyzing data in Fig. 5-12*D*, obtained during a sequence of random occlusions and releases of the pulmonary artery and venae cavae so as to break up the sample-based multicollinearity in the independent variables, analyzed after also centering the data. The variance inflation factors, are all now quite small, indicating that the combination of modifying the experiment to break up the sample-based multicollinearity and centering the independent variables permit accurate estimates of the parameters in the regression model.

Before we go on to more advanced multicollinearity diagnostics and solutions, it is worth reflecting on what we have done so far with the analysis of the interaction between the two sides of the heart. We began by formulating a regression model based on what we knew about the function of the heart and a preliminary look at the data. We then discovered we had severe multicollinearity when we tried to fit the regression model [Eq. (5.3)] to data obtained from only a pulmonary artery constriction. This multicollinearity resulted in P_{ed} being associated with a variance inflation factor of 18,762. We discovered that this multicollinearity came from two sources: the structure of Eq. (5.3) and the data. We showed how we could reduce the multicollinearity by centering and the variance inflation factor for P_{ed} decreased to 16.8. Then, we showed that it was possible to redesign the experiment to collect data using a combination of constrictions of the pulmonary artery and the veins bringing blood back to the heart to break up the sample-based multicollinearity. The combination of centering and additional data reduced the variance inflation factor for left ventricular end-diastolic pressure to a very acceptable level of 1.5 (remember, 1.0 is the lowest possible value for a variance inflation factor). Thus, because there were two distinct sources of multicollinearity, it took a combination of strategies to solve the problem.

Of course, it is possible that such a combination of strategies will not always be applicable or will not sufficiently reduce the multicollinearity. Therefore, more esoteric methods of dealing with multicollinearity have been developed.

USING PRINCIPAL COMPONENTS TO DIAGNOSE AND TREAT MULTICOLLINEARITY*

The problems associated with multicollinearity arise because the values of the independent variables tend to change together, which makes it difficult to attribute changes in the dependent variable uniquely to each independent variable. As we have already seen, this situation graphically boils down to one in which the values of the independent variables are "bunched" in a way that does not provide a broad base of support for the regression plane (Fig. 5-1C). When the values of the independent variables are "spread out," they provide a broad base to support the regression plane, so we can estimate the coefficients that describe this plane with greater precision (Fig. 5-1A). Thus far, we have used variance inflation factors, which are essentially the multiple correlations between each independent variable and the others, to diagnose multicollinearity. Once the variance inflation factors identify variables with high multicollinearity, we used the auxiliary regression equations to identify the structure of these multicollinearities. It is possible to replace this two-step procedure with a single so-called *principal components analysis* that identifies the number and nature of serious multicollinearities.

Principal components analysis essentially involves identifying the lines in the multidimensional space spanned by the independent variables which describe the relationships among the independent variables. Deriving the expressions for principal components analysis is done using the matrix representation of the linear regression equation. Each principal component associated with the matrix of independent variables defines one of these directions, known as an *eigenvector*, and an associated number, known as an *eigenvalue*, which can be used to define whether or not a multicollinearity exists. Essentially we seek to determine the axes of an ellipse (in as many dimensions as there are independent variables) that enclose the independent variables and determine how long these axes are. When the values of the independent variables are broadly distributed, all the axes of this ellipse will be about the same

---

*This section requires the use of matrix notation for linear regression. It can be skipped without compromising the understanding of the remaining material in this book.

length and multicollinearity will not be a problem. In contrast, when one or more of the axes is much shorter than the others, the independent variables exhibit multicollinearity. Figure 5-14 illustrates these two situations for two independent variables. The eigenvectors and eigenvalues associated with the independent variables represent the direction of these axes and their lengths, respectively.

Before we continue, we must address the issue of how the independent variables are scaled. All the results we have obtained so far are independent of the scales used to quantify the independent and dependent variables.* In other words, if we measured cell sizes in miles instead of micrometers or weights in stones instead of micrograms, and then used the data in a regression analysis, the specific numbers would come out different depending on the choice of units, but all the correlation coefficients, F and t test statistics, and results of hypothesis tests would be the same. Moreover, the numerical differences in the regression coefficients would simply reflect the differences in units used, and the results obtained with one set of units could be converted to the other by simply applying the appropriate conversion factors. Likewise, the zero point on the scales of measurement can be changed without fundamentally affecting the resulting regression equation; these shifts in the zero points are simply absorbed into a compensating change in the constant term in the regression equation (as happens when we center data).

The situation with principal components analysis is different. The results depend on both the zero points and the scales of measurement. The intuitive justification for this fact follows from considering the problem of defining the ellipse described above. We wish to compare the lengths of the different axes; in order to make such comparisons, all the independent variables need to be measured on a common scale. It would make no sense to say that a mass axis 65 g long was longer than a length axis that was 54 cm long. To make such comparisons, we need to *standardize* all the variables so that they contain *no units*. By doing so we make it possible to compare the spread in different directions among the independent variables.

Standardized Variables, Standardized Regression, and the Correlation Matrix

We will standardize all the variables by converting the raw observations to unitless *standardized deviates* by subtracting the mean of each variable (i.e., centering each variable as discussed above) and dividing by the

*For a more detailed discussion of the issue of scaling and scale invariance, see S. Weisberg, *Applied Linear Regression* (2nd ed.), New York, Wiley, 1985, pp. 185–186.

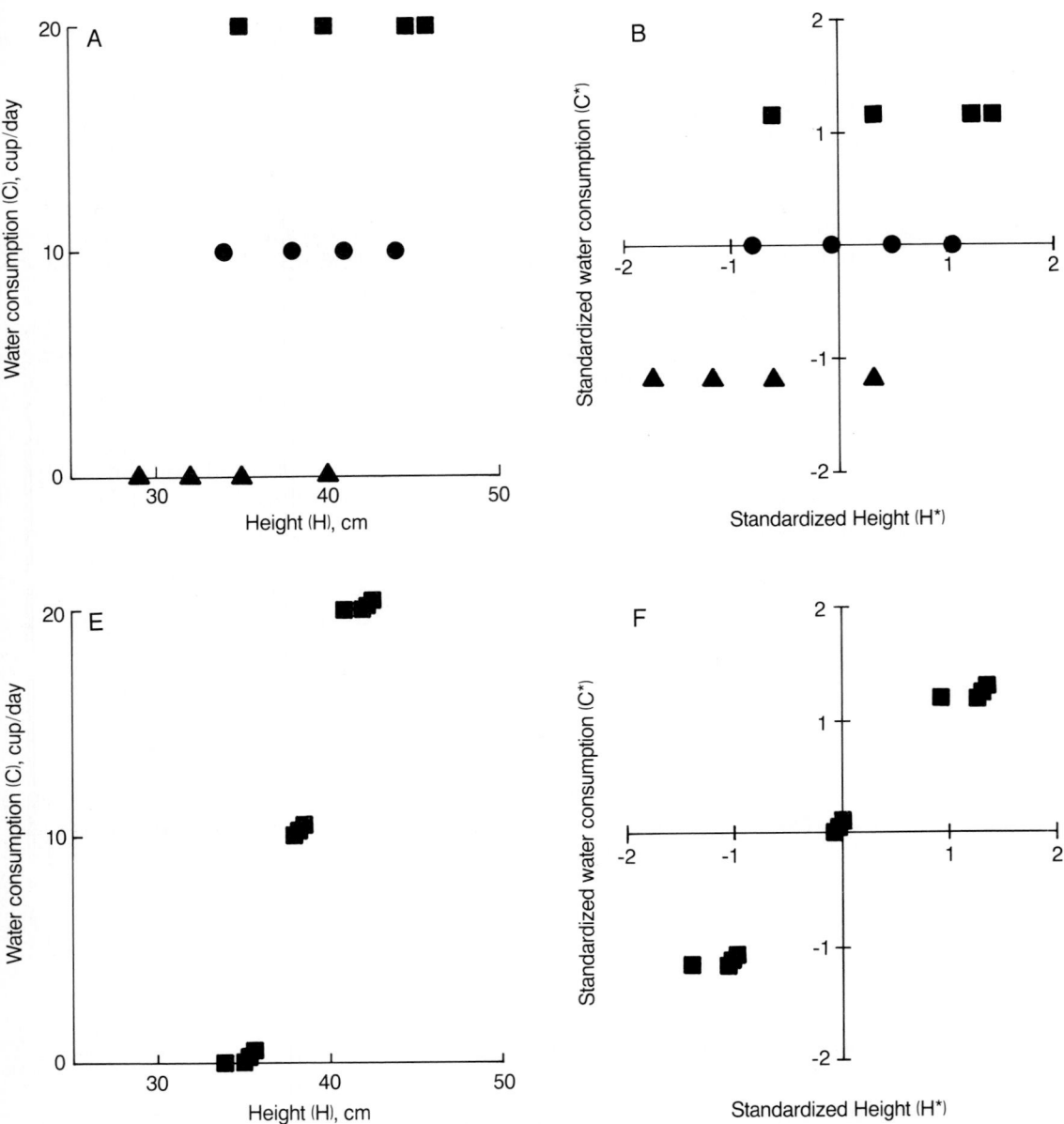

FIGURE 5-14 This figure illustrates the interpretation of the eigenvalues and eigenvectors of the correlation matrix of the independent variables and how these quantities are used to diagnose multicollinearity. The top set of panels (A–D) presents the values of the independent variables (height H and water consumption C) from our first study of the effects of height and water consumption on the weight of Martians presented in Fig. 1-2 (and reproduced in Figs. 3-1 and 5-1). A. The observed values of the independent variables H and C. Examining these points, there is little evidence of multicollinearity; the values are broadly

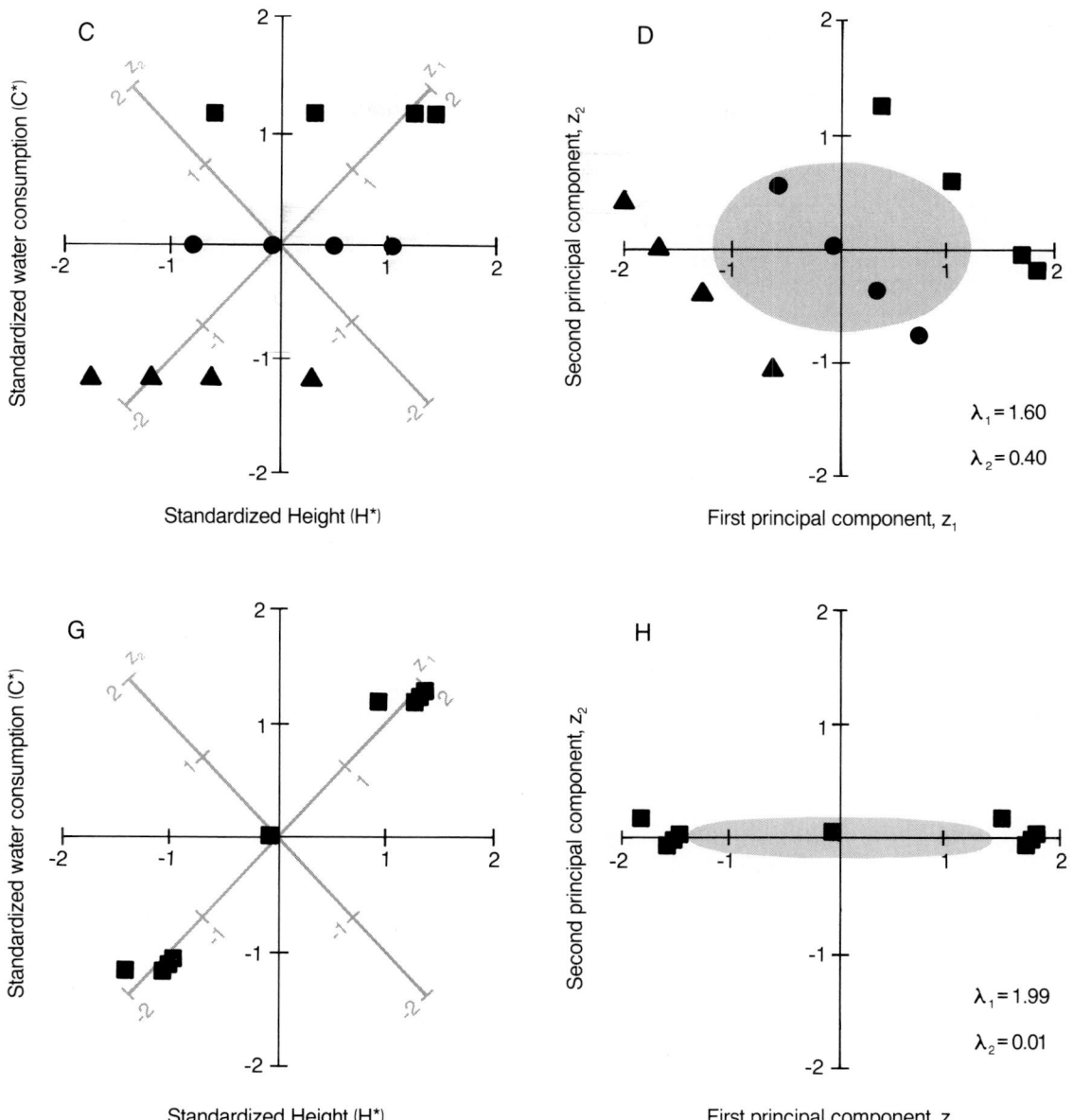

FIGURE 5-14 (*continued*)   distributed and provide a wide "base" to support the regression plane. *B*. This panel shows the data after standardizing the variables by subtracting the mean and dividing by the standard deviation to obtain the dimensionless variables H^* and C^*. *C*. The first principal component of the standardized data is the direction along which there is the maximum variance and the second principal component is perpendicular to it. (If there had been more than two

(*continued on page 220*)

standard deviation of that variable. Specifically, for each of the k independent variables, we define

$$x_1^\star = \frac{x_1 - \overline{X}_1}{s_{x_1}}$$

$$x_2^\star = \frac{x_2 - \overline{X}_2}{s_{x_2}}$$

$$\vdots$$

$$x_k^\star = \frac{x_k - \overline{X}_k}{s_{x_k}}$$

where $\overline{X}_i$ and s_{x_i} are the mean and standard deviation of the observed values of X_i, respectively. By definition, all of the standardized variables $x_i^\star$ have mean 0 and standard deviation 1, and are unitless. It will simplify matters to also standardize the dependent variable in a similar manner according to

$$y^\star = \frac{y - \overline{Y}}{s_y} \tag{5.10}$$

where $\overline{Y}$ and s_y are the mean and standard deviation of the dependent variable, respectively. This standardization is equivalent to changing the coordinate system for describing the data to one whose origin is on the

---

FIGURE 5-14 (*continued*) independent variables, the second principal component would be in the direction with the greatest variance perpendicular to the first principal component and so on.) The directions of the principal components are the eigenvectors of the correlation matrix of the independent variables. *D*. Expressing the standardized data in terms of the principal components corresponds to simply rotating the coordinates to make the axes correspond to the eigenvectors of the correlation matrix. The eigenvalues associated with each eigenvector are a measure of the variance along the corresponding principal components. The light ellipse has axes equal to the lengths of the two eigenvalues, $\lambda_1 = 1.60$ and $\lambda_2 = .40$. (Note that the eigenvalues add up to the number of independent variables, 2.) The fact that the eigenvalues are not too different indicates that multicollinearity is not a problem in these data. The bottom set of panels (*E–H*) presents the data from our second study of Martian height, water consumption, and weight presented in Fig. 5-1, when serious multicollinearity is present. Each panel corresponds to the one above it. In this second study the data cluster along the first principal component, so there is a large difference in the eigenvalues ($\lambda_1 = 1.99$ and $\lambda_2 = .01$) and multicollinearity is a serious problem. Note that the "base" for the regression plane is now long and narrow rather than wide, making it hard to come up with reliable estimates for the regression parameters.

mean of all the variables (i.e., centering) and adjusting the scale so that all the variables have a standard deviation equal to 1. Figure 5-14 shows the values of the independent variables from the two studies of Martian weight, height, and water consumption (from Fig. 5-1*A* and *C*) before and after this change of coordinates. (The $y^\star$ axis is coming out of the page.) Note that the shape of the ellipses containing the observed values of the independent variables has changed because of the rescaling.

Next, use the definitions of the standardized dependent and independent variables to rewrite the multiple regression equation

$$\hat{y} = b_0 + b_1 x_1 + b_2 x_2 + \cdots + b_k x_k \tag{5.11}$$

as

$$\hat{y}^\star = b_0^\star + b_1^\star x_1^\star + b_2^\star x_2^\star + \cdots + b_k^\star x_k^\star \tag{5.12}$$

where the regression coefficients in the rewritten equation $b_i^\star$ are known as *standardized regression coefficients.*[*] The standardized regression coefficients are dimensionless and related to the original regression coefficients according to

$$b_i = \frac{s_y}{s_{x_i}} b_i^\star \tag{5.13}$$

Recall that the regression plane must pass through the mean of the dependent variable and the means of all the independent variables. All these means are zero by construction, so, from Eq. (5.12)

$$0 = b_0^\star + b_1^\star \cdot 0 + b_2^\star \cdot 0 + \cdots + b_k^\star \cdot 0 = b_0^\star$$

and the intercept in the standardized regression equation vanishes, leaving

$$\hat{y}^\star = b_1^\star x_1^\star + b_2^\star x_2^\star + \cdots + b_k^\star x_k^\star$$

We will conduct our principal components analysis using this standardized regression equation.

As before, it is possible to fit the data using least-squares parameter estimation. Let $\mathbf{y}^\star$ be the vector of values of the standardized dependent variable and $\mathbf{X}^\star$ be the design matrix $\mathbf{X}$ of standardized values of the independent variables (with each variable in a column of $\mathbf{X}^\star$), and $\mathbf{b}^\star$ be the vector of regression coefficients. These definitions are exactly analogous to those we used in Chap. 3 (and throughout the book) to present the regression equations in matrix form, except now the constant intercept term is zero and we do not need to allow for it in the data matrix

---

[*]The standardized regression coefficients are often denoted as β_i, but we will use the notation $b_i^\star$ to avoid confusion with the parameters in the equation for the plane of means [Eq. (3.13)].

and, thus, it has no column of ones. In matrix notation, the regression equation is

$$\hat{\mathbf{y}}^{\star} = \mathbf{X}^{\star}\mathbf{b}^{\star} \tag{5.14}$$

where the least-squares estimate of the standardized regression parameters is

$$\mathbf{b}^{\star} = (\mathbf{X}^{\star T}\mathbf{X}^{\star})^{-1}\mathbf{X}^{\star T}\mathbf{y}^{\star}$$

The results of this regression analysis can be directly converted back to the original variables using Eqs. (5.10) and (5.13). All the results of the statistical hypothesis tests are identical. Only the form of the equations is different.[*]

It can be shown that the matrix $\mathbf{R} = \mathbf{X}^{\star T}\mathbf{X}^{\star} = [r_{ij}]$ is a $k \times k$ matrix whose elements are the correlations between all possible pairs of the independent variables.[†] Specifically, the *correlation matrix* is

$$\mathbf{R} = \mathbf{X}^{\star T}\mathbf{X}^{\star} = [r_{ij}] = \begin{bmatrix} 1 & r_{12} & r_{13} & \cdots & r_{1k} \\ r_{21} & 1 & r_{23} & \cdots & r_{2k} \\ \vdots & & & & \\ r_{k1} & r_{k2} & r_{k3} & \cdots & 1 \end{bmatrix}$$

The *ij*th element of R, r_{ij}, is the correlation between the two independent variables x_i and x_j (or $x_i^{\star}$ and $x_j^{\star}$). Because the correlation of x_i with x_j equals the correlation of x_j with x_i, $r_{ij} = r_{ji}$ and the correlation matrix is symmetric. The correlation matrix contains *all* the information about the interrelationships among the standardized independent variables. We will therefore use it to identify and characterize any multicollinearities that are present.[‡]

Principal Components of the Correlation Matrix

It is possible to do a rotation of the coordinate axes used to describe the standardized independent variables in a way that will define a new set of mutually perpendicular directions so that the first direction is aligned

---

[*]In fact, computer programs used for conducting regression analysis almost always convert the variables to their standardized forms before estimating the regression coefficients, and then convert the results back to the original variables because doing so provides more uniform scaling among all the variables and improves the numerical accuracy of the computations.

[†]For a derivation of this result, see J. Neter, W. Wasserman, and M. H. Kutner, *Applied Linear Regression Models*, Homewood, Ill., Irwin, 1983, pp. 380–381.

[‡]It can also be shown that the diagonal elements of $\mathbf{R}^{-1}$ are the variance inflation factors.

along the direction of maximum variation in the standardized variables; the second direction is aligned along the direction of the next greatest amount of variation in the independent variables, subject to the condition that it has to be perpendicular to the first direction; the third direction oriented to maximize the remaining variation, subject to the condition that it is perpendicular to the other two directions; and so forth. Figure 5-14 shows such a rotation of coordinates for the two samples we collected on Mars. Note that in both examples, the first axis is aligned along the direction of most variation in the independent variables and the second axis is perpendicular to it. (There are two axes because there are two independent variables.) In the first example, when there is no severe multicollinearity, the data points are spread about equally far along both axes (Fig. 5-14D). In the second example, when there is severe multicollinearity, the points tend to cluster along the first axis, with relatively little spread perpendicular to it (Fig. 5-14H).

This rotation of coordinates is represented by the matrix $\mathbf{P}$, whose columns represent the eigenvectors of $\mathbf{R}$. In the new coordinate system, the correlation matrix is represented by the diagonal matrix

$$\mathbf{P}^T\mathbf{R}\mathbf{P} = \mathbf{P}^T(\mathbf{X}^{\star T}\mathbf{X}^\star)\mathbf{P} = \lambda = \begin{bmatrix} \lambda_1 & \cdots & 0 \\ 0 & \lambda_2 & \cdots & 0 \\ \vdots & & & \\ 0 & 0 & \cdots & \lambda_k \end{bmatrix}$$

where the eigenvalue λ_i corresponds to the ith eigenvector of $\mathbf{R}$, which is in the ith column, $\mathbf{p}_i$, of the rotation matrix $\mathbf{P}$. If there are k independent variables, there will be k eigenvalues and k eigenvectors. By convention, the eigenvalues (and associated eigenvectors) are numbered in descending order of magnitude of the eigenvalues, with λ_1 the largest and λ_k the smallest. All the eigenvalues must be greater than or equal to zero.

The elements of the vector

$$\mathbf{p}_i = \begin{bmatrix} p_{1i} \\ p_{2i} \\ \vdots \\ p_{ki} \end{bmatrix}$$

define the direction of the ith eigenvector. The elements of the ith eigenvector define the ith *principal component* associated with the data, and the associated eigenvalue is a measure of how widely scattered the data are about the ith rotated coordinate axis.

The k new principal component variables z_i are related to the k independent variables $x_i^\star$ measured in the original standardized coordinate system according to

$$z_1 = p_{11}x_1^* + p_{12}x_2^* + p_{13}x_3^* + \cdots + p_{1k}x_k^*$$

$$z_2 = p_{21}x_1^* + p_{22}x_2^* + p_{23}x_3^* + \cdots + p_{2k}x_k^*$$

$$z_3 = p_{31}x_1^* + p_{32}x_2^* + p_{33}x_3^* + \cdots + p_{3k}x_k^*$$

$$\vdots$$

$$z_k = p_{k1}x_1^* + p_{k2}x_2^* + p_{k3}x_3^* + \cdots + p_{kk}x_k^*$$

Thus, the elements of the eigenvectors describe how each original standardized independent variable contributes to each new variable defined in the principal component coordinate system. A small magnitude value of p_{ij} means that original variable x_i^* contributes little information to defining new variable z_i. As we will now see, this information can be used to define the multicollinearities that exist in the data.

Let $\mathbf{Z}$ be the matrix of the data on the independent variables expressed in the principal components coordinate system, where each column of $\mathbf{Z}$ corresponds to one of the new independent variables and each row corresponds to one observation. (This definition is exactly analogous to the one we used to define the standardized data matrix of independent variables $\mathbf{X}^*$). In matrix notation, we can represent the rotation of coordinates defined by the equations above as

$$\mathbf{Z} = \mathbf{X}^*\mathbf{P} \tag{5.15}$$

The ith column of the $\mathbf{Z}$ matrix is the collection of all values of the new variable z_i. It is given by

$$\mathbf{z}_i = \mathbf{X}^*\mathbf{p}_i$$

because $\mathbf{p}_i$ is the ith column of the eigenvector matrix $\mathbf{P}$. Because the origin of the new coordinate system is at the mean of all the variables, $\bar{z}_i = 0$, and we can measure the variation in z_i with

$$\Sigma(z_i - \bar{z}_i)^2 = \Sigma z_i^2 = \mathbf{z}_i^T\mathbf{z}_i$$

where the summation is over all the n data points. Use the last three equations to obtain

$$\Sigma z_i^2 = \mathbf{z}_i^T\mathbf{z}_i = (\mathbf{X}^*\mathbf{p}_i)^T(\mathbf{X}^*\mathbf{p}_i) = \mathbf{p}_i^T(\mathbf{X}^{*T}\mathbf{X}^*)\mathbf{p}_i = \lambda_i$$

Thus, *the eigenvalue λ_i quantifies the variation about the ith principal component axis.* A large value of λ_i means that there is substantial variation in the direction of this axis, whereas a value of λ_i near zero means that there is little variation along principal component i.

Moreover, it can be shown that the sum of the eigenvalues must equal k, the number of independent variables. If the independent variables are uncorrelated (statistically independent), so that they contain no redundant information, all the data will be spread equally about all the principal components, there will not be any multicollinearity, and all the eigenvalues will equal 1.0. In contrast, if there is a perfect multicollinearity, the associated eigenvalue will be 0. More likely, if there is serious (but not perfect) multicollinearity, one or more eigenvalues will be very small and there will be little variation in the standardized independent variables in the direction of the associated principal component axes. Because the eigenvalues have to add up to the number of independent variables, other eigenvalues will be larger than 1. Each small eigenvalue is associated with a serious multicollinearity. The associated eigenvector defines the relationships between the collinear variables (in standardized form).

There are no precise guidelines for what "small" eigenvalues are. One way to assess the relative magnitudes of the eigenvalues is to compute the *condition number:*

$$\phi = \frac{\lambda_{max}}{\lambda_{min}} = \frac{\lambda_1}{\lambda_k}$$

or, more generally, a *condition index* for each principal component:

$$\phi_i = \frac{\lambda_1}{\lambda_i}$$

Condition numbers above about 100 suggest serious multicollinearity.

Principal Components to Diagnose Multicollinearity on Mars

Armed with all this esoteric knowledge about principal components, we can go back and more formally assess the multicollinearity in our second set of Martian weight, height, and water consumption data (Fig. 5-1C). The two eigenvalues are $\lambda_1 = 1.9916$ and $\lambda_2 = .0084$. Thus, almost all of the variability in the original height and water consumption independent variables is conveyed by the first principal component. Specifically, this component accounts for $(1.9916/2.0) \times 100$ percent $= 99.6$ percent of the total variability in height and water consumption. The condition number $\lambda_1/\lambda_2 = 237$ exceeds the suggested cutoff value of 100. This large condition number provides additional evidence, from a different perspective than we had before, that there is a severe multicollinearity between height and water consumption in this example.

In Fig. 5-14H, the ellipse is much longer than it is wide (i.e., $\lambda_1 \gg \lambda_2$). That is, most information specified by the values of H and C is

```
Eigenanalysis of the Correlation Matrix

Eigenvalue    3.7441     0.2533     0.0026     0.0000
Proportion    0.936      0.063      0.001      0.000
Cumulative    0.936      0.999      1.000      1.000


Variable         PC1        PC2        PC3        PC4
Ped           -0.517      0.020     -0.104      0.850
Dr             0.482     -0.714     -0.438      0.256
Ped2          -0.516     -0.032     -0.754     -0.406
PedDr         -0.484     -0.699      0.479     -0.219
```

FIGURE 5-15 Principal components analysis of the correlation matrix of the ventricular interaction data obtained from a simple pulmonary artery constriction. Because there are four independent variables, there are four eigenvalues and eigenvectors. There is one very small eigenvalue, $\lambda_4 < .0001$. This small eigenvalue contrasts with the largest eigenvalue, $\lambda_1 = 3.7441$. Thus, these eigenvalues do not come close to the desired situation where each approximately equals 1; this reflects the serious multicollinearity in these data.

concentrated along the major axis of the ellipse (z_1), whereas almost no information about H and C is contained in the direction of the minor axis (z_2). This is a geometric way to say that H and C are highly correlated. Thus, in the context of this two-variable example, the problem of multicollinearity can be thought of as occurring when the joint distribution of two predictor variables is much longer than it is wide.

Principal Components and the Heart

When there are more than two independent variables, principal components can be used to evaluate multicollinearities in much the same way we used auxiliary regressions. We can look at the eigenvectors and eigenvalues of the correlation matrix of these data to get another perspective on the extent and nature of the multicollinearity (see Fig. 5-15). The advantage of using principal components is largely technical: They are easier to obtain using statistical software.

We saw above that when fitting the data from a pulmonary artery constriction to Eq. (5.3), the variance inflation factor for P_{ed} was 18,762 and that the auxiliary regression for P_{ed} indicated that all the independent variables in Eq. (5.3) were involved in the multicollinearity. Because there are $k = 4$ independent variables, the eigenvalues sum to 4. Instead of the desired situation where the eigenvalues are near 1, we see that $\lambda_1 = 3.7441$, and thus the first principal component accounts for almost 94 percent of the combined variability in the independent variables. In fact, the first 2 eigenvectors account for 99.9 percent of the variability in the independent variables. On the other extreme the smallest eigenvalue λ_4 equals .00004, and thus accounts for virtually none of the variability (1/1000 of 1 percent). The condition number for this eigenvalue, λ_1/λ_4, is 93,603, far above the threshold for concern of 100. The second smallest eigenvalue λ_3 is only .0026, and has a condition index of 1440. Very small eigenvalues (in this instance there are two) identify the important multicollinearities, and the independent variables involved in these multicollinearities can be identified by examining the associated eigenvectors.

For example, the eigenvector associated with λ_4 (Fig. 5-15) is

$$\mathbf{p}_4 = \begin{bmatrix} .850 \\ -.406 \\ .256 \\ -.219 \end{bmatrix}$$

Because of the small eigenvalue λ_4, we set this vector equal to zero and rewrite it in scalar form as a linear function of the standardized independent variables,

$$0 \approx .85P_{ed}^{\star} - .406P_{ed}^{2\star} + .256D_R^{\star} - .219P_{ed}^{\star}D_R^{\star}$$

The variables most involved in the associated multicollinearity are those which have large coefficients in this equation. Although there are no formal rules for what constitutes a large coefficient, most people consider any coefficient with an absolute value greater than about .1 to be "large." All the coefficients in this eigenvector are large by this definition, so this eigenvector identifies a four-variable multicollinearity involving all independent variables.

We can similarly evaluate $\mathbf{p}_3$, which also identifies a multicollinearity (because λ_3 is very small). In scalar form, we can write $\mathbf{p}_3$ as

$$0 \approx -.104P_{ed}^{\star} - .754P_{ed}^{2\star} - .438D_R^{\star} + .479P_{ed}^{\star}D_R^{\star}$$

If we consider the coefficient for $P_{ed}^{\star}$ to be small enough to ignore, this principal component identifies a three-variable multicollinearity involving all the independent variables except $P_{ed}^{\star}$.

These two components, $\mathbf{p}_3$ and $\mathbf{p}_4$, tell us slightly different stories about the nature of the multicollinearity. $\mathbf{p}_4$ indicates a multicollinearity involving all variables: it identifies both the sample-based portion of the multicollinearity (between $P_{ed}^{\star}$ and $D_R^{\star}$) as well as the structural multicollinearity between these two variables and $P_{ed}^{2\star}$ and $P_{ed}D_R^{\star}$. In contrast, $\mathbf{p}_3$ identifies only the structural aspects of the multicollinearity ($P_{ed}^{\star}$ does not play a big role in this three-variable multicollinearity).

Principal Components Regression

Using principal components regression to compensate for unavoidable multicollinearities follows directly from the use of principal components to diagnose multicollinearity. The idea is quite simple: The principal components contain the same information as the original variables. Therefore, we can rewrite the original regression equation [Eq. (5.12)] in terms of the new variables z_i, the principal components. These new variables have the benefit of being defined in a way so that they are *orthogonal* — mutually uncorrelated — so that there is no redundant information between them. If some of the principal components do not contain any information about the dependent variable (i.e., $\lambda_i = 0$ for component i), then they can simply be deleted from the regression equation without losing any information. Deleting these variables will, however, reduce the condition number of the correlation matrix and eliminate the multicollinearity. In fact, perfect multicollinearities rarely occur in practice, so we remove the principal components associated with small eigenvalues (i.e., $\lambda_i \approx 0$) on the grounds that they supply little unique information about the response variable. The method of least squares is then used to estimate the parameters in the transformed data space created by deleting some principal components of the original data. Finally, these parameter estimates are transformed back into the original coordinate system so that they can be interpreted in terms of the original variables.

To derive the equation for principal components regression, we simply rewrite the regression equation for the standardized variables in terms of the principal component variables.

The eigenvector matrix $\mathbf{P}$ represents the rotation of coordinates between the standardized variables and principal component variables. By convention each eigenvector $\mathbf{p}_i$ is normalized to have a length of 1. All matrices that represent rotations of coordinates, such as $\mathbf{P}$, have the property that $\mathbf{P}^T$ represents the so-called inverse rotation from the new coordinates back to the original coordinates. Thus, if we rotate one way, then reverse the change, we arrive back at the coordinates we started with. In matrix notation, we write this statement as $\mathbf{P}^T\mathbf{P} = \mathbf{P}\mathbf{P}^T = \mathbf{I}$ where $\mathbf{I}$ is the identity matrix. (Note that this result also means that $\mathbf{P}^T = \mathbf{P}^{-1}$.) Thus, from Eq. (5.14)

$$\hat{\mathbf{y}}^\star = \mathbf{X}^\star\mathbf{b}^\star = \mathbf{X}^\star\mathbf{I}\mathbf{b}^\star = \mathbf{X}^\star(\mathbf{P}\mathbf{P}^T)\mathbf{b}^\star = (\mathbf{X}^\star\mathbf{P})(\mathbf{P}^T\mathbf{b}^\star)$$

But, from Eq. (5.15), $\mathbf{X}^\star\mathbf{P} = \mathbf{Z}$, the independent variable data matrix in terms of the principal components. Let

$$\mathbf{a} = \mathbf{P}^T\mathbf{b}^\star = \begin{bmatrix} a_1 \\ a_2 \\ \cdot \\ \cdot \\ \cdot \\ a_k \end{bmatrix} \tag{5.16}$$

be the vector of regression coefficients in terms of the principal components. Thus, we can rewrite the regression equation as

$$\hat{\mathbf{y}}^\star = \mathbf{Za}$$

Or, in scalar form,

$$\hat{y}^\star = a_1 z_1 + a_2 z_2 + a_3 z_3 + \cdots + a_k z_k \tag{5.17}$$

Because $\mathbf{P}^T = \mathbf{P}^{-1}$, we can rewrite Eq. (5.16) as

$$\mathbf{a} = \mathbf{P}^T \mathbf{b}^\star = \mathbf{P}^{-1}\mathbf{b}^\star$$

so that

$$\mathbf{b}^\star = \mathbf{Pa} \tag{5.18}$$

Up to this point, we have not changed anything except the way in which we express the regression equation. The differences between Eqs. (5.11), (5.12), and (5.17) are purely cosmetic. You can conduct the analysis based on any of these equations and obtain *exactly the same results*, including any multicollinearities.

The thing that is different now is the fact that we have defined the new independent variables, the principal components, so that they have three desirable properties:

- They are orthogonal, so that there is no redundancy of information. As a result there is no multicollinearity between the principal components, and the sums of squares associated with each principal component in the regression are independent of the others.

- The eigenvalue associated with each principal component is a measure of how much additional independent information it adds to describing all the variability within the independent variables. The principal components are ordered according to descending information.

- The eigenvalues must add up to k, the number of independent variables. This fact permits us to compare the values of the eigenvalues associated with the different principal components to see if any of the components contribute nothing (in which case the associated eigenvalue equals 0) or very little (in which case the associated eigenvalue is very small) to the total variability within the independent variables.

We can take advantage of these properties to reduce the multicollinearity by simply deleting any principal components with small eigenvalues from the computation of the regression equation, thus eliminating the associated multicollinearities. This step is equivalent to setting the regression coefficients associated with these principal components to 0 in the vector of coefficients **a** to obtain the modified coefficient vector **a'** in which the last one or more coefficients is zero. This step removes the principal components from computation of the final regression equation. For example, if we delete the last two principal components from the regression equation, we obtain

$$\mathbf{a'} = \begin{bmatrix} a_1 \\ a_2 \\ \vdots \\ \vdots \\ a_{k-2} \\ 0 \\ 0 \end{bmatrix}$$

After setting these principal components to zero, we simply convert the resulting regression back to the original variables. Following Eq. (5.18),

$$\mathbf{b}^{*\prime} = \mathbf{Pa'}$$

and, from Eq. (5.13),

$$b'_i = \frac{s_y}{s_{x_i}} b_i^{*\prime}$$

The intercept follows from Eq. (5.11). Note that, while one or more principal components is deleted, these deletions do not translate into deletions of the original variables in the problem; in general, each component is a weighted sum of all of the original variables.

Once you have the regression equation back in terms of the original variables, you can also compute the residual sum of squares between the resulting regression equation and the observed values of the dependent variable. Most computer programs that compute principal components regressions show the results of these regressions as each principal component is entered. By examining the eigenvalues associated with each principal component and the associated reduction in the residual sum of squares, it becomes possible to decide how many principal components to include in the analysis. You delete the components associated with very small eigenvalues so long as this does not cause a large increase in the residual sum of squares.*

*There are only a few commercial computer programs to do principal components *regression*. Depending on the selection criteria, different components may enter into the final regression computations, and this could lead to the deletion of components associated with intermediate, or even large, eigenvalues. In terms of the justification for principal components regression, it only makes sense to delete components associated with the very smallest eigenvalues. We recommend that, if necessary, you force the program to enter all components sequentially. The resulting output lets you see how the regression coefficients change as each component is added, up to, and including, the regression with all components added, which is the ordinary least-squares solution.

More Principal Components on Mars

At the beginning of this chapter, we repeated our study of how weight depended on height and water consumption in a sample of Martians. We computed the regression equation and obtained coefficients much different than we found in our first study (back in Chap. 3). Furthermore, none of the regression coefficients was significant, even though R^2 was .70 and was highly significant ($P = .004$). As we went through our discussion of the effects of multicollinearity, we saw that the diagnostics we developed indicated there was a severe problem with multicollinearity. Because we did not analyze these data until we got back to earth, we are stuck with them. Therefore, we will use principal components regression to try to obtain more precise parameter estimates. The principal components regression output for these data (Fig. 5-1C) is in Fig. 5-16.

Because there are only two independent variables, there are only two eigenvectors and eigenvalues. As we saw above, the eigenvalues are $\lambda_1 = 1.9916$ and $\lambda_2 = .0084$ (remember that the sum of eigenvalues equals the number of independent variables). These eigenvalues result in a condition number of $\lambda_1/\lambda_2 = 237$, which is larger than our recommended cutoff value of 100. In fact, almost all of the information about height and water consumption is carried by the first principal component (or eigenvector); it accounts for 99.6 percent of the total variability in the independent variables. Thus, based on our discussion of principal components regression, it seems reasonable to delete z_2, the principal component associated with the small eigenvalue λ_2, which results in the regression equation

$$\hat{W} = -2.05 + .31 \text{ g/cm } H + .11 \text{ g/(cups/day) } C$$

The R^2 for this regression analysis was reduced by only .01 to .69, and all regression coefficients are very similar to what we obtained with our first analysis in Chap. 3 [Eq. (3.1)]. Thus, principal components regression has allowed us to obtain much more believable parameter estimates in the face of data containing a serious multicollinearity.

The Catch

On the face of it, principal components regression seems to be an ideal solution to the problem of multicollinearity. One need not bother with designing better sampling strategies or eliminating structural multicollinearities; just submit the data to a principal components regression analysis and let the computer take care of the rest.

The problem with this simplistic view is that there is no formal justification (beyond the case where an eigenvalue is exactly zero) for removing any of the principal components. By doing so, we impose an external constraint, or restriction, on the results of the regression anal-

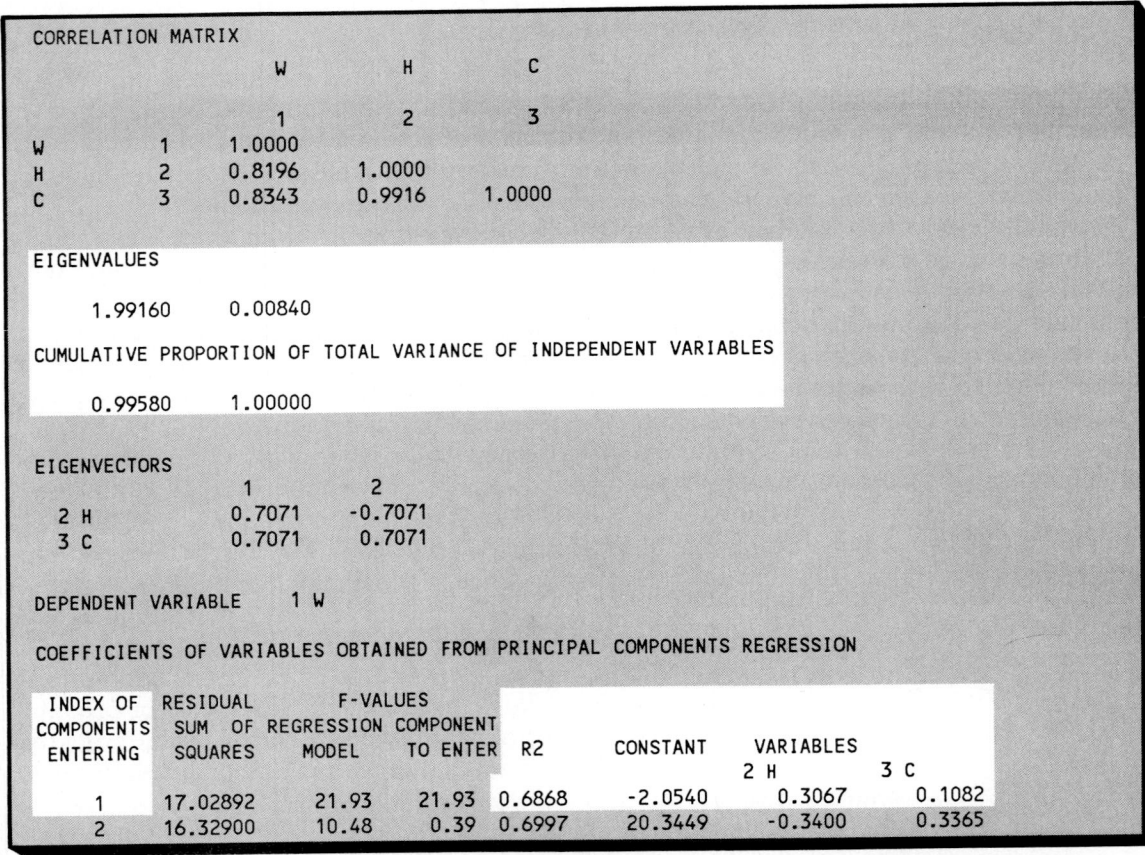

```
CORRELATION MATRIX
                     W          H          C

                     1          2          3
W          1      1.0000
H          2      0.8196     1.0000
C          3      0.8343     0.9916     1.0000


EIGENVALUES

    1.99160        0.00840

CUMULATIVE PROPORTION OF TOTAL VARIANCE OF INDEPENDENT VARIABLES

    0.99580        1.00000


EIGENVECTORS
                     1          2
   2 H             0.7071    -0.7071
   3 C             0.7071     0.7071


DEPENDENT VARIABLE      1 W

COEFFICIENTS OF VARIABLES OBTAINED FROM PRINCIPAL COMPONENTS REGRESSION
```

INDEX OF COMPONENTS ENTERING	RESIDUAL SUM OF SQUARES	F-VALUES REGRESSION MODEL	F-VALUES COMPONENT TO ENTER	R2	CONSTANT	VARIABLES 2 H	3 C
1	17.02892	21.93	21.93	0.6868	-2.0540	0.3067	0.1082
2	16.32900	10.48	0.39	0.6997	20.3449	-0.3400	0.3365

FIGURE 5-16 Principal components regression analysis of the data on Martian height, weight, and water consumption shown in Fig. 5-1C. The first principal component accounts for 99.58 percent of the total variance in the independent variables.

ysis in order to get around the harmful effects of the multicollinearities in the independent variables.* Although introducing the restrictions that

*There are other types of restricted least-squares procedures that are also useful in mitigating multicollinearity. See G. G. Judge, W. E. Griffiths, R. C. Hill, and T.-C. Lee, *Theory and Practice of Econometrics*, New York, Wiley, 1980, pp. 468–497, for an overview of many methods. A commonly used procedure is *ridge regression*. For a discussion of ridge regression, see A. E. Hoerl and R. W. Kennard, "Ridge Regression: Biased Estimation for Nonorthogonal Problems," *Technometrics* 12:55–67, 1970; D. W. Marquardt and R. D. Snee, "Ridge Regression in Practice," *Am. Stat.* 20:3–20, 1975; R. H. Myers, *Classical and Modern Regression Analysis with Applications*, Boston, Duxbury Press, 1986, pp. 243–262; G. Smith and F. Campbell, "A Critique of Some Ridge Regression Methods," *J. Am. Stat. Asoc.* 75:74–81, 1980 (many commentary articles follow this primary article).

some of the principal components are forced to be 0 reduces the variance of the parameter estimates, it also introduces an unknown *bias* in the estimates of the regression parameters. To the extent that the assumed restrictions are, in fact, true, this bias will be small or nonexistent.*

---

*In using a restricted regression procedure, one hopes to get a favorable trade-off where the introduction of bias is more than compensated for by a reduction in variance. To evaluate this trade-off, we can think about the mean squared error of the parameter estimates (MSE) which quantifies the two types of error in estimating a parameter: the random sampling uncertainty (the variance) and the bias (systematic error). Specifically, MSE = variance + bias^2. We know that the parameter estimates we obtain are unlikely to be exactly equal to the true (and unknown) parameter, and we are used to quantifying uncertainty only in terms of the standard error of the parameter estimates. The reason we concentrate on this is that the ordinary (unrestricted) least-squares estimator of a parameter is unbiased (so the bias = 0) when the regression model is correctly specified (see Chap. 6 for further discussion of bias and model specification). Thus, for the ordinary case, MSE = variance. In the face of multicollinearity, the variances are inflated and, thus, MSE is inflated. Placing restrictions on a regression solution always reduces the variance in the parameter estimates, but, unless the restrictions are true, there will be bias. What we desire with a restricted least-squares regression, then, is a reduction in variance which more than offsets the introduction of bias so that their sum, MSE = variance + bias^2, decreases. If the MSE decreases, the parameter estimates from the restricted regression are "better" than those obtained from ordinary least quares with multicollinearity.

One can gain some confidence that the specific sample data support the restrictions implied in principal components (and ridge) regression by formally testing to see if the MSE of the restricted regression procedure is less than the MSE of the ordinary least-squares procedure. These formal tests for MSE superiority require the construction of F tests based on the changes in sums of squares caused by the restrictions. For unbiased parameter estimation in ordinary least squares, we are used to using a central F distribution. The tests for restrictions in least squares require a noncentral F distribution. (A noncentral F distribution looks like a central F distribution, but the peak of the distribution is shifted to a higher value of F, with the amount of shift determined by the noncentrality parameter.) The appropriate F tables can be found in J. Goodnight and T. D. Wallace, "Operational Techniques and Tables for Making Weak MSE Tests for Restrictions in Regressions," *Econometrica* 40:699–709, 1972, and T. D. Wallace and C. E. Toro-Vizcarrondo, "Tables for the Mean Square Error Test for Exact Linear Restrictions in Regression," *J. Am. Stat. Assoc.* 64:1649–1663, 1969. For further general discussion of testing restrictions in least-squares regression, see R. C. Hill, T. B. Fomby, and S. R. Johnson, "Component Selection Norms for Principal Components Regression," *Commun. Stat.* A6:309–334, 1977; R. C. Mittelhammer, D. L. Young, D. Tasanasanta, and J. T. Donnelly, "Mitigating the Effects of Multicollinearity Using Exact and Stochastic Restrictions: The Case of an Aggregate Agricultural Function in Thailand," *Am. J. Agric. Econ.* 62:199–210, 1980.

However, the *restrictions implied when using a restricted regression procedure, such as principal components, to combat multicollinearity are usually arbitrary.* Therefore, the usefulness of the results of principal components analysis and other restricted regression procedures depends on a willingness to accept arbitrary restrictions placed on the data analysis. This arbitrariness lies at the root of most objections to such procedures.

Another problem is that restricted regression procedures make it difficult to assess statistical significance. The introduction of bias means that the sampling distribution of the regression parameters no longer follows the usual t distribution. Thus comparison of t statistics with a standard t distribution will usually be biased toward accepting parameter estimates as statistically significantly different from zero when they are not, but little is known about the size of the bias. An investigator can arbitrarily choose a more stringent significance level (e.g., $\alpha = 0.01$), or one can accept, as significant, parameter estimates with t statistics likely to have $P > 0.05$.

Recapitulation

In sum, principal components derived from the standardized variables provide a direct approach to diagnosing multicollinearity and identifying the variables that contain similar information. Sometimes you can use this information to design new experimental interventions to break up the multicollinearity or to identify structural multicollinearities that can be reduced by centering the independent variables. Often, however, you are simply stuck with the multicollinearity, either because of limitations on the sampling process (such as data from questionnaires or existing patterns of hospital use) or because you cannot exert sufficient experimental control. Sometimes the results of the principal components analysis will identify one or more independent variables that can be deleted to eliminate the multicollinearity without seriously damaging your ability to describe the relationship between the dependent variable and the remaining independent variables. Other times, deleting collinear independent variables cannot be done without compromising the goals of the study. In such situations, principal components regression provides a useful, if ad hoc, solution to mitigate the effects of serious multicollinearity, especially when ordinary least-squares regression produces parameter estimates so imprecise as to be meaningless. Principal components regression provides a powerful tool for analyzing such data. Like all powerful tools, however, it must be used with understanding and care after

solutions based on ordinary (unrestricted) least squares have been ex-
hausted.

SUMMARY

Part of doing a thorough analysis of a set of data is to check the multicollinearity diagnostics, beginning with the variance inflation factor.
When the variance inflation factor is greater than 10, the data warrant
further investigation to see whether or not multicollinearity among the
independent variables results in imprecise or misleading values of the
regression coefficients.

When multicollinearity is a problem, ordinary least-squares multiple
linear regression can give unreliable results for the regression coefficients
because of the redundant information in the independent variables. Thus,
when the desired result of the regression analysis is estimating the parameters of a model or identifying the structure of a model, ordinary
least-squares multiple linear regression is clearly inadequate in the face
of multicollinearity.

When the problem is structural multicollinearity caused by transforming one or more of the original predictor variables (e.g., polynomial
or cross-products), the multicollinearity can sometimes be effectively
dealt with by substituting a different transformation of predictor (or response) variables. Such structural multicollinearities may also be effectively dealt with by simply centering the native variables before forming
polynomial and interaction (cross-product) terms.

The best way to deal with sample-based multicollinearity is to avoid
it from the outset by designing the experiment to minimize it, or failing
that, to collect additional data over a broader region of the data space to
break up the harmful multicollinearities. To be forewarned is to be fore-
armed, so the understanding of multicollinearity we have tried to provide
should give the investigator some idea of what is needed to design better
experiments so that multicollinearity can be minimized. There may even
be occasions when one must use the results of an evaluation of multicollinearity (particularly the auxiliary regressions or principal components) to redesign an experiment.

In the event that analyzing data containing harmful multicollinearities is unavoidable, the simplest solution to the problem of multicollinearity is to delete one or more of the collinear variables. However, it may
not be possible to select a variable to delete objectively, and, particularly
if more than two predictor variables are involved in a multicollinearity,
it may be impossible to avoid the problem without deleting important
predictor variables. Thus this approach should be used only when the

goal of the regression analysis is predicting a response using data that have correlations among the predictor variables similar to those in the data originally used to obtain the regression parameter estimates.

When obtaining additional data, deleting or substituting predictor variables, or centering data are impossible or ineffective, other ad hoc methods such as principal components regression are available. However, these methods are controversial. The major objection is that the arbitrariness of the restrictions implied by the restricted estimation methods makes the resulting parameter estimates as useless as those obtained with ordinary least squares applied to collinear data. Nevertheless, there is a large literature supporting the careful application of these methods as useful aids to our quantitative understanding of systems that are difficult to manipulate experimentally. To assure careful use of restricted regression methods, one should choose definite guidelines for evaluating the severity of multicollinearity and a definite set of selection criteria (e.g., for the principal components to delete). One should consistently abide by these rules and criteria, even if the results still do not conform to prior expectations. It is inappropriate to hunt through data searching for the results that support some preconceived notation of the biological process under investigation.

Up to this point, we have been talking about fitting regression models where we were pretty sure we knew which variables should be in the model and we just wanted to get estimates of the parameters or predict new responses. However, sometimes we do not have a good idea which variables should be included in a regression model, and we want to use regression methods to help us decide what the important variables are. Under such circumstances, where we have many potential variables we want to screen, multicollinearity can be very important. In the next chapter we will develop ways to help us decide what a good model is and indicate where multicollinearity may cause problems in making those decisions.

PROBLEMS

5.1   *A.* Is multicollinearity a problem in the study of breast cancer and involuntary smoking and diet in Prob. 3.17, excluding the interaction? *B.* Including the interaction? The data are in Table D-2, Appendix D.

5.2   In Table D-6, Appendix D, T_1 and T_2 are the two independent variables for the multiple regression analysis of Prob. 3.12. Using only correlation analysis (no multiple regressions allowed), calculate the variance inflation factors for T_1 and T_2.

5.3 For the multiple regression analysis of Prob. 3.12, is there any evidence that multicollinearity is a problem?

5.4 In Chap. 3, we fit the data on heat exchange in gray seals in Fig. 2-11 (and Table C-1, Appendix C) using the quadratic function

$$\hat{C}_b = C_0 + ST_a + S_2T_a^2$$

This equation describes the data better than a simple linear regression but contains a potential structural multicollinearity due to the presence of both T_a and T_a^2. Is there a serious multicollinearity? Explain your answer.

5.5 In Prob. 3.8, you used multiple linear regression analysis to conclude that urinary calcium U_{Ca} was significantly related to dietary calcium D_{Ca} and urinary sodium U_{Na}, but not to dietary protein D_p or glomerular filtration rate G_{fr}. This conclusion assumes that there is no serious multicollinearity. Evaluate the multicollinearity in the set of four independent variables used in that analysis. (The data are in Table D-5, Appendix D.)

5.6 In Prob. 4.3, you analyzed the relationship between ammonium ion A and urinary anion gap U_{ag}. There was a suggestion of a quadratic relationship. A. Evaluate the extent of the structural multicollinearity between U_{ag} and U_{ag}^2 in this model. B. What effect does centering the data have in this analysis and why?

5.7 In Prob. 3.11, you used regression and dummy variables to see if two different antibiotic dosing schedules affected the relationship between efficacy and the effective levels of drug in the blood. A. Reanalyze these data, including in your analysis a complete regression diagnostic workup. B. If the linear model seems inappropriate, propose another model. C. If multicollinearity is a problem, do an evaluation of the severity of multicollinearity. If necessary to counter the effects of multicollinearity, reestimate the parameters, using an appropriate method.

5.8 Here are the eigenvalues of a correlation matrix of a set of independent variables: $\lambda_1 = 4.231$, $\lambda_2 = 1.279$, $\lambda_3 = .395$, $\lambda_4 = .084$, $\lambda_5 = .009$, and $\lambda_6 = .002$. A. How many independent variables are there? B. Calculate the condition index of this matrix. C. Is there a problem with multicollinearity?

5.9 Compare and interpret the eigenvalues and condition indices of the correlation matrix of the native and centered independent variables in the regression model analyzed in Prob. 5.7.

5.10 Use principal components regression to obtain parameter estimates for the regression model analyzed in Prob. 5.7. A. What is the condition index of the correlation matrix? B. How many principal components are you justified in omitting? Report the regression equation. *C.* Is there a better way to deal with the multicollinearity in these data?

5.11 Evaluate the extent of multicollinearity when using a quadratic equation to fit the data of Table D-1, Appendix D, relating sedation S to blood cortisol level C during Valium administration. Reanalyze the data, if necessary, using an appropriate method. If you reanalyze the data, discuss any changes in your conclusions.

5.12 Draw graphs illustrating the effect of centering on the relationship between C and C^2 for the data in Table D-1, Appendix D.

Selecting the "Best" Regression Model

Our discussion of regression analysis to this point has been based on the premise that we have correctly identified all the relevant independent variables. Given these independent variables, we have concentrated on investigating whether it was necessary to transform these variables or to consider interaction terms, evaluate data points for undue influence (in Chap. 4), or resolve ambiguities arising out of the fact that some of the variables contained redundant information (in Chap. 5). It turns out that, in addition to such analyses of data using a predefined model, multiple regression analysis can be used as a tool to screen large numbers of potential independent variables to select that subset of them which make up the "best" regression model. As a general principle, we wish to identify the simplest model with the smallest number of independent variables that will describe the data adequately.

Procedures known as *all possible subsets regression* and *stepwise regression* permit you to use the data to guide you in the formulation of the regression model. *All possible subsets regression* involves forming all possible combinations of the potential independent variables, computing the regressions associated with them, and then selecting the one with the best characteristics. *Stepwise regression* is a procedure for sequentially entering independent variables one at a time in a regression equation in the order that most improves the regression equation's predictive ability or removing them when doing so does not significantly degrade its predictive ability. These methods are particularly useful for screening data sets in which there are many independent variables in order to identify a smaller subset of variables that determine the value of a dependent variable.

Although these methods can be very helpful, there are two important caveats: First, the list of candidate independent variables to be screened must include all the variables that actually predict the dependent variable; and, second, there is no single criterion that will always be the best measure of the "best" regression equation. In short, these methods, although very powerful, require considerable thought and care to produce meaningful results.

SO WHAT DO YOU DO?

The problem of specifying the correct regression model has several elements, only some of which can be addressed directly via statistical calculations. The first, and most important, element follows not from statistical calculations but from knowledge of the substantive topic under study: *You need to carefully consider what you know — from both theory and experience — about the system under study to select the potentially important variables for study.* Once you have selected these variables, statistical methods can help you decide whether the model is adequately specified, whether you have left out important variables, or whether you have included extraneous or redundant variables. Completing this analysis requires a combination of studying the *residual diagnostics* for an adequate functional form of the regression equation (discussed in Chap. 4), studying the *multicollinearity diagnostics* for evidence of model overspecification (discussed in Chap. 5), and using the *variable selection procedures* (discussed in this chapter).

WHAT HAPPENS WHEN THE REGRESSION EQUATION CONTAINS THE WRONG VARIABLES?

Recall that we began our discussion of regression analysis by assuming that the value of the dependent variable y at any given values of the k independent variables $x_1, x_2, x_3, \ldots, x_k$ is

$$y = \beta_0 + \beta_1 x_1 + \beta_2 x_2 + \beta_3 x_3 + \cdots + \beta_k x_k + \epsilon$$

We estimated the parameters β_i in this equation by computing the regression coefficients b_i in the regression equation

$$\hat{y} = b_0 + b_1 x_1 + b_2 x_2 + b_3 x_3 + \cdots + b_k x_k$$

that minimized the sum of squared deviations between the observed and predicted values of the dependent variables. So far we have tacitly assumed that these equations contain all relevant independent variables necessary to describe the observed value of the dependent variable accurately at any combination of values of the independent variables. Although the values of the regression coefficients estimated from different random samples drawn from the population will generally differ from the true value of the regression parameter, the estimated values will cluster around the true value and the mean of all possible values of b_i will be β_i. In more formal terms, we say that b_i (the statistic) is an *un-*

biased estimator of the β_i (the population parameter) when the expected value of b_i (the mean of the values of b_i computed from all possible random samples from the population) equals β_i. In addition, because we compute the regression coefficients to minimize the sum of squared deviations between the observed and predicted values, it is a *minimum variance estimate*, which means that $s_{y|x}$, the standard error of the estimate, is an unbiased estimator of the underlying population variability about the plane of means $\sigma_{y|x}$.*

But what happens when the independent variables included in the regression equation are not the ones that actually determine the dependent variable? To answer this question, we first need to consider the four possible relationships between the set of independent variables in the regression equation and the set of variables that actually determine the dependent variable:

- The regression equation contains all relevant independent variables, including necessary interaction terms and transformations, and no redundant or extraneous variables, in which case the model is *correctly specified.*
- The regression equation is missing one or more important independent variables, in which case the model is *underspecified.*
- The regression equation contains two or more *extraneous variables* that are not related to the dependent variables or other independent variables.
- The regression equation contains one or more redundant independent variables, in which case the model is *overspecified.*

These different possibilities have different effects on the results of the regression analysis.

When the model is *correctly specified*, it contains all independent variables necessary to predict the mean value of the dependent variable and the regression coefficients are unbiased estimates of the parameters in the model. As a result, the predicted value of the dependent variable, at any specified location defined by the independent variables, is an unbiased estimator of the true value of the dependent variable on the plane of means defined by the regression equation. This is the situation we have taken for granted up to this point.

When the regression model is *underspecified*, it lacks important independent variables, and the computed values of the regression coefficients and the standard error of the estimate are no longer unbiased estimators of the corresponding parameters in the true equation that

---

*For a mathematical derivation of these results, see R. H. Myers, *Classical and Modern Regression with Applications*, Boston, Duxbury Press, 1986, pp. 102, 111–114.

determines the dependent variable. This bias appears in three ways. First, the values of the regression coefficients in the model will be distorted as the computations try to compensate for the missing information contained in the missing independent variables. Second, as a direct consequence of the first effect, the predicted values of the dependent variable are also biased. There is no way to know the magnitude or even the direction of these biases. Third, and finally, leaving out important independent variables means that the estimate of the standard error of the estimate from the regression analysis $s_{y|x}$ will be biased upward by an amount that depends on how much information the ignored independent variables contain* and thus will tend to overestimate the true variation around the regression plane $\sigma_{y|x}$.

To illustrate the effects of model underspecification, let us return to our first study of Martian height, weight, and water consumption in Chap. 1. Weight depends on both height and water consumption, so the fully specified regression model is

$$\hat{W} = b_0 + b_H H + b_C C \tag{6.1}$$

Based on the data we collected, we concluded that the relationship between Martian weight and height and water consumption is [from Eq. (1.2)]

$$\hat{W} = -1.2 \text{ g} + .28 \text{ g/cm } H + .11 \text{ g/cup } C$$

where $\hat{W}$ is the predicted weight at any observed height H and water consumption C. Recall that we initially failed to take into account the fact that weight depended on water consumption as well as height. Thus, we underspecified the regression model by omitting the important independent variable water consumption and used

$$\hat{W} = b_0 + b_H H \tag{6.2}$$

When we simply regressed weight on height, we obtained the regression equation [Eq. (1.1)]

$$\hat{W} = -4.1 \text{ g} + .39 \text{ g/cm } H$$

Figure 6-1 reproduces the data and regression lines for these two equations from Fig. 1-2. The more negative intercept and more positive coefficient associated with H in this underspecified model reflect the biases associated with underspecification. Note also that the standard error of the estimate increased from .13 g in the correctly specified model to

---

*See R. H. Myers, *Classical and Modern Regression with Applications*, Boston, Duxbury Press, 1986, pp. 102–103 and 334–335 for a mathematical derivation of the bias in $s_{y|x}$.

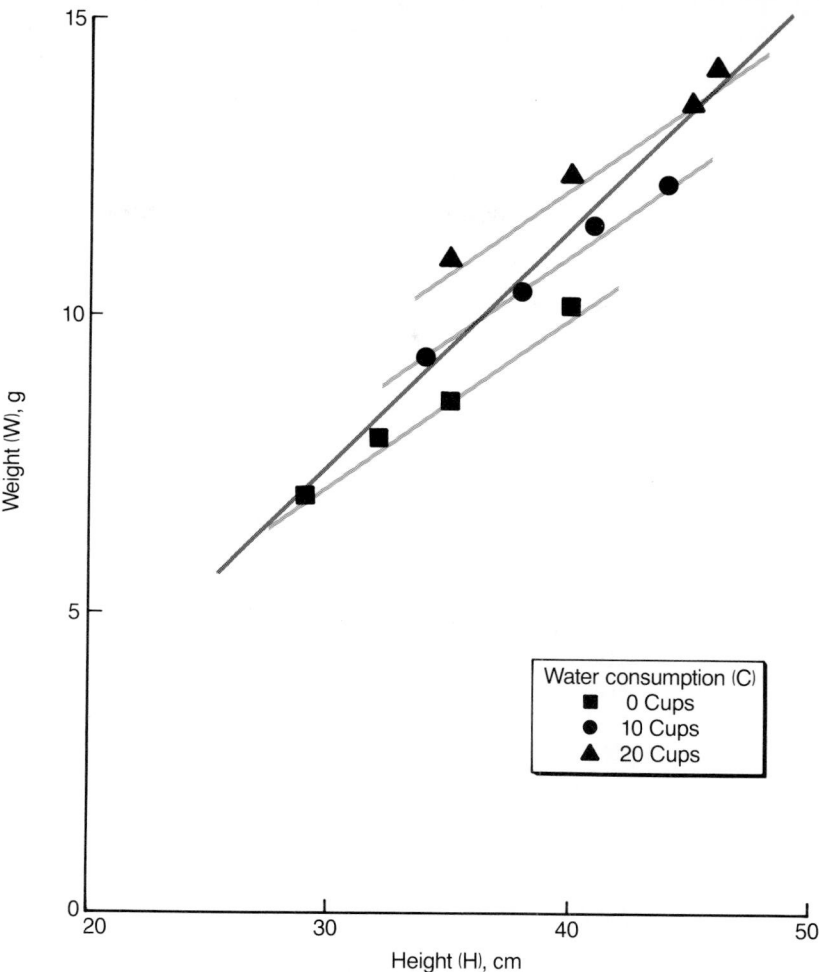

FIGURE 6-1 Data on Martian height, weight, and water consumption from Fig. 1-2. The dark line shows the results of regressing Martian weight on height, ignoring the fact that water consumption is an important determinant of Martian weight [Eq. (6.2)], whereas the light lines show the result of the full regression model, including both height and water consumption [Eq. (6.1)]. Leaving the effect of water comsumption out of the regression model leads to an overestimation of the effect of height, i.e., the slope of the line computed using only height as an independent variable is steeper than the slopes of the three parallel lines that result when we include the effect of water consumption, as well as height, on weight. Leaving the effect of water consumption out of the regression model introduces a *bias* in the estimates of the effects of height. Bias is also introduced into the intercept term of the regression equation. These effects are due to the fact that the regression model is trying to compensate for the missing information concerning the effects of water consumption on height. Although it is possible to prove that there will be biases introduced in the estimates of the coefficient associated with height when water consumption is left out of the model, there is no way *a priori* to know the exact nature of these biases.

.81 g in the underspecified model that omitted the effects of water consumption.

Likewise, had we omitted height from our regression model and treated weight as though it just depended on water consumption, we would have used the regression model

$$\hat{W} = b_0 + b_C C \tag{6.3}$$

Fitting the data in Table 1-1 to this regression equation yields

$$\hat{W} = 8.56 \text{ g} + .22 \text{ g/cup } C$$

in which the two regression coefficients also differ substantially from those obtained from the complete model. The estimated effect of drinking canal water on weight is twice what we estimated from the complete model. In addition, the standard error of the estimate is increased to 1.29 g, nearly 10 times the value obtained from the complete model. Table 6-1 summarizes in more detail the results of analyzing our data on Martian weight using these three different regression models. Note that all three regressions yield statistically significant values for the regression coefficients and reasonable values of R^2, but that the values of the regression coefficients (and the conclusions one would draw from them) are very different for the two underspecified models than for the full model.

Given these problems associated with model underspecification, why not simply include every possible independent variable, then simply concentrate on the ones associated with coefficients that are statistically significantly different from zero?

We could use this strategy if we were sure that any extra variables we added were *extraneous*, i.e., unrelated to both the dependent variable and the other independent variables. We would lose little by including extra variables beyond the cost of collecting the data and a few degrees of freedom in the hypothesis tests associated with having extra variables in the equation. The extraneous variables will (within sampling variation) be associated with regression coefficients that are not significantly different from zero, and the regression computations will lead us to conclude that they are extraneous. Indeed, this situation prevailed in the study of β-blockers to control hypertension in diabetics we presented in Chap. 3. Because these extraneous variables contribute nothing to predicting the dependent variable, they do not affect the other regression coefficients or the residual variance.

Unfortunately, situations in which variables are totally extraneous are rare in practice. Because it costs effort and money to collect and process data, people generally include independent variables for which there is some reason based on experience or theory to think that they may affect the dependent variable. Thus, the more likely outcome

TABLE 6-1 Comparison of Regression Models Relating Martian Weight to Height, Water Consumption, or Both

Regression Model $\hat{W} =$	$s_{y\|x}$	Intercept		Height		Water Consumption		R^2
		b_0	s_{b_0}	b_H	s_{b_H}	b_C	s_{b_C}	
$b_0 + b_H H + b_C C$ (6.1)	.131	−1.22	.32	.28	.01	.11	.01	.997
$b_0 + b_H H$ (6.2)	.808	−4.14	1.75	.39	.05	—	—	.880
$b_0 + b_C c$ (6.3)	1.285	8.56	.59	—	—	.22	.05	.696

of adding too many independent variables is that some of the variables will be redundant, in which case the model will be *overspecified* and we will have the problems associated with multicollinearity discussed in Chap. 5. The regression coefficients and the standard error of the estimate are unbiased, so that the resulting equation can be used to describe the data, as well as for interpolation and, with due caution, prediction or extrapolation. However, the uncertainty associated with the regression coefficients (quantified with the standard errors) is inflated, and, thus, one cannot draw meaningful conclusions about the values of the regression coefficients. The problems associated with model overspecification are important because multiple regression analysis is often used to identify important variables and quantify their effects on some dependent variable. Hence, the uncertainty introduced by overspecification undermines the entire goal of doing the analysis in the first place.

WHAT DOES "BEST" MEAN?

In essence, the problem of selecting the best regression model boils down to finding the middle ground between an underspecified model, which yields biased estimates of the regression parameters and a high residual variance, and an overspecified model, which yields unbiased but imprecise estimates of the regression parameters. For statistical computations, such as variable selection procedures, to guide us in this process, we need to quantitatively define how good a regression model is. There are several such criteria, all of which are reasonable, that one could use to identify the "best" model from a collection of candidate models. However, because these different criteria quantify different aspects of the regression problem, they can lead to different choices for the best set of variables. You often have to consider several of these criteria simultaneously and, combined with your knowledge of the subject under consideration, select the most sensible model.

We will now discuss these different measures of "bestness" and see how to use them as part of formal variable selection procedures. These criteria include two familiar statistics — the coefficient of determination R^2 and the standard error of the estimate $s_{y|x}$, which measure how closely the regression equation agrees with the observations — and three new statistics: the *adjusted R^2*; the *PRESS statistic*, which estimates how well the regression model will predict new data; and C_p, which estimates the combined effects of residual variation plus the effects of bias due to model misspecification.

The Coefficient of Determination R^2

The coefficient of determination R^2 is a nondimensional measure of how well a regression model describes a set of data: R^2 equals 0 when taking the information in the independent variables into account does not improve your ability to predict the dependent variable at all, and R^2 equals 1 when it is possible to predict perfectly the dependent variable from the information in the independent variables. Recall (from Chap. 3) that because

$$R^2 = \frac{SS_{\text{reg}}}{SS_{\text{tot}}} = 1 - \frac{SS_{\text{res}}}{SS_{\text{tot}}} \tag{6.4}$$

people say that R^2 is the fraction of the variance in the dependent variable that the regression model "explains." Thus, one possible criterion is to find the model that maximizes R^2.

The problem with this criterion is that R^2 always increases as more variables are added to the regression equation, even if these new variables add little new independent information. Indeed, it is possible — even common — to add additional variables that produce modest increases in R^2 along with serious multicollinearities. Therefore, R^2 by itself is not a very good criterion for the best regression model.

However, the *increase* in R^2 that occurs when a variable is added to the regression model, denoted ΔR^2, can provide useful insight into how much additional predictive information one gains by adding the independent variable in question to the regression model. Frequently, ΔR^2 will be used in combination with other criteria to determine when adding additional variables does not improve the model's descriptive ability enough to warrant problems that come with it, such as an increase in multicollinearity.

The Adjusted R^2

As we just noted, R^2 is not a particularly good index of bestness because including another independent variable (thus, removing a degree of freedom) will always decrease SS_{res} and, thus, will always increase R^2, even if that variable adds little or no useful information about the dependent variable. To take this possibility into account, an adjusted R^2, denoted R^2_{adj}, has been proposed. Although similar to R^2, and interpreted in much the same way, R^2_{adj} makes you pay a penalty for adding additional variables. The penalty is in the form of taking into account the loss of degrees of freedom when additional variables are added. R^2_{adj} is computed by substituting MS_{res} and MS_{tot} for SS_{res} and SS_{tot} in Eq. (6.4):

$$R^2_{adj} = 1 - \frac{MS_{res}}{MS_{tot}} = 1 - \frac{SS_{res}/(n - k - 1)}{SS_{tot}/(n - 1)}$$

where k is the number of independent variables in the regression equation. Adding another independent variable increases k and reduces S_{res}, but this reduction must more than compensate for the lost degree of freedom (because $n - k - 1$ falls) for R^2_{adj} to increase. In fact, because R^2_{adj} is an approximately unbiased estimator of the true population R^2, it is possible for R^2_{adj} to be *negative*. For example, if the true population R^2 were 0, we would expect an unbiased estimator of this true R^2 to have an expected value of 0 and that the estimates would cluster about zero and be equally likely to be above or below zero. Thus, an unbiased estimator of the population R^2 would be expected to occasionally take on a value less than zero; the usual R^2 defined by Eq. (6.4) cannot do this. [*] According to this criterion, the best regression model is the one with the largest R^2_{adj}.

The Standard Error of the Estimate $s_{y|x}$

The standard error of the estimate is our measure of the variation of the members of the underlying population about the plane of means. It follows directly from the residual sum of squares that one minimizes when estimating the regression coefficients. Its square, known as the *residual variance*, or *residual mean square*, plays a crucial role in all hypothesis testing related to linear regression. From Eq. (3.15),

$$s^2_{y|x} = \frac{SS_{res}}{n - k - 1} = MS_{res}$$

---

[*] For further discussion, see J. Neter, W. Wasserman, and M. H. Kutner, *Applied Linear Regression Models*, Homewood, Ill., Irwin, 1983, pp. 423–424.

where n is the number of observations and k is the number of independent variables in the regression equation. As we just noted when discussing R^2_{adj}, although adding more independent variables reduces SS_{res} (unless the new variables are totally extraneous) and so increases R^2, it is possible for the residual variance to increase as unnecessary variables are added because the denominator in the equation for $s^2_{y|x}$ falls faster than the numerator. Thus, one could take the model with the smallest standard error of the estimate as the best model.

Because $s^2_{y|x}$ equals MS_{res}, which is the numerator for R^2_{adj}, the model with the highest R^2_{adj} will have the lowest $s_{y|x}$. There is no question that such a model will have the smallest prediction errors when used to predict or interpolate within the data. There is no guarantee, however, that it will not be overspecified and contain serious multicollinearities that will make it difficult to use the values of the regression coefficients to draw conclusions about the system under study. As with R^2 and R^2_{adj}, sometimes we will be as interested in how large a reduction in $s_{y|x}$ is associated with adding a variable into the regression model as we will be in the absolute value of the standard error of the estimate.

Independent Validations of the Model with New Data

We generally do research and statistical analysis to reach general conclusions that will apply beyond the specific set of observations used to reach the conclusions. Therefore, it makes sense to think of the best regression model as the one that best decribes the relationship between independent and dependent variables for a *set of data other than that used to compute the regression coefficients.*

Such tests against independent observations are particularly revealing because they avoid the biases inherent in using the same set of data to both compute the regression coefficients and test it. The least-squares algorithm for estimating the regression coefficients (developed in Chap. 3) produces the best estimate of the regression model *for the specific data used to compute the regression coefficients.* Because the regression coefficients are optimized for one specific set of data, the fit to any other set of data will generally not be as good. A good regression model with reliably estimated parameters will, however, yield a reasonably good description of a new set of data.[*]

---

[*]For further discussion of the desirability of independent model validation, see R. H. Myers, *Classical and Modern Regression with Applications*, Boston, Duxbury Press, 1986, pp. 103–111; and C. Cobelli, E. R. Carson, L. Finkelstein, and M. S. Leaning, "Validation of Simple and Complex Models in Physiology and Medicine," *Am. J. Physiol.* 246:R259–R266, 1984.

Ideally, one should collect more than one set of data (or, equivalently, divide the available data into two different sets), one to estimate the coefficients in one or more regression equations and another set to test the equations. Having estimated the coefficients with the first set, we would then compute the sum of squared residuals for the second data set using the prediction equation and coefficients obtained independently from the first data set. The model with the smallest residuals for predicting the new data would be deemed the best. Having selected the best form of the model, one could then go back and reestimate the parameters in the model using *all* the data to obtain estimates based on a larger sample size and hence with greater precision.

The principal problem with this approach is that it requires obtaining a lot of extra data. In many research projects the data are expensive to collect, or other constraints (such as the difficulty of recruiting people into clinical studies) limit the total sample size, so people are reluctant to split the data set. Moreover, for this approach to be feasible, one generally needs a relatively large data set to achieve a reasonable number of degrees of freedom for the parameter estimates and hypothesis tests associated with interpreting the regression equation. Meeting these requirements is often not practical. When it is possible to divide the data into fitting and test subsets, it greatly improves the confidence one can have in the results.*

The Predicted Residual Error Sum of Squares, PRESS

Although it is often not possible to test a regression model against an independent data set, we can approximate an independent test of a model using a procedure similar to that which we used to develop the Cook's distance statistic in Chap. 4. We used Cook's distance to quantify the influence of each data point on the estimated values of the regression coefficients by deleting that data point, then estimating the regression coefficients without it. The Cook's distance statistic is simply a normalized measure of how different the resulting regression coefficients were from the values computed from all the data. Using similar logic we can derive a measure of how well a given regression model predicts data other than that used to fit it.

Suppose that we simply remove the first data point from the data set before computing the regression coefficients. We can then use this single point as a test data set and see how well the regression equation, esti-

---

*For an example of where it was possible to follow this procedure, see B. K. Slinker and S. A. Glantz, "Beat-to-Beat Regulation of Left Ventricular Function in the Intact Cardiovascular System," *Am. J. Physiol.* 256:R962–R975, 1989.

mated without it, predicts its value. Specifically, let $\hat{y}_{-1}$ be the predicted value of the dependent variable for the first data point using the regression equation computed from the rest of the data. The observed value of the dependent variable for the first point is Y_i, so the *prediction error* or *PRESS residual* is

$$e_{-1} = Y_1 - \hat{y}_{-1}$$

Now, repeat this process, withholding the second data point from the computation of the regression coefficients. The prediction error will be

$$e_{-2} = Y_2 - \hat{y}_{-2}$$

In general, if we remove the ith data point from the data set, estimate the regression coefficients from the remaining data, and then predict the value of the dependent variable for this ith point, the prediction error will be

$$e_{-i} = Y_i - \hat{y}_{-i}$$

The accumulation of these single prediction errors forms a sort of poor person's set of independent test data for the ability of the regression model to predict values of data points not used to estimate the coefficients in the regression equation.[*]

Finally, we compute the *PRESS statistic* as an overall measure of how well the candidate regression model predicts new data:

$$\text{PRESS} = \Sigma(Y_i - \hat{y}_{-i})^2 = \Sigma\, e_{-i}^2$$

We can compare several different candidate regression models by comparing the associated values of the PRESS statistic. The model with the smallest value could be judged best on the basis of predictive ability.

Bias Due to Model Underspecification and C_p

In our discussion of the Martian data above, we have already seen that model underspecification introduces biases in the parameter estimates and predicted values of the dependent variable because of the omission of information necessary to predict the dependent variable. We now develop a statistic C_p, which estimates the magnitude of the biases intro-

---

[*]The specific values of the regression coefficients will be different for each PRESS residual, but if none of the data points is exerting undue influence on the results of the computations, there should not be too much difference from point to point. As a result, the PRESS residuals can also be used as diagnostics to locate influential points; points with large values of PRESS residuals are potentially influential. In fact, the PRESS residual associated with the ith point is directly related to the raw residual e_i and leverage h_{ii} defined in Chap. 4, according to $e_{-i} = e_i/(1 - h_{ii})$.

duced in the estimates of the dependent variable by leaving variables out of the regression model. C_p is particularly useful for screening large numbers of regressions obtained in all possible subsets regression.

To motivate C_p we need to further investigate what it means for a regression equation to be unbiased or biased. To do so we need to recall that the *expected value* of a statistic is the value one would obtain by averaging all possible values of the statistic that can be obtained from a population. For example, the expected value of a variable x, denoted $E(x)$, is just the mean of all values of x in the population,

$$E(x) = \mu_x$$

Note that the expected value of a statistic is a characteristic of the population rather than a specific sample. *We say that a statistic is an unbiased estimator of a population parameter when its expected value equals the population parameter.* For example, the sample mean $\overline{X}$ is an unbiased estimator of the population mean μ_x because

$$E(\overline{X}) = \mu_x$$

In a population described by a linear regression, the mean value of all possible values of the dependent variable at any given set of values of the independent variables is a point on the regression plane,

$$E(y_i) = \mu_{y|x_i}$$

and the predicted value of the dependent variable $\hat{y}_i$ is an unbiased estimator of the plane of means because (Fig. 6-2)

$$E(\hat{y}_i) = \mu_{y|x_i}$$

Furthermore, the standard deviation of the distribution of all possible predicted values of the dependent variable will be $\sigma_{\hat{y}_i}$.

If there is no bias, both expected values $E(y_i)$ and $E(\hat{y}_i)$ equal $\mu_{y|x_i}$, so

$$E(\hat{y}_i) - E(y_i) = 0$$

If the difference between the expected value of the predicted value of the dependent variable and the expected value of the dependent variable is not zero, there is a bias in the prediction obtained from the regression model.

Now, suppose that we fit the data with an underspecified regression model, which will result in the introduction of bias into the predicted value of the dependent variable at the ith point. This bias will be

$$B_i = E(\hat{y}_i) - E(y_i)$$

where $\hat{y}_i$ refers to the predicted value of the ith data point using the underspecified regression model. Because of the bias, the variance in the

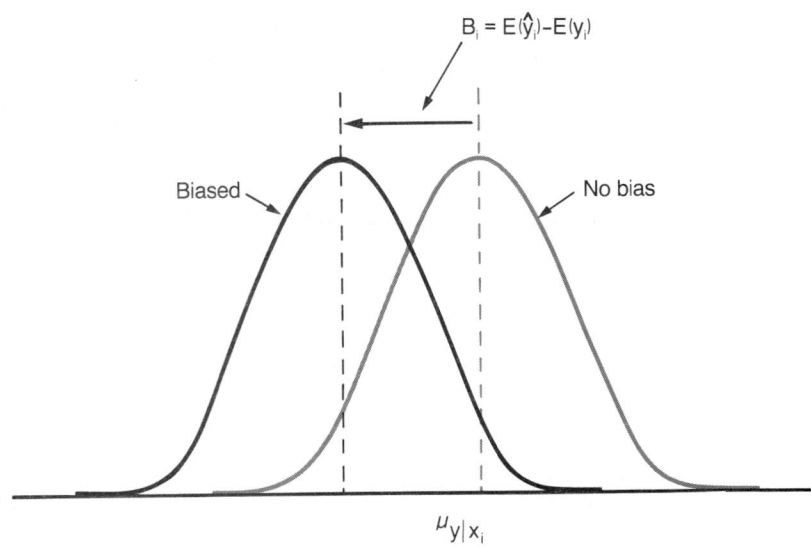

FIGURE 6-2 Bias in the estimates of the regression model parameters arises when important terms are left out of the regression equation. Accordingly, there will be bias in the values of the dependent variable estimated from the regression equation. When the model is correctly specified, the distribution of estimates of the dependent variable at any given values of the independent variables will be normally distributed with a mean equal to the true mean value of the population of values of the dependent variable at the given combination of the independent variables. In contrast, when one or more important determinants of the dependent variable is left out of the regression model, the distribution of predicted values will no longer have a mean equal to the true population mean of the values of the dependent variable at those values of the independent variables.

predicted values at this point is no longer simply that due to random sampling variation, $\sigma_{\hat{y}_i}^2$. We also have to take into account the variance associated with this bias B_i^2. To obtain a measure of the total variation associated with the n data points, we sum these two variance components over all the n data points and divide by the variation about the plane of means $\sigma_{y|x}^2$:

$$\Gamma_p = \frac{1}{\sigma_{y|x}^2} \{\Sigma \sigma_{\hat{y}_i}^2 + \Sigma[E(\hat{y}_i) - E(y_i)]^2\}$$

where the subscript p denotes the number of parameters (including the intercept) in the regression model. (Because p includes the intercept, there are $p - 1 = k$ independent variables in the regression equation.) Moreover, if there is no bias, all the terms in the second summation in this equation are zero and Γ_p achieves its smallest value. It can be shown[*] that this smallest value is

$$\Gamma_p = \frac{1}{\sigma_{y|x}^2} \left(\Sigma \sigma_{\hat{y}_i}^2 + 0 \right) = p$$

Thus, Γ_p would seem to be a good measure of the bias in an underspecified model. The best model is simply the one with the smallest value of Γ_p. We even know the theoretical minimum value: p.

If one knows the true variance about the plane of means $\sigma_{y|x}^2$, one can estimate Γ_p with

$$C_p = p + \frac{(s_p^2 - \sigma_{y|x}^2)(n - p)}{\sigma_{y|x}^2}$$

where s_p^2 is the standard error of the estimate associated with fitting the reduced model containing only p parameters ($p - 1$ independent variables plus the intercept) to the data. Thus, one simply screens candidate regression models, selecting the one in which C_p is closest to p because this model has the lowest bias.

The problem, of course, is that you do not know $\sigma_{y|x}^2$ because you do not know the true population parameters. In fact, the goal of the entire analysis is to use C_p to help find the best regression model. People deal with this problem by simply using the standard error of the estimate from the candidate regression model that includes all k of the independent variables under consideration, s_{k+1} as an estimate of $\sigma_{y|x}$. *This practice assumes that there are no biases in the full model,* an assumption which may or may not be valid, but which cannot be tested without additional information. It also ensures that the value of C_p for the full model (which includes all k independent variables and for which $p = k + 1$) will equal $k + 1$. In the event that all the models (except the full model) are associated with large values of C_p, you should suspect that some important independent variable has been left out of the analysis.

As a practical matter, this assumption does not severely restrict the use of C_p because, when one designs a regression model, one usually takes care to see that all relevant variables are included in order to obtain unbiased estimates. In other words, one would not knowingly leave out relevant information. The problem, of course, comes when you unknowingly leave out relevant independent variables. Neither C_p nor any other diagnostic statistic can warn you that you forgot something important.

---

[*]For a proof of this result, together with a more rigorous derivation of C_p, see R. H. Myers, *Classical and Modern Regression with Applications*, Boston, Duxbury Press, 1986, pp. 114–116.

Only knowledge of the substantive topic at hand and careful thinking can help reduce that possibility.

Because of the assumption that the full model contains all relevant variables, you should not use C_p as a strict measure of how good a candidate regression model is and blindly take the model with C_p closest to p. Rather, you should use C_p to screen models and select the few with C_p values near p — which indicates small biases under the assumption that all relevant variables are included in the full model — to identify a few models for further investigation. In general, one seeks the simplest model (usually the one with the fewest independent variables) that produces an acceptable C_p because models with fewer parameters are often simpler to interpret and use and are less prone to multicollinearity than are models with more variables. In fact, one of the primary practical uses of C_p is to screen large numbers of possible models when there are many potential independent variables in order to identify the important variables for further study.

But What Is "Best"?

The six measures of how good a regression model is — R^2, R^2_{adj}, $s_{y|x}$, prediction of new data, PRESS, and C_p — all measure related, but different, aspects of the regression model.[*]

The coefficient of determination R^2 is a standardized measure of how much of the variance in the dependent variable is "explained" by the independent variables in the regression model. Clearly, large values of R^2 indicate better agreement between the model and data, but R^2 will always increase as more variables are added to the model, so it does not represent a simple criterion to be maximized or minimized. The increases in R^2 as variables are added to the model (or reduction in R^2 as they are deleted) can provide insight into just how much additional information each variable contains, taking into account the information in the other variables. The standard error of the estimate $s_{y|x}$ directly measures how accurately the regression equation predicts the dependent variable. Unlike R^2, $s_{y|x}$ can reach a minimum before all the variables have been added to the equation. The similar R^2_{adj} depends on $s_{y|x}$, but moves in the opposite direction; as the prediction error goes down, R^2_{adj} goes up. R^2_{adj} is better than R^2 in the sense that it takes into account the number of independent variables and makes you pay a penalty if the reduction in degrees of freedom outpaces the reduction in the residual

---

[*]For a discussion of the specific interrelations between these (and other) measures of how "good" a regression equation is, see R. R. Hocking, "The Analysis and Selection of Variables in Linear Regression," *Biometrics* 32:1–49, 1976.

sum of squares. If the sole purpose of the regression analysis is predicting new observations or interpolating within existing data, minimizing the standard error of the estimate (or, equivalently, maximizing R^2_{adj}) is a good way to find the best regression equation.

Testing the model's ability to predict new data or using the related PRESS statistic are also good measures of the model's predictive ability and sensitivity to the specific data set used to estimate the regression coefficients. Selecting the model that will minimize one of these quantities will increase your confidence about the generalizability of the results.

Finally, C_p provides a way for selecting a complete enough subset of candidate independent variables to avoid bias in the estimates of the regression coefficients or predictions of the model. Models with values of C_p near p will increase your confidence that the results are unbiased — so long as you have not left any important variables out of the list of potential independent variables.

All these "bestness" criteria measure qualities that are desirable in a good regression model. Occasionally, a single regression equation will produce the optimum values of all these measures at once, and then you can be confident that it is the best regression model in terms of these criteria. Unfortunately, this happy situation rarely arises, and these different diagnostics identify different models. In addition, in applied science, people are often seeking insight into the system under study and would like to draw conclusions from the values associated with the regression coefficients. None of these measures of how good a regression model is provide any protection against model overspecification and the associated multicollinearity-induced problems with the regression coefficients. Nevertheless, taken together, these criteria do identify a small subset of all the different regression models that could be constructed from the independent variables you have selected. You can then study these few models in greater detail, taking into account the goals of the study, other knowledge of the system under study, and the methods discussed in Chaps. 4 and 5, to make an intelligent judgment as to what the best model is for the problem at hand.

SELECTING VARIABLES WITH ALL POSSIBLE SUBSETS REGRESSION

The most straightforward — and many would say best — approach to finding which regression model is best, given a set of candidate independent variables, is to compute the regressions associated with all the models obtained by all possible combinations of the candidate independent variables. For example, if there are three possible independent variables, one would fit the data with $2^3 = 8$ different regression models:

$$\hat{y} = a_0$$

$$\hat{y} = b_0 + b_1 x_1$$

$$\hat{y} = c_0 + c_2 x_2$$

$$\hat{y} = d_0 + d_3 x_3$$

$$\hat{y} = e_0 + e_1 x_1 + e_2 x_2$$

$$\hat{y} = f_0 + f_1 x_1 + f_3 x_3$$

$$\hat{y} = g_0 + g_2 x_2 + g_3 x_3$$

$$\hat{y} = h_0 + h_1 x_1 + h_2 x_2 + h_3 x_3$$

in which $a_i, b_i, \ldots, h_i$ are the coefficients in the different regression models. The different models can then be compared using the criteria just discussed and the "best" model selected.

The obvious problem associated with this approach is that the number of possible equations grows quickly with the number of independent variables. There are 2^k different possible regression models with k candidate independent variables. For example, if there are 15 candidate independent variables, there are $2^{15} = 32{,}768$ possible subset regression models. Fortunately, doing these computations is only a matter of waiting for the computer to finish the dirty work. To avoid being buried in computer printout, the programs that do such computations take whatever criterion you specify (or whatever is built into the program, generally C_p) and find the best few regression equations with each number of independent variables.* The programs generally report the results of most or all of the diagnostics discussed above for the few best equations, so you can use this information to determine which sets of independent variables warrant further investigation, particularly for tests of multi-collinearity.

What Determines an Athlete's Time in a Triathlon?

The triathlon, which is an endurance event that combines running, bicycling, and swimming, is one of the most demanding athletic events. Because these different types of exercise use different muscle groups and require different kinds of training, and because different people are intrinsically more or less well suited for each type of exercise than others, it would be of interest to determine which factors predict overall performance in the triathlon.

*In fact, the programs do not have to compute all 2^k equations. Criteria such as C_p make it possible to identify combinations of independent variables that will have large values of C_p without actually computing all the regressions.

Data that could be used to answer this question were reported by Kohrt and her colleagues,[*] who studied men training for a half-triathlon (1.2-mile swim, 56-mile bike ride, and 13.1-mile run) over a 6-week period. They observed the athletes' age, weight, and number of years of triathlon experience and measured the number of kilometers per week that they ran, biked, or swam as they prepared to compete. They also measured their maximum oxygen consumption $\dot{V}O_{2\ max}$ (in mL/min/kg) while running on a treadmill, pedaling on a fixed cycle, and swimming on a tether as a measure of these athletes' exercise capacity. They also measured the total time required to complete the actual half-triathlon. Thus, they have measurements of triathlon performance and measurements of the athletes' gross physical characteristics, training (specific amounts for each event, plus total triathlon experience), and exercise capacity as potential predictors of performance (the data are in Table C-13, Appendix C). The questions we wish to ask are: What are the important determinants of the athletes' final time when they competed in the triathlon, and what can we conclude about the role these different variables are playing? Are the changes primarily due to who the athletes are (how old they are, how much they weigh, their amount of experience); how much they train (perhaps also reflected in their total triathlon experience); or their physiological makeup, as reflected in maximum oxygen consumption?

We will address this question by submitting the nine potential independent variables to an all possible subsets analysis, to see which ones produce the best regression equation with the total time (in minutes) as the dependent variable. The full regression equation is

$$\hat{t} = b_0 + b_A A + b_W W + b_E E + b_{T_R} T_R + b_{T_B} T_B + b_{T_S} T_S + b_{V_R} V_R + b_{V_B} V_B + b_{V_S} V_S$$

in which the dependent variable t is the time for the half-triathlon (in minutes) and the nine candidate independent variables are

A = age, years

W = weight, kg

E = experience running triathlons, years

T_R = running training, km/week

T_B = bicycling training, km/week

T_S = swimming training, km/week

[*]W. M. Kohrt, D. W. Morgan, B. Bates, and J. S. Skinner, "Physiological Responses of Triathletes to Maximal Swimming, Cycling, and Running," *Med. Sci. Sports Exerc.* 19:51–55, 1987.

V_R = maximum oxygen consumption while running, mL/min/kg

V_B = maximum oxygen consumption while biking, mL/min/kg

V_S = maximum oxygen consumption while swimming, mL/min/kg

There are $2^9 = 512$ different possible combinations of these nine independent variables.

Figure 6-3 shows the results of the all possible subsets regression analysis. Even though this program ranks the models (we asked the program to list the four best models of each size) using the R^2 criterion, we want to focus on C_p. All the models with fewer than five independent variables in the regression equation have large values of C_p, indicating considerable underspecification bias, which means that important independent variables are not included in the analysis.

One model with five variables ($p = 5 + 1 = 6$) has a value of $C_p = 5.3 \approx 6 = p$. This model contains the independent variables age A, experience E, running training T_R, bike training T_B, and oxygen consumption while running V_R. It has a coefficient of determination of $R^2 = .793$ and a standard error of the estimate of 22.6 min (compared with an average total time for the triathlon of 343 min). This model appears promising. It reasonably predicts the dependent variable, and C_p does not detect any serious biases compared with the full model.

We further investigate this model by conducting a multiple linear regression to obtain the regression coefficients and the associated variance inflation factors. Figure 6-4 shows the results of this analysis. The first thing to note is that all the variance inflation factors are small (between 1.3 and 3.4), so there does not appear to be a problem with multicollinearity to complicate the interpretation of the regression coefficients. The resulting regression equation is

$$\hat{t} = 516 \text{ min} + 3.5 \text{ min/year } A - 20.8 \text{ min/year } E + .796 \text{ min/km/week } T_R \quad (6.5)$$
$$- .24 \text{ min/km/week) } T_B - 4.09 \text{ min per mL/min/kg } V_R$$

All the coefficients, except perhaps b_{T_R}, have the expected sign. Total time in the triathlon increases as the participant is older and decreases with experience, amount of bicycle training, and maximum oxygen consumption during running. The one surprising result is that the total time seems to increase with amount of running done during training.

In general, we would expect that performance would increase (time would decrease) as the amount of training increases, so the coefficients related to training should be negative. In general, the regression results support this expectation. The single largest impact on performance is experience, with total time decreasing almost 21 min for every year of experience. (This may be a cumulative training effect, but one needs to be careful in interpreting it as such, because it is possible that it relates

Best Subsets Regression of T

Vars	R-sq	Adj. R-sq	C-p	s	A	W	E	T r	T b	T s	V r	V b	V s
1	48.3	47.5	84.4	34.482						X			
1	41.8	40.9	102.6	36.578							X		
1	35.6	34.5	120.2	38.497								X	
1	24.2	23.0	152.2	41.760				X					
2	63.8	62.6	42.8	29.081				X		X			
2	58.8	57.4	57.0	31.050						X	X		
2	58.2	56.8	58.7	31.273			X			X			
2	57.4	56.0	60.8	31.551		X				X			
3	71.3	69.9	23.8	26.117	X	X				X			
3	71.2	69.7	24.1	26.177		X			X	X			
3	69.3	67.8	29.2	26.988		X				X	X		
3	68.2	66.6	32.5	27.496		X				X		X	
4	76.2	74.6	12.0	23.987	X	X		X		X			
4	75.1	73.5	14.9	24.500	X	X				X	X		
4	74.6	72.9	16.4	24.766		X				X	X	X	
4	72.7	70.9	21.8	25.674		X		X	X	X			
5	79.3	77.5	5.3	22.573	X		X	X	X		X		
5	76.9	75.0	11.9	23.801	X	X					X	X	X
5	76.8	74.8	12.4	23.893	X	X				X	X	X	
5	76.4	74.4	13.4	24.078	X	X		X			X	X	
6	79.5	77.4	6.7	22.631	X			X	X	X	X	X	
6	79.5	77.3	6.8	22.651	X			X	X	X	X	X	
6	79.3	77.2	7.1	22.714	X			X	X	X	X		X
6	79.3	77.1	7.3	22.765	X	X	X	X	X		X		
7	80.1	77.6	7.0	22.506	X			X	X	X	X	X	X
7	79.7	77.2	8.1	22.721	X		X	X	X		X	X	X
7	79.6	77.0	8.5	22.796	X			X	X	X	X	X	X
7	79.5	77.0	8.6	22.815	X	X	X	X	X		X	X	
8	80.4	77.6	8.2	22.535	X		X	X	X	X	X	X	X
8	80.1	77.3	8.9	22.687	X	X	X	X	X	X	X	X	
8	79.7	76.8	10.1	22.913	X	X	X	X	X		X	X	X
8	79.6	76.7	10.3	22.964	X	X	X	X	X	X	X		X
9	80.4	77.2	10.0	22.699	X	X	X	X	X	X	X	X	X

FIGURE 6-3 Results of all possible subsets regression for the analysis of the data on triathlon performance. The four best models in terms of R^2 are displayed for each number of independent variables. The simplest model, in terms of fewest independent variables, that yields a value of C_p near $p = k + 1$, where k is the number of independent variables in the equation, is the first model containing the five variables, age, experience, running training time, bicycling training time, and running oxygen consumption. For the full model with 9 variables (the last line in the printout) $C_p = p$. This situation will always arise because the computation assumes that the full model specified contains all relevant variables.

```
The regression equation is
T = 516 + 3.53 A - 20.8 E + 0.796 Tr - 0.242 Tb - 4.09 Vr

Predictor      Coef      Stdev     t-ratio       p        VIF
Constant      516.10     54.51       9.47      0.000
A              3.5335     0.8188      4.32      0.000      1.9
E            -20.752      3.141      -6.61      0.000      1.9
Tr             0.7958     0.2689      2.96      0.004      3.4
Tb            -0.24185    0.05154    -4.69      0.000      2.9
Vr            -4.0886     0.5490     -7.45      0.000      1.3

s = 22.57      R-sq = 79.3%      R-sq(adj) = 77.5%
```

Figure 6-4 Details of the regression analysis of the model identified for predicting triathlon performance using the all possible subsets regression in Fig. 6-3. The model gives a good prediction of the triathlon time with an $R^2 = .793$ and a standard error of 22.6 min. All of the independent variables contribute significantly to predicting the triathlon time, and the variance inflation factors are all small, indicating no problem with multicollinearity.

in part to psychological factors unrelated to training.) In addition, T_B, the amount of training on the bicycle, also supports our expectation that increased training decreases time. However, T_S, the amount of swimming training, does not appear to affect performance, and T_R, the amount of running training actually had the opposite effect—performance went down as the amount of running training went up. Because of the low variance inflation factors associated with these variables, we do not have any reason to believe that a multicollinearity is affecting these results. Thus, we have a difficult result to explain.

Of the physiological variables related to oxygen consumption, only V_R, the maximal oxygen consumption while running, is significantly related to overall performance. As the intrinsic ability to exercise when running increases, triathlon time decreases. This makes sense. However, neither of the other oxygen consumptions for biking and swimming seems to be related to performance. Age, which might be loosely considered a physiological variable, negatively influences performance. The older triathletes had higher times.

If we take these results at face value, it appears that certain types of training—specifically, biking—are more important for achieving good overall performance than the others. On the other hand, it looks as though the maximum oxygen consumption attained while running is the major physiological influence: neither of the other oxygen consumption measurements appears in our regression model. It appears that the best triathletes are those who are intrinsically good runners, but who emphasize training on the bike. In addition, everything else being equal, experience counts a lot.

Figure 6-3 also shows that there are several other models, all involving six or more independent variables, which appear reasonable in terms of C_p, R^2, and the standard error of the estimate. Despite the fact that these other regression models contain more variables than the one with five independent variables, none of them performs substantially better in terms of prediction, measured by either R^2 or the standard error of the estimate. The highest R^2 (for the model containing all nine independent variables) is .804, only .011 above that associated with the model we have selected, and the smallest standard error of the estimate (associated with the first model with seven independent variables) is 22.506 min, only .067 min (4 s) smaller than the value associated with the five-variable model. Thus, using the principle that the fewer independent variables the better, the five-variable model appears to be the best.

To double-check, let us examine the regression using the full model. The results of this computation, presented in Fig. 6-5, reveal that the four variables not included in the five-variable model do not contribute significantly to predicting the total time needed to complete the triathlon. Because there is no evidence of multicollinearity in the full model, we can be confident in dropping these four variables from the final model.

```
The regression equation is
T = 486 + 3.41 A + 0.347 W - 21.4 E + 0.702 Tr - 0.173 Tb - 1.37 Ts - 3.36 Vr
    - 1.38 Vb + 0.893 Vs
```

Predictor	Coef	Stdev	t-ratio	p	VIF
Constant	486.3	114.5	4.25	0.000	
A	3.410	1.091	3.13	0.003	3.4
W	0.3470	0.7862	0.44	0.661	2.2
E	-21.424	3.697	-5.80	0.000	2.6
Tr	0.7025	0.2771	2.54	0.014	3.6
Tb	-0.17251	0.06920	-2.49	0.016	5.2
Ts	-1.3727	0.9566	-1.43	0.157	3.4
Vr	-3.3550	0.8338	-4.02	0.000	2.9
Vb	-1.3845	0.9098	-1.52	0.134	3.2
Vs	0.8934	0.9217	0.97	0.337	3.2

```
s = 22.70      R-sq = 80.4%      R-sq(adj) = 77.2%
```

FIGURE 6-5 To further investigate the value of the model used to predict triathlon performance identified in Figs. 6-3 and 6-4, we analyzed the full model containing all the potential independent variables. Adding the remaining four variables only increased the R^2 from .793 to .804 and actually increased the standard error of the estimate from 22.57 to 22.70 min. In addition, none of the variables not included in the analysis reported in Fig. 6-4 contributed significantly to predicting the athlete's performance in the triathlon.

SEQUENTIAL VARIABLE SELECTION TECHNIQUES

Although thorough in exhausting all the possible models, all possible subsets regression can produce models with serious multicollinearity because none of the criteria used to identify the best regression model take into account the effects of multicollinearity. An alternative approach which often avoids producing a regression model with serious multicollinearity among the independent variables involves the sequential selection of candidate dependent variables, based on how much independent information each one contains about the dependent variable, allowing for the information contained in the variables already in the regression equation. These *sequential variable selection techniques* are algorithms for systematically building a regression equation by adding or removing one variable at a time, based on the incremental change in R^2 (or, equivalently, SS_{res}). There are three related strategies for sequential variable selection: *forward selection, backward elimination,* and *stepwise regression.*

These techniques are particularly good for screening large numbers of possible independent variables, and they generally produce regression models without serious multicollinearity.[*] The problem with sequential variable selection procedures is that one can obtain different regression models by changing the inclusion and exclusion criteria. Moreover, these three different searching algorithms can yield three different regression models. Nevertheless, these methods are extremely useful, particularly when combined with all possible subsets regression.

Forward Selection

The forward selection algorithm begins with no independent variables in the regression equation. Each candidate independent variable is tested to see how much it would reduce the residual sum of squares if it were included in the equation, and the one that causes the greatest reduction is added to the regression equation. Next, each remaining independent variable is tested to see if its inclusion in the equation will significantly reduce the residual sum of squares further, given the other variables already in the equation. At each step, the variable that produces the largest incremental reduction in the residual sum of squares is added to

---

[*]This is not to say that a sequential variable technique will *never* yield multicollinear results. It is possible to set the entry and exclusion criteria so that multicollinearity will result; however, one normally does not do this. More important, the presence of serious multicollinearities in the set of candidate independent variables affects the order of entry (or removal) of candidate variables from the equation and can lead to differences in the final set of variables chosen, depending on both the algorithm used and small changes in the data.

the equation, with the steps repeated until none of the remaining candidate independent variables significantly reduces the residual sum of squares.

We base our formal criterion for whether or not an additional variable significantly reduces the residual sum of squares on the F test for the *incremental sum of squares* that we introduced in Chap. 3. Recall that the incremental sums of squares is the reduction in the residual sum of squares that occurs when a variable is added to the regression equation after taking into account the effects of the variables already in the equation. Specifically, let $SS_{x_j|x_1,\ldots,x_{j-1}}$ be the reduction in the residual sum of squares obtained by fitting the regression equation containing independent variables $x_1, x_2, \ldots, x_j$ compared to the residual sum of squares obtained by fitting the regression equation to the data with the independent variables $x_1, x_2, \ldots, x_{j-1}$. In other words, $SS_{x_j|x_1,\ldots,x_{j-1}}$ is the reduction in the residual sum of squares associated with adding the new independent variable x_j to the regression model. *At each step we add the independent variable to the equation that maximizes the incremental sum of squares.*

We test whether this incremental reduction in the residual sum of squares is large enough to be considered significant by computing the F statistic [after Eq. (3.17)]:

$$F = \frac{\text{mean square associated with adding } x_j \text{ given that } x_1, \ldots, x_{j-1} \text{ are already in the regression equation}}{\text{mean square residual for equation containing } x_1, \ldots, x_j}$$

There is 1 degree of freedom associated with $SS_{x_j|x_1,\ldots,x_{j-1}}$, so $MS_{x_j|x_1,\ldots,x_{j-1}} = SS_{x_j|x_1,\ldots,x_{j-1}}/1$, and we can compute

$$F = \frac{MS_{x_j|x_1,\ldots,x_{j-1}}}{MS_{\text{res}|x_1,\ldots,x_j}} \tag{6.6}$$

Most forward selection algorithms denote this ratio F_{enter} for each variable not in the equation and select the variable with the largest F_{enter} for addition to the equation on the next step. This criterion is equivalent to including the variable with the largest incremental sum of squares because it maximizes the numerator and minimizes the denominator in Eq. (6.6). New variables are added to the equation until F_{enter} falls below some predetermined cutoff value, denoted F_{in}.

Rather than doing a formal hypothesis test on F_{enter} using the associated degrees of freedom, F_{in} is usually specified as a fixed cutoff value. Formal hypothesis tests are misleading when conducting sequential variable searches because there are so many possible tests to be conducted at each step. The larger the value of F_{in}, the fewer variables will enter the equation. Most people use values of F_{in} around 4 because this value approximates the critical values of the F distribution for $\alpha = .05$, when the number of denominator degrees of freedom (which is approximately equal to the sample size when there are many more data points than independent variables in the regression equation) is large (see Table E-2, Appendix E). If you do not specify F_{in}, the computer program you use will use a default value. These default values vary from program to program and, as just noted, will affect the results. It is important to know (and report) what value is used for F_{in}, whether you explicitly specify it or use the program's default value.

The appeal of the forward selection algorithm is that it begins with a "clean slate" when selecting variables for the regression equation. It also begins by identifying the "most important" (in the sense of best predictor) independent variable. There are, however, some difficulties with forward selection. First, because the denominator in the equation for F_{enter} in Eq. (6.6) only includes the effects of variables already included in the candidate equation (not the effects of variables $x_{j+1}, \ldots, x_k$) on step j the estimated residual mean square will be larger for the earlier steps than later when more variables are taken into account. This computation may make the selection process less sensitive to the effects of additional independent variables than desirable and, as a result, the forward inclusion process may stop too soon. In addition, once a variable is in the equation, it may become redundant with some combination of new variables that are entered on subsequent steps.

Backward Elimination

An alternative to forward selection which addresses these concerns is the *backward elimination* procedure. With this algorithm the regression model begins with all the variables in the equation. The variable that accounts for the smallest decrease in residual variance is removed if doing so will not significantly increase the residual variance. This action is equivalent to removing the variable that causes the smallest reduction in R^2. Next, the remaining variables in the equation are tested to see if any additional variables can be removed without significantly increasing the residual variance. At each step, the variable that produces the smallest increase in residual variance is then removed, then the process repeated, until there are no variables remaining in the equation that could be removed without significantly increasing the residual variance. In other words, backward elimination proceeds like forward selection, but backwards.

Like forward selection, we use the incremental sum of squares $SS_{x_j|x_1,\ldots,x_{j-1}}$, and the associated mean square to compute the F ratio in Eq. (6.6). In contrast with forward selection where the incremental sum of squares was a measure of the decrease in the residual sum of squares associated with entering a variable, in backward elimination the incremental sum of squares is a measure of the *increase* in the residual sum of squares associated with *removing* the variable. As a result, the associated F statistic is now denoted F_{remove} and is a measure of the effect of removing the variable in question on the overall quality of the regression model. At each step, we identify the variable with the smallest F_{remove} and remove it from the regression equation until there are no variables left with values of F_{remove} below some specified cutoff value, denoted F_{out}.

The same comments we just made about F_{in} apply to F_{out}. The smaller the value of F_{out}, the more variables will remain in the final equation. Be sure you know what value is being used by the program if you do not specify F_{out} explicitly.

The major benefit of backward elimination is that, because it starts with the smallest possible mean square residual in the denominator of Eq. (6.6), it avoids the problem where forward selection occasionally stops too soon. However, this small denominator of Eq. (6.6) leads to a major problem of backward elimination: it may stop too soon and result in a model that has extraneous variables. In addition, a variable may be deleted as unnecessary because it is collinear with two or more of the other variables, some of which are deleted on subsequent steps, leaving the model underspecified.

Stepwise Regression

Because of the more or less complementary strengths and weaknesses of forward selection and backward elimination, these two methods have been combined into a procedure called *stepwise regression*. The regression equation begins with no independent variables; then variables are entered as in forward selection (the value with the largest F_{enter} is added). After each variable is added, the algorithm does a backward elimination on the set of variables included to that point to see if any variable, in combination with the other variables in the equation, makes one (or more) of the independent variables already in the equation, redundant. If redundant variables are detected (F_{remove} below F_{out}), they are deleted and then the algorithm begins stepping forward again. The process stops when no variables not in the equation would significantly reduce the mean square residual (no variable not in the equation has a value of $F_{\text{enter}} \geq F_{\text{in}}$) and no variables in the equation could be removed without significantly increasing the mean square residual (i.e., no variable in the equation has a value of $F_{\text{remove}} \leq F_{\text{out}}$).

The primary benefit of stepwise regression is that it combines the best features of both forward selection and backward elimination. The first variable is the strongest predictor of the dependent variable, and variables can be removed from the equation if they become redundant with the combined information of several variables entered on subsequent steps. In fact, it is not unusual for variables to be added, then deleted, then added again, as the algorithm searches for the final combination of independent variables. However, because stepwise regression is basically a modified forward selection it may stop too soon, just like forward selection.

As with forward selection and backward elimination, the behavior of the algorithm will depend on the values of F_{in} and F_{out} you specify.* We generally set $F_{in} = F_{out}$, but some people prefer to make F_{out} slightly smaller than F_{in} (say, 3.9 and 4.0, respectively) to introduce a small bias to keep a variable in the equation once is has been added.[†]

Because variables can be removed as well as entered more than once, it is possible for a stepwise regression to take many steps before it finally stops. (This situation contrasts with the forward and backward methods, which can only take as many steps as there are candidate independent variables.) Many programs have a default limit to the number of steps the algorithm will take before it stops. If you encounter this limit, the model may not be complete. In fact, when the stepwise regression takes many, many steps it is a good indication that something is wrong with the selection of candidate independent variables.

Of the three techniques, stepwise regression is the most popular.[‡]

Interpreting the Results of Sequential Variable Selection

People tend to overinterpret the results of sequential variable selection procedures, in particular the order of entry of variables into the regression equation. Although it is reasonable to think of the first variable as the "most important," in the sense of it being the one with the highest simple correlation with the dependent variable, one needs to be very careful in drawing any conclusions about the order of entry (or associated increase in the coefficient of determination ΔR^2) associated with any variable, particularly when there are multicollinearities in the full data set. The changes in R^2 as additional variables are added can, however, provide additional useful information about how many variables are needed in the final model. It is not uncommon for sequential variable

---

*In fact, a stepwise regression can be made to behave like a forward inclusion by setting $F_{out} = 0$ or a backward elimination by setting F_{in} to a very large number and forcing all variables into the equation at the start.

[†]Some computer programs require that F_{out} be less than F_{in}.

[‡]A similar method, which is somewhat of a hybrid between stepwise selection and all possible subsets regression, is called MAXR. See R. H. Myers, *Classical and Modern Regression with Applications*, Boston, Duxbury, 1986, pp. 122–126.

selection methods to include variables that have detectable, but very small, contributions to predicting the dependent variable. These effects may be small enough to ignore. Such judgments depend on knowledge of the substantive problem at hand; they are not statistical issues.

The one exception to this caveat is when you *force* the order of entry by specifying it based on prior knowledge of the system. Then the incremental sum of squares and ΔR^2 do provide some insight into the independent contribution of each factor in the context of the other independent variables already in the regression equation.*

Another Look at the Triathlon

Let us now reanalyze the data we discussed earlier in this chapter on the determinants of total time for completing a triathlon. Figures 6-6 and 6-7 present the results of submitting the data on total time for running

STEPWISE REGRESSION OF	T	ON	9 PREDICTORS, WITH N =	65	
STEP	1	2	3	4	5
CONSTANT	687.9	709.7	704.1	532.8	516.1
Vr	-5.68	-5.20	-4.82	-3.96	-4.09
T-RATIO	-7.67	-8.24	-8.37	-6.81	-7.45
Tb		-0.203	-0.187	-0.128	-0.242
T-RATIO		-5.15	-5.24	-3.51	-4.69
E			-10.7	-16.9	-20.8
T-RATIO			-3.94	-5.56	-6.61
A				3.03	3.53
T-RATIO				3.56	4.32
Tr					0.80
T-RATIO					2.96
S	34.5	29.1	26.2	24.0	22.6
R-SQ	48.31	63.82	71.15	76.17	79.25

FIGURE 6-6 Sequential variable selection analysis of the triathlon performance data using the forward selection method. This procedure yields the same five-variable regression model as we obtained using the all possible subset regressions analysis reported in Fig. 6-3. In this example, stepwise regression yields an identical printout.

*We will use precisely this procedure in Chaps. 7, 8, and 9 when we use multiple regression models to handle missing data or incomplete designs in analysis of variance.

```
STEPWISE REGRESSION OF     T    ON  9 PREDICTORS, WITH N =    65
```

	STEP	1	2	3	4	5
	CONSTANT	486.3	522.5	565.4	539.0	516.1
A		3.41	3.41	2.88	3.32	3.53
	T-RATIO	3.13	3.15	3.13	3.86	4.32
W		0.35				
	T-RATIO	0.44				
E		-21.4	-20.8	-19.8	-20.3	-20.8
	T-RATIO	-5.80	-6.13	-6.18	-6.38	-6.61
Tr		0.70	0.71	0.73	0.80	0.80
	T-RATIO	2.54	2.57	2.68	2.96	2.96
Tb		-0.173	-0.181	-0.185	-0.238	-0.242
	T-RATIO	-2.49	-2.74	-2.82	-4.58	-4.69
Ts		-1.37	-1.19	-1.08		
	T-RATIO	-1.43	-1.39	-1.28		
Vr		-3.36	-3.53	-3.44	-3.74	-4.09
	T-RATIO	-4.02	-4.82	-4.75	-5.40	-7.45
Vb		-1.38	-1.38	-1.15	-0.65	
	T-RATIO	-1.52	-1.52	-1.32	-0.83	
Vs		0.89	0.84			
	T-RATIO	0.97	0.92			
S		22.7	22.5	22.5	22.6	22.6
R-SQ		80.44	80.37	80.08	79.50	79.25

FIGURE 6-7 Sequential variable selection analysis of the triathlon performance data using the backward elimination method. Backward elimination yields the same model as forward selection and the all possible subsets regression. The concordance of all these methods increases the confidence we have in this regression model.

the triathlon to stepwise, forward selection, and backward elimination variable selection procedures. All three procedures produced the same regression model we selected using the all possible regressions procedure (compare with Fig. 6-3). The fact that all these procedures produced the same results, combined with the fact that the same model was a good candidate based on criteria such as C_p in the all possible subsets regression analysis, greatly increases the confidence we can have in the final regression equation given by Eq. (6.5).

In fact, the combination of sequential variable procedures and all possible subsets regression analysis can be a very effective tool in deriving a final regression model when there are a large number of candidate independent variables. As before, it is still important to check for multicollinearity and examine the other regression diagnostics before settling on a final regression model.

SUMMARY

All possible subsets regression and sequential variable selection techniques can help build useful regression models when you are confronted with many potential independent variables from which to build a model. Unfortunately, you cannot submit your data to one of these procedures and let the computer blithely select the best model.

The first reason for caution is that there is no single numerical criterion for what the best regression model is. If the goal is simply prediction of new data, the model with the highest coefficient of determination R^2 or smallest standard error of the estimate $s_{y|x}$ would be judged best. Such models, however, often contain independent variables that exhibit serious multicollinearity, and this multicollinearity muddies the water in trying to draw conclusions about the actual effects of the different independent variables on the dependent variable. Avoiding this multicollinearity usually means selecting a model with fewer variables, and doing so raises the specter of producing an underspecified model that leaves out important variables. As a result, unknown biases can be introduced in the estimates of the regression coefficients associated with the independent variables remaining in the model. The C_p statistic provides some guidance concerning whether or not substantial biases are likely to be present in the model. Selecting the most sensible model usually requires taking all these factors into account.

All possible subsets regression is, by definition, the most thorough way to investigate the different possible regression models that can be constructed from the independent variables under consideration. However, this approach often produces many reasonable models that will require further study. Some of these models will contain obvious problems — such as those associated with multicollinearity — and can be eliminated from consideration, particularly if the additional predictive ability one acquires by adding the last few independent variables is small.

Sequential variable selection algorithms are a rational way to screen a large number of independent variables to build a regression equation. They do not guarantee the best equation using any of the criteria we developed in the first part of this chapter. These procedures were developed in the 1960s when the computational power was developed to do the arithmetic that was necessary, but when it was prohibitively expensive to do all possible subsets regression. Since then the cost of computing has dropped and better algorithms for computing all possible subsets regressions (when you only need the best few models with each number of independent variables) have become available. As a result, the consensus in the statistical community is now that all possible subsets regression is the preferable way for using the data to search for the best model.

Sequential variable selection procedures still remain a useful tool, particularly because they are less likely to develop a model with serious multicollinearities because the offending variables would not enter or would be removed from the equation, depending on the specific procedure you use. Sequential variable selection procedures are particularly useful when there are a very large number (tens or hundreds) of potential independent variables, a situation where all possible subsets regression might be too expensive or unruly. Sequential regression techniques also provide a good adjunct to all possible subsets regression to provide another view of what a good regression model is and help make choices among the several plausible models that may emerge from an all possible subsets regression analysis.

Finally, there is nothing in any of these techniques that can compensate for a poor selection of candidate independent variables. Variable selection techniques can be powerful tools for identifying the important variables in a regression model, but they are not foolproof. They are best viewed as guides for using independent scientific judgment for formulating sensible regression models.

PROBLEMS

6.1 Calculate R^2_{adj} from the computer output in Fig. 3-8.

6.2 For the regression analysis of Prob. 3.12, what is the incremental sum of squares associated with adding T_2 to the regression equation already containing T_1?

6.3 In Prob. 3.17, we concluded that involuntary smoking was associated with an increased risk of female breast cancer, allowing for the effects of dietary animal fat. Are both variables necessary? The data are in Table D-2, Appendix D.

6.4 Reanalyze the data from the study of urinary calcium during parenteral nutrition (Probs. 2.6, 3.8, and 5.5; data in Table D-5, Appendix D) using variable selection methods. Of the four potential variables, which ones are selected? Do different selection methods yield different results?

6.5 Data analysts are often confronted with a set of a few measured independent variables and are to choose the "best" predictive equation. Not infrequently, such an analysis consists of taking the measured variables, their pairwise cross-products, and their squares, throwing the whole lot into a computer, and using a variable selection method (usually stepwise regression) to select the best model. Using the data for the urinary calcium study in Probs. 2.6, 3.8, 5.5, and 6.4 (data in Table D-5, Appendix D), do such an analysis using several variable selection methods. Do the methods agree? Is there a problem with multicollinearity, and, if so, to what extent can it explain problems with variable selection? Does centering help? Is this "throw a whole bunch of things in the computer and stir" approach useful?

CHAPTER

SEVEN

One-Way Analysis of Variance

People often wish to compare the effects of several different treatments on an outcome variable. Sometimes the different treatments are applied to different individuals, and sometimes all the experimental subjects receive all the treatments. Although commonly analyzed statistically with multiple t tests, such data should be analyzed using an analysis of variance, followed by an appropriate multiple comparisons procedure to isolate treatment effects.[*] Analyzing a single set of data with multiple t tests both decreases the power of the individual tests and increases the risk of reporting a false positive. In this chapter we will analyze so-called *completely randomized experiments*, in which one observation is collected on each experimental subject, selected at random from the population of interest.[†] Every treatment is applied to a different individual, so some of the variability in the data is due to the treatments and some of the variability is due to random variation among the experimental subjects. The purpose of an analysis of variance is to discern whether the differences associated with the treatments are large enough, compared to the underlying random variability, to warrant concluding that the treatments had an effect. This chapter develops *one-way*, or *single factor*, analysis of variance when we wish to compare several different treatments, say the effects of several drugs on blood pressure. We will generalize these ideas in Chap. 8 to include *two-way*, or *two factor*, designs, when the experimental treatments are compared after taking into account some other factor, such as the effect of a drug on blood pressure, accounting for whether the people taking the drug are male or female.

---

[*]For a detailed discussion of this issue, see S. Glantz, *Primer of Biostatistics* (2nd ed.), New York, McGraw-Hill, 1987, Chaps. 3 and 4.

[†]We will analyze *repeated measures* experiments, in which each experimental subject is observed under several different treatments, in Chap. 9.

We will approach analysis of variance as a special case of multiple regression. This approach has the advantage of allowing us to transfer all we know about what constitutes a good regression model to our evaluation of whether we have a good analysis of variance model. In addition, the multiple linear regression approach yields estimates of the size of the treatment effects—not just an indication of whether or not such effects are present—and directly provides many of the pairwise multiple comparisons. Finally, when we consider more complicated analysis of variance designs in Chaps. 8 and 9, a regression approach will make it easier to understand how these analyses work and how to analyze experiments in which some observations are missing. In fact, it is in these situations that the regression approach to analysis of variance is particularly useful.

Because analysis of variance and linear regression are generally presented as distinct statistical methods in most introductory texts, it may seem strange to assert that one can formulate analysis of variance problems as equivalent regression problems. After all, analysis of variance is presented as a technique for testing differences between mean values of a variable of interest in the presence of several different (and distinct) treatments, such as before and after administration of a drug, whereas linear regression is presented as a way to estimate a continuous linear relationship between a dependent variable and one or more independent variables. We will use *dummy variables* to encode the treatments an individual subject received, and then use them as independent variables in a regression equation to obtain results identical to an analysis of variance on the same data. In fact, one can prove mathematically that the analysis of variance and regression are simply different ways of computing the same results.

USING A *t* TEST TO COMPARE TWO GROUPS*

Student's *t* test is widely used to test the null hypothesis that two samples were drawn from populations with the same mean, under the assumption that both underlying populations are normally distributed with the same standard deviation. The *t* test is the most widely used test in biomedical research.

The *t* statistic for the difference of two means is the ratio of the estimated difference to the standard error of this difference:

$$t = \frac{\overline{X}_1 - \overline{X}_2}{s_{\overline{X}_1 - \overline{X}_2}}$$

*For a more detailed discussion of the *t* test, see S. Glantz, *Primer of Biostatistics* (2nd ed.), New York, McGraw-Hill, 1987, pp. 64–91.

in which $\overline{X}_1$ and $\overline{X}_2$ are the observed mean values of the two treatment groups (samples) and $s_{\overline{X}_1 - \overline{X}_2}$ is the standard error of their difference, $\overline{X}_1 - \overline{X}_2$. If the observed difference between the two means is large compared with the uncertainty in the estimate of this difference (quantified by the standard error of the difference $s_{\overline{X}_1 - \overline{X}_2}$), t will be a big number and we reject the null hypothesis that both samples were drawn from the same population.

The definition of a "big" value of t depends on the total sample size, embodied in the degrees of freedom parameter, $v = n_1 + n_2 - 2$, where n_1 and n_2 are the number of subjects in treatment groups 1 and 2, respectively. When the value of t associated with the data exceeds t_α, the (two-tailed) critical value that determines the α percent most extreme values of the t distribution with v degrees of freedom, we conclude that the treatment had an effect ($P < \alpha$).

If n_1 equals n_2, the standard error of the difference of the means is

$$s_{\overline{X}_1 - \overline{X}_2} = \sqrt{s_{\overline{X}_1}^2 + s_{\overline{X}_2}^2}$$

where $s_{\overline{X}_1}^2$ and $s_{\overline{X}_2}^2$ are the standard errors of the two group means $\overline{X}_1$ and $\overline{X}_2$. When n_1 does not equal n_2, $s_{\overline{X}_1 - \overline{X}_2}$ is computed from the so-called pooled estimate of the population variance according to

$$s_{\overline{X}_1 - \overline{X}_2} = \sqrt{\frac{s^2}{n_1} + \frac{s^2}{n_2}}$$

where

$$s^2 = \frac{(n_1 - 1)s_1^2 + (n_2 - 1)s_2^2}{n_1 + n_2 - 2}$$

in which s_1 and s_2 are the standard deviations of the two samples.

Does Secondhand Tobacco Smoke Nauseate Martians?

The t test can be formulated as a simple linear regression problem, with a dummy independent variable. Before presenting this formulation, let us solve a problem using a traditional t test. We will then re-solve the same problem using a regression formulation.

Figure 7-1A shows more of the data collected during the study of the effects of secondhand tobacco smoke on Martians we started in Chap. 1. In this study we exposed one sample of Martians to steam (a placebo) and the other sample to secondhand smoke (a drug), and then measured the level of nausea they experienced. The question is: Is the level of nausea significantly different among the Martians in the two groups, i.e., does secondhand smoke nauseate Martians?

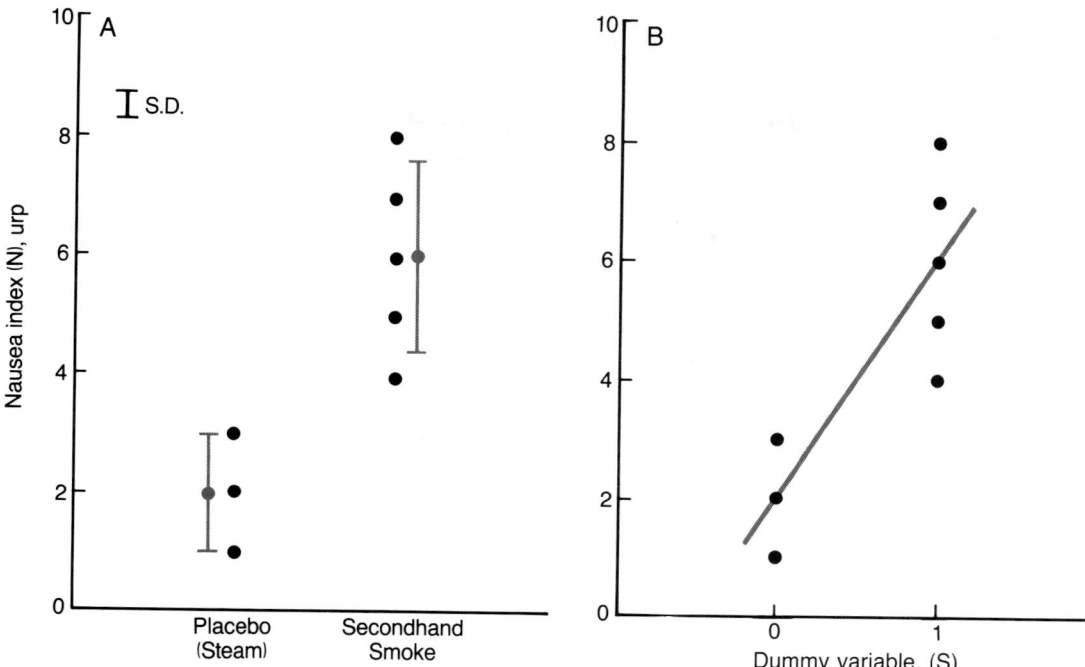

FIGURE 7-1 *A.* A traditional way to present and analyze data from an experiment in which the level of nausea was measured in two groups of Martians, one exposed to secondhand smoke and one exposed to a steam placebo, would be to plot the observed values under each of the two experimental conditions and use a *t* statistic to test the hypothesis that the means are not different. *B.* An alternative approach would be to define a dummy variable *S* to be equal to 0 for those Martians exposed to the steam placebo and 1 for those exposed to secondhand smoke. We then conduct a linear regression analysis with the dummy variable *S* as the independent variable and the nausea level *N* as the dependent variable. Both these approaches yield identical *t* statistics, degrees of freedom, and *P* values.

We can answer this question using a *t* test to compare the mean nausea level in the two treatment groups:

$$t = \frac{\overline{N}_S - \overline{N}_P}{s_{\overline{N}_S - \overline{N}_P}}$$

where $\overline{N}_P$ is the mean nausea level in the placebo (*P*) group and $\overline{N}_S$ is the mean nausea level in the secondhand smoke (*S*) group. To compute this *t* test, we first compute the mean nausea observed in the placebo group according to

$$\overline{N}_P = \frac{\Sigma N_{P_i}}{n_P} = \frac{1 + 2 + 3}{3} = 2 \text{ urp}$$

in which N_{P_i} is the nausea level observed in subject i within the placebo group P. Likewise, the mean nausea level of the Martians exposed to secondhand smoke is

$$\overline{N}_S = \frac{\Sigma N_{S_i}}{n_S} = \frac{4 + 5 + 6 + 7 + 8}{5} = 6 \text{ urp}$$

Because the samples have different sizes, we compute the standard error of the difference in mean nausea using the pooled estimate of the population variance. The standard deviations within each group are

$$s_P = \sqrt{\frac{\Sigma(N_{P_i} - \overline{N}_P)^2}{n_P - 1}}$$

$$= \sqrt{\frac{(1 - 2)^2 + (2 - 2)^2 + (3 - 2)^2}{3 - 1}} = 1.00 \text{ urp}$$

and

$$s_S = \sqrt{\frac{\Sigma(N_{S_i} - \overline{N}_S)^2}{n_S - 1}}$$

$$= \sqrt{\frac{(4 - 6)^2 + (5 - 6)^2 + (6 - 6)^2 + (7 - 6)^2 + (8 - 6)^2}{5 - 1}}$$

$$= 1.58 \text{ urp}$$

The pooled variance estimate is

$$s^2 = \frac{(n_S - 1)s_S^2 + (n_P - 1)s_P^2}{n_P + n_S - 2} = \frac{(5 - 1)1.58^2 + (3 - 1)1.00^2}{5 + 3 - 2} = 2.00 \text{ urp}^2$$

and the standard error of the mean difference is

$$s_{\overline{N}_S - \overline{N}_P} = \sqrt{\frac{s^2}{n_S} + \frac{s^2}{n_P}} = \sqrt{\frac{2.00}{5} + \frac{2.00}{3}} = 1.03 \text{ urp}$$

Finally, the t statistic to test the null hypothesis that breathing secondhand smoke does not affect the mean level of nausea is

$$t = \frac{\overline{N}_S - \overline{N}_P}{s_{\overline{N}_S - \overline{N}_P}} = \frac{6 - 2}{1.03} = 3.873$$

The magnitude of this value of t exceeds 3.707, the critical value of t that defines the 1 percent most extreme values of the t distribution with $\nu = n_P + n_S - 2 = 3 + 5 - 2 = 6$ degrees of freedom. Therefore, we conclude that the secondhand smoke produced significantly more nausea than the steam placebo ($P < .01$).

Using Linear Regression to Compare Two Groups

Figure 7-1*B* shows an alternative way to think about the data we just analyzed with a traditional *t* test. At first glance, it looks just like Fig. 7-1*A*, but there is an important difference. We have now represented the data on a graph with all the observations under placebo conditions associated with a value of $S = 0$ and all the observations associated with secondhand smoke plotted with values of $S = 1$. The new variable S encodes which group a given individual belongs to; it is just another example of a dummy variable. Viewed this way, these data could be thought of as the set of (S, N) data points in Table 7-1.

Now let us use linear regression to fit the straight line

$$\hat{N} = b_0 + b_1 S$$

through these (S, N) data points, where S is the independent variable. Figure 7-2 shows the results of these computations. The intercept is $b_0 = 2$ urp and the slope is $b_1 = 4$ urp, so the regression equation is

$$\hat{N} = 2 \text{ urp} + 4 \text{ urp } S$$

When the data are from the placebo group, $S = 0$ and the regression equation is

$$\hat{N}(0) = b_0 + b_1 \cdot 0 = b_0 = 2 \text{ urp}$$

and b_0 *is an estimate of the mean value of nausea for the placebo group,* $\overline{N}_p$. Likewise, when the data are from the involuntary smoking group, $S = 1$ and the regression equation is

$$\hat{N}(1) = b_0 + b_1 \cdot 1 = b_0 + b_1 = 2 \text{ urp} + 4 \text{ urp} = 6 \text{ urp} \qquad (7.1)$$

TABLE 7-1 Martian Nausea as a Function of Exposure to Secondhand Smoke

Nausea Index N, urp	Smoke Exposure S	
1	0	
2	0	Steam (placebo)
3	0	
4	1	
5	1	
6	1	Secondhand smoke (drug)
7	1	
8	1	

```
The regression equation is
N = 2.00 + 4.00 D

Predictor        Coef       Stdev     t-ratio        p
Constant       2.0000      0.8165        2.45    0.050
D               4.000       1.033        3.87    0.008

s = 1.414        R-sq = 71.4%      R-sq(adj) = 66.7%

Analysis of Variance

SOURCE       DF          SS         MS          F        p
Regression    1      30.000     30.000      15.00    0.008
Error         6      12.000      2.000
Total         7      42.000
```

FIGURE 7-2 Results of regression analysis for the data in Fig. 7-1B.

and $(b_0 + b_1)$ is an estimate of the mean level of nausea of the involuntary smoking group $\overline{N}_S$. Note that in addition to being the slope of the linear regression, b_1 *represents the size of secondhand smoke's effect* (4 urp = $\overline{N}_P - \overline{N}_S$, the numerator of the t test above), in that it is the average change in N for a one-unit change in the dummy variable S, which is defined to change from 0 to 1 when we move from the placebo to the involuntary smoking group.

To test whether the slope (which equals the change in mean nausea between the two groups) b_1 is significantly different from zero, use the standard error of the slope $s_{b_1} = 1.033$ urp (from Fig. 7-2) to compute the t statistic just as in Chaps. 2 and 3:

$$t = \frac{b_1}{s_{b_1}} = \frac{4}{1.033} = 3.873$$

We compare this value of t with the critical value for $v = n - 2 = 8 - 2 = 6$ degrees of freedom for $\alpha = .01$, 3.707, and conclude that the slope is statistically significantly different from zero ($P < .01$).

The value of t, the degrees of freedom, the associated P values, and our conclusion are *identical* to those obtained with the t test on the group means. Any t test can be conducted as a linear regression using a dummy independent variable which takes on the values 0 and 1 to encode whether a given experimental subject is in one treatment group or the other.

Reflecting on this example, however, should convince you that using a linear regression is doing it the hard way. We presented this example only to illustrate the close relationship between linear regression using dummy variables and the *t* test. We will generalize these ideas to analysis of variance in which there are several treatment groups by simply creating more dummy variables and using multiple regression. Whereas we gained nothing by using linear regression with dummy variables to do a *t* test for the difference between two groups, we will reap considerable benefits when we formulate a more general analysis of variance to test for the difference between three or more group means. When there are more than two experimental factors,[*] the regression approach will allow us to handle unbalanced designs (when there are different numbers of subjects in the different treatment groups) and missing data better than in a traditional analysis of variance.

THE BASICS OF ONE-WAY ANALYSIS OF VARIANCE

Analysis of variance is the multigroup generalization of the *t* test. Like the *t* test, analysis of variance rests on the assumptions that the samples are randomly drawn from normally distributed populations with the same standard deviations. The question is: Are the differences in the observed means of the different treatment (sample) groups likely to simply reflect random variation associated with the sampling process, or are these differences in the group means too large to be consistent with this so-called null hypothesis? If the differences between the group means are larger than one would expect from the underlying population variability, estimated with the standard deviations within the treatment groups, we reject the null hypothesis and conclude that at least one of the treatment groups has a different mean than the others.[†]

We use the *F* statistic to conduct an analysis of variance. This statistic is the ratio of an estimate of the population variance computed from the means of the different treatment groups assuming that all the treatment groups were drawn from the same population (i.e., that the treatment had no effect) to an estimate of the population variance computed from the variance within each treatment group. When *F* is a big number, we will conclude that at least one of the treatments had an effect different from the others.

---

[*]These designs will be discussed in Chaps. 8 and 9.

[†]For a detailed discussion of traditional analysis of variance, see S. A. Glantz, *Primer of Biostatistics* (2nd ed.), New York, McGraw-Hill, 1987, Chaps. 3 and 9. The discussion here follows pp. 34–37 of *Primer of Biostatistics.*

Suppose there are four treatment groups, each of size n, with mean values $\overline{X}_1$, $\overline{X}_2$, $\overline{X}_3$, and $\overline{X}_4$ and standard deviations s_1, s_2, s_3, and s_4.[*] We will use this information to obtain two estimates of the variance of the underlying population σ_ϵ^2, assuming that the treatments had no effect. Given this assumption, each sample standard deviation is an estimate of σ_ϵ, so we can obtain an estimate of σ_ϵ^2 from within the treatment groups by simply averaging s_1^2, s_2^2, s_3^2, and s_4^2 to obtain

$$s_{\text{wit}}^2 = \frac{s_1^2 + s_2^2 + s_3^2 + s_4^2}{4}$$

s_{wit}^2 is known as the *within groups variance*. This estimate of σ_{wit}^2 does not depend on the mean values of the different treatment groups.

Next, we estimate the population variance from the means of the four samples. Because we have hypothesized that all four samples were drawn from the same population, the standard deviation of all four sample means will approximate the standard error of the mean (which is the standard deviation of the distribution of all possible means of samples of size n drawn from the population in question). Recall that the standard error of the mean $\sigma_{\overline{X}}$ is related to the population standard deviation σ_ϵ and the sample size n according to

$$\sigma_{\overline{X}} = \frac{\sigma_\epsilon}{\sqrt{n}}$$

Therefore, the population variance is related to the standard error of the mean according to

$$\sigma_\epsilon^2 = n\sigma_{\overline{X}}^2$$

We use this relationship to estimate the population variance σ_ϵ^2 from the variability between the sample means using

$$s_{\text{bet}}^2 = ns_{\overline{X}}^2$$

where $s_{\overline{X}}$ is the standard deviation of the observed means of the four treatment groups $\overline{X}_1$, $\overline{X}_2$, $\overline{X}_3$, and $\overline{X}_4$, and s_{bet}^2 is the so-called *between groups variance*.

[*]The computations that follow assume equal sample sizes. Equal sample sizes are not required for one-way analysis of variance, but the computations are not as intuitive. In general, one requires equal sample sizes for more complex (e.g., two-way) analyses unless one uses the techniques presented in Chap. 8. The equations for one-way analysis of variance with unequal sample sizes can be found in S. A. Glantz, *Primer of Biostatistics* (2nd ed.), New York, McGraw-Hill, 1987, pp. 352–353. In this book we will let the computer handle the necessary adjustments in the computations for one-way analysis of variance with unequal sample sizes.

If the hypothesis that all four samples were drawn from the same population (i.e., if the treatments had no effect) is true, then these two different estimates of the underlying population variance will be similar and the ratio

$$F = \frac{s_{\text{bet}}^2}{s_{\text{wit}}^2} \tag{7.2}$$

will be approximately 1. If, on the other hand, there is more variation between the means of the four samples than one would expect based on the variation observed within the treatment groups, F will be much larger than 1 and we conclude that the treatments produced differences in the variable of interest. Thus, we compute F to see if the s_{bet}^2 is "large" relative to the s_{wit}^2, and then compare this value with the critical value F_α (in Table E-2, Appendix E) for $v_n = k - 1$ numerator and $v_d = k(n - 1)$ denominator degrees of freedom, where k is the number of treatment groups; in this example, $k = 4$. If the observed value of F exceeds F_α, we conclude that at least one of the treatment (group) means is different from the others $(P < \alpha)$.

In traditional analysis of variance nomenclature, these variance estimates are generally called *mean squares* and denoted MS rather than s^2. The variance estimate obtained from the different group means is denoted MS_{bet}, and the variance estimate obtained from within the different treatment groups is denoted MS_{wit}. MS_{bet} is sometimes called the *treatment* or *between treatments* mean square. MS_{wit} is sometimes called the *residual, within groups,* or *error* mean square, and denoted MS_{res} or MS_{err}.

With this notation, the F statistic in Eq. (7.2) is written

$$F = \frac{MS_{\text{bet}}}{MS_{\text{wit}}} \tag{7.3}$$

Expected Mean Squares

We have already introduced the idea of the expected value of a statistic, which is the average of all possible values of the statistic. When we discussed expected values previously, we used them to see if a sample statistic (like a regression coefficient) was an unbiased estimator of the corresponding population parameter. Now, however, we will use expected values as a guide to understanding how F statistics are constructed. In fact, we have already implicitly used expected values in arguing that, if the treatments had no effect, the expected value of the F statistic would be around 1.0. We now formalize this argument. Understanding the expected values of the mean squares that comprise the numerator and denominator of F is particularly important for understanding complicated analyses of variance, particularly those involving more than one experimental factor or repeated measures that we will discuss in Chaps. 8 and 9.

We just finished discussing MS_{bet} and MS_{wit} and why the F test computed as their ratio made sense in terms of the null hypothesis: If there is no difference between the group means, both MS_{bet} and MS_{wit} are estimates of the same underlying random population variance σ_ϵ^2. In terms of expected values, we would write

$$E(MS_{\text{bet}}) = \sigma_\epsilon^2$$

and

$$E(MS_{\text{wit}}) = \sigma_\epsilon^2$$

so, if the null hypothesis is true, we would expect F to have a value near

$$F = \frac{E(MS_{\text{bet}})}{E(MS_{\text{wit}})} = \frac{\sigma_\epsilon^2}{\sigma_\epsilon^2} = 1 \tag{7.4}$$

In contrast, if the null hypothesis is not true and there is a difference between the group means, only MS_{wit} is an estimate of the underlying population variance. MS_{bet} now includes two components: a component due to the underlying population variance (the same σ_ϵ^2 that MS_{wit} estimates) and a component due to the variation between the group means. In terms of expected values, we can restate this general result as follows. The expected value of MS_{bet} is

$$E(MS_{\text{bet}}) = \sigma_\epsilon^2 + \sigma_{\text{bet}}^2$$

and the expected value of MS_{wit} is

$$E(MS_{\text{wit}}) = \sigma_\epsilon^2$$

Hence, we would expect the F statistic to have a value around

$$F = \frac{\sigma_\epsilon^2 + \sigma_{\text{bet}}^2}{\sigma_\epsilon^2} \tag{7.5}$$

Note that the F statistic is constructed such that the expected value of the mean square in the numerator differs from the expected value of the mean square in the denominator by only one component, the variance due to the treatment or grouping factor which we are interested in testing. When the null hypothesis is true, $\sigma_{\text{bet}}^2 = 0$ and $F = 1$, consistent with Eq. (7.4). This is simply a restatement, in terms of expected values, of our argument above that if the null hypothesis is true, MS_{bet} and MS_{wit} are both estimates of the underlying population variance σ_ϵ^2 and F should be about 1. If there is a difference between the group means, σ_{bet}^2 will be greater than zero and F should exceed 1. Hence, if we obtain large values of F, we will conclude that it is unlikely that σ_{bet}^2 is zero and assert that at least one treatment group mean differs from the others.

Expected mean squares for more complicated analyses of variance will not be this simple, but the concept is the same. We will partition the total variability in the sample into components that depend on the design of the experiment. Each resulting mean square will have an expected value that is the sum of one or more variances due to our partitioning, and the more complicated the design, the more variance estimates appear in the expected mean squares. However, the construction of F statistics is the same: The numerator of the F statistic is the mean square whose expected value includes the variance due to the factor of interest, and the denominator of the F statistic is the mean square whose expected value includes all the components that appear in the numerator, *except* the one due to the factor of interest. Viewed in this way, an F statistic tests the null hypothesis that the variance due to the treatment or grouping factor of interest equals zero.

Using Linear Regression to Do Analysis of Variance with Two Groups

The F statistic we computed above as the ratio of MS_{bet} to MS_{wit} should remind you of the F we calculated to test whether the slope of a regression line is significantly different from zero in Eq. (2.18), with MS_{reg} playing a role analogous to MS_{bet} and MS_{res} playing a role analogous to MS_{wit}. In fact, just as it is possible to derive the same value of t via either a traditional t test or a regression analysis, we can also derive this value of F from a regression analysis. We will illustrate this fact using our study of secondhand smoke and nausea of Martians.

Table 7-2 presents the results of analyzing the data in Fig. 7-1 (and Table 7-1) using a one-way analysis of variance. Comparing the results in Table 7-2 with those in Fig. 7-2 reveals that the MS_{reg}, which is the average sum of squared difference between the regression line and the mean nausea experienced by *all* Martians $\overline{N}$, is actually an estimate of

TABLE 7-2 Analysis of Variance for Martian Nausea and Exposure to Secondhand Smoke

Source	Degrees of Freedom	Sum of Squares	Mean Square	F
S	1	30.00	30.00	15.00
Error	6	12.00	2.00	
Total	7	42.00		

the variation between the treatment groups; MS_{reg} is identical to MS_{bet}, the mean square associated with the treatment, calculated in a traditional analysis of variance. Both values are 30 urp². Likewise, MS_{res}, which is the normalized sum of squared differences between the data points and the regression line (in this case the regression line simply connects the means of the two groups of sample data points) is identical to the MS_{wit} of a traditional analysis of variance. Both values are 2 urp². Thus, the F test of a regression equation (using a properly defined dummy variable to encode the treatment effect) and the F test of a traditional one-way analysis of variance that tests the null hypothesis of equality between group means both yield

$$F = \frac{s_{bet}^2}{s_{wit}^2} = \frac{MS_{bet}}{MS_{wit}} = \frac{MS_{reg}}{MS_{res}} = \frac{30 \text{ urp}^2}{2 \text{ urp}^2} = 15$$

This introductory example contained only two treatment groups because it is much easier to visualize the correspondence between regression and analysis of variance when using only two groups. (Ordinarily, these data would be analyzed using a t test.) To use linear regression to conduct an analysis of variance with more than two groups, we simply define additional dummy variables to encode the additional sample groups and use multiple linear regression instead of simple linear regression. In the following chapters we will see that by proper coding of dummy variables we can handle not only one-way analysis of variance (where there are several levels of single experimental condition or factor), but also multiple factor analysis of variance (where there are two or more levels of two or more factors), and repeated measures analysis of variance (where each subject receives multiple treatments).

USING LINEAR REGRESSION TO DO ONE-WAY ANALYSIS OF VARIANCE WITH ANY NUMBER OF TREATMENTS

To see how the ideas presented above generalize, suppose we had three treatment groups: a placebo (control) group, a group that received drug 1, and a group that received drug 2. To encode these three groups, define two dummy variables as

$$D_1 = \begin{cases} 1 \text{ if drug 1} \\ 0 \text{ otherwise} \end{cases}$$

and

$$D_2 = \begin{cases} 1 \text{ if drug 2} \\ 0 \text{ otherwise} \end{cases}$$

These two dummy variables encode all three treatment groups: for the placebo group $D_1 = 0$ and $D_2 = 0$; for the drug 1 group $D_1 = 1$ and $D_2 = 0$; and for the drug 2 group $D_1 = 0$ and $D_2 = 1$.

The multiple linear regression equation relating the response variable y to the two predictor variables D_1 and D_2 is

$$\hat{y} = b_0 + b_1 D_1 + b_2 D_2 \tag{7.6}$$

For the placebo group, Eq. (7.6) reduces to

$$\hat{y} = b_0 + b_1 \cdot 0 + b_2 \cdot 0 = b_0$$

So, as we saw above when using simple linear regression to do a t test, the intercept b_0 is an estimate of the mean of the reference group (the group for which all the dummy variables are zero). In this example, when the data are from drug 1, $D_1 = 1$ and $D_2 = 0$ so that Eq. (7.6) reduces to

$$\hat{y} = b_0 + b_1 \cdot 1 + b_2 \cdot 0 = b_0 + b_1$$

and, as we saw for Eq. (7.1), b_1 is an estimate of the difference between the control and drug 1 group means. Likewise, when the data are from drug 2, $D_1 = 0$ and $D_2 = 1$, so Eq. (7.6) reduces to

$$\hat{y} = b_0 + b_1 \cdot 0 + b_2 \cdot 1 = b_0 + b_2$$

and b_2 is an estimate of the difference between the control and the drug 2 group means. Thus, we can also use the results of this regression equation to quantify differences between the drug groups and the placebo group, as well as test hypotheses about these differences.

As before, we partition the total variability about the mean value of the dependent variable into a component due to the regression fit (SS_{reg}) and a component due to scatter in the data (SS_{res}). As with simple linear regression, the overall goodness of fit of the entire multiple regression equation to the data is assessed by an F statistic computed as the ratio of the mean squares produced by dividing each of these sums of squares by the corresponding degrees of freedom. If the resulting F exceeds the appropriate critical value, it indicates that there is a significant difference among the treatment groups.

This approach immediately generalizes to more than three treatment groups by simply adding more dummy variables, with the total number of dummy variables being one fewer than the number of treatment groups (which equals the numerator degrees of freedom for the F statistic). As already illustrated, you should select the treatment group for which all dummy variables equal zero to be the control or reference group, because then the values of the coefficients in the regression equation will provide estimates of the magnitude of the difference between the mean response of each of the different treatments and the control group mean. Specifically, the procedure for formulating a one-way analysis of variance problem with k treatment groups as a multiple linear regression problem is

1. Select a treatment group to be the control or reference group, and denote it group k.
2. Define $k - 1$ dummy variables according to

$$D_i = \begin{cases} 1 \text{ if group } i & i < k \\ 0 \text{ otherwise} \end{cases}$$

    This type of coding is called *reference cell coding.*
3. Fit the observed data Y_i with the multiple linear regression equation:

$$\hat{y} = b_0 + b_1 D_1 + b_2 D_2 + \cdots + b_{k-1} D_{k-1}$$

4. Use the resulting sums of squares and mean squares to test whether the regression equation significantly reduces the unexplained variance in y.

The coefficient b_0 is an estimate of the mean value of y for the control or reference group. The other coefficients quantify the average change of each treatment group from the reference group. The associated standard errors provide the information necessary to do the multiple comparison tests (discussed later in this chapter) to isolate which treatment group means differ from the others.

Hormones and Depression

Depression in adults has been linked to abnormalities in the so-called hypothalamic-pituitary-adrenal axis which links a region of the brain (the hypothalamus, which is involved in regulating emotional states) to hormones that are produced outside the brain. To gain a better understanding of how abnormalities in this system might affect whether or not a person will suffer depression, Price and his colleagues[*] wanted to learn how a specific kind of nerve cell in the brain affects the adrenal gland's production of the hormone cortisol. They measured cortisol levels in the blood of healthy adults and adults diagnosed with depression. The depressed individuals were further divided into two diagnostic groups: those with melancholic depression and those with the less severe non-melancholic depression. To begin their study, they wanted to see if there was any difference in the baseline (that is, before they gave a drug) levels

[*]L. H. Price, D. S. Charney, A. L. Rubin, and F. R. Heninger, "α_2-Adrenergic Receptor Function in Depression: The Cortisol Response to Yohimbine," *Arch. Gen. Psychiatry* 43:849–858, 1986.

of cortisol in these three groups. Because we have more than two group means to compare, we begin by analyzing these data using a one-way analysis of variance.

Table 7-3 shows the data and Table 7-4 presents the results of a traditional one-way analysis of variance. The mean square due to the between-groups effect MS_{bet} is 82.3 $(\mu g/dL)^2$, and the within-groups (or error) mean square MS_{wit} is 12.5 $(\mu g/dL)^2$. (Figure 7-3 shows a plot of these data as they would often appear in a journal.) We compare these two estimates of the population variance with

$$F = \frac{MS_{bet}}{MS_{wit}} = \frac{82.3 \ (\mu g/dL)^2}{12.5 \ (\mu g/dL)^2} = 6.6 \tag{7.7}$$

TABLE 7-3 Cortisol Levels ($\mu g/dL$) in Healthy and Depressed Individuals

		Depressed	
	Healthy	Nonmelancholic	Melancholic
	2.5	5.4	8.1
	7.2	7.8	9.5
	8.9	8.0	9.8
	9.3	9.3	12.2
	9.9	9.7	12.3
	10.3	11.1	12.5
	11.6	11.6	13.3
	14.8	12.0	17.5
	4.5	12.8	24.3
	7.0	13.1	10.1
	8.5	15.8	11.8
	9.3	7.5	9.8
	9.8	7.9	12.1
	10.3	7.6	12.5
	11.6	9.4	12.5
	11.7	9.6	13.4
		11.3	16.1
		11.6	25.2
		11.8	
		12.6	
		13.2	
		16.3	
Mean	9.2	10.7	13.5
SD	2.9	2.8	4.7
n	16	22	18

Table 7-4 Traditional Analysis of Variance Table for Cortisol and Depression Data of Table 7-3

Source	Degrees of Freedom	Sum of Squares	Mean Square	F	P
Between groups	2	164.67	82.3	6.612	.003
Within groups (error)	53	660.02	12.5		
Total	55	824.69			

This value of F exceeds the critical value of F for $P < .01$ with 2 numerator degrees of freedom (number of groups − 1) and 53 denominator degrees of freedom (total number of patients − number of groups), 5.03 (interpolating in Table E-2, Appendix E), so we reject the null hypothesis that the mean cortisol levels are the same in all three groups of people

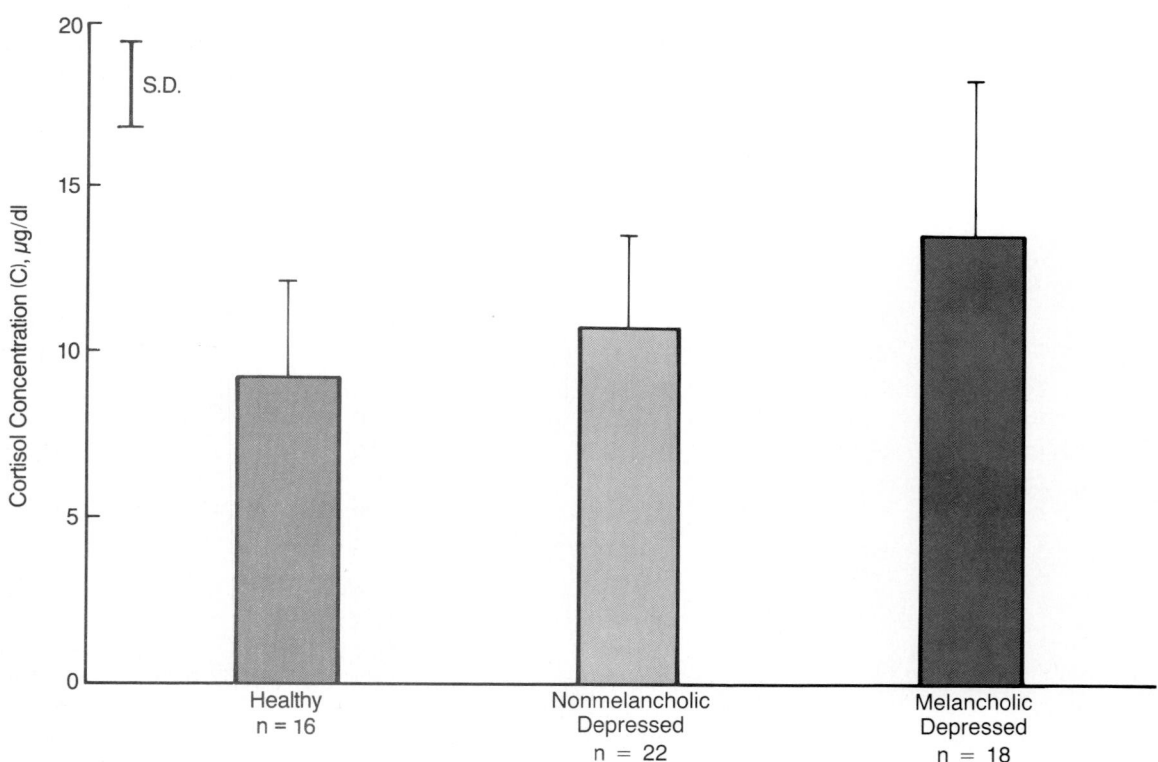

Figure 7-3 Cortisol levels measured in healthy individuals and nonmelancholy depressed and melancholy depressed patients.

in favor of the alternative that at least one group has a different mean cortisol level than the others (the exact P value reported by the computer program is .003). Hence, we conclude that different mental states are associated with different baseline cortisol levels. We will do *multiple comparison tests* later to identify what the differences are.

Now, let us do this one-way analysis of variance with multiple linear regression and dummy variables. We have three groups of patients to compare, so we need to define two dummy variables to identify the three groups. We take the healthy people as our reference group and define

$$D_N = \begin{cases} 1 \text{ if nonmelancholic depression} \\ 0 \text{ otherwise} \end{cases}$$

$$D_M = \begin{cases} 1 \text{ if melancholic depression} \\ 0 \text{ otherwise} \end{cases}$$

If a patient has no depression, $D_N = 0$ and $D_M = 0$. For patients with nonmelancholic depression, $D_N = 1$ and $D_M = 0$; for patients with melancholic depression, $D_N = 0$ and $D_M = 1$. Table 7-5 presents the same cortisol data as Table 7-3 together with the associated values of D_N and D_M as they would be submitted for multiple linear regression analysis. The multiple linear regression equation used to do the analysis of variance is

$$\hat{C} = b_0 + b_N D_N + b_M D_M \tag{7.8}$$

where $\hat{C}$ is the predicted cortisol concentration and D_N and D_M are defined above.

The results of the multiple linear regression on these data appear in Fig. 7-4. First, let us compare the traditional analysis of variance shown in Table 7-4 with the analysis of variance portion of this multiple linear regression output. The mean squares due to the regression model fit is 82.3 $(\mu g/dL)^2$, which is identical to the MS_{bet} in the traditional analysis of variance. Likewise, the residual (or error) mean squares in the regression model is 12.5 $(\mu g/dL)^2$, which is also identical to the MS_{wit} in the traditional analysis of variance. Thus, the F statistic computed from this regression model

$$F = \frac{MS_{reg}}{MS_{res}} = \frac{82.3 \ (\mu g/dL)^2}{12.5 \ (\mu g/dL)^2} = 6.6$$

is identical to that obtained in a traditional analysis of variance [compare with Eq. (7.7)]. The degrees of freedom associated with the numerator and denominator of this F test statistic are the same as before (2 for MS_{reg} and 53 for MS_{res}), so the critical value of F for $\alpha = .01$ remains 5.03 and we again reject the null hypothesis that the means of all three groups are the same ($P < .01$).

TABLE 7-5 Cortisol Levels (µg/dL) from Table 7-3 with Dummy Variables to Encode the Different Study Groups

Cortisol C, µg/dL	D_N	D_M
HEALTHY INDIVIDUALS		
2.5	0	0
7.2	0	0
8.9	0	0
9.3	0	0
9.9	0	0
10.3	0	0
11.6	0	0
14.8	0	0
4.5	0	0
7.0	0	0
8.5	0	0
9.3	0	0
9.8	0	0
10.3	0	0
11.6	0	0
11.7	0	0
NONMELANCHOLIC DEPRESSED INDIVIDUALS		
5.4	1	0
7.8	1	0
8.0	1	0
9.3	1	0
9.7	1	0
11.1	1	0
11.6	1	0
12.0	1	0
12.8	1	0
13.1	1	0
15.8	1	0
7.5	1	0
7.9	1	0
7.6	1	0
9.4	1	0
9.6	1	0
11.3	1	0
11.6	1	0
11.8	1	0
12.6	1	0
13.2	1	0
16.3	1	0

TABLE 7-5 (*continued*)   **Cortisol Levels (μg/dL) from Table 7-3 with Dummy Variables to Encode the Different Study Groups**

Cortisol C, μg/dL	D_N	D_M
MELANCHOLIC DEPRESSED INDIVIDUALS		
8.1	0	1
9.5	0	1
9.8	0	1
12.2	0	1
12.3	0	1
12.5	0	1
13.3	0	1
17.5	0	1
24.3	0	1
10.1	0	1
11.8	0	1
9.8	0	1
12.1	0	1
12.5	0	1
12.5	0	1
13.4	0	1
16.1	0	1
25.2	0	1

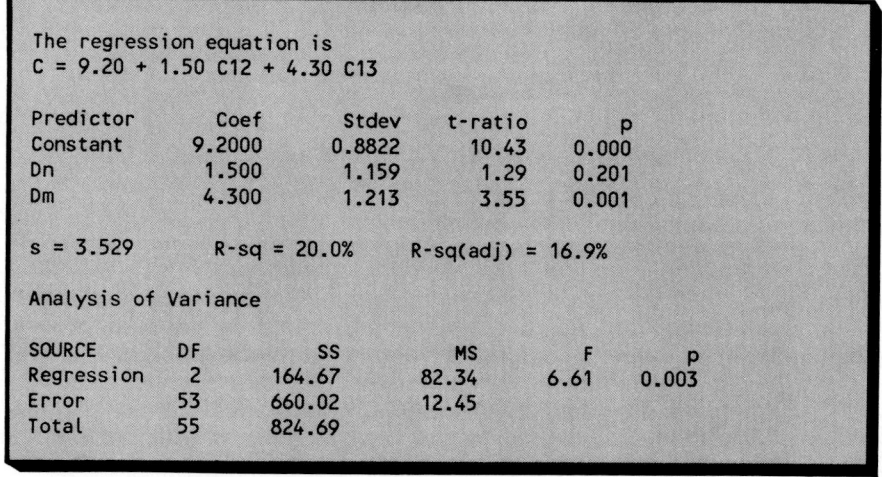

```
The regression equation is
C = 9.20 + 1.50 C12 + 4.30 C13

Predictor        Coef       Stdev     t-ratio        p
Constant       9.2000      0.8822       10.43    0.000
Dn              1.500       1.159        1.29    0.201
Dm              4.300       1.213        3.55    0.001

s = 3.529      R-sq = 20.0%     R-sq(adj) = 16.9%

Analysis of Variance

SOURCE        DF         SS         MS        F        p
Regression     2      164.67      82.34     6.61    0.003
Error         53      660.02      12.45
Total         55      824.69
```

FIGURE 7-4 Regression implementation of a one-way analysis of variance for the data shown in Fig. 7-3.

Now, let us examine the information contained in the regression coefficients b_0, b_N, and b_M. When the data are from healthy subjects, both D_N and D_M are 0 and Eq. (7.8) reduces to

$$\hat{C} = b_0 + b_N \cdot 0 + b_M \cdot 0 = b_0 = 9.2 \ \mu\text{g/dL}$$

Thus, b_0 is the mean cortisol level of the group of healthy subjects. When the data are from nonmelancholic depressed subjects, D_N is 1, D_M is 0, and Eq. (7.8) reduces to

$$\hat{C} = b_0 + b_N \cdot 1 + b_M \cdot 0 = 9.2 + 1.5 \cdot 1 + 4.3 \cdot 0 = 10.7 \ \mu\text{g/dL}$$

Thus, the mean cortisol level of the group of nonmelancholic depressed subjects is the sum of b_0 and b_N. In other words, b_N is the *difference* between the mean cortisol level of healthy subjects and the mean cortisol level of nonmelancholic depressed subjects. Similarly, b_M is the difference between the mean cortisol level of the healthy subjects and the mean cortisol level of the melancholic depressed subjects:

$$\hat{C} = b_0 + b_N \cdot 0 + b_M \cdot 1 = 9.2 + 1.5 \cdot 0 + 4.3 \cdot 1 = 13.5 \ \mu\text{g/dL}$$

Each of these parameter estimates in the multiple regression equation has an associated standard error. We can use these standard errors to test whether the coefficients b_N and b_M are significantly different from zero with t statistics. On first blush, these t tests would appear to test whether the mean cortisol level of the depressed individuals is significantly different from the mean of the controls. One can indeed conduct such tests, but, because they are related to each other, we must consider the fact that these comparisons represent a *family of related tests* and we must use *multiple comparison procedures*.

MULTIPLE COMPARISON TESTING

If an analysis of variance (whether done with traditional methods or as a linear regression) yields a statistically significant value of F, we conclude that at least one of the group means differs from the others by more than would be expected by sampling variation. The question still remains as to which groups are different. There are a variety of methods, called *multiple comparison procedures*, that can be used to help answer this question. All are essentially based on the t test but include appropriate corrections for the fact that we are comparing more than one pair of means.*

Before we develop these multiple comparison procedures, we first need to think about why it matters that we are comparing more than

---

*For a more detailed introduction to multiple comparison testing, see S. Glantz, *Primer of Biostatistics* (2nd ed.), New York, McGraw-Hill, 1987, pp. 85–96.

one pair of means. In the example above, there are three possible comparisons to be made. We could simply barge ahead and do three t tests (healthy vs. nonmelancholic depressed, healthy vs. melancholic depressed, and nonmelancholic depressed vs. melancholic depressed), but this practice is incorrect because the true probability of erroneously concluding that any two of these groups differ from each other is actually higher than the nominal level, say 5 percent, used when looking up the critical cutoff value of the t statistic in Table E-1, Appendix E, for each individual pairwise comparison.

To understand why, suppose that the value of the t statistic computed in one of the three possible comparisons is in the most extreme 5 percent of the values that would occur if there was really no difference in cortisol production; so we assert that there is a difference in cortisol production between the two groups in question with $P < .05$. This situation means that there is a 5 percent risk of erroneously rejecting the null hypothesis of no treatment effect and reaching a false-positive conclusion. In other words, in the long run we are willing to accept the fact that 1 statement in 20 will be wrong. Therefore, when we test healthy vs. nonmelancholic depressed, we can expect erroneously to assert a difference 5 percent of the time. Similarly, when testing healthy vs. melancholic depressed, we expect erroneously to assert a difference 5 percent of the time, and, likewise, when testing nonmelancholic depressed vs. melancholic depressed, we expect erroneously to assert a difference 5 percent each time. Therefore, when considering the three tests together, we expect to erroneously conclude that at least one pair of groups differ about 5 percent + 5 percent + 5 percent = 15 percent of the time, even if, in reality, there is no difference in baseline cortisol concentration. If there are not too many comparisons, simply adding the P values obtained in multiple tests produces a realistic (and conservative) estimate of the true P value for the set of comparisons.

In the example above, there were three t tests, so the effective P value was about $3(.05) = .15$, or 15 percent. When comparing four groups, there are six possible t tests (1 vs. 2, 1 vs. 3, 1 vs. 4, 2 vs. 3, 2 vs. 4, 3 vs. 4), so if you conclude that there is a difference and report $P < .05$, the effective P value is about $6(.05) = .30$; there is about a 30 percent chance of at least one incorrect statement when you conclude that the treatments had an effect if you fail to account for the multiple comparisons!*

---

*In this discussion, the total probability α_T of erroneously rejecting a null hypothesis (the effective P value) is estimated as $\alpha_T = k\alpha$, where α is the desired error rate of each test and k is the number of tests. This expression overestimates the true value of $\alpha_T = 1 - (1 - \alpha)^k$. For example, when $\alpha = .05$ and $k = 3$, $\alpha_T = 0.14$ (compared to $3 \cdot .05 = .15$), and when $k = 6$, $\alpha_T = 0.27$ (compared to $6 \cdot .05 = .30$). As k increases, the approximate α_T begins to seriously overestimate the true α_T.

Our goal is to control the total probability of erroneously rejecting the null hypothesis of no differences between group means. The general approach is to first perform an analysis of variance to see whether anything appears different, then use a multiple-comparison procedure to isolate the treatment or treatments producing the different results. A *multiple comparison procedure* controls the error rate of a related *family of tests*, rather than simply controlling the error rate of the individual tests. We will discuss three multiple comparison procedures, the *Bonferroni t test*, the *Student-Newman-Keuls test*, and *Dunnett's test*.

The Bonferroni t Test

The Bonferroni t test is the simplest multiple comparison procedure. It follows directly from the discussion we just had regarding the accumulation of error with multiple t tests. In the last section, we saw that if one analyzes a set of data with three t tests, each using the 5 percent critical value, there is about a $3(5) = 15$ percent chance of obtaining a false positive. This result is a special case of the *Bonferroni inequality*, which states that if k statistical tests are performed at the α level, the total probability of obtaining at least one false-positive result α_T does not exceed $k\alpha$:

$$\alpha_T \leq k\alpha$$

α_T is the error rate we want to control. From the Bonferroni inequality,

$$\frac{\alpha_T}{k} \leq \alpha$$

Thus, if we require the value of t associated with each pairwise comparison to exceed the critical value of t corresponding to $\alpha = \alpha_T/k$ before concluding that there is a significant difference between the two groups in question, the error rate for the family of all comparisons will be at most α_T. This procedure is called the *Bonferroni t test* because it is based on the Bonferroni inequality.

For example, if we wished to compare the three mean cortisol levels in the depression example above with Bonferroni t tests to keep the total probability of making at most one false-positive statement to less than 5 percent, we would compare each of the values of t computed for the three individual comparisons to the critical value of t corresponding to $\alpha = .05/3 = .016 = 1.6$ percent for each of the individual comparisons. If, on the other hand, we only wished to compare each of the depressed groups with the control healthy group, then there are only $k = 2$ comparisons, so we would require the value of t associated with each of these two comparisons to exceed the critical value for $\alpha = .05/2 = .025 = 2.5$ percent to achieve a family error rate of 5 percent. The number of degrees of freedom is that associated with the error or residual.

Comparisons between the reference group and the treatment groups are given directly by the t statistics for the regression coefficients. Pairwise comparisons other than between the control group and the other treatment groups can also be obtained from the results of the multiple regression. However, to do so one needs to do some additional computations to obtain the appropriate values of t. In general, to compare groups i and j, we compute a t test statistic according to

$$t = \frac{b_i - b_j}{\sqrt{MS_{res}(1/n_i + 1/n_j)}} \tag{7.9}$$

where b_i and b_j are the regression coefficients which estimate how much groups i and j differ from the reference group (which corresponds to all dummy variables being equal to zero) and n_i and n_j are the number of individuals in groups i and j, respectively. MS_{res} is the residual mean square from the regression equation, and the number of degrees of freedom for the test is the number of degrees of freedom associated with this mean square. For multiple comparison testing in which all possible pairwise comparisons are of potential interest, one may use this equation with an appropriate Bonferroni correction.

More on Hormones and Depression

We are now in a position to use the results of the multiple regression analysis of the data on cortisol concentrations and depression to test whether or not either of the depressed groups had cortisol levels significantly different from those of the healthy group. We will address these questions by testing whether the regression coefficients b_N and b_M in Eq. (7.8) are significantly different from zero. If either of these coefficients is significantly different from zero, it will indicate that the mean cortisol concentration in the corresponding group of people is significantly different from the mean cortisol concentration observed in the healthy group.

We want to control the overall P value for this family of comparisons at .05, so we will require that each individual comparison have an associated value of t that exceeds the critical value corresponding to $0.05/2 = 0.025$ before concluding that there is a significant difference from the healthy group. According to Fig. 7-4, there are 53 degrees of freedom associated with the residual mean square in the regression equation, so we use a critical value of t of 2.31 (interpolating in Table E-1, Appendix E).

The results of the multiple regression analysis of the cortisol data show that the average change in cortisol concentration from control for nonmelancholic depressed people is $b_N = 1.5$ µg/dL with an associated standard error of $s_{b_N} = 1.16$ µg/dL. We test whether this coefficient is significantly different from zero by computing

$$t = \frac{b_N}{s_{b_N}} = \frac{1.5 \ \mu g/dL}{1.16 \ \mu g/dL} = 1.29 \tag{7.10}$$

(This value of t also appears in the computer output in Fig. 7-4.) This value of t does not exceed the critical value of 2.31, so we conclude that cortisol concentrations are not significantly elevated in nonmelancholic depressed individuals compared with healthy individuals.

Likewise, Fig. 7-4 shows that the mean difference in cortisol concentration for melancholic depressed individuals compared to healthy individuals is $b_M = 4.3 \ \mu g/dL$ with a standard error of $s_{b_M} = 1.21 \ \mu g/dL$. We test whether this coefficient is significantly different from zero by computing

$$t = \frac{b_M}{s_{b_M}} = \frac{4.3 \ \mu g/dL}{1.21 \ \mu g/dL} = 3.55 \tag{7.11}$$

which exceeds the critical value of 2.31, so we conclude that cortisol levels are significantly increased over healthy individuals in melancholic depressed individuals. The overall risk of a false-positive statement (type I error) in the family consisting of these two tests is 5 percent.

Suppose we had wished to do all possible comparisons among the different groups of people rather than simple comparisons against the healthy group. There are three such comparisons: healthy vs. nonmelancholic depressed, healthy vs. melancholic depressed, and nonmelancholic depressed vs. melancholic depressed. Because there are now three comparisons in the family rather than two, the Bonferroni correction requires that each individual test be significant at the $.05/3 = 0.016$ level before reporting a significant difference to keep the overall error rate at 5 percent. (The number of degrees of freedom remains the same at 53.) By interpolating in Table E-1, Appendix E, we find the critical value of t is now 2.49.

We still use Eqs. (7.10) and (7.11) to compute t statistics for the two comparisons against the healthy group and we reach the same conclusions: the t of 1.29 from comparing healthy to nonmelancholic depressed individuals does not exceed the critical value of 2.49, and the t of 3.55 from comparing healthy to melancholic depressed does exceed this critical value. To compare the cortisol concentrations in the nonmelancholic depressed and melancholic depressed groups, use the results in Table 7-3 and Fig. 7-4 with Eq. (7.9) to compute

$$t = \frac{b_M - b_N}{\sqrt{MS_{res}(1/n_M + 1/n_N)}} = \frac{4.300 - 1.500}{\sqrt{12.45(1/18 + 1/22)}} = 2.50 \tag{7.12}$$

This value exceeds the critical value of 2.49, so we conclude that these two groups of people have significantly different cortisol concentrations in their blood. The results of this family of three tests lead us to conclude that there are two subgroups of individuals. The healthy and nonmelancholic depressed individuals have similar levels of cortisol, whereas the melancholic depressed individuals have higher levels of cortisol (Fig. 7-3).

Thus, in terms of the original clinical question, severe melancholic depression shows evidence of abnormality in cortisol production by the adrenal gland. Less severely affected nonmelancholic depressed individuals do not show this abnormal cortisol level. Thus, severe depression is associated with an abnormal regulation of the adrenal gland. Further studies by Price and his coworkers showed that the abnormal cortisol production by the adrenal gland was due to an abnormally low inhibitory effect of a certain type of nerve cell in the brain.

The Bonferroni correction to the t test works reasonably well when there are only a few groups to compare, but as the number of comparisons k increases above about 10, the value of t required to conclude that a difference exists becomes much larger than it really needs to be and the method becomes very conservative.* Other multiple comparison procedures, such as Dunnett's test and the Student-Newman-Keuls test (discussed below), are less conservative. All, however, are similar to the Bonferroni test in that they are essentially modifications of the t test to account for the fact that we are making multiple comparisons.

Dunnett's Test for Multiple Comparisons against a Single Control Group

The principal problem with the Bonferroni t test is the fact that it is based on the Bonferroni inequality, which provides an *upper bound* for the family error rate. Because the critical values of t for the individual comparisons are based on this upper bound rather than the actual compounded risk of a false positive (type I error), the test is conservative. Dunnett's test is an alternative for comparing several groups to a single control group. Dunnett's test statistic q' is computed exactly the same way as the Bonferroni t test statistic, but with a more sophisticated mathematical model of the way the error accumulates in order to derive the associated table of critical values for hypothesis testing (Table E-3, Appendix E). Like the t test, the critical values for Dunnett's test depend on α_T, the total risk of erroneously reporting at least one false-positive conclusion from all the comparisons made, and the number of degrees of freedom associated with the residual (or error) term in the regression (or analysis of variance). In addition, the critical value depends on a parameter p, the number of means spanned in the comparison. Dunnett's

---

*The probability of a false-negative (type II) error increases.

test has the advantage of automatically accounting for the multiple comparisons and being less conservative than the Bonferroni t test.

The procedure for Dunnett's test is to first rank the means in ascending order. For example, if we had 6 groups, we first would rank them in order with group 1 having the smallest mean and group 6 having the largest. Group 3 is the control group (Fig. 7-5). Next, compare the control group mean (group 3) with the means below it, starting with the largest difference (group 3 vs. group 1) and working toward the smallest difference (group 3 vs. group 2). Next, repeat this procedure to compare the control group mean with the means above it, again testing the largest difference first (group 3 vs. group 6) and working down to the smallest

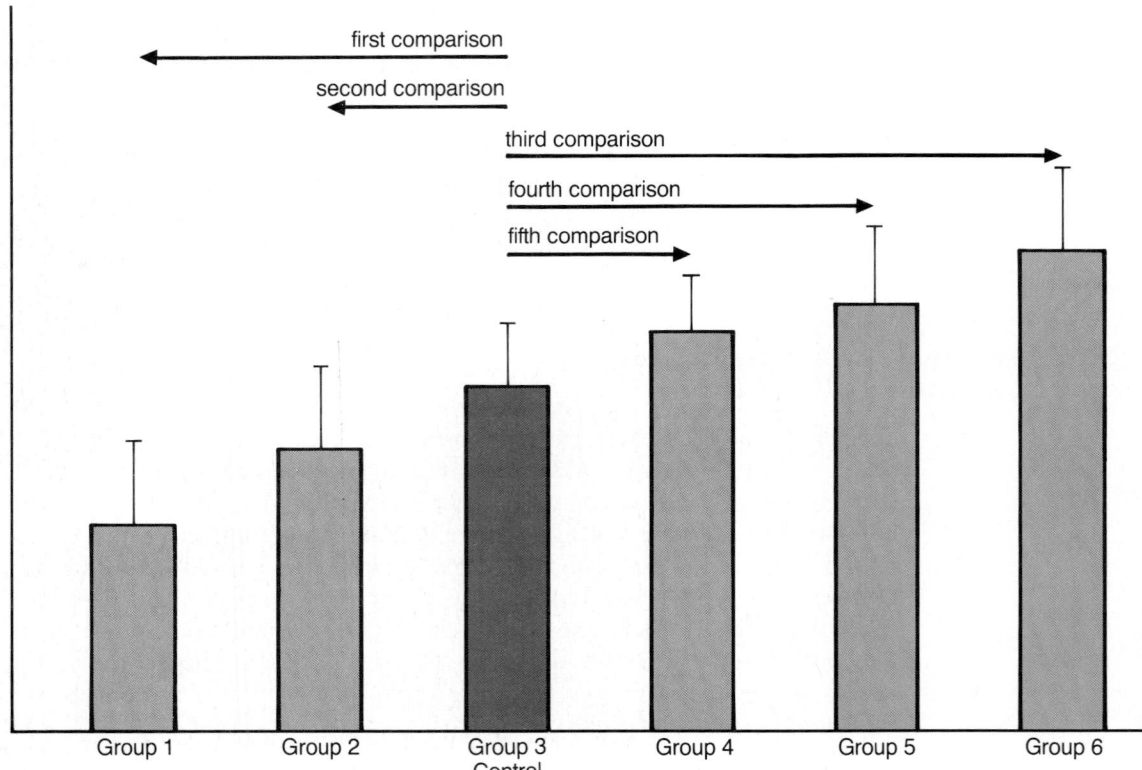

FIGURE 7-5 Schematic illustration of multiple comparisons against a single control group such as one would do with Dunnett's test. The groups are first arranged in order of increasing value of the mean. Then we begin with the comparison of the control group (group 3) with the group that is most different below the control (group 1), then the next closest group (group 2). After all of the comparisons below control are completed, we make the comparisons above the control group, again beginning with the largest difference.

(group 3 vs. group 4). An additional rule is that if we do not reject the null hypothesis of no difference between two means (say, group 3 vs. group 1), we conclude that no difference exists between the control mean and other means enclosed by the two (here, group 3 vs. group 2) and *do not explicitly test* for such differences. For each comparison, the parameter p equals the number of means being compared. In this example, six groups are being compared, so $p = 6$ (Fig. 7-5).

Dunnett's test is easy to conduct with the results of a multiple regression formulation of an analysis of variance, so long as all the dummy variables equal zero under control conditions. With this reference cell coding, the individual regression coefficients associated with the different treatments quantify the average differences between the treatment group means and the control (or reference) group mean. Therefore, $t = b_i/s_{b_i} = q'$, and we only need to compare these values with an appropriate table of critical values for q'. If at any point we do not reject the null hypothesis in question, we do not test for smaller differences enclosed by two means for which we conclude there is no difference, and we assume that all smaller differences are also not significant.

To use Dunnett's test to test whether the two depressed groups in the above example had significantly different cortisol concentrations from the healthy group, we rank the group means obtained from Table 7-3 from smallest to largest together with the values of t from the regression output in Fig. 7-4 (Table 7-6) and then compare the healthy group with the most distant of the other means, the next most distant, and so on. The first comparison is the healthy group with the largest mean above it, melancholic depressed. The difference between the mean of the melancholic depressed, 9.2 μg/dL, and the mean of the healthy individuals, 13.5 μg/dL, is 4.3 μg/dL. Note that this value is equal to the value of b_M in Eq. (7.8), as it should be. From Fig. 7-4 the associated t (which equals q') value is 3.55. There are six group means being compared, so the p parameter for Dunnett's test is 6. Interpolating in Table E-3, Appendix E, the critical value of q' for $\alpha = .05$ with 53 degrees of freedom (the denominator degrees of freedom for the F test) and $p = 6$ is 2.60. The observed value of q' exceeds 2.60, so we conclude that there is a statistically

TABLE 7-6 Dunnett's Test Results for Comparing Cortisol Levels (μg/dL) in Depression Groups to the Healthy Group

Comparison	Difference in Means	q'	p	$q'_{.05}$	$P < .05$?
Melancholic vs. healthy	$13.5 - 9.2 = 4.3$	3.55	6	2.60	Yes
Nonmelancholic vs. healthy	$10.7 - 9.2 = 1.5$	1.29		2.60	No

significant difference between the mean cortisol level in healthy subjects and melancholic depressed subjects ($P < .05$). Because we rejected the null hypothesis for this test, we proceed with the test of the next smaller mean difference, between the healthy group and nonmelancholic depressed groups. This difference, 1.5 μg/dL, is given by b_M, and the q' value is identical to the t value associated with b_M, 1.29. Because 1.29 is less than 2.60, we cannot reject the null hypothesis that the mean cortisol levels in the healthy and nonmelancholic depressed individuals are the same. Thus, we obtain the same conclusion as we did with Bonferroni t tests.

The Student-Newman-Keuls Test for All Pairwise Comparisons

The *Student-Newman-Keuls (SNK) test* is a procedure for doing *all* pairwise comparisons between several experimental groups.[*] Like Dunnett's test, the SNK test is similar in spirit to a t test but uses a more sophisticated mathematical model of the underlying sampling process to determine the critical values of the SNK test statistic q. This more sophisticated model gives rise to a less conservative estimate of the total true probability of erroneously concluding a difference exists α_T than does the Bonferroni t test.[†]

The first step in the analysis is to complete an analysis of variance on all the data to test the global null hypothesis that all the samples were drawn from a single population. If this test yields a significant value of F, arrange all the means in increasing order and compute the SNK test statistic q according to

$$q = \frac{\overline{X}_A - \overline{X}_B}{\sqrt{(MS_{res}/2)(1/n_A + 1/n_B)}} \qquad (7.13)$$

[*]For a discussion of the Student-Newman-Keuls test in the context of a traditional analysis of variance, see S. Glantz, *Primer of Biostatistics* (2nd ed.), New York, McGraw-Hill, 1987, pp. 91–96.

[†]In fact, we had to modify Price et al.'s data on cortisol and depression to obtain the unambiguous results we did with Bonferroni t tests for all pairwise comparisons earlier in this chapter. Had we used the actual data, the comparison between the two depressed groups would not have reached statistical significance, giving rise to the ambiguous result that the healthy group was significantly different from the melancholic depressed but not nonmelancholic depressed individuals while both depressed groups were not significantly different from each other. This situation developed because of the conservatism of the Bonferroni t test. Sometimes such ambiguities are unavoidable and require one to exercise judgment in interpreting the results of multiple comparisons.

where $\overline{X}_A$ and $\overline{X}_B$ are the two means being compared, MS_{res} is the variance within the treatment groups from the analysis of variance, and n_A and n_B are the sample sizes from the two groups being compared.

The value of the q test statistic follows easily from the results of a multiple regression formulation of an analysis of variance. Comparing this equation with Eq. (7.9) reveals that

$$q = \sqrt{2}\, t$$

As with Dunnett's test, the critical value of q depends on α_T, the total risk of erroneously asserting a difference for all comparisons combined, the denominator degrees of freedom from the analysis of variance, and a parameter p, which is the number of means which the comparison spans. (Table E-4, Appendix E, is a table of critical values of q.) For example, when comparing the largest and smallest of four means, $p = 4$; when comparing the second smallest and smallest means, $p = 2$.

As with Dunnett's test, the conclusions reached by SNK testing depend on the order in which the pairwise comparisons are made. After ranking the means in ascending order, one first compares the largest mean with the smallest, then the largest with the second smallest, and so on, until the largest has been compared with the second largest. Then one compares the second largest with the smallest, the second largest with the next smallest, and so forth. For example, after ranking four means in ascending order, the sequence of comparisons should be 4 vs. 1, 4 vs. 2, 4 vs. 3, 3 vs. 1, 3 vs. 2, and, finally, 2 vs. 1. If no significant difference exists between two means, we conclude that no difference exists between any means enclosed by the two and do not test them. Thus, in the preceding example, if we had failed to find a significant difference between means 3 and 1, we would not have tested for a difference between means 3 and 2 or means 2 and 1.

Table 7-7 shows the results of a SNK test for the data on cortisol and depression in Table 7-3. We have already computed the t statistics for the three comparisons (healthy vs. nonmelancholic depressed, healthy vs. melancholic depressed, and nonmelancholic vs. melancholic depressed) in Eqs. (7.10), (7.11), and (7.12), so we simply multiply these

TABLE 7-7 Student-Newman-Keuls Test Results for All Possible Comparisons of Mean Cortisol (μg/dL) among Healthy and Depressed Groups

Comparison	Difference in Means	q'	p	$q'_{.05}$	$P < .05$?
Melancholic vs. healthy	$13.5 - 9.2 = 4.3$	5.02	3	3.41	Yes
Nonmelancholic vs. melancholic	$13.5 - 10.7 = 2.8$	3.53	2	2.84	Yes
Nonmelancholic vs. healthy	$10.7 - 9.2 = 1.5$	1.82	2	2.84	No

values by $\sqrt{2}$ to obtain the corresponding values of q. For example, for the comparison of healthy vs. melancholic depressed (the first comparison we make because it is the greatest difference),

$$q = \sqrt{2}\, t = \sqrt{2} \cdot 3.55 = 5.02$$

There are 53 degrees of freedom and $p = 3$ because this comparison spans 3 means, so, interpolating in Table E-4, Appendix E, the critical value of q for $\alpha_T = 0.05$ is 3.41. The value of q associated with this comparison exceeds this critical value, so we conclude that there is a significant difference between the melancholic depressed and healthy groups and proceed to do the next comparison of the melancholic depressed to the nonmelancholic depressed group. As shown in Table 7-7, this comparison yields a value of $q = 3.53$, which exceeds the critical value of 2.84 for 53 degrees of freedom and $p = 2$, so we conclude that this difference is significant. Finally, we compare the nonmelancholic depressed group to the healthy group and find that there is no significant difference between the cortisol levels in these two groups. Thus, as before, we conclude that the melancholic depressed individuals have, on the average, higher concentrations of cortisol in their blood than do the healthy or nonmelancholic depressed groups.

What Is a Family?

We have devoted considerable energy to developing multiple comparison procedures to control the family error rate, the false-positive (type I) error rate for a set of related tests. The question remains, however: What constitutes a family?

At one extreme is the situation in which each individual test is considered to be a family, in which case there is no multiple comparisons problem because we control the false-positive rate for each test separately. We have already rejected this approach because the risks of false positives for the individual tests rapidly compound to create a situation in which the overall risk of a false-positive statement becomes unacceptably high.

At the other extreme is the situation in which one could construe *all* statistical hypothesis tests one conducted during his or her entire career as a family of tests. If you take this posture, you would become tremendously conservative with regard to rejecting the null hypothesis of no treatment effect on each individual test. Although this would keep you from making false-positive errors, this tremendous conservatism would make the risk of a false-negative (type II) error skyrocket.

Clearly, these extreme positions are both unacceptable, and we are still left with the question: What is a reasonable rule for considering a set of tests as being a family? This is an issue related to the judgment and sensibility of the individual investigator, rather than a decision which can be derived based on statistical criteria. Like most other authors, *we will consider a family to be a set of related tests done on a single variable.* Thus, in the context of analysis of variance, the individual F tests for the different treatment effects will be considered members of separate families, but the multiple comparisons procedures done to isolate the origins of significant variability between the different treatment effects will be considered a family of related tests.

Although the issue of controlling family error rates arises most directly in analysis of variance, one is left with the question of whether or not one ought to control the family error rate for a set of statements made about individual regression coefficients in multiple regression. With the exception of the use of regression implementations of analysis of variance, we have done individual tests of hypotheses (t tests) on the separate coefficients in a multiple regression model and treated each one as an independent test. Specifically, we have not controlled for the fact that when we are analyzing a multiple regression equation with, say, three independent variables, we do not correct the t tests done on the individual regression coefficients for the fact that we are doing multiple t tests. This approach is consistent with the philosophy outlined above because we are treating hypothesis tests concerning each variable as a separate family.

Diet, Drugs, and Atherosclerosis

Atherosclerosis is a disorder in which cholesterol, among other things, builds up in the walls of blood vessels, damaging the vessels and restricting blood flow. If vessels are damaged badly enough, they become blocked and can cause a heart attack or stroke. Sievers and his colleagues[*] wanted to determine if the progression of atherosclerosis (and, thus, blood vessel damage) could be slowed by changing diet or giving the calcium channel-blocking drug, verapamil, which could theoretically affect cholesterol incorporation into the blood vessel wall. They measured the percentage of the rabbits' aortas covered by atherosclerotic plaque under controlled high-cholesterol diet conditions and under various combinations of high-cholesterol or normal diet and use of verapamil with the goal of determining whether the progression of atherosclerosis could be halted by normalizing diet or giving verapamil after the atherosclerosis had started.

All rabbits received a high-cholesterol diet for 12 weeks and then were divided into treatment groups including

[*]R. E. Sievers, T. Rasheed, J. Garrett, S. Blumlein, and W. W. Parmley, "Verapamil and Diet Halt Progression of Atherosclerosis in Cholesterol Fed Rabbits," *Cardiovasc. Drugs Therapy* 1:65–69, 1987.

1. Immediate study, i.e., no further treatment after 12 weeks ($n = 8$).
2. High-cholesterol diet for 12 more weeks ($n = 5$).
3. Normal diet for 12 more weeks ($n = 6$).
4. Normal diet with verapamil for 12 more weeks ($n = 6$).
5. High-cholesterol diet with verapamil for 12 more weeks ($n = 5$).

We ask the question: Is there any difference in the amount of plaque observed in the treatment groups compared to that seen after only 12 weeks of cholesterol diet (group 1)? Figure 7-6 shows these data (also in Table C-14, Appendix C).

Because there are five groups, we need to define four dummy variables. We take group 1 as the reference (control) group, and let

$$G_2 = \begin{cases} 1 \text{ if group 2} \\ 0 \text{ otherwise} \end{cases}$$

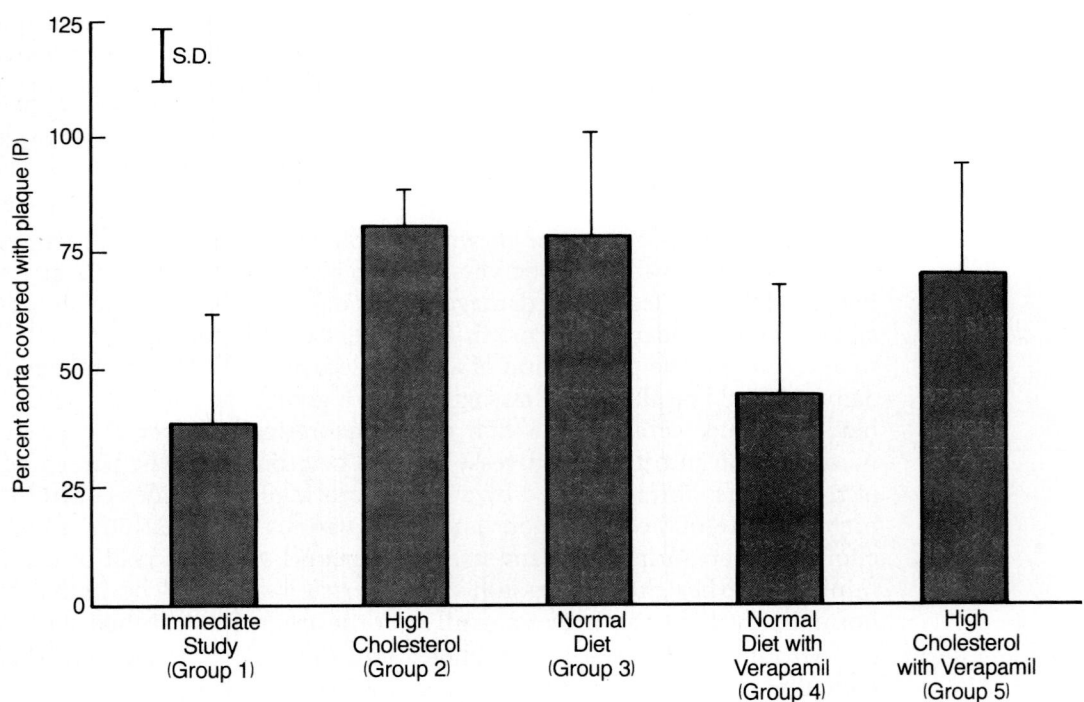

FIGURE 7-6 Percent of aorta covered with atherosclerotic plaque in rabbits fed different diets, some of which were given the calcium channel-blocking drug, verapamil.

$$G_3 = \begin{cases} 1 \text{ if group 3} \\ 0 \text{ otherwise} \end{cases}$$

$$G_4 = \begin{cases} 1 \text{ if group 4} \\ 0 \text{ otherwise} \end{cases}$$

$$G_5 = \begin{cases} 1 \text{ if group 5} \\ 0 \text{ otherwise} \end{cases}$$

or, in more compact notation,

$$G_i = \begin{cases} 1 \text{ if group } i \quad i = 2, \ldots, 5 \\ 0 \text{ otherwise} \end{cases}$$

All four dummy variables equal 0 if the data are from group 1. (These dummy variables are included in the data set in Table C-14, Appendix C.)

We then write the following multiple regression equation to do the analysis of variance:

$$\hat{P} = b_0 + b_2G_2 + b_3G_3 + b_4G_4 + b_5G_5$$

in which $\hat{P}$ is the estimated percentage of the aorta covered by plaque; b_0 is an estimate of the mean percentage of the aorta covered by plaque in the control group (group 1); and b_2, b_3, b_4, and b_5 are estimates of the difference in the percentage of the aorta covered by plaque between the control group mean and each of the four treatment group means.

The results of this multiple linear regression appear in Fig. 7-7. The F value for testing the regression model, which is equivalent to testing the null hypothesis that there is no difference in group means, is 5.32 with 4 numerator degrees of freedom and 25 denominator degrees of freedom, which is highly significant ($P = .003$). We conclude that there is at least one difference between these group means.

By examining the coefficients b_0 and b_2 through b_5, we can obtain estimates of the difference in means of each treatment group compared to the reference group and the significance of these differences. For example, b_0 is 39 percent, which indicates that the mean plaque coverage of the aorta for the control group (group 1) is 39 percent. b_2 is 42 percent, which indicates that the mean aortic plaque coverage in group 2 is 42 percent higher than the mean of the reference group (group 1). The mean of group 2 is $b_0 + b_2 = 39$ percent $+ 42$ percent $= 81$ percent.

To use Dunnett's test for comparison of treatment means with a control group, we first rank the means in ascending order: group 1, group 4, group 5, group 3, group 2. We start by testing the largest difference,

```
The regression equation is
P = 38.6 + 42.3 G2 + 39.9 G3 + 6.1 G4 + 32.2 G5

Predictor         Coef       Stdev      t-ratio          p
Constant        38.638       7.616         5.07      0.000
G2               42.28       12.28         3.44      0.002
G3               39.90       11.63         3.43      0.002
G4                6.13       11.63         0.53      0.603
G5               32.18       12.28         2.62      0.015


s = 21.54        R-sq = 46.0%      R-sq(adj) = 37.3%


Analysis of Variance

SOURCE          DF          SS           MS          F          p
Regression       4       9875.8       2468.9       5.32      0.003
Error           25      11601.2        464.0
Total           29      21476.9
```

FIGURE 7-7 Regression implementation of one-way analysis of variance of the data in Fig. 7-6.

group 2 vs. group 1 (control), by comparing the t statistic associated with each regression coefficient with the critical value of Dunnett's q' (Table 7-8). For this comparison, we use the value of t associated with b_2, the difference between group 2 and the control group. The value of 3.44 (from Fig. 7-7) exceeds the critical value of q' for 25 degrees of freedom and $p = 5$, 2.61, so we conclude that there is a significant difference between group 1 and group 2 ($P < .05$).

We continue by comparing the next largest difference — in this case group 3 vs. group 1 — then the next, and so on. Table 7-8 presents all these comparisons and their associated q' values. This analysis reveals that groups 3 and 5 also are significantly different from group 1; only group 4 is not significantly different from group 1.

TABLE 7-8 Dunnett's Test for Comparison of Percentage of Aorta Covered with Plaque in Treatment Groups Compared with the Control Group Mean

Comparison	Difference in Means	q'	p	$q_{.05}$	$P < .05$?
Group 2 vs. group 1	$80.9 - 38.6 = 42.3$	3.44	5	2.61	Yes
Group 3 vs. group 1	$78.5 - 38.6 = 39.9$	3.43	4	2.51	Yes
Group 5 vs. group 1	$70.8 - 38.6 = 32.2$	2.62	3	2.35	Yes
Group 4 vs. group 1	$44.7 - 38.6 = 6.1$	.53	2	2.06	No

This one run of a regression program, followed by a multiple comparisons using Dunnett's test, has given us all the information we need to answer the original question about whether the calcium channel-blocking drug, verapamil, could halt the progression of atherosclerosis. All rabbits were fed high-cholesterol diets for 12 weeks to establish moderate atherosclerosis. (Group 1 was studied to determine this level.) Group 2 was continued on a high-cholesterol diet to provide a second reference state for the effects of diet and verapamil. Groups 3 (normalized diet only) and 5 (verapamil only) showed significantly higher atherosclerosis than the reference group at 12 weeks, indicating that neither treatment alone halted the progression of atherosclerosis. However, group 4 (which received both normalized diet and verapamil) had atherosclerosis that was not significantly higher than the 12 week reference group, indicating that the progression of atherosclerosis was halted (or greatly slowed) when the combination of normalized diet and verpamil was used. Thus, the combination of treatments, but neither treatment alone, was effective in slowing the damage to blood vessels caused by high cholesterol.

TESTING THE ASSUMPTIONS IN ANALYSIS OF VARIANCE

The primary assumptions for analysis of variance are that the population members within each treatment group are normally distributed with the same variance. These assumptions are the counterparts of those we evaluated for multiple regression in Chap. 4. The fact that an analysis of variance problem can be treated as a regression problem with appropriately defined dummy variables should make it seem reasonable that we can use the same approaches to testing the data for outliers and model misspecification. In contrast to the situation for multiple regression, we have several observations under the same circumstances (i.e., in each treatment group) for analysis of variance, so we can approach the problem of testing the assumptions more directly that we did in multiple regression. Finally, because analysis of variance is a form of linear regression, we can use the same techniques discussed in Chap. 4 for linear regression to identify outliers and leverage points that warrant further consideration.

As with linear regression in general, the first step in evaluating whether a set of data meet the assumptions of analysis of variance is to examine a plot of the residuals. In analysis of variance there is not a continuously varying independent variable, only different treatment groups. Therefore, we simply plot the residuals as a function of the different treatment groups. A great deal can be learned from examining these plots as well as the variances within each of the groups. Likewise, the residuals can be studied by preparing a normal probability plot of the residuals to test whether or not they are consistent with the assumption of normally distributed members of the population.

Analysis of variance is a robust method, which means that it can tolerate deviations from the equal variance assumption without introducing major errors in the results, particularly when the data are balanced (the sample sizes in each treatment group are equal). Although statisticians are vague about just how large a deviation from the equal variance assumption is acceptable, as a rule of thumb, a factor of 3 or 4 difference in the residual variances within the groups is probably large enough to investigate further. If the equal variance assumption appears to be violated, you can transform the dependent variable (using one of the transformations in Table 4-7) or revise the model (perhaps by introducing interaction terms). If these steps fail to stabilize the variances, there are a variety of nonparametric techniques for conducting an analysis of variance based on ranks, which avoid the need for assuming equal variances or a normally distributed population.*

Formal Tests of Homogeneity of Variance

Bartlett's test is the most widely used test of homogeneity of variance because it was one of the first tests developed and it does not require equal sample sizes. However, it is not very highly recommended because, although it has reasonable power when it can be safely assumed that the normality assumption holds, it performs badly when the normality assumption does not hold.[†] Furthermore, it is complex to calculate and not

---

*For a discussion of analysis of variance based on ranks, see S. Glantz, *Primer of Biostatistics* (2nd ed.), New York, McGraw-Hill, 1987, pp. 310–326. Rank-based methods have the advantage of avoiding the normality and equal variance assumptions but generally require balanced designs and no missing data.

[†]Most of the formal tests for homogeneity of variance are very sensitive to deviations from the normality assumption. This poor performance of tests for homogeneity of variance coupled with the robustness of analysis of variance has led some statisticians to recommend that formal tests for homogeneity of variance should not be used before an analysis of variance. For further discussion of Bartlett's test, see J. H. Zar, *Biostatistical Analysis* (2nd ed.), Englewood Cliffs, NJ, Prentice-Hall, 1984, pp. 181–183, and B. J. Winer, *Statistical Principles in Experimental Design* (2nd ed.), New York, McGraw-Hill, 1971, pp. 208–210. Most of the common alternatives to Bartlett's test are more intuitive, although they suffer from the same lack of robustness as Bartlett's test and some, like Hartley's test, are even less useful because they require equal sample sizes. For further discussion of Hartley's test, see B. J. Winer, *Statistical Principles in Experimental Design* (2nd ed.), New York, McGraw-Hill, 1971, pp. 206–209; and L. Ott, *An Introduction to Statistical Methods and Data Analysis* (3rd ed.), Boston, PWS-Kent, 1988, pp. 415–416.

very intuitive. Better behaved, but less common, tests have been developed. The best of these tests is a modified Levene test, which we will call the *Levene median test.*[*] The Levene median test is not as sensitive to deviations from the assumption of normality, and it is only slightly less powerful than Bartlett's test.

The rationale for the *Levene median test* is as follows. The raw data points within each group are reexpressed as *absolute* deviations from the group median

$$Z_{ij} = |X_{ij} - \tilde{X}_i|$$

where X_{ij} is the jth data point in the ith group and $\tilde{X}_i$ is the median of the ith group. Like the process of subtracting the mean, this transformation is also known as centering because it moves the center of the distribution of data points to zero and reexpresses each value as a deviation from the center of the sample distribution. (The original Levene test centered using the group mean $\overline{X}_i$, but modifying it to use $\tilde{X}_i$ improves its performance.) We test to see if the means of the Z_{ij} values, $\overline{Z}_i$, for each group are different in the different groups by using an analysis of variance on the transformed data. If the F statistic associated with this analysis of variance of the Z_{ij} is statistically significant, we conclude that the equal variance assumption is violated. When the sample size is an odd number and small (less than 20 for the group in question), the Levene median test is too likely to lead to the conclusion that the variances are unequal. This peculiarity disappears if the median values (i.e., $Z_{ij} = 0$) are omitted from those groups in which the sample size is odd and less than 20 before doing the analysis of variance.

More on Diet, Drugs, and Atherosclerosis

Earlier in this chapter, we studied how lowering the cholesterol in the diets of rabbits and administering verapamil affected the progression of atherosclerosis in rabbit arteries. At that time, we concluded that only the group treated with verapamil and normal diet showed evidence of halting the progression of atherosclerosis. However, we did not assess the assumption of homogeneity of variance.

[*]W. J. Conover, M. E. Johnson, and M. M. Johnson, "A Comparative Study of Test for Homogeneity of Variances, with Applications to the Outer Continental Shelf Bidding Data," *Technometrics* 23:351–361, 1981. They recommended the Levene median test based on their assessment of the relative performance of more than 50 different tests of homogeneity of variances. For further discussion of the relative merits of different tests for homogeneity of variances, see G. A. Milliken and D. E. Johnson, *Analysis of Messy Data: Volume 1: Designed Experiments,* New York, Van Nostrand Reinhold, 1984, pp. 18–19.

We have seen what a violation of this assumption looks like for linear regression, and it seems reasonable to examine the homogeneity of variance assumption using similar methods. First, we examine this assumption by plotting the residuals from the regression implementation of analysis of variance vs. group number (Fig. 7-8). Although the band of residual does not have a shape that obviously suggests a violation of this assumption, there is one group (group 2) that has a smaller variance than the other four.

To formally test whether this difference is severe enough to constitute evidence for unequal variances, we will use the Levene median test. First, we transform the raw data (from Table C-14, Appendix C) into the absolute deviates from the median in Table 7-9C. For example, the median percentage of the aorta covered by plaque in group 1 is 33.7 percent, so the absolute deviation from the group median for the first data point in this group, 22 percent, is

$$Z_{11} = |X_{11} - \tilde{X}_1| = |22 - 33.7| = |-11.7| = 11.7$$

All other data points are similarly transformed. Because all sample sizes are less than 20, we need to also delete any deviate value that is equal to zero in any group with an odd sample size. This situation occurs in groups 2 and 5, so we delete the zeroes before proceeding. We then do an ordinary analysis of variance on the resulting data (Table 7-9C); the output is in Fig. 7-9. F is .78 with 4 and 23 degrees of freedom. The associated P value is .55, so we conclude that there is no significant difference in the variability within each group ($P > .55$). Thus, we have no evidence that these data violated the assumption of homogeneity of variances.

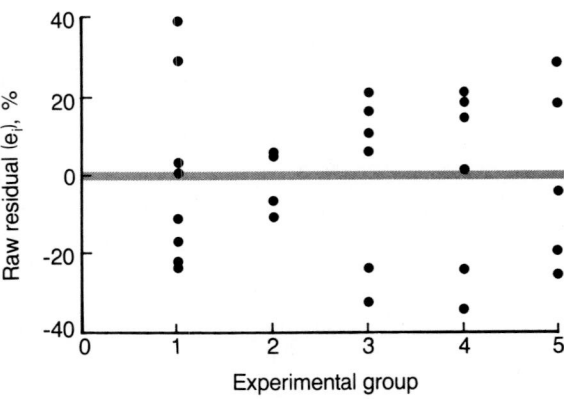

FIGURE 7-8 Residual plot from the regression implementation of the analysis of variance in Fig. 7-7. There is no systematic pattern to the residuals. Group 2 appears to have a smaller variance than the other groups, but this visual impression is not confirmed by the Levene median test, which shows no statistically significant heterogeneity of variances.

TABLE 7-9 Absolute Values of Centered (to the Median) Percentages of Aortic Plaque in Cholesterol-Fed Rabbits for the Levene Median Test of Homogeneity of Variances in Analysis of Variance

	Group 1	Group 2	Group 3	Group 4	Group 5
A. RAW PERCENTAGE OF AORTA COVERED BY PLAQUE					
	22.0	85.7	46.4	66.4	45.7
	14.9	87.0	100.0	63.9	67.1
	16.8	74.6	89.6	46.5	89.6
	68.0	86.7	95.4	10.8	100.0
	78.0	70.6	85.0	21.0	51.7
	42.0		54.8	60.0	
	39.4				
	28.0				
Median	33.7	85.7	87.3	53.25	67.1
B. DATA CENTERED TO THE GROUP MEDIANS					
	−11.7	0.0	−40.9	13.15	−21.4
	−18.8	1.3	12.7	10.65	0.0
	−16.9	−11.1	2.3	−6.75	22.5
	34.3	1.0	8.1	−42.45	32.9
	44.3	−15.1	−2.3	−32.25	−15.4
	8.3		−32.5	6.75	
	5.7				
	−5.7				

C. ABSOLUTE VALUES OF CENTERED DATA WITH MEDIANS DELETED IN GROUPS WITH ODD SAMPLE SIZE

	Z_1	Z_2	Z_3	Z_4	Z_5
	11.7	—	40.9	13.15	21.4
	18.8	1.3	12.7	10.65	—
	16.9	11.1	2.3	6.75	22.5
	34.3	1.0	8.1	42.45	32.9
	44.3	15.1	2.3	32.25	15.4
	8.3		32.5	6.75	
	5.7				
	5.7				

```
ANALYSIS OF VARIANCE
SOURCE       DF        SS        MS        F        p
FACTOR        4       567       142      0.78     0.549
ERROR        23      4173       181
TOTAL        27      4740
                                    INDIVIDUAL 95 PCT CI'S FOR MEAN
                                    BASED ON POOLED STDEV
LEVEL         N      MEAN      STDEV   ------+---------+---------+---------+
Group1 Z      8     18.21     14.12                 (-------*-------)
Group2 Z      4      7.13      7.09    (-----------*-----------)
Group3 Z      6     16.47     16.37            (---------*-------)
Group4 Z      6     18.67     15.03          (---------*--------)
Group5 Z      4     23.05      7.27               (---------*-----------)
                                        ------+---------+---------+---------+
POOLED STDEV =      13.47                  0        12        24        36
```

FIGURE 7-9 The analysis of variance computed as part of the Levene median test for equality of variance using the data on atherosclerosis in Figs. 7-6, 7-7, and 7-8.

SUMMARY

One of the advantages of doing analysis of variance with multiple regression is that it is an extremely flexible approach. As we will see in the next two chapters, depending on the experimental design, one can incorporate various features into a complicated model that will be more easily understood and interpreted if handled in a multiple regression form.

This ability to generalize easily to complex multi-way experimental designs and the flexibility of the regression model approach to analyzing multivariate data are not without limits. One must always keep in mind that the multiple linear regression approach to analysis of variance makes the same assumptions that traditional approaches do. Specifically, analysis of variance assumes that the population from which the data were drawn is normally distributed within each treatment group with the same variances. Another assumption is that the effects associated with the different experimental factors (and their interactions) are *additive*. This assumption, which is immediately evident by examining the regression equation, is often overlooked in traditional analysis of variance computations. There is nothing in the multiple regression approach that allows one to violate these assumptions. Thus, these assumptions should be evaluated in the same manner as in any regression problem. If these assumptions can be reasonably expected to hold, analysis of variance using multiple linear regression is a powerful tool for analyzing real data.

PROBLEMS

7.1 Recently, Yoshida and coworkers[*] noted that a patient with myotonic dystrophy (a neuromuscular disease) had elevated blood calcium levels. A search of the clinical literature revealed other case reports that suggested a problem with the parathyroid glands in patients with myotonic dystrophy. They hypothesized that patients with myotonic dystrophy also have hyperparathyroidism, a syndrome caused by an overactive parathyroid gland. To test this hypothesis, they measured the body's response to a calcium challenge in patients with myotonic dystrophy, patients with other types of dystrophy (nonmyotonic dystrophy), and normal subjects. One of the biochemical variables they measured was the amount of cyclic AMP (cAMP) excreted by the kidneys (the data are in Table D-18, Appendix D). Elevated renal cAMP excretion would be consistent with hyperparathyroidism. Is there any evidence that cAMP excretion is elevated in myotonic dystrophic subjects?

7.2 Evaluate the equality of variance assumption for the analysis of Prob. 7.1 (the data are in Table D-18, Appendix D). If there is a problem, try to stabilize the variances. If successful in stabilizing the variances, reanalyze the data. Do your conclusions change? If not successful, what do you do next?

7.3 Evaluate the normality assumption of the data used in the analysis of Prob. 7.1. If the data are not normally distributed, what can you do?

7.4 Sports medicine workers often want to determine the body composition (density and percentage of fat) of athletes. The typical way this is done is to weigh the individual in a tub of water to determine the amount of water displaced. For paraplegic athletes, weighing in water is technically difficult and stressful. Because body composition in normal athletes can be predicted using a variety of body measurements (for example, neck size, biceps circumference, weight, height, chest size), Bulbulian and his coworkers[†] wanted to see if a similar approach could be used to determine the body composition of paraplegic athletes. The equations used to predict body composition from a variety of measurements are different for different body types; for example, ectomorphs—skinny athletes such

[*]H. Yoshida, H. Oshima, E. Saito, and M. Kinoshita, "Hyperparathyroidism in Patients with Myotonic Dystrophy," *J. Clin. Endocrinol. Metab.* 67:488–492, 1988.

[†]R. Bulbulian, R. E. Johnson, J. J. Gruber, and B. Darabos, "Body Composition in Paraplegic Male Athletes," *Med. Sci. Sports Exerc.* 19:195–201, 1987.

as long-distance runners—require a different equation than meso-morphs—beefy athletes such as fullbacks—of similar height. There-fore, as part of their evaluation, these investigators wanted to determine which body type was predominant in paraplegic athletes. They hypothesized that the increased upper body development of paraplegics would make them have the characteristics of meso-morphs (the data for biceps circumference are in Table D-19, Appendix D). Is there any evidence that the biceps muscle circumference is different in the two body types (ectomorph or mesomorph) and paraplegics? If so, use appropriate pairwise comparisons to pinpoint the differences.

7.5 Evaluate the assumption of homogeneity of variances in the data analyzed in Prob. 7.4 (the data are in Table D-19, Appendix D). If necessary, reanalyze the data and discuss any differences from the original conclusions.

7.6 Saliva contains many enzymes and proteins that serve a variety of biological functions, including initiating digestion and controlling the mouth's environment. The latter may be important in the development of tooth decay. The protein content of saliva seems to be regulated by sympathetic nerves, a component of the autonomic nervous system. β_1-receptors for catecholamines released by sympathetic nerves have been shown to be important, but little is known about the role of β_2-receptors. To better define the relative importance of these two receptor types in regulating salivary proteins, particularly those proteins that contain large amounts of the amino acid proline, Johnson and Cortez[*] gave five groups of rats one of four drugs or a saline injection (control). The four drugs were isoproterenol, a nonselective β-stimulant that affects both types of receptors; dobutamine, a stimulant selective for β_1-receptors; ter-butaline, a stimulant selective for β_2-receptors; and metoprolol, a blocker selective for β_1-receptors. They then measured the proline-rich protein content of the saliva produced by one of the salivary glands (the data are in Table D-20, Appendix D). Which receptor type, if any, seems to control the content of proline-rich proteins in rat saliva?

7.7 Evaluate the assumption of homogeneity of variances in the data analyzed in Prob. 7.6 (the data are in Table D-20, Appendix D). If necessary, reanalyze the data and discuss any effect on the interpretation of the data.

[*]D. A. Johnson and J. E. Cortez, "Chronic Treatment with Beta Adrenergic Agonists and Antagonists Alters the Composition of Proteins in Rat Parotid Saliva," *J. Dent. Res.* 67:1103–1108, 1988.

7.8 Use the Student-Newman-Keuls test to do all possible comparisons of the data analyzed in Prob. 7.6. Interpret the results in terms of the physiological question asked in Prob. 7.6.

7.9 What is the critical value of t to which each t statistic is compared if you use Bonferroni t tests to do all possible comparisons of the data analyzed in Probs. 7.6 and 7.8?

7.10 Compute all possible comparisons among group means computed in Prob. 7.6 using Bonferroni t tests. Interpret your results.

Two-Way Analysis of Variance

The experiments we analyzed in Chap. 7 were ones in which the subjects were divided according to a *single factor*, such as mental status or diet. Although many experiments can be analyzed using that design, there are also times when one wishes to divide the experimental subjects into groups according to *two factors*. For example, in Chap. 7 we tested the hypothesis that circulating cortisol levels differed among normal, nonmelancholy depressed, and melancholy depressed people. Now, suppose that we wanted to investigate whether or not the results depended on the sex of the person being studied. In this case we have a *two-factor*, or *two-way*, analysis of variance problem in which each individual experimental subject is classified according to two factors (Table 8-1). Although fundamentally the same as the single-factor analysis of variance discussed in Chap. 7, the two-factor analysis of variance provides a different perspective on the data.

TABLE 8-1 Schematic Diagram of Design of Hypothetical Two-Factor Analysis of Variance for Depression Effect on Cortisol Level, Also Taking into Account the Sex of the Individuals

	Diagnosis		
Sex	Healthy	Nonmelancholy Depressed	Melancholy Depressed
Male			
Female			

First, two-way analysis of variance allows us to use the same data to test *three* hypotheses about the data we collected. We can test whether or not the dependent variable changes significantly with each of the two factors (while taking into account the effects of the other factor) as well as test whether or not the effects of each of the two factors are the same regardless of the level of the other one. This third hypothesis is whether or not there is a significant *interaction effect* between the two factors. In terms of the hypothetical study of the dependence of cortisol concentration on mental state and sex, these three hypotheses would be

1. Cortisol concentration does not depend on mental state (the same hypothesis as before).
2. Cortisol concentration does not depend on sex.
3. The effect of mental state on cortisol does not depend on sex.

Second, all things being equal, a two-factor analysis of variance is more sensitive than multiple single-factor analyses of variance looking for similar effects. When one sets up an experiment or observational study according to the analysis of variance paradigm, the experimental subjects are assigned at random to the different treatment groups, so the only difference between the different groups is the treatment (or, in an observational study, the presence of a certain characteristic of interest). The randomization process is designed to eliminate systematic effects of other potentially important characteristics of the experimental subjects by randomly distributing these characteristics across all the sample groups. It is this assumption of randomness that allows us to compare the different group means by examining the ratio of the variance estimated from looking between the group means to the variance estimated from looking within the groups. If the resulting F statistic is large, we conclude that there is a statistically significant difference among the treatment groups.

There are two ways in which the value of F can become large. First, and most obvious, if the size of the treatment effect can be increased, the variation between the group means increases (compared with the variation within the groups), so that the numerator of F increases. Second, and less obvious, if the amount of variation within the various sample groups can be reduced, the denominator of F decreases. This residual variation is a reflection of the random variation within the treatment groups. Part of this variation is the underlying random variation between subjects — including biological variation — and part of this variation is due to the fact that there are other factors that could affect the dependent variable which are not being considered in the experimental design. In a two-way design, we explicitly take into account one of these other factors, and the net result is a reduction in the residual variation, and thus a reduction in the denominator of F. This situation is exactly parallel to adding another independent variable to a regression equation to improve its ability to describe the data.

In sum, two-way analysis of variance permits you to consider simultaneously the effects of two different factors on a variable of interest and test for interaction between these two factors. Because we are using more information to classify individual experimental subjects, we reduce the residual variation and produce a more sensitive test of the hypotheses that each of the factors under study had a significant effect on the dependent variable of interest than we would obtain with the analogous one-way (single-factor) analysis of variance.

We will again take a regression approach to analyses of variance. The reasons for so doing include the ease of interpretability that led us to use this approach in Chap. 7, and additional benefits that result when there are missing data. Traditional analysis of variance with two or more treatment factors generally requires that each sample group be the same size and that there be no missing data.[*] Unfortunately, this requirement is rarely satisfied in practice, and investigators are often confronted with the problem of how to analyze data that do not strictly fit the traditional analysis of variance paradigm. Confronted with this problem, people will sometimes estimate values for the missing data, delete all the observations for subjects with incomplete data, or fall back on pairwise analysis of the data using multiple t tests. Filling in missing data involves "making up" points, whereas deleting subjects with incomplete data means ignoring potentially important information which is often gathered at considerable expense. Deleting data also reduces the power of the statistical analysis to detect true effects of the treatment. As we noted in Chap. 7, analyzing a single set of data with multiple t tests both decreases the power of the individual tests and increases the risk of reporting a false positive.

One can avoid these pitfalls by recasting the analysis of variance as a multiple linear regression problem through the use of dummy variables.

---

[*]Traditional analysis of variance computations can also be completed in some other restricted conditions, such as when the unequal numbers of observations in the different cells vary in proportion, so that the number of observations in any cell n_{ij} is

$$n_{ij} = \frac{(\text{number of observations in row } i) \times (\text{number of observations in column } j)}{N}$$

where N is the total number of observations in all cells. For further discussion of proportional replication, see B. J. Winer, *Statistical Principles in Experimental Design* (2nd ed.), New York, McGraw-Hill, 1971, pp. 419–422, and J. H. Zar, *Biostatistical Analysis* (2nd ed.), Englewood Cliffs, N.J., Prentice-Hall, 1984, pp. 214–216.

When the sample sizes for each treatment are the same (a so-called balanced experimental design) and there are no missing data, the results of a traditional analysis of variance and the corresponding multiple regression problem are identical. When the sample sizes are unequal or there are missing data, one can use a regression formulation to analyze data that cannot be handled in some traditional analysis of variance paradigms.

PERSONALITY ASSESSMENT AND FAKING HIGH GENDER IDENTIFICATION

Clinical psychologists have developed several ways to assess an individual's personality traits from answers given to a series of carefully selected questions. One such instrument is the California Psychological Inventory, which consists of 15 scales designed to give insight into some facet of a respondent's behavior and personality and 3 additional validity scales designed to detect individuals trying to fake certain personality traits. Even though these validity scales are present, Montross and his coworkers[*] wondered if the overall assessment of personality with this instrument could be affected by faking strong gender identification—men responding in a very "macho" manner and women responding in a very "feminine" manner—without being detected by the validity scales.

To evaluate the effect of faking strong gender identification, they administered the California Psychological Inventory to 16 men and 16 women. Half of each sex were given standard instructions, whereas the other half were instructed to role-play so that they answered the questions in the way they thought an extremely masculine man (male subjects) or extremely feminine woman (female subjects) would answer. Thus the study design is a two-factor experiment in which one factor is gender (male or female) and the other factor is role playing (present or absent). Data for 1 of the 15 personality assessment scales, the S_p scale, appear in Table 8-2. We can use these data to test three hypotheses:

1. There is no difference in S_p score between men and women.
2. There is no difference in S_p score between people instructed to answer in a very masculine or feminine manner and those taking the test normally.
3. There is no interaction between gender and role playing; the difference in S_p score between role players and controls is the same for men and women.

[*]J. F. Montross, F. Neas, C. L. Smith, and J. H. Hensley, "The Effect of Role-Playing High Gender Identification on the California Psychological Inventory," *J. Clin. Psychol.* 44:160–164, 1988.

TABLE 8-2 Data for Effect of Role-Playing High Gender Identification on the S_p Scale of the California Psychological Inventory in Males and Females

	Instructions for Taking Test	
	Honest Answers	Role Playing
	MALE	
	64.0	37.7
	75.6	53.5
	60.6	33.9
	69.3	78.6
	63.7	46.0
	53.3	38.7
	55.7	65.8
	70.4	68.4
Mean	64.1	52.8
SD	7.6	16.5
	FEMALE	
	41.9	25.6
	55.0	23.1
	32.1	32.8
	50.1	43.5
	52.1	12.2
	56.6	35.4
	51.8	28.0
	51.7	41.9
Mean	48.9	30.3
SD	8.1	10.3

Just as we could analyze a traditional one-way analysis of variance with a multiple linear regression model with appropriately coded dummy variables, we can do this two-way analysis of variance as a multiple linear regression problem. In this example, we have two factors — gender and role playing. We use what we know about dummy variables and generalize from the one-way analysis of variance example to define two dummy variables. The first dummy variable encodes the two levels of the gender effect (i.e., male and female):

$$G = \begin{cases} 0 \text{ if male} \\ 1 \text{ if female} \end{cases}$$

TABLE 8-3 Schematic Diagram of Two-Factor Analysis of Variance for Role-Playing Data Showing Reference Cell Dummy Variable Codes

Gender	Test Instructions	
	Honest Answers	Role Playing
Male	$G = 0$ $R = 0$	$G = 0$ $R = 1$
Female	$G = 1$ $R = 0$	$G = 1$ $R = 1$

and the other dummy variable codes for the two levels of the role-play effect (i.e., role-play or no role-play):

$$R = \begin{cases} 0 \text{ if no role-play} \\ 1 \text{ if role-play} \end{cases}$$

In addition to these two dummy variables to quantify the *main effects* due to the primary experimental factors, we also need to include a variable to quantify the *interaction effect* described above.[*] We do this by including the product of G and R in the regression model,

$$\hat{S} = b_0 + b_G G + b_R R + b_{GR} GR \tag{8.1}$$

in which $\hat{S}$ is the predicted S_p score for a given combination of experimental factors (i.e., dummy variables). Table 8-3 shows the values of the dummy variables associated with each cell in the experimental design.

To see how to interpret the dummy variables, let us consider some of the possible combinations of experimental factors. For men who are not role-playing, $G = 0$ and $R = 0$, so Eq. (8.1) reduces to

$$\hat{S} = b_0 + b_G \cdot 0 + b_R \cdot 0 + b_{GR} \cdot 0 \cdot 0 = b_0$$

and thus b_0 is the mean response of men who are not role-playing. (This is analogous to the reference group in the one-way analysis of variance.) For women who are not role-playing, $G = 1$ and $R = 0$, so Eq. (8.1) reduces to

$$\hat{S} = b_0 + b_G \cdot 1 + b_R \cdot 0 + b_{GR} \cdot 1 \cdot 0 = b_0 + b_G$$

[*]You must have at least two observations per cell to test for interaction. If there is only one observation per cell, the analysis proceeds similarly, but without the interaction term in the regression equation.

Hence b_G is the mean difference in score between women and men who do not role-play. Similar substitutions of the dummy variables reveal that b_R is the mean difference between men who do not role-play and those who do.

Here b_{GR} is the interaction effect, i.e., the incremental effect of being both female and role playing. To see this, we write Eq. (8.1) for role-playing $(R = 1)$ men $(G = 0)$

$$\hat{S} = b_0 + b_G \cdot 0 + b_R \cdot 1 + b_{GR} \cdot 0 \cdot 1 = b_0 + b_R$$

and role-playing $(R = 1)$ women $(G = 1)$

$$\hat{S} = b_0 + b_G \cdot 1 + b_R \cdot 1 + b_{GR} \cdot 1 \cdot 1 = b_0 + b_G + b_R + b_{GR}$$

If there is no interaction, b_{GR} is zero, the effect of gender b_G simply adds to the effect of role-playing b_R, and we conclude that the effects of gender and role playing were simply additive and independent of each other. If, on the other hand, the interaction term is not zero, we conclude that the effects of the two factors are not simply additive, and an additional factor, b_{GR}, needs to be considered. The interpretation of two-way (and other multifactorial analyses of variance) is simpler in the absence of significant interaction effects because it means that the effects of the main experimental factors are independent of each other and simply add.

These data—including the dummy variables as defined above—as they would be submitted for multiple linear regression analysis to do this analysis of variance are shown in Table 8-4. Because there are only two independent variables in this example $(G$ and $R)$, we can plot the data in three dimensions, with the observations concentrated at the values of the corresponding dummy variables (Fig. 8-1A).* Conducting the analysis of variance is equivalent to finding the plane that passes through these points.

The results of the regression analysis are in Fig. 8-2, and the corresponding regression plane is indicated in Fig. 8-1B. As was true for one-way analysis of variance, we can substitute the dummy variables associated with each data cell (Table 8-3) into Eq. (8.1) to show that various sums of the b_i are estimates of the S_p means for each cell. These combinations are shown in Table 8-5. As expected, the appropriate sums of parameter estimates (from Fig. 8-2) equal the various cell mean scores.† Verifying this agreement is a good way to check that you have properly coded the dummy variables.

---

*If there were more than two dummy variables, we could not draw such a picture in three dimensions, but the idea would be the same in a higher dimension space.

†This equality will only be present if all interaction terms are included in the regression model.

TABLE 8-4 Personality Scores and Reference Cell Dummy Variable Codes for Two-Way Analysis of Variance (Refer to Table 8-3)

S_p	G	R	GR	
		MALE		
64.0	0	0	0	
75.6	0	0	0	
60.6	0	0	0	
69.3	0	0	0	Honest
63.7	0	0	0	
53.3	0	0	0	
55.7	0	0	0	
70.4	0	0	0	
37.7	0	1	0	
53.5	0	1	0	
33.9	0	1	0	
78.6	0	1	0	Role-play
46.0	0	1	0	
38.7	0	1	0	
65.8	0	1	0	
68.4	0	1	0	
		FEMALE		
41.9	1	0	0	
55.0	1	0	0	
32.1	1	0	0	
50.1	1	0	0	Honest
52.1	1	0	0	
56.6	1	0	0	
51.8	1	0	0	
51.7	1	0	0	
25.6	1	1	1	
23.1	1	1	1	
32.8	1	1	1	
43.5	1	1	1	Role-play
12.2	1	1	1	
35.4	1	1	1	
28.0	1	1	1	
41.9	1	1	1	

The tests for significant factor and interaction effects require identifying the correct sums of squares due to the various factor and interaction dummy variables, computing the corresponding factor mean

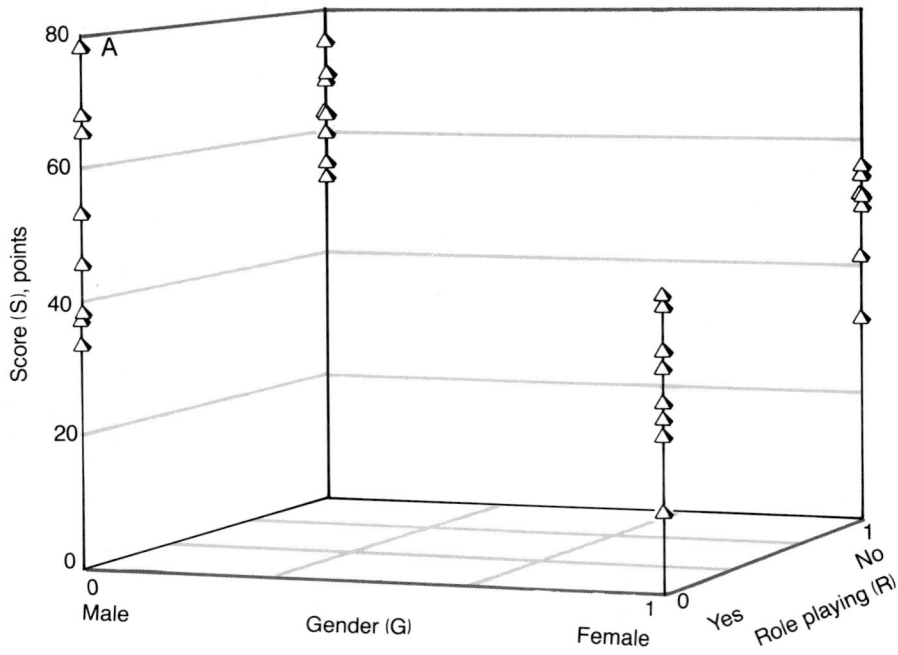

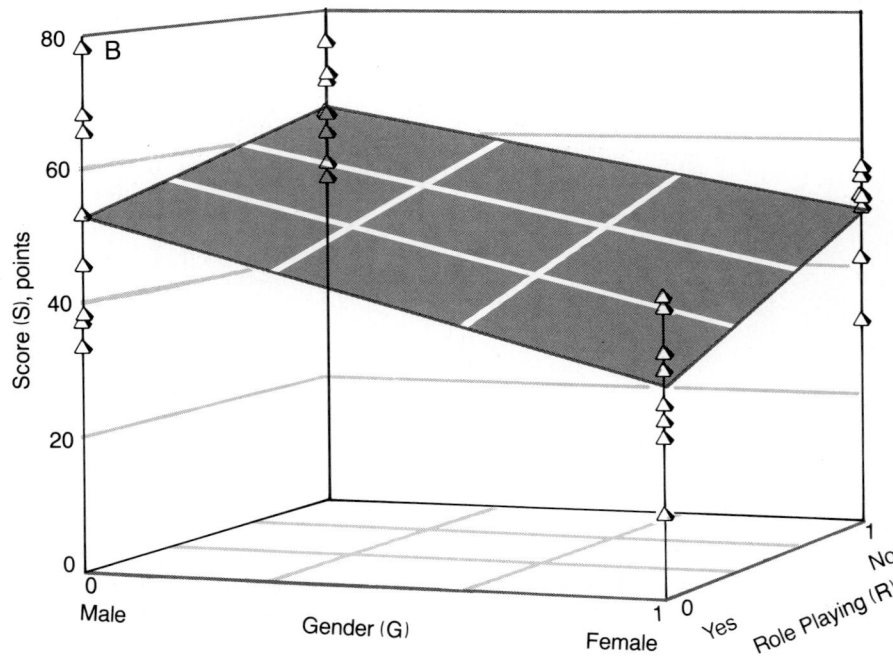

```
DEP VARIABLE: S        Personality Score

                                    ANALYSIS OF VARIANCE

                                 SUM OF          MEAN
                    SOURCE   DF   SQUARES        SQUARE      F VALUE      PROB>F

                    MODEL     3   4728.90125   1576.30042    12.571       0.0001
                    ERROR    28   3510.90750   125.38955357
                    C TOTAL  31   8239.80875

                       ROOT MSE      11.19775     R-SQUARE       0.5739
                       DEP MEAN      49.03125     ADJ R-SQ       0.5283
                       C.V.          22.83798

                                    PARAMETER ESTIMATES

              PARAMETER    STANDARD    T FOR H0:                              VARIANCE   VARIABLE
VARIABLE  DF  ESTIMATE     ERROR       PARAMETER=0   PROB > |T|   TYPE I SS   INFLATION  LABEL

INTERCEP  1   64.07500000  3.95900167   16.185       0.0001    76930.03125         0    INTERCEPT
G         1   -15.1625     5.59887385   -2.708       0.0114     2838.81125   2.00000000 dummy variable = 1 if female
R         1   -11.25       5.59887385   -2.009       0.0542     1782.04500   2.00000000 dummy variable = 1 if role play
GR        1   -7.35        7.91800333   -0.928       0.3612      108.04500000 3.00000000 interaction of G and R
```

FIGURE 8-2 Regression implementation of a two-way analysis of variance for the data in Fig. 8-1. Computations done using reference cell coding.

square by dividing the sum of squares by the appropriate degrees of freedom, then computing F as the factor (or interaction) mean square divided by the error mean square.

From Fig. 8-2, the sum of squares due to the gender is simply the sum of squares attributable to the dummy variable G. Each variable in a linear regression is associated with 1 degree of freedom, and there is only one dummy variable associated with gender; so, from Fig. 8-2,

$$MS_G = \frac{SS_G}{DF_G} = \frac{2838.8}{1} = 2838.8$$

◀ **FIGURE 8-1** (A) Personality score assessments for males and females who did and did not role-play can be plotted in three dimensions, just like we did for regression problems in Chap. 3, with the gender of the experimental subject as one dimension and whether or not they are role-playing as another dimension. The only difference between this situation and the multiple regression analyses discussed in earlier chapters is that the independent variables are dummy variables. (B) Conducting a two-way analysis of variance on these data involves simply fitting the best plane (in the sense of minimizing the residual sum of squares) to these data. If there is a significant effect of gender on the test score, the plane will have a slope in the G direction. If there is a significant effect of role playing on the score, the plane will have a slope in the R direction. If there is an interaction between gender and role playing, the plane will be twisted just as it was when there was an interaction in ordinary linear regression. Note that the number of dimensions depends on the number of dummy variables. We were able to draw this plot because there are only two levels, and hence only one dummy variable, for each of two experimental factors.

TABLE 8-5 Personality Score Cell Means Computed from Regression Coefficients

Gender	Test Instructions	
	Honest Answers	Role Playing
Male	b_0 $= 64.1$	$b_0 + b_R$ $= 64.1 - 11.3 = 52.8$
Female	$b_0 + b_G = 64.1 - 15.2 = 48.9$	$b_0 + b_G + b_R + b_{GR} = 64.1 - 15.2 - 11.3 - 7.4$ $= 30.2$

and

$$MS_{res} = \frac{SS_{res}}{DF_{res}} = \frac{3510.9}{28} = 125.4$$

Use these results to test the hypothesis that the gender of the respondent does not affect the S_p score by computing

$$F = \frac{MS_G}{MS_{res}} = \frac{2838.8}{125.4} = 22.64$$

This F statistic, with 1 numerator and 28 denominator degrees of freedom, exceeds the critical value of F for $\alpha = .01$, 7.64, and we conclude that men and women have different S_p scores $(P < .01)$.

We use exactly the same procedure to test whether role playing has a significant effect on the S_p score. We quantify the variability between the role-playing and non-role-playing groups with the mean square associated with the variable R. From Fig. 8-2,

$$MS_R = \frac{SS_R}{DF_R} = \frac{1782.0}{1} = 1782.0$$

and test whether this variation is large by comparing it to the variability within the groups (i.e., cells) with

$$F = \frac{MS_R}{MS_{res}} = \frac{1782.0}{125.4} = 14.21$$

This value also exceeds the critical value of F, 7.64, and we conclude that role playing significantly affects the S_p score $(P < .01)$.

So far we have concluded that both gender and role playing affect the S_p score of the psychological inventory. The remaining question is: Are these changes independent or does the change in score with role playing depend on gender? To answer this question, we examine the variation associated with the interaction term GR. Using the numbers in Fig. 8-2, we estimate this variation as

$$MS_{GR} = \frac{SS_{GR}}{DF_{GR}} = \frac{108.0}{1} = 108.0$$

and test whether it is large by comparing this variance estimate with MS_{res} by computing

$$F = \frac{MS_{GR}}{MS_{res}} = \frac{108.0}{125.4} = .86$$

which does not even approach the critical value of F for $\alpha = .05$ with 1 and 28 degrees of freedom, 4.20. Therefore, we conclude that there is not a significant interaction effect. In terms of Fig. 8-1B, these results mean that the regression surface is a plane with slopes in both the G and R directions, but that there is no twist in the regression surface. This is the same concept of interaction introduced in Chap. 3, where, in the presence of interaction, the regression surface was twisted because the effect of one independent variable depended on the value of the other.

This analysis leads to the conclusions that the average test scores are significantly different for men and women, that the average test score is affected by role playing, and that these two effects are independent (do not interact). In addition to permitting us to reach these qualitative conclusions, the values of the regression parameters provide a quantitative estimate of the size of the effects of gender and role playing on test scores on this personality scale. b_0 indicates that the average score for men who are not role playing is 64 points; b_G indicates that, on the average, women score 15 points below men when answering honestly, and, most important in terms of the question being studied, b_R indicates that the score drops by an average of 11 points for men when they role-play. Thus, the faking of high gender identification does affect personality assessment by the S_p scale. Furthermore, using other data, these authors found that the validity scales were unaffected, so such faking might very well go undetected.

Table 8-6 shows a traditional analysis of variance table for these data. Comparing the sums of squares, degrees of freedom, mean squares, and F values reveals that they are exactly the same as we achieved using a multiple regression formulation.

The process just described probably seems the long way around to testing whether or not gender and role playing have any effect on personality assessment. After all, we could have reached this same conclusion by simply using a t statistic to test whether or not the coefficients b_G, b_R, and b_{GR} were significantly different from zero. This situation will exist whenever there are only two levels of the factor under study, and

TABLE 8-6 **Traditional Analysis of Variance Table from Analysis of Role-Playing Gender Identification and Its Effects on Personality Assessment**

Source	Degrees of Freedom	Sum of Squares	Mean Square	F	P
Gender (G)	1	2838.8	2838.8	22.64	.001
Role-play (R)	1	1782.0	1782.0	14.21	.001
Interaction (G × R)	1	108.0	108.0	.86	.361
Error	28	3510.9	125.4		
Total	31	8239.8			

so a single dummy variable encodes that factor. When there are more than two levels of a factor, one must conduct the F test using all the variation associated with all the dummy variables used to encode that factor; simply looking at the individual regression coefficients will not work. The next example will illustrate this point, but before we go on to more complicated analyses of variance, we need a new way to code dummy variables.

AN ALTERNATIVE APPROACH FOR CODING DUMMY VARIABLES

So far we have defined dummy variables to encode the different treatment groups in ways that led to the regression coefficients being equal to estimates of the differences between the different treatment groups and the reference (or control) group. This procedure, known as *reference cell coding*, is not the only way that one can define dummy variables. Often, one may desire to encode the dummy variables so that the coefficients of the dummy variables in the regression model quantify *deviations of each treatment group from the overall mean* response rather than as deviations from the mean of a reference group. In fact, not only is this so-called *effects coding* desirable, but it is also helpful for ensuring that you always compute the correct sums of squares when interaction is included in the model. *When there are missing data, effects coding is essential for obtaining the correct sums of squares.*

As before, we define $k - 1$ dummy variables to encode k levels of the factor of interest. Rather than defining the dummy variables as being 0 or 1, however, we define

$$E_1 = \begin{cases} 1 \text{ if group 1} \\ -1 \text{ if group } k \\ 0 \text{ otherwise} \end{cases}$$

$$E_2 = \begin{cases} 1 \text{ if group 2} \\ -1 \text{ if group } k \\ 0 \text{ otherwise} \end{cases}$$

$$E_3 = \begin{cases} 1 \text{ if group 3} \\ -1 \text{ if group } k \\ 0 \text{ otherwise} \end{cases}$$

and so on, up to E_{k-1} for group $k - 1$.

The associated regression equation for a one-way analysis of variance is

$$\hat{y} = e_0 + e_1 E_1 + e_2 E_2 + e_3 E_3 + \cdots + e_{k-1} E_{k-1}$$

where e_0 is the mean value of the dependent variable y observed under all conditions, e_1 estimates the difference between the mean of group 1 and the overall mean, e_2 estimates the difference between group 2 and the overall average, and so forth, through e_{k-1}. Because all the dummy variables are set to -1 for group k, the estimate of the difference of group k from the overall mean is

$$e_k = -\sum_{i=1}^{k-1} e_i$$

An Alternative Approach to Personality

To illustrate this alternative coding, let us recode the dummy variables identifying the treatment groups in the study of the effect of role-playing gender identification on personality assessment using this new coding scheme. As before, we define two dummy variables; one to encode the two genders and one to code whether or not the subjects role-played or answered the questions honestly:

$$G = \begin{cases} 1 \text{ if male} \\ -1 \text{ if female} \end{cases}$$

$$R = \begin{cases} 1 \text{ if no role playing} \\ -1 \text{ if role playing} \end{cases}$$

Table 8-7 shows these new dummy variables; compare it to Table 8-4. Because there are only two levels of each factor, none of the groups has a code of 0.

We obtain the necessary sums of squares as before by fitting these data with the regression equation

$$\hat{S}_p = e_0 + e_G G + e_R R + e_{GR} GR \tag{8.2}$$

TABLE 8-7 Personality Scores and Dummy Variable Codes for Two-Factor Analysis of Variance Recoded Using Effects Coding (Compare to Table 8-4)

S_p	G	R	GR	
		MALE		
64.0	1	1	1	
75.6	1	1	1	
60.6	1	1	1	
69.3	1	1	1	
63.7	1	1	1	Honest
53.3	1	1	1	
55.7	1	1	1	
70.4	1	1	1	
37.7	1	-1	-1	
53.5	1	-1	-1	
33.9	1	-1	-1	
78.6	1	-1	-1	
46.0	1	-1	-1	Role-play
38.7	1	-1	-1	
65.8	1	-1	-1	
68.4	1	-1	-1	
		FEMALE		
41.9	-1	1	-1	
55.0	-1	1	-1	
32.1	-1	1	-1	
50.1	-1	1	-1	
52.1	-1	1	-1	Honest
56.6	-1	1	-1	
51.8	-1	1	-1	
51.7	-1	1	-1	
25.6	-1	-1	1	
23.1	-1	-1	1	
32.8	-1	-1	1	
43.5	-1	-1	1	
12.2	-1	-1	1	Role-play
35.4	-1	-1	1	
28.0	-1	-1	1	
41.9	-1	-1	1	

Figure 8-3 shows the results of fitting this equation to the data in Table 8-7. The total sum of squares, residual sum of squares, degrees of freedom, and multiple correlation coefficient are the same as we obtained before (compare with the results in Fig. 8-2). This result should not sur-

```
DEP VARIABLE: S        Personality Score


                              ANALYSIS OF VARIANCE

                                  SUM OF          MEAN
                   SOURCE    DF    SQUARES         SQUARE       F VALUE      PROB>F

                   MODEL     3    4728.90125     1576.30042     12.571       0.0001
                   ERROR    28    3510.90750     125.38955357
                   C TOTAL  31    8239.80875

                   ROOT MSE      11.19775      R-SQUARE      0.5739
                   DEP MEAN      49.03125      ADJ R-SQ      0.5283
                   C.V.          22.83798


                              PARAMETER ESTIMATES

                 PARAMETER    STANDARD    T FOR HO:                          VARIANCE   VARIABLE
VARIABLE  DF     ESTIMATE      ERROR    PARAMETER=0    PROB > |T|   TYPE I SS INFLATION  LABEL

INTERCEP  1    49.03125000   1.97950083    24.770      0.0001    76930.03125        0   INTERCEPT
G1        1     9.41875000   1.97950083     4.758      0.0001     2838.81125 1.00000000 dummy variable = -1 if female
R1        1     7.46250000   1.97950083     3.770      0.0008     1782.04500 1.00000000 dummy variable = -1 if role play
GR1       1    -1.8375       1.97950083    -0.928      0.3612      108.04500000 1.00000000 interaction of G1 and R1
```

FIGURE 8-3 Analysis of the data in Fig. 8-1 using effects coding. Although the coefficients associated with individual terms are different from those computed using reference cell coding in Fig. 8-2, the degrees of freedom and sums of squares are the same. In contrast to the result obtained using reference cell coding, all the variance inflation factors are equal to 1, indicating no multicollinearity or shared information between any of the dummy variables.

prise you, because we are simply taking a slightly different approach to describing the same data.

Because we defined the dummy variables differently, however, the values of the regression coefficients and their associated standard errors and sums of squares are different than we obtained before. With the new coding procedure, the intercept e_0 is 49.0, indicating that the mean of *all* personality scores was 49 points. The value of $e_G = 9.4$ indicates that the mean personality score in men was *above* the overall mean by 9.4 points because the men were coded with $+1$, whereas the mean personality score for women was *below* the overall mean by 9.4 points because the women were coded with -1. Likewise, the value of $e_R = 7.5$ indicates that the mean for non-role players was above the overall mean by 7.5 points, whereas the mean for role players was below the overall mean by 7.5 points. (The standard errors in Fig. 8-3 provide estimates of the precision of these differences.)

These coefficients contain the same information about differences between the groups that we found in our original coding scheme. For example, we can use the results in Fig. 8-3 to estimate that the mean personality score for men $(G = +1)$ who role-played $(R = -1)$ is

$$e_0 + e_G G - e_R R - e_{GR} GR = 49 + 9.4(+1) + 7.5(-1) + (-1.8)(+1)(-1) = 52.7$$

which is the estimate we obtained before (see Table 8-5).

As before, we test the hypothesis that there is no significant difference between men and women by comparing the mean square associated with the dummy variable G to the residual mean square. From the mean squares shown in Fig. 8-3, we compute

$$F = \frac{MS_G}{MS_{res}} = \frac{2838.8}{125.4} = 22.64$$

with 1 and 28 degrees of freedom to test this hypothesis. This is exactly the same value we found before, and so we conclude that there is a statistically significant difference in personality scores between men and women. Similarly, we compute

$$F = \frac{MS_R}{MS_{res}} = \frac{1782.0}{125.4} = 14.21$$

to test the null hypothesis of no difference between the role-play and non-role-play groups, and

$$F = \frac{MS_{GR}}{MS_{res}} = \frac{108.0}{125.4} = .86$$

to test the null hypothesis of no interaction. As before, each of these F values is associated with 1 and 28 degrees of freedom, and we conclude that there are significant effects of gender and role playing, but that there is no interaction — the effects of gender and role playing are independent.

In general, this coding of the dummy variables does not lead to the kind of straightforward multiple comparison testing as did our original coding scheme, because there is no reference cell. On the other hand, when we code two-way (or higher-order) designs with reference cell coding, the straightforward correspondence between the changes of a cell mean from the reference cell mean and the regression parameters breaks down anyway (see Table 8-5), because some cell means are estimated by the sum of more than two regression coefficients. Multiple comparisons can be computed directly on the group means (computed directly or from the e_i) using MS_{res} as before.

Why Does It Matter How We Code the Dummy Variables?

Given the fact that we obtained identical results with effects coding (1, 0, −1) and reference cell coding (0, 1), one is left with the question of why we bothered to introduce this alternative coding scheme. Sometimes one is interested in deviations of each group from the overall mean, particularly when there is no natural control (or reference group). For one-way analyses of variance, it makes no difference which coding you use, but we recommend reference cell coding because of the ease of interpreting the parameter estimates as changes from the reference cell mean. For anything other than simple one-way analysis of variance, however, it does matter which coding scheme is used: In general, the more complicated the analysis, the more reasons there are to use effects coding $(1, 0, -1)$ rather than reference cell coding $(0, 1)$.

In particular, effects coding will be important when encoding between-subjects differences in repeated-measures designs (discussed in Chap. 9) where the same subjects are observed before and after one or more treatments, and for a two-way (or higher-order) analysis of variance that includes interaction. The reason effects coding is so important in these settings is that it produces *sets* of dummy variables that are orthogonal (independent) when there is interaction in the model, whereas reference cell coding does not. If there is an interaction and we use reference coding, we introduce a multicollinearity, with all the attendant uncertainties in the parameter estimates and associated sums of squares.*

To see this multicollinearity, we use the methods developed in Chap. 5 to compare the principal components of the set of reference cell dummy variable codes in Table 8-4 with the principal components of the set of effects dummy variable codes in Table 8-7.[†] The three principal components for the effects dummy variables have eigenvalues of 1.0, 1.0, and 1.0, exactly what we would expect from orthogonal independent variables (see Table 8-8). The corresponding eigenvectors also illustrate the orthogonality: All the coefficients in the eigenvectors, except one associated with each dummy variable, are zero, indicating that no dummy variable contains information in common with another. In contrast, the principal components of the reference cell dummy variables in Table 8-8 show evidence of nonorthogonality: The eigenvalues are not equal, and the third one is small, only .184, with a condition index $\lambda_1/\lambda_3 = 1.817/.184 = 9.9$. All the coefficients in the corresponding eigenvectors are nonzero, indicating that each of the three dummy variables shares information with the others. This nonorthogonality can also be seen in the variance inflation factors shown in Fig. 8-2, which are 2 or 3

---

*Effects coding of dummy variables assures only that each set of dummy variable codes for one factor is orthogonal to the set of dummy variables for the other factor and the interactions. Thus, it is important to keep the sets of dummy variable codes together in the regression equation. There are other ways to code dummy variables, so that all the dummy variables are orthogonal to one another and their interactions and the sets need not be kept together. In general it is simpler to keep the sets of variables together than to try to use a different type of coding. For example, see A. L. Edwards, *Multiple Regression and the Analysis of Variance and Covariance* (2nd ed.), New York, Freeman, 1985, pp. 103–113.

[†]If you did not cover principal components in Chap. 5, skip the rest of this paragraph.

TABLE 8-8 Principal Components of Reference Cell Dummy Variable Codes and Effects Dummy Variable Codes for the Gender and Personality Score Data

	Principal Component		
	1	2	2
EFFECTS CODING (1, −1)			
Eigenvalues	1.000	1.000	1.000
Eigenvectors:			
Variable			
G	0.000	0.000	1.000
R	0.000	1.000	0.000
GR	1.000	0.000	0.000
REFERENCE CELL CODING (0, 1)			
Eigenvalues	1.817	1.000	0.184
Eigenvectors:			
Variable			
G	−0.500	0.707	−0.500
R	−0.500	−0.707	−0.500
GR	−0.707	0.000	0.707

for the reference cell coding, indicating a mild multicollinearity. In contrast, the variance inflation factors associated with the effects coded dummy variables are all 1.0, the smallest possible value (Fig. 8-3), indicating no multicollinearity.

Given that reference coding induces multicollinearity when interaction is included in the model, we now must ask: Does it matter? The variance inflation factors are much less than 10, and the condition index is much less than 100. Thus, if this were an ordinary multiple regression, we would not be concerned about multicollinearity—in fact, we would be delighted that it was so mild. However, this is not an ordinary regression analysis. To do analysis of variance with regression models, we use the sequential sums of squares associated with the dummy variables to compute F statistics to test our hypotheses about the main effects and interaction. *In the face of multicollinearity, no matter how small, these sequential sums of squares associated with the dummy variables depend on the order the variables are entered into the regression equation.* Hence, if we used

$$\hat{S}_p = e_0 + e_{GR}GR + e_G G + e_R R \tag{8.3}$$

TABLE 8-9 Comparison of Sums of Squares Using Reference Cell and Effects Coding of Dummy Variables (Compare to Fig. 8-2)

Source	SS	DF	MS
REFERENCE CELL CODING IN EQ. (8.3)			
G	485.1	1	485.1
R	506.2	1	506.2
GR	3737.5	1	3737.5
Residual	3510.9	28	125.4
EFFECTS CODING IN EQ. (8.3)			
G	2838.8	1	2838.8
R	1782.0	1	1782.0
GR	108.0	1	108.0
Residual	3510.9	28	125.4

instead of Eq. (8.2) to analyze the data in Table 8-4 (reference cell codes), we would get different sums of squares (compare Table 8-9 with Fig. 8-2). Not only are they different, *they are wrong*. However, if we used this equation to reanalyze the data in Table 8-7 (effects codes), we obtain sums of squares identical to those shown in Figs. 8-2 and 8-3 (Table 8-9).* If we do not include interaction in the model, it does not matter which type of coding we use. This is because the multicollinearity arises when we form the cross-products between dummy variables needed for coding interaction effects.

So, what does all of this mean? *When you are analyzing data with a two-way (or higher-order) analysis of variance using dummy variables and multiple regression, and you include interaction effects, you should use effects coding for the dummy variables to ensure that you obtain the correct sums of squares regardless of the order in which you enter the dummy variables into the regression equation.* If you do not include interactions, the coding does not matter. However, we rarely have enough prior information about the system we are studying to justify

*We obtained the correct result in the earlier analysis using reference cell coding because we entered the interaction dummy variables last. Because the main effects dummy variables are orthogonal to one another, their order of entry does not matter as long as the interactions are always entered into the model after the main effects (as long as there are no missing data). To ensure there are no mistakes, it is simpler to get in the habit of using effects coding, even when there are no missing data.

excluding interactions, so we recommend using effects coding for any-thing other than simple one-way analysis of variance.[*]

The Kidney, Sodium, and High Blood Pressure

Sodium (Na) plays an important role in the genesis of high blood pressure, and the kidney is the principal organ that regulates the amount of sodium in the body. The kidney contains an enzyme, Na-K-ATPase, which is essential for maintaining proper sodium levels. If this enzyme does not function properly, high blood pressure may result. The activity of this enzyme has been studied in whole kidney extracts, even though the kidney is known to contain many functionally distinct sites. To see whether any particular site of Na-K-ATPase activity was abnormal in hypertension, or whether this enzyme was uniformly abnormal throughout the kidney, Garg and his colleagues[†] studied Na-K-ATPase activity at different sites along the nephrons of normal rats and specially bred rats which spontaneously develop hypertension.

We specify the presence of hypertension (strain of rat) as the first experimental factor and the nephron site—the distal collecting tubule (DCT), cortical collecting duct (CCD), or outer medullary collecting duct (OMCD)—as the second experimental factor. Thus, there are two levels of the strain of rat factor and three levels of the nephron site factor. The data are shown in Table 8-10. We can use these data to test three hypotheses:

1. There is no difference in Na-K-ATPase between normal and hypertensive rats.
2. There is no difference in Na-K-ATPase between different sites in the kidney.
3. There is no interaction between the presence of hypertension and the site in the kidney in terms of Na-K-ATPase; the presence or absence of hypertension has the same effect on the Na-K-ATPase regardless of site within the kidney.

Like the previous example, this is a two-way analysis of variance, only now we have more than two levels of one of the factors. To encode the two levels of the hypertension factor, we use effects coding to define

---

[*]Of course, you can also use effects coding for a one-way analysis of variance. Doing so simply changes the interpretation of the regression coefficients.

[†]L. C. Garg, N. Narang, and S. McArdle, "Na-K-ATPase in Nephron Segments of Rats Developing Spontaneous Hypertension," *Am. J. Physiol* 249:F863–F869, 1985.

TABLE 8-10 Data (by Cell) from Study of Na-K-ATPase Activity [pmol/(min·mm)] as a Function of Nephron Site and Hypertension

Nephron Site	Strain of Rat	
	Normal	Hypertensive
DCT	62 73 58 66	44 49 46 37
CCD	15 31 19 35	8 36 11 18
OMCD	7 7 9 17	19 7 15 4

one dummy variable:

$$H = \begin{cases} 1 \text{ if normal} \\ -1 \text{ if hypertensive} \end{cases}$$

For the three levels (sites) of the nephron factor, we define two dummy variables as

$$N_1 = \begin{cases} 1 \text{ if distal collecting tubule (DCT)} \\ 0 \text{ if cortical collecting duct (CCD)} \\ -1 \text{ if outer medullary collecting duct (OMCD)} \end{cases}$$

and

$$N_2 = \begin{cases} 0 \text{ if DCT} \\ 1 \text{ if CCD} \\ -1 \text{ if OMCD} \end{cases}$$

These three dummy variables completely identify all cells of data. Table 8-11 shows the dummy variables assigned to each cell in the data table (Table 8-10). For example, the data from the CCD site of the normal rats is coded as $H = 1$, $N_1 = 0$, and $N_2 = 1$.

TABLE 8-11 Schematic of Analysis of Variance for Na-K-ATPase Activity Data Showing Effects Encoding of Dummy Variables

Nephron Site	Strain of Rat	
	Normal	Hypertensive
DCT	$H = 1$ $N_1 = 1$ $N_2 = 0$	$H = -1$ $N_1 = 1$ $N_2 = 0$
CCD	$H = 1$ $N_1 = 0$ $N_2 = 1$	$H = -1$ $N_1 = 0$ $N_2 = 1$
OMCD	$H = 1$ $N_1 = -1$ $N_2 = -1$	$H = -1$ $N_1 = -1$ $N_2 = -1$

The complete data set as it would be submitted to a multiple regression program, including the associated dummy variables, is shown in Table 8-12. We have also included two additional variables, HN_1 and HN_2, to account for possible interactions among the two experimental factors (hypertension and nephron site). They allow for the possibility that the influences of hypertension and nephron site are not independent, but rather that the effect of one factor depends on the level of the other factor.

The multiple regression model for a two-way analysis of variance on the data in Table 8-12 is specified by letting Na-K-ATPase activity (A) equal a linear function of the hypertension factor plus the nephron factor plus the interaction of hypertension and nephron factors:

$$\hat{A} = b_0 + b_H H + b_{N_1} N_1 + b_{N_2} N_2 + b_{HN_1} HN_1 + b_{HN_2} HN_2 \qquad (8.4)$$

where the intercept b_0 is an estimate of the overall mean A for all nephron sites in both strains of rat, b_H estimates the difference between the overall mean and the mean of normal rats $(-b_H$ is the difference between the overall mean and the mean of hypertensive rats), b_{N_1} estimates the difference between the overall mean and the mean of all DCT nephron sites, b_{N_2} estimates the difference between the overall mean and the mean of all CCD nephron sites, and $-(b_{N_1} + b_{N_2})$ estimates the difference between the overall mean and the mean of all OMCD sites. Figure 8-4 shows the results of this regression analysis.

TABLE 8-12 Rat Kidney Na-K-ATPase Activity [pmol/(min·mm)] and Effects Coded Dummy Variables for Hypertension and Nephron Site

Na-K-ATPase A	H	N_1	N_2	HN_1	HN_2	
NORMAL						
62	1	1	0	1	0	DCT
73	1	1	0	1	0	
58	1	1	0	1	0	
66	1	1	0	1	0	
15	1	0	1	0	1	CCD
31	1	0	1	0	1	
19	1	0	1	0	1	
35	1	0	1	0	1	
7	1	-1	-1	-1	-1	OMCD
7	1	-1	-1	-1	-1	
9	1	-1	-1	-1	-1	
17	1	-1	-1	-1	-1	
HYPERTENSIVE						
44	-1	1	0	-1	0	DCT
49	-1	1	0	-1	0	
46	-1	1	0	-1	0	
37	-1	1	0	-1	0	
8	-1	0	1	0	-1	CCD
36	-1	0	1	0	-1	
11	-1	0	1	0	-1	
18	-1	0	1	0	-1	
19	-1	-1	-1	1	1	OMCD
7	-1	-1	-1	1	1	
15	-1	-1	-1	1	1	
4	-1	-1	-1	1	1	

The sum of squares due to the hypertension factor is simply the sum of squares attributable to the dummy variable H. Each variable in a linear regression is associated with 1 degree of freedom, and there is only one dummy variable associated with the presence or absence of hypertension; so, from Fig. 8-4,

$$MS_H = \frac{SS_H}{DF_H} = \frac{459.38}{1} = 459.38 \text{ [pmol/(min·mm)]}^2$$

```
DEP VARIABLE: A          Sodium-potassium ATPase

                                    ANALYSIS OF VARIANCE

                                SUM OF         MEAN
                    SOURCE   DF  SQUARES        SQUARE      F VALUE      PROB>F

                    MODEL     5  9242.37500    1848.47500   28.727       0.0001
                    ERROR    18  1158.25000    64.34722222
                    C TOTAL  23  10400.62500

                        ROOT MSE    8.021672    R-SQUARE    0.8886
                        DEP MEAN     28.875     ADJ R-SQ    0.8577
                        C.V.        27.78068

                                  PARAMETER ESTIMATES

              PARAMETER      STANDARD    T FOR H0:                            VARIABLE
VARIABLE  DF  ESTIMATE       ERROR    PARAMETER=0   PROB > |T|   TYPE I SS    LABEL

INTERCEP  1   28.87500000   1.63741695   17.634      0.0001    20010.37500   INTERCEPT
H         1    4.37500000   1.63741695    2.672      0.0156      459.37500000 dummy variable =-1 if hypertensive
N1        1   25.50000000   2.31565725   11.012      0.0001     7656.25000   dummy variable = 1 if DCT group
N2        1   -7.25         2.31565725   -3.131      0.0058      630.75000000 dummy variable = 1 if CCD group
HN1       1    6.00000000   2.31565725    2.591      0.0184      484.00000000 interaction of H and N1
HN2       1   -1            2.31565725   -0.432      0.6710       12.00000000 interaction of H and N2
```

FIGURE 8-4 Regression implementation of analysis of variance of the data on sodium potassium ATPase activity in different nephron sites in normal and hypertensive rats.

and

$$MS_{res} = \frac{SS_{res}}{DF_{res}} = \frac{1158.25}{18} = 64.35 \; [\text{pmol}/(\text{min}\cdot\text{mm})]^2$$

Use these results to test the hypothesis that the presence of hypertension has no effect on Na-K-ATPase activity by computing

$$F = \frac{MS_H}{MS_{res}} = \frac{459.38}{64.35} = 7.14$$

This value of F exceeds the critical value of F for $\alpha = .05$ and 1 and 18 degrees of freedom, 4.41, and so we conclude there is a significant difference in the Na-K-ATPase activity between the two strains of rats ($P < .05$).

To test whether or not there is a significant effect of nephron site on Na-K-ATPase activity, we need to compare the total variation between sites with MS_{res}. Because there are two dummy variables (N_1 and N_2) that account for the nephron factor, we add the sums of squares due to N_1 and N_2 and divide this sum by 2 degrees of freedom (one for N_1 plus one for N_2) to obtain the nephron mean square,

$$MS_N = \frac{SS_N}{DF_N} = \frac{SS_{N_1} + SS_{N_2}}{DF_{N_1} + DF_{N_2}} = \frac{7656.25 + 630.75}{1 + 1} = 4143.50 \; [\text{pmol}/(\text{min}\cdot\text{mm})]^2$$

We test the hypothesis that there is no significant difference among the nephron sites by comparing this variance estimate with the residual variance estimate $MS_{res} = 64.35$ [pmol/(min·mm)]^2:

$$F = \frac{MS_N}{MS_{res}} = \frac{4143.50}{64.35} = 64.39$$

This value of F far exceeds the critical value of F for $\alpha = .01$ with 2 and 18 degrees of freedom, 6.01, and so we conclude that there is a significant difference in Na-K-ATPase activity among nephron sites ($P < .01$).

What Do Interactions Tell Us?

So far we have concluded that the Na-K-ATPase activity differs in normal and hypertensive rats (controlling for the nephron site) and that the Na-K-ATPase activity differs among nephrons taken from different sites. These two statements followed from testing the *main effects*, i.e., whether there are differences associated with the different factors used to classify the experimental subjects. In addition to testing for main effects, we can also test for *interactions*, i.e., whether the main effects act independently or whether they depend on each other. The presence of significant interaction indicates that the difference in Na-K-ATPase activity between normal and hypertensive rats depends on the nephron site. If there is no significant interaction, we would conclude that the difference in Na-K-ATPase activity between normal and hypertensive rats is the same among all sites and vice versa.

To test for significant interaction between hypertension and nephron site, we compute the interaction mean square MS_{HN} and compare it with the variance estimate from within the groups, MS_{res}. There are two variables in the regression equation, HN_1 and HN_2, that describe the interaction. As with the nephron factor, we compute the sum of squared deviations associated with the interaction by adding together the sums of squares associated with each of these variables, then divide by their total degrees of freedom to get the interaction mean square.

$$MS_{HN} = \frac{SS_{HN}}{DF_{HN}} = \frac{SS_{HN_1} + SS_{HN_2}}{DF_{HN_1} + DF_{HN_2}} = \frac{484 + 12}{1 + 1} = 248 \text{ [pmol/(min·mm)]}^2$$

We then compute

$$F = \frac{MS_{HN}}{MS_{res}} = \frac{248}{64.35} = 3.85$$

This value of F exceeds the critical value of F for $\alpha = .05$ and 2 and 18 degrees of freedom, 3.55, and so we conclude that there is a significant interaction between the presence of hypertension and nephron site ($P <$

.05). This significant interaction means that not only is the Na-K-ATPase activity generally lower at all sites in the hypertensive rats than in the normal rats, but it is also lower at some nephron sites than at others. Hence, the abnormalities in the kidney that may contribute to hypertension are not uniformly distributed. Rather, some sites are more abnormal than others. This conclusion may provide important clues about why hypertension develops in these rats.

In the analysis of these data we have treated the interaction effect just as we would have in an ordinary multiple regression analysis—it is simply another effect that indicates the main effects are not independent. We make no distinction between the interaction and the main effects when we test the hypotheses for these different effects. In this example, we found significant interaction in addition to significant main effects for hypertension and nephron site. We interpreted this as meaning

1. Na-K-ATPase activity differs in normal and hypertensive rats.
2. Na-K-ATPase activity differs among the different nephron sites.
3. The differences in Na-K-ATPase activity at the different nephron sites depend on whether the rats were normal or hypertensive.

This outcome is reasonably straightforward to interpret, especially if you plot the cell means to get a picture of the outcome as shown in Fig. 8-5H. There are two lines connecting the nephron site means, one for normal rats and one for hypertensive rats. The separation between these lines reflects the significant main effect due to hypertension. The fact that these lines have a slope reflects the main effect due to the nephron site. Finally, the fact that the lines are not parallel reflects the interaction. (Compare this figure with Figs. 3-17, 3-20, and 3-21.) In such two-dimensional data plots interaction appears as nonparallel lines. The same interpretation holds for interaction in analysis of variance as can be seen in Fig. 8-5H, where the line for the hypertensive rats does not parallel the line for the normal rats. Thus, interaction effects are not particularly hard to interpret if you think about them in terms of regression.

There are many other possible outcomes from a two-way analysis of variance with interaction, several of which are shown in the other panels of Fig. 8-5. Some of these are more complex, and some, like interaction with no main effects (Fig. 8-5E), are harder to interpret. In general, the main effects in the outcomes diagrammed in panels A through D, which do not involve interaction, are easier to interpret than the main effects in those outcomes that do involve interaction (panels E through H).

The added difficulty in interpreting main effects in the presence of significant interaction has led many statisticians to follow a different hypothesis testing procedure than the one we have suggested. They suggest that one test for interaction *first*. If there is significant interaction,

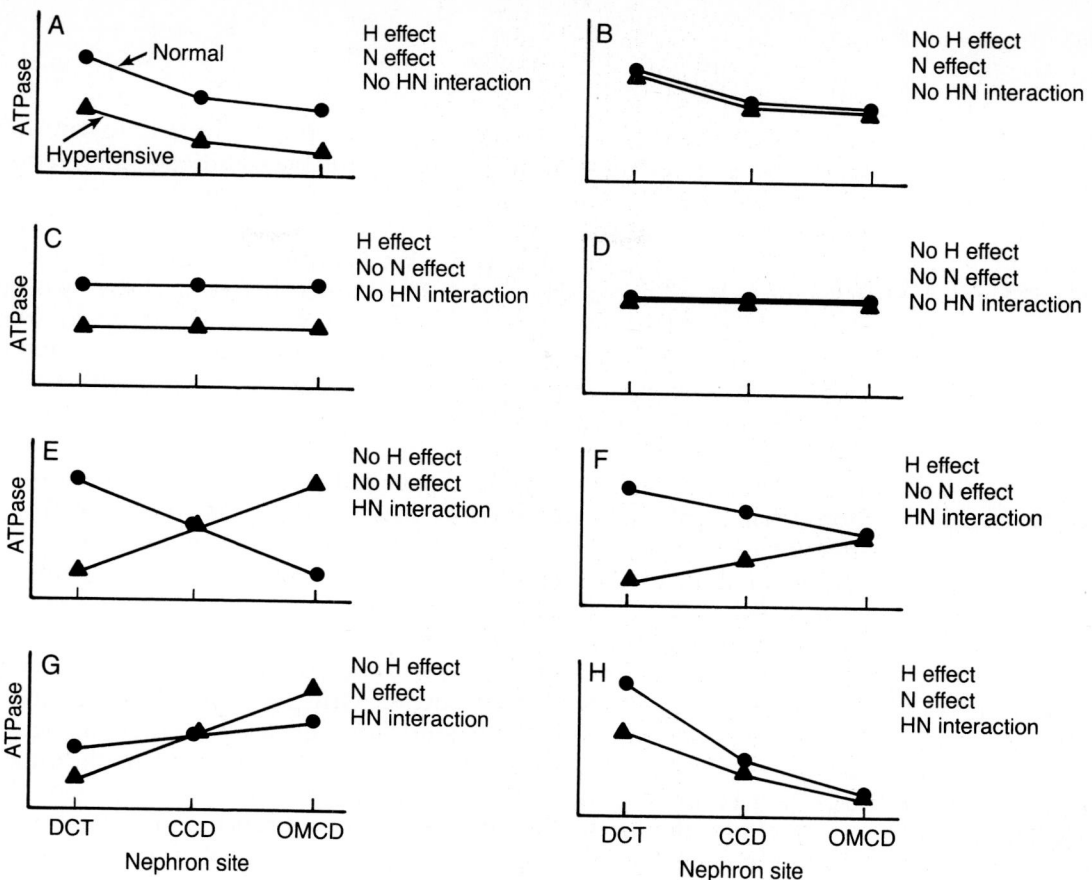

FIGURE 8-5 This schematic geometric interpretation illustrates how the pattern of mean sodium potassium ATPase activities of the different nephron site and hypertensive groups changes for several different combinations of significant and not significant main effects and interaction effects. The results obtained from the analysis of the data in Fig. 8-4 are shown in panel *H*, where the significant nephron site effect is reflected in a slope in the direction of the nephron site axis, the significant hypertension effect is reflected in a separation between the two lines, and the significant interaction is reflected by the nonparallel separation of the two lines. Panels *A* through *G* illustrate other possible combinations of significant and nonsignificant nephron site, hypertension, and interaction effects. The key to interpreting these schematics is to relate these pictures to the hypotheses being tested in the analysis of variance. For example, consider panel *E*, in which the two lines cross symmetrically. The test for significant nephron site differences asks the question: Is there a difference in sodium potassium ATPase activity *averaged across both hypertension groups?* Although each of the lines has a slope, the imaginary line that is the *average* of the two lines does not, and thus there is no significant nephron effect. Similarly, when *averaged across all nephron sites,* there is no displacement between the two lines, indicating no significant hypertension effect. The fact that the two lines are not parallel indicates there is a significant interaction. Similar thinking will permit you to interpret the remaining situations.

do not test the main effects hypotheses, but proceed directly to multiple comparisons of all the cell means. If there is no significant interaction, then delete the interactions, reestimate the model, and then test for the main effects and follow this up with multiple comparisons, if desired.

There is no theoretical justification for this approach, compared to the approach we have followed.* The choice between these two approaches is philosophical: If you see analysis of variance, as we do, from a regression point of view, interactions are simply another component of the model. They may complicate interpretation of the main effects, but multiple comparisons can be used to sort out the complete picture. We make no distinction between main effects and interaction when testing hypotheses. On the other hand, if you see analysis of variance from a more traditional point of view, interactions mean that the main effects are not easily interpretable—some would say uninterpretable—so you do not even bother to test main effects hypotheses if you find a significant interaction.

The approach of reestimating the main effects after deleting interaction when the interaction is found to be not significant is equivalent to pooling the interaction mean square with the residual mean square. In the same way we treated nonsignificant regression coefficients, we do not recommend reestimation after deleting a component of the model following failure to reject the null hypothesis for that component. For example, had the main effect of hypertension been found to be not significant, no one would suggest reestimating the model after deleting this term. Thus, we recommend that you simply test the hypotheses that interest you on an equal footing, not making distinctions between interaction and main effects hypotheses. If you do decide to test the interaction effect first, we find no justification for reestimating the model without interaction if it is found to be not significant. In either case, a complete analysis includes the analysis of variance, a graphical presentation of the data as in Fig. 8-5, and the multiple comparisons that make sense in terms of the experimental goals.

Multiple Comparisons in Two-Way Analysis of Variance

We handle multiple comparisons in two-way analysis of variance in much the same way we did for single-factor analysis of variance, using Bonferroni t tests, the Student-Newman-Keuls test, and Dunnett's test. These multiple comparison procedures control the error rate of a family

---

*In fact, there are probably better theoretical reasons for not following the practice of pooling the nonsignificant interaction mean square with the residual mean square. See S. R. Searle, *Linear Models for Unbalanced Data*, New York, Wiley, 1987, pp. 106–107.

of related tests rather than the error rates of the individual tests. Because there are two main effects rather than one, we have two choices for formulating the analysis. The question is how to define a family of pairwise comparisons. In a one-way analysis of variance the answer to this question was obvious: A family was the collection of all pairwise comparisons across the different treatments (levels of the experimental factor). Because we are now dealing with two experimental factors, there are two different ways we could define the family of tests to be considered in each set of multiple comparisons. How one defines a family of comparisons will affect how the pairwise tests are conducted.

One definition of a family is the set of all pairwise comparisons among the different levels of each experimental factor, *treating comparisons within each factor as a separate family*. In the study of Na-K-ATPase activity at different kidney sites in normal and hypertensive rats we just analyzed, we would do one set of multiple comparisons to control the family error rate among all pairwise comparisons among the two different classes of rats (normal vs. hypertensive) and another set of multiple comparisons to control the family error rate among all pairwise comparisons between the three different kidney sites. The other definition of a family is the set of all pairwise comparisons among all the treatments, *treating all possible pairwise comparisons as a single family*. There is nothing special about either of these ways to define the family of tests whose error rate is to be controlled. Which definition you choose depends on what information you want from the analysis and whether or not there is significant interaction.

On one hand, if you want to compare the different levels of the two factors separately, then it makes sense to define each set of comparisons as a family. On the other hand, when you are interested in all possible comparisons among the different treatment combinations, then all the treatments should be considered members of a single family. When there is a significant interaction effect in the analysis of variance, you should generally treat all possible pairwise comparisons as a single family. Doing so will make it possible to isolate where the interactions play a role. In the absence of a significant interaction effect, one generally treats the effects of the two factors as two separate families.

We will now illustrate how to conduct the multiple comparisons using the two definitions of the family of tests just described.

More on the Kidney, Sodium, and High Blood Pressure

In the study of rat kidney Na-K-ATPase activity we just analyzed, we found significant differences in activity between the normal and hypertensive rats, significant differences in activity between sites within the kidney, and a significant interaction effect between hypertension and nephron site (Figs. 8-4 and 8-5*H*). Because we found a significant interaction between the two main experimental factors, it makes sense to consider all pairwise comparisons between the mean values of Na-K-ATPase activity observed under all 6 treatments (2 rat strains × 3 kidney sites; see Table 8-10) as a single family of 15 pairwise comparisons.

We are interested in all possible comparisons, so we will use the Student-Newman-Keuls test. We first rank all the cell means from smallest to largest, then compare the smallest with the largest. If this difference is significant, we compare the largest to the second smallest, and so forth, until a test fails to reach significance. The q test statistics are computed just as we did for the one-way analysis of variance [compare with Eq. (7.13)]:

$$q = \frac{\overline{A}_{i_1 j_1} - \overline{A}_{i_2 j_2}}{\sqrt{\dfrac{MS_{res}}{2}\left(\dfrac{1}{n_{i_1 j_1}} + \dfrac{1}{n_{i_2 j_2}}\right)}} \tag{8.5}$$

where $\overline{A}_{i_1 j_1}$ and $\overline{A}_{i_2 j_2}$ are the means of the two cells in rows i_1 and i_2 and columns j_1 and j_2, respectively; $n_{i_1 j_1}$ and $n_{i_2 j_2}$ are the number of observations used to compute the respective cell means, and MS_{res} is the residual mean square from the analysis of variance.

From Fig. 8-4, $MS_{res} = 64.35$ with 18 degrees of freedom. We use this information, together with the means of each cell and associated sample size to conduct the multiple comparisons in Table 8-13. Figure 8-6 summarizes the results of these tests (compare with Fig. 8-5*H*). These pairwise comparisons reveal that the CCD and OMCD sites are not significantly different from each other in both normal and hypertensive rats, but they are significantly different from the DCT sites. The DCT sites from normal and hypertensive rats are also different. Thus, the Na-K-ATPase activity for the CCD and OMCD sites does not appear to be different or to change in response to hypertension.* In contrast, the Na-K-ATPase at the DCT sites differs from the other sites and also from each other when hypertension is present. Hence, the DCT site appears to be the crucial factor in the kidney's Na-K-ATPase response to hypertension.

Although the practical goal of these multiple comparisons is to un-

---

*Note that the horizontal line on Fig. 8-6 connects the treatment groups that are *not* statistically significantly different, rather than indicating the ones that are significantly different. This notation is standard in multiple comparisons because it focuses the reader's attention on the similarities. Most writers in biomedical research do it the other way around, indicating the significant differences. While widely applied, this convention makes the results harder to follow than the standard convention we follow in Fig. 8-6.

Table 8-13 Student-Newman-Keuls Tests for All Possible Comparisons among All Cell Mean Na-K-ATPase Activities

Comparison	Difference in Means	q	p	$q_{.05}$	$P < .05?$
DCT normal vs. OMCD normal	$65 - 10 = 55$	13.65	6	4.495	Yes
DCT normal vs. OMCD hypertensive	$65 - 11 = 54$	13.34	5	4.277	Yes
DCT normal vs. CCD hypertensive	$65 - 18 = 47$	11.66	4	3.997	Yes
DCT normal vs. CCD normal	$65 - 25 = 40$	9.91	3	3.609	Yes
DCT normal vs. DCT hypertensive	$65 - 44 = 21$	5.17	2	2.971	Yes
DCT hypertensive vs. OMCD normal	$44 - 10 = 34$	8.48	5	4.277	Yes
DCT hypertensive vs. OMCD hypertensive	$44 - 11 = 33$	8.17	4	3.997	Yes
DCT hypertensive vs. CCD hypertensive	$44 - 18 = 26$	6.48	3	3.609	Yes
DCT hypertensive vs. CCD normal	$44 - 25 = 19$	4.74	2	2.971	Yes
CCD normal vs. OMCD normal	$25 - 10 = 15$	3.74	4	3.997	No
CCD normal vs. OMCD hypertensive	$25 - 11 = 14$	3.43	3	Do not test	
CCD normal vs. CCD hypertensive	$25 - 18 = 7$	1.75	2	Do not test	
CCD hypertensive vs. OMCD normal	$18 - 10 = 8$	1.99	3	3.609	No
CCD hypertensive vs. OMCD hypertensive	$18 - 11 = 7$	1.68	2	Do not test	
OMCD hypertensive vs. OMCD normal	$11 - 10 = 1$	.31	2	Do not test	

derstand how the kidney changes in response to hypertension, they also provide some insight into what is happening in the statistical analysis itself. These data exhibited significant effects associated with both experimental factors as well as a significant interaction. The results of the multiple comparison tests reveal that the significant main effect of the site that was detected is due to the DCT site being different from the other two sites. Similarly, the significant main effect of hypertension is at the DCT site only. The fact that these differences appear only at one combination of rat type and nephron site led to the significant interaction effect.

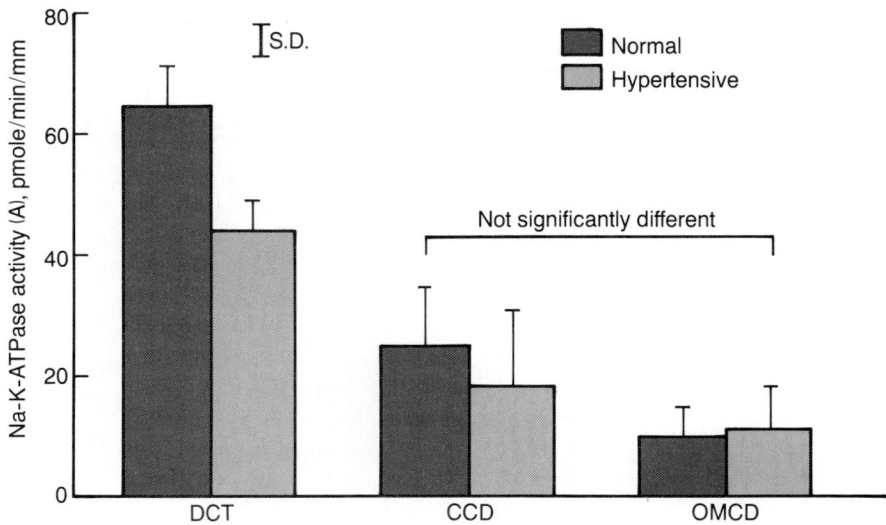

FIGURE 8-6 A summary of the results of the analysis of sodium potassium ATPase activity at different nephron sites in normal and hypertensive rats, including the results of all possible pairwise comparisons conducted with the Student-Newman-Keuls test. These tests reveal that nephrons taken from the distal collecting tubule (DCT) in normal and hypertensive rats differed from each other and from the nephrons taken at all other sites in both normal and hypertensive rats. There were no significant differences among nephrons taken from the other two sites (CCD and OMCD) in either normal or hypertensive rats. Note that we have indicated the differences that are *not significant* rather than the ones that are. This procedure generally makes it easier to illustrate the various subgroups among the data than to indicate the differences that *are* significant.

This approach of treating all possible comparisons as a single family makes most sense if there is a significant interaction effect. Had the interaction effect not been significant, we would have no evidence that the differences in Na-K-ATPase activity between sites depends on the strain of rat or, conversely, that the difference between rats depends on site. Thus, it would make sense to conduct multiple comparisons among the mean responses of different levels of the experimental factors one factor at a time.

We will now use the same data we just discussed to illustrate how to conduct the multiple comparisons *treating the comparisons for each experimental factor separately*. In this scheme of multiple comparisons testing, we define two different families of tests: one family consisting of tests among levels of the first factor (hypertension) and one family consisting of tests among levels of the second factor (nephron site). Within each of these families, we can use Bonferroni t tests, Dunnett's test, or the Student-Newman-Keuls test, whichever is most appropriate.

The procedure is similar to what we just did, with one important exception: *The tests are no longer between the means of each cell, but rather between the mean values at each level of the factor used to define the family of tests.*

For illustration, suppose we had decided to consider the comparisons across kidney sites as our family of comparisons. These comparisons ignore any differences between rat type, so we first average Na-K-ATPase activity across both kinds of rats at each of the three sites (Table 8-14). We are asking the question: Is the Na-K-ATPase activity at the sites, averaged over both kinds of rats, different?

As before, we rank the group means in ascending order, then begin with the test for the largest difference, and proceed until no significant differences are found. The test statistic is computed as

$$q = \frac{\overline{A}_{j_1} - \overline{A}_{j_2}}{\sqrt{\frac{MS_{res}}{2}\left(\frac{1}{n_{j_1}} + \frac{1}{n_{j_2}}\right)}} \tag{8.6}$$

$\overline{A}_{j_1}$ and $\overline{A}_{j_2}$ are the mean Na-K-ATPase activity values observed at each nephron site, averaged over both rat strains, and n_{j_1} and n_{j_2} are the number of observations *used to compute the mean Na-K-ATPase activity at each site*, not the number of observations per cell. The first comparison is between DCT and OMCD nephrons. We first average Na-K-ATPase activity for all rats at the DCT site. This value is

$$\overline{A}_{DCT} = \frac{(62 + 73 + 58 + 66) + (44 + 49 + 46 + 37)}{4 + 4} = \frac{435}{8} = 54 \text{ pmol/(min·mm)}$$

Note that we averaged across the first row in Table 8-10, without worrying about which column (strain of rat) the observations came from. This mean of 54 pmol/(min·mm) is based on $n_{DCT} = 8$ observations. Likewise, we average all $n_{OMCD} = 8$ observations in the last row of Table 8-10 to obtain $\overline{A}_{OMCD} = 11$ pmol/(min·mm). From Fig. 8-4 $MS_{res} = 64.35$

TABLE 8-14 Student-Newman-Keuls Tests for All Possible Comparisons among the Three Nephron Site Mean Na-K-ATPase Activities (Averaged across Both Rat Strains), pmol/(min·mm)

Comparison	Difference in Means	q	p	$q_{.05}$	$P < .05$?
DCT vs. OMCD	$54 - 11 = 43$	15.43	3	3.609	Yes
DCT vs. CCD	$54 - 23 = 31$	11.55	2	2.971	Yes
CCD vs. OMCD	$23 - 11 = 12$	3.88	2	2.971	Yes

$[pmol/(min \cdot mm)]^2$, so the Student-Newman-Keuls statistic for this comparison is

$$q = \frac{54 - 11}{\sqrt{\dfrac{64.35}{2}\left(\dfrac{1}{8} + \dfrac{1}{8}\right)}} = 15.43$$

which exceeds the critical value of q for $p = 3$ and $\alpha = .05$, 3.609; so we conclude that there is a statistically significant difference in Na-K-ATPase activity at these two sites $(P < .05)$.

Similarly we find statistically significant differences between DCT and CCD $(P < .05)$ and between CCD and OMCD $(P < .05)$ (Table 8-14), and conclude that all three kidney sites exhibit different Na-K-ATPase activities. This conclusion is different from the one we reached when using the all-possible-comparisons approach above. When we did those 15 pairwise comparisons (Table 8-13) we found that the OMCD and CCD sites were not different, whereas the family of 3 pairwise comparisons we just did between OMCD, DCT, and CCD (Table 8-14) showed that the OMCD and CCD groups were, on the average, different. This discrepancy in conclusions illustrates the relative lack of power of doing a finer dissection of the data with all possible pairwise comparisons.

We could conduct another family of similar tests to compare the different rat types, averaged over sites. However, in this example this is not necessary because there are only two strains of rat and thus only one comparison, which we already made with the F statistic, and so we already know that hypertension affects Na-K-ATPase activity.

In sum, you can use any of the multiple comparison procedures we have developed to analyze two-way designs. If there is a significant interaction effect, it makes most sense to consider the family of all possible pairwise comparisons as the family of tests and the analysis proceeds in much the same way as with multiple comparisons following a one-way analysis of variance. If there is not a significant interaction, it makes most sense to treat the pairwise comparisons within each treatment factor as a separate family; you compare the mean values observed at each level of the factor of interest averaged across all levels of the other factor.

UNBALANCED DATA

So far our entire discussion of two-way analysis of variance has presumed a *balanced* design, in which there are the same number of observations under each combination of experimental conditions (i.e., in each cell of the data table). Classic two-way analysis of variance was developed for balanced designs. In one-way analysis of variance, the sample sizes of the different groups do not have to be equal; the design does not have to be balanced. If there are unequal sample sizes, the arithmetic is a little harder, but the hypothesis testing is unaffected. However, when dealing with two-way (or higher-order) analyses of variance, we have to be more careful about how we handle missing data.

When the data are balanced, the sums of squares associated with the different experimental factors all add up to the total sum of squares and the hypothesis tests can be based on F values computed from a single run of either a regression or analysis of variance computer program. This situation no longer holds for *unbalanced designs* when there are different numbers of observations under each combination of experimental conditions, either because of missing data or the way the experiment was designed. Some statistical software requires balanced designs for conducting a two-way (or higher-order) analysis of variance. *Other software will run and produce familiar looking F and P values that are not testing the hypotheses you want.* Although one can use a regression approach to analysis of variance to analyze such data, it is important to formulate this regression analysis correctly to obtain sums of squares and mean squares that lead to F tests (and P values) that actually test meaningful hypotheses. The problems of analyzing unbalanced data provide ample proof (if you need any more) that just because the computer gives you a number does not mean the number is worth anything.

There are many reasons why you can be faced with analyzing unbalanced designs: data are lost due to technical problems, experiments are not planned with an eye toward how the data will be analyzed, or data from an observational study are grouped after data collection and thus sample sizes are unequal. Usually these factors will not result in the complete loss of data for one or more treatment combinations, but occasionally you may end up with one or more empty cells. The primary benefit of a regression approach to analysis of variance when there are missing data is that it always permits analysis of such unbalanced designs, if you take care to compute the correct F statistic. This benefit is important because many computer software packages do not allow you to analyze unbalanced designs, but almost all of them have multiple regression commands. An additional benefit of the regression approach is that it makes it easier to understand what the numbers mean, even if you are able to analyze your data directly using more sophisticated analysis of variance or general linear models software.

When we have missing data in two-way analyses of variance, the basic problem is our old nemesis, multicollinearity. As we discussed above for balanced data, the sets of effects coded dummy variables representing the different treatments and their interactions are orthogonal and the sums of squares associated with each factor (i.e., with each set of dummy variables) do not depend on the order in which the sets of dummy variables are entered in the regression equation. In contrast, when there are missing data, the dummy variables are no longer orthogonal, so the sums of squares associated with each factor (i.e., with each set of dummy variables) depend on which variables have already been entered into the regression equation, just as when we discussed multicollinearity in ordinary multiple regression. To construct hypothesis tests for the different factors, we base the associated F statistics on the *marginal sums of squares* associated with each factor. In other words, we need to compute the regression analysis several times, putting the dummy variables representing each factor of interest into the regression equation *last* each time.

Because of the multicollinearity among the independent (dummy) variables, there is shared information between these variables and so the sums of squares associated with each variable will depend on the order of entry. As a result, *the sum of squares associated with each experimental factor will not add up to the total sum of squares when there are unbalanced or missing data.*

There are two very different situations involving missing data: the situation where the cells do not have the same number of observations, but there is at least one observation in each cell, and the much more complicated situation where there are one or more empty cells. When there is at least one observation per cell, it is possible, using the marginal sums of squares, to compute F statistics to test the same hypotheses concerning main effects and interactions we have discussed so far. In contrast, when there are one or more empty cells, it is no longer possible to make meaningful tests for interaction between the main effects. Depending on the pattern of missing cells, it may be possible to test for main effects if you are willing to *assume* that there is no interaction. We now discuss each of these two cases.

ALL CELLS FILLED, BUT SOME CELLS HAVE MISSING OBSERVATIONS

If the number of observations in each cell is the same, the main and interaction effects are independent and the total sum of squares equals the sum of the factor, interaction, and error sums of squares. Under these circumstances, the hypotheses we test about each main effect are independent of hypotheses about other main effects and the interaction effects, and we can compute the necessary sums of squares with a single regression analysis. However, if the number of observations in each cell is not the same, the main and interaction effects are interdependent, and we must obtain the *marginal* sum of squares associated with each factor after *all the other factors have already been included in the model.* Recall that the marginal sum of squares for a variable equals the incremental sum of squares for that variable when it is entered into the equation last. In terms of a regression approach to analysis of variance, the marginal sum of squares due to a factor is the sum of squares for the set of dummy variables associated with that factor when those dummy variables are entered into the model last, after all other dummy variables. (When the design is balanced, the incremental sum of squares equals the marginal sum of squares, regardless of order of entry, as long as you use effects coding.) You then use these marginal sums of squares and the associated degrees of freedom to compute F statistics, as before.

This procedure requires us to fit the regression equation to the data three times to conduct a two-way analysis of variance with missing data, once with the set of dummy variables encoding the first factor entered last, after all the other dummy variables, once with the set of dummy variables encoding the second factor entered last, and once with the interaction terms entered last. Next, add up the incremental sum of squares for the last factor, which equals the marginal sum of squares for that factor. Then, use these sums of squares and associated degrees of freedom to compute the marginal mean squares associated with that factor.* The residual sum of squares, degrees of freedom, and mean square are the same for all three of these regressions. Finally, use these mean squares to compute F statistics, just as in balanced designs.

The Case of the Missing Kidneys

Suppose four data points from the study of Na-K-ATPase activity at different sites in kidneys of normal and hypertensive rats (Table 8-10) were missing, yielding the data set in Table 8-15. Otherwise, these data are identical to the example we just finished. To begin, we define the same dummy variable (H) to encode the presence of hypertension, the same two dummy variables (N_1 and N_2) to encode the three different nephron sites, and compute the interaction terms HN_1 and HN_2.

To obtain the correct main effect sums of squares for the hypertension factor, we must do a regression analysis to obtain the marginal sum of squares for the H dummy variable, which equals the incremental sums of squares due to the hypertension *after* taking into account nephron sites and the interaction. To do this we rearrange Eq. (8.4) to obtain

$$\hat{A} = b_0 + \underbrace{b_{N_1}N_1 + b_{N_2}N_2}_{\text{Nephron site}} + \underbrace{b_{HN_1}HN_1 + b_{HN_2}HN_2}_{\text{Interaction}} + \underbrace{b_H H}_{\text{Hypertension}}$$

---

*Because the dummy variables representing the different experimental effects are not orthogonal, the sum of the main effects, interaction, and residual marginal sums of squares does not equal the total sum of squares.

TABLE 8-15 Data (by Cell) from Study of Na-K-ATPase Activity as a Function of Nephron Site and Hypertension; Some Data Are Missing (Compare to Table 8-10), pmol/(min·mm)

Nephron Site	Strain of Rate	
	Normal	Hypertensive
DCT	58 66	44 49 46
CCD	15 31 19 35	8 11 18
OMCD	7 7 9 17	19 7 15 4

Figure 8-7A shows the results of this analysis. The sum of squares associated with the presence or absence of hypertension is the sum of squares associated with the dummy variable H, 382.7 [pmol/(min·mm)]^2. There is 1 degree of freedom associated with this sum of squares, so

$$MS_H = \frac{SS_H}{DF_H} = \frac{382.7}{1} = 382.7 \text{ [pmol/(min·mm)]}^2$$

The residual sum of squares is 582.1 [pmol/(min·mm)]^2 with 14 degrees of freedom, so

$$MS_{\text{res}} = \frac{SS_{\text{res}}}{DF_{\text{res}}} = \frac{582.1}{14} = 41.6 \text{ [pmol/(min·mm)]}^2$$

Finally, we test the hypothesis that there is no difference in Na-K-ATPase activity in kidneys from normal and hypertensive rats by computing

$$F = \frac{MS_H}{MS_{\text{res}}} = \frac{382.7}{41.6} = 9.2$$

This value of F exceeds 8.86, the critical value of the F distribution with 1 numerator and 14 denominator degrees of freedom and $\alpha = .01$, so we conclude that the presence of hypertension significantly affects Na-K-ATPase activity ($P < .01$).

A

DEP VARIABLE: A   Sodium-potassium ATPase

ANALYSIS OF VARIANCE

SOURCE	DF	SUM OF SQUARES	MEAN SQUARE	F VALUE	PROB>F
MODEL	5	6229.66667	1245.93333	29.967	0.0001
ERROR	14	582.08333333	41.57738095		
C TOTAL	19	6811.75000			

ROOT MSE	6.448052	R-SQUARE	0.9145	
DEP MEAN	24.25	ADJ R-SQ	0.8840	
C.V.	26.58991			

PARAMETER ESTIMATES

VARIABLE	DF	PARAMETER ESTIMATE	STANDARD ERROR	T FOR H0: PARAMETER=0	PROB > |T|	TYPE I SS	VARIABLE LABEL
INTERCEP	1	27.81944444	1.48782074	18.698	0.0001	11761.25000	INTERCEPT
N1	1	26.34722222	2.25852356	11.666	0.0001	5010.00498	dummy variable = 1 if DCT group
N2	1	-9.15278	2.05784675	-4.448	0.0006	646.95573421	dummy variable = 1 if CCD group
HN1	1	3.31944444	2.25852356	1.470	0.1637	141.77812500	interaction of H and N1
HN2	1	1.81944444	2.05784675	0.884	0.3915	48.22855202	interaction of H and N2
H	1	4.51388889	1.48782074	3.034	0.0089	382.69927536	dummy variable =-1 if hypertensive

B

DEP VARIABLE: A   Sodium-potassium ATPase

ANALYSIS OF VARIANCE

SOURCE	DF	SUM OF SQUARES	MEAN SQUARE	F VALUE	PROB>F
MODEL	5	6229.66667	1245.93333	29.967	0.0001
ERROR	14	582.08333333	41.57738095		
C TOTAL	19	6811.75000			

ROOT MSE	6.448052	R-SQUARE	0.9145	
DEP MEAN	24.25	ADJ R-SQ	0.8840	
C.V.	26.58991			

PARAMETER ESTIMATES

VARIABLE	DF	PARAMETER ESTIMATE	STANDARD ERROR	T FOR H0: PARAMETER=0	PROB > |T|	TYPE I SS	VARIABLE LABEL
INTERCEP	1	27.81944444	1.48782074	18.698	0.0001	11761.25000	INTERCEPT
.H	1	4.51388889	1.48782074	3.034	0.0089	92.45000000	dummy variable =-1 if hypertensive
HN1	1	3.31944444	2.25852356	1.470	0.1637	34.27920000	interaction of H and N1
HN2	1	1.81944444	2.05784675	0.884	0.3915	108.70830000	interaction of H and N2
N1	1	26.34722222	2.25852356	11.666	0.0001	5171.72727	dummy variable = 1 if DCT group
N2	1	-9.15278	2.05784675	-4.448	0.0006	822.50189394	dummy variable = 1 if CCD group

FIGURE 8-7 In order to use regression to analyze a two-way analysis of variance with missing data, it is necessary to repeat the regression analysis three times, once with the dummy variables corresponding to each main effect entered last and once with the interaction term entered last in order to obtain the marginal sum of squares associated with each of these effects. (Recall that the marginal sum of squares is the incremental sum of squares associated with a variable or group of variables entered into the regression equation *last.*) The marginal sum of squares is a measure of the unique information contained in the independent variable after taking into account all other information included in the other variables. (*A*) To test for the effect of hypertension, we enter dummy variable *H* last. (*continued*)

```
C

DEP VARIABLE: A          Sodium-potassium ATPase

                                          ANALYSIS OF VARIANCE

                                    SUM OF          MEAN
                        SOURCE   DF  SQUARES         SQUARE        F VALUE     PROB>F

                        MODEL     5  6229.66667   1245.93333      29.967      0.0001
                        ERROR    14  582.08333333   41.57738095
                        C TOTAL  19  6811.75000

                        ROOT MSE      6.448052    R-SQUARE        0.9145
                        DEP MEAN         24.25    ADJ R-SQ        0.8840
                        C.V.          26.58991

                                          PARAMETER ESTIMATES

                   PARAMETER    STANDARD     T FOR H0:                               VARIABLE
VARIABLE   DF       ESTIMATE       ERROR    PARAMETER=0    PROB > |T|    TYPE I SS    LABEL

INTERCEP    1    27.81944444   1.48782074       18.698        0.0001   11761.25000   INTERCEPT
H           1     4.51388889   1.48782074        3.034        0.0089      92.45000000 dummy variable =-1 if hypertensive
N1          1    26.34722222   2.25852356       11.666        0.0001    5116.67280   dummy variable = 1 if DCT group
N2          1    -9.15278      2.05784675       -4.448        0.0006     741.89609535 dummy variable = 1 if CCD group
HN1         1     3.31944444   2.25852356        1.470        0.1637     246.14587738 interaction of H and N1
HN2         1     1.81944444   2.05784675        0.884        0.3915      32.50189394 interaction of H and N2
```

Figure 8-7 (*continued*) (*B*) To test for the effect of nephron site, we repeat the regression with the dummy variables N_1 and N_2 last and add the sums of squares associated with each of these variables together to obtain the total sum of squares associated with the nephron site factor in the analysis. (*C*) Finally, to obtain the sum of squares associated with the hypertension by nephron interaction, we repeat the regression a third time with the dummy variables HN_1 and HN_2 entered last.

To test whether there are differences in Na-K-ATPase activity among the nephron sites, we repeat our analysis using the same general regression model as before, but with the set of dummy variables that encode the nephron site entered into the equation last to obtain

$$\hat{A} = b_0 + \underbrace{b_{HN_1}HN_1 + b_{HN_2}HN_2}_{\text{Interaction}} + \underbrace{b_H H}_{\text{Hypertension}} + \underbrace{b_{N_1}N_1 + b_{N_2}N_2}_{\text{Nephron site}}$$

Note that the two dummy variables that encode the nephron site, N_1 and N_2, are entered last as a *set*. You do not compute two regression equations, one with N_1 last and one with N_2 last, but rather put N_1 and N_2 last as a set. (Whether you put N_1 or N_2 last in the set will affect the incremental sum of squares associated with each variable, but will not affect the total sum of squares associated with the set of the two variables N_1 and N_2.) From Fig. 8-7*B*,

$$MS_N = \frac{SS_N}{DF_N} = \frac{SS_{N_1} + SS_{N_2}}{DF_{N_1} + DF_{N_2}} = \frac{5171.7 + 822.5}{1 + 1} = 2997.1 \; [\text{pmol}/(\text{min·mm})]^2$$

Note that the residual sum of squares, degrees of freedom, and mean square are the same as in Fig. 8-7A. We compute

$$F = \frac{MS_N}{MS_{res}} = \frac{2997.1}{41.6} = 72.05$$

which indicates that there is a significant difference in Na-K-ATPase activity among nephron sites, controlling for the presence or absence of hypertension.

Finally, we test for interaction between hypertension and kidney site by fitting the regression equation with the set of interaction dummy variables entered last:

$$\hat{A} = b_0 + \underbrace{b_H H}_{\text{Hypertension}} + \underbrace{b_{N_1} N_1 + b_{N_2} N_2}_{\text{Nephron site}} + \underbrace{b_{HN_1} HN_1 + b_{HN_2} HN_2}_{\text{Interaction}}$$

As with the nephron site dummy variables above, the interaction dummy variables are entered last as a set; the order within the set does not affect the total sum of squares associated with the set of the two variables. From Fig. 8-7C,

$$MS_{HN} = \frac{SS_{HN}}{DF_{HN}} = \frac{SS_{HN_1} + SS_{HN_2}}{DF_{HN_1} + DF_{HN_2}} = \frac{246.1 + 32.5}{1 + 1} = 139.3 \ [\text{pmol}/(\text{min} \cdot \text{mm})]^2$$

Again, MS_{res} has not changed, so we test for a significant interaction effect with

$$F = \frac{MS_{HN}}{MS_{res}} = \frac{139.3}{41.6} = 3.35$$

which does not exceed 3.74, the critical value of F for $\alpha = .05$ and 2 and 14 degrees of freedom. Thus, we conclude that there is a marginally insignificant interaction between hypertension and nephron site ($P \approx .07$). These conclusions are basically what we obtained for the full data set, except with the missing data, there is not much evidence of an interaction.

Summary of the Procedure

To recapitulate, the procedure for two-way analysis of variance with unbalanced data and no empty cells is

1. For the a levels of factor A, define $a - 1$ dummy variables using effects coding (1, 0, −1).
2. For the b levels of factor B, define $b - 1$ dummy variables using effects coding (1, 0, −1).

3. Form the $(a - 1)(b - 1)$ dummy variables that represent the interaction of the two main effects by computing the products of the $a - 1$ dummy variables associated with factor A and the $b - 1$ dummy variables associated with factor B.

4. Fit a regression model containing all the dummy variables with the set of $a - 1$ factor A dummy variables entered into the equation last.

5. Add up the incremental sums of squares associated with the set of $a - 1$ factor A dummy variables. This sum is the marginal sum of squares associated with factor A, SS_A.

6. Compute MS_A by dividing SS_A by the associated degrees of freedom, which equals the number of dummy variables used to encode factor A, $a - 1$.

7. Repeat steps 4 through 6 with the set of $b - 1$ factor B dummy variables entered into the equation last to obtain MS_B.

8. Repeat steps 4 through 6 with the set of $(a - 1)(b - 1)$ interaction dummy variables entered into the equation last to obtain MS_{AB}.

9. Use the residual mean square MS_{res} obtained from any of the regressions and each mean square computed in steps 6, 7, and 8, to compute the F statistics to test each main effect and the interaction, just as we did with balanced data.

Recall that when we discussed multicollinearity in Chap. 3, we used a similar procedure to obtain the marginal sum of squares associated with *each variable* in the regression equation. The procedure we follow here is slightly different in that we want the marginal sum of squares associated with *each set of dummy variables* that represent each factor in the analysis of variance. Just as some computer programs have an option for printing the marginal sum of squares associated with each variable, some have the capability to print the marginal sum of squares associated with each set of variables corresponding to an analysis of variance factor. This procedure generalizes directly to more than two factors. For example, a three-way analysis of variance involving factors A, B, and C, and all interactions (AB, AC, BC, and ABC) would require seven regression analyses of the data, one with the factor A dummy variables entered last, one with the factor B dummy variables entered last, and so forth.*

The easy way to analyze unbalanced data is to use statistical software

---

*In such higher-order designs, people often pool the third and higher-order interactions with the residual because it is virtually impossible to interpret the high-order interactions. In this example, if you were to eliminate the ABC interaction through such pooling, you would only need to do six regressions, with the A, B, C, AB, AC, and BC sets of dummy variables entered last, in turn.

GENERAL LINEAR MODELS PROCEDURE							
DEPENDENT VARIABLE: A		Sodium-potassium ATPase					
SOURCE	DF	SUM OF SQUARES	MEAN SQUARE	F VALUE	PR > F	R-SQUARE	C.V.
MODEL	5	6229.66666667	1245.93333333	29.97	0.0001	0.914547	26.5899
ERROR	14	582.08333333	41.57738095		ROOT MSE		A MEAN
CORRECTED TOTAL	19	6811.75000000			6.44805249		24.25000000

SOURCE	DF	TYPE I SS	F VALUE	PR > F	DF	TYPE III SS	F VALUE	PR > F
H	1	92.45000000	2.22	0.1581	1	382.69927536	9.20	0.0089
NGROUP	2	5858.56889535	70.45	0.0001	2	5994.22916667	72.09	0.0001
H*NGROUP	2	278.64777132	3.35	0.0647	2	278.64777132	3.35	0.0647

FIGURE 8-8 In some computer programs it is possible to obtain the marginal sums of squares in a single run, rather than having to run three separate regressions as we did in Fig. 8-7. The SAS Type III sum of squares in the general linear models (GLM) procedure is the marginal sum of squares. This figure shows an analysis of the same data as in Fig. 8-7. Compare the Type III sums of squares in this figure with the corresponding sums of squares from Fig. 8-7. The Type I sum of squares in the SAS GLM procedure is the incremental sum of squares. Note that these Type I sum of squares values correspond to those in Fig. 8-7*C* where the hypertension factor was entered first, the nephron factor second, and the interaction third. When there are missing data, the incremental sums of squares (Type I in SAS GLM) and the marginal sum of squares (Type III in SAS GLM) are not equal. When there are no missing data and the dummy variables have effects coding, the marginal and incremental sums of squares are equal. (Note that this notation is specific to SAS GLM procedure; generally it is not a nomenclature used for incremental and marginal sums of squares in all programs or, for that matter, even within SAS. The numbers labeled Type II sums of squares in the GLM and REG procedures are different. For a discussion of these issues for all the programs illustrated in this book, see Appendix B.)

that computes the correct sum of squares on one pass, as shown in Fig. 8-8, which shows the result of analyzing the data in Table 8-15 with a general linear models program. The sums of squares labeled "type III"* and the associated *F* statistics are the marginal sums of squares obtained from the repeated regressions for the hypertension effect (Fig. 8-7*A*), the

---

*The SAS procedure GLM is a common way to compute these sums of squares, which are labeled type III sums of squares in the SAS computer output. This terminology is sometimes carried over into the general statistical literature, but has no special meaning—it is specific to SAS. For more detail on how to get BMDP, SAS, SPSS, and Minitab to give you the appropriate marginal sums of squares when dealing with missing data in multifactor analysis of variance, see Appendix B. You need to be very careful that you know what sums of squares the computer program you are using is actually computing!

nephron effect (Fig. 8-7*B*), and interaction (Fig. 8-7*C*). If you have access to such software, obtaining the correct marginal sums of squares is simplified. If not, the repeated regression procedure we have outlined is a small price to pay to avoid deleting entire cases to make cell sizes equal when some data are missing.

In summary, when cell sample sizes are not equal, you must take care to base the hypothesis testing on the marginal sums of squares associated with the different factors in the model. The regression approach to analysis of variance is very useful for analyzing these data sets because you can obtain the marginal sums of squares by simply repeating the regression computations with each effect (main effects and interactions) entered after all the others. The effects are entered in sequence: Factor *A* after factor *B* and interaction for one regression, factor *B* after factor *A* and interaction for a second regression, and, finally, interaction after factors *A* and *B*. The factor *A* *F* statistic is computed from the regression analysis that entered factor *A* last, the factor *B* *F* statistic is computed from the regression analysis that entered factor *B* last, and the interaction *F* statistic is computed from the regression analysis that entered interaction last.

What If You Use the Wrong Sum of Squares?

We have now encountered three different ways to compute the sum of squares associated with a factor in analysis of variance. The first way is the *incremental* sum of squares, which is the reduction of the residual sum of squares that occurs when a variable is added to the regression equation after the variables above it are in the equation. The second way is the *marginal sum of squares associated with entering an individual variable* in the regression equation given that *all* the other variables are in the equation, and the third way is the *marginal sum of squares associated with a set of dummy variables that encode an experimental factor* that are entered after all the other dummy variables. Various computer programs print some or all of these sums of squares, often with precious little indication of how they were computed in the output.[*] When the data are balanced, all these sums of squares are equal, so it does not matter which type of sum of squares one uses. In contrast, when the data are not balanced, the sums of squares computed using these different strategies can be very different and lead you to different conclusions if you use them to construct *F* statistics. In each case, the *F* statistic will test *some* hypothesis, but there is no guarantee that the

---

[*]In fact, there are other strategies for computing sums of squares. For example, the SAS procedure GLM computes type II sum of squares by putting the variable of interest second to last followed by all interaction terms that involve the term of interest.

hypothesis will be the one you are interested in or will even be interpretable.[*]

For example, suppose we had simply used our original regression model, given by Eq. (8.4), to analyze the data in Table 8-15, without worrying about properly computing the marginal sums of squares. The computer output in Fig. 8-7C (hypertension factor first, nephron site next, interaction last) is what we would obtain. Because the interaction dummy variables were entered last, the associated sums of squares are correct, and we used them above to draw the correct conclusion that there was a marginally insignificant interaction effect. In contrast, because the nephron site dummy variables were not entered after all the other variables, the incremental sum of squares associated with the nephron factor does not equal its marginal sum of squares, and the F value we compute for the nephron site,

$$F = \frac{MS_N}{MS_{res}} = \frac{(SS_{N_1} + SS_{N_2})/2}{MS_{res}} = \frac{(5116.7 + 741.9)/2}{41.6} = 70.4$$

is incorrect. Similarly, the F value computed for the hypertension effect,

$$F = \frac{MS_H}{MS_{res}} = \frac{SS_H/1}{MS_{res}} = \frac{92.5}{41.6} = 2.22$$

is incorrect.

From these F statistics we conclude that there are significant differences in Na-K-ATPase activity with nephron site $(P < .01)$, that there is no difference in Na-K-ATPase activity between the two kinds of rats $(P > .05)$, and that there is a marginally insignificant interaction between nephron site and hypertension factor (rat strain) $(P \approx .07)$. The only correct hypothesis test was the one for interaction. The others are incorrect because they are not based on the marginal sums of squares associated with the main effects. The incorrect F of 70.4 for the nephron site effect still leads us to the same conclusion we reached using the correct F computed from the marginal sums of squares for N_1 and N_2 in Fig. 8-7B, that there is a significant nephron site effect. However, the incorrect F of 2.22 for the hypertension effect is much smaller than the correct value of 9.2 computed from the marginal sum of squares in Fig. 8-7A and leads us to conclude that there is not a significant hypertension effect $(P \approx .16)$. Had we not followed the procedure of obtaining the marginal

---

[*]For a discussion of the actual hypotheses that F test statistics based on these different sums of squares actually test, see S. R. Searle, *Linear Models for Unbalanced Data*, New York, Wiley, 1987, Chap. 4, "The 2-Way Crossed Classification with All-Cells-Filled Data: Cell Means Models," and Chap. 12, "Comments on Computing Packages"; and G. A. Milliken and D. E. Johnson, *Analysis of Messy Data: Volume 1; Designed Experiments*, New York, Van Nostrand Reinhold, 1984, Chaps. 9 and 10.

sums of squares for each effect, we would have reached the wrong conclusion about the difference in Na-K-ATPase activity between normal and hypertensive rats.

Multiple Comparisons with Missing Data

We use the same techniques for doing multiple comparisons — Bonferroni t tests, the Student-Newman-Keuls test, and Dunnett's test — as for balanced designs. If these tests are conducted comparing individual cells (as, say, one might do in the presence of interaction), we proceed exactly as we did in Chap. 7, when comparing treatment means in a one-way analysis of variance with unequal sample sizes [with Eq. (7.9)]. The only difference is that you use the MS_{res} and associated degrees of freedom from the two-way analysis of variance. In contrast, when you wish to do multiple comparisons across different levels of one factor (i.e., across the rows or columns of a two-way table), the computations are a bit more involved because it is necessary to take into account the different numbers of observations used to compute the means within the individual cells as well as in each column or row.

The need for this adjustment is best illustrated with an example, so let us return to the data on Na-K-ATPase and kidneys we have been discussing. Examining Table 8-15 reveals that there are two, three, or four observations in each of the cells. Because there are no empty cells, we can use these observations to estimate the mean value of Na-K-ATPase activity that occurs under each combination of experimental conditions. For example, the mean observed Na-K-ATPase activity of normal rats at the DCT site is

$$\overline{A}_{DCT,N} = \frac{58 + 66}{2} = 62.0 \ \text{pmol/(min·mm)}$$

and for hypertensive rats at the DCT site, it is

$$\overline{A}_{DCT,H} = \frac{44 + 49 + 46}{3} = 46.3 \ \text{pmol/(min·mm)}$$

Table 8-16 shows these cell means, together with the means of the other cells. As noted above, doing multiple comparisons among these cell means simply requires taking into account the differences in sample size with Eq. (7.9).

The situation in estimating the average values across the individual rows or columns, however, is a bit more complicated. The question is: How should we estimate the mean of any row (or column)? There are two possible strategies.

TABLE 8-16 Estimated Cell Means, Row Means, Column Means, and Sample Sizes for Rat Nephron Na-K-ATPase Activity Data Shown in Table 8-15, pmol/(min·mm)

Nephron Site	Strain of Rat		Row Means
	Normal	Hypertensive	
DCT	Mean 62.0 n 2	Mean 46.3 n 3	54.2
CCD	Mean 25.0 n 4	Mean 12.3 n 3	18.7
OMCD	Mean 10.0 n 4	Mean 11.3 n 4	10.7
Column Means	32.3	23.3	

The first strategy is to simply average all the observations in that row, without regard for the column, to obtain an estimate of the mean response in that row. For example, using this procedure, we would estimate the mean Na-K-ATPase activity for all DCT nephrons (regardless of whether the rat was normal or hypertensive) as

$$\frac{(58 + 66) + (44 + 49 + 46)}{2 + 3} = \frac{263}{5} = 52.6 \text{ pmol/(min·mm)}$$

Using this strategy, the row mean represents a weighted average of the means of the individual cells, with the weights depending on the number of observations within each cell. Thus, a cell with many missing points does not contribute as much to the row average as a cell with no missing observations.

The second strategy is to begin not with the raw observations, but with the cell means, because each of these means is our best estimate of the value of Na-K-ATPase activity within that cell. To obtain an estimate of the row mean, we average the cell means, without regard for the number of observations used to derive each cell mean. With this strategy, the mean Na-K-ATPase activity for DCT nephrons is

$$\overline{A}_{\text{DCT}} = \frac{\overline{A}_{\text{DCT},N} + \overline{A}_{\text{DCT},H}}{2} = \frac{62.0 + 46.3}{2} = 54.2 \text{ pmol/(min·mm)}$$

which differs from the estimate we obtained using the first strategy. Because we wish to test the hypothesis that the row means are not different (as opposed to the hypothesis that there is a difference in row means, weighted by the sample sizes), we will estimate the row (or column) means based on the cell means when conducting multiple comparisons in the face of unbalanced designs.* Table 8-16 shows the row and column means computed from the cell means, together with the number of observations in each cell used to estimate the respective means. We will use these values for our multiple comparison testing.

The differences in sample size within the cells do, however, enter into the computation of the multiple comparison test statistic because these numbers affect the precision with which we can estimate the differences between different row (or column) means. The test proceeds as before. We first rank the row means from smallest to largest, then compare the largest difference first, and work toward smaller differences. The test statistic for the Bonferroni t test for differences in row means is

$$t = \frac{\overline{X}_{r_1} - \overline{X}_{r_2}}{\frac{1}{c}\sqrt{MS_{res}\left(\sum\frac{1}{n_{rc_1}} + \sum\frac{1}{n_{rc_2}}\right)}}$$

where there are c columns and r rows and the summations in the denominator are across the columns of the rows being compared, MS_{res} is the residual mean square from the two-way analysis of variance, and the number of degrees of freedom is that associated with MS_{res}. Dunnett's statistic q' is identical to t (we just use a different table of critical values), whereas Student-Newman-Keuls statistic q is computed analogously to t and is equal to $\sqrt{2}\, t$.

For example, to compare the DCT and OMCD nephron Na-K-ATPase activity for the data in Tables 8-15 and 8-16, we compute

$$t = \frac{54.2 - 10.7}{\frac{1}{2}\sqrt{41.6\left[\left(\frac{1}{2} + \frac{1}{3}\right) + \left(\frac{1}{4} + \frac{1}{4}\right)\right]}} = 11.68$$

There are a total of three possible comparisons within the nephron site factor (DCT vs. OMCD, DCT vs. CCD, and CCD vs. OMCD), so to reject the hypothesis of no difference with a family error rate of $\alpha_T = .05$, we must conduct each test at the $.05/3 = .016$ level of significance. For 14 degrees of freedom, this specification means that the critical value of t is 2.77 (interpolating in Table E-1, Appendix E). Because the value of t associated with this comparison exceeds this critical value, we conclude that there is a significant difference and go on to the next desired comparison.

*This procedure requires that there not be any empty cells. We will discuss the situation when there are empty cells in the following section.

ONE OR MORE CELLS EMPTY

The basic idea behind a multi-way analysis of variance is that it allows you to combine information obtained under a variety of combinations of two or more treatment factors, say the drug being given to treat high blood pressure and the sex of the recipient. By using a two-way design, you can test hypotheses about the effects of the drug "controlling for" differences in sex. The traditional analysis of variance paradigm requires that the number of observations in each cell be the same (or, in some cases, proportional to each other). In the last section, we saw how to use a regression implementation of analysis of variance to relax the requirement for balanced data and handle unbalanced designs as well as missing data. We now turn our attention to the extreme case in which one or more of the cells in a two-way analysis of variance is empty. Empty cells greatly complicate the analysis of such data, because it is no longer possible to make any statements about the mean of the members of the population who fall in the empty cells. We simply have no observations in the cell to average.

In terms of the regression model, where there are empty cells it is not possible to estimate all the coefficients in the regression model because some of the needed information is absent; in terms of the model selection procedures we discussed in Chap. 6, the model is overspecified. As with other overspecified models, we must reduce the number of parameters in the model in order to estimate parameters and draw conclusions about the data. None of the approaches for dealing with this problem is ideal; you are best off striving to avoid empty cells in two-way (and higher) analyses of variance.

If you are stuck with an empty cell, what should you do? There are three strategies:

- Abandon the two-way design and treat the data as a one-way analysis of variance, with the different treatments representing the combinations of the two-factor treatments for which you have data. If you follow this strategy, you can conduct the analysis of variance and multiple comparisons just as we did in single-factor designs in Chap. 7. This approach allows you to draw conclusions about those cells for which you have data, and is often the most sensible, particularly if there are many empty cells.

- Divide the two-way table into subtables in which the subtables contain no empty cells, then conduct two-way analyses on the subtables. This approach has the benefit of maintaining the two-way design but limits the hypotheses you can test and the nature of the interactions you can study. This approach generally leads to more confusion than enlightenment and should generally be avoided.

- *Assume* that there is no interaction between the two factors and test for the main effects. If the assumption of no interaction is true, this approach has the benefit of reducing the number of parameters in the model and can, under certain conditions (discussed below), permit testing for the main effects. If the assumption of interaction is not correct, the model will be misspecified and the results biased. We will devote the remainder of this section to discussing this strategy.

Before we discuss the specifics of a no-interaction model, it is important to emphasize that some statistical packages contain routines that calculate F (and other) statistics in the presence of empty cells. As with the case of unbalanced designs, there are many ways in which such computations can be done, and the actual hypothesis tests that these calculations are performing are often not of practical interest or even interpretable. Even more than in the unbalanced designs discussed in the last section, you should not be lulled into a false sense of security simply because these packages produce lots of numbers.

As noted above, the first requirement for conducting a two-way analysis of main effects is to *assume* that there is no interaction. This assumption eliminates the dummy variables associated with the interaction and so reduces the number of parameters that need to be estimated. This situation reduces the likelihood that the model will be overspecified and makes it possible to estimate the mean value of the population within empty cells based on the values of the surrounding cells. While greatly facilitating the analysis, this assumption must be made carefully because if it is wrong, the results of the statistical hypothesis test will be unreliable.

The second requirement is that the pattern of empty cells be arranged in such a way that permits using the cells that do contain observations to estimate the mean values within the empty cells. This requirement is that the cells be *connected*. While the definition of connected data allows for multi-way analyses of variance, as a practical matter there is a simple geometric definition of connected data that will suffice for most practical problems when there are not a large number of missing cells.[*] The definition of *geometrically connected* data is

> *Data of a two-way design are geometrically connected when the filled cells of the data grid can be joined by a continuous line, con-*

---

[*]For a general discussion of connected data, see S. R. Searle, *Linear Models for Unbalanced Data*, New York, Wiley, 1987, pp. 139–145; or G. A. Milliken and D. E. Johnson, *Analysis of Messy Data: Volume 1: Designed Experiments*, New York, Van Nostrand Reinhold, 1984, pp. 105–106.

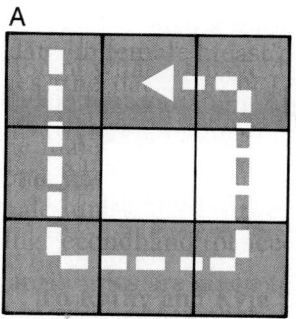 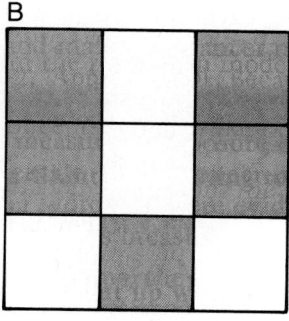

FIGURE 8-9 Schematic representative of geometrically connected data in a two-way design. Shaded cells contain data and blank cells are empty. When there is a completely empty cell in a two-way experimental design, it is still possible to conduct a two-way analysis of variance if you can reasonably *assume* that there is no interaction between the two experimental factors and if the data are *connected*. Although there are less restrictive definitions of connectedness, for all practical purposes one can determine if data are connected by testing whether or not they are *geometrically connected*. Data in a two-way table are connected if it is possible to draw a single line, consisting only of horizontal and vertical segments, through all cells containing data while making all directional changes within a cell that contains data. Hence, the pattern of observations in panel *A* is connected. The pattern of missing data in panel *B* is such that it is not connected; even if you change the order of the columns, there is no way to draw a single line through all the cells containing data (shaded) and to make turns only in cells that are not empty.

sisting solely of horizontal and vertical segments, that changes direction only in filled cells.

The rows and columns of the cells can be rearranged to meet the requirement of geometrically connected data. Figure 8-9 shows examples of connected and disconnected data. If the data are connected and you are willing to assume no interactions, then an analysis of variance can be conducted just as was done in the last section, using the marginal sums of squares.*

---

*If you must include interaction or your data are not connected, it is possible to test some hypotheses that allow you to get a partial look at interactions. However, these hypotheses are rarely unique—they depend on the pattern of missing cells and the computer algorithms that select the hypotheses to be tested—and their interpretation and justification is well beyond the scope of this book. Further explanation of hypothesis testing under such circumstances can be found in S. R. Searle, *Linear Models for Unbalanced Data*, New York, Wiley, 1987, pp. 158–165; and G. A. Milliken and D. E. Johnson, *Analysis of Messy Data: Volume 1: Designed Experiments*, New York, Van Nostrand Reinhold, 1984, Chaps. 13 and 14.

Multiple Comparisons with Empty Cells

If we assume no interactions and the data are connected, it is possible to estimate the mean values of the empty cells from the cells that contain data. This estimation is possible because, in the absence of interaction, the differences in estimated cell means (from the regression model) between any two rows (or columns) will be the same for any columns (or rows). For example, consider a hypothetical two-factor example with two levels of each factor and one missing cell. In the absence of an interaction term, the regression equation to describe this situation is

$$\hat{y} = b_0 + b_R R + b_C C$$

where R and C are dummy variables coded with effects coding, so

$$R = \begin{cases} -1 \text{ for the first row} \\ +1 \text{ for the second row} \end{cases}$$

and

$$C = \begin{cases} -1 \text{ for the first column} \\ +1 \text{ for the second column} \end{cases}$$

According to the regression equation, the predicated cell means are

$$\hat{y}_{11} = b_0 - b_R - b_C$$
$$\hat{y}_{12} = b_0 - b_R + b_C$$
$$\hat{y}_{21} = b_0 + b_R - b_C$$
$$\hat{y}_{22} = b_0 + b_R + b_C$$

where $\hat{y}_{ij}$ is the predicted mean for the cell in row i and column j. We can use these relationships to express any of the cell means in terms of the others. For example,[*]

[*]Equation (8.7) is determined from inspection of the relationships between the cell means and the regression coefficients. To see this, consider

$$\hat{y}_{12} + \hat{y}_{21} = (b_0 - b_R + b_C) + (b_0 + b_R - b_C)$$
$$= b_0 + b_0 - b_R + b_R + b_C - b_C$$
$$= 2b_0$$

then subtract $\hat{y}_{22}$ to obtain

$$(\hat{y}_{12} + \hat{y}_{21}) - \hat{y}_{22} = 2b_0 - (b_0 + b_R + b_C)$$
$$= (2b_0 - b_0) - b_R - b_C = b_0 - b_R - b_C$$
$$= \hat{y}_{11}$$

$$\hat{y}_{11} = \hat{y}_{12} + \hat{y}_{21} - \hat{y}_{22} \tag{8.7}$$

This computation immediately generalizes to any other missing cell for any two-factor design, so long as the cells are connected. The specific formulas are different, depending on which cells are missing, but the computations still boil down to making use of the fact that the difference in mean value between any two rows (or columns) is the same for all columns (or rows).

We can use this information to construct multiple comparisons tests among the rows and columns similar to those we have conducted before, even though some cells may be empty. Recall that each of these test statistics is of the form (give or take a $\sqrt{2}$, depending on the test)

$$t = \frac{\hat{y}_i - \hat{y}_j}{s_{\hat{y}_i - \hat{y}_j}} \tag{8.8}$$

We only need to come up with estimates of the numerator and denominators of this equation.

Suppose we wish to compare two rows of an experiment in which there are three levels of the other (column) factor, and there is an empty cell so that we cannot directly observe $\hat{y}_{11}$. Following the principle developed above that we want to base the multiple comparisons on the averages (across all columns) of the cell means, we wish to compare

$$\hat{y}_1 = \frac{1}{2}(\hat{y}_{11} + \hat{y}_{12})$$

with

$$\hat{y}_3 = \frac{1}{2}(\hat{y}_{31} + \hat{y}_{32})$$

so the numerator of Eq. (8.8) is

$$\hat{y}_1 - \hat{y}_3 = \frac{1}{2}(\hat{y}_{11} + \hat{y}_{12}) - \frac{1}{2}(\hat{y}_{31} + \hat{y}_{32})$$

The only problem with this is that we do not have any values in the top left cell with which to estimate $\hat{y}_{11}$. We can, however, use the other cell means, together with the assumption of no interaction to replace $\hat{y}_{11}$ in this equation with the known cell means using Eq. (8.7) to obtain

$$\hat{y}_1 - \hat{y}_3 = \frac{1}{2}[(\hat{y}_{12} + \hat{y}_{21} - \hat{y}_{22}) + \hat{y}_{12}] - \frac{1}{2}(\hat{y}_{31} + \hat{y}_{32})$$

$$= \hat{y}_{12} + \frac{1}{2}\hat{y}_{21} - \frac{1}{2}\hat{y}_{22} - \frac{1}{2}\hat{y}_{31} - \frac{1}{2}\hat{y}_{32}$$

and we have an expression for the numerator of Eq. (8.8) in terms of things we can observe, i.e., in terms of cells in which we have observations.

To obtain the denominator of Eq. (8.8), we make use of a general theoretical result for the standard error of a sum of means from an analysis of variance.* Suppose we have a weighted sum (or difference) of means in a two-way analysis of variance, $\sum c_{ij}\hat{y}_{ij}$, where the c_{ij} are constants. The standard error of this sum is

$$s_{\sum c_{ij}\hat{y}_{ij}} = \sqrt{MS_{\text{res}} \sum \frac{c_{ij}^2}{n_{ij}}} \tag{8.9}$$

where n_{ij} is the number of observations in cell ij. We can use this equation to estimate the denominator of Eq. (8.8). From the expression for $\hat{y}_1 - \hat{y}_3$ above, we have

$$c_{12} = 1$$

$$c_{21} = \frac{1}{2}$$

$$c_{22} = -\frac{1}{2}$$

$$c_{31} = -\frac{1}{2}$$

$$c_{32} = -\frac{1}{2}$$

Thus, we can use Eq. (8.9), together with the MS_{res} from the analysis of variance and the sample sizes from the filled cells to obtain the denominator for the multiple comparison test statistics.

This procedure is obviously more complicated than those we have used for experiments with no empty cells; the results need to be computed on a case-by-case basis. It is a good idea to do these computations by hand unless you are *absolutely certain* that you know how the computer package you are using is doing them, if it will do them at all.

We now illustrate these computations by returning to the kidney Na-K-ATPase activity example one last time, with even fewer data than before.

---

*For further discussion of this result, see G. A. Milliken and D. E. Johnson, *Analysis of Messy Data: Volume 1: Designed Experiments*, New York, Van Nostrand Reinhold, 1984, pp. 188–189.

More on the Missing Kidney

Suppose that the upper-left-hand cell shown in Table 8-15 was completely empty instead of only missing two points, yielding the data set in Table 8-17. In this situation, with only one missing cell, it is easy to demonstrate that the cells are connected, so we can do a two-way analysis of variance to test for the main effects (hypertension and nephron site) *if we are willing to assume that there is no interaction between the two main effects.* We will make this assumption for the sake of argument here, knowing that it is a dangerous one, because we have already demonstrated significant interaction between the nephron site and presence or absence of hypertension when we analyzed the complete data set.

The regression model for Na-K-ATPase activity, without interaction, is

$$\hat{A} = b_0 + b_H H + b_{N_1} N_1 + b_{N_2} N_2 \tag{8.10}$$

where the dummy variables are defined with effects coding as before. To obtain the marginal sums of squares for the nephron site, we fit the equation with the nephron site dummy variables N_1 and N_2 entered last (Fig. 8-10A), and to obtain the marginal sum of squares for the hyperten-

TABLE 8-17 Data (by Cell) from Study of Na-K-ATPase Activity [pmol/(min·mm)] of Nephron Site and Hypertension; Some Data Are Missing (Compare to Table 8-15) and There Is One Empty Cell

Nephron Site	Strain of Rat	
	Normal	Hypertensive
DCT		44 49 46
CCD	15 31 19 35	8 11 18
OMCD	7 7 9 17	19 7 15 4

A

DEP VARIABLE: A   Sodium-potassium ATPase

ANALYSIS OF VARIANCE

SOURCE	DF	SUM OF SQUARES	MEAN SQUARE	F VALUE	PROB>F
MODEL	3	2884.08547	961.36182336	18.466	0.0001
ERROR	14	728.85897436	52.06135531		
C TOTAL	17	3612.94444			

ROOT MSE	7.215356	R-SQUARE	0.7983	
DEP MEAN	20.05556	ADJ R-SQ	0.7550	
C.V.	35.97684			

PARAMETER ESTIMATES

VARIABLE	DF	PARAMETER ESTIMATE	STANDARD ERROR	T FOR H0: PARAMETER=0	PROB > |T|	TYPE I SS	VARIABLE LABEL
INTERCEP	1	26.24893162	1.94002444	13.530	0.0001	7240.05556	INTERCEPT
H	1	2.58653846	1.87193207	1.382	0.1887	94.04444444	dummy variable =-1 if hypertensive
N1	1	22.67094017	3.32416301	6.820	0.0001	2398.33820	dummy variable = 1 if DCT group
N2	1	-7.04701	2.56911877	-2.743	0.0159	391.70282339	dummy variable = 1 if CCD group

B

DEP VARIABLE: A   Sodium-potassium ATPase

ANALYSIS OF VARIANCE

SOURCE	DF	SUM OF SQUARES	MEAN SQUARE	F VALUE	PROB>F
MODEL	3	2884.08547	961.36182336	18.466	0.0001
ERROR	14	728.85897436	52.06135531		
C TOTAL	17	3612.94444			

ROOT MSE	7.215356	R-SQUARE	0.7983	
DEP MEAN	20.05556	ADJ R-SQ	0.7550	
C.V.	35.97684			

PARAMETER ESTIMATES

VARIABLE	DF	PARAMETER ESTIMATE	STANDARD ERROR	T FOR H0: PARAMETER=0	PROB > |T|	TYPE I SS	VARIABLE LABEL
INTERCEP	1	26.24893162	1.94002444	13.530	0.0001	7240.05556	INTERCEPT
N1	1	22.67094017	3.32416301	6.820	0.0001	2476.47046	dummy variable = 1 if DCT group
N2	1	-7.04701	2.56911877	-2.743	0.0159	308.21803606	dummy variable = 1 if CCD group
H	1	2.58653846	1.87193207	1.382	0.1887	99.39697802	dummy variable =-1 if hypertensive

FIGURE 8-10 Analysis of the data on sodium potassium ATPase activity in kidneys from normal and hypertensive rats with the nephrons obtained from different sites in the presence of empty cells (data in Table 8-17). As before, the analysis needs to be based on the marginal sums of squares. However, because there is a missing cell, we are assuming no interaction; thus, we only have to run the regression analysis twice, once putting each factor in last, in order to obtain the necessary sums of squares. (A) To test for a significant effect of nephron site, we entered the dummy variables N_1 and N_2 last. (B) To test for a significant hypertension effect, we entered the dummy variable H last.

sion effect, we fit the data with the regression model with the hypertension dummy variable H entered last (Fig. 8-10B):

$$\hat{A} = b_0 + b_{N_1}N_1 + b_{N_2}N_2 + b_H H$$

From Fig. 8-10A,

$$MS_N = \frac{SS_{N_1} + SS_{N_2}}{DF_{N_1} + DF_{N_2}} = \frac{2398.3 + 391.7}{1 + 1} = 1395 \; [\text{pmol}/(\text{min·mm})]^2$$

and from Fig. 8-10B,

$$MS_H = \frac{SS_H}{DF_H} = \frac{99.4}{1} = 99.4 \; [\text{pmol}/(\text{min·mm})]^2$$

In both cases, the residual mean square is $MS_{\text{res}} = 52.06$ $[\text{pmol}/(\text{min·mm})]^2$ with $DF_{\text{res}} = 14$ degrees of freedom. We test for a significant effect of hypertension by computing

$$F = \frac{MS_H}{MS_{\text{res}}} = \frac{99.4}{52.06} = 1.91$$

and conclude that there is not a significant hypertension effect. Likewise, we test for differences in Na-K-ATPase activity between nephron sites with

$$F = \frac{MS_N}{MS_{\text{res}}} = \frac{1395}{52.06} = 26.8$$

and conclude that there is a significant difference in Na-K-ATPase activity between different nephron sites ($P < .0001$).

The significant nephron site effect is consistent with what we found before when we analyzed the full data set. (The computer output for the full data set is in Fig. 8-4.) However, based on these data with the missing cell, we do not even come close to concluding that there is a significant hypertension effect, whereas with the full data set, we found a highly significant effect. This difference in conclusions regarding the hypertension effect is worth exploring in detail, because it gets to the heart of the problems of assuming no interaction and having to estimate empty cell means.

The regression equation for these data (Fig. 8-10) allows us to estimate what the cell mean might have been for the empty cell (the DCT nephron site in normal rats). This equation is

$$\hat{A} = 26.249 + 2.587H + 22.671N_1 - 7.047N_2$$

To estimate any cell mean, we simply substitute the dummy variable codes for that treatment combination and solve for $\hat{A}$. For example, the

DCT site in normal rats has codes $H = +1$, $N_1 = +1$, and $N_2 = 0$. Substituting these values into the regression equation, we obtain

$$\hat{A} = 26.249 + 2.587(+1) + 22.671(+1) - 7.047(0) = 51.5 \text{ pmol/(min·mm)}$$

The other cell means are estimated similarly and are shown in Table 8-18. As noted above, these are the best estimates of the true cell means in the face of a missing cell, but because we are assuming no interaction, they do not equal the means you would actually calculate from the observations in the cells.

From the full data set shown in Table 8-10, our best estimate of the normal rat DCT cell mean is the average of the four observations in the cell, $(62 + 73 + 58 + 66)/4 = 64.8$ pmol/(min·mm). Thus, by being forced to exclude interaction, and by chance missing all the observations in a cell at the site that showed the biggest difference between normal and hypertensive rats in the full data set, we have an incomplete look at the data. Thus, we greatly underestimate the true magnitude of Na-K-ATPase activity at this site. This incorrect conclusion about the difference between normal and hypertensive rats is a reflection of the model misspecification bias induced by leaving out the interaction effect. Another reflection of the misspecification is the magnitude of the mean square residual, 52.06 [pmol/(min·mm)]^2 with 14 degrees of freedom. If we were to treat the five cells that contain data as a one-way analysis of variance with five groups, we would estimate a MS_{wit} of 42.31 [pmol/(min·mm)]^2 with 13 degrees of freedom, much more like the 41.6 [pmol/(min·mm)]^2 with 14 degrees of freedom obtained above for the data in Table 8-15. Thus, much of the variability that was attributable to

TABLE 8-18 Estimated Cell Means, Row Means, Column Means, and Observed Sample Sizes for Rat Nephron Na-K-ATPase Activity [pmol/(min·mm)] Data with an Empty Cell

Nephron Site	Strain of Rat		Row Means
	Normal	Hypertensive	
DCT	Mean 51.5 n 0	Mean 46.3 n 3	48.9
CCD	Mean 21.8 n 4	Mean 16.6 n 3	19.2
OMCD	Mean 13.2 n 4	Mean 8.1 4	10.7
Column means	28.8	23.7	

interaction in earlier analyses is now absorbed into the residual variation, thus inflating it.

Thus, one should be extremely cautious in drawing conclusions when there are empty cells in the data. The only way to analyze such data in the original multi-way analysis of variance format is to leave out interactions and run the risk of misspecifying the model. In the absence of clear evidence that interactions are not important, it would probably be better to abandon the multifactor design, which has already lost much of its appeal by excluding interactions, and resort to simpler one-way analyses of variance, in this example with five groups: DCT hypertensive, CCD hypertensive, OMCD hypertensive, CCD normal, and OMCD normal.

Multiple Comparisons for the Missing Kidney

Following this two-way analysis of variance without interaction when there are empty (but connected) cells, you may want to do multiple comparisons. The general method for comparing any two row or column means was introduced above. Now that we have completed the analysis of variance for these data, we can demonstrate how one uses Eq. (8.8) to compute the appropriate t tests. The procedures for the Student-Newman-Keuls and Dunnett's tests are analogous.

As before, the F statistic for the hypertension effect obviates the need to compare the columns in Table 8-18—there is only one comparison, which we have already made with the F statistic. There are three possible comparisons to be made between the rows (nephron sites). For example, to compare the DCT and OMCD sites, we need to compute, following Eq. (8.8),

$$t = \frac{\hat{A}_{\text{DCT}} - \hat{A}_{\text{OMCD}}}{s_{\hat{A}_{\text{DCT}} - \hat{A}_{\text{OMCD}}}}$$

The numerator is obtained from the row means computed in Table 8-18:

$$\hat{A}_{\text{DCT}} - \hat{A}_{\text{OMCD}} = \frac{1}{2}(\hat{A}_{\text{DCT},N} + \hat{A}_{\text{DCT},H}) - \frac{1}{2}(\hat{A}_{\text{OMCD},N} + \hat{A}_{\text{OMCD},H})$$

where, for example, $\hat{A}_{\text{DCT},N}$ is the estimated mean for the DCT site in normal rats. Because we did not actually make any observations in the DCT site in normal rats, we need to estimate the cell mean $\hat{A}_{\text{DCT},N}$ indirectly using the other known cell means. We estimate the empty cell mean, using *all* the other cell means from filled cells,

$$\hat{A}_{\text{DCT},N} = \hat{A}_{\text{DCT},H} + \frac{1}{2}\hat{A}_{\text{CCD},N} - \frac{1}{2}\hat{A}_{\text{CCD},H} + \frac{1}{2}\hat{A}_{\text{OMCD},N} - \frac{1}{2}\hat{A}_{\text{OMCD},H}$$

We substitute this into the expression for $\hat{A}_{DCT} - \hat{A}_{OMCD}$ to obtain

$$\hat{A}_{DCT} - \hat{A}_{OMCD} = \frac{1}{2}\left(\hat{A}_{DCT,H} + \frac{1}{2}\hat{A}_{CCD,N} - \frac{1}{2}\hat{A}_{CCD,H} + \hat{A}_{DCT,H} + \frac{1}{2}\hat{A}_{OMCD,N}\right.$$
$$\left. - \frac{1}{2}\hat{A}_{OMCD,H}\right) - \frac{1}{2}\left(\hat{A}_{OMCD,N} + \hat{A}_{OMCD,H}\right)$$

$$= \hat{A}_{DCT,H} + \frac{1}{4}\hat{A}_{CCD,N} - \frac{1}{4}\hat{A}_{CCD,H} - \frac{1}{4}\hat{A}_{OMCD,N} - \frac{3}{4}\hat{A}_{OMCD,H}$$

The denominator is obtained following Eq. (8.9) as the sum of the weights c_{ij} and their corresponding sample sizes n_{ij} for those cells involved in the numerator of the t test. In computing the t statistic, we can only use information we actually observed. Thus, we do not base the c_{ij} and n_{ij} on the empty cell, but from the three filled cells from which the empty cell mean was estimated. Thus, the weights c_{ij} are $+1$, $+\frac{1}{2}$, $-\frac{1}{2}$, $-\frac{1}{2}$, and $-\frac{1}{2}$. The corresponding sample sizes n_{ij} are 3, 4, 3, 4, and 4, and

$$\sum \frac{c_{ij}^2}{n_{ij}} = \frac{1^2}{3} + \frac{(1/4)^2}{4} + \frac{(-1/4)^2}{3} + \frac{(-1/4)^2}{4} + \frac{(-3/4)^2}{4} = .562$$

We substitute this value along with the MS_{res}, into Eq. (8.7) to obtain the denominator for the t statistics,

$$s_{\hat{A}_{DCT} - \hat{A}_{OMCD}} = \sqrt{MS_{res} \sum \frac{c_{ij}^2}{n_{ij}}} = \sqrt{52.06(.562)} = 5.233$$

Finally, the t statistic to compare the DCT and OMCD nephrons is

$$t = \frac{48.9 - 10.7}{5.233} = 7.30$$

Because we will make three hypothesis tests for all possible nephron site comparisons, we select $\alpha = .05/3 = .016$ to keep the family error rate at .05. The observed value of t exceeds the critical value of t for $\alpha = .016$ and 14 degrees of freedom, 2.77 (interpolating in Table E-1, Appendix E), so we conclude that Na-K-ATPase activity at the DCT site differs significantly from the OMCD site ($P < .05$).

The comparison of DCT vs. CCD follows similarly, whereas, the comparison of CCD vs. OMCD follows the procedure given on pages 362–364.

Recapitulation

When there are empty cells, it is still possible to draw conclusions about data collected in a two-way design. One approach is to abandon the two-way design and simply conduct a single-factor analysis of variance between the different treatment groups. Alternatively, if the cells are connected and if it is reasonable to *assume* that there is no interaction between the experimental factors, it is possible to conduct an analysis of variance, making use of effects coding of the treatments and computing the marginal sums of squares. These computations can be done using two regression analyses, putting the dummy variables associated with the factor of interest last, or with one of the general linear model computer programs discussed earlier in this chapter. In any event, you need to be sure that you are obtaining the correct sums of squares to test the hypotheses of interest. It is always better to avoid empty cells.

SUMMARY

When we analyze data that can be grouped according to two (or more) factors, we use two-way (or higher-order) analysis of variance. This approach allows us to test hypotheses about each main effect and also to test whether the two effects interact. We use the same concepts as introduced for one-way analysis of variance in Chap. 7, and all of the same assumptions apply. We can test these assumptions using the methods introduced in Chaps. 4 and 7. We introduced effects coding (1, 0, -1) because the presence of interaction makes the regression implementation of analysis of variance sensitive to the order of entry of effects into the model if reference cell codes (0, 1) are used.

Unlike one-way analysis of variance, two-way analysis of variance is affected by missing data. When some observations are missing, but all treatment combinations (cells) have some data, we need to consider the order of entry of effects into the model (regression implementation or traditional), adjusting each effect (including interactions) for all other effects. We can do this using regression by repeating the regressions, each time with a different effect entered last to obtain the correct marginal sums of squares when there are missing data. Sophisticated analysis of variance and general linear models computer software are available to directly obtain the marginal sums of squares. Be warned, however: Just because you can coax a computer program into giving you sums of squares and F statistics does not mean that they are correct. If incorrect, you will not be testing the hypothesis you think you are, and even if you knew what hypothesis you were testing, it probably would not be of interest anyway.

When data are missing such that one or more cells are empty, the problems are worse. If you can reasonably assume no interaction and if the data cells are connected, then the marginal sums of squares can be used to correctly test main effects. However, you risk model misspecification by omitting interaction and may be worse off for having tried to force a particular analysis on unwilling data. In many cases, it is preferable to abandon the factorial aspect of the analysis and focus on simpler analyses, such as dropping from a two-way to a one-way analysis of variance.

All of the methods we presented for two-way analyses of variance generalize to more than two factors. However, as more factors are added, it becomes more difficult to interpret the results, especially the complex interactions that may result.*

PROBLEMS

8.1   Dietary fat is known to promote certain types of cancer. One of the ways certain fats may facilitate cancer is through a class of reactive chemicals derived from the fat, arachidonic acid, called n-6 fatty acids. Another class of fats, the so-called n-3 fatty acids that are found in high quantities in fish oil, may be less active in promoting tumors. Because n-3 acids can replace n-6 acids in cell membrane phospholipids, it may be possible to modify tumorigenesis by modifying the type of fat in the diet. In order to compare the relative tumorigenicity of the n-6 fatty acid, arachidonic acid (AA), with that of the n-3 fatty acid, eicosapentanoic acid (EPA), Belury and her coworkers[†] used a well-known tumor induction model where the chemical 12-O-tetradecanoylphorbol-13-acetate (TPA) is added to a cell culture to promote the carcinogenic activity of a suspected carcinogen. If either AA or EPA is carcinogenic, they would be expected to increase the DNA synthesis in skin cells grown in cell culture in the presence of TPA. These investigators measured the specific activity of DNA synthesis under four conditions, AA, EPA, AA with TPA, and EPA with TPA (the data are in Table D-21, Appendix D). Is there any evidence that AA and EPA have different carcinogenic potential?

8.2   Evaluate the assumption of homogeneity of variance for Prob. 8.1.

8.3 γ-Aminobutyric acid (GABA) is a nerve transmitter in the brain that has many functions that are not clearly understood. One such activity is control of secretion of the sex hormone prolactin from the pituitary gland. Control of prolactin secretion by other nerve transmitters, such as dopamine, is thought to be exerted through cyclic nucleotides, cAMP and cGMP. Therefore, Apud and his coworkers[‡]

---

*For further discussion of higher-order (factorial) analyses of variance, including designs with nested (heirarchical) factors, see B. J. Winer, *Statistical Principles in Experimental Design* (2nd ed.), New York, McGraw-Hill, 1971, Chaps. 5, 6, 8, and 9.

[†]M. A. Belury, K. E. Patrick, M. Locniskar, and S. M. Fischer, "Eicosapentaenoic and Arachidonic Acid: Comparison of Metabolism and Activity in Murine Epidermal Cells," *Lipids* 24:423–429, 1989.

[‡]J. A. Apud, D. Cocchi, V. Locatelli, C. Masotto, E. E. Muller, and G. Racagni, "Review: Biochemical and Functional Aspects on the Control of Prolactin Release by the Hypothalamo-Pituitary GABAergic System," *Psychoneuroendocrinology* 14:3–17, 1989.

measured cGMP levels in two sites—one in the cerebellum of the central nervous system and one in the anterior pituitary gland—in response to three different GABA stimulants, ethanolamine-*O*-sulfate (EOS), aminooxyacetic acid (AAOA), and muscimol (M) (the data are in Table D-22, Appendix D). Is there evidence that activation of the GABAergic nervous system with these three agonists affects cGMP?

8.4 Suppose that you were missing three data points from the study analyzed in Prob. 8.3, one observation from the AOAA group, and two observations from the M group in samples taken from the pituitary. (The modified data are in Table D-23, Appendix D.) *A.* Analyze the modified data. *B.* Do any of the conclusions you reached in Prob. 8.3 change?

8.5 Suppose no data were obtained for cGMP levels in the pituitary gland when M was given, so there is a missing cell. *A.* Modify the data in Table D-22, Appendix D, accordingly and repeat the analysis of Prob. 8.3. *B.* Are any of the conclusions based on this incomplete data set different from those reached in Prob. 8.3?

8.6 Nurse practitioner is a title granted to individuals with widely different training: from a short, continuing-education program to a 2-year program leading to a master's degree. There is agreement among nursing professionals that it would be good to narrow the applicability of the title so that it referred to a standard type of training. However, there are no data to support the selection of any one type of training as better than another. Therefore, Glascock and her coworkers[*] sought to determine if education (master's vs. nonmaster's) made a difference in the type or quantity of nursing activity engaged in by pediatric nurse practitioners. One of the things they measured was the total number of nursing-related assessment activities performed. Because the nurse practitioners who had master's degrees also tended to have more experience, these investigators did a two-way analysis of variance to see if educational background influenced assessment activities, while controlling for the confounding factor of experience. They divided experience into two categories: those with 7 years experience or less and those with more than 7 years experience (the data are in Table D-24, Appendix D). Is there evidence that the number of nursing assessment activities depends on the level of education?

8.7 Evaluate the assumption of homogeneity of variance in the data of Prob. 8.6 (Table D-24, Appendix D).

[*]J. Glascock, C. Webster-Stratton, and A. M. McCarthy, "Infant and Preschool Well-Child Care: Master's- and Nonmaster's-Prepared Pediatric Nurse Practitioners," *Nurs. Res.* 34:39–43, 1985.

8.8 One approach to handling data in which there are empty cells in two-way and higher-order analyses of variance is to fall back on a one-way analysis of the remaining cells. Do a one-way analysis of variance on the data in the five remaining cells of the kidney Na-K-ATPase activity data in Table 8-17. Follow this analysis by an appropriate multiple comparisons testing of pairwise differences between the means. How do your conclusions compare with what we found previously?

8.9 Many chemicals known as polycyclic aromatic hydrocarbons cause cancer. One of the best known of these is benzo(a)pyrine, which has many sources, including second-hand tobacco smoke. A related chemical, 7, H-dibenzo(c,g)carbazole is also found in tobacco smoke. However, these two polycyclic hydrocarbons seem to cause far different types of cancer: benzo(a)pyrine seems to cause largely epithelial cancers, whereas 7, H-dibenzo(c,g)carbazole seems to cause mostly cancers in the parenchyma of the internal organs. In this respect, 7, H-dibenzo(c,g)carbazole is much more like a completely different class of chemicals, the aromatic amines, one of which is 2-acetylaminofluorene. However, because the carcinogenicity of these different chemicals has been studied in a variety of animals and with a wide variety of methods, it is not clear how the carcinogenicity of 7, H-dibenzo(c,g)carbazole compares to other aromatic carcinogens. To clarify the relative carcinogenicity of these three chemicals, Schurdak and Randerath[*] compared the ability of these three aromatic carcinogens, administered by three different routes, topically (on the skin), orally, or subcutaneously, to cause damage to DNA in four different tissues from mice, liver, lung, kidney, and skin (the data are in Table D-25, Appendix D). *A.* Is there evidence that the different carcinogens have different effects? *B.* If so, is there evidence that the different carcinogens have different mechanisms of action?

8.10 Repeat the analysis of Prob. 8.9, leaving out the three-way interaction term. *A.* What effect does this omission have on the statistics and statistical conclusions? *B.* What effect does this omission have on your ability to interpret the biological questions asked?

---

[*]M. E. Schurdak and K. Randerath, "Effects of Route of Administration on Tissue Distribution of DNA Adducts in Mice: Comparison of 7, H-Dibenzo(c,g)carbazole, Benzo(a)pyrine, and 2-Acetylaminofluorene," *Cancer Res.* 49:2633–2638, 1989.

Repeated Measures

The procedures for testing hypotheses discussed in Chaps. 7 and 8 apply to experiments in which the control and treatment groups contain *different* subjects (individuals). It is often possible to design experiments in which *each* experimental subject can be observed *before and after* one or more treatments. Such experiments gain sensitivity because they make it possible to measure how the treatment *affects each individual*. When the control and treatment groups consist of different individuals, the changes brought about by the treatment may be masked by variability between experimental subjects. This chapter shows how to analyze experiments in which each subject is observed repeatedly under different experimental conditions. In this chapter we will use dummy variables to account for different subjects, in much the same way we used dummy variables in previous chapters to account for different experimental conditions. We will start by showing how subject dummy variables can be used to account for the between-subjects variability in a linear regression problem where data from several subjects are pooled to obtain one regression equation, thus allowing more precise estimates of regression coefficients. Then we will show how subject dummy variables can be used to account for between-subjects variability in analyses of variance when there is more than one treatment per subject in so-called *repeated-measures analysis of variance*.

ACCOUNTING FOR BETWEEN-SUBJECTS VARIABILITY IN LINEAR REGRESSION

In our treatment of regression to this point, we have assumed that we obtained one data point from each experimental subject. How should we handle the situation in which we obtain several points from each subject?

One strategy would be simply to treat all the points as before and submit them to a linear regression as if they were collected in different subjects. There are two potential disadvantages of this approach. First, the between-subjects differences are pooled into the residual variance, which reduces the precision of the parameter estimates and may even obscure a dependence on the independent variable. Second, such pooling may make it appear that there is a relationship between the dependent and independent variables if different experimental subjects have different values of the independent and dependent variables even though there is no relationship between the dependent and independent variables *within* each individual subject. How can we deal with this problem?

One could simply fit the data from each subject separately, then examine the regression coefficients. This strategy has two major disadvantages: First, it greatly reduces the number of points used for each individual regression and so reduces the power of the analysis. Second, it does not really test the hypothesis of interest, namely, *is there a relationship between the dependent and independent variables, taking into account the between-subjects differences?* By using dummy variables to encode the different subjects, we can pool all the data together to estimate a single regression equation and at the same time account for the between-subjects variability and thus increase the precision with which we estimate the associated regression coefficients.

The basic approach we will use is the same one we used to account for the effect of exposure to secondhand tobacco smoke on the relationship between weight and height in Martians (see Figs. 3-7 and 3-8) where we used a dummy variable to account for a parallel shift between the line describing Martians not exposed to smoke and that describing Martians who were exposed to smoke. In the same way, we can estimate an overall regression line while letting each subject have its own line that is shifted up or down (in parallel fashion) from this overall line. The between-subjects deviation from the overall line is accounted for by the dummy variables, with the net effect of reducing the MS_{res} and $s_{y|x}$ and increasing the precision with which we estimate the regression coefficients.

Measuring Heart Size with a Catheter

Heart muscle has the characteristic that it develops more force during a contraction if it is stretched to a longer length prior to the beginning of the contraction. This property, called the Frank-Starling mechanism, plays an important role in permitting the heart to adapt to changing demands on a beat-to-beat basis. When the filling of the heart increases (say, because you begin exercising and blood pooled in your venous system is forced into circulation), the heart gets bigger during the filling part of the cardiac cycle (known as diastole) and stretches the heart muscle in the wall to a longer length. Because the muscle has been stretched to a greater length, it contracts harder and expels the increased volume of blood into the aorta and out to the body. As a result, the volume of the heart—particularly the volume of the left ventricle which pumps blood into the aorta—is of crucial importance in determining how well the heart is working as a pump. Measuring left ventricular volume is important both to studies of cardiac physiology and to clinical management of patients with heart disease.

It is difficult to measure left ventricular volume, particularly if you want to obtain a continuous measure of volume over more than a few beats. One promising technique involves threading a special catheter backward through an artery into the heart so that it passes through the aortic valve at the top of the left ventricle and down the middle of the "barrel" of the ventricle to the apex. This catheter has a series of electrodes on it that establishes an electrical field within the blood in the ventricular cavity and produces a voltage signal that is proportional to the volume of blood in the left ventricle. This signal provides a continuous measure of left ventricular volume for as long as it is feasible to leave the catheter in place. Unfortunately, the heart is not made of rubber (or any other insulator), so some of the electrical field passes out of the blood inside the ventricle and adds an offset term, called the *parallel conductance volume* V_P, to the voltage due to the blood in the left ventricular cavity. If this offset is constant, it can be measured and subtracted out from the total volume signal measured by the catheter V_T to obtain the left ventricular cavity volume V_L:

$$V_L = V_T - V_P$$

If V_P is not actually constant, then correcting for this offset term is much more complicated and the catheter becomes more difficult to use.

Boltwood and coworkers[*] investigated the question of whether or not the parallel conductance volume remains constant over a wide range of loading conditions on the heart. They anesthetized dogs, placed the catheter in the left ventricle through an artery, much as it would be used in humans, then partially constricted various arteries and veins (with balloons placed in them) to change the loading conditions on the heart and obtain a wide range of cardiac volumes. Figure 9-1A shows the result of plotting the parallel conductance volume V_P against the total volume

[*]C. M. Boltwood, R. Appleyard, and S. A. Glantz, "Left Ventricular Volume Measurement by Conductance Catheter in Intact Dogs: The Parallel Conductance Volume Increases with End-Systolic Volume," *Circulation* 80:1360–1377, 1989.

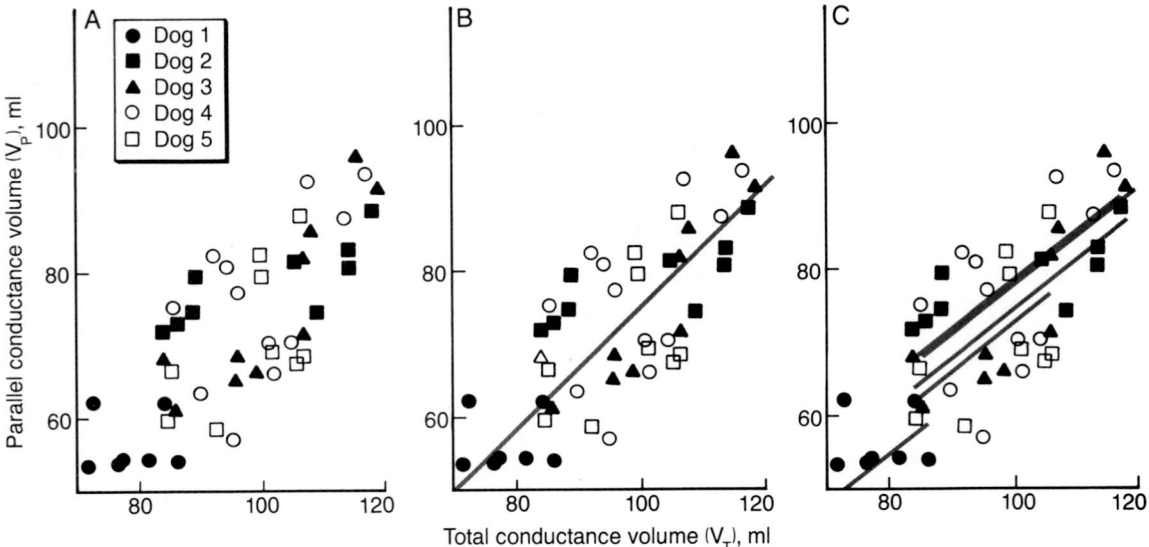

FIGURE 9-1 (*A*) Parallel conductance volume plotted against total left ventricular conductance volume measured in several dogs suggests that as total conductance volume signal increases, so does parallel conductance. (*B*) This impression is confirmed by fitting a regression line through these data. The problem with simply pooling the data from several animals, however, is that different animals are different sizes and so have different size hearts and could reasonably be expected to have both different total conductance volumes and parallel conductance volumes. Hence, pooling the data from several animals could introduce an artifact which would make it *appear* that parallel conductance volume changes within a given dog even when it does not. (*C*) To allow for between-dog differences we introduce dummy variables that permit the line relating parallel conductance volume to total volume to shift up and down (i.e., change its intercept) for different dogs. Allowing for the between-dog differences results in a different estimate of the slope of the regression line than pooling all the data but does not affect the conclusion that parallel conductance volume depends on total volume.

signal produced by the catheter V_T for data collected in five dogs. Simply inspecting this plot reveals that V_P is not constant. We analyze these data using linear regression and the equation

$$\hat{V}_P = a_0 + a_V V_T$$

to fit the data shown in Fig. 9-1*A*. The results confirm this impression; the slope a_V equals .82 with a standard error of .09 and is significantly different from zero ($P < .001$). The regression equation is

$$\hat{V}_P = -7 \text{ mL} + .82 V_T$$

Figure 9-1*B* shows this line superimposed on the data.

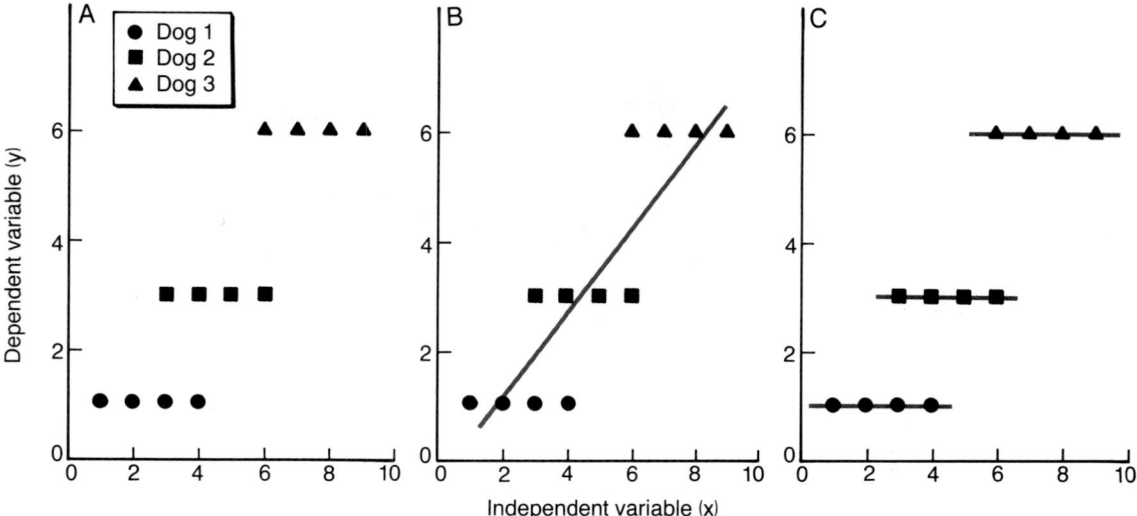

FIGURE 9-2 A simplified example of the potential problem encountered with the data in Fig. 9-1. (*A*) Suppose we have collected data on a dependent variable *y* as a function of an independent variable *x* in three dogs. (*B*) If we subject these data to a simple regression analysis without accounting for the between-dog differences, we conclude that there is a significant relationship between *y* and *x*. (*C*) In contrast, if we allow for between-dog differences, we find that the slope of the relationship between *y* and *x* within a given dog is zero; *y* does not change with *x* within any given animal. There appeared to be a slope in panel B because the simple regression model we used did not account for between-dog differences and, therefore, was underspecified and led to biased estimates of the slope and intercept.

There is, however, a problem with this straightforward analysis. The data in Fig. 9-1 come from five different dogs with different-sized hearts, and it is not unreasonable to assume that the parallel conductance volume would vary from dog to dog, with larger dogs having larger hearts and larger parallel conductance volumes. Specifically, we might have a situation like that shown in Fig. 9-2*A*, which shows observations of some dependent variable *y* collected in three different subjects (dogs). Inspecting this plot clearly shows that *y* does not depend on *x* *within a given subject*, but that there are changes across the three subjects. If we ignore the fact that there are different subjects, there appears to be a relationship between *y* and *x* (Fig. 9-2*B*). Hence, if we look at each experimental subject one at a time (Fig. 9-2*C*), there is no relationship between *y* and *x* (i.e., the slope of each regression line is zero). In our study of parallel conductance volume, we are interested in changes within a given dog and need to account for differences between dogs to be sure that our conclusion that V_P depends on V_T is not just an artifact due to the fact that some dogs are big and some are little.

```
The regression equation is
y = - 0.396 + 0.772 x

Predictor      Coef      Stdev     t-ratio       p
Constant     -0.3959    0.7129      -0.56     0.591
x             0.7716    0.1328       5.81     0.000

s = 1.076       R-sq = 77.2%    R-sq(adj) = 74.9%
```

FIGURE 9-3 Regression analysis of the data in Fig. 9-2B. The x coefficient (slope) is significantly different from zero.

We can use dummy variables to account for these between-dog differences, so that the slope in the regression equation is an estimate of the dependence of parallel conductance volume on total volume signal *after allowing for differences between different dogs*. Before doing so, however, we will illustrate this process using the hypothetical data in Fig. 9-2. If we submit these data to a linear regression with the model

$$\hat{y} = a_0 + a_x x$$

we obtain the results in Fig. 9-3. The slope is .77, with a standard error of .13, which is significantly different from zero ($P < .001$).

Now, define dummy variables to account for the between-subjects differences. There are three subjects, so we need two dummy variables, which we define using effects coding as

$$D_1 = \begin{cases} 1 \text{ if subject 1} \\ 0 \text{ if subject 2} \\ -1 \text{ if subject 3} \end{cases}$$

and

$$D_2 = \begin{cases} 0 \text{ if subject 1} \\ 1 \text{ if subject 2} \\ -1 \text{ if subject 3} \end{cases}$$

and use the regression model

$$\hat{y} = b_0 + b_x x + b_1 D_1 + b_2 D_2 \tag{9.1}$$

Table 9-1 shows the data in Fig. 9-2, including these dummy variables. For data from subject 1, the regression equation becomes

$$\hat{y} = b_0 + b_x x + b_1 \cdot 1 + b_2 \cdot 0 = (b_0 + b_1) + b_x x$$

TABLE 9-1 Hypothetical Data from Three Dogs to Illustrate Repeated Measures in Linear Regression

Dog	x	y	D_1	D_2
1	1	1	1	0
1	2	1	1	0
1	3	1	1	0
1	4	1	1	0
2	3	3	0	1
2	4	3	0	1
2	5	3	0	1
2	6	3	0	1
3	6	6	-1	-1
3	7	6	-1	-1
3	8	6	-1	-1
3	9	6	-1	-1

For subject 2, it becomes

$$\hat{y} = b_0 + b_x x + b_1 \cdot 0 + b_2 \cdot 1 = (b_0 + b_2) + b_x x$$

and for subject 3, it becomes

$$\hat{y} = b_0 + b_x x + b_1 \cdot (-1) + b_2 \cdot (-1) = [b_0 - (b_1 + b_2)] + b_x x$$

Thus, fitting the data with Eq. (9.1) is equivalent to fitting each subject with equations with different intercepts *but a common slope*.

Figures 9-2C and 9-4 show the results of fitting this equation to the data. After allowing for between-subjects differences, the slope is now estimated to be 0. b_0 is estimated to be 3.3, the overall average for all subjects. b_1 is -2.3, indicating that the response of subject 1 is 2.3 units below the average, and b_2 is $-.3$, indicating that subject 2 is .3 units below the average. From the last equation, subject 3 is $-(b_1 + b_2) = -(-2.3 - .3) = 2.6$ units above the average. By including dummy variables to allow for between-subjects differences, we were able to estimate the effects of x on y within each subject.*

---

*We could have accomplished the same goal using reference cell (0, 1) coding for between-subjects differences, but the interpretation of the intercept term would have been different. Here, it represents the average intercept for all subjects; with the 0, 1 coding it would represent the intercept for the individual with both D_1 and D_2 coded 0.

```
The regression equation is
y = 3.33 +0.000000 x - 2.33 D1 - 0.333 D2

Predictor        Coef       Stdev     t-ratio          p
Constant       3.33333     0.00000       *             *
x           0.00000000  0.00000000       *             *
D1            -2.33333     0.00000       *             *
D2           -0.333333    0.000000       *             *

s = 0              R-sq = 100.0%    R-sq(adj) = 100.0%
```

FIGURE 9-4 Regression analysis of the data in Fig. 9-2C, allowing for between-dog differences. The x coefficient (slope) equals zero.

We are now ready to analyze our data on the parallel conductance volume, taking into account the differences between dogs. Because there are five dogs, we define four dummy variables, D_1, D_2, D_3, and D_4, according to

$$D_i = \begin{cases} 1 \text{ if dog } i & i < 5 \\ -1 \text{ if dog } 5 \\ 0 \text{ otherwise} \end{cases}$$

Table 9-2 shows the data, together with the values of these dummy variables. Figure 9-5 shows the results of analyzing these data with the regression equation

$$\hat{V}_P = b_0 + b_1 D_1 + b_2 D_2 + b_3 D_3 + b_4 D_4 + b_V V_T = b_0 + \Sigma b_i D_i + b_V V_T$$

In contrast to the situation we observed in the hypothetical example of Fig. 9-2, the coefficient b_V associated with the total volume is still sig-

```
The regression equation is
Vp = 8.2 - 6.15 D1 + 4.00 D2 + 0.39 D3 + 3.37 D4 + 0.659 Vt

Predictor        Coef       Stdev     t-ratio          p
Constant        8.20       10.92        0.75        0.457
D1             -6.145       3.115       -1.97        0.055
D2              3.995       2.406        1.66        0.104
D3              0.393       2.336        0.17        0.867
D4              3.369       2.138        1.58        0.122
Vt              0.6595      0.1135       5.81        0.000

s = 7.924      R-sq = 69.6%    R-sq(adj) = 66.2%
```

FIGURE 9-5 Regression analysis of the conductance catheter data in Fig. 9-1.

TABLE 9-2 Left Ventricular Volume Measured by Conductance Catheter V_T and Parallel Conductance Volume V_P Measured in Five Dogs

Dog	V_T, mL	V_P, mL	D_1	D_2	D_3	D_4
1	81.7	54.3	1	0	0	0
1	84.3	62.0	1	0	0	0
1	72.8	62.3	1	0	0	0
1	71.7	47.3	1	0	0	0
1	76.7	53.6	1	0	0	0
1	75.8	38.0	1	0	0	0
1	77.3	54.2	1	0	0	0
1	86.3	54.0	1	0	0	0
1	71.8	53.4	1	0	0	0
2	105.0	81.5	0	1	0	0
2	113.6	80.8	0	1	0	0
2	108.7	74.5	0	1	0	0
2	83.9	71.9	0	1	0	0
2	89.0	79.5	0	1	0	0
2	86.1	73.0	0	1	0	0
2	88.7	74.7	0	1	0	0
2	117.6	88.6	0	1	0	0
2	113.9	83.1	0	1	0	0
3	95.5	65.0	0	0	1	0
3	95.7	68.3	0	0	1	0
3	84.0	67.9	0	0	1	0
3	85.8	61.0	0	0	1	0
3	98.8	66.0	0	0	1	0
3	106.2	81.8	0	0	1	0
3	106.4	71.4	0	0	1	0
3	115.0	96.0	0	0	1	0
3	118.5	91.4	0	0	1	0
3	107.7	85.6	0	0	1	0
4	113.1	87.5	0	0	0	1
4	116.5	93.6	0	0	0	1
4	100.8	70.4	0	0	0	1
4	101.5	66.1	0	0	0	1
4	120.8	101.4	0	0	0	1
4	95.0	57.0	0	0	0	1
4	91.9	82.5	0	0	0	1
4	94.0	80.9	0	0	0	1
4	107.2	92.7	0	0	0	1
4	104.4	70.4	0	0	0	1
4	85.4	75.2	0	0	0	1
4	89.8	63.5	0	0	0	1
4	95.9	77.3	0	0	0	1

(continued on page 390)

TABLE 9-2 (*continued*)   Left Ventricular Volume Measured by Conductance Catheter V_T and Parallel Conductance Volume V_P Measured in Five Dogs

Dog	V_T, mL	V_P, mL	D_1	D_2	D_3	D_4
5	99.5	79.4	-1	-1	-1	-1
5	99.2	82.5	-1	-1	-1	-1
5	106.1	87.9	-1	-1	-1	-1
5	85.2	66.4	-1	-1	-1	-1
5	106.3	68.4	-1	-1	-1	-1
5	84.6	59.5	-1	-1	-1	-1
5	92.1	58.6	-1	-1	-1	-1
5	101.2	69.2	-1	-1	-1	-1
5	105.4	67.5	-1	-1	-1	-1

nificantly different from zero ($b_V = .66 \pm .11$, $P < .001$). This result indicates that the parallel conductance volume changes within a given dog, with the parallel conductance volume increasing, on the average, by .66 mL for each 1-mL increase in the total volume signal. $b_0 = 8.2$ mL, so the average regression line for all dogs is*

$$\hat{V}_P = 8.2 \text{ mL} + .66 V_T$$

Thus, these data show that the parallel conductance volume cannot be assumed constant and point to the need for further development of this technique for measuring ventricular volume before it can be used reliably to measure absolute volume. Compared to the results of computing the regression equation ignoring the fact that we had different dogs, this slope is smaller (.66 vs. .82) as is $s_{y|x}$ (7.9 vs. 8.2 mL).

Designs such as this in which one has repeated observations in each experimental subject are very common in biomedical research, particularly in clinical studies. This example illustrates how to account for such repeated observations in the context of a simple linear regression problem. These procedures can be generalized directly to more complex models.

ACCOUNTING FOR BETWEEN-SUBJECTS VARIABILITY IN ANALYSIS OF VARIANCE

Experiments in which each subject receives more than one treatment are called *repeated-measures designs* and are analyzed using *repeated-*

---

*The coefficients associated with the D_i variables provide information about how much each dog varies from this average, which does not particularly concern us at this point.

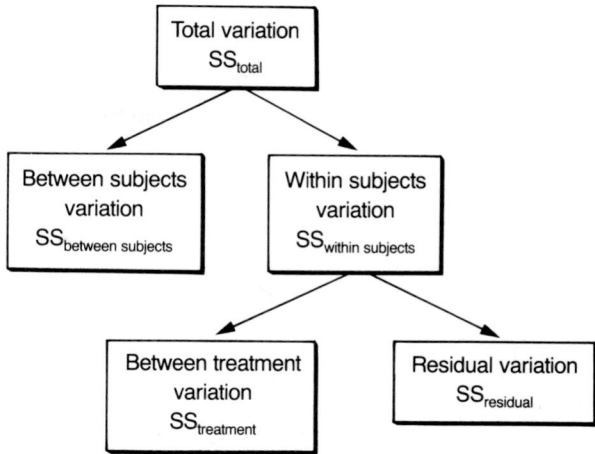

FIGURE 9-6 Partitioning of the sums of squares for a single-factor, repeated-measures analysis of variance with no missing data. We first divide the variation into a between-subjects component that arises because of random differences between subjects and a within-subjects component. This procedure allows us to "discard" the variation between subjects and concentrate on the variation within experimental subjects to test hypotheses about the different treatment effects. The degrees of freedom partition in the same way.

*measures analysis of variance.** Repeated-measures analysis of variance is the generalization of a paired t test in the same sense that ordinary analysis of variance is a generalization of an unpaired t test. Repeated-measures analysis of variance permits testing hypotheses about any number of treatments whose effects are measured repeatedly in the same subjects. We will explicitly separate the total variability in the observations into three components: variability between experimental subjects, variability in each individual subject's response, and variability due to the treatments (Fig. 9-6). Then, depending on the experimental design, each of these components may be further subdivided into effects of interest for hypothesis testing. Repeated-measures designs are generally more powerful than ordinary designs because the variability between subjects (e.g., people, animals, or cell cultures) can be isolated and the analysis can then focus more precisely on the factor effects.

ONE-WAY REPEATED-MEASURES ANALYSIS OF VARIANCE

The simplest repeated-measures design is one in which each experimental subject is observed under several different levels of a single experimental factor. We will analyze this situation by first taking into account the differences in the mean response of each subject and removing the *between-subjects variability* associated with these differences. We will then concentrate on the differences *within each subject* that can be attributed to the experimental treatments and compare this within-subjects variability due to the treatments with the residual within-subjects

*For an introductory discussion of repeated-measures analysis of variance, see S. Glantz, *Primer of Biostatistics* (2nd ed.), McGraw-Hill, New York, 1987, Chap. 9.

variability. We will illustrate how to formulate and interpret such an analysis with a specific example. In this example there are no missing data; had there been missing data, the analysis would still proceed in the same manner, with the caveat that you must *put the subjects dummy variables in first.* If you do not, you will obtain incorrect sums of squares for the factor dummy variables because of the multicollinearity caused by the missing data.

Hormones and Food

When a meal is eaten, it must be digested to make the nutrients available to the body. This digestion is the result of a coordinated series of actions by the stomach, intestines, and hormones. The changes in body activity that accompany digestion can be grossly evaluated by the change in the rate of metabolism of the body; metabolism increases as food is digested. Diamond and LeBlanc[*] reasoned that the events that initiate digestion fall into two distinct classes: those events associated with the act of eating (e.g., chewing, taste, smell) and those events associated with the stomach filling with food. To see if these two different classes of initiators had different effects on body function after eating, they studied animals that first ate normally, then ate a meal that was diverted so that it did not enter the stomach (bypassing the factors associated with the stomach filling with food), and then, finally, received food directly into the stomach (bypassing the factors associated with the act of eating). They then measured the metabolic rate of the body, in milliliters of oxygen consumed per kilogram of body weight during the first 45 min after the meal, in each animal under each treatment condition; metabolic rate was measured in *each dog* under *each experimental condition.* The data from this early digestive phase are shown in Fig. 9-7A and Table 9-3. The question is: Is there any difference in the metabolic rate between the three treatment groups, taking into account the fact that each animal received each treatment?

We begin the regression implementation of this repeated-measures analysis of variance by defining dummy variables to account for the fact that different subjects may have different mean responses, independent of the treatment effects. In this example, we have data from six dogs ($n = 6$), so we must define five dummy variables to account for between-subjects differences. We will use the 0, 1, -1 coding scheme for the same reason we used it above when we analyzed the behavior of the catheter used to measure volume in the heart, that is, to make the coefficients

[*]P. Diamond and J. LeBlanc, "Hormonal Control of Postprandial Thermogenesis in Dogs," *Am. J. Physiol.* 253:E521–E529, 1987.

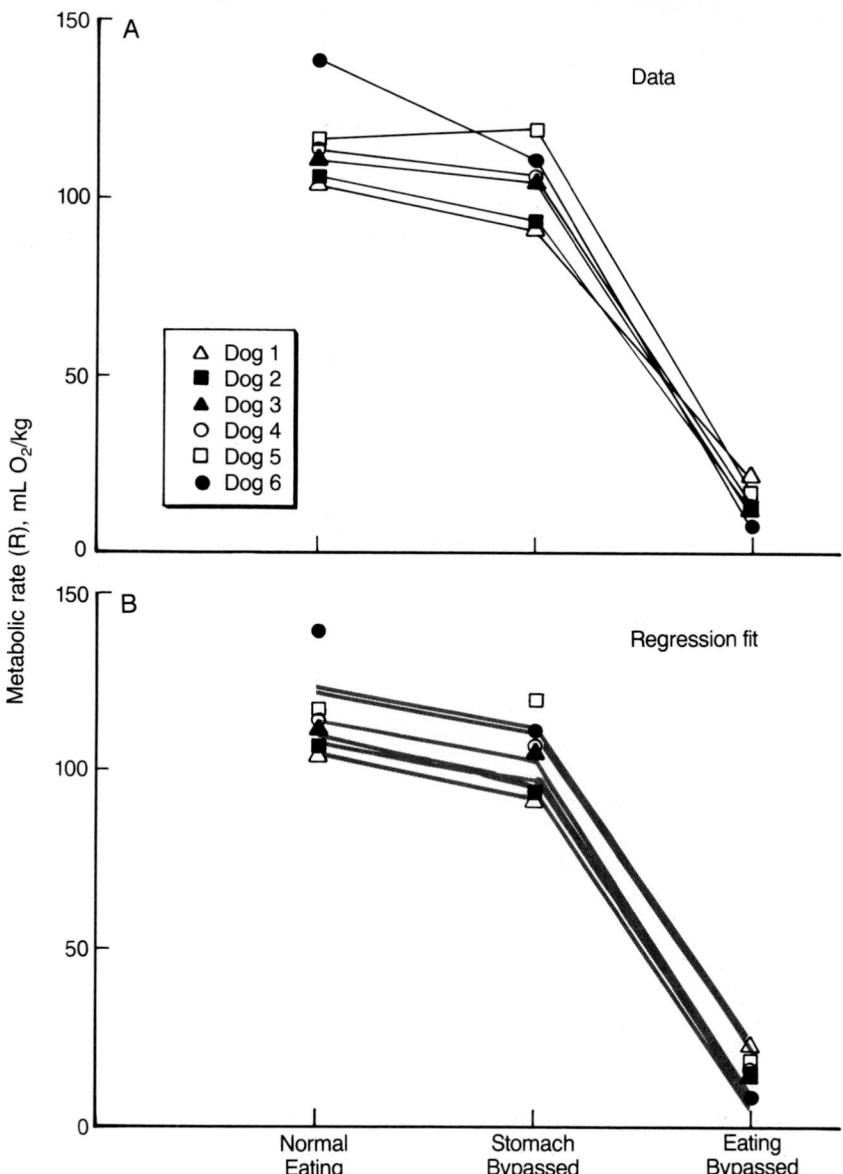

FIGURE 9-7 Data collected from six dogs in an experiment in which they were allowed to eat normally, then allowed to eat but with the food bypassing the stomach, and, finally, had food placed directly into the stomach to bypass the act of eating. Each animal was observed under each experimental condition, so this is a single-factor, repeated-measures analysis of variance. (*A*) The measured metabolic rates from each dog are connected by lines. (*B*) Results of a regression implementation of a one-way, repeated-measures analysis of variance on the data in panel A. The lines indicate the values *predicted* from the regression equation for each treatment. The differences between the predicted and observed values for each point are the residuals that comprise the residual sum of squares.

TABLE 9-3 Metabolic Rate (mL O_2/kg) in Dogs Early after Three Different Methods of Eating

Dog	Normal Eating	Stomach Bypassed	Eating Bypassed
1	104	91	22
2	106	94	14
3	111	105	14
4	114	106	15
5	117	120	18
6	139	111	8

associated with the dummy variable for each dog represent the deviation of that dog's average response from the overall average response of all dogs. Specifically, we define five dummy variables (one fewer than the number of dogs), A_1, A_2, A_3, A_4, and A_5, according to

$$A_i = \begin{cases} 1 \text{ if animal } i \quad i < 6 \\ -1 \text{ if animal } 6 \\ 0 \text{ otherwise} \end{cases}$$

For example, the value of the regression coefficient associated with A_2 estimates the amount by which the mean metabolic rate in dog 2 differs from the mean metabolic rate of all dogs averaged over all the treatments.[*]

We quantify the treatment effects the same way we would in an ordinary one-way analysis of variance by defining two dummy variables to code the three treatments. Because this is a one-way design, you can use either effects coding or reference cell coding. As before, in some situations, you may want to use effects codes so that each coefficient is the difference between the overall mean and the effect coded by the dummy variable. More often, however, it makes more sense to specify the dummy variables in balanced one-way designs using reference cell coding.[†] Specifically, we define

$$E_E = \begin{cases} 1 \text{ if dogs eat with stomach bypassed} \\ 0 \text{ otherwise} \end{cases}$$

[*]The deviation from the overall mean response for the nth dog, dog 6, is $A_n = A_6 = -\Sigma A_i$. A_6 does not appear in the regression equation.

[†]When we go on to discuss two-way designs and the problems of unbalanced data, we will need to use effects coding. We prefer to use reference cell coding here because the regression coefficients are more easily interpretable.

and

$$E_S = \begin{cases} 1 \text{ if food placed in stomach without eating} \\ 0 \text{ otherwise} \end{cases}$$

Thus, normal eating has codes E_E and $E_S = 0$, eating with bypass of the stomach has codes $E_E = 1$ and $E_S = 0$, and bypassing eating by putting food directly into the stomach has codes $E_E = 0$ and $E_S = 1$. Table 9-4 shows how the dummy variable codes correspond to the data for all the dogs under each of the three experimental conditions.

We now fit these data to a multiple regression model that incorporates these dummy variables to account for the eating factor and the fact that we have made multiple measurements in the same animals:

$$\hat{R} = b_0 + \Sigma a_i A_i + e_E E_E + e_S E_S$$

where $\hat{R}$ is the predicted metabolic rate and the other variables are as defined above. b_0 is the mean metabolic rate (in mL O_2/kg) under control

TABLE 9-4 Metabolic Rate and Reference Cell Dummy Variables to Encode Three Different Methods of Eating and Subjects

Dog	Metabolic Rate R, mL O_2/kg	Animal					Eating Mode	
		A_1	A_2	A_3	A_4	A_5	E_E	E_S
1	104	1	0	0	0	0	0	0
2	106	0	1	0	0	0	0	0
3	111	0	0	1	0	0	0	0
4	114	0	0	0	1	0	0	0
5	117	0	0	0	0	1	0	0
6	139	−1	−1	−1	−1	−1	0	0
1	91	1	0	0	0	0	1	0
2	94	0	1	0	0	0	1	0
3	105	0	0	1	0	0	1	0
4	106	0	0	0	1	0	1	0
5	120	0	0	0	0	1	1	0
6	111	−1	−1	−1	−1	−1	1	0
1	22	1	0	0	0	0	0	1
2	14	0	1	0	0	0	0	1
3	14	0	0	1	0	0	0	1
4	15	0	0	0	1	0	0	1
5	18	0	0	0	0	1	0	1
6	8	−1	−1	−1	−1	−1	0	1

conditions for all dogs. Each a_i is the difference between the mean metabolic rate of dog i and the mean of all dogs. e_E is the average difference in metabolic rate between the control metabolic rate and that observed when the dog eats without the food entering the stomach, and e_S is the difference in metabolic rate between the control metabolic rate and that observed when food is placed directly in the stomach without the dog actually eating it.

One should enter the subject dummy variables into the model first, followed by the experimental factor dummy variables (in this case E_E and E_S).* There are two reasons for this convention. First, it is conceptually closer to what repeated-measures analysis of variance is doing: accounting for the between-subjects differences, then focusing the analysis on the treatment effects within each subject. Second, as discussed in Chap. 8, the order in which variables are entered in the regression equation affects the incremental sums of squares and mean squares when there are missing data. Although, by definition, all repeated-measures designs start out balanced, it is possible to end up with unbalanced designs because of missing data due to technical difficulties during data collection or loss of experimental subjects (particularly in clinical trials). The same considerations we discussed previously for missing data in two-way analysis of variance apply here as well. However, because there is no interaction, it does not matter whether you use reference cell or effects coding, and one can always be assured of computing the correct F value for testing the hypothesis of no treatment effect if the subject dummy variables are entered into the model first. As before, the order of entry will not affect the computations when there are no missing data.

Figure 9-8 shows the computer output from the multiple linear regression analysis, and Fig. 9-7B shows a plot of both the data points and the predicted values from the regression equation for the individual dogs. From these figures, $b_0 = 115$ mL O_2/kg, which indicates that the average metabolic rate under control conditions is 115 mL O_2/kg. $e_E = -11$ mL O_2/kg, which indicates that the metabolism decreases by 11 mL O_2/kg from control when the stomach is bypassed, and $e_S = -100$ mL O_2/kg, which indicates that metabolism decreases by 100 mL O_2/kg when eating is bypassed and the food is placed directly in the stomach.

---

*We have written the regression equation to explicitly indicate that the between-subjects dummy variables are entered first. People often write the equation with the between-subjects dummy variables last to focus attention on the dummy variables used to quantify the treatment effects which are actually under study. Nevertheless, the between-subjects dummy variables should always be entered into the actual calculation first.

```
DEP VARIABLE: R          metabolic rate after eating

                                      ANALYSIS OF VARIANCE

                              SUM OF          MEAN
          SOURCE     DF       SQUARES        SQUARE      F VALUE      PROB>F

          MODEL       7    36761.38889    5251.62698     57.570      0.0001
          ERROR      10      912.22222222   91.22222222
          C TOTAL    17    37673.61111

                     ROOT MSE       9.551033     R-SQUARE     0.9758
                     DEP MEAN      78.27778      ADJ R-SQ     0.9588
                     C.V.          12.20146

                                     PARAMETER ESTIMATES

                    PARAMETER      STANDARD     T FOR H0:
VARIABLE   DF        ESTIMATE        ERROR     PARAMETER=0    PROB > |T|    TYPE I SS        VARIABLE
                                                                                            LABEL

INTERCEP   1      115.16666667    3.89919270     29.536        0.0001    110293.38889   INTERCEPT
A1         1       -5.94444       5.03383613     -1.181        0.2650       280.16666667  dummy variable =1 if subject 1
A2         1       -6.94444       5.03383613     -1.380        0.1978       122.72222222  dummy variable =1 if subject 2
A3         1       -1.61111       5.03383613     -0.320        0.7555         0.02777778  dummy variable =1 if subject 3
A4         1        0.05555556    5.03383613      0.011        0.9914         7.35000000  dummy variable =1 if subject 4
A5         1        6.72222222    5.03383613      1.335        0.2113       162.67777778  dummy variable =1 if subject 5
EE         1      -10.6667        5.51429120     -1.934        0.0818      6188.44444     dummy var. = 1 if bypass stomach
ES         1     -100            5.51429120    -18.135        0.0001     30000.00000     dummy var. = 1 if direct to stomach
```

FIGURE 9-8 Regression implementation of the repeated-measures analysis of variance of the data in Fig. 9-7. The dummy variables A_1, A_2, . . . , and A_5 are entered first with the dummy variables corresponding to the treatment effects E_E and E_S entered last. In regression implementations of repeated-measures analysis of variance, the dummy variables representing the between-subjects differences are *always* entered first.

To test whether these differences between the treatments are greater than can be reasonably attributed to sampling variability, we compute an F statistic to compare the variability within the subjects due to the treatments with the residual variability within subjects, after accounting for the differences between subjects. To obtain the treatment sum of squares, we add the sums of squares associated with the two treatment variables (E_E and E_S), then divide this sum by 2 degrees of freedom (one for each dummy variable) to obtain our estimate of variability associated with the treatments:

$$MS_{\text{treat}} = \frac{SS_E + SS_S}{DF_E + DF_S} = \frac{6188.4 + 30{,}000}{1 + 1} = \frac{36{,}188.4}{2} = 18{,}094.2 \ (\text{mL O}_2/\text{kg})^2$$

The residual variation within subjects is $MS_{\text{res}} = 91.2 \ (\text{mL O}_2/\text{kg})^2$, with 10 degrees of freedom, so

$$F = \frac{18{,}094.2}{91.2} = 198.4$$

This value greatly exceeds the critical value of F for 2 numerator and 10 denominator degrees of freedom and a 1 percent risk of a false-positive error ($\alpha = .01$), 7.56, so we reject the null hypothesis of no treatment effect and conclude that at least one treatment produces a different metabolic rate from the others ($P < .01$).

Comparison with Simple Analysis of Variance

The output in Fig. 9-8 can also be used to illustrate how a repeated-measures design relates to a standard (completely randomized) analysis of variance. Each subject dummy variable contributes a certain amount to the total sums of squares, which is a reflection of the fact that different dogs (subjects) have different mean responses, independent of the presence or absence of treatment effects. The sum of these contributions equals the between-subjects sum of squares. Recall that the reason repeated-measures designs are often useful is that they remove this between-subjects variation from the analysis, thus reducing the residual (or error) sum of squares. In this case, the residual sum of squares SS_{res} is 912.2 (mL O_2/kg)^2 with 10 degrees of freedom, so MS_{res} is $912.2/10 = 91.2$ (mL O_2/kg)^2. The between-subjects sum of squares is the total of the sums of squares associated with each of the five dummy variables A_i, $280.2 + 122.7 + 0 + 7.4 + 162.7 = 573.0$ (mL O_2/kg)^2. Had we not used a repeated-measures design, this between-subjects variation would have been folded into the residual sum of squares, raising SS_{res} from 912.2 to 1485.2 (mL O_2/kg)^2. Because this sum of squares includes the effects of the five between-subjects dummy variables as well as the residual variation, it is associated with $5 + 10 = 15$ degrees of freedom. This means that without taking into account the repeated measures, MS_{res} would have increased from $912.2/10 = 91.2$ (mL O_2/kg)^2 to $1485.2/15 = 99$ (mL O_2/kg)^2, a 9 percent increase. Thus, in the presence of a constant treatment effect (quantified by MS_{treat}), the value of $F = MS_{treat}/MS_{res}$ used to test for a significant treatment effect would be reduced by 9 percent. The loss of power associated with this reduction in the value of F is partially offset by the increase in the number of degrees of freedom associated with the residual (from 10 to 15). Thus, the critical value of F (for $\alpha = .05$ and 2 numerator degrees of freedom) falls from 4.10 to 3.68, or by about 9 percent. In this example, the between-dogs variability is small compared to the total variability, so the repeated-measures design does not appear to be of great benefit—the reduced mean square error when taking into account the between-subjects differences is just about offset by the loss of degrees of freedom.

This conclusion omits one important fact. The study in question required 6 dogs, with each dog being studied 3 times. To achieve the same result using a completely randomized design, we would have needed 18 dogs, 6 for each of the 3 treatment conditions. Thus, *we were able to reduce the number of experimental subjects by two-thirds with a repeated-measures design.* This consideration is particularly important in biomedical research where experimental subjects—be they animals for laboratory studies or humans for clinical studies—are often difficult to obtain. In addition, ethical considerations require that experiments use only the minimum number of animal or human subjects consistent with good scientific practice. Moreover, in clinical studies of rare disorders, it is often impossible to locate and recruit large numbers of patients.

The result in our example, where the modest reduction in residual variability is just offset by the corresponding reduction in residual degrees of freedom, is also more likely to occur in laboratory studies where the investigator can obtain a more homogeneous study population than in clinical studies of people—particularly sick people—where the underlying biological variability will generally be higher. In general, a repeated-measures design increases the power of the experiments and reduces the necessary sample size compared to a standard design by explicitly taking the differences between experimental subjects into account in the analysis.

Multiple Comparisons in Repeated-Measures Analysis of Variance

To answer further questions about which treatments lead to different metabolic rates, we can use any of the multiple comparisons procedures we have already presented. The physiological question of interest—which component of meal ingestion induces the early digestive response—can be answered with a Dunnett's test for multiple comparisons against the control group.

The way we have set up the regression problem, with the normal eating (control) group having both E_E and $E_S = 0$ means that the t values associated with e_E (the mean difference between control conditions and eating with the food diverted from the stomach) and e_S (the mean difference between control conditions and placing the food directly in the stomach) directly give us the q' values we need for Dunnett's test.

From Fig. 9-8, the mean control metabolism is $b_0 = 115$ mL O_2/kg. The largest difference from control is $e_S = -100$ mL O_2/kg, with an associated $t = q' = -18.14$, which greatly exceeds 3.53, the critical value of q' for $\alpha = 0.01$, $p = 3$ (number of means in the list of ranked means spanned by the two means being compared), and 10 degrees of freedom (from Table E-3, Appendix E). Because this difference was significantly different from zero, we go on to test the next largest (and last) difference. This comparison is between normal eating and eating without the food entering the stomach. Eating without allowing the food to enter the stomach is associated with an average reduction in metabolism of $e_E = -11$ mL O_2/kg, with an associated value of $t = q' = -1.93$. This value falls below the critical value of 2.23 ($p = 2$, $\alpha = 0.05$, and 10 degrees of freedom), so we conclude that metabolism does not differ significantly from control in this condition. These two comparisons lead to the conclusion that metabolic rate in the early digestive period is significantly lower than control when chewing, smelling, and tasting food are bypassed, and it is these acts which initiate the early phase of digestion after eating a meal.

Had we wished to conduct all possible comparisons, we would have used Bonferroni t tests or the Student-Newman-Keuls test as outlined in Chap. 7. The value of MS_{res} and its associated degrees of freedom (91.2 and 10, respectively, in this example) would be used to compute the test statistics and determine the associated critical values.

Recapitulation

We have shown how making repeated measurements in the same individuals changes the way we partition the variability in a one-way analysis of variance. We can use either reference cell coding or effects coding, and the order of entry of the sets of treatment dummy variable codes and the subject dummy variable codes does not matter when the data are balanced. However, order of entry does matter when the data are unbalanced, and the treatment effect sum of squares must be the marginal sum of squares for the set of treatment dummy variables. Thus, we recommend that you get in the habit of always entering the treatment dummy variables after the subject dummy variables, so that you will obtain the correct sum of squares for the treatment dummy variables when there are missing data.

We will use similar partitioning of the total sum of squares into between-subjects and within-subjects sums of squares when we consider two-way repeated-measures analysis of variance. However, we have two very different situations to consider. The first situation is when there are two factors, but we have made repeated measurements over only one of the factors. The second situation is when we make repeated measurements over both of the factors. The total sums of squares partition very differently in these two situations.

ASSUMPTIONS UNDERLYING REPEATED-MEASURES ANALYSIS OF VARIANCE

In a repeated-measures design we take into account the fact that different experimental subjects can have different mean responses because of between-subjects differences (as opposed to treatment effects). Doing so requires that we make assumptions in addition to those we already made to develop the test procedures for completely randomized designs in Chaps. 7 and 8.

The *treatment effects* are *fixed* in the underlying mathematical model of the experiment. We assume that the treatments exert a fixed (but unknown) effect on the mean response of the dependent variable, and we use the data to estimate this fixed effect. In the regression implementation of analysis of variance, the regression coefficients associated with the treatment dummy variables provide the estimates of these fixed effects. As with earlier regression and analysis of variance models, we assume that

- The random components of the observations under any given set of experimental factors and for any given individual subject are normally distributed about the mean values with the same variance, regardless of the specific experimental conditions.
- The random components of the different observations are independent of each other.

As with simple and multiple linear regression and analysis of variance, we can test these assumptions by examining the residuals between the predicted and observed values of the data points.

In a repeated-measures design there is another random element, the experimental subjects who are selected for the study. Because we observe each experimental subject under all treatments, we can take differences between subjects into account by incorporating dummy variables into our regression model that allow different subjects to have different mean responses, independent of the effects of the treatments on the means. While these dummy variables look like the dummy variables we defined to describe the treatment effects, there is an important theoretical difference: the *between-subjects effects* are *random*. Whereas the differences in mean response between any particular set of subjects selected for study are fixed, the specific means depend on which specific subjects happen to be selected at random from the population under study. This situation contrasts with that of the fixed treatment effects, which are theoretically constant values, independent of the specific individuals selected at random for study.

In completely randomized designs the random effects associated with the fact that different subjects who received different treatments had different mean responses independent of the treatment are simply lumped in with the random elements of the response and experimental error to form the residual variance we used in hypothesis testing. In a repeated-measures design, we explicitly allow for differences in the mean responses of the different subjects. When we estimate the regression coefficients associated with the between-subjects dummy variables, we are estimating the differences in mean response *for those individuals who happened to be selected for study.* Because these values are samples from a larger population, we must make additional assumptions about the nature of the underlying population. Specifically, we assume:

- The effects of different subjects add to the effects of the treatments.
- The mean responses of all subjects (independent of treatment) are normally distributed with zero mean and a standard deviation that is independent of the treatment effects.
- The mean responses of different subjects are independent of each other and of the random components associated with the treatment effects.

These additional assumptions affect repeated-measures analysis of variance in two ways, compared with the completely randomized designs we presented in Chaps. 7 and 8.

First, they affect the construction of the hypothesis tests (*F* tests) in repeated-measures analysis of variance, because the random elements due to the experimental subjects must be taken into account in computing the residual variation and the variation between the different treatments. In the one-way repeated-measures analysis of variance, the between-subjects effects simply add to the treatment effects, so the *F* test for a treatment effect looks just like what we did in Chap. 7. However, as we will see when we consider two-factor designs, the random element associated with the between-subjects effects interacts with the fixed treatment effects in a way that affects the construction of the *F* tests.

Second, these assumptions, combined with the assumptions about the treatment effects, lead to the condition that the correlation between any pair of observations on a given subject be the same and that the correlations on pairs of observations between subjects be zero. This condition is called *compound symmetry*. When compound symmetry exists, the *F* statistic we construct for repeated-measures analysis of variance follows the theoretical *F* distribution and produces accurate estimates of the risk of erroneously rejecting the null hypothesis of no treatment effect.* When compound symmetry does not exist, the *F* values do not follow the theoretical *F* distribution and the results will be biased toward rejecting the null hypothesis—the *P* values will be smaller than the data justify.

---

*This condition is only important for testing the factors upon which there are repeated measures. In multifactor experimental designs in which some factors do not have repeated measures, these restrictions do not come into play for those factors that do not involve repeated measures.

Two similar techniques, known as the *Huynh-Feldt* and *Greenhouse-Geisser corrections*, have been developed to correct for this effect. Both these corrections involve adjusting the degrees of freedom used to derive the critical value of F downward, so as to increase the value of F required to conclude there is a statistically significant treatment effect. Both the numerator and denominator degrees of freedom are multiplied by an adjustment factor ε, which is derived from the data.[*] ε equals 1 when the assumption of compound symmetry is satisfied and falls towards 0 with greater deviations from compound symmetry. The difference between the two corrections is how ε is defined, with the Huynh-Feldt definition leading to larger (i.e., less conservative) values of ε than the Greenhouse-Geisser definition. Some computer programs report one or both values of ε, and some even display P values that have been corrected for ε. If the ε values are near 1, it increases your confidence in the results. Unfortunately, these two definitions of ε can produce widely varying values, and there is no consensus on which is preferable. It is known, however, that the Greenhouse-Geiser correction is very conservative. Moreover, although the condition of compound symmetry is a sufficient condition for the F distribution to be accurate, it is not necessary.[†] In other words, it is possible that these corrections may not even be necessary. We are left with the question: How restrictive is the assumption of compound symmetry and how can we tell when a test is appropriate? Unfortunately, there is no easy answer to this question because there is no simple, generally accepted test for compound symmetry.

The general conditions (called the Huynh-Feldt conditions) under which the F test is valid are often not met because observations on the same subject taken closely in time are more highly correlated than are observations taken farther apart in time. For instance, in evaluating several different drugs in the same patients and trying to obtain subjective measures of how the patient "feels," they may tend to give higher (or lower) ratings for treatments at the end of the sequence than at the beginning because of a placebo effect. Residual effects of earlier treatments may also affect subsequent treatments, giving rise to a so-called carryover effect.

From a practical standpoint, the best thing to do is try to design the experiment to see that the data meet the assumptions that underlie the analysis. Make certain that there is sufficient time between applications

---

[*]For details on the computation, see H. Huynh and L. S. Feldt, "Conditions under Which Mean Square Ratios in Repeated Measurement Designs Have Fixed F Distribution," *J. Am. Stat. Assoc.* 65:1582–1589, 1970; and S. W. Greenhouse and S. Geisser, "On Methods in the Analysis of Profile Data," *Psychometrika* 24:95–112, 1959.

[†]For a discussion of some of these other conditions, see B. J. Winer, *Statistical Principles in Experimental Design* (2nd ed.), New York, McGraw-Hill, 1971, pp. 282–283.

of the treatment to allow washout (or elimination) of the previous treatment to reduce carryover and to make certain that the design is applied in only those situations where the disease (or other underlying biological state) is relatively stable, so that following treatment and washout, each patient (or other experimental subject) is essentially the same as prior to receiving treatment. Randomization of the treatment orders for each subject independently will make it more reasonable to analyze data as if the error terms are independent. It may be desirable at times to balance the order of treatment presentations and sometimes even the number of times each treatment is preceded by any other treatment. Although it is often possible to follow this prescription, there are also times, such as when one follows a patient over time during some treatment regimen, when it is not. In any event, you should consider how closely the experimental protocol met the assumptions underlying the analysis when deciding how confident you can be in the results.

TWO-FACTOR ANALYSIS OF VARIANCE WITH REPEATED MEASURES ON ONE FACTOR

In a one-way analysis of variance with repeated measures, each individual in the study received each treatment. We then accounted for the underlying differences between subjects and compared the variability within subjects that could be attributed to the treatments with the residual variation within subjects. Such a design is commonly used to study the same individuals under a variety of experimental conditions or over time following some intervention, such as giving a drug. It is not always possible to use this design, particularly in clinical studies of drugs or other treatments. For example, suppose we wished to compare the effectiveness of different modes of postsurgical intensive care on clinical outcome. You can only do the surgery once on each patient, so the people receiving the different follow-up treatments would have to be different. We would, however, be able to make repeated observations on each individual in the experiment (over time). Thus, we would do a two-factor (intensive care mode and time) analysis of variance with repeated measures on one factor (time). This design is one of the most common ones used in clinical research.

Table 9-5 shows the general form of such an experiment. In this design there are two factors under study, A and B, with A having two levels and B having three levels. There are a total of eight experimental subjects, with half assigned to each of the two groups under treatment A. All eight subjects are observed under all three levels of factor B. For the example above, factor A would be the mode of intensive care, with four patients receiving each mode, and factor B would be time after the operation. Thus, each experimental group under factor A (mode of intensive care) has different subjects, so this experimental grouping factor

TABLE 9-5 Schematic of Experimental Design for Two-Way Repeated-Measures Analysis of Variance (Factors A and B) with Repeated Measures on Factor B

Factor A	Subject	B_1	B_2	B_3
A_1	S_1 S_2 S_3 S_4			
A_2	S_5 S_6 S_7 S_8			

involves *no repeated measures*. On the other hand, individual subjects within each factor A experimental group have *repeated measurements* made on them over all levels of factor B (time). Because different experimental subjects receive each of the different levels of the factor A treatment, tests to see if there are significant differences between the different levels of factor A will be similar to those we conducted in Chap. 8. In contrast, because each subject is observed under all levels of factor B, the test for significant differences among the different levels of factor B will resemble that which we developed earlier in this chapter.

Partitioning the Variability

As with all analyses of variance, we will compute F statistics to compare the variation between the different treatment groups with an estimate of the underlying variation within the treatment groups. Up to this point, we always used the residual variation left over after taking all the treatment effects into account as the underlying variation to put in the denominator of the F statistic. In two-way analysis of variance with repeated measures on one factor, this is no longer true. To understand why this happens, we will examine how the variability in the data can be attributed to the different elements of the experimental design.*

*We will present a qualitative and intuitive justification of the origins of the F statistics used in the text. For a formal mathematical derivation of these results, see B. J. Winer, *Statistical Principles in Experimental Design* (2nd ed.), New York, McGraw-Hill, 1971, Chap. 7, "Multifactor Experiments Having Repeated Measures on the Same Elements."

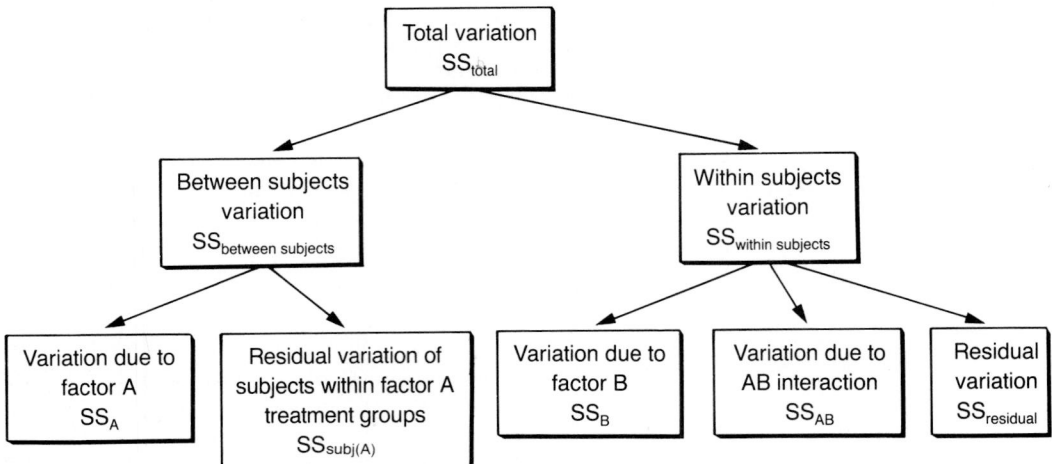

FIGURE 9-9 Partitioning of the total variation in a two-way, repeated-measures analysis of variance with repeated measures on one factor (factor B). As before, we start by dividing the total variation into components according to differences between subjects and variation within subjects. However, unlike the other repeated-measures designs we discuss, the between-subjects variation contains the information about variability according to the non-repeated-measures factor, in this example, factor A. The other component of the between-subjects variation is due to the fact that not all subjects are alike. Because there are different subjects in each level of factor A, the between-subjects variation is said to be nested within factor A and the corresponding sum of squares is denoted $SS_{subj(A)}$. The within-subjects variation is partitioned into components according to the effect of the repeated-measures factor, factor B, the A by B interaction, and a residual. The degrees of freedom partition in the same way.

As with all analyses of variance, we begin with the total sum of squared deviations of the observations about the mean of all the data SS_{tot} defined by Eq. (2.15),

$$SS_{tot} = \Sigma(Y - \overline{Y})^2$$

as our measure of variability. We will now partition this sum of squares (variation) into components associated with the different elements of the experimental design. Figure 9-9 shows this partitioning.

As with a one-way repeated-measures design, we first divide the variation into that between the different experimental subjects and that within subjects. In the one-way design, all subjects received all treatments, so the between-subjects variation did not contain any information insofar as hypothesis testing was concerned. We simply removed it to reduce the residual variation and concentrated on changes within the subjects (Fig. 9-6). However, now the different subjects received different treatments under factor A, so the between-subjects variation contains the information about the effect of factor A. Thus, after partitioning the total variation into components between and within subjects, we will further partition the between-subjects component into the variation associated with factor A and the residual variation between subjects within each of the factor A treatment groups (Fig. 9-9). These measures of variability are analogous to the treatment sum of squares and residual sum of squares in a completely randomized analysis of variance.

All subjects received all levels of factor B, and so, as with a one-way repeated-measures analysis of variance, the within-subjects variation has a component due to factor B (the treatment). In addition, because each subject receives some combination of both factor A and factor B, we can attribute some of the variation within subjects to the interaction between these two factors. The remaining variation within the subjects is the residual (Fig. 9-9). The variability due to factor B and the residual variation are analogous to the treatment sum of squares and residual sum of squares in a one-way repeated-measures analysis of variance (Fig. 9-6).

We next need to construct F statistics that compare the variability attributed to the treatment with the residual variability "within" that treatment.

Testing the Nonrepeated-Measures Factor

We will now formalize the computations for the experimental factor which is applied to different individuals, factor A in Fig. 9-9. In terms of our hypothetical example above, this factor would be mode of intensive care. Different individuals receive each level of factor A: In Table 9-5, subjects S_1, S_2, S_3, and S_4 receive one treatment (level A_1 of factor A) and subjects S_5, S_6, S_7, and S_8 receive a different treatment (level A_2 of factor A).

In one-way repeated-measures analysis of variance, we were only interested in $SS_{\text{bet subj}}$ because it allowed us to remove the inherent variability between subjects from the analysis so that we could focus more clearly on the treatment effect as changes within subjects. However, in this two-way design with repeated measures on factor B, but not factor A, $SS_{\text{bet subj}}$ has two components: SS_A and $SS_{\text{subj}(A)}$. SS_A is the variation due to the different treatments; some subjects received treatment A_1 while other subjects received treatment A_2. $SS_{\text{subj}(A)}$ (read "the sum of squares for subjects in A") is the variation due to the fact that not all subjects are alike and is our estimate of the residual variation within the treatment groups of factor A. We will test whether the variation associated with the different levels of factor A, SS_A, is "large" by comparing it with the residual variation of subjects within the factor A treatment groups $SS_{\text{subj}(A)}$.

Before we can construct the appropriate F statistics, we must convert these sums of squares into appropriate estimates of the underlying population variance by dividing by the associated degrees of freedom to obtain the associated mean squares.[*] The degrees of freedom associated with factor A are $DF_A = a - 1$, where there are a levels of factor A. The degrees of freedom associated with the residual variation among subjects within factor A are $DF_{\text{subj}(A)} = a(n - 1)$, where there are n different subjects within each of the a levels of factor A. For the example in Table 9-5, $a = 2$ and $n = 4$, so $DF_A = 2 - 1 = 1$ and $DF_{\text{subj}(A)} = 2(4 - 1) = 6$. Thus,

$$MS_A = \frac{SS_A}{DF_A}$$

and, for subjects in A,

$$MS_{\text{subj}(A)} = \frac{SS_{\text{subj}(A)}}{DF_{\text{subj}(A)}}$$

Finally, we compute the F statistic:

$$F = \frac{MS_A}{MS_{\text{subj}(A)}}$$

to test the null hypothesis that the means of the treatment groups under factor A are not different. This value of F is compared with the critical value of F with DF_A numerator and $DF_{\text{subj}(A)}$ denominator degrees of freedom.

Testing the Repeated-Measures Factor

Every subject received every level of factor B, so we will construct our test for whether different levels of factor B produced different average responses by examining the variation within subjects. The partitioning of $SS_{\text{wit subj}}$ in this experimental design is similar to that used for a one-way repeated-measures analysis of variance. There is a component due to the effect of the treatment (factor B) SS_B and a residual or error component SS_{res}. However, because this is a two-factor design, there is also the possibility of an interaction between the two experimental factors that form an additional component of $SS_{\text{wit subj}}$, SS_{AB} (Fig. 9-9). Thus, we partition $SS_{\text{wit subj}}$ into

$$SS_{\text{wit subj}} = SS_B + SS_{AB} + SS_{\text{res}}$$

[*]The degrees of freedom presented here assume fully balanced designs with no missing data. This assumption is not necessary when one formulates the analysis as a multiple regression problem and uses the degrees of freedom from the regression computations. We will illustrate this procedure in detail below.

Each of these sums of squares has an associated number of degrees of freedom. Specifically, $DF_B = b - 1$, $DF_{AB} = (a - 1)(b - 1)$, and $DF_{res} = a(b - 1)(n - 1)$, where there are a levels of factor A, b levels of factor B, and n subjects in each level of factor A.

To test whether the variability between the treatment groups is large compared with the residual variation, we must convert each of these sums of squares to an appropriate variance estimate by dividing by the associated degrees of freedom. Specifically,

$$MS_B = \frac{SS_B}{DF_B}$$

$$MS_{AB} = \frac{SS_{AB}}{DF_{AB}}$$

and

$$MS_{res} = \frac{SS_{res}}{DF_{res}}$$

Finally, we test for a significant factor B effect by comparing the variance estimate obtained from between the levels of factor B, MS_B, with the residual variance estimate within subjects, just like we did for single-factor repeated-measures analysis of variance:

$$F = \frac{MS_B}{MS_{res}}$$

We compare this value of F with the critical value for DF_B numerator and DF_{res} denominator degrees of freedom. If the value of F associated with the data exceeds this critical value, we conclude that there is more variation between the factor B treatment levels than would be expected based on the residual variability within the subjects and conclude that there are significant differences between the different levels of factor B.

We can also test for a significant interaction effect, just as we did in ordinary two-way analysis of variance, by computing

$$F = \frac{MS_{AB}}{MS_{res}}$$

and comparing it with the critical value of F with DF_{AB} numerator and DF_{res} denominator degrees of freedom.

Table 9-6 summarizes these computations for this two-way analysis of variance for two factors with repeated measures on one factor. We will now use a specific example to illustrate how to formulate this analysis as a multiple regression problem and derive the necessary sums of squares and mean squares to complete these computations.

TABLE 9-6 Analysis of Variance Table for Two-Factor Experiment (*a* Levels of Factor *A* and *b* Levels of Factor *B*) with Repeated Measures on Factor *B*

Source	SS	DF	MS	F
BETWEEN SUBJECTS				
Factor A	SS_A	$(a - 1)$	$SS_A/(a - 1)$	$MS_A/MS_{subj(A)}$
Subjects in A	$SS_{subj(A)}$	$a(n - 1)$	$SS_{subj(A)}/[a(n - 1)]$	
WITHIN SUBJECTS				
Factor B	SS_B	$(b - 1)$	$SS_B/(b - 1)$	MS_B/MS_{res}
$A \times B$ interaction	SS_{AB}	$(a - 1)(b - 1)$	$SS_{AB}/[(a - 1)(b - 1)]$	MS_{AB}/MS_{res}
Residual	SS_{res}	$a(b - 1)(n - 1)$	$SS_{res}/[a(b - 1)(n - 1)]$	

Is Alcoholism Associated with a History of Childhood Aggression?

Antisocial personality disorder (ASP), which is diagnosed from a history of childhood aggressive behavior and violence, has been implicated in the development of alcohol- and drug-related problems. Because of this association of childhood aggression and alcoholism, Jaffee and his co-workers[*] speculated that, because alcoholics as a group tend to be more aggressive and violent when drinking than do nonalcoholics, alcoholics diagnosed as having ASP might be more prone to aggressive and violent behavior when drinking than alcoholics without a diagnosis of ASP. They also wanted to see if drinking affected other factors—feelings of anger, depression, sociability, and sense of well-being—and if these factors could be related to differences in aggressive behavior between antisocial personality alcoholics (ASP alcoholics) and non-antisocial personality alcoholics (non-ASP alcoholics).

They developed a questionnaire to gage feelings of aggression and hostility and used the answers to compute scores for feelings of aggression, anger, depression, sociability, and well-being, with higher scores indicating stronger feelings. They administered this questionnaire to alcoholics under two different conditions: when they were sober and after they had been drinking. They studied both ASP and non-ASP alcoholics and examined the results of their questionnaire to see if there were differences in their feelings, both when they were sober and when they were drinking. This study is a two-factor design with one factor being drinking

[*]J. H. Jaffe, T. F. Babor, and D. H. Fishbein, "Alcoholics, Aggression and Antisocial Personality," *J. Stud. Alcohol* 49:211–218, 1988.

state (sober vs. drinking) and the other factor being antisocial personality (ASP vs. non-ASP). Because each subject was studied while sober and while drinking, the drinking factor is a repeated-measures factor. Obviously, the group of ASP alcoholics consisted of different individuals than the group of non-ASP alcoholics, so this factor does not have repeated measures.

Table 9-7 shows the design of this study, together with values of the aggression score for 16 subjects (8 ASP and 8 non-ASP alcoholics). Figure 9-10 shows the mean scores for these data, as well as the other scores they computed for depression, well-being, sociability, and anger. We will work through the analysis of the aggression score in detail, then present and discuss the analysis of the anger and depression scores. Because the experimental design is the same for all the scores, we only need to formulate one model for the analysis. We will then apply the same analysis to each set of scores.

The coding of dummy variables necessary to implement a multiple regression solution to this problem uses the same rules we used for ordinary two-way analysis of variance. We use *effects coding* to define

TABLE 9-7 Data for Study of Aggressiveness before and after Drinking in Alcoholics Diagnosed as either ASP or Non-ASP

Personality Type	Subject	Drinking Status	
		Sober	Drinking
Non-ASP	1	.81	.59
	2	.91	1.04
	3	.98	1.11
	4	1.08	1.13
	5	1.10	1.15
	6	1.16	1.16
	7	1.19	1.25
	8	1.44	1.70
ASP	9	.72	.83
	10	.82	.99
	11	.89	1.17
	12	1.01	1.24
	13	1.10	1.33
	14	1.14	1.47
	15	1.24	1.59
	16	1.34	1.73

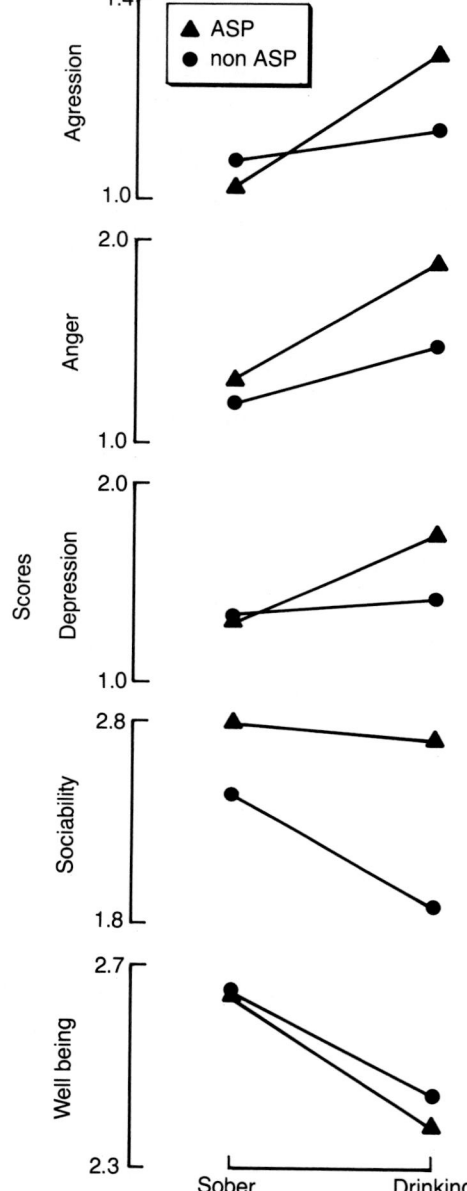

FIGURE 9-10 Personality traits as a function of personality type (antisocial personality or non-antisocial personality) and alcohol consumption (sober or drinking) in alcoholics. Each subject answers questions while sober and then while drinking. Thus, drinking status (alcohol consumption) is a repeated-measures factor. The two personality types consist of groups of different individuals. Thus, personality type is not a repeated-measures factor. Based on their answers to the questions, the subjects were given scores for aggression (panel *A*), anger (panel *B*), depression (panel *C*), sociability (panel *D*), and well-being (panel *E*). *(Modified from Figure 1 of J. H. Jaffee, T. F. Babor, and D. H. Fishbein: "Alcoholics, Aggression, and Anti-social Personality," J. Stud. Alcohol 49:211–218, 1988.)*

dummy variables to encode the experimental factors (both main effects and, because it is a two-factor design, the interaction effect) and, because it involves repeated measures, we also encode subject dummy variables to account for the fact that there are repeated measures on one of the factors.

There are two experimental factors, drinking status and antisocial personality, each with two levels. Thus, we need one dummy variable to code each factor:

$$D = \begin{cases} -1 \text{ if drinking} \\ 1 \text{ if sober} \end{cases}$$

and

$$A = \begin{cases} -1 \text{ if ASP alcoholic} \\ 1 \text{ if non-ASP alcoholic} \end{cases}$$

The interaction between drinking state and personality is simply DA.

There are 16 experimental subjects, so it would seem that we will require $16 - 1 = 15$ dummy variables to encode the between-subjects differences. However, because the subjects are grouped (or *nested*) under one of the experimental factors (different ASP diagnoses), *we must encode the between-subjects variability separately within each diagnostic group.*

The reason for this procedure is that we can account for repeated measures *within* each level of the personality factor (ASP or non-ASP alcoholic) but not *across* different levels of this factor. Thus, the between-subjects differences within each level of the experimental factor must sum to zero so that the differences in mean response between the two levels of the experimental factor will be reflected in the coefficient associated with the dummy variable A.* Thus, instead of defining 15 subject dummy variables to code for the 16 subjects, we define 7 dummy variables for the 8 ASP alcoholic subjects and 7 more for the 8 non-ASP alcoholics, for a total of 14 subject dummy variables. Specifically,

---

*The precise technical reason for this procedure is the following: Personality type (ASP or non-ASP) is the non-repeated-measures experimental factor in the design, which we denoted as factor A in our general discussion earlier in this section. We must define our dummy variables so that we can compute the sum of squares of subjects within the levels of factor A, $SS_{subj(A)}$. $SS_{subj(A)}$ has two components due to the fact that the subjects are grouped (or nested) under different levels of factor A:

$$SS_{subj(A)} = SS_{subj(A_1)} + SS_{subj(A_2)}$$

where $SS_{subj(A_1)}$ is the sum of squared deviations of the different subject means around the overall mean value observed of factor A under level 1 and $SS_{subj(A_2)}$ is the sum of squared deviations of the different subject means around the overall mean value observed of factor A under level 2. The between-subjects dummy variables must be defined to obtain these sums of squares. The two components of $SS_{subj(A)}$ are independent of each other. Therefore, we must define the dummy variables to encode the between-subjects differences in a way that keeps the between-subjects effects associated with levels A_1 and A_2 independent of each other. For a further discussion of nesting and hierarchical designs in analysis of variance, see B. J. Winer, *Statistical Principles in Experimental Design*, New York, McGraw-Hill, 1971, pp. 359–366.

$$P_{\text{non-ASP}_i} = \begin{cases} 1 \text{ if person } i \text{ in non-ASP group} & i < 8 \\ -1 \text{ if person 8 in non-ASP group} \\ 0 \text{ otherwise} \end{cases}$$

and

$$P_{\text{ASP}_i} = \begin{cases} 1 \text{ if person } i \text{ in ASP group} & i < 8 \\ -1 \text{ if person 8 in ASP group} \\ 0 \text{ otherwise} \end{cases}$$

Now that we have defined all the necessary dummy variables, we can write the regression model for this analysis:

$$\hat{S} = b_0 + \sum_{i=1}^{7} b_{\text{non-}A_i} P_{\text{non-ASP}_i} + \sum_{i=1}^{7} b_{A_i} P_{\text{ASP}_i} + b_D D + b_A A + b_{DA} DA$$

where $\hat{S}$, the predicted score, is any of the hostility and aggression scores derived from the questions the subjects answered. With this formulation, b_0 is the overall mean score for all subjects, b_D is the change from the overall mean score due to drinking, b_A is the change from the overall mean score due to ASP diagnosis, and b_{DA} represents the interaction between drinking and ASP characteristics.

The data for the aggression score and the associated dummy variables appear in Table 9-8. Figure 9-11 presents the results of the multiple linear regression fit to these data. We now use these results to obtain the sums of squares and mean squares we need to test the hypotheses that:

1. There is no difference in mean aggressiveness score associated with drinking, after controlling for personality.
2. There is no difference in mean aggressiveness score associated with ASP diagnosis, after controlling for drinking.
3. The effects of drinking and ASP diagnosis on aggressiveness score are independent and additive; they do not interact.

Because there are only two levels of the drinking factor, there is only one dummy variable, and the sum of squares for drinking is simply the sum of squares associated with the variable D. There is 1 degree of freedom associated with D, so

$$MS_D = \frac{SS_D}{DF_D} = \frac{0.2032}{1} = 0.2032$$

Likewise, for personality type,

$$MS_A = \frac{SS_A}{DF_A} = \frac{0.0205}{1} = 0.0205 \tag{9.2}$$

and for interaction between drinking and personality type,

TABLE 9-8 Aggression Score S (points) while Sober ($D = 1$) or Drinking ($D = -1$) in Alcoholic Subjects Diagnosed as either ASP ($A = -1$) or Non-ASP ($A = 1$)

S	D	A	$P_{\text{non-ASP}_i}$							P_{ASP_i}						
			1	2	3	4	5	6	7	1	2	3	4	5	6	7
.81	1	1	1	0	0	0	0	0	0	0	0	0	0	0	0	0
.91	1	1	0	1	0	0	0	0	0	0	0	0	0	0	0	0
.98	1	1	0	0	1	0	0	0	0	0	0	0	0	0	0	0
1.08	1	1	0	0	0	1	0	0	0	0	0	0	0	0	0	0
1.10	1	1	0	0	0	0	1	0	0	0	0	0	0	0	0	0
1.16	1	1	0	0	0	0	0	1	0	0	0	0	0	0	0	0
1.19	1	1	0	0	0	0	0	0	1	0	0	0	0	0	0	0
1.44	1	1	-1	-1	-1	-1	-1	-1	-1	0	0	0	0	0	0	0
.72	1	-1	0	0	0	0	0	0	0	1	0	0	0	0	0	0
.82	1	-1	0	0	0	0	0	0	0	0	1	0	0	0	0	0
.89	1	-1	0	0	0	0	0	0	0	0	0	1	0	0	0	0
1.01	1	-1	0	0	0	0	0	0	0	0	0	0	1	0	0	0
1.10	1	-1	0	0	0	0	0	0	0	0	0	0	0	1	0	0
1.14	1	-1	0	0	0	0	0	0	0	0	0	0	0	0	1	0
1.24	1	-1	0	0	0	0	0	0	0	0	0	0	0	0	0	1
1.34	1	-1	0	0	0	0	0	0	0	-1	-1	-1	-1	-1	-1	-1
.59	-1	1	1	0	0	0	0	0	0	0	0	0	0	0	0	0
1.04	-1	1	0	1	0	0	0	0	0	0	0	0	0	0	0	0
1.11	-1	1	0	0	1	0	0	0	0	0	0	0	0	0	0	0
1.13	-1	1	0	0	0	1	0	0	0	0	0	0	0	0	0	0
1.15	-1	1	0	0	0	0	1	0	0	0	0	0	0	0	0	0
1.16	-1	1	0	0	0	0	0	1	0	0	0	0	0	0	0	0
1.25	-1	1	0	0	0	0	0	0	1	0	0	0	0	0	0	0
1.70	-1	1	-1	-1	-1	-1	-1	-1	-1	0	0	0	0	0	0	0
.83	-1	-1	0	0	0	0	0	0	0	1	0	0	0	0	0	0
.99	-1	-1	0	0	0	0	0	0	0	0	1	0	0	0	0	0
1.17	-1	-1	0	0	0	0	0	0	0	0	0	1	0	0	0	0
1.24	-1	-1	0	0	0	0	0	0	0	0	0	0	1	0	0	0
1.33	-1	-1	0	0	0	0	0	0	0	0	0	0	0	1	0	0
1.47	-1	-1	0	0	0	0	0	0	0	0	0	0	0	0	1	0
1.59	-1	-1	0	0	0	0	0	0	0	0	0	0	0	0	0	1
1.73	-1	-1	0	0	0	0	0	0	0	-1	-1	-1	-1	-1	-1	-1

$$MS_{DA} = \frac{SS_{DA}}{DF_{DA}} = \frac{0.08303}{1} = 0.08303$$

Had any of these factors had more than two levels, there would have been more than one dummy variable per factor, and we would have added the sums of squares and degrees of freedom for all the dummy variables associated with each factor to obtain the appropriate mean squares.

```
DEP VARIABLE: S        Aggression score

                                    ANALYSIS OF VARIANCE

                                 SUM OF          MEAN
                SOURCE    DF      SQUARES        SQUARE      F VALUE     PROB>F

                MODEL     17    2.06262812    0.12133107     17.454      0.0001
                ERROR     14    0.09731875    0.006951339
                C TOTAL   31    2.15994687

                    ROOT MSE    0.08337469    R-SQUARE      0.9549
                    DEP MEAN    1.137813      ADJ R-SQ      0.9002
                    C.V.        7.32763

                                    PARAMETER ESTIMATES

            PARAMETER      STANDARD     T FOR HO:                              VARIABLE
VARIABLE  DF  ESTIMATE       ERROR    PARAMETER=0    PROB > |T|   TYPE I SS     LABEL

INTERCEP  1   1.13781250   0.0147387     77.199       0.0001    41.42775313   INTERCEPT
PASP1     1    -0.4125     0.05514718    -7.480        0.0001     0.75690000   subject dummy for first ASP subject
PASP2     1    -0.1375     0.05514718    -2.493        0.0258     0.03413333
PASP3     1    -0.0675     0.05514718    -1.224        0.2412     0.002016667
PASP4     1    -0.0075     0.05514718    -0.136        0.8938     0.00169
PASP5     1     0.0125     0.05514718     0.227        0.8240     0.003526667
PASP6     1     0.0475     0.05514718     0.861        0.4036     0.009219048
PASP7     1   0.10750000   0.05514718     1.949        0.0716     0.02641429   subject dummy for seventh ASP subject
PNONASP1  1   -0.388125    0.05514718    -7.038        0.0001     0.57760000   dummy for first non-ASP subject
PNONASP2  1   -0.258125    0.05514718    -4.681        0.0004     0.08333333
PNONASP3  1   -0.133125    0.05514718    -2.414        0.0301     0.002604167
PNONASP4  1   -0.038125    0.05514718    -0.691        0.5007     0.0065025
PNONASP5  1    0.051875    0.05514718     0.941        0.3628     0.033135
PNONASP6  1   0.14187500   0.05514718     2.573        0.0221     0.07381071
PNONASP7  1   0.25187500   0.05514718     4.567        0.0004     0.14500804   dummy for seventh non-ASP subject
A         1   -0.0253125   0.0147387     -1.717        0.1079     0.02050312   Dummy variable = −1 if ASP alcoholic
D         1   -0.0796875   0.0147387     -5.407        0.0001     0.20320312   Dummy variance = −1 if drinking
DA        1   0.0509375    0.0147387      3.456        0.0039     0.08302813   Interaction of D and A
```

FIGURE 9-11 Results of regression implementation of the two-way analysis of variance with repeated measures on one factor for the aggression scores shown in Fig. 9-10A. The subject dummy variables are entered first, followed by the main effects of antisocial personality diagnosis A, drinking D, and their interaction DA.

We estimate the variance of subjects within the two personality groups by adding the sums of squares associated with the 14 dummy variables used to encode the 16 different subjects:

$$SS_{\text{subj}(A)} = SS_{\text{subj}(ASP)} + SS_{\text{subj}(\text{non-ASP})}$$

$$= \sum_{i=1}^{7} SS_{P_{ASP_i}} + \sum_{i=1}^{7} SS_{P_{\text{non-ASP}_i}}$$

$$= (0.75690 + 0.03413 + 0.00202 + 0.00169$$

$$+ 0.00353 + 0.00922 + 0.02641) \tag{9.3}$$

$$+ (0.57760 + 0.08333 + 0.00260 + 0.00650$$

$$+ 0.03314 + 0.07381 + 0.14501)$$

$$= 1.7559$$

Each of these 14 dummy variables has 1 associated degree of freedom, so

$$DF_{\text{subj}(A)} = \sum_{i=1}^{7} DF_{P_{\text{ASP}_i}} + \sum_{i=1}^{7} DF_{P_{\text{non-ASP}_i}}$$

$$= (1 + 1 + 1 + 1 + 1 + 1 + 1)$$

$$+ (1 + 1 + 1 + 1 + 1 + 1 + 1) = 14$$

Therefore,

$$MS_{\text{subj}(A)} = \frac{SS_{\text{subj}(A)}}{DF_{\text{subj}(A)}} = \frac{1.7559}{14} = 0.12542$$

Finally, as before, we obtain $MS_{\text{res}} = .00695$ (with 14 degrees of freedom) directly from the regression output. We now have all the quantities we need to test the three hypotheses listed above.

We first test whether there is any change in aggressiveness score associated with drinking. We have repeated measures on aggressiveness score at both levels of drinking (sober and drinking), so we test for differences with

$$F = \frac{MS_D}{MS_{\text{res}}} = \frac{0.20320}{0.00695} = 29.24$$

This value greatly exceeds 8.86, the critical value of F for $\alpha = .01$ with 1 numerator and 14 denominator degrees of freedom. Hence, we conclude that drinking significantly changes aggressiveness ($P < .01$). The negative value of $b_D = -0.08$ indicates that, because drinking is coded with $D = -1$, drinking increases aggressiveness ($-1 \cdot -.08 = +.08$).

We next test whether there is any change in aggressiveness score associated with ASP vs. non-ASP traits. There are no repeated measures on this factor, so we test this hypothesis by computing

$$F = \frac{MS_A}{MS_{\text{subj}(A)}} = \frac{0.02050}{0.12542} = 0.16$$

This value does not even approach statistical significance, so we conclude that the presence or absence of ASP traits does not affect the aggressiveness score.

Finally, we test for interaction between drinking and personality traits with

$$F = \frac{MS_{DA}}{MS_{\text{res}}} = \frac{0.08303}{0.00695} = 11.95$$

This value of F is also associated with 1 numerator and 14 denominator degrees of freedom, so we conclude that there is a significant interaction between drinking and personality traits in determining the aggressiveness score ($P < .01$). This significant interaction between ASP diagnosis and drinking indicates that alcoholics diagnosed as ASP reported more aggression when drinking than did non-ASP alcoholics.

Figure 9-10, which shows the mean values of aggressiveness score under the different experimental conditions, clearly illustrates this interaction. Had there been no interaction, the changes in aggressiveness score with drinking would have been the same for both personality types; the lines would have been parallel. Thus, we conclude that alcoholics with ASP disorder are significantly more prone to aggressive behavior when drinking than are alcoholics without ASP, which supports the original hypothesis.

Figure 9-12 presents the results of a similar analysis for anger scores on the questionnaire (the data are in Table C-15, Appendix C.) The design of the experiment is identical to that just described, as are the computations of the sums of squares, mean squares, and F statistics. Table 9-9 presents the results of this analysis: All three hypothesis tests proved to be statistically significant. Figure 9-10 illustrates these effects: The two

```
DEP VARIABLE: S       Anger score

                         ANALYSIS OF VARIANCE

                             SUM OF        MEAN
          SOURCE    DF      SQUARES       SQUARE     F VALUE    PROB>F

          MODEL     17    3.28211250   0.19306544    36.681     0.0001
          ERROR     14    0.0736875    0.005263393
          C TOTAL   31    3.35580000

                  ROOT MSE    0.07254924    R-SQUARE     0.9780
                  DEP MEAN         1.47     ADJ R-SQ     0.9514
                  C.V.         4.935323

                         PARAMETER ESTIMATES

                   PARAMETER    STANDARD    T FOR HO:                             VARIABLE
VARIABLE   DF      ESTIMATE      ERROR    PARAMETER=0   PROB > |T|   TYPE I SS     LABEL

INTERCEP   1     1.47000000   0.01282502    114.620       0.0001    69.14880000   INTERCEPT
PASP1      1    -0.300625     0.04798681     -6.265       0.0001     0.34222500   subject dummy for first ASP subject
PASP2      1    -0.160625     0.04798681     -3.347       0.0048     0.03100833
PASP3      1    -0.085625     0.04798681     -1.784       0.0960     0.001066667
PASP4      1    -0.050625     0.04798681     -1.055       0.3093     0.00036
PASP5      1     0.054375     0.04798681      1.133       0.2762     0.022815
PASP6      1     0.10437500   0.04798681      2.175       0.0473     0.03729643
PASP7      1     0.15437500   0.04798681      3.217       0.0062     0.05447232   subject dummy for seventh ASP subject
PNONASP1   1    -0.324375     0.04798681     -6.760       0.0001     0.49702500   dummy for first non-ASP subject
PNONASP2   1    -0.139375     0.04798681     -2.904       0.0115     0.03740833
PNONASP3   1    -0.079375     0.04798681     -1.654       0.1203     0.004004167
PNONASP4   1    -0.034375     0.04798681     -0.716       0.4856     0.0000625
PNONASP5   1    -0.024375     0.04798681     -0.508       0.6194     0.000375
PNONASP6   1     0.040625     0.04798681      0.847       0.4115     0.01029643
PNONASP7   1     0.18062500   0.04798681      3.764       0.0021     0.07457232   dummy for seventh non-ASP subject
A          1    -0.129375     0.01282502    -10.088       0.0001     0.53561250   Dummy variable = -1 if ASP alcoholic
D          1    -0.213125     0.01282502    -16.618       0.0001     1.45351250   Dummy variance = -1 if drinking
DA         1     0.075        0.01282502      5.848       0.0001     0.18000000   Interaction of D and A
```

FIGURE 9-12 Results of regression implementation of the two-way analysis of variance with repeated measures on one factor for the anger scores shown in Fig. 9-10B.

TABLE 9-9 Analysis of Variance Table for Anger Scores while Drinking

Source	SS	DF	MS	F
BETWEEN SUBJECTS				
A	.5356	1	.5356	6.74
Subj(A)	1.1130	14	.0795	
WITHIN SUBJECTS				
D	1.4535	1	1.4535	276.16
AD	.1800	1	.18	34.20
Residual	.0737	14	.00526	

lines are separated but not parallel. They are separated because of the significant main effect due to personality trait, have a slope due to the drinking effect, and are not parallel because of the significant interaction effect. Hence, we conclude that anger changed significantly with the presence of ASP trait or drinking and that there was an interaction between these two factors, so that alcoholics with ASPs became more angry when they drank than did people without this personality trait.

Figure 9-13 and Table 9-10 present the results of the analysis of the depression score (the data are in Table C-16, Appendix C). None of the F tests for the depression score indicate any statistically significant effect, so we conclude that the depression score is unaffected by drinking or ASP diagnosis.* Thus, the differences in mean values presented for this variable in Fig. 9-10 appear to be due to sampling variability.

*Jaffe and his coworkers had a much larger sample size for the actual study (77 subjects) and identified a significant interaction effect for this variable.

TABLE 9-10 Analysis of Variance Table for Depression Scores while Drinking

Source	SS	DF	MS	F
BETWEEN SUBJECTS				
A	.1554	1	.1554	.92
Subj(A)	2.3717	14	.1694	
WITHIN SUBJECTS				
D	.5592	1	.5592	1.38
AD	.2503	1	.2503	.62
Residual	5.6712	14	.4051	

DEP VARIABLE: S   Depression score

ANALYSIS OF VARIANCE

SOURCE	DF	SUM OF SQUARES	MEAN SQUARE	F VALUE	PROB>F
MODEL	17	3.33655313	0.19626783	0.485	0.9214
ERROR	14	5.67121875	0.40508705		
C TOTAL	31	9.00777188			

ROOT MSE	0.6364645	R-SQUARE	0.3704
DEP MEAN	1.445938	ADJ R-SQ	-0.3941
C.V.	44.01743		

PARAMETER ESTIMATES

VARIABLE	DF	PARAMETER ESTIMATE	STANDARD ERROR	T FOR H0: PARAMETER=0	PROB > |T|	TYPE I SS	VARIABLE LABEL
INTERCEP	1	1.44593750	0.11251209	12.851	0.0001	66.90352812	INTERCEPT
PASP1	1	0.00875	0.42098169	0.021	0.9837	0.009025	subject dummy for first ASP subject
PASP2	1	0.29875000	0.42098169	0.710	0.4896	0.07840833	
PASP3	1	0.00875	0.42098169	0.021	0.9837	0.02470417	
PASP4	1	-0.19625	0.42098169	-0.466	0.6483	0.14520250	
PASP5	1	0.46375000	0.42098169	1.102	0.2892	0.29260167	
PASP6	1	-0.26625	0.42098169	-0.632	0.5373	0.24862976	
PASP7	1	-0.42125	0.42098169	-1.001	0.3340	0.40560357	subject dummy for seventh ASP subject
PNONASP1	1	0.13937500	0.42098169	0.331	0.7455	0.11222500	dummy for first non-ASP subject
PNONASP2	1	-0.060625	0.42098169	-0.144	0.8875	0.18007500	
PNONASP3	1	-0.565625	0.42098169	-1.344	0.2005	0.84375000	
PNONASP4	1	0.059375	0.42098169	0.141	0.8898	0.00625	
PNONASP5	1	0.069375	0.42098169	0.165	0.8715	0.006	
PNONASP6	1	-0.075625	0.42098169	-0.180	0.8600	0.01547143	
PNONASP7	1	-0.040625	0.42098169	-0.097	0.9245	0.003772321	dummy for seventh non-ASP subject
A	1	-0.0696875	0.11251209	-0.619	0.5456	0.15540312	Dummy variable = -1 if ASP alcoholic
D	1	-0.132188	0.11251209	-1.175	0.2596	0.55915313	Dummy variance = -1 if drinking
DA	1	0.0884375	0.11251209	0.786	0.4450	0.25027813	Interaction of D and A

FIGURE 9-13 Results of regression implementation of the two-way analysis of variance with repeated measures on one factor for the depression scores shown in Fig. 9-10C.

The other scores are left as problems, but we will finish this example by stating that Jaffee and his colleagues concluded that drinking induces so-called negative affective states (as indicated by less sociability and less feeling of well-being), and that these states, when coupled with a history of ASP might reliably predict those alcoholics who are most prone to drinking-induced aggressive behavior or violence.

Multiple Comparisons

In the current example there are only two levels of each of the experimental factors, so there is no need to conduct multiple comparisons on either of the factors when the analysis of variance reveals a significant factor effect. When there are more than two levels of a factor, you can conduct multiple comparisons among different means within either factor using any of the methods we have developed so far. To do so, simply use the same mean square and associated degrees of freedom as was used in the denominator of the F test for that factor.[*]

MISSING DATA IN TWO-FACTOR ANALYSIS OF VARIANCE WITH REPEATED MEASURES ON ONE FACTOR

All of the problems we encountered with missing data and the resultant nonorthogonality of the dummy variables when we used ordinary two-way analysis of variance arise when we have repeated measures. Thus, at first blush, it would seem that we could simply use the strategy of obtaining the marginal sums of squares for each effect by adjusting each effect for all other effects in the model. This is true to a point: We do, indeed, have to obtain marginal sums of squares for each effect of interest, and all of our subsequent discussion assumes we have obtained the correct sums of squares. Unfortunately, when there are repeated measures, the missing data cause another problem: The way in which the F statistics are computed must be adjusted for distortions created by missing data. To understand why this happens and how to deal with it, we must consider the origins of the F statistic in more detail.

Expected Mean Squares and Random Factors

Although we have developed the F statistics for two-way analysis of variance with repeated measures on one factor from qualitative arguments about the partitioning of the sums of squares, as we have seen for other analyses of variance, it is possible to derive the F statistics formally by considering the expected mean squares for each effect in the model. These expected mean squares consist of sums of the different sources of variance in the observations. The F statistics are computed such that we effectively test whether or not the source of variance due to a particular effect is zero: The numerator contains a sum of variances that includes the variance due to the factor to be tested, and the denominator contains all the variances in the numerator except the one being tested. The same development can be prepared for two-way repeated-measures analysis of variance with repeated measures on one of the factors.

The expected mean squares are more complicated than in a completely randomized design. The complication arises because of the additional *random effect* due to the subjects. The subjects consist of a

[*]The fact that the mean squares associated with the two different factors are different complicates the task of doing multiple comparisons among *all* treatment means. To do so, one must compute an "average" mean square to use in the denominator of the t, q, or q' test statistics. For details on how to do these computations, see B. J. Winer, *Statistical Principles in Experimental Design*, New York, McGraw-Hill, 1971, pp. 528–532.

(hopefully) random sample of all possible subjects. There is nothing special about the ones in a study; they just happened to be the ones chosen from the population of interest. For completely randomized designs, we have been dealing with *fixed effects* which represent factors we specifically chose to study. Factors are said to be fixed when the levels of the factor are not meant to be representative of a much larger population of interest, but rather were chosen as the only levels of interest for the purposes of the experiment—they are not randomly chosen and the study is not meant to be extrapolated beyond the chosen levels.

The repeated-measures analysis of variance model contains the fixed factors of interest *and* an additional random factor, which is the subjects. Models which contain both fixed effects and random effects are called *mixed models*. For example, the simple one-way analysis of variance with repeated measures we considered above is equivalent to a two-way mixed model which has one fixed factor—the different ways of getting food into the stomach—and one random factor—the different dogs. Similarly, the two-way repeated-measures analysis of variance with repeated measures on one factor we have been considering is equivalent to a three-way mixed-model analysis of variance with two fixed factors—drinking and ASP diagnosis—and one random factor—the subjects, who in this model are *nested* within the levels of the ASP factor. The idea of nesting is just another way to say that the different levels of the ASP factor contain different subjects. Any given type of repeated-measures analysis of variance can be recast as a higher order mixed-model analysis of variance where we have an explicit subjects factor which is random.

The expected mean squares of mixed models have more terms in them than models containing only fixed effects, but the concept of forming the F statistics is the same. For the general two-way analysis of variance with repeated measures on one of the factors with a levels of factor A and b levels of factor B, with subjects nested within the levels of factor A, the expected mean squares are*

---

*For a more thorough treatment of the expected mean squares for this experimental design, see B. J. Winer, *Statistical Principles in Experimental Design* (2nd ed.), New York, McGraw-Hill, 1971, pp. 518–521. We state the expected mean squares specifically for the design in question and in so doing do not consider some of their more general features. In particular, what we call σ^2_{res} actually consists of two components:

$$\sigma^2_{\text{res}} = \sigma^2_\epsilon + \sigma^2_{\text{subj}(A) \times B}$$

where σ^2_ϵ is the underlying population variance and $\sigma^2_{\text{subj}(A) \times B}$ is the interaction between factor B and the subjects nested in factor A (this interaction arises because different subjects respond differently over the levels of B). When we are considering the usual repeated-measures experiment, there is only one observation made on each subject at each combination of levels of factors A and B. Thus, we cannot estimate σ^2_ϵ and $\sigma^2_{\text{subj}(A) \times B}$ separately.

$$E(MS_A) = \sigma^2_{res} + b\sigma^2_{subj(A)} + \sigma^2_A$$

$$E(MS_{subj(A)}) = \sigma^2_{res} + b\sigma^2_{subj(A)}$$

$$E(MS_B) = \sigma^2_{res} + \sigma^2_B$$

$$E(MS_{AB}) = \sigma^2_{res} + \sigma^2_{AB}$$

$$E(MS_{res}) = \sigma^2_{res}$$

The mean squares used to compute F for each hypothesis test follow from the same rule we have previously applied: The numerator contains the variance due to the effect we desire to test, and the denominator contains all the variances in the numerator *except* the one to be tested. For example, to test for a significant factor A effect we estimate

$$F = \frac{\sigma^2_{res} + b\sigma^2_{subj(A)} + \sigma^2_A}{\sigma^2_{res} + b\sigma^2_{subj(A)}}$$

with

$$F = \frac{MS_A}{MS_{subj(A)}}$$

Likewise, to test for a significant factor B effect we estimate

$$F = \frac{\sigma^2_{res} + \sigma^2_B}{\sigma^2_{res}}$$

with

$$F = \frac{MS_B}{MS_{res}}$$

Finally, to test for a significant interaction, we estimate

$$F = \frac{\sigma^2_{res} + \sigma^2_{AB}}{\sigma^2_{res}}$$

with

$$F = \frac{MS_{AB}}{MS_{res}}$$

In each case, if the treatment of interest has zero effect, the last term in the numerator will be zero, and F will be expected to be 1. In contrast, if the treatment has an effect, the last term in the numerator will be greater than zero and F will be expected to exceed 1.

Figure 9-14 shows the computer output from an analysis of the data in Table 9-8 using a traditional mixed-model analysis of variance on the fixed effects of drinking and ASP diagnosis and the random effect of subjects nested within ASP diagnosis. The sums of squares and mean

```
                            GENERAL LINEAR MODELS PROCEDURE

DEPENDENT VARIABLE: S        Aggression score

SOURCE              DF      SUM OF SQUARES      MEAN SQUARE      F VALUE      PR > F      R-SQUARE         C.V.

MODEL               17         2.06262813        0.12133107        17.45      0.0001      0.954944       7.3276

ERROR               14         0.09731875        0.00695134               ROOT MSE                     S MEAN

CORRECTED TOTAL     31         2.15994687                            0.08337469                       1.13781250


SOURCE              DF      TYPE I SS     F VALUE     PR > F      DF      TYPE III SS     F VALUE     PR > F
A                    1      0.02050312       2.95     0.1079       1      0.02050312         2.95     0.1079
D                    1      0.20320313      29.23     0.0001       1      0.20320313        29.23     0.0001
A*D                  1      0.08302812      11.94     0.0039       1      0.08302812        11.94     0.0039
SUBJECT(A)          14      1.75589375      18.04     0.0001      14      1.75589375        18.04     0.0001
```

FIGURE 9-14 Analysis of aggression score as a function of antisocial personality and drinking, computed using the SAS general linear models (GLM) procedure. Because there are no missing data, the incremental sums of squares (labeled Type I *SS*) equal the marginal sums of squares (labeled Type III *SS*). Compare with the regression analysis of the same data presented in Fig. 9-11.

squares are identical to what we computed using the multiple regression approach and dummy variables.* The expected mean squares for these data, shown in Table 9-11, are what we would expect based on the general

TABLE 9-11 Expected Mean Squares for Aggression Score in Alcoholics while Sober and Drinking

---

NO MISSING DATA

$$E(MS_A) = \sigma_{res}^2 + 2\sigma_{subj(A)}^2 + \sigma_A^2$$
$$E(MS_{subj(A)}) = \sigma_{res}^2 + 2\sigma_{subj(A)}^2$$
$$E(MS_D) = \sigma_{res}^2 + \sigma_D^2$$
$$E(MS_{AD}) = \sigma_{res}^2 + \sigma_{AD}^2$$
$$E(MS_{res}) = \sigma_{res}^2$$

MISSING DATA

$$E(MS_A) = \sigma_{res}^2 + 1.659\sigma_{subj(A)}^2 + \sigma_A^2$$
$$E(MS_{subj(A)}) = \sigma_{res}^2 + 1.769\sigma_{subj(A)}^2$$
$$E(MS_D) = \sigma_{res}^2 + \sigma_D^2$$
$$E(MS_{AD}) = \sigma_{res}^2 + \sigma_{AD}^2$$
$$E(MS_{res}) = \sigma_{res}^2$$

---

*Because we did not explicitly tell this program what to use for the denominator of the *F* statistics, it assumed that it was using MS_{res}, and so the *F* statistic for the ASP effect is incorrect and needs to be calculated by hand from the sums of squares and degrees of freedom in the output.

discussion above (there are $b = 2$ levels of factor B, so $b\sigma^2_{\text{subj}(A)} = 2\sigma^2_{\text{subj}(A)}$).*

As before, these expected mean squares are just a more formal statement of the way we have been partitioning the total variability in the observations. However, unlike the experiments we have been considering up until now, the expected mean squares will play an important role in how we handle missing data in two-way repeated-measures analyses of variance.

What Happens to the Expected Mean Squares When There Are Missing Data?

The expected mean squares we have been discussing change when there are missing data in a way that distorts the F test statistics we have been using. As a result, these F statistics no longer test the hypotheses you want and can lead to erroneous conclusions if you are not aware of this fact. We now explore these distortions and illustrate how to take them into account and obtain F statistics that actually test the hypotheses we want.

As just described, when there are no missing data, we test for the main effect of factor A by examining the ratio of the mean square asso-

---

*Computer programs generally report the expected mean squares from the analysis of mixed models using very general notation that includes separate components for each random effect in the model (including σ^2_ϵ) and a general function of some combination of fixed effects, usually given as $Q[\cdot]$, where the notation $Q[\cdot]$ refers to a function known as the *quadratic form* of the fixed effects and the • notation refers to some list of fixed effects in the model. Do not get hung up on the general $Q[\cdot]$ notation; the principles of deciding which mean squares to put in the F statistic are the same. For example, in the general model we have been discussing, the computer output might show

$$E(MS_A) = \text{VAR[error]} + b\text{VAR[subj}(A)] + Q[A, AB]$$

$$E(MS_{\text{subj}(A)}) = \text{VAR[error]} + b\text{VAR[subj}(A)]$$

$$E(MS_B) = \text{VAR[error]} + Q[B, AB]$$

$$E(MS_{AB}) = \text{VAR[error]} + Q[AB]$$

$$E(MS_{\text{res}}) = \text{VAR[error]}$$

Although the $E(MS_A)$ has in it a term $Q[A, AB]$ which also includes the interaction, when the specific assumptions underlying the model are applied to the mathematics underlying the analysis, the interaction portion goes to zero and $Q[A, AB]$ reduces to what we have called σ^2_A. Similar considerations apply to $E(MS_B)$. See G. A. Milliken and D. E. Johnson, *Analysis of Messy Data: Volume 1: Designed Experiments*, New York, Van Nostrand Reinhold, 1984, pp. 216–231 for a more thorough discussion of this topic.

ciated with factor A and the mean square associated with subjects in A because both the numerator and denominator used in this comparison contain $\sigma^2_{res} + b\sigma^2_{subj(A)}$, where σ^2_{res} is the residual variation, $\sigma^2_{subj(A)}$ is the variation of experimental subjects within each level of factor A, and b is the number of levels of factor B. The numerator also contains the term σ^2_A, the variation due to the effect of factor A. When factor A has an effect, σ^2_A will exceed zero and the F ratio will be expected to be greater than 1. This test rests on the fact that the term $\sigma^2_{res} + b\sigma^2_{subj(A)}$ appears in the expressions for the expected mean squares in both the numerator and denominator.

When there are missing data, the coefficient associated with $\sigma^2_{subj(A)}$ changes, depending on the pattern of missing data, and the expected mean squares we have used to construct the F test statistic for factor A become[*]

$$E(MS_A) = \sigma^2_{res} + b'\sigma^2_{subj(A)} + \sigma^2_A \tag{9.4}$$

$$E(MS_{subj(A)}) = \sigma^2_{res} + b''\sigma^2_{subj(A)} \tag{9.5}$$

where $b' \neq b'' \neq b$. The precise values of b' and b'' depend on the pattern of missing data. Computation of the values of b' and b'' can be quite involved, but the results are available from some computer packages that do analysis of variance if you request that the expected mean squares be printed. Hence, the F test statistic we have been using to test for factor A main effects becomes

$$F = \frac{E(MS_A)}{E(MS_{subj(A)})} = \frac{\sigma^2_{res} + b'\sigma^2_{subj(A)} + \sigma^2_A}{\sigma^2_{res} + b''\sigma^2_{subj(A)}} \tag{9.6}$$

Because $b' \neq b''$, the denominator differs from the numerator by more than just the additive term σ^2_A, so this F test statistic *does not test the hypothesis of no factor A treatment effect*. It tests some other strange hypothesis that is of no practical interest.

Following the rules outlined above for determining the numerator and denominator of F, the proper F statistic to test for an effect of factor A would be

$$F = \frac{\sigma^2_{res} + b'\sigma^2_{subj(A)} + \sigma^2_A}{\sigma^2_{res} + b'\sigma^2_{subj(A)}} \tag{9.7}$$

In short, we need to adjust the F test statistic we compute from our data to take into account the fact that $b \neq b' \neq b''$ in the presence of missing

---

[*]The theory underlying this procedure is developed in G. A. Milliken and D. E. Johnson, *Analysis of Messy Data: Volume 1: Designed Experiments*, New York, Van Nostrand Reinhold, 1984, Chap. 28, "Analyzing Split-Plot and Certain Repeated Measures Experiments with Unbalanced and Missing Data."

data so that the F statistic will have the property that the numerator and denominator differ only by the variance associated with the treatment effect being tested, σ_A^2. We can obtain the information we need to do this correction from computer programs that print out the formulas for the expected mean squares and compute an appropriate F test statistic.

To obtain the F given in Eq. (9.7), first solve Eq. (9.5) for $\sigma_{\text{subj}(A)}^2$:

$$\sigma_{\text{subj}(A)}^2 = \frac{E(MS_{\text{subj}(A)}) - \sigma_{\text{res}}^2}{b''}$$

But, $E(MS_{\text{res}}) = \sigma_{\text{res}}^2$, so

$$\sigma_{\text{subj}(A)}^2 = \frac{E(MS_{\text{subj}(A)}) - E(MS_{\text{res}})}{b''}$$

Substitute from Eq. (9.4) and the results above into Eq. (9.7) to obtain

$$F = \frac{E(MS_A)}{E(MS_{\text{res}}) + b'[E(MS_{\text{subj}(A)}) - E(MS_{\text{res}})]/b''}$$

Rearranging,

$$F = \frac{E(MS_A)}{(1 - b'/b'')E(MS_{\text{res}}) + (b'/b'')E(MS_{\text{subj}(A)})}$$

Replace the expected mean squares with the mean squares estimated from the data to obtain

$$F = \frac{MS_A}{(1 - b'/b'')MS_{\text{res}} + (b'/b'')MS_{\text{subj}(A)}} \tag{9.8}$$

This test statistic follows an F distribution with $(a - 1)$ numerator degrees of freedom and approximate denominator degrees of freedom given by

$$\frac{[b'MS_{\text{subj}(A)} + (b'' - b')MS_{\text{res}}]^2}{[b'MS_{\text{subj}(A)}]^2/DF_{\text{subj}(A)} + [(b'' - b')MS_{\text{res}}]^2/DF_{\text{res}}} \tag{9.9}$$

Hence, we can use the reported mean squares and degrees of freedom, together with the values of b' and b'' obtained from the expected mean squares reported by a computer program, to compute F and the associated degrees of freedom when there are missing data in repeated-measures analysis of variance.

If there are only a few missing points, $b' \approx b'' \approx b$ and Eq. (9.8) approximates the ordinary F. Thus, if the missing data are such that the expected mean squares for random effects are not affected very much, the F computed directly from the estimated mean squares will not be much in error. In general, the more missing data, the greater the effect of this adjustment. However, given that we know there will be an error when there are missing data, we recommend that you always take the time to compute the proper F and its associated approximate denominator degrees of freedom according to Eqs. (9.8) and (9.9). We will now work an example to show you how to implement this procedure.

More on Drinking and Antisocial Personality

Table 9-12 shows the same set of data shown in Table 9-7, but missing 5 observations: subjects 3 and 5 in the non-ASP group and subject 10 in the ASP group did not complete the questionnaire when they were sober, and subjects 5 and 14 in the ASP group did not complete the questionnaire when drunk. We will now analyze these data with a general linear models routine,* declaring the subjects with an ASP diagnosis effect as random and issuing the necessary commands to obtain the correct marginal sums

TABLE 9-12 Data for Study of Aggressiveness before and after Drinking in Alcoholics Diagnosed as either ASP or Non-ASP; Missing Four Observations

Personality Type	Subject	Drinking Status	
		Sober	Drinking
Non-ASP	1	.81	.59
	2	.91	1.04
	3		1.11
	4	1.08	1.13
	5		
	6	1.16	1.16
	7	1.19	1.25
	8	1.44	1.70
ASP	9	.72	.83
	10		.99
	11	.89	1.17
	12	1.01	1.24
	13	1.10	1.33
	14	1.14	
	15	1.24	1.59
	16	1.34	1.73

*We could obtain the marginal sums of squares with repeated regression analyses, with each of the sets of dummy variables associated with a factor of interest entered last for one of the repetitions, just as we did for the ordinary two-way analysis of variance in Chap. 7. However, we also need the estimated coefficients of the variances due to random effects (subjects) in the expected mean squares.

```
                              GENERAL LINEAR MODELS PROCEDURE

DEPENDENT VARIABLE: S          Aggression score
```

SOURCE	DF	SUM OF SQUARES	MEAN SQUARE	F VALUE	PR > F	R-SQUARE	C.V.
MODEL	16	1.83331019	0.11458189	13.06	0.0001	0.954326	8.1875
ERROR	10	0.08774167	0.00877417		ROOT MSE		S MEAN
CORRECTED TOTAL	26	1.92105185			0.09367052		1.14407407

SOURCE	DF	TYPE I SS	F VALUE	PR > F	DF	TYPE III SS	F VALUE	PR > F
A	1	0.01361669	1.55	0.2413	1	0.00823253	0.94	0.3556
D	1	0.10851174	12.37	0.0056	1	0.14570417	16.61	0.0022
A*D	1	0.04521152	5.15	0.0466	1	0.07150417	8.15	0.0171
SUBJECT(A)	13	1.66597024	14.61	0.0001	13	1.66597024	14.61	0.0001

FIGURE 9-15 When there are missing data, the marginal sums of squares used to compute the hypothesis tests in repeated-measures analysis of variance with repeated measures on one factor (labeled Type III SS) are no longer equal to the incremental sums of squares (labeled Type I SS), except for the last factor entered in the regression model, in this case subjects within A [labeled Subject (A)]. This example shows the results of analyzing the aggression scores when there are some missing data, as shown in Table 9-12.

of squares for each effect and a printout of the expected mean squares so that we can obtain the values of b' and b''. The analysis of variance results are shown in Fig. 9-15, and the expected mean squares are shown in the lower half of Table 9-11. From the analysis of variance table, MS_A = .0082, $MS_{subj(A)}$ = .128, and MS_{res} = .0088. From the expected mean squares we see that b' = 1.659 and b'' = 1.769. Substitute these values into Eq. (9.8) to obtain

$$F = \frac{.0082}{(1 - 1.659/1.769).0088 + (1.659/1.769).128} = .07$$

With numerator degrees of freedom $(a - 1)$ = 1 and approximate denominator degrees of freedom given according to Eq. (9.9)

$$\frac{[1.659(.128) + (1.769 - 1.659).0088]^2}{[1.659(.128)]^2/13 + [(1.769 - 1.659).0088]^2/10} = 13.1$$

This result compares to the unadjusted value of F equal to .0082/.128 = .064 with 1 and 13 degrees of freedom.* The approximate F and the unadjusted F are in such close agreement because b' was not too different

---

*Notice that the F statistic listed in the computer output is .94. This is because, unless this computer program (SAS procedure GLM) is told otherwise, it computes all F statistics as though the factors were fixed, even though you have told it the subjects factor is random. It does not check to see what the appropriate denominator mean square is and simply calculates F with MS_{res} in the denominator.

from b'' (1.659 vs. 1.769). Because the approximate degrees of freedom computed in this manner will often not equal whole integers, you can either round *down* to the nearest whole integer or interpolate in Table E-2, Appendix E, to get an exact critical value for the approximate degrees of freedom.

We next obtain the F statistics for the remaining factors in the model, drinking status and the interaction between drinking and ASP diagnosis, whose expected mean squares involve only σ^2_{res} and fixed effects. The expected mean squares involved in testing these fixed factors do not depend on the random element associated with the repeated-measures aspect of the experimental design $\sigma^2_{subj(A)}$, so the missing data do not affect the construction of the F statistics used to test for these effects. We compute

$$F = \frac{MS_D}{MS_{res}} = \frac{.1457}{.0088} = 16.6$$

to test for an effect of drinking on aggression scores, and

$$F = \frac{MS_{AD}}{MS_{res}} = \frac{.0715}{.0088} = 8.1$$

to test for an interaction between ASP diagnosis and drinking on aggression scores. (These F values are the same as those shown in the computer output in Fig. 9-15.) Thus, we conclude that there is no main effect due to ASP diagnosis, but that drinking changes aggression scores: Both personality types become more aggressive when they drink. The interaction between drinking and ASP diagnosis indicates that drinking brings out a difference in aggressiveness between ASP and non-ASP alcoholics.

Multiple Comparisons

When there are missing data, multiple comparisons are complicated. In general, we can test for pairwise differences between any two row or column means using the procedures discussed for the empty cell situation in Chap. 8. The reason we have to consider this an empty cell situation is that, because there is only one observation per subject per treatment combination, missing any observation leaves an empty cell where that observation should have been. In addition, for testing across the fixed effect in this model, whose F statistic involves the random subjects effect, we must also take into account the difference in b' and b'', just as we had to adjust for this difference when computing the F statistic. This adjustment takes the form of computing the approximate degrees of freedom for the t statistic analogously to the procedure we just used to compute the approximate denominator degrees of freedom for F'. In this example, there are only two levels of the personality diagnosis effect, so we do not have any additional comparisons to make and we are off the hook.*

TWO-WAY ANALYSIS OF VARIANCE WITH REPEATED MEASURES ON BOTH FACTORS

In the example we just finished, there were two experimental factors (drinking status and personality type) with repeated measures on one of the factors (drinking status). Hence, the same subjects were observed when sober or drinking, but different subjects comprised the ASP and non-ASP groups. Because one of the factors was a personality trait, we obviously had to have different individuals in each of the two levels of the personality factor. However, there are occasions when it is possible to observe the same individuals under all possible combinations of the two treatments. This situation is known as a two-factor design with repeated measures on *both* factors. Table 9-13 shows a schematic presentation of this design. Comparing Table 9-13 with Table 9-5 illustrates the difference between this design and the one we just finished discussing: The same experimental subjects are exposed to both levels of factor A.

TABLE 9-13 Schematic of Experimental Design for Two-Way Repeated-Measures Analysis of Variance (Factors A and B) with Repeated Measures on Both Factors

		Factor B		
Factor A	Subjects	B_1	B_2	B_3
A_1	S_1 S_2 S_3 S_4			
A_2	S_1 S_2 S_3 S_4			

*For details of computing these pairwise comparisons, including examples, see G. A. Milliken and D. E. Johnson, *Analysis of Messy Data: Volume 1: Designed Experiments*, New York, Van Nostrand Reinhold, 1984, Chap. 28, "Analyzing Split-Plot and Certain Repeated Measures Experiments with Unbalanced and Missing Data."

As with all analysis of variance techniques, we will need to construct an F statistic to quantify how the variability between the means associated with the different levels of an experimental factor compares to the residual variation within the treatment groups. In nonrepeated-measures completely randomized designs, we used the random population variation within the treatment groups as the standard against which to decide that the variation associated with the treatments was "large." However, as with two-factor repeated-measures analysis of variance with repeated measures on only one factor, the presence of repeated measures complicates this situation, and we must consider the fact that subjects were observed repeatedly when deciding if the variation associated with the treatments is large enough to warrant rejecting the null hypothesis of no treatment effect.

Partitioning the Variability

As always, we use the total sum of squared deviations of the observations about the grand mean SS_{tot} as our measure of total variability in the data. As with the other repeated-measures designs, we first partition the variability into a component reflecting between-subjects variability $SS_{bet\ subj}$ and a component reflecting within-subjects variability $SS_{wit\ subj}$ (Fig. 9-16). The between-subjects variation arises because different subjects have different mean responses, independent of any treatment effects. Like the single-factor repeated-measures analysis of variance, all subjects receive all treatments, so the between-subjects variation contains no information about the treatment effects and we simply remove it from the analysis to reduce the residual variation and make it easier to detect treatment effects. This situation contrasts with the two-factor design with repeated measures on one factor, where the between-subjects variation contained information about one of the treatment effects because different subjects received different levels of that treatment. (Compare Fig. 9-16 with Figs. 9-6 and 9-9.)

Like the single-factor repeated-measures analysis of variance, we subdivide the variation within experimental subjects into components associated with the treatments in order to construct our hypothesis tests. The fact that we have repeated observations on all levels of all factors for all subjects leads to another difference with two-factor analysis of variance with repeated measures on only one factor: We can estimate the interaction between different subjects and different treatments as well as the interaction of the treatments with each other. We first subdivide the variation within subjects into three components, one associated with each of the two experimental factors and one associated with the interaction between the two experimental factors. Because of the interaction between the subjects and experimental factors, we can further subdivide each of these three elements into two parts.

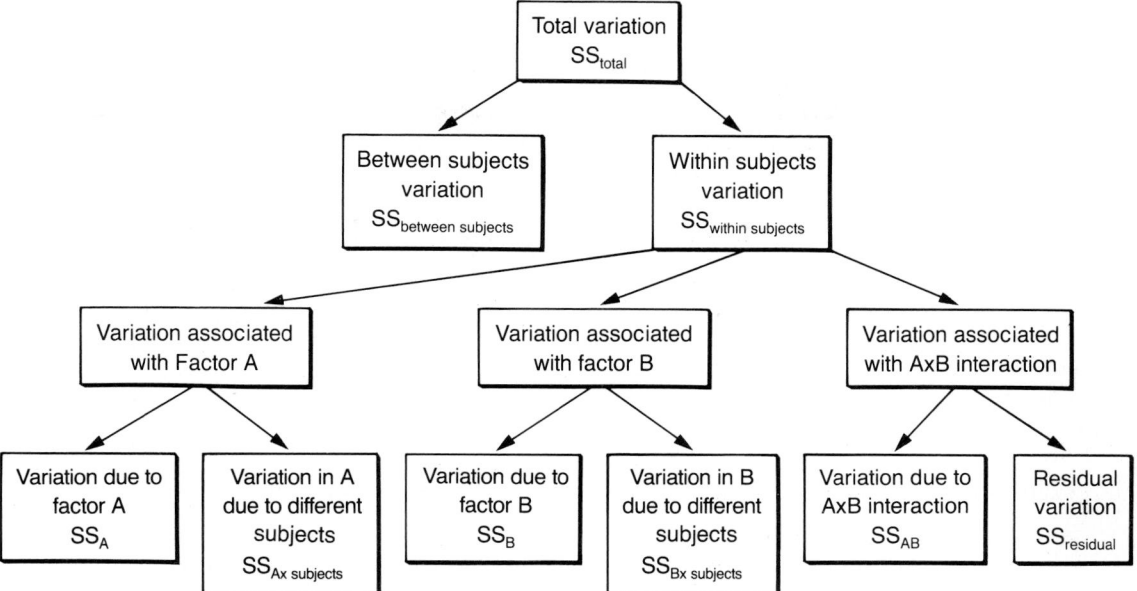

FIGURE 9-16 Partitioning of sums of squares for two-way analysis with repeated measures on both factors. As before, we first divide the total variability into between- and within-subjects components. However, in contrast to the situation where only one factor had repeated measures, we do a partitioning similar to that done in a one-way design. After "removing" the between-subjects variation, the within-subjects variation is further broken up into variations associated with each main effect, the interaction of the main effects, and subjects by factor interactions in order to construct the hypothesis tests. The degrees of freedom partition in the same way.

For example, consider factor A in Fig. 9-16. Part of the variation associated with factor A is due to the differences in mean value of the dependent variable across the different levels of A, computed without regard for the level of factor B and thus averaged across all levels of factor B and *all experimental subjects.* But different experimental subjects have different mean responses, independent of the effects of factor A. Because we have observations on all subjects during all treatments, we can estimate the difference in the response at any level of A for each subject averaged across all levels of B *for each experimental subject.* This difference simply represents the interaction between different subjects and different levels of factor A. We will test whether there are significant differences in the mean responses to different levels of factor A by comparing the variation due to the different levels of A with the variation at different levels of A we observe among different experimental subjects, the factor A by subjects interaction.

The variability associated with the different mean responses at different levels of A is measured by the sum of squared deviations of the factor A means about the grand mean of all the data, SS_A. The variability in the mean response of each subject around the mean value at each level of A (averaged across all levels of B) is the factor A by subjects interaction, $SS_{A \times \text{subj}}$. As always, before we can compare these two measures of variability, we need to convert them to variance estimates (mean squares) by dividing the sums of squares by the associated degrees of freedom. Specifically,

$$MS_A = \frac{SS_A}{DF_A}$$

and

$$MS_{A \times \text{subj}} = \frac{SS_{A \times \text{subj}}}{DF_{A \times \text{subj}}}$$

in which $DF_A = a - 1$ and $DF_{A \times \text{subj}} = (a - 1)(n - 1)$, where a is the number of levels of factor A and n is the number of experimental subjects. If the variability due to the change caused by exposing each subject to the levels of factor A, MS_A, is large compared to the variability in response of the subjects to the factor A treatments, $MS_{A \times \text{subj}}$, then we will conclude that it is unlikely that the change we observed was due to chance and that there is a statistically significant effect due to factor A. As before, we compute an F statistic to test the null hypothesis as

$$F = \frac{MS_A}{MS_{A \times \text{subj}}}$$

and compare this value with the critical value of the F distribution with DF_A numerator and $DF_{A \times \text{subj}}$ denominator degrees of freedom.

The second component of $SS_{\text{with subj}}$ arises from looking within subjects across the levels of the second factor (factor B) and partitions in exactly the same way as factor A into two components: SS_B is due to the variability in mean values associated with the different levels of factor B (averaged across all levels of factor A and all subjects), and $SS_{B \times \text{subj}}$ is due to the variability introduced because different subjects respond differently to the effects of factor B (averaged over all levels of factor A), the factor B by subjects interaction. Just as we computed an F ratio to test for a significant factor A effect, we compute

$$F = \frac{MS_B}{MS_{B \times \text{subj}}} = \frac{SS_B / DF_B}{SS_{B \times \text{subj}} / DF_{B \times \text{subj}}}$$

to see if the variability associated with factor B is large relative to the variability in responses to different subjects within levels of factor B.

$DF_B = (b - 1)$ and $DF_{B \times subj} = (b - 1)(n - 1)$ where there are b levels of factor B and n subjects. This value of F is compared to the theoretical value of the F distribution under the null hypothesis with DF_B and $DF_{B \times subj}$ numerator and denominator degrees of freedom to test for a significant effect of factor B.

Finally, the remaining components of $SS_{with\ subj}$ are a sum of squares due to a factor A by factor B interaction SS_{AB} and a residual sum of squares SS_{res}. We can test for significant interaction between factors A and B by computing

$$F = \frac{MS_{AB}}{MS_{res}} = \frac{SS_{AB}/DF_{AB}}{SS_{res}/DF_{res}}$$

in which $DF_{AB} = (a - 1)(b - 1)$ and $DF_{res} = (a - 1)(b - 1)(n - 1)$. Although this partitioning of sums of squares may look different from what we did for the two main effects (factors A and B), it is not. SS_{res} equals the sum of squares associated with the interaction between factor A by factor B and the different experimental subjects $SS_{A \times B \times subj}$. When there is only one observation on each subject at each treatment combination (there is no replication), $SS_{A \times B \times subj}$ is the only component of variability left when partitioning SS_{tot}, and so it is more convenient to refer to it as SS_{res}.

We now turn to a specific example to illustrate these ideas. In thinking about this example, pay special attention to the regression equation and the meaning of the various terms in it. Doing so should make the meaning of the various components of the sum of squares we have been discussing more concrete.

Candy, Chewing Gum, and Tooth Decay

When you eat candy, you fill your mouth with sugars. Then, bacteria that are normally present in your mouth ferment these sugars and cause rapid acidification (decrease in pH) of the plaque on your teeth. This low pH presumably contributes to tooth decay, cavity formation, and gum inflammation. Saliva has natural buffers (compounds that limit the change in pH) that reduce mouth acidity after eating candy. The easiest way to stimulate saliva production is to chew gum (sugarless, of course). Igarashi and his colleagues[*] reasoned that if the natural buffers in saliva were beneficial because they raised plaque pH and helped protect against tooth decay and gum disease, additional benefit could be gained by chewing gum that also contained the additional buffer sodium bicarbonate.

[*]K. Igarashi, I. K. Lee, and C. F. Schachtele, "Effect of Chewing Gum Containing Sodium Bicarbonate on Human Interproximal Plaque pH," *J. Dent. Res.* 67:531–535, 1988.

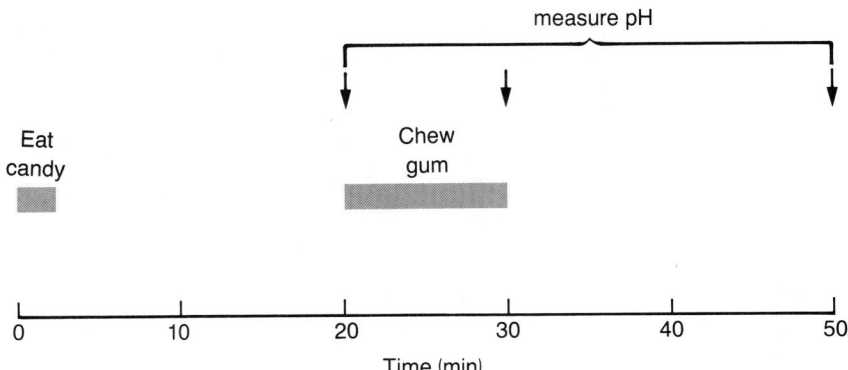

FIGURE 9-17 Schematic representation of the design of the study of mouth pH and the response to chewing different kinds of gum after eating candy. This is a repeated-measures experiment because each experimental subject is observed under all experimental conditions.

To test this reasoning, they placed pH-measuring electrodes in eight volunteers' mouths to measure plaque pH after eating candy and chewing two different kinds of gum (a standard sugarless gum and a gum containing sodium bicarbonate). The experiment consisted of eating a piece of candy, waiting 20 min, then chewing the test gum for 10 min (Fig. 9-17). Igarashi and his colleagues measured plaque pH 20 min after eating the candy (immediately before chewing the gum), 30 min after eating the candy (immediately after chewing the gum), and 50 min after eating the candy (20 min after chewing the gum). All eight subjects chewed both types of gum. Four people were randomly selected to chew the standard sugarless gum first; the remaining four chewed the bicarbonate gum first. This randomization was designed to avoid biases that might be associated with the order in which the two test gums were chewed. In addition, the two sets of data were collected one day apart, to give time for any effects of the candy or gum chewing from the previous day to dissipate and reduce carry-over effects.

Table 9-14 shows the data from this study. Every one of the eight people (subjects) chewed both the sugarless and bicarbonate gum and had his or her mouth pH measured at three times (20, 30, and 50 min after eating the candy). There are two factors: type of gum (standard or bicarbonate) and time (20, 30, and 50 min). Because all subjects chewed both kinds of gum and were measured at all three times, there are repeated measures on both factors.

We approach the regression implementation of this analysis just like we did before. We need to define dummy variables to quantify the different treatment effects and different experimental subjects. We also need to include all the interaction terms. The magnitudes of the regression coefficients and associated sums of squares and degrees of freedom will

TABLE 9-14 Tooth Plaque pH Measured 20, 30, and 50 min after Eating Candy and Chewing One of Two Gums, a Standard Sugarless Gum and a Sodium Bicarbonate–Containing Gum

Gum	Subject	Time after Eating Candy, min		
		20	30	50
Standard	1	4.6	4.4	4.1
	2	4.2	4.7	4.7
	3	3.9	5.0	5.7
	4	4.6	5.2	5.6
	5	4.6	5.1	5.7
	6	4.7	5.3	5.3
	7	4.9	5.6	5.9
	8	5.0	6.0	5.9
Bicarbonate	1	3.8	5.3	5.1
	2	4.1	5.8	5.1
	3	4.2	5.8	5.6
	4	4.2	6.9	5.6
	5	4.3	6.2	5.8
	6	4.4	6.3	5.9
	7	4.9	6.0	5.9
	8	4.5	7.2	6.2

provide the information we need to estimate the sizes of the various treatment effects and do hypothesis tests.

As always, we need one less dummy variable than there are experimental conditions. Therefore, we use effects coding to define a single dummy variable to encode the two gums:

$$G = \begin{cases} -1 \text{ if bicarbonate in gum} \\ 1 \text{ if standard sugarless gum} \end{cases}$$

Likewise, two dummy variables encode the three time points:

$$T_{20} = \begin{cases} 1 \text{ if 20 min after eating candy} \\ 0 \text{ if 30 min after eating candy} \\ -1 \text{ if 50 min after eating candy} \end{cases}$$

and

$$T_{30} = \begin{cases} 1 \text{ if 30 min after eating candy} \\ 0 \text{ if 20 min after eating candy} \\ -1 \text{ if 50 min after eating candy} \end{cases}$$

Because all subjects received all treatment combinations, we do not have to segregate the between-subjects effects as we did when we had repeated measures on only one of the two factors. Hence, we define seven dummy variables to describe the differences between the eight subjects:

$$S_i = \begin{cases} 1 \text{ if subject } i \quad i < 8 \\ -1 \text{ if subject } 8 \\ 0 \text{ otherwise} \end{cases}$$

We are now ready to write the regression equation to implement the analysis of variance using these dummy variables. This regression equation includes the dummy variables defined above, as well as the cross-product terms $(GT_{20}, GT_{30}, S_iG, S_iT_{20},$ and $S_iT_{30}, i = 1, \ldots, 7)$ necessary to account for the various interaction effects between the experimental effects and the different subjects. There are a total of 33 dummy variables: the 3 defined to quantify the treatment effects, the 7 defined to allow for between-subjects differences, and 23 interactions, which represent products between the other variables. The regression equation is

$$\hat{pH} = b_0 + b_G G + b_{T_{20}} T_{20} + b_{T_{30}} T_{30} + b_{GT_{20}} GT_{20} + b_{GT_{30}} GT_{30}$$

$$+ \sum_{i=1}^{7} b_i S_i + \sum_{i=1}^{7} b_{S_iG} S_iG + \sum_{i=1}^{7} b_{S_iT_{20}} S_iT_{20} + \sum_{i=1}^{7} b_{S_iT_{30}} S_iT_{30}$$

in which $\hat{pH}$ is the predicted pH for any given subject and set of experimental conditions. Table 9-15 shows the data as they will be submitted for regression analysis.

Before continuing, let us rewrite this equation and discuss what the various coefficients mean. This discussion will lay the foundation for the computations we will do below as well as help illustrate the logic for the partitioning of sums of squares in Fig. 9-16.

$$\hat{pH} = b_0 + \sum_{i=1}^{7} b_i S_i$$

$$+ \left(b_G G + \sum_{i=1}^{7} b_{S_iG} S_iG \right)$$

$$+ \left[(b_{T_{20}} T_{20} + b_{T_{30}} T_{30}) + \left(\sum_{i=1}^{7} b_{S_iT_{20}} S_iT_{20} + \sum_{i=1}^{7} b_{S_iT_{30}} S_iT_{30} \right) \right]$$

$$+ (b_{GT_{20}} GT_{20} + b_{GT_{30}} GT_{30}) \tag{9.10}$$

The first line in this equation contains the regression constant b_0 and the between-subjects coefficients b_i. b_0 estimates the overall mean pH. The seven b_i coefficients estimate the average difference between

TABLE 9-15 Plaque pH Values from Gum Study with Effects Dummy Variable Codes for Time Factor (T_{20} and T_{30}), Gum Type (G), and Subjects (S_i)

pH	G	T_{20}	T_{30}	S_1	S_2	S_3	S_4	S_5	S_6	S_7
4.6	1	1	0	1	0	0	0	0	0	0
4.2	1	1	0	0	1	0	0	0	0	0
3.9	1	1	0	0	0	1	0	0	0	0
4.6	1	1	0	0	0	0	1	0	0	0
4.6	1	1	0	0	0	0	0	1	0	0
4.7	1	1	0	0	0	0	0	0	1	0
4.9	1	1	0	0	0	0	0	0	0	1
5.0	1	1	0	-1	-1	-1	-1	-1	-1	-1
4.4	1	0	1	1	0	0	0	0	0	0
4.7	1	0	1	0	1	0	0	0	0	0
5.0	1	0	1	0	0	1	0	0	0	0
5.2	1	0	1	0	0	0	1	0	0	0
5.1	1	0	1	0	0	0	0	1	0	0
5.3	1	0	1	0	0	0	0	0	1	0
5.6	1	0	1	0	0	0	0	0	0	1
6.0	1	0	1	-1	-1	-1	-1	-1	-1	-1
4.1	1	-1	-1	1	0	0	0	0	0	0
4.7	1	-1	-1	0	1	0	0	0	0	0
5.7	1	-1	-1	0	0	1	0	0	0	0
5.6	1	-1	-1	0	0	0	1	0	0	0
5.7	1	-1	-1	0	0	0	0	1	0	0
5.3	1	-1	-1	0	0	0	0	0	1	0
5.9	1	-1	-1	0	0	0	0	0	0	1
5.9	1	-1	-1	-1	-1	-1	-1	-1	-1	-1
3.8	-1	1	0	1	0	0	0	0	0	0
4.1	-1	1	0	0	1	0	0	0	0	0
4.2	-1	1	0	0	0	1	0	0	0	0
4.2	-1	1	0	0	0	0	1	0	0	0
4.3	-1	1	0	0	0	0	0	1	0	0
4.4	-1	1	0	0	0	0	0	0	1	0
4.9	-1	1	0	0	0	0	0	0	0	1
4.5	-1	1	0	-1	-1	-1	-1	-1	-1	-1
5.3	-1	0	1	1	0	0	0	0	0	0
5.8	-1	0	1	0	1	0	0	0	0	0
5.8	-1	0	1	0	0	1	0	0	0	0
6.9	-1	0	1	0	0	0	1	0	0	0
6.2	-1	0	1	0	0	0	0	1	0	0
6.3	-1	0	1	0	0	0	0	0	1	0
6.0	-1	0	1	0	0	0	0	0	0	1
7.2	-1	0	1	-1	-1	-1	-1	-1	-1	-1
5.1	-1	-1	-1	1	0	0	0	0	0	0
5.1	-1	-1	-1	0	1	0	0	0	0	0
5.6	-1	-1	-1	0	0	1	0	0	0	0

(continued on page 440)

TABLE 9-15 (*continued*) **Plaque pH Values from Gum Study with Effects Dummy Variable Codes for Time Factor (T_{20} and T_{30}), Gum Type (G), and Subjects (S_i)**

pH	G	T_{20}	T_{30}	S_1	S_2	S_3	S_4	S_5	S_6	S_7
5.6	-1	-1	-1	0	0	0	1	0	0	0
5.8	-1	-1	-1	0	0	0	0	1	0	0
5.9	-1	-1	-1	0	0	0	0	0	1	0
5.9	-1	-1	-1	0	0	0	0	0	0	1
6.2	-1	-1	-1	-1	-1	-1	-1	-1	-1	-1

the mean pH of all subjects and the response of subject i. The sum of squares associated with the b_i is the between-subjects sum of squares in Fig. 9-16.

The second line in Eq. (9.10) contains the terms related to the type of gum and the interaction of gum with the subjects dummy variables. The sum of squares associated with the terms on this line quantify the total variation associated with the gum experimental factor (analogous to factor A in Fig. 9-16). We can divide this sum of squares into two components, one associated with the average effect of the treatment and another associated with the differences in the mean responses of different experimental subjects. b_G estimates the change in pH from the overall mean associated with chewing the bicarbonate gum, averaged across all the subjects and all the times. The sum of squares associated with this dummy variable equals the treatment sum of squares associated with the different forms of chewing gum SS_G (which is analogous to SS_A in Fig. 9-16). The interaction term S_iG is 1 for subject i when he or she is chewing the standard gum (or -1 for subject 8) and 0 otherwise. Hence, the coefficient b_{S_iG} is an estimate of how much subject i's response to chewing the standard gum deviates from the average of all subjects chewing standard gum. The sum of the sums of squares and degrees of freedom associated with these seven dummy variables equals the interaction sum of squares between the different gums and the different subjects, $SS_{G \times \text{subj}}$ (which is analogous to $SS_{A \times \text{subj}}$ in Fig. 9-16).

The third line in Eq. (9.10) provides analogous information for the time effect. Because there are three times, there are two dummy variables (T_{20} and T_{30}) to quantify the main time effects and 14 dummy variables to quantify the interaction of time with the different experimental subjects. The sums of squares associated with the terms on this line represent all the variations associated with time (analogous to factor B in Fig. 9-16). The coefficient $b_{T_{20}}$ estimates how much the pH, 20 min after eating the candy and averaged over both kinds of gum and all people, deviates from the overall mean pH. $b_{T_{30}}$ provides the same information for 30 min after eating the candy. The sum of squares associated with these two terms equals the variation due to the time factor SS_T (analogous to SS_B in Fig. 9-16). As before, the values of the interaction coefficients $b_{S_iT_{20}}$ and $b_{S_iT_{30}}$ quantify how much the mean pH of subject i deviates from the overall mean of all subjects at times 20 and 30 min, respectively. The sums of squares associated with these variables yield the interaction sum of squares $SS_{T \times subj}$ (analogous to $SS_{B \times subj}$ in Fig. 9-16).

Finally, the last line in Eq. (9.10) quantifies the interaction between gum and time, given by $b_{GT_{20}}$ and $b_{GT_{30}}$. The sum of squares associated with $b_{GT_{20}}$ and $b_{GT_{30}}$ is the interaction sum of squares $SS_{G \times T}$ (analogous to $SS_{A \times B}$ in Fig. 9-16). This last line does not include the interaction between the gum by time interaction and the individual subjects because we are considering that source of variation to be the residual sum of squares SS_{res}, which is what is left of the total sum of squares after taking into account the sources of variability due to gum, gum by subjects, time, time by subjects, and gum by time. (If we had included the gum by time by subjects interactions, we would have added another 14 dummy variables and had a fully specified regression equation in which the number of parameters would have equaled the number of observations. This model would perfectly describe the data with no residual error.)

Figure 9-18 shows the regression output from fitting Eq. (9.10) to the data in Tables 9-14 and 9-15. Figure 9-19 shows the mean values of pH observed under each combination of experimental conditions, together with the values of pH predicted by the regression equation.

To test whether there are significant differences between the two gums, over time, and for interactions between gum and time, we extract the sums of squares needed to compute the F ratios from the regression output in Fig. 9-18. The sum of squares associated with G is $SS_G = 1.47$ and is associated with 1 degree of freedom, so the mean square due to the two gums is

$$MS_G = \frac{SS_G}{DF_G} = \frac{1.47}{1} = 1.47 \tag{9.11}$$

The mean square due to the gum by subjects interaction is obtained by adding up the sums of squares associated with the seven interaction variables b_{S_iG}, each with 1 degree of freedom,

$$MS_{G \times subj} = \frac{SS_{G \times subj}}{DF_{G \times subj}} = \frac{\Sigma SS_{S_iG}}{\Sigma DF_{S_iG}}$$

$$= \frac{\begin{matrix}.00083 + .01361 + .00347 + .00408 \\ + .00939 + .0048 + .08048\end{matrix}}{1 + 1 + 1 + 1 + 1 + 1 + 1} \tag{9.12}$$

$$= \frac{.1167}{7} = .0167$$

```
DEP VARIABLE: PH        plaque pH

                                    ANALYSIS OF VARIANCE

                                 SUM OF          MEAN
                  SOURCE    DF   SQUARES        SQUARE      F VALUE      PROB>F

                  MODEL     33   28.12458333   0.85226010    9.665      0.0001
                  ERROR     14    1.23458333   0.08818452
                  C TOTAL   47   29.35916667

                     ROOT MSE      0.2969588    R-SQUARE     0.9579
                     DEP MEAN      5.204167     ADJ R-SQ     0.8588
                     C.V.          5.706174

                                   PARAMETER ESTIMATES
```

VARIABLE	DF	PARAMETER ESTIMATE	STANDARD ERROR	T FOR H0: PARAMETER=0	PROB > |T|	TYPE I SS	VARIABLE LABEL
INTERCEP	1	5.20416667	0.04286231	121.416	0.0001	1300.00083	INTERCEPT
S1	1	-0.654167	0.11340301	-5.769	0.0001	4.68750000	dummy variable =1 if subject 1
S2	1	-0.4375	0.11340301	-3.858	0.0017	0.66694444	dummy variable =1 if subject 2
S3	1	-0.170833	0.11340301	-1.506	0.1542	0.0001388889	dummy variable =1 if subject 3
S4	1	0.14583333	0.11340301	1.286	0.2193	0.46875000	dummy variable =1 if subject 4
S5	1	0.07916667	0.11340301	0.698	0.4965	0.16805556	dummy variable =1 if subject 5
S6	1	0.11250000	0.11340301	0.992	0.3380	0.17813492	dummy variable =1 if subject 6
S7	1	0.32916667	0.11340301	2.903	0.0116	0.74297619	dummy variable =1 if subject 7
GS1	1	-0.00833333	0.11340301	-0.073	0.9425	0.0008333333	interaction of gum and subject1
GS2	1	-0.0583333	0.11340301	-0.514	0.6150	0.01361111	
GS3	1	0.008333333	0.11340301	0.073	0.9425	0.003472222	
GS4	1	-0.0416667	0.11340301	-0.367	0.7188	0.004083333	
GS5	1	0.025	0.11340301	0.220	0.8287	0.009388889	
GS6	1	-0.0416667	0.11340301	-0.367	0.7188	0.004801587	
GS7	1	0.10833333	0.11340301	0.955	0.3556	0.08047619	
G	1	-0.175	0.04286231	-4.083	0.0011	1.47000000	dummy variable =-1 if bicarb gum
T20	1	-0.772917	0.06061646	-12.751	0.0001	9.24500000	dummy variable = 1 if 20 min
T30	1	0.47083333	0.06061646	7.767	0.0001	5.32041667	dummy variable = 1 if 30 min
GT20	1	0.30625000	0.06061646	5.052	0.0002	0.60500000	interaction of gum and T20
GT30	1	-0.3375	0.06061646	-5.568	0.0001	2.73375000	interaction of gum and T30
T20S1	1	0.42291667	0.16037608	2.637	0.0195	0.40500000	interaction of T20 and Subject1
T20S2	1	0.15625000	0.16037608	0.974	0.3465	0.006666667	
T20S3	1	-0.210417	0.16037608	-1.312	0.2106	0.46020833	
T20S4	1	-0.177083	0.16037608	-1.104	0.2881	0.028125	
T20S5	1	-0.0604167	0.16037608	-0.377	0.7120	0.05208333	
T20S6	1	0.00625	0.16037608	0.039	0.9695	0.001488095	
T20S7	1	0.13958333	0.16037608	0.870	0.3988	0.006428571	
T30S1	1	-0.170833	0.16037608	-1.065	0.3048	0.37500000	
T30S2	1	0.0125	0.16037608	0.078	0.9390	0.008888889	
T30S3	1	-0.104167	0.16037608	-0.650	0.5265	0.05840278	
T30S4	1	0.22916667	0.16037608	1.429	0.1750	0.10837500	
T30S5	1	-0.104167	0.16037608	-0.650	0.5265	0.06669444	
T30S6	1	0.0125	0.16037608	0.078	0.9390	0.0009722222	
T30S7	1	-0.204167	0.16037608	-1.273	0.2237	0.14291667	interaction of T30 and Subject7

FIGURE 9-18 Results of a regression implementation of a two-way, repeated-measures analysis of variance for the study of changes in plaque pH following chewing different kinds of gum. The dummy variables for the treatment effects are entered *after* the dummy variables which describe between-subjects differences.

To test whether there are significant differences in mean pH between the two gums, we compute

$$F = \frac{MS_G}{MS_{G \times \text{subj}}} = \frac{1.47}{.0167} = 88.02 \tag{9.13}$$

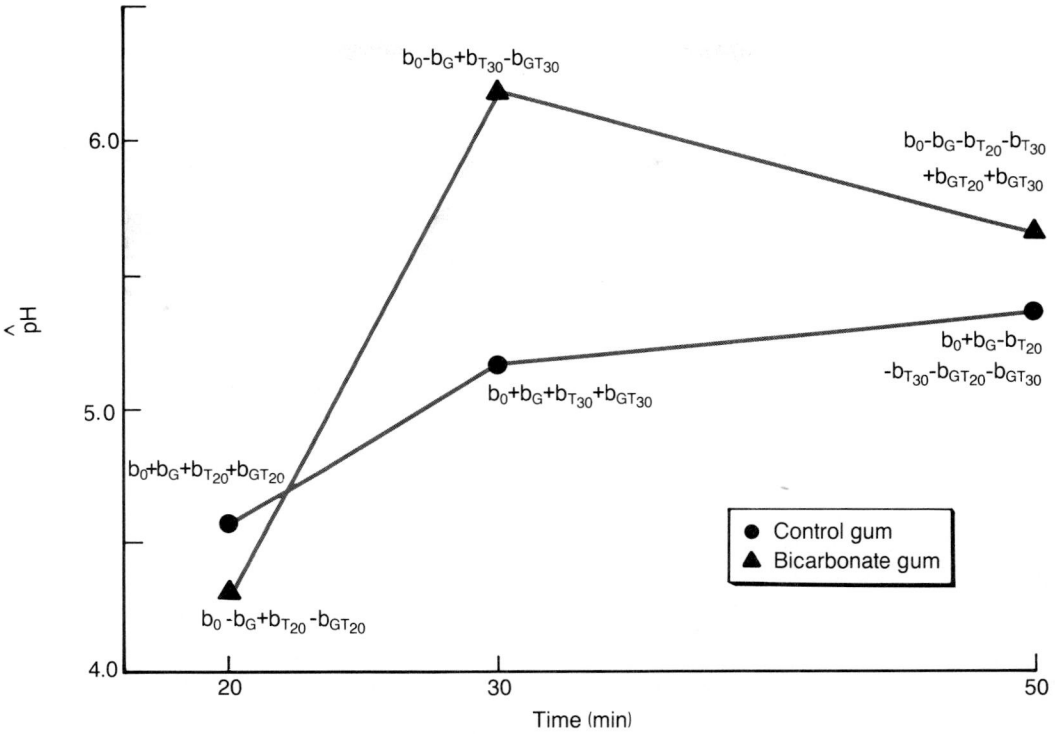

FIGURE 9-19 The predicted value of mouth pH under each set of experimental conditions can be computed from the regression model for the analysis of variance by substituting in the appropriate values of the dummy variables.

This F ratio exceeds the critical value of F with 1 and 7 degrees of freedom for $\alpha = 0.01$, 12.25, so we conclude that there is a significant difference in plaque pH when chewing bicarbonate gum compared to chewing standard sugarless gum.

We do similar computations using the dummy variables associated with the time variables T_{20} and T_{30} and the interactions between these two variables and the experimental subjects, $S_i T_{20}$ and $S_i T_{30}$, to test for significant differences over time:

$$MS_T = \frac{SS_T}{DF_T} = \frac{SS_{T_{20}} + SS_{T_{30}}}{DF_{T_{20}} + DF_{T_{30}}} \tag{9.14}$$

Use these sums of squares from the regression output in Fig. 9-18 to obtain

$$MS_T = \frac{9.245 + 5.32042}{1 + 1} = \frac{14.56542}{2} = 7.2827$$

Likewise,

$$MS_{T \times \text{subj}} = \frac{SS_{T \times \text{subj}}}{DF_{T \times \text{subj}}} = \frac{\Sigma SS_{S_i T_{20}} + \Sigma SS_{S_i T_{30}}}{\Sigma DF_{S_i T_{20}} + \Sigma DF_{S_i T_{30}}}$$

$$= \frac{\begin{matrix}(.40500 + .00667 + .46021 + .02813 \\ + .05208 + .00149 + .00643) \\ + (.37500 + .00889 + .05840 + .10838 \\ + .06669 + .00097 + .14292)\end{matrix}}{\begin{matrix}(1 + 1 + 1 + 1 + 1 + 1 + 1) \\ + (1 + 1 + 1 + 1 + 1 + 1 + 1)\end{matrix}} \qquad (9.15)$$

$$= \frac{1.7213}{14} = .123$$

To test for significant differences over time, we compute

$$F = \frac{MS_T}{MS_{T \times \text{subj}}} = \frac{7.2827}{.123} = 59.21$$

This value of F exceeds the critical value of F for 2 and 14 degrees of freedom and $\alpha = .01$, 6.51, and so we conclude that there was a significant change in pH with time after starting gum chewing.

Finally, we test for a significant interaction effect between gum chewing and time (i.e., test whether the pH response over time after chewing standard gum parallels the response to chewing gum to which bicarbonate has been added) by computing

$$MS_{GT} = \frac{SS_{GT}}{DF_{GT}} = \frac{SS_{GT_{20}} + SS_{GT_{30}}}{DF_{GT_{20}} + DF_{GT_{30}}} = \frac{.605 + 2.73375}{1 + 1} = \frac{3.33875}{2} = 1.6694 \qquad (9.16)$$

The residual mean square is

$$MS_{\text{res}} = \frac{SS_{\text{res}}}{DF_{\text{res}}} = \frac{1.23458}{14} = .08818$$

and the F statistic to test for interaction between time and gum is

$$F = \frac{MS_{GT}}{MS_{\text{res}}} = \frac{1.6694}{.08818} = 18.93$$

This value of F exceeds the critical value of F with 2 numerator and 14 denominator degrees of freedom with $\alpha = .01$, 6.51, so we conclude that there is a significant interaction effect between gum and time. Figure 9-19 reflects this interaction; the time courses of change in pH for the two gums do not parallel one another.

We conclude that pH changes after chewing gum (significant time effect), that chewing bicarbonate gum raises plaque pH more than chewing standard gum (significant gum effect), and that pH rises faster over time when chewing bicarbonate gum than when chewing standard gum (significant gum by time interaction effect). Hence it is possible to augment the natural buffers in saliva by adding bicarbonate to chewing gum. These results suggest that chewing bicarbonate gum changes plaque pH in a way that may help prevent cavities and gum disease.

On reflection, the fact that we found a significant interaction effect is not surprising. The model we used [Eq. (9.10)] assumes that the effects of the different gums and time are additive at all times, particularly at time zero. In other words, in the absence of an interaction between gum and time, we assume that the changes in pH over time are the same for both gums (so the lines in Fig. 9-19 would be parallel). This may be a reasonable assumption at times 30 and 50 min, after the people started chewing the gum, but it is not reasonable at time 20 min, which is *before* the people started chewing the gum. Because the order in which people chewed the gum was randomized, there should not have been any difference between the two gum groups at time 20 min. This requirement, combined with the fact that there were differences in pH after chewing the different gums, led to the significant interaction effect.

This situation—in which all people are observed during a control state before some intervention—is very common in clinical research. When approached as a straightforward repeated-measures analysis of variance, this fact often leads to a significant interaction effect, particularly when one of the factors is time, as it is in this study.

Multiple Comparisons

In this example there are only two levels of the bicarbonate factor, so there is no need to conduct multiple comparisons on this factor. When there are more than two levels of a factor, such as with the time factor in this example, you can conduct multiple comparisons among different means within either factor using any of the methods we have developed so far. To do so, simply use the same mean square and associated degrees of freedom as was used in the denominator of the F test for that factor.[*]

---

[*]The fact that the mean squares associated with the denominators of the F tests for the two different factors are different complicates the task of doing multiple comparisons on *all* treatment means. As was true with two-way repeated-measures analysis of variance with repeated measures on one factor, one must compute an "average" mean square to use in the denominator of the t, q, or q' test statistics. For details on how to do these computations, see B. J. Winer, *Statistical Principles in Experimental Design*, New York, McGraw-Hill, 1971, pp. 528–532.

MISSING DATA IN REPEATED MEASURES
ON BOTH OF TWO FACTORS

The problems of missing data in this two-way design with repeated measures on both factors are identical to those we previously encountered for the two-way design with repeated measures on only one factor. Any F statistic computed from expected mean squares that includes a variance due to a random effect (other than σ_{res}^2) will be incorrect when there are missing data because the missing data cause the weights of the variances due to the random effect to be unequal. In this design, we have two such F statistics, one to test for a factor A effect and one to test for a factor B effect (when there were repeated measures on only factor B, only the F statistic for factor A was affected). To understand how we need to adjust the computations of the associated F test statistics, we again need to consider why these test statistics are constructed as they are in terms of the expected mean squares.

Expected Mean Squares

Two-way analysis of variance with repeated measures on both factors is equivalent to a three-way analysis of variance with two fixed factors (A and B) and one random factor (the subjects). In contrast to when we had repeated measures on only one of the two factors, there is no nesting of the subjects within another factor in this design. The expected mean squares consist of sums of variances attributable to the fixed and random effects, and the variances associated with the random effects have coefficients relating to the number of levels of the fixed factors.

The essential features of the expected mean squares for the general two-way analysis of variance with repeated measures on both factors are

$$E(MS_A) = \sigma_{res}^2 + b\sigma_{A \times subj}^2 + \sigma_A^2$$

$$E(MS_{A \times subj}) = \sigma_{res}^2 + b\sigma_{A \times subj}^2$$

$$E(MS_B) = \sigma_{res}^2 + a\sigma_{B \times subj}^2 + \sigma_B^2$$

$$E(MS_{B \times subj}) = \sigma_{res}^2 + a\sigma_{B \times subj}^2$$

$$E(MS_{AB}) = \sigma_{res}^2 + \sigma_{AB}^2$$

$$E(MS_{res}) = \sigma_{res}^2$$

where a is the number of levels of factor A and b is the number of levels of factor B. As always, we compute an F statistic for testing a given factor effect by dividing the expected mean square containing the variance due to that effect with the expected mean square that contains everything in the numerator except the variance due to the effect being tested. For example, in terms of expected mean squares, the F statistic for factor A is

$$F = \frac{\sigma^2_{\text{res}} + b\sigma^2_{A \times \text{subj}} + \sigma^2_A}{\sigma^2_{\text{res}} + b\sigma^2_{A \times \text{subj}}} = \frac{E(MS_A)}{E(MS_{A \times \text{subj}})}$$

the F statistic for factor B is

$$F = \frac{\sigma^2_{\text{res}} + a\sigma^2_{B \times \text{subj}} + \sigma^2_B}{\sigma^2_{\text{res}} + a\sigma^2_{B \times \text{subj}}} = \frac{E(MS_B)}{E(MS_{B \times \text{subj}})}$$

and, finally, the F statistic for the factor A by factor B interaction is

$$F = \frac{\sigma^2_{\text{res}} + \sigma^2_{AB}}{\sigma^2_{\text{res}}} = \frac{E(MS_{AB})}{E(MS_{\text{res}})}$$

Both of the F statistics for the main effects of factors A and B involve variances attributable to the random factor, subjects. Variance due to subjects does not enter into the expected mean squares directly, but through the subjects by factor interactions. In a mixed-model analysis of variance, the interaction of a fixed factor and a random factor is considered random.

Figure 9-20 shows the computer output from an analysis of the data in Table 9-15 using a traditional mixed-model analysis of variance on the fixed effects of gum (analogous to factor A) and time (analogous to factor

GENERAL LINEAR MODELS PROCEDURE

DEPENDENT VARIABLE: PH   plaque pH

SOURCE	DF	SUM OF SQUARES	MEAN SQUARE	F VALUE	PR > F	R-SQUARE	C.V.
MODEL	33	28.12458333	0.85226010	9.66	0.0001	0.957949	5.7062
ERROR	14	1.23458333	0.08818452		ROOT MSE		PH MEAN
CORRECTED TOTAL	47	29.35916667			0.29695879		5.20416667

SOURCE	DF	TYPE I SS	F VALUE	PR > F	DF	TYPE III SS	F VALUE	PR > F
G	1	1.47000000	16.67	0.0011	1	1.47000000	16.67	0.0011
T	2	14.56541667	82.58	0.0001	2	14.56541667	82.58	0.0001
G*T	2	3.33875000	18.93	0.0001	2	3.33875000	18.93	0.0001
G*SUBJ	14	7.02916667	5.69	0.0012	7	0.11666667	0.19	0.9831
T*SUBJ	14	1.72125000	1.39	0.2711	14	1.72125000	1.39	0.2711
SUBJ	0	0.00000000	.	.	7	6.91250000	11.20	0.0001

FIGURE 9-20 Analysis of the repeated-measures study of plaque pH after chewing different kinds of gum computed using the SAS GLM procedure. Because there are no missing observations, the marginal sums of squares (labeled Type III SS) equal the incremental sums of squares (labeled Type I SS). Compare the regression implementation of the analysis of these same data shown in Fig. 9-18.

TABLE 9-16 Expected Mean Squares for Plaque pH and Gum Chewing Data

NO MISSING DATA

$$E(MS_G) = \sigma^2_{res} + 3\sigma^2_{G \times subj} + \sigma^2_G$$
$$E(MS_{G \times subj}) = \sigma^2_{res} + 3\sigma^2_{G \times subj}$$
$$E(MS_T) = \sigma^2_{res} + 2\sigma^2_{T \times subj} + \sigma^2_T$$
$$E(MS_{T \times subj}) = \sigma^2_{res} + 2\sigma^2_{T \times subj}$$
$$E(MS_{GT}) = \sigma^2_{res} + \sigma^2_{GT}$$
$$E(MS_{res}) = \sigma^2_{res}$$

MISSING DATA

$$E(MS_G) = \sigma^2_{res} + 2.333\sigma^2_{G \times subj} + \sigma^2_G$$
$$E(MS_{G \times subj}) = \sigma^2_{res} + 2.714\sigma^2_{G \times subj}$$
$$E(MS_T) = \sigma^2_{res} + 1.750\sigma^2_{T \times subj} + \sigma^2_T$$
$$E(MS_{T \times subj}) = \sigma^2_{res} + 1.857\sigma^2_{T \times subj}$$
$$E(MS_{GT}) = \sigma^2_{res} + \sigma^2_{GT}$$
$$E(MS_{res}) = \sigma^2_{res}$$

B), and the random effect, subjects. The sums of squares and degrees of freedom are identical to those we computed above from the output shown in Fig. 9-18. The expected mean squares for these data, shown in Table 9-16, are what we expect from the discussion above.

What Happens to the Expected Mean Squares When There Are Missing Data?

When there are missing data, the expected mean squares involving the random effects of subjects become

$$E(MS_A) = \sigma^2_{res} + b'\sigma^2_{A \times subj} + \sigma^2_A$$

$$E(MS_{A \times subj}) = \sigma^2_{res} + b''\sigma^2_{A \times subj}$$

where $b' \neq b'' \neq b$, and

$$E(MS_B) = \sigma^2_{res} + a'\sigma^2_{B \times subj} + \sigma^2_B$$

$$E(MS_{B \times subj}) = \sigma^2_{res} + a''\sigma^2_{B \times subj}$$

where $a' \neq a'' \neq a$. Hence, when we try to compute the F statistics developed above for factors A and B, the denominators are incorrect; they do not match the numerators because $a' \neq a''$ and $b' \neq b''$. Thus, we need to compute adjusted F statistics and approximate denominator degrees of freedom analogous to Eqs. (9.8) and (9.9).

Following the procedure developed above when we had repeated measures on only one of the two factors, we test for factor A, by computing

$$F = \frac{MS_A}{(1 - b'/b'')MS_{\text{res}} + (b'/b'')MS_{A \times \text{subj}}} \tag{9.17}$$

which follows an approximate F distribution with $(a - 1)$ numerator degrees of freedom and approximate denominator degrees of freedom given by

$$\frac{[b'MS_{A \times \text{subj}} + (b'' - b')MS_{\text{res}}]^2}{(b'MS_{A \times \text{subj}})^2/DF_{A \times \text{subj}} + [(b'' - b')MS_{\text{res}}]^2/DF_{\text{res}}} \tag{9.18}$$

where DF_{res} is the residual degrees of freedom and $DF_{A \times \text{subj}}$ is the subjects by factor A degrees of freedom.

To test for factor B, we compute

$$F = \frac{MS_B}{(1 - a'/a'')MS_{\text{res}} + (a'/a'')MS_{B \times \text{subj}}} \tag{9.19}$$

which follows an approximate F distribution with $(b - 1)$ numerator degrees of freedom and approximate denominator degrees of freedom given by

$$\frac{[a'MS_{B \times \text{subj}} + (a'' - a')MS_{\text{res}}]^2}{(a'MS_{B \times \text{subj}})^2/DF_{B \times \text{subj}} + [(a'' - a')MS_{\text{res}}]^2/DF_{\text{res}}} \tag{9.20}$$

where $DF_{B \times \text{subj}}$ is the subjects by factor B degrees of freedom.

Because the $A \times B$ interaction expected mean square does not involve a variance due to the random subjects factor, its F statistic is unaffected by the coefficients of the variance term arising from the random subjects effect and is simply given, as before, by

$$F = \frac{MS_{AB}}{MS_{\text{res}}} \tag{9.21}$$

As when there were repeated measures on only one of the two factors, we need to obtain the marginal sums of squares for each effect and use a computer program (usually a general linear models procedure) that will display the values of the expected mean squares computed from the unbalanced data set. We can then use the mean squares, the values of the coefficients of the random subjects effects in the expected mean squares, and Eqs. (9.17) through (9.21) to compute the necessary F statistics to test the main effects and interaction null hypothesis.

More on Chewing Gum

Table 9-17 shows the same set of data shown in Table 9-14, but missing two observations: subject 4 was unable to complete the protocol during the regular gum chewing phase of the study. We will now analyze these data with a general linear models routine, declaring the subjects (and interactions involving the subjects) to be random, and issuing the necessary commands to obtain the correct marginal sums of squares for each effect and a printout of the expected mean squares so that we can obtain the values of a', a'', b', and b'' needed to compute the approximate F statistics for gum (factor A) and time (factor B) effects. The analysis of variance results are shown in Fig. 9-21, and the expected mean squares are shown in the lower half of Table 9-16.

To compute the approximate F statistic for testing for the gum effect, we compute $MS_G = SS_G/DF_G = .9465/1 = .9465$, $MS_{res} = .0773$, and $MS_{G \times subj} = SS_{G \times subj}/DF_{G \times subj} = .1156/7 = .0165$, from the marginal sums of squares (labeled "type III" in Fig. 9-21) and associated degrees of

TABLE 9-17 Tooth Plaque pH Measured 20, 30, and 50 min after Eating Candy and Chewing One of Two Gums, a Standard Sugarless Gum and a Sodium Bicarbonate–Containing Gum; Some Data Are Missing

Gum	Subject	Time after Eating Candy, min		
		20	30	50
Standard	1	4.6	4.4	4.1
	2	4.2	4.7	4.7
	3	3.9	5.0	5.7
	4	4.6		
	5	4.6	5.1	5.7
	6	4.7	5.3	5.3
	7	4.9	5.6	5.9
	8	5.0	6.0	5.9
Bicarbonate	1	3.8	5.3	5.1
	2	4.1	5.8	5.1
	3	4.2	5.8	5.6
	4	4.2	6.9	5.6
	5	4.3	6.2	5.8
	6	4.4	6.3	5.9
	7	4.9	6.0	5.9
	8	4.5	7.2	6.2

```
                           GENERAL LINEAR MODELS PROCEDURE

DEPENDENT VARIABLE: PH        plaque pH

SOURCE                DF      SUM OF SQUARES       MEAN SQUARE      F VALUE        PR > F        R-SQUARE          C.V.

MODEL                 33        28.27103520         0.85669804       11.08        0.0001        0.968215        5.3526

ERROR                 12         0.92809524         0.07734127                    ROOT MSE                     PH MEAN

CORRECTED TOTAL       45        29.19913043                                      0.27810298                 5.19565217


SOURCE                DF          TYPE I SS     F VALUE      PR > F      DF        TYPE III SS     F VALUE      PR > F
G                      1         1.69000165       21.85      0.0005       1        0.94651124       12.24      0.0044
T                      2        14.35844468       92.83      0.0001       2       13.45752976       87.00      0.0001
G*T                    2         3.23175554       20.89      0.0001       2        2.40190476       15.53      0.0005
G*SUBJ                14         6.99357143        6.46      0.0013       7        0.11556548        0.21      0.9753
T*SUBJ                14         1.99726190        1.84      0.1471      14        1.99726190        1.84      0.1471
SUBJ                   0         0.00000000         .          .         7        6.91675595       12.78      0.0001
```

FIGURE 9-21 Analysis of the study of chewing gum and plaque pH when there are missing data (as shown in Table 9-17). The marginal sum of squares (labeled Type III SS) no longer equals the incremental sum of squares (labeled Type I SS). In this case of missing data, the hypothesis tests should be based on the marginal sums of squares.

freedom, and substitute these mean squares and values of $b' = 2.333$ and $b'' = 2.714$ into Eq. (9.17) to obtain

$$F = \frac{.9465}{(1 - 2.333/2.714).0773 + (2.333/2.714).0165} = 37.81$$

with numerator degrees of freedom $(a - 1) = 1$ and approximate denominator degrees of freedom given by Eq. (9.18):

$$\frac{[2.333(.0165) + (2.714 - 2.333).0773]^2}{[2.333(.0165)]^2/7 + [(2.714 - 2.333).0773]^2/12} = 16.3$$

This F of 37.81, with 1 and 16 degrees of freedom,[*] exceeds the critical value of F for $\alpha = .01$, 8.53, so we conclude that there is a significant effect of gum on plaque pH.

Similarly, to compute the approximate F statistic for testing for the time effect, we compute $MS_T = SS_T/DF_T = 13.458/2 = 6.729$, $MS_{res} = .0773$, and $MS_{T \times subj} = SS_{T \times subj}/DF_{T \times subj} = 1.9973/14 = .1427$, from the marginal sums of squares (labeled "type III" in Fig. 9-21) and associated

---

[*]We rounded the degrees of freedom *down*. If you wish to be more precise, and less conservative, you can interpolate in the table of critical values of F to obtain a more exact critical value.

degrees of freedom, and substitute these mean squares and values of $a' = 1.750$ and $a'' = 1.857$ into Eq. (9.19) to obtain

$$F = \frac{6.729}{(1 - 1.750/1.857).0773 + (1.750/1.857).1427} = 48.43$$

with numerator degrees of freedom $(b - 1) = 2$ and approximate denominator degrees of freedom given by Eq. (9.20):

$$\frac{[1.750(.1427) + (1.857 - 1.750).0773]^2}{[1.750(.1427)]^2/14 + [(1.857 - 1.750).0773]^2/12} = 14.9$$

This approximate F of 48.43, with 2 and 14 degrees of freedom, exceeds the critical value of F for $\alpha = .01$, 6.51, and so we conclude that there is a significant effect of gum on plaque pH.

The F statistic for testing gum by time interaction, which can be computed from Eq. (9.21) or read directly from the computer printout (under the "type III" sums of squares entry), is 15.53 with 2 and 12 degrees of freedom $(P = .0005)$.

These conclusions are the same ones we reached before when there were no missing data.

Multiple Comparisons

Multiple comparisons are handled as we described above for the missing data situation in two-way analysis of variance with repeated measures on only one of the factors.

EQUALITY OF VARIANCES REVISITED

Earlier in this chapter we discussed the assumptions underlying repeated-measures analysis of variance. They are more complicated than if there are no repeated measures, but one of the essential assumptions common to all analyses of variance is that the underlying variances in the cells be equal, so-called homogeneity of variances. In Chap. 7 we showed how to use the modified Levene median test to test the null hypothesis that the variances were equal for the rabbit cholesterol and plaque data. In that example, even though the standard deviation of one group was much smaller than those of the other four groups, we had insufficient evidence to reject the null hypothesis of equal variances and concluded that we had not violated the homogeneity of variance assumption. Now, we will work another example to illustrate the situation where the equal variance assumption is violated, how to fix the problem, and the difference it makes in the conclusions drawn from the data.

How Similar Are the Mechanical Properties of the Whole Heart to Heart Muscle?

The heart, which pumps blood to the lungs and the body, consists of muscular chambers whose walls are constructed from complexly oriented and intertwined bundles of heart muscle. To study the mechanical properties of the contraction process in heart muscle, people often remove a small piece of muscle so that the complex architecture of muscle in the wall of the whole heart is avoided. Presumably, the mechanical properties observed in these isolated muscle preparations relate more directly to the properties of the muscle itself. However, it would be interesting to know if the complex arrangement of heart muscle in the walls of the heart chambers somehow modifies the expression of basic heart muscle contraction as observed in an isolated piece of muscle.

To answer this question, Campbell and his colleagues* studied hearts that had been removed from ferrets so that a balloon could be placed into the main pumping chamber, the left ventricle. This balloon allowed fine control of the volume in the left ventricle. In addition, they made the heart go into tetanus, which is a maximum sustained contraction of the muscle. Using this preparation allows study of the basic behavior of the heart muscle in the whole heart. The tetanus means that the muscle is maximally contracted, so that when the volume is suddenly changed by either infusing or withdrawing about 15 to 20 percent of the volume of fluid in the balloon, the resulting changes in pressure reflect the interactions of the proteins in the muscle that are causing the contraction. The time course of the pressure change relates to the rate of the molecular events at the protein level. These experiments were designed to be analogous to classic isolated heart muscle studies where a tiny piece of heart muscle is held at a constant length, tetanized, and then subjected to sudden small stretches or releases from the intial fixed length (analogous to holding volume constant in the whole heart). The change in force (analogous to pressure in the whole heart) that follows the sudden change in length reflects the properties of the proteins that cause contraction, actin and myosin.

Previous findings from isolated heart muscle indicate that there are two phases to the force response to a sudden length change: one fast and one slow. Furthermore, the time course of the force response differs when length is suddenly increased (corresponding to a volume infusion in the whole heart) compared to the response when length is suddenly de-

*K. B. Campbell, A. R. Rahimi, D. L. Bell, R. D. Kirkpatrick, and J. A. Ringo, "Pressure Response to Quick Volume Changes in Tetanized Isolated Ferret Hearts," *Am. J. Physiol.* 257:H38-H46, 1989.

creased. The specific goals of the experiments described by Campbell and colleagues were to see if a similar two-phase response of pressure to a sudden volume change occurred in analogous experiments in the whole heart and if the pressure response also depended on whether volume was infused or withdrawn. Thus, they have a two-way experimental design with one factor being speed (fast phase vs. slow phase, quantified with time constants) and the other being volume change (infusion vs. withdrawal). The experiment was conducted so that there were repeated measures on both factors.

In terms of the physiological questions of interest, they wanted to test whether the apparent difference in speed of the fast and slow phases was significant, and, more importantly, whether or not infusion was different from withdrawal. This latter question would be answered affirmatively if either the main effect for volume or the volume by speed interaction was significant.

The means and standard deviations of the four treatment combinations are shown in Fig. 9-22, and the individual observations from seven ferret hearts are shown in Table 9-18. We analyzed these data using a traditional mixed-model analysis of variance with the speed and volume factors being fixed, and a random subjects factor. The computer output

TABLE 9-18 Time Constants (ms) of Fast and Slow Pressure Responses to Sudden Volume Withdrawal of Infusion in Tetanized Isolated Ferret Left Ventricles

Volume Change	Heart	Speed	
		Fast	Slow
Infusion	1	8	183
	2	6	68
	3	8	119
	4	13	147
	5	4	102
	6	5	174
	7	6	89
Withdrawal	1	23	123
	2	7	74
	3	14	177
	4	25	79
	5	18	40
	6	7	196
	7	33	65

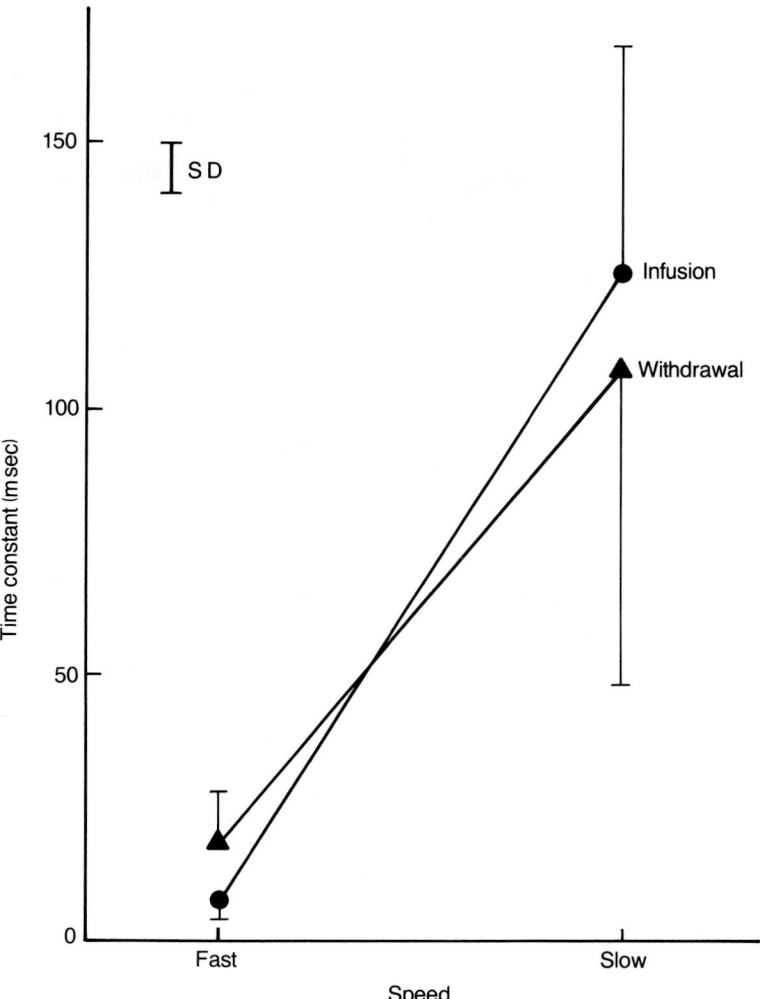

FIGURE 9-22 Time constants of the two-phase pressure response to sudden volume infusions or withdrawals in tetanized, isolated ferret hearts. One phase of the pressure response was characterized by a fast time constant, and the other phase was characterized by a slow time constant. Note that the standard deviations in the data for the slow responses are much greater than for the fast responses. This difference makes it important to test the equal variance assumption of the analysis of variance to ensure that the test produces reliable results.

is shown in Fig. 9-23. There is a statistically significant difference in the speed of the two phases ($F = 33.5$; $P < .001$). However, neither the main effect of volume ($F = .19$; $P = .68$) nor the volume by speed interaction ($F = 2.07$; $P = .2$) is significant. Thus, it would seem that, unlike, iso-

```
Factor          Type Levels Values
Volume         fixed    2    0    1
Speed          fixed    2    0    1
subject       random    7    1    2    3    4    5    6    7

Analysis of Variance for Timecon

Source            DF        SS        MS       F      P
Volume             1      91.5      91.5    0.19  0.681
Speed              1   75928.1   75928.1   33.48  0.001
subject            6   12247.5    2041.2     *
Volume*Speed       1    1520.8    1520.8    2.07  0.200
Volume*subject     6    2951.0     491.8    0.67  0.680
Speed*subject      6   13605.6    2267.6    3.09  0.098
Error              6    4400.2     733.4
Total             27  110744.7
```

FIGURE 9-23 Repeated-measures analysis of variance of the data in Fig. 9-22.

lated heart muscle, the whole heart does not show a differential response to infusion vs. withdrawal. Thus, something about the complexity with which muscle is arranged in the wall of the heart appears to modify the fundamental muscle behavior.

However, Fig. 9-22 shows that the standard deviations of the fast-phase speed were much smaller (by an order of magnitude) than those of the slow phases. Thus, we may be violating the equal variance assumption. To formally test whether we are violating this assumption, we will use the Levene median test to compare the standard deviations of the four cells of data shown in Table 9-18. Following the procedure we outlined in Chap. 7, the seven observations in each cell (upper third of Table 9-19) are centered to the median of each cell. These centered observations are shown in the middle third of Table 9-19. Next, we take the absolute values of the centered values. Finally, because the sample size is odd for each group and less than 20, we delete the centered value corresponding to the median, as shown in the bottom third of Table 9-19. To conduct the Levene test, we analyze the data in the bottom third of Table 9-19 using an ordinary one-way analysis of variance. The computer output shown in Fig. 9-24 shows that the F statistic is 6.0 with an associated P value of .004. Thus, we reject the null hypothesis of equality of variances and conclude that we have violated the equal variance assumption.

In Chap. 4 we introduced several transformations commonly used to stabilize (or equalize) variances. For data like these, in which the variance increases as the dependent variable increases, the logarithm transformation is commonly used, so we will transform the data in Table 9-19 by taking their natural logarithms.

TABLE 9-19 Steps in the Levene Median Test for Homogeneity of Variances in the Raw Ventricle Time Constant Data

	Fast		Slow	
	Infuse	Withdraw	Infuse	Withdraw
	RAW DATA			
	8	23	183	123
	6	7	68	74
	8	14	119	177
	13	25	147	79
	4	18	102	40
	5	7	174	196
	6	33	89	65
Median	6	18	119	79
	DATA CENTERED TO THE MEDIAN OF EACH COLUMN			
	2	5	64	44
	0	−11	−51	−5
	2	−4	0	98
	7	7	28	0
	−2	0	−17	−39
	−1	−11	55	117
	−0*	15	−30	−14
	ABSOLUTE VALUES OF CENTERED DATA, WITH MEDIANS DELETED			
	2	5	64	44
	2	11	51	5
	7	4	28	98
	2	7	17	39
	1	11	55	117
	0	15	30	14

*This value is actually a small negative number which rounds to zero. Because it is not actually zero, it is retained in the calculations, which are based on the actual values.

Did this transformation rectify the problem of unequal variances? To answer this question, we again apply the Levene median test. Following the procedure outlined above, we center the transformed data, take their absolute values, and delete all observations equal to zero. These data are then analyzed using ordinary one-way analysis of variance to test for

```
ANALYSIS OF VARIANCE
SOURCE     DF       SS        MS        F        p
FACTOR      3     10825      3608     6.00    0.004
ERROR      20     12030       601
TOTAL      23     22855

                                   INDIVIDUAL 95 PCT CI'S FOR MEAN
                                   BASED ON POOLED STDEV
  LEVEL     N      MEAN      STDEV  -------+---------+---------+---------
Zwithfas     6      8.95      4.24     (------*------)
Zwithslo     6     52.82     45.15                        (------*------)
Zinfufas     6      2.03      2.30  (------*------)
Zinfuslo     6     40.83     18.55                  (------*------)
                                   -------+---------+---------+---------
POOLED STDEV =    24.53                  0        30        60
```

FIGURE 9-24 Results of analysis of variance computation for the Levene median test for unequal variances in the time constants from the tetanized heart study shown in Fig. 9-22. Because this analysis of variance detects significant differences in variability around each group median (as expressed by the absolute difference of each data point from its group median), we conclude that the assumption of equal variances is violated.

equality of variances. The computer output from this analysis is in Fig. 9-25, and shows that F is 1.5 and is associated with a P value of .25. Thus, we do not reject the null hypothesis of equality of variances and conclude our transformation of the data had the desired effect of stabilizing the variances in these data.

We now repeat the three-way mixed-model analysis of this experiment, this time using the logarithm transformed data. The computer output from this analysis of variance is in Fig. 9-26. There is a significant effect of speed ($F = 110$; $P < .001$), a significant effect of volume change ($F = 12.7$; $P = .012$), and a significant interaction between the speed of the response and the method of volume change ($F = 9.2$; $P = .023$). This result is remarkably different from the result obtained from the raw data (Fig. 9-23), which showed only a significant speed effect.

In terms of the physiology, these two sets of results could not be more different. Whereas before we concluded that the way heart muscle is assembled to form the walls of the heart somehow changed how it expressed its elemental mechanical behavior, now we conclude that the complex arrangement of the heart muscle in the wall of the heart does not greatly modify the expression of heart muscle mechanics in terms of whole ventricle mechanics.

The lesson is clear. Check the assumptions behind the analysis before you accept the results.

```
ANALYSIS OF VARIANCE
SOURCE      DF       SS       MS         F        p
FACTOR       3     0.3271   0.1090     1.50    0.246
ERROR       20     1.4556   0.0728
TOTAL       23     1.7827

                                    INDIVIDUAL 95 PCT CI'S FOR MEAN
                                    BASED ON POOLED STDEV
LEVEL        N      MEAN     STDEV  -------+---------+---------+-------
Zlnwifas     6     0.5523   0.3035            (-----------*----------)
Zlnwislo     6     0.5166   0.3391           (-----------*---------)
Zlninfas     6     0.2758   0.2488  (-----------*----------)
Zlninslo     6     0.3372   0.1489      (-----------*----------)
                                    -------+---------+---------+-------
POOLED STDEV =     0.2698             0.20      0.40      0.60
```

FIGURE 9-25 Results of the Levene median test for unequal variances in the time constants from the tetanized heart study after doing a logarithmic transformation of the data. The fact that the F test associated with the analysis of variance of the centered, transformed data is no longer significant indicates that the logarithmic transformation has had the desired effect of stabilizing the variances so that the assumption of equal variance required for analysis of variance is no longer violated.

```
Factor             Type Levels Values
Volume             fixed    2     0    1
Speed              fixed    2     0    1
subject            random   7     1    2    3    4    5    6    7

Analysis of Variance for lnTcon
```

Source	DF	SS	MS	F	P
Volume	1	0.6492	0.6492	12.67	0.012
Speed	1	37.8160	37.8160	110.02	0.000
subject	6	1.9791	0.3299	*	
Volume*Speed	1	2.0699	2.0699	9.22	0.023
Volume*subject	6	0.3075	0.0513	0.23	0.952
Speed*subject	6	2.0622	0.3437	1.53	0.309
Error	6	1.3468	0.2245		
Total	27	46.2308	1.7123		

FIGURE 9-26 Results of two-way, repeated-measures analysis of variance for the data from the study of quick volume infusions and withdrawals in tetanized hearts after variance stabilization using a logarithmic transformation. In contrast to the result shown in Fig. 9-23, after variance stabilization both the main effects and the interaction are highly significant, leading us to much different physiological conclusions than we obtained before making this adjustment.

SUMMARY

Repeated-measures designs are common in biomedical research, especially clinical trials of new drug and other therapies. Repeated-measures designs benefit from accounting for differences between experimental subjects, which generally reduces the residual variation in the experiment and so increases the sensitivity of the experiments to detect treatment effects. This reduction in residual variance is offset to some extent by a reduction in the associated degrees of freedom (which means that the critical value of F required to reject the null hypothesis of no treatment effect increases), but this tradeoff is usually worth making, particularly in the presence of significant biological variability in the individual experimental subjects. Perhaps more important, repeated-measures designs greatly reduce the number of experimental subjects necessary for a study. These benefits do have a price. Repeated-measures analysis of variance requires making additional assumptions about the underlying population and requires care in the design of experiments to avoid carryover effects between the different treatments. Nevertheless, repeated-measures designs are particularly important in clinical and animal studies where the number of available subjects may be limited.

PROBLEMS

9.1   In this chapter we analyzed the metabolic rate response early after different methods of eating in dogs. The investigators also analyzed the metabolic rate later after the different types of eating. These data are in Table D-26, Appendix D. *A.* Analyze them the same way we did the early metabolic rate response. *B.* Are any of the conclusions different from before? *C.* Putting the results of these two separate analyses together, what can you say about the physiology of eating in terms of the control of the early and late phases of digestion?

9.2   In this chapter, and in Prob. 9.1, we analyzed the metabolic responses early (data in Table 9-3) and late (data in Table D-26, Appendix D) after a meal that dogs ate normally, ate but had their stomachs bypassed, or did not eat, but had food placed directly into their stomachs. Combine these two data sets and use a two-way analysis to reanalyze the data. Does your answer agree with the conclusions reached by considering the results of the two separate one-way analyses? Which method is preferable?

9.3   Bacterial contamination of surgical patients can occur from bacteria that contaminate the scrub suits worn by surgical personnel. During surgery these scrub suits are covered by surgical gowns. However, because this effort is not perfect, one presumably could gain an advantage if the contamination of scrub suits was reduced. One of the sources of the contamination of scrub suits is bacteria picked up when personnel leave the operating room, for example, to go to lunch. Traditional practice is to wear a covergown over the scrub suits so as to minimize contamination while outside the operating room. No data support the use of covergowns, which are expensive. Therefore, Copp and her coworkers[*] sought to obtain evidence on the effectiveness of covergowns and also to test related hypotheses about various strategies to reduce scrub suit contamination in the operating room. One of their hypotheses was that a fresh scrub suit put on after lunch break was more effective in reducing contamination than changing to street clothes for lunch and protecting the scrub suit in a clean wrap in a locker until it was put on for the afternoon shift. These two strategies were followed by two groups of 19 operating room personnel whose scrub suits were swabbed on the shoulder for bacterial cultures at four times during the day: start of shift, immediately before lunch, on return from lunch, and at the end of the shift (the data are in Table D-27, Appendix D). Is there evidence to support their hypothesis? Include a graph of the data.

9.4 Physical training causes adaptations in the metabolism of muscles, one of which is a switch in the energy sources utilized. To study the switch in energy metabolism as a result of exercise, Hood and Terjung[†] measured the contribution of the branched-chain amino acid, leucine, to muscle energy utilization in skeletal muscle bundles isolated from two groups of rats: one physically trained and one not trained. They measured the contribution of leucine to the total energy consumption of the muscle (percentage of total energy used) with the muscle at rest, and then during two frequencies of stimulation (15/min and 45/min) to simulate two levels of acute exercise (the data are in Table D-28, Appendix D). Is there evidence to support the hypothesis that the percentage contribution of leucine to energy needs during exercise in trained muscle is lower than the percentage contribution in untrained (control muscle)?

9.5 In this chapter, we analyzed three different personality scores — aggression, anger, and depression — in two groups of alcoholics, those with antisocial personality disorder and those without, when

---

[*]G. Copp, C. B. Mailhot, M. Zalar, L. Slezak, and A. J. Copp, "Covergowns and the Control of Operating Room Contamination." *Nurs. Res.* 35:263–268, 1986.

[†]D. A. Hood and R. L. Terjung, "Effect of Endurance Training on Leucine Metabolism in Perfused Rat Skeletal Muscle," *Am. J. Physiol.* 253:E648–E656, 1987.

they were sober and when they were drinking. That study included an assessment of two other personality traits, a sociability score and a score for feeling of well-being (these data are in Tables D-29 and D-30, Appendix D). *A.* What effects do antisocial personality and drinking have on sociability? *B.* What effects do antisocial personality and drinking have on feelings of well-being?

9.6 Suppose that some subjects gave incomplete responses to the sociability assessment analyzed in Prob. 9.5*A* (the data are in Table D-31, Appendix D). *A.* Reanalyze the effect of antisocial personality diagnosis and drinking on sociability when these data are missing. *B.* Do any conclusions reached in Prob. 9.5 change?

9.7 Strenuous exercise to which one is not accustomed can lead to muscle damage that is perceived as soreness in the afflicted muscles. Some evidence suggests that repeating a bout of exercise as much as 2 or 3 weeks later, without any intervening exercise, causes less damage and, thus, less soreness. To examine this adaptation in more detail, Triffletti and his coworkers[*] measured muscle soreness at 6, 18, and 24 hours after a bout of unaccustomed exercise in six subjects who each underwent four different bouts of exercise spaced 1 week apart (the data are in Table D-32, Appendix D). They hypothesized that there would be less muscle soreness after bouts 2, 3, or 4 than there was after bout 1. In addition, they hypothesized that the usual increase in soreness with time immediately following a bout of exercise would be attenuated after bouts 2, 3, or 4, compared to after the first bout. Is there any evidence to support these hypotheses?

9.8 Suppose some of the data analyzed in Prob. 9.7 were missing as shown in Table D-33, Appendix D. *A.* Analyze these missing data using the same analysis as in Prob. 9.7. *B.* Are any of the conclusions different from those reached in Prob. 9.7?

9.9 To study the physiology of the stomach and intestines, it is often necessary to measure the blood flow in a localized area in the wall of these organs. One standard method is to measure the washout (clearance) of a known amount of hydrogen gas. This technique is difficult. More recently, an easier method has been developed based on measuring the Doppler shift in laser light scattered from flowing blood cells. One problem with laser Doppler flow measurement is

[*]P. Triffletti, P. E. Litchfield, P. M. Clarkson, and W. C. Byrnes, "Creatine Kinase and Muscle Soreness after Repeated Isometric Exercise," *Med. Sci. Sports Exer.* 20:242–248, 1988.

determining a calibration factor so that absolute flow can be computed. To assess the feasibility of computing a calibration factor, Gana and coworkers[*] compared laser Doppler L to hydrogen clearance H estimates of gastric flow in four dogs (the data are in Table D-34, Appendix D). Are the relationships between L and H different in the different dogs?

9.10 Recompute the regression analysis in Prob. 9.9 without including dummy variables to account for the between-dogs variability. What effect does this have on R^2 and $s_{y|x}$?

[*]T. J. Gana, R. Huhlewych, and J. Koo, "Focal Gastric Mucosal Blood Flow by Laser-Doppler and Hydrogen Gas Clearance: A Comparative Study," *J. Surg. Res.* 43:337–343, 1987.

Nonlinear Regression

So far we have concentrated on describing data with models in which the effects of the different independent variables on the dependent variable could be represented as a weighted sum of the independent variables. The weights in this sum are the regression coefficients we have learned to estimate. In the simplest case, the dependent variable changed in proportion to each of the independent variables, but it was possible to describe many nonlinear relationships by transforming one or more of the variables (such as by taking a logarithm or a power of one or more of the variables), while still maintaining the fact that the dependent variable remained a weighted sum of the independent variables. These models are *linear in the regression parameters*. The fact that the model is linear in the regression parameters means that the problem of solving for the set of parameter estimates (coefficients) that minimizes the sum of squared residuals between the observed and predicted values of the dependent variable boils down to the problem of solving a set of simultaneous linear algebraic equations in the unknown regression coefficients. Although the arithmetic can be messy, we have computers to handle it. The important point is that there is a unique solution to this problem and the coefficients can be computed for any model from any set of data. Moreover, because the coefficients are themselves weighted sums (more precisely, linear combinations) of the observations, statistical theory demonstrates that the sampling distributions of the regression coefficients follow the t distribution, so that we can conduct hypothesis tests and compute confidence intervals for these coefficients and for the regression model as a whole. Indeed, these topics have occupied us for most of this book. These techniques are powerful tools for describing complex biological, clinical, economic, social, and physical systems and gaining insight into the important variables that define these systems, despite the restriction that the model be linear in the regression parameters.

There are, however, many instances in which it is not desirable — or even possible — to describe a system of interest with a linear regression model. For example, predictive theories are commonplace in the physical sciences and are beginning to appear in the life and health sciences. Rather than being based on a purely empirical description of the process under study, such theories are derived from fundamental assumptions about the underlying structure of a problem (often embodied in one or more differential equations), and manipulated to produce theoretical relationships between observable variables of interest. For example, in studies of the distribution of drugs within the body — pharmacokinetics — one often wishes to describe or predict the concentration of drugs over time after giving an injection or a pill. If one assumes that the drug is distributed in and exchanged between several compartments within the body, the resulting concentration of the drug in the blood will be described by a sum of exponential functions of time, in which the associated coefficients and exponents depend on the characteristics of the compartments and how the drug moves between them. Such descriptions can be extremely useful for describing the drug's behavior, and so people often want to estimate these parameters from measurements of drug concentration in experimental subjects or patients. Doing so requires fitting the theoretical equation to the observations, much as we have done so far in this book. The problem is that, in contrast to what we have done so far, the parameters enter into the equation describing the concentration in a *nonlinear* way, so we cannot simply apply the formulas for estimating the parameters in the model that we developed for linear regression.

In addition to the desire to fit a theory-based model to a set of data, sometimes linear functions of the independent variables (or their transformed values) cannot be made to describe a set of observations adequately or the transformations required become hard to interpret. For example, exponential growth and decay processes are common in biology (and elsewhere) where the rate of change of a process depends on its magnitude (Fig. 10-1). In the case of an exponential decay, the process might be described by the equation

$$c(t) = c_0 e^{-t/\tau} \tag{10.1}$$

where c_0 and τ are parameters. The parameter τ, known as the *time constant*, is a measure of how long c takes to fall to $1/e = 37$ percent of its original value at time $t = 0$, c_0. These parameters are closely related to the underlying differential equations which govern the system and can

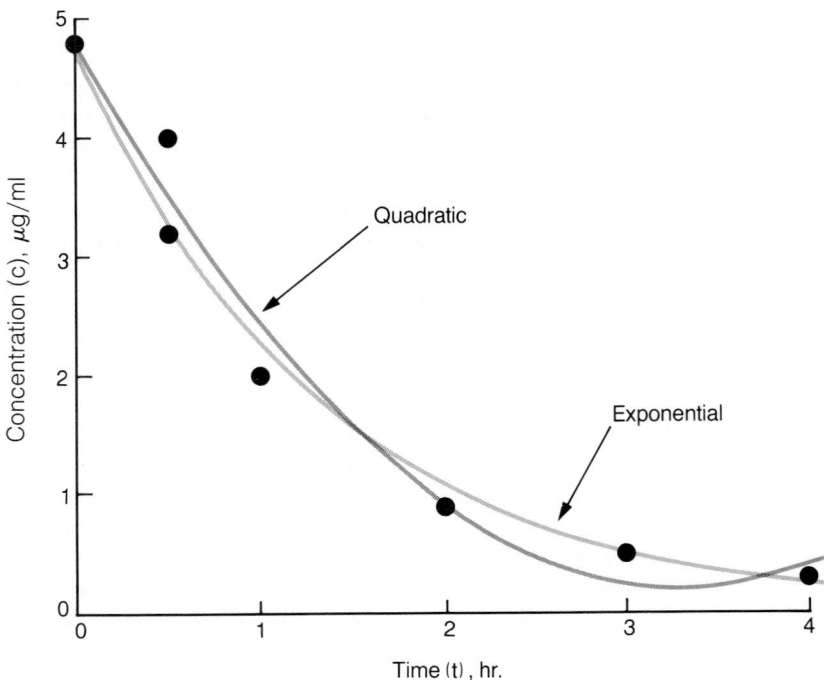

FIGURE 10-1 It is possible to describe reasonably well a falling nonlinear relationship between concentration and time with both a polynomial and an exponential function of time. The exponential function is generally preferable, however, because it arises naturally out of theoretical models based on first principles of pharmacokinetics and also only contains two parameters (c_0 and τ) compared to three parameters in the quadratic equation (d_0, d_1, and d_2). The disadvantage of the exponential function is that it is a nonlinear function which may require a nonlinear least-squares parameter estimation technique, whereas the quadratic equation can always be treated as a linear regression problem.

be interpreted in terms of the system's properties.* We could also describe this relationship with the polynomial

$$c(t) = d_0 + d_1 t + d_2 t^2 \tag{10.2}$$

Figure 10-1 shows that this functional form also describes the data reasonably well. As we have already seen, Eq. (10.2) is linear in the regression parameters (d_0, d_1, and d_2), so we can use the methods we already know to compute the parameters and test hypotheses about them. The problem is that, in contrast to the parameters c_0 and τ in Eq. (10.1), the

*For a discussion of the relationship between the time constants and the underlying differential equations which describe a process, see S. A. Glantz, *Mathematics for Biomedical Applications*, Berkeley, UC Press, 1979.

parameters d_0, d_1, and d_2 in Eq. (10.2) have no clear interpretation. Thus, a second reason to deal with a nonlinear model is to obtain parameters to describe the problem which are more easily interpretable than in some other model. In addition, if we explicitly allow for the nonlinear structure, we can describe the process with two rather than three parameters. This reduction in the number of parameters simplifies the process of estimating and interpreting the results and also conserves the error degrees of freedom and leads to more precise parameter estimates.

But wait. In Chap. 4 we showed that we could transform Eq. (10.1) into one that was linear in the model parameters by simply taking the natural logarithm of both sides to obtain

$$\ln c(t) = \ln c_0 - \frac{t}{\tau}$$

or

$$\ln c(t) = C_0 + C_t t$$

in which $C_0 = \ln c_0$ and $C_t = -1/\tau$, and simply use linear regression methods to estimate the parameters C_0 and C_t in the transformed data, then convert these parameter estimates back to the original valuables with $c_0 = e^{C_0}$ and $\tau = -1/C_t$. Indeed, as discussed in Chap. 4, this procedure often not only resolves difficulties connected with the nonlinear shape of the curve, but also stabilizes the variance of the residuals and so provides better estimates of the parameters.

There are, however, two potential difficulties. First, as we discussed in Chap. 4, transforming the dependent variable implicitly changes the presumed nature of the random component of the measurement from one that is additive to one that is multiplicative. When the error structure is multiplicative (as it often is in exponential processes), this transformation of the error structure is desirable because it makes the mathematical model of the process more closely match reality. There are many instances, however, in which one wants to maintain the structure of additive errors, in which case the logarithm transformation, while straightening out the function and converting it to a linear regression problem, complicates the interpretation of the residuals or violates the equal variance assumption. Second, we are often interested in not only the value of a regression coefficient, but also some measure of how precise this estimate is. The associated standard error provides this estimate of precision. If the model is transformed to a new set of variables and regression parameters—a process known as *reparameterization*—the standard errors produced by the regression analysis apply to the new parameters, not the original ones of interest.

In sum, there are several reasons why one might wish to fit a nonlinear regression model to a set of data:

- The equations may have some basis in theory, in which case the regression parameters can provide more insight into the underlying processes than a simple empirical fit based on, say, a polynomial.
- A nonlinear equation may provide a more sensible equation for an empirical fit, often with fewer parameters than one that is linear in the model parameters.
- Transforming a nonlinear equation into one that is linear in the parameters, while simplifying the curve-fitting process, may require making unrealistic assumptions about the nature of the error term (reflected in the fact that the residuals no longer are consistent with the equal variance assumption); doing so also loses information about the standard errors of the native parameters of the problem.

And, of course, some problems contain nonlinearities which cannot be eliminated by any transformation. All these reasons often make it necessary or desirable to deal directly with a nonlinear regression model.

Although we will concentrate on examples in which the independent variables are continuous (such as time), there is no reason that the independent variables in nonlinear regression need to be continuous. Indeed, we can make the same use of dummy variables to indicate the presence or absence of a condition or to code for between-subjects differences (in a repeated-measures design) as we did in linear regression.

The problems with nonlinear regression are that, in contrast to the situation that prevails for linear regression models, we cannot compute the best estimates of the regression parameters with simple algebra and that the statistical properties of these estimates are not well characterized. Indeed, every nonlinear problem has its own character. Despite these problems, however, there are well-developed methods for nonlinear parameter estimation. It is important to understand these methods to obtain reasonable and reliable results from them. In particular, keep in mind what we do *not* know about the estimates as well as what we do know. Nonlinear parameter estimation is perhaps the best example of the fact that just because a computer produces a result, it does not mean that it is correct or reasonable.

MARTIAN MOODS

In our continuing effort to understand life on Mars, we return again, this time to conduct a study to see how many Martians think that the time to invade earth has arrived. We find that 17.3 percent of the Martians we surveyed think that it would be a good idea, if for no reason other than stopping pesky academics from periodically visiting to study them. To see how this pattern is changing with time, we conduct daily surveys for

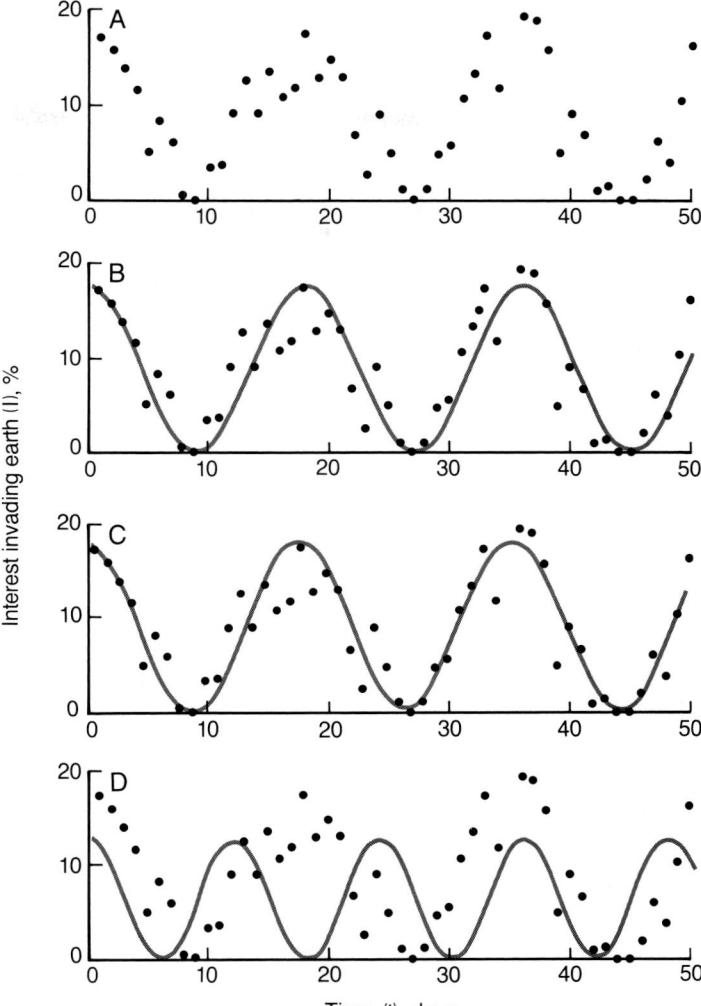

FIGURE 10-2 (*A*) Martian interest in invading earth as a function of time. (*B*) Best-fitting function using Eq. (10.3) with the parameters identified using the grid search (see Table 10-1) superimposed on the observed data. (*C*) Best-fitting function using Eq. (10.3) with the parameters identified using a gradient method and good first guesses for the parameters superimposed on the observed data. (*D*) Best-fitting function using Eq. (10.3) with the parameters identified using bad first guesses for the parameters superimposed on the observed data.

the next 49 days. Figure 10-2*A* shows the results of these surveys. Examining this plot reveals that interest in invading earth varies periodically with time. The question is, What is the peak level of interest and how rapidly does it vary?

Simply examining the data in Fig. 10-2*A* reveals that no simple linear function of time will describe this relationship, and none of the variable transformations we have discussed will linearize these data. Instead, we decide to fit these data with the nonlinear function

$$\hat{I}(t) = I_{\max} \frac{\cos(2\pi t/T) + 1}{2} \tag{10.3}$$

where $\hat{I}(t)$ is the predicted interest in invading earth (as a proportion of the Martians we interview who think it is a good idea), I_{max} is the maximum level of interest in invasion, and T is the period of the variation in interest. We want to fit this function to the data in Fig. 10-2A (and Table C-18, Appendix C), to estimate these parameters.

As before, we seek the values of the parameters in the regression model that minimize the sum of squared residuals between the predicted and observed values of interest.

$$SS_{res} = \Sigma\,[\,I(t) - \hat{I}(t)\,]^2 = \Sigma\left[\,I(t) - I_{max}\,\frac{\cos(2\pi t/T) + 1}{2}\,\right]^2 \tag{10.4}$$

As with linear regression, this quantity will be a minimum when the partial derivatives of SS_{res} with respect to the two parameters I_{max} and T equal zero. Solving the resulting nonlinear equations directly is a very difficult problem for which no closed form exists; therefore we require a numerical solution. This situation contrasts markedly with linear models, where it was a matter of simple algebra to find the set of regression coefficients that minimized SS_{res}.

Grid Searches

Rather than try to solve these unpleasant equations, let us take a more straightforward strategy; we will do a *grid search* in which we simply fix I_{max} and T at a variety of values, compute the residual sum of squares SS_{res}, then pick the values of I_{max} and T that produce the smallest value of SS_{res}. Simply looking at the data in Fig. 10-2A shows that the maximum level of interest in invading earth is somewhere around 20 percent and the period with which this interest varies is around 18 days. Table 10-1 shows the results of computing SS_{res} with values around these *initial guesses*. Examining this table reveals that values of I_{max} = 18 percent and T = 18 days lead to the smallest SS_{res}, 462 percent^2, *among the values we checked*. Figure 10-2B shows the fit obtained by substituting these values of I_{max} and T into Eq. (10.3); the line goes through the data. Thus, we can assert that the best-fitting curve to these data is characterized by these values.

Doing a grid search is feasible because we have a good idea of the true values of the regression parameters and could search in the neighborhood of these values. There are, however, several problems with a grid search that make it impractical for most situations:

- The parameter estimates are only approximations of those that minimize the residual sum of squares; the true parameter estimates almost always lie between values used in the grid search.

TABLE 10-1 Residual Sums of Squares (percent²) for Grid Search for Parameter Estimates of Martian Moods Function

T, days	I_{max}, percent										
	10	12	14	16	18	20	22	24	26	28	30
10	3033	2630	3722	2121	1362	2473	3550	3583	3354	3079	2894
12	2981	2515	3788	1909	934	2297	3392	3668	3337	2974	2760
14	3079	2554	3996	1854	641	2267	3594	3911	3461	2999	2758
16	3327	2749	4347	1956	484	2383	3956	4312	3724	3152	2888
18	3725	3100	4841	2215	462	2646	4478	4873	4128	3435	3149
20	4274	3606	5479	2632	576	3054	5160	5592	4672	3847	3542
22	4972	4267	6259	3205	827	3608	6001	6470	5357	4388	4067
24	5820	5084	7182	3936	1212	4308	7003	7506	6181	5059	4724
26	6818	6056	8249	4824	1734	5155	8165	8702	7146	5859	5512
28	7966	7184	9458	5870	2391	6147	9487	10,055	8521	6788	6433
30	9265	8468	10,810	7073	3184	7285	10,969	11,568	9496	7846	7485

- The results of the search depend strongly on having a good initial guess of the true parameter values; one often lacks this knowledge. If the grid is established over a range of values that do not contain the true values of the parameters to be estimated, there is nothing in the approach that will warn you.

- This method requires extensive computation, particularly if you establish a fine grid to get accurate estimates of the parameters. This inefficiency in the computation costs time and, often, money.

- There is no way to estimate the precision of the parameter estimates (i.e., associate standard errors with the parameter estimates).

- These problems become even more acute as the number of parameters increases, because the number of different combinations of the parameters, and thus points on the grid, increases rapidly.

We clearly need a better way to identify the best estimates of the parameters in Eq. (10.3).

FINDING THE BOTTOM OF THE BOWL

To develop a more efficient approach to finding the values of the parameters in the regression model that minimize the residual sum of squares, let us plot the values we found in our grid search. Figure 10-3 shows a plot of SS_{res} vs. the values of I_{max} and T used to compute SS_{res} using

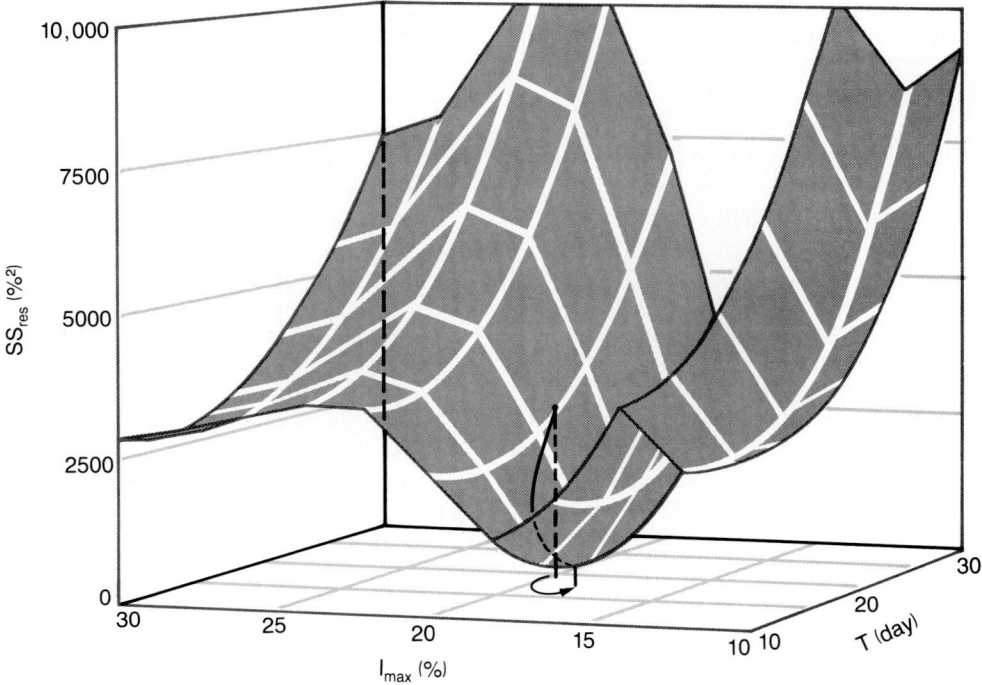

FIGURE 10-3 Residual sum of squares as a function of the parameters, maximum interest in invading earth I_{max}, and a period of variation of interest T. The best estimates of I_{max} and T are those values which correspond to the smallest residual sum of squares, i.e., the lowest point on the surface. The method of steepest descent involves making successive estimates of the parameters I_{max} and T that produce the most rapid reductions in the residual sum of squares SS_{res}. The figure illustrates the trajectory down the error surface beginning with first guesses of $I_{max} = 20$ percent and $T = 20$ days. The line on the error surface shows the values of SS_{res} associated with successive iterations as SS_{res} is approached. The arrow in the I_{max}-T plane shows the corresponding estimates of I_{max} and T on successive iterations from the first guess to the best estimate of $I_{max} = 18.4$ percent and $T = 17.6$ days. Note that there are local minimums on both the left and right of the surface. Bad first guesses for parameter values might cause an algorithm to converge on the local minimum, as was the case in Fig. 10-2D, which corresponds to the local minimum on the right side of this figure.

Eq. (10.4). Note that this plot is different from all those we have considered before; instead of plotting the value of the dependent variable I against the value of the independent variable t, we have plotted the residual sum of squares SS_{res} against the possible values of the two parameters in the regression equation (I_{max} and T). Because SS_{res} depends on the regression equation, the values of the parameters, and the specific data set used to compute SS_{res}, the precise shape of the surface in Fig.

10-3 will depend on the specific data set used to do the fit. The values of I_{max} and T that lead to the best fit (or, equivalently, the smallest residual sum of squares), are those that define the lowest point on this surface. The problem, then, is one of finding "the bottom of the bowl."

Examining the surface in Fig. 10-3 reveals that the minimum sum of squared residuals corresponds to a deep crater in the surface. We want to find the bottom of that crater. We will explore three strategies: *steepest descent, Gauss-Newton,* and *Marquardt's method,* which is a combination of the first two methods. All three methods are *iterative,* which means they involve making successive estimates of the regression parameters based on the current estimate until we cannot locate values for the parameters that will further reduce the residual sum of squares. Like a grid search, these methods require many evaluations of SS_{res} for different values of the parameters. By following a specific strategy, however, we can greatly reduce the number of such estimates needed, and so the computations are much more efficient than a grid search. In addition, the values of the estimates are not constrained to fall at specific points on a grid. Most of these algorithms should converge to an estimate of the regression coefficients in 5 to 10 steps. We will examine each approach in qualitative terms, then develop the corresponding mathematical embodiment of the methods.

The Method of Steepest Descent

The idea behind the *method of steepest descent* or *gradient method* is very simple: Find the point on the SS_{res} surface corresponding to the initial guess of the regression parameters, then move a step down the surface in the steepest direction computed from the derivatives of SS_{res} with respect to the regression parameters, reevaluate the derivatives, and take another step, until there are no further reductions in SS_{res} (Fig. 10-4).

Although the method of steepest descent moves quickly down toward the minimum residual sum of squares, it often has trouble converging, because it bounces back and forth across the bottom of the bowl. To deal with this problem, most algorithms that use the method of steepest descent have a provision for reducing the step size (usually cutting it in half) if SS_{res} increases on any step. Thus, the step size often decreases as the algorithm converges (Fig. 10-4C). The search terminates when the estimate reaches the bottom of the bowl.

Figure 10-3 illustrates the path of successive estimates obtained by a steepest descent algorithm for estimating the parameters I_{max} and T to fit the data on Martian interest in invading earth, beginning with the initial guess of $I_{max} = 20$ percent and $T = 20$ days. Figure 10-5 also shows the corresponding computer printout, which presents the successive estimates of these parameters as the computer applied the steepest

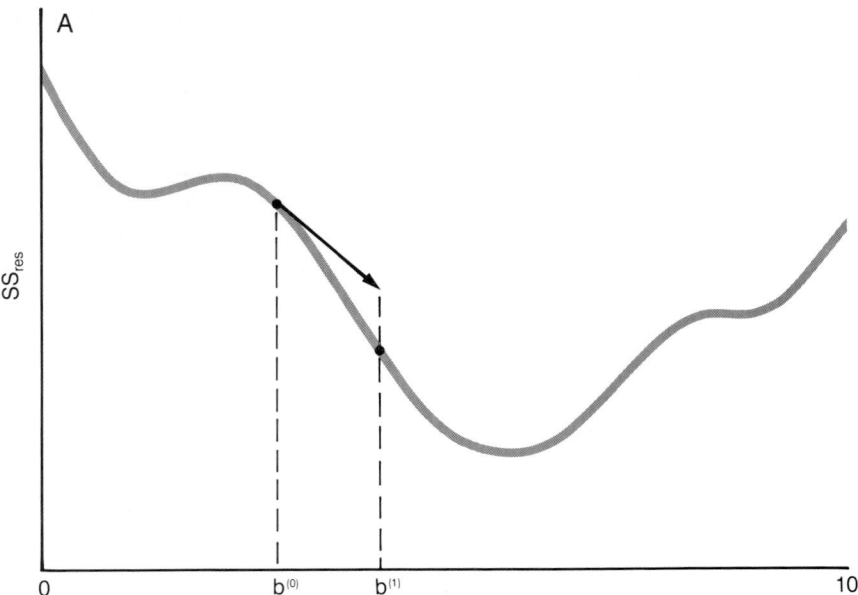

FIGURE 10-4 In the method of steepest descent, one begins with a first guess of the regression parameter (or parameters), $b^{(0)}$, and estimates the shape of the error surface locally, then takes a step down the gradient in the direction of steepest descent, i.e., most rapidly falling SS_{res}, and continues stepping until reaching the minimum of SS_{res}. If the step size is fixed, as in panels A and B, this method moves toward the minimum point quickly but then often has a hard time converging because of a tendency to overshoot the "bottom of the bowl," as in panel B. One deals with this problem by implementing an algorithm which reduces the step size by one-half if SS_{res} does not decrease on subsequent steps, as shown in panel C.

descent method. The first column shows the iteration number; the next two columns, the estimates of I_{max} and T; and the final column, the corresponding values of SS_{res}. The initial guess corresponds to step (or iteration) 0 and is associated with a residual sum of squares of 3054 percent^2. After the first step (indicated by the line with a dot under the line corresponding to iteration 0), the estimates of I_{max} and T have been modified to 15.0 percent and -7.8 days (an impossible value), respectively, and SS_{res} has actually *increased* to 3512 percent^2. This situation occurred because the step size was too big, and the method of steepest descent "overshot" the values of I_{max} and T that would minimize the residual sum of squares (analogous to the situation in Fig. 10-4B). Because SS_{res} increased, the program cuts the step size in half, and tries again, this time with estimates of $I_{max} = 17.5$ percent and $T = 6.1$ days (on the next line of the output). These estimates correspond to a residual sum of squares of 3176 percent^2, which is better than before, but still higher

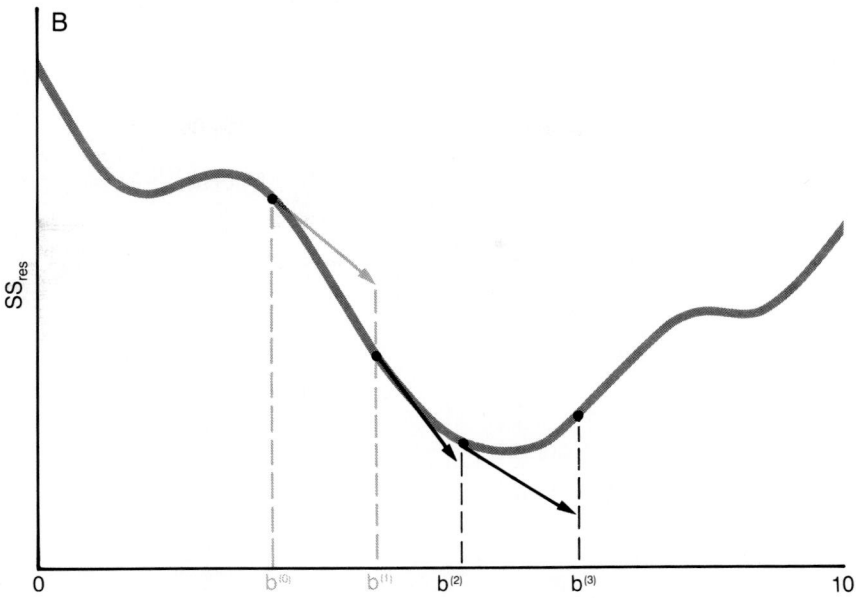

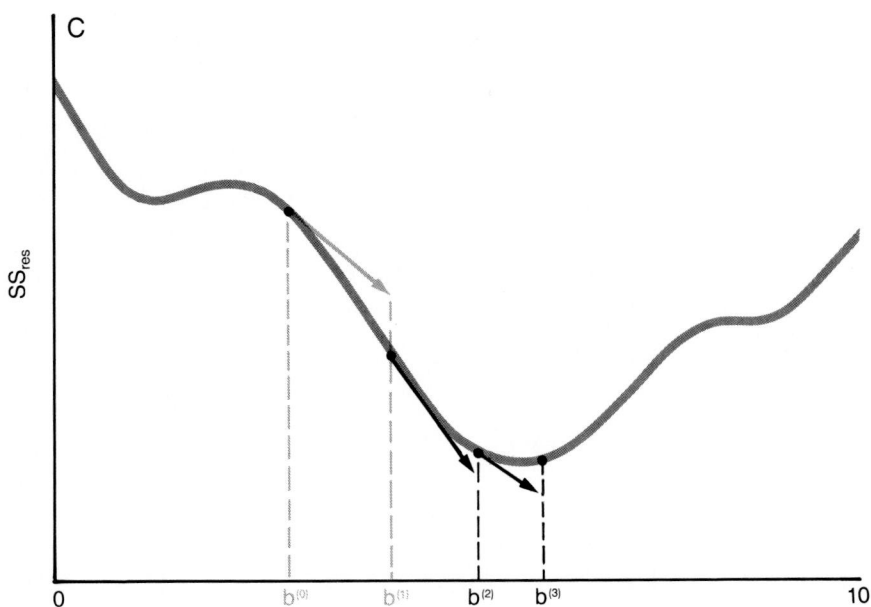

FIGURE 10-4 (*continued*)

```
          NON-LINEAR LEAST SQUARES ITERATIVE PHASE

      DEPENDENT VARIABLE: I      METHOD: GRADIENT

ITERATION        IMAX            T          RESIDUAL SS

    0      20.000000000    20.000000000    3054.395917347
    .      15.009290905    -7.840489       3511.976480400
    .      17.504645452     6.079755545    3176.425036735
    .      18.752322726    13.039877773    3842.482827310
    1      19.376161363    16.519938886    1476.300622809
    .      17.074867931    41.878331156    3695.557106847
    .      18.225514647    29.199135021    3265.140838357
    .      18.800838005    22.859536954    5123.082077075
    .      19.088499684    19.689737920    2448.576073178
    2      19.232330524    18.104838403     581.470288195243
    .      16.608510092    -8.177918       3575.424779278
    .      17.920420308     4.963460267    3817.707932216
    .      18.576375416    11.534149335    3599.073788190
    .      18.904352970    14.819493869    4595.664621589
    .      19.068341747    16.462166136    1529.671156378
    3      19.150336135    17.283502270     517.716001976980
    .      15.972453173    42.883421892    3522.817001115
    .      17.561394654    30.083462081    3074.453203558
    .      18.355865395    23.683482175    5034.349902940
    .      18.753100765    20.483492222    3353.948307828
    .      18.951718450    18.883497246    1361.287555470
    .      19.051027293    18.083499758     555.581937447747
    4      19.100681714    17.683501014     409.113391909493
```

```
   13      18.537299135    17.590362712     380.778677792725
    .       5.139413641    39.351543376    3581.291519516
    .      11.838356388    28.470953044    2917.240369808
    .      15.187827761    23.030657878    4120.152335614
    .      16.862563448    20.310510295    2805.509025648
    .      17.699931292    18.950436504    1333.720247915
    .      18.118615213    18.270399608     648.006922419481
    .      18.327957174    17.930381160     438.720485387043
    .      18.432628154    17.760371936     388.172295628141
   14      18.484963645    17.675367324     378.795861662684
    .       6.027221767    -4.661086       3693.543827697
    .      12.256092706     6.507140537    3050.578433987
    .      15.370528175    12.091253931    2649.598916119
    .      16.927745910    14.883310628    3973.026546507
    .      17.706354777    16.279338976    1661.421218882
    .      18.095659211    16.977353150     698.363852453513
    .      18.290311428    17.326360237     443.568841047413
    .      18.387637536    17.500863781     387.062414229037
   15      18.436300590    17.588115553     376.951036323348
```

NOTE: CONVERGENCE CRITERION MET.

```
   NON-LINEAR LEAST SQUARES SUMMARY STATISTICS     DEPENDENT VARIABLE I

      SOURCE            DF  SUM OF SQUARES    MEAN SQUARE

      REGRESSION         2   5165.3989637     2582.6994818
      RESIDUAL          48    376.9510363        7.8531466
      UNCORRECTED TOTAL 50   5542.3500000

      (CORRECTED TOTAL) 49   1776.9658000
```

PARAMETER	ESTIMATE	ASYMPTOTIC STD. ERROR	ASYMPTOTIC 95 % CONFIDENCE INTERVAL	
			LOWER	UPPER
IMAX	18.43630059	0.67714471913	17.074813222	19.797787959
T	17.58811555	0.09840870134	17.390252083	17.785979023

```
      ASYMPTOTIC CORRELATION MATRIX OF THE PARAMETERS

            CORR            IMAX               T

            IMAX           1.0000            0.0813
            T              0.0813            1.0000
```

than the value associated with iteration 0. The program cuts the step size in half again, and again (the next two lines in the output), before taking a small enough step that SS_{res} falls below 3054 percent^2, to 1476 percent^2. The corresponding estimates of $I_{max} = 19.4$ percent and $T = 16.5$ days become the starting point for the next iteration, iteration number 1. After 15 iterations, the program finally converged to estimates of $I_{max} = 18.4$ percent and $T = 17.6$ days, with $SS_{res} = 377$ percent^2 (Fig. 10-2C).

As long as the first guess is somewhere in the deepest crater, the method of steepest descent will move down toward the point corresponding to the smallest residual sum of squares. If, however, the first guess is not in the deepest crater, the methods of steepest descent (and all the other methods) will move down to a *relative minimum,* which will give the smallest SS_{res} for values locally around the first guess, but not the best fit in terms of the *global minimum.* The problem of converging to a relative minimum is always present when doing nonlinear regression, especially because one rarely has a good idea of what the SS_{res} surface actually looks like.

For example, suppose we had started with a first guess of $I_{max} = 16$ percent and $T = 12$ days in our Martian problem. Figure 10-3 shows that the algorithm would converge to estimates of $I_{max} = 13$ percent and $T = 12$ days with $SS_{res} = 2489$ percent^2. Comparing this residual sum of squares with the value of 377 percent^2 that we obtained before (in Fig. 10-5) reveals that we have converged to a local minimum, because of the shallow trough paralleling the deeper global minimum. We should use the results of the first analysis, because SS_{res} is smaller. The problem is that if you run only one analysis, how can you be sure that you have found the global minimum SS_{res}?

Most important, *always look at a plot of the regression equation superimposed on the data.** If you have converged to a local minimum,

---

*It happens that nonlinear regression problems often have a single independent variable (such as time), which makes examining these plots possible. If there are two or more independent variables, examine the residual plots, as we did for linear regression in Chap. 4.

◀ **FIGURE 10-5** Results of the steepest descent (gradient) method for computing the parameter estimates from the data in Fig. 10-2A. Each dot under the "iteration" heading indicates one step as the algorithm searches for a decrease in SS_{res}. Several of the intermediate steps have been omitted. Once the algorithm has converged, it is possible to estimate approximate standard errors and construct approximate confidence intervals (or, equivalently, t tests) for the coefficients in the model. It is also possible to construct an F test for the overall regression fit.

the resulting fit will almost never look very good. For example, Fig. 10-2 shows plots of Eq. (10.3) with the values of I_{max} and T from the two analyses we just did. Note that the parameter estimates obtained from the second set of first guesses produces a curve that obviously is not a good description of the data (Fig. 10-2D). In the event that a plot of the predicted function looks suspicious, run the program again with first guesses of the model parameters that are very different from the ones you used before to see if the program converges to a different, more reasonable, set of estimates. In fact, if there is any uncertainty that the first guess is reasonably close to the true parameter values, it is a good idea to run the program several times with varying first guesses to make sure that it always converges to the same point.*

The Gauss-Newton Method

It should be clear by now that the shape of the error surface is very important, and that the only time we can be sure that an iterative non-linear least-squares estimation will always converge to the correct values of the parameter estimates is when there are no local minima, i.e., when the error surface is a simple bowl. This situation always exists in linear regression, where the error surface is a (multidimensional) parabolic bowl. In fact, if the regression model is linear in the regression parameters, we can simply compute the location of the bottom of the bowl and immediately jump to it. The Gauss-Newton method takes advantage of this fact by matching the shape of the error surface in the region of our current estimate of the regression parameters with the surface for a linear regression problem that has the same local characteristics. (This process is analogous to approximating a curved line over a limited range with a straight line that is tangent to the curve at the point of interest.) The error surface for this associated linear regression problem is a well-behaved paraboloid, so we can jump to the bottom of its bowl to obtain our next estimate of the regression parameters. We repeat this process until the algorithm converges.

To illustrate this process, suppose we have a problem in which we are trying to estimate a single parameter β with the statistic b. We can plot SS_{res} against b on a simple graph (Fig. 10-6A). We begin with an initial guess of $b^{(0)} = 2$, where the superscript "(0)" indicates that it is

---

*In fact, several computer programs have the capability to do an automatic grid search (using coarse values of the parameters) to find the best first guess before beginning the iterative process.

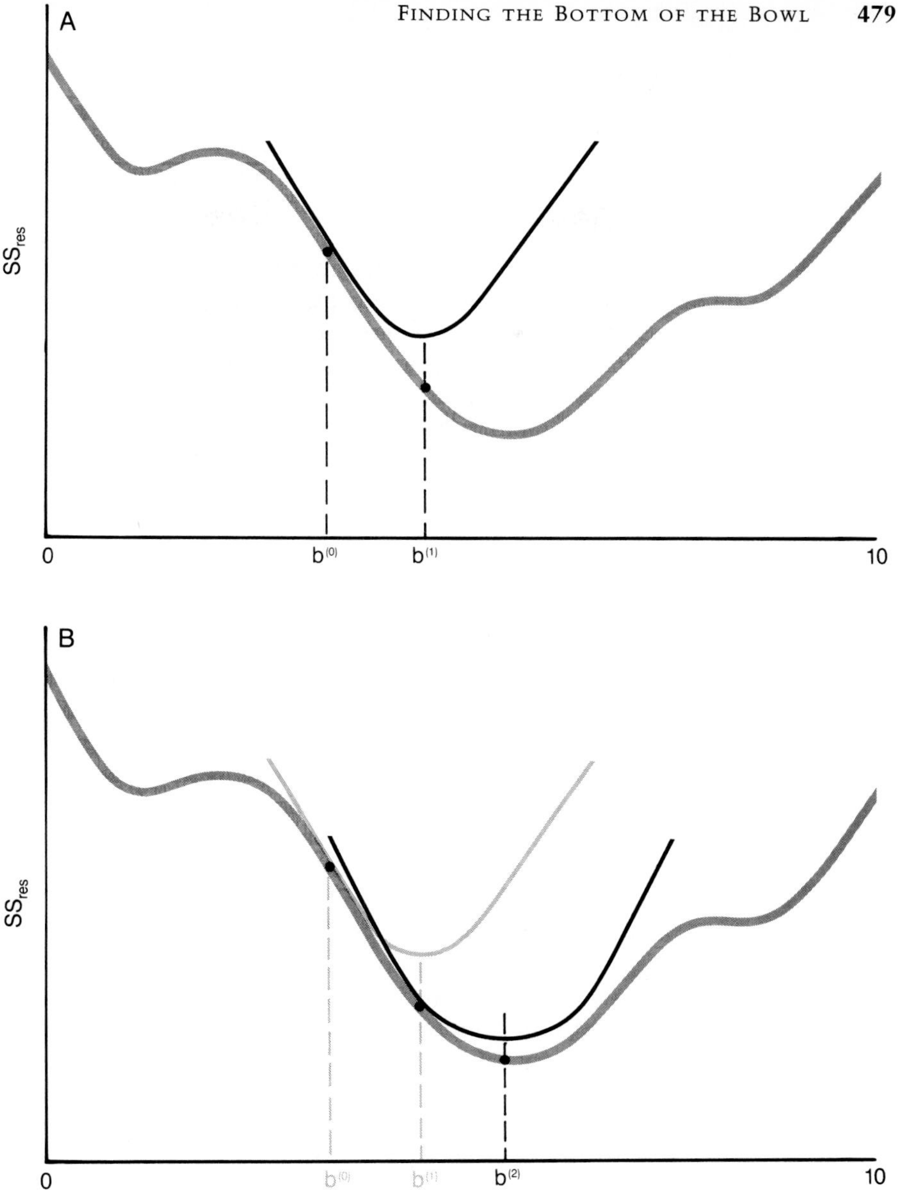

FIGURE 10-6 A schematic description of the Gauss-Newton method. At each step the algorithm estimates the shape of the error surface based on its local characteristics and then computes the parameter estimates based on the linear regression problem that would have the same local shape as the error surface. Because the error surface in a linear regression is a well-behaved parabolic bowl, the algorithm can analytically compute the value of the regression parameters corresponding to the minimum of the linear approximation of the error surface and then uses this new value to locate the next point on the error surface and fit a new local linear regression approximation. This process repeats until the method converges.

the initial value of b. We next solve the corresponding linear regression problem whose error surface has the same shape (we will define more precisely what we mean by "same shape" in the following section) as does the error surface for the nonlinear function when $b = 4$. Next, we jump to the bottom of the well-behaved bowl the associated linear regression problem defines to find our next estimate of $b^{(1)} = 5$, then repeat the entire process over again, this time finding the associated linear regression problem that matches the shape of the error surface at $b^{(1)} = 5$ (Fig. 10-6B) to find the next estimate of $b^{(2)} = 6$. We repeat this process until subsequent steps produce no further reductions in SS_{res}, i.e., the method has converged on the final estimate.

This process directly generalizes to more dimensions, with the parabola in Fig. 10-6 becoming a parabolic bowl.

There are two potential benefits to the Gauss-Newton approach. First, by converting the estimation problem into an associated linear regression problem at each step, the changes in the parameter estimates may be larger than with a steepest descent algorithm, so the Gauss-Newton algorithm may converge with fewer iterations than a steepest descent algorithm, particularly if the error surface for the nonlinear function is reasonably well behaved (i.e., more or less bowl shaped). Second, because the nonlinear problem is approximated with a linear problem, we can use the results to derive approximate standard errors of the parameter estimates once the algorithm has converged.

The major problem associated with the Gauss-Newton method is that it essentially involves making an extrapolation outside the region in which you currently have information about the error surface (at the current step), and these extrapolations can be very inaccurate. Sometimes the algorithm can converge very slowly, or even *diverge*, and produce parameter estimates that lead to *increases* in SS_{res} on subsequent steps. These problems arise when the changes in parameter estimates with each step are too big. Most computer programs that do Gauss-Newton estimations check to make sure that the algorithm is not diverging; if the SS_{res} increases on any step, they halve the increments in each of the parameter estimates, then try again, and continue halving the increments until SS_{res} decreases. This modification to the pure Gauss-Newton method significantly improves its performance.

Figure 10-7 shows the results of fitting Eq. (10.3) to the Martian data again, this time using the Gauss-Newton method. The program converged to estimates of $I_{max} = 17.4$ percent and $T = 17.6$ days after only 6 iterations, without any necessity to halve the step size. The corresponding residual sum of squares SS_{res} is 356 percent^2. Note that when the algorithm stopped iterating, the changes in the residual sum of squares was very small on the last iteration. This small change increases the

NON-LINEAR LEAST SQUARES ITERATIVE PHASE

DEPENDENT VARIABLE: I   METHOD: GAUSS-NEWTON

ITERATION	IMAX	T	RESIDUAL SS
0	20.000000000	20.000000000	3054.395917347
1	14.026631135	18.796095749	1179.794948809
2	16.462061213	17.383686918	404.116185505712
3	17.314201285	17.628330277	355.794455387420
4	17.360232086	17.618687846	355.684558514581
5	17.360327718	17.618769957	355.684553671097
6	17.360327508	17.618769527	355.684553670966

NOTE: CONVERGENCE CRITERION MET.

NON-LINEAR LEAST SQUARES SUMMARY STATISTICS   DEPENDENT VARIABLE I

SOURCE	DF	SUM OF SQUARES	MEAN SQUARE
REGRESSION	2	5186.6654463	2593.3327232
RESIDUAL	48	355.6845537	7.4100949
UNCORRECTED TOTAL	50	5542.3500000	
(CORRECTED TOTAL)	49	1776.9658000	

PARAMETER	ESTIMATE	ASYMPTOTIC STD. ERROR	ASYMPTOTIC 95 % CONFIDENCE INTERVAL LOWER	UPPER
IMAX	17.36032751	0.65815073472	16.037030011	18.683625005
T	17.61876953	0.10183828069	17.414010442	17.823528611

ASYMPTOTIC CORRELATION MATRIX OF THE PARAMETERS

CORR	IMAX	T
IMAX	1.0000	0.0772
T	0.0772	1.0000

FIGURE 10-7 Results of the Gauss-Newton method for fitting the data in Fig. 10-2*A*. In contrast to the results of the steepest descent algorithm in Fig. 10-5, the Gauss-Newton method converged after only six steps.

confidence you can have in the results.* For this problem, the Gauss-Newton method is clearly superior to the steepest descent method.

As with the steepest descent method, it is important to have a good first guess of the regression parameters. Otherwise, the Gauss-Newton method may converge to a local minimum, just like steepest descent.

Marquardt's Method

The steepest descent and Gauss-Newton methods have complementing strengths and weaknesses. Assuming a reasonable first guess at the parameters, the steepest descent algorithm will converge toward the minimum SS_{res}, even on relatively "bumpy" error surfaces, but has difficulty converging to the final estimates once it gets close. The Gauss-Newton method will move toward the minimum SS_{res} more quickly on well-behaved error surfaces and is particularly good near the minimum SS_{res} point, where the error surface is usually well approximated by the paraboloid the Gauss-Newton method uses. Marquardt's method is a combination of these two methods that takes advantage of the best characteristics of both. The precise relationship between the Marquardt method and the other two will become obvious when we present its mathematical formulation in the next section, but we can describe it in qualitative terms without the actual mathematics.

Marquardt's method uses a parameter λ which sets the balance between the steepest descent and Gauss-Newton methods. When λ is very small (say, less than about 10^{-8}), Marquardt's method is equivalent to the Gauss-Newton method. When λ is very large (say, greater than about 10^4; the specific value depends on the problem), Marquardt's method is equivalent to steepest descent, with step sizes of $1/\lambda$. The algorithm proceeds as follows: On each step compute the increment in the parameter estimates with λ set to a small number, say 10^{-8}, then compute SS_{res} for the new parameter estimates. If SS_{res} is smaller than the previous step, simply go on to the next step. If SS_{res} does not fall, increase λ by a factor of 10 and repeat the process again until SS_{res} falls. This process is equivalent to gradually shifting from a Gauss-Newton procedure to a steepest descent procedure, with progressively smaller step sizes. λ is

---

*This result compares with a change from 387.06 to 376.95 percent^2 on the last two steps of the steepest descent results in Fig. 10-5. The reason for this difference is that we used a less strict convergence criterion of 5×10^{-3} for the steepest descent, compared with 10^{-8} for the Gauss-Newton method because the steepest descent method never converged when the convergence criterion was set to 10^{-8}. In this example we used SAS procedure NLIN, for which convergence occurs when $(SS_{res}^{(i-1)} - SS_{res}^{(i)})/(SS_{res} + 10^{-6}) < 10^{-8}$, where i is the ith iteration.

reset to 10^{-8} at the beginning of each step. The value of λ tells you what the algorithm is doing.

The results of fitting our Martian data with the Marquardt method are identical to those we obtained with the Gauss-Newton method (Fig. 10-7) because the Marquardt algorithm did not have to increase λ.

As with the steepest descent and Gauss-Newton methods upon which it is based, Marquardt's method requires that you provide it a good first guess of the regression parameters to avoid converging to a relative minimum.

Where Do You Get a Good First Guess?

All nonlinear regression methods require a good first guess to find the global minimum in the residual sum of squares and the associated best estimates of the regression parameters. There are several ways to make these first guesses. Often you can estimate the parameters based on prior experience. Sometimes simply examining the data will give you an idea of the right numbers. For example, in our study of Martian interest in invading earth, simple examination of Fig. 10-2A shows that the maximum interest $I_{\max}$ is around 20 percent and the period of the variation T is around 20 days. There are times when you can simplify the nonlinear equation by linearizing it or dropping some terms, and the simplified equation can be used to compute a guess of the parameters using simple algebraic manipulations. Finally, you can always conduct a coarse grid search to find reasonable starting values.

How Do You Tell You Are at the Bottom of the Bowl?

The question of how to tell you are at the bottom of the bowl is really two questions: (1) When should the computer stop tying to find values of the regression parameters that will further reduce the residual sum of squares? and, (2) How can you be sure that the program has not stopped at a local minimum?

The first question deals with the technical issue of the *convergence criterion* used to stop the program, and the second deals with a question of judging whether or not the results are reasonable. There are several different convergence criteria that can be used to tell a nonlinear regression program to stop. The most common criterion is the stop when SS_{res} stops decreasing on subsequent steps, with even small step sizes. This limit can be specified in absolute units or in terms of a fractional change between steps. Occasionally, rather than using the SS_{res} to define the convergence criterion, the programs stop when the parameter estimates themselves stop changing on subsequent steps. All programs also have a maximum number of permissible steps to protect against the program running forever if the computations diverge. *Never accept the parameter estimates from a nonlinear regression computation when the program terminated because of too many iterations.*

Most programs provide default values of the convergence criterion; be sure you know and approve of the one your program is using. If you wish to change the convergence criterion, all programs permit you to override the default value in one way or another. Although the convergence criterion should be set small enough to ensure that no further improvement is possible in SS_{res} (at least locally), be careful not to set the convergence criterion too small. For example, if you set it to zero, the program will never stop, simply because of numerical round-off errors, and may actually diverge. Examining the residual sum of squares for comparable problems will give you a good idea of what a reasonable convergence criterion would be.

There are also some numerical issues that can be important in getting accurate results from nonlinear regression analysis that can affect convergence. The programs work best when the shape of the error surface is reasonably symmetrical along all the directions defined by the different regression parameters. In other words, it is better to have a nice round crater than a deep but narrow canyon with a poorly defined minimum point along one of the directions. The reason for this is that the calculations are done to a finite number of digits of accuracy in the computer, and large differences in the scales on which the parameters are measured can cause problems. The solution to this problem is to see that the variables are scaled so that they are all of a comparable order of magnitude, or at least not more than 1 or 2 orders of magnitude different. Often simply rescaling a problem will help solve convergence problems.

Another possible source of numerical problems arises from the fact that all these methods require calculating the derivatives of the regression function in order to define the increments in the dependent variables on each step. Some programs require that you provide formulas for these derivatives; others estimate the derivatives numerically themselves. Although this capability is a convenience, it can be a source of small, but troublesome errors in the computation that both slow the computations and introduce errors. If you are having convergence problems, see if you can provide formulas for these derivatives.* Doing so will make the computations faster and more accurate and may solve convergence problems.

---

*Another source of convergence problems is incorrectly specifying the derivatives. Sometimes the algorithm will converge, albeit slowly or to a relative minimum, even if the derivatives are incorrectly specified.

The more difficult problem is that of convincing yourself that the program has not settled into a local minimum. There are several indicators of possible difficulty:

- The regression function with the parameter values the program produced does not describe the data in a reasonable way. This procedure is the single best test of the regression parameters. *Always examine a plot of the raw data with the final regression equation.*

- The final estimates of the regression parameters are unreasonable values. Often you will have some idea of the magnitude or sign of the regression parameters (for example, a mass could not be negative). If the parameter values are unrealistic, you have probably located a relative minimum in the error surface.

- The algorithm had a difficult time converging. While the problems associated with conducting a nonlinear regression are involved, the algorithms are reasonably reliable for well-formulated problems and will often converge in under 5 to 10 iterations. If the algorithm takes many more steps than that, be suspicious. In addition, if the estimates of the individual parameters do not vary smoothly from step to step after the first few steps, be suspicious.

If any of these conditions exist, you should be concerned. Carefully check the regression model for appropriateness (including the specification of the derivatives) and check the data for errors. Start the algorithms from several widely different starting values and see if you can find a better (both in the sense of smaller residual sum of squares and also fit to the data) set of regression coefficients. If these problems persist, it is good evidence that there is something wrong with the regression model you are using. Often reformulating the model with fewer parameters will solve the problem.

MATHEMATICAL DEVELOPMENT OF NONLINEAR REGRESSION ALGORITHMS

We now turn our attention to converting the qualitative ideas outlined in the last section into the mathematical relationships that are actually used to compute the nonlinear regressions that we just described. For the sake of simplicity, we will begin with the simplest case of fitting a nonlinear function of a single independent variable x containing a single parameter β:

$$y = f(x, \beta) \tag{10.5}$$

to a set of observations. We will then generalize these results to the multivariate case of a nonlinear function of k independent variables containing p parameters.[*] We will also show how the approximations used to derive the Gauss-Newton method lead directly to estimates of the standard errors of the regression parameters in nonlinear regression.

For the more general case, we use matrix notation to write

$$y = f(\mathbf{x}, \boldsymbol{\beta})$$

in which, as before, each observation $\mathbf{x}$ contains k independent variables,

$$\mathbf{x} = [x_1 \; x_2 \; \ldots \; x_k]$$

and the parameters given by

$$\boldsymbol{\beta} = \begin{bmatrix} \beta_1 \\ \beta_2 \\ \cdot \\ \cdot \\ \cdot \\ \beta_p \end{bmatrix}$$

We seek the estimate of $\boldsymbol{\beta}$ which minimizes SS_{res}; this estimate is

$$\mathbf{b} = \begin{bmatrix} b_1 \\ b_2 \\ \cdot \\ \cdot \\ \cdot \\ b_p \end{bmatrix}$$

in which each element b_i of $\mathbf{b}$ is the best estimate of the corresponding element β_i of $\boldsymbol{\beta}$. In either case, we seek the estimate (b or $\mathbf{b}$) of the regression parameters (β or $\boldsymbol{\beta}$) that minimizes the residual sum of squares

$$SS_{res} = \Sigma[Y_i - \hat{y}_i]^2 = \Sigma[Y_i - f(\mathbf{x}_i, \mathbf{b})]^2 \tag{10.6}$$

Because the values of the dependent and independent variables are specified for any specific problem, SS_{res} is a function of the value of the parameter or parameters. In each case we begin with a first guess for the parameters, then use the characteristics of the error surface in the neighborhood of that point to compute the changes in the parameters (increments) to go to the next step. This process is repeated until the convergence criterion is met.

---

[*]In linear regression (with an intercept) the number of parameters is always one more than the number of independent variables (to allow for the intercept in a multiple regression equation), so $p = k + 1$. This situation need not occur in nonlinear regression.

The Method of Steepest Descent

Figure 10-4 shows a plot of the residual sum of squares as a function of the single parameter β in Eq. (10.5). The initial guess is $b^{(0)}$. The slope of the relationship between SS_{res} and β at the point $\beta = b^{(0)}$ is

$$\frac{\partial SS_{res}}{\partial \beta}\Bigg|_{\beta = b^{(0)}} = -2\Sigma[Y_i - f(X_i, \beta)]\frac{\partial f(X_i, \beta)}{\partial \beta}\Bigg|_{\beta = b^{(0)}} \qquad (10.7)$$

where the vertical lines indicate that the derivatives are evaluated at the point $\beta = b^{(0)}$. The term in brackets is just the residual for the ith point computed using the current parameter estimates,

$$e_i^{(0)} = Y_i - f(X_i, b^{(0)}) \qquad (10.8)$$

The partial derivative, evaluated using the current estimate of the parameter β at each value of the independent variable, is

$$D(X_i, b^{(0)}) = \frac{\partial f(X_i, \beta)}{\partial \beta}\Bigg|_{\beta = b^{(0)}} \qquad (10.9)$$

It may be computed from an analytical form you (or the program) provide, or may be estimated numerically by the program. As discussed above, it is usually preferable to provide the analytical form for the derivative.

We take a step of length $h/2$ "down" the gradient defined by this equation by adjusting our parameter estimate with

$$b^{(1)} = b^{(0)} - 2h\frac{\partial SS_{res}}{\partial \beta}\Bigg|_{\beta = b^{(0)}}$$

Substitute from Eqs. (10.7) through (10.9) into this equation to obtain

$$b^{(1)} = b^{(0)} + h\,\Sigma\,D(X_i, b^{(0)})\,e^{(0)}$$

This is the equation for the method of steepest descent. In general, for the mth iteration,

$$b^{(m+1)} = b^{(m)} + h\,\Sigma\,D(X_i, b^{(m)})\,e^{(m)}$$

For the general case of k variables and p parameters, we define the matrix of partial derivatives, evaluated with the parameter values at the mth step

$$\mathbf{D}^{(m)} = [D_{ij}^{(m)}] = \left[\frac{\partial f(\mathbf{x}_i, \boldsymbol{\beta})}{\partial \beta_j}\Bigg|_{\boldsymbol{\beta} = \mathbf{b}^{(m)}}\right]$$

and the vector of residuals computed using the parameter estimates at the mth step

$$\mathbf{e}^{(m)} = \begin{bmatrix} Y_1 - f(\mathbf{x}_1, \mathbf{b}^{(m)}) \\ Y_2 - f(\mathbf{x}_2, \mathbf{b}^{(m)}) \\ \vdots \\ Y_3 - f(\mathbf{x}_3, \mathbf{b}^{(m)}) \end{bmatrix}$$

In terms of this more general nomenclature, the estimate of the p parameters $\boldsymbol{\beta}$ on the $(m + 1)$th step is

$$\mathbf{b}^{(m+1)} = \mathbf{b}^{(m)} + h\,\mathbf{D}^{(m)T}\,\mathbf{e}^{(m)} \tag{10.10}$$

In sum, the method of steepest descent proceeds as follows:

1. Select a first guess for the parameters $\beta_1, \beta_2, \ldots, \beta_p$ that comprise $\boldsymbol{\beta}$ and an initial step size h for the $(m = 0)$ step.
2. For the $(m + 1)$th step use Eq. (10.10) to estimate a new set of parameters and compute the associated residual sum of squares $SS_{res}^{(m+1)}$ with Eq. (10.6).
3. If $SS_{res}^{(m+1)} \geq SS_{res}^{(m)}$, the fit is not improved, so cut the step size in half, then go back to step 2.
4. If $SS_{res}^{(m+1)} < SS_{res}^{(m)}$, the fit is improved, so check to see if the convergence criterion has been met. If so, stop. If not, reset the step size back to its original value and go back to step 2.

The Gauss-Newton Method

As already described, the Gauss-Newton method works by converting the problem of computing the nonlinear regression problem to a series of linear regression problems. To do this, we replace the nonlinear function we are trying to fit with a linear approximation in the region of the parameter estimates, then solve the associated linear regression problem to come up with the revised estimates of the regression parameters. As with steepest descent, we repeat this process until we meet some specified convergence criterion. As before, we will begin with the simple case of a nonlinear equation with one parameter and one independent variable.

Just as we used the derivative of the SS_{res} function to obtain a linear approximation to derive the method of steepest descent, we can approximate the nonlinear function $y = f(x, \beta)$ in the region near $\beta = b^{(0)}$ with

$$y = f(x, \beta) \approx f(x, b^{(0)}) + (\beta - b^{(0)}) \left. \frac{\partial f(x, \beta)}{\partial \beta} \right|_{\beta = b^{(0)}} \tag{10.11}$$

(Figure 10-6 shows this approximation.) Therefore,

$$f(x, \beta) - f(x, b^{(0)}) \approx (\beta - b^{(0)}) \left. \frac{\partial f(x, \beta)}{\partial \beta} \right|_{\beta = b^{(0)}} \tag{10.12}$$

For the ith observation, we approximate $f(x_i, \beta)$ with the observed value Y_i, let $\Delta b^{(0)}$ be the (unknown) distance from the current estimate of $b^{(0)}$ to the true value of the regression parameter β, $\beta - b^{(0)}$, and use the notation in Eq. (10.9) for the derivative to obtain

$$Y_i - f(X_i, b^{(0)}) = e_i^{(0)} = \Delta b^{(0)} D(X_i, b^{(0)}) \tag{10.13}$$

We can now treat estimating Δb in this equation as a linear regression problem (with no intercept), given the other terms, which can be computed from the data. We then compute the next estimate of the parameter β with

$$b^{(1)} = b^{(0)} + \Delta b^{(0)}$$

and repeat this process until the method converges.

When there are p parameters and k independent variables, we replace Eq. (10.11) with

$$y = f(\mathbf{x}, \boldsymbol{\beta}) \approx f(\mathbf{x}, \mathbf{b}^{(0)}) + (\beta_1 - b_1^{(0)}) \left. \frac{\partial f(\mathbf{x}, \boldsymbol{\beta})}{\partial \beta_1} \right|_{\boldsymbol{\beta} = \mathbf{b}^{(0)}} + (\beta_2 - b_2^{(0)}) \left. \frac{\partial f(\mathbf{x}, \boldsymbol{\beta})}{\partial \beta_2} \right|_{\boldsymbol{\beta} = \mathbf{b}^{(0)}}$$

$$+ \cdots + (\beta_p - b_p^{(0)}) \left. \frac{\partial f(\mathbf{x}, \boldsymbol{\beta})}{\partial \beta_p} \right|_{\boldsymbol{\beta} = \mathbf{b}^{(0)}}$$

Using exactly the same substitutions as in Eqs. (10.11) through (10.13) above, we obtain, for the ith observation,

$$e_i^{(0)} = \Delta b_1^{(0)} D_{i1}^{(0)} + \Delta b_2^{(0)} D_{i2}^{(0)} + \cdots + \Delta b_p^{(0)} D_{ip}^{(0)}$$

But this equation is simply a linear regression problem in which the residuals e_i are the dependent variables, the derivatives $D_{i1}, \ldots, D_{ip}$ are the independent variables, and the increments in the regression parameter estimates in the original problem, $\Delta b_1^{(0)}, \Delta b_2^{(0)}, \ldots, \Delta b_p^{(0)}$, are the coefficients to be estimated. In matrix notation the equation becomes

$$\mathbf{e}^{(0)} = \mathbf{D}^{(0)} \, \Delta \mathbf{b}^{(0)} \tag{10.14}$$

where

$$\Delta \mathbf{b}^{(0)} = \begin{bmatrix} \Delta b_1^{(0)} \\ \Delta b_2^{(0)} \\ \vdots \\ \Delta b_p^{(0)} \end{bmatrix}$$

Equation (10.14) is just a linear regression problem in matrix notation with $\mathbf{e}$ playing the role of the dependent variable, $\mathbf{D}$ playing the role of the data matrix of independent variables $\mathbf{X}$, and $\Delta\mathbf{b}$ playing the role of the regression coefficients $\mathbf{b}$. From Eq. (3.18), the best least-squares estimate of Δb is

$$\Delta\mathbf{b}^{(0)} = (\mathbf{D}^{(0)T}\mathbf{D}^{(0)})^{-1}\,\mathbf{D}^{(0)T}\,\mathbf{e}^{(0)}$$

Finally, we compute the next estimate of the parameters in the original regression equation with

$$\mathbf{b}^{(1)} = \mathbf{b}^{(0)} + \Delta\mathbf{b}^{(0)}$$

In general, for the $(m + 1)$th step

$$\mathbf{b}^{(m+1)} = \mathbf{b}^{(m)} + h(\mathbf{D}^{(m)T}\mathbf{D}^{(m)})^{-1}\,\mathbf{D}^{(m)T}\,\mathbf{e}^{(m)} \tag{10.15}$$

where we have added the parameter h to allow for variable step sizes.

In sum, the Gauss-Newton method proceeds as follows:

1. Select a first guess $\mathbf{b}^{(0)}$ for the parameters $\beta_1, \beta_2, \ldots, \beta_p$ that comprise $\boldsymbol{\beta}$ and set the initial step size h to 1 for the $m = 0$th step.

2. For the $(m + 1)$th step use Eq. (10.15) to estimate a new set of parameters and compute the associated residual sum of squares $SS_{\text{res}}^{(m+1)}$ with Eq. (10.6).

3. If $SS_{\text{res}}^{(m+1)} \geq SS_{\text{res}}^{(m)}$, the fit is not improved, so cut the step size in half, then go back to step 2.

4. If $SS_{\text{res}}^{(m+1)} < SS_{\text{res}}^{(m)}$, the fit is improved, so check to see if the convergence criterion has been met. If so, stop. If not, reset the step size back to its original value and go back to step 2.

Note that this procedure is quite similar to that used in steepest descent. The difference lies in the way in which we change the parameter estimates from step to step.

Marquardt's Method

Marquardt's method is a combination of the steepest descent and Gauss-Newton methods. In Marquardt's method, one begins with a first guess $\mathbf{b}^{(0)}$, then uses it to estimate the regression parameters on the next step using the formula

$$\mathbf{b}^{(m+1)} = \mathbf{b}^{(m)} + (\mathbf{D}^{(m)T}\mathbf{D}^{(m)} + \lambda\mathbf{I})^{-1}\,\mathbf{D}^{(m)T}\mathbf{e}^{(m)} \tag{10.16}$$

in which λ is an adjustable parameter and I is a $p \times p$ identity matrix. To see the relationship between Marquardt's method and the other two methods, let us consider two extreme cases. First, if $\lambda = 0$, this equation becomes

$$\mathbf{b}^{(m+1)} = \mathbf{b}^{(m)} + (\mathbf{D}^{(m)T} \mathbf{D}^{(m)} + 0\mathbf{I})^{-1}\mathbf{D}^{(m)T}\mathbf{e}^{(m)} = \mathbf{b}^{(m)} + (\mathbf{D}^{(m)T}\mathbf{D}^{(m)})^{-1}\mathbf{D}^{(m)T}\mathbf{e}^{(m)} \quad (10.17)$$

which is identical to Eq. (10.15), with $h = 1$, the rule for the Gauss-Newton method. Second, suppose that λ is a large number. If it is, the elements of $\lambda\mathbf{I}$ will dominate those of $\mathbf{D}^{(m)T}\mathbf{D}^{(m)}$ in the term in parentheses in Eq. (10.16), and then $\mathbf{D}^{(m)T}\mathbf{D}^{(m)} + \lambda\mathbf{I} \approx \lambda\mathbf{I}$ and Eq. (10.16) becomes

$$\mathbf{b}^{(m+1)} \approx \mathbf{b}^{(m)} + (\lambda\mathbf{I})^{-1}\mathbf{D}^{(m)T}\mathbf{e}^{(m)} = \mathbf{b}^{(m)} + \left(\frac{1}{\lambda}\right) \mathbf{D}^{(m)T}\mathbf{e}^{(m)}$$

because the identity matrix is its own inverse (i.e., $\mathbf{I} = \mathbf{I}^{-1}$) and the identity matrix times any matrix leaves it unchanged. Comparing this result with Eq. (10.10) reveals that it is simply a steepest descent algorithm, with the step size h equal to $1/\lambda$. Thus, for small values of λ, the Marquardt method effectively follows the Gauss-Newton algorithm, and as λ increases, the behavior of the Marquardt method becomes similar to a steepest descent algorithm, with progressively smaller step sizes.

The implementation of the Marquardt method is very similar to the other two methods we have already discussed:

1. Select a first guess $\mathbf{b}^{(0)}$ for the parameters $\beta_1, \beta_2, \ldots, \beta_p$ that comprise $\boldsymbol{\beta}$ and set the initial value of λ to a very small number, say 10^{-8}, for the $m = 0$th step.

2. For the $(m + 1)$th step use Eq. (10.16) to estimate a new set of parameters and compute the associated residual sum of squares $SS_{res}^{(m+1)}$ with Eq. (10.6).

3. If $SS_{res}^{(m+1)} \geq SS_{res}^{(m)}$, the fit is not improved, so multiply λ by 10, then go back to step 2.

4. If $SS_{res}^{(m+1)} < SS_{res}^{(m)}$, the fit is improved, so check to see if the convergence criterion has been met. If so, stop. If not, reset λ back to its original value and go back to step 2.

Because it combines the desirable properties of both the steepest descent and Gauss-Newton methods, the Marquardt method is widely used to do parameter estimation in nonlinear regression.

HYPOTHESIS TESTING IN NONLINEAR REGRESSION

If the sample of data points used in the nonlinear regression analysis is drawn from a population that is normally distributed about the true relationship given by $y = f(\mathbf{x}, \boldsymbol{\beta})$ with mean 0 and a constant variance, independent of the value of the independent variables $x_1, x_2, \ldots, x_k$, it is possible to estimate the standard errors associated with the regression coefficients and conduct other hypothesis tests similar to those we used to analyze the results of linear regression. The fact that the Gauss-Newton method reduces the nonlinear regression problem to an approximate linear regression problem permits us to estimate the standard errors associated with the final regression coefficients. These estimates are theoretically accurate only as the sample size becomes large, so the resulting standard errors and associated statistical hypothesis tests should be viewed as only approximate.

We can estimate the standard deviation of the distribution of population members around the nonlinear function $f(\mathbf{x}, \boldsymbol{\beta})$ with the standard error of the estimate computed from the residual sum of squares obtained once the nonlinear regression has converged:

$$s_{y|x} = \sqrt{\frac{SS_{res}}{n - p}} = \sqrt{MS_{res}}$$

As with linear regression, the standard error of the estimate is a good measure of the overall predictive value of the regression equation.[*]

We can also conduct an approximate test for the overall fit of the regression equation by computing an F statistic defined exactly like we used for linear regression:

$$F = \frac{MS_{reg}}{MS_{res}}$$

where, as before,

$$MS_{reg} = \frac{SS_{reg}}{DF_{reg}} = \frac{SS_{tot} - SS_{res}}{p}$$

Note that the degrees of freedom depend on the number of *parameters*, not the number of independent variables. This F statistic is compared with the critical value for p numerator and $n - p$ denominator degrees of freedom.

We can obtain approximate estimates for the standard errors of the regression coefficients by taking the square roots of the diagonal elements of the matrix

$$\mathbf{s}_b^2 = s_{y|x}^2 (\mathbf{D}^T \mathbf{D})^{-1}$$

---

[*]The most common measure of goodness of fit in nonlinear regression is the standard error of the estimate $s_{y|x}$. Sometimes people also compute a correlation coefficient or coefficient of determination. In most instances, you will have to calculate R^2 yourself from the sums of squares reported by the computer. Some programs do not report the total sum of squares, so you will have to calculate it from the standard deviation of the dependent variable with

$$SS_{tot} = (n - 1)s_Y^2$$

You can then use this SS_{tot} with SS_{res} to compute R^2.

where **D** is the matrix of partial derivatives evaluated after the algorithm has converged.[*] These standard errors can then be used to construct an approximate test of whether or not the associated regression coefficients are significantly different from zero by computing the t statistics

$$t = \frac{b_i}{s_{b_i}}$$

and comparing the result with the t distribution with $n - p$ degrees of freedom. These standard errors can also be used to construct approximate confidence intervals for the regression coefficients.[†] Because the actual distribution of the parameters and standard errors only approximates those present in linear regression, you should be cautious in interpreting the results of these hypothesis tests, particularly when the results are near the borderline of statistical significance.

REGRESSION DIAGNOSTICS IN NONLINEAR REGRESSION

Before drawing any conclusions about the results of a nonlinear regression, one should carefully examine how good the fit of the equation is to the data as well as a set of residual plots to see that there are no obvious systematic deviations between the regression model and the data and that there are no violations of the equal variance assumption.

In addition, because the **D** matrix evaluated at the solution to the parameter estimation problem plays a role analogous to the **X** design matrix in linear regression, it is possible to derive all the regression diagnostics we developed in Chap. 4, including standardized and Studentized residuals, leverage values, and Cook's distance.[‡] As the sample size increases, the properties of these statistics approach those associated with linear regression problems. We need not be too concerned about the precise statistical properties of the regression diagnostics, however, because

[*]Note that **D** is playing the role of the design matrix **X** in linear regression.

[†]Several computer packages report the 95 percent confidence intervals for the regression coefficients as well as their standard errors of the estimate. Specifically, the $(1 - \alpha)$ confidence interval for the regression coefficient β_i is

$$b_i - t_\alpha s_{b_i} < \beta_i < b_i + t_\alpha s_{b_i}$$

where t_α is the two-tail critical value for the t distribution with $n - p$ degrees of freedom. Confidence intervals can also be used for hypothesis testing: if the $(1 - \alpha)$ confidence interval does *not* include 0, then the coefficient is significantly different from zero with $P < \alpha$. For more discussion of confidence intervals, see S. A. Glantz, *Primer of Biostatistics* (2nd ed.), New York, McGraw-Hill, 1987, Chap. 7, "Confidence Intervals."

[‡]These diagnostics are derived from $\mathbf{D}^T\mathbf{D}$ rather than $\mathbf{X}^T\mathbf{X}$.

we will use them to identify problems with the data or the regression model that warrant further investigation, rather than for formal hypothesis testing.

As with linear regression, the regression diagnostics can be used to identify erroneous data points or improve the formulation of the model by modifying the regression model.

Multicollinearity can occur in nonlinear regression problems just as in linear regression problems, and it produces the same ambiguities for interpreting the values of the regression coefficients. There are two ways to diagnose multicollinearity in nonlinear regression problems, again based on the **D** matrix after the problem has converged. Occasionally the program will report the variance inflation factor, or the tolerance, which is the reciprocal of the variance inflation factor. When available, you can use the variance inflation factor or tolerance to identify potential multicollinearities. In addition, most programs print the correlation matrix of the regression parameter estimates. Large values of the correlation — above about .90 in magnitude — between any two parameter estimates indicate that these estimates both depend on the same information in the independent variables, and so multicollinearity is a problem.

The solutions to multicollinearity problems in nonlinear regression are the same as we described in Chap. 5 for linear regression: Get more data to break up the multicollinearity, or eliminate some of the parameters from the regression model to make it easier to estimate the remaining parameters based on the available information.

EXPERIMENTING WITH DRUGS

Carnitine is a ubiquitous compound manufactured by the body and is particularly abundant in muscle cells, where it is essential for helping many long-chain fatty acid compounds get into mitochondria, where the energy in these fats is made available for muscle contraction. It is known that there can be deficiencies in carnitine when the body cannot make enough of it. Because of the reports of carnitine deficiency and the fact that it is so important in supporting muscle contraction, it is being investigated as a potential therapy for certain kinds of muscle disease. However, how the body responds to externally administered carnitine is not well characterized. This lack of information is important because the rational administration of any drug requires knowing how blood levels relate to the route of administration (a pill or an injection) and the dose. Therefore, Harper and coworkers* sought to study how humans responded to different doses of carnitine.

*P. Harper, C.-E. Elwin, and G. Cederblad, "Pharmacokinetics of Intravenous and Oral Bolus Doses of L-Carnitine in Healthy Subjects," *Eur. J. Clin. Pharmacol.* 35:555–562, 1988.

Like many drugs, the concentration of carnitine in the blood plasma falls following a biexponential function of time after an injection of the drug at time $t = 0$:

$$c(t) = Ae^{-t/\tau_1} + Be^{-t/\tau_2} \tag{10.18}$$

where $c(t)$ is the blood concentration of the drug and t is the time after giving the drug. The four parameters A, B, τ_1, and τ_2 describe the nonlinear relationship between carnitine and time and relate to the physiological characteristics of the way in which the carnitine is distributed within and removed from the body.[*] Harper and coworkers used this equation to describe the response of the body to different doses of carnitine, then compared these values to characterize how people respond to different doses. They were particularly interested in the values of the time constants τ_1 and τ_2.[†] The first step in this process is to describe the data from a single dose in a single individual to obtain the parameters in Eq. (10.18).

In order to fit these data, shown in Fig. 10-8, to Eq. (10.18), we need to use nonlinear regression. As we noted above, most computer programs that do nonlinear regression require that you specify the equation and the partial derivatives of the equation with respect to each of the parameters in the equation. Because these are the most generally reliable and efficient types of programs, we will illustrate their use. In this case, we enter Eq. (10.18), and the partial derivatives of $c(t)$ with respect to the four regression parameters A, B, τ_1, and τ_2:

$$\frac{\partial c}{\partial A} = e^{-t/\tau_1}$$

$$\frac{\partial c}{\partial B} = e^{-t/\tau_2}$$

[*]The study of drug distribution and clearance is known as *pharmacokinetics*. For a discussion of pharmacokinetic models for drug distribution and how the parameters in these models relate to the parameters in Eq. (10.18), see S. Glantz, *Mathematics for Biomedical Applications*, Berkeley, University of California Press, 1979, pp. 27–30, or J. A. Jaquez, *Compartmental Analysis in Biology and Medicine*, Ann Arbor, University of Michigan Press, 1985.

[†]Harper and coworkers actually used the equation

$$c(t) = Ae^{-\alpha t} + Be^{-\beta t}$$

This equation is equivalent to Eq. (10.18) by letting $\alpha = 1/\tau_1$ and $\beta = 1/\tau_2$. We can estimate τ_1 and τ_2 directly using Eq. (10.18) or indirectly by calculating them from the estimates of α and β in the equation above. However, because we want to think in terms of time constants, we should estimate them directly so that the standard errors and resultant hypothesis tests will follow directly from the parameter estimation. Had we estimated α and β, the standard errors would apply to those parameters, not the time constants in which we were really interested.

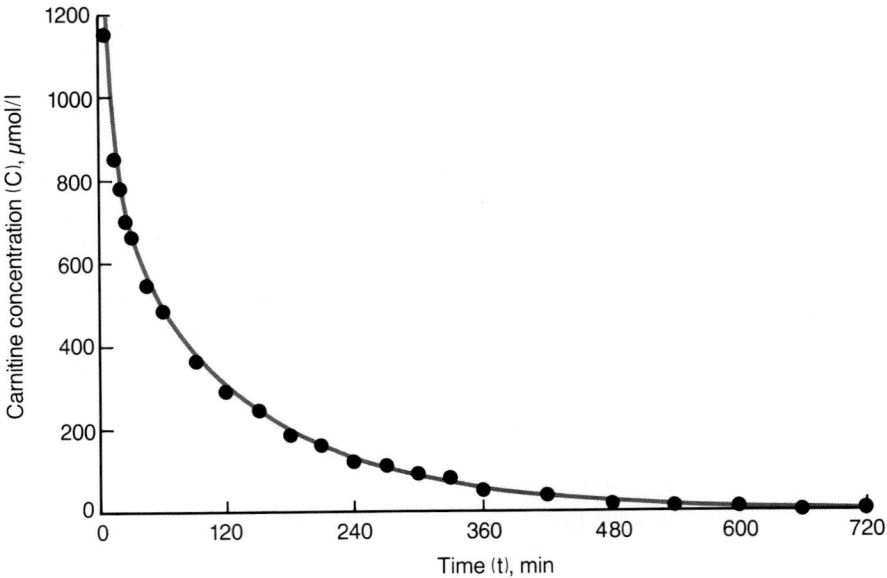

FIGURE 10-8 Carnitine concentration measured over time following an intravenous injection. These data are fitted with the sum of two exponentials given by Eq. (10.18). (*Modified from Figure 1 of P. Harper, C.-E. Elwin, and G. Cederblad: "Pharmacokinetics of Intravenous and Oral Bolus Doses of L-Carnitine in Healthy Subjects," Eur. J. Clin. Pharmacol. 35:555–562, 1988.*)

$$\frac{\partial c}{\partial \tau_1} = \frac{tAe^{-t/\tau_1}}{\tau_1^2}$$

$$\frac{\partial c}{\partial \tau_2} = \frac{tBe^{-t/\tau_2}}{\tau_2^2}$$

The results of Marquardt's method to fit these data (in Table C-19, Appendix C) to Eq. (10.18) are in Fig. 10-9. Before we examine the parameter estimates and influence diagnostics, we should examine the information supplied at each step of the iterative process used to estimate the parameters. It took 11 steps for the Marquardt algorithm to converge. After a little hunting around for the right direction to proceed on the first two steps (there were two or three step-size halvings needed to find the direction of a decrease in the residual sum of squares), the algorithm

FIGURE 10-9 Results of nonlinear least-squares fit to the data of Fig. 10-8 using the Marquardt algorithm. The residual and influence diagnostics computed from the approximate linear regression problem computed at the convergence point are also included.

```
ITERATION    RESIDUAL
NO. HALVING  SUM OF SQ    A          B          tau1       tau2

  0    0   ***********  600.00000  300.00000  100.00000  300.00000
  1    3   ***********  431.78857  492.67867   69.40136  193.64954
  2    2   72101.76405  136.12181  848.89167   15.25870  128.63712
  3    0   45175.94900  651.91563  716.49935    8.14820  136.76185
  4    0    3915.24429  567.12646  749.17349   12.64074  133.43888
  5    0    1255.86240  652.68167  721.79549   13.61766  137.42127
  6    0    1244.07325  653.98122  719.08576   13.63973  138.06859
  7    0    1244.06514  654.02718  718.95335   13.64602  138.10014
  8    0    1244.06503  654.02907  718.93648   13.64696  138.10319
  9    0    1244.06502  654.02921  718.93421   13.64709  138.10358
 10    0    1244.06502  654.02923  718.93390   13.64711  138.10364
 11    0    1244.06502  654.02923  718.93386   13.64712  138.10364
```

ITERATION   11 HAS THE SMALLEST RESIDUAL SUM OF SQUARES(SUBJECT TO CONSTRAINTS,
IF ANY).
REMAINING CALCULATIONS ARE BASED ON THE RESULTS OF THIS ITERATION.

ASYMPTOTIC CORRELATION MATRIX OF THE PARAMETERS

```
                 A          B        tau1      tau2

                 1          2          3         4
A       1     1.0000
B       2    -0.4007     1.0000
tau1    3    -0.0484    -0.8469     1.0000
tau2    4     0.4095    -0.8722     0.6652    1.0000
```

RESIDUAL MEAN SQUARE   65.4771

DEGREES OF FREEDOM   19

THE RESIDUAL MEAN SQUARE,   65.477   , IS USED IN COMPUTING
STANDARD DEVIATIONS.

PARAMETER	ESTIMATE	ASYMPTOTIC STANDARD DEVIATION	TOLERANCE
A	654.029230	17.855398	0.3000834880
B	718.933855	15.129389	0.0517034410
tau1	13.647115	0.794851	0.0946753399
tau2	138.103645	3.070895	0.2147066629

CASE NO. LABEL	PREDICTED c(t)	STD DEV OF PRED VALUE	OBSERVED c(t)	RESIDUAL	COOK DISTANCE	WEIGHT * DIAG HAT MAT	STANDARDIZED RESIDUAL	t
1	1146.77	7.95153	1150.00	3.22960	32.5565	0.965631	2.15290	5.00000
2	862.835	4.77468	850.000	-12.8349	0.515433	0.348176	-1.96463	15.0000
3	773.061	4.43901	780.000	6.93895	0.113213	0.300942	1.02563	20.0000
4	704.607	3.95540	700.000	-4.60692	0.334292E-01	0.238941	-0.652614	25.0000
5	651.154	3.64649	660.000	8.84639	0.955454E-01	0.203077	1.22465	30.0000
6	543.198	4.13031	545.000	1.80247	0.591066E-02	0.260541	0.259040	45.0000
7	473.651	4.56098	485.000	11.3488	0.335612	0.317707	1.69793	60.0000
8	375.579	3.86476	365.000	-10.5791	0.163605	0.228116	-1.48808	90.0000
9	301.625	3.09315	290.000	-11.6254	0.103416	0.146121	-1.55477	120.000
10	242.663	2.86556	245.000	2.33692	0.341868E-02	0.125410	0.308814	150.000
11	195.275	2.88916	185.000	-10.2746	0.674971E-01	0.127483	-1.35935	180.000
12	157.146	2.91840	160.000	2.85434	0.534689E-02	0.130077	0.378199	210.000
13	126.462	2.87698	120.000	-6.46230	0.264115E-01	0.126411	-0.854454	240.000
14	101.770	2.76259	110.000	8.22994	0.386215E-01	0.116558	1.08209	270.000
15	81.8991	2.59358	90.0000	8.10092	0.319730E-01	0.102733	1.05689	300.000
16	65.9080	2.39048	80.0000	14.0920	0.794321E-01	0.872732E-01	1.82288	330.000
17	53.0392	2.17061	50.0000	-3.03921	0.294653E-02	0.719573E-01	-0.389881	360.000
18	34.3491	1.72891	40.0000	5.65092	0.611122E-02	0.456513E-01	0.714860	420.000
19	22.2450	1.33075	20.0000	-2.24504	0.549817E-03	0.270462E-01	-0.281276	480.000
20	14.4063	0.999203	15.0000	0.593744	0.211648E-04	0.152482E-01	0.739420E-01	540.000
21	9.32973	0.736355	10.0000	0.670268	0.144430E-04	0.828105E-02	0.831782E-01	600.000
22	6.04209	0.534802	5.00000	-1.04209	0.182709E-04	0.436814E-02	-0.129066	660.000
23	3.91296	0.383917	5.00000	1.08704	0.102020E-04	0.225105E-02	0.134490	720.000

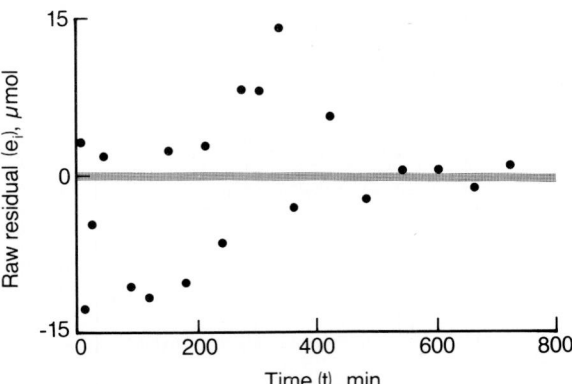

FIGURE 10-10 Residual plot corresponding to the analysis of Fig. 10-9. The residual pattern shows no problems.

quickly jumped to near the bottom of the bowl (steps 3, 4, and 5) and then fine-tuned the parameter estimates in steps 6 through 11.

The parameters A and τ_1 describe the fast decaying exponential, and B and τ_2 describe the slow phase of the concentration change.* The estimates of A and B are 654 and 719 μmol/L, with standard errors of 18 and 15 μmol/L, respectively. The estimates of τ_1 and τ_2 are 14 and 138 min, with standard errors of .8 and 3.1 min, respectively. In all cases the standard errors are small compared with the parameter estimates, which indicates that we have obtained quite precise estimates of the underlying population parameters for this individual. We could also use these standard errors to compute t test statistics to test whether or not the coefficients were significantly different from zero, but such a test is not really relevant to the question at hand.

Figure 10-10 shows a residual plot for these data, and Fig. 10-11 shows a normal probability plot of the residuals. Neither of these plots reveals any problem with the regression model. Had these plots shown an inconsistency with the equal variance or normality assumptions, we would have tried the variance stabilizing transformations on the dependent variable discussed in Chap. 4.

Figure 10-9 shows the diagnostics for this regression. There is only one Studentized residual r_i worth noting, 2.1529 (associated with the carnitine concentration at 5 min), but it is not too large. However, this same point has a large leverage value, .9656, compared with the suggested cutoff value of twice the average value, $2(p/n) = 2(4/23) = 2(.1739) = .3478$, and a very large Cook's distance of 32.5565, compared with the

*People often refer to the fast phase as the "distribution phase" and the slow phase as the "clearance (removal) phase." In fact, the magnitudes and time constants of both phases of the concentration vs. time curve depend on how the drug is distributed in and removed from the body.

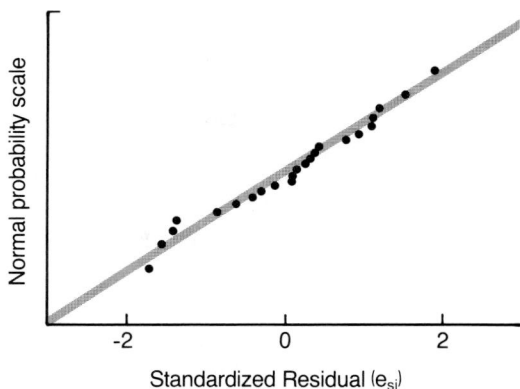

FIGURE 10-11 Normal probability plot of the residuals corresponding to the analysis of Fig. 10-9. Because the normal probability plot follows a straight line, the residuals appear to meet the assumption of normality.

suggested cutoff value of 4, indicating that this point has a large effect on the parameter estimates.[*] We have no reason to doubt the validity of this data point, so we accept the results of the analysis.

The fact that the first data point has such a large effect on the parameter estimates is typical in problems that involve fitting multiexponential functions to data. The fast time constant τ_1 is 14 min, which indicates that the process it represents has decayed to $1/e = .37$ of its original value after one time constant, or 14 min. The first data point is at 5 min, at which time the contribution of the first term in Eq. (10.18) has fallen to $e^{-5 \text{ min}/14 \text{ min}} = .70$ of its original value. In addition, the second point is measured at 15 min, after one time constant has already elapsed. Because the exponential is changing so quickly, we effectively have very little data upon which to base the estimates of the parameters in the model associated with the first exponential term in Eq. (10.18), which accounts for the high Cook's distance associated with this point. When fitting exponential models to data, such as commonly appear in pharmacokinetics, it is important to obtain several points during the interval in which the fast time constant is affecting the results. The example we have discussed here is right at the margin of that required to obtain reliable results.

Likewise, it is important that data be collected over a period several times longer than the long time constant. The slow time constant τ_2 is 138 min and we have data collected out to 720 min. Therefore, there is no problem in reliably estimating this time constant. Had we only collected data for 240 min, we would have had a problem reliably estimating τ_2 and B. The problems associated with obtaining adequate sampling times become even more critical when there are more than two expo-

---

[*]If we reestimate the parameters omitting this one point, A decreases to 520 μmol/L, B decreases to 685 μmol/L, τ_1 increases to 19 min, and τ_2 increases to 143 min.

nentials in the regression model. For example, one could not reliably fit a regression equation containing more than two exponentials to these data.

The asymptotic correlation matrix of the parameters suggests that there is a multicollinearity problem involving the parameter B and the time constants. In addition, the tolerance associated with B is below .1, which corresponds to a variance inflation factor greater than 10, which also suggests a multicollinearity problem. The other tolerances (or the corresponding variance inflation factors) are not too large, however, so we have little evidence of a serious multicollinearity and, thus, accept the results of this analysis. We would have avoided these problems had we had an observation after 1 or 2 min and 10 min, as well as 5 min and 15 min, to reduce the dependence of the results on the first observation, obtained at 5 min.

As we noted above, the analysis we present here is just the fitting of a single carnitine concentration vs. time curve after a single dose. Harper and colleagues repeated this experiment at two different doses in six different subjects, then analyzed the resulting pharmacokinetic parameters. Using the estimated model parameters, they concluded that elimination of the carnitine from the body occurred in a dose-dependent fashion and was not saturated. Thus, the body's ability to get rid of carnitine at the levels studied was not limiting, and doses corresponding to the high blood level studied will be no problem in terms of elimination from the body.

KEEPING BLOOD PRESSURE UNDER CONTROL

Blood pressure is tightly regulated by the body via a coordinated series of adjustments mediated by the nervous system. In terms of the hydraulics of blood flow, the basic pressure control problem is to adjust the flow into blood vessels or the resistance of blood vessels to keep the pressure within a narrow range. This is a classic feedback problem which requires the central nervous system to know what the blood pressure is, compare this pressure with the desired pressure (or set point), then use the difference between actual and desired pressure to generate an appropriate signal to change the blood flow or arterial resistance to move the actual pressure closer to the desired pressure. Thus, there must be transducers that measure blood pressure so that the brain knows what it is. Various sites of such transducers have been identified: a principal one is the carotid sinus, a small widening in the carotid artery of the neck near the base of the skull. The wall of the carotid sinus contains stretch-sensitive elements that generate nerve impulses at a frequency proportional to the amount of stretch. Increased blood pressure in the carotid sinus increases the stretch on these receptors and thus increases the frequency of discharge of the nerve leading from the carotid sinus to the brain.

High blood pressure, or hypertension, is a prolonged elevation of blood pressure. Hypertension has many long-term adverse health effects, including heart disease, kidney damage, and stroke. If the body has such well-regulated blood pressure, how can there be a sustained increase? Simply speaking, this situation can only persist if the pressure control system is somehow reset to maintain the blood pressure at the higher level. To investigate the mechanisms of this resetting, Koushanpour and Behnia* studied how carotid sinus nerve activity responded to changes in carotid sinus pressure in normal dogs and dogs that had been made hypertensive by restricting blood flow to their kidneys. They expressed nerve activity as a percentage of the maximal activity reached during a given experiment. They controlled carotid sinus pressure by closing all entrances and exits from the carotid sinus and infusing or withdrawing blood to change the pressure over a range from about 0 to about 250 mmHg. The data from normal and hypertensive dogs appear in Fig. 10-12. In both cases, nerve activity is low at low blood pressures, then increases to a plateau as blood pressure increases. The question is: Are

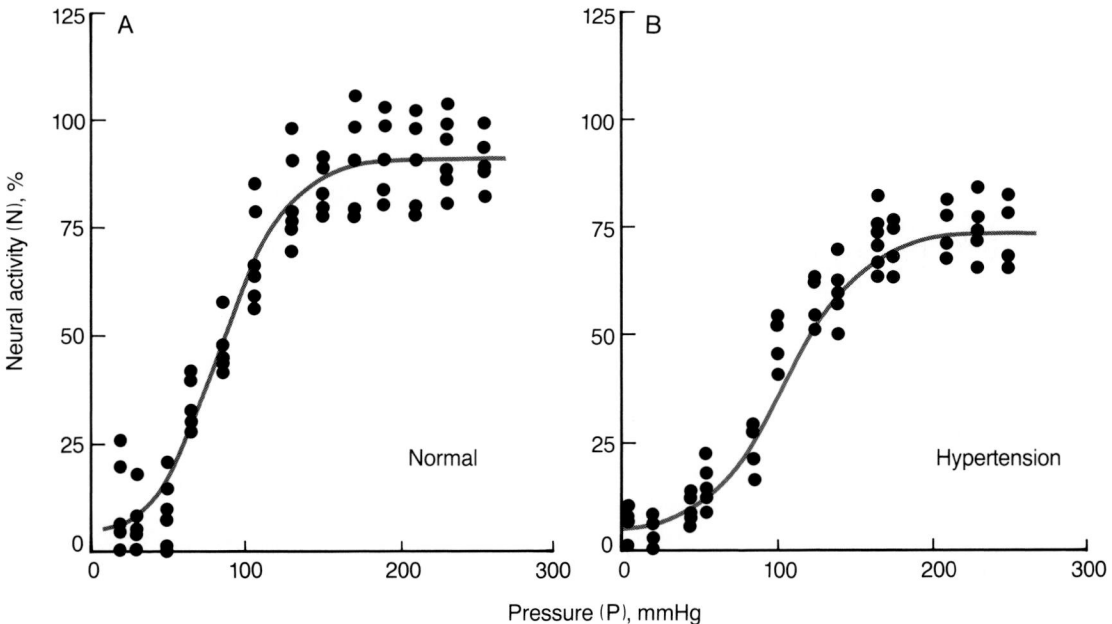

FIGURE 10-12 Normalized carotid sinus nerve activity as a function of carotid sinus blood pressure in normal (A) and hypertensive (B) dogs. The lines indicate the best fit to the logistic function given by Eq. (10.19).

*E. Koushanpour and R. Behnia, "Partition of Carotid Baroreceptor Response in Two-Kidney Renal Hypertensive Dogs," *Am. J. Physiol.* 253:R568–R575, 1987.

the responses of the carotid sinus to changes in blood pressure different in the normal and hypertensive dogs, and, if so, how? In particular, is there any evidence that the pressure control system has been reset?

The type of sigmoidal nonlinear relationship in Fig. 10-12 is common in biology, when a process increases to a point, then saturates. It is common in neural control problems as well as many problems relating to growth processes. The *logistic equation** often describes such processes:

$$\hat{N} = N_{\min} + \frac{N_{\max} - N_{\min}}{1 + e^{(C + DP)}} \tag{10.19}$$

where $\hat{N}$ is the predicted normalized level of neural activity, $N_{\min}$ is the minimum value of neural activity that is asymptotically approached as blood pressure P decreases, $N_{\max}$ is the maximum value of neural activity that is asymptotically approached as blood pressure P increases, and C and D are parameters that define the shape of the curve, as N increases from $N_{\min}$ to $N_{\max}$ as P increases.

Koushanpour and Behnia fit their data to the logistic equation and examined the resulting parameters to see how the carotid sinus response differed in the normal and hypertensive dogs. The results of fitting the data (in Table C-20, Appendix C) from the normal animals appear in Fig. 10-13. The algorithm had only a little trouble getting started—it took two halvings to take the first step—but then it only took two or three steps to get close and then fine-tuned the parameter estimates on steps 4 through 8. It converged to a reasonable fit, with a standard error of the estimate of 9 percent (computed as the square root of MS_{res}), which is acceptable, compared to the average level of neural activity around 50 percent. Examining the plot of this equation with the estimated parameters (Fig. 10-12*A*), as well as the residual plots in Fig. 10-14, reveals that there are no problems with the data or regression model, so we can consider these four numbers a reasonable parameterization of the data. The parameter estimates given in Fig. 10-13, $N_{\min} = 1$ percent (which is not significantly different from zero) and $N_{\max} = 91$ percent, appear reasonable from an inspection of this plot. The values of the shape parameters, $C = 3.8$ and $D = -.046$/mmHg are more difficult to interpret in terms of the physiology in question.

Figure 10-12*B* also shows the results of fitting the data collected in the hypertensive dogs (in Table C-21, Appendix C) with Eq. (10.19). As before, the fit is good. Table 10-2 summarizes the results of both fits. Examining the numbers in this table suggests that the only difference between the patterns of neural activity in the carotid sinus from normal and hypertensive dogs follows from the fact that the maximum level of

---

*The logistic equation also provides the basis for regression analysis when the dependent variable is qualitative in Chap. 11.

NON-LINEAR LEAST SQUARES ITERATIVE PHASE

DEPENDENT VARIABLE: N   METHOD: MARQUARDT

ITERATION	NMAX	NMIN	C	D	RESIDUAL SS
0	90.000000000	5.000000000	4.000000000	-0.1	55321.656405505
.	83.037001216	-30.29804	-2.440293	0.08868335	726605.643368731
.	82.188764905	1.481310483	1.569154578	0.015814767	355894.762374835
1	81.723422436	18.884508195	4.475566408	-0.04440145	14796.156799812
2	90.617883659	2.876106495	3.098829658	-0.04138559	7677.529903180
3	90.638302326	3.588740546	3.897048371	-0.04669728	6094.982484168
4	90.676038515	1.173329022	3.756980673	-0.04577434	6057.636959765
5	90.645627852	1.374062201	3.796323120	-0.0461487	6057.310598504
6	90.645625794	1.336033766	3.791953169	-0.04611465	6057.305087687
7	90.645086067	1.341962212	3.792817610	-0.04612264	6057.304966948
8	90.645100615	1.341112222	3.792707552	-0.04612173	6057.304964198

NOTE: CONVERGENCE CRITERION MET.

NON-LINEAR LEAST SQUARES SUMMARY STATISTICS   DEPENDENT VARIABLE N

SOURCE	DF	SUM OF SQUARES	MEAN SQUARE
REGRESSION	4	378497.45504	94624.36376
RESIDUAL	74	6057.30496	81.85547
UNCORRECTED TOTAL	78	384554.76000	
(CORRECTED TOTAL)	77	91037.95179	

PARAMETER	ESTIMATE	ASYMPTOTIC STD. ERROR	ASYMPTOTIC 95 % CONFIDENCE INTERVAL LOWER	UPPER
NMAX	90.64510061	1.7411733812	87.175727350	94.114473880
NMIN	1.34111222	5.0238777927	-8.669210408	11.351434852
C	3.79270755	0.6934334415	2.411007448	5.174407657
D	-0.04612173	0.0071209144	-0.060310504	-0.031932963

ASYMPTOTIC CORRELATION MATRIX OF THE PARAMETERS

CORR	NMAX	NMIN	C	D
NMAX	1.0000	-0.3007	-0.4031	0.4981
NMIN	-0.3007	1.0000	0.8674	-0.7860
C	-0.4031	0.8674	1.0000	-0.9743
D	0.4981	-0.7860	-0.9743	1.0000

FIGURE 10-13 Results of fitting the data of Fig. 10-12A to Eq. (10.18) using the Marquardt method.

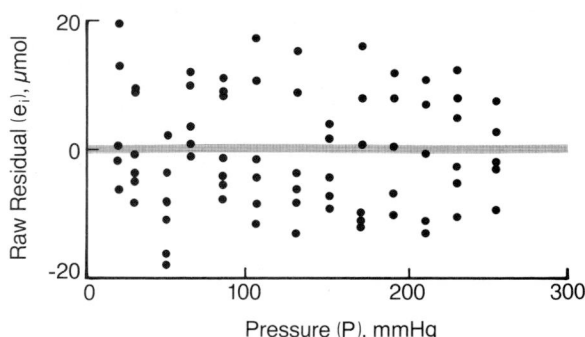

FIGURE 10-14 The residual plot corresponding to the analysis of Fig. 10-13 shows no systematic deviations in the residuals.

TABLE 10-2 Parameter Estimates Describing Baroreceptor Control of Blood Pressure via Logistic Function in Normal and Hypertensive Dogs; Usual Parameterization of the Model

Parameter	Parameter Estimate $\pm$ SE		t	P
	Normal Dogs	Hypertensive Dogs		
N_{max}, %	90.6 ± 1.7	72.7 ± 1.4	8.14	<.001
N_{min}, %	1.3 ± 5.0	3.3 ± 2.2	$-.35$	>.50
C	$3.8 \pm .7$	$4.2 \pm .5$	$-.46$	>.50
D, mmHg^{-1}	$-.05 \pm .01$	$-.04 \pm .01$	$-.35$	>.50

neural output N_{max} falls from 91 to 73 percent, with the other characteristics of the response unchanged.

We can test this impression formally by computing an approximate t statistic using the reported values of the parameter estimates and their standard errors. Recall that the t statistic is just the ratio of the parameter estimate to its standard error, so we have

$$t = \frac{N_{max_N} - N_{max_H}}{s_{N_{max_N} - N_{max_H}}}$$

in which the subscript N stands for normal and the subscript H stands for hypertensive. The standard error of the difference of the two regression parameters is just the square root of the sum of the squared standard errors associated with the separate parameter estimates:*

$$s_{N_{max\ N} - N_{max_H}} = \sqrt{s_{N_{max_N}}^2 + s_{N_{max_H}}^2}$$

We compare this value of the t statistic with the critical value of the t distribution with $n_N + n_H - 8$ degrees of freedom, where n_N and n_H are

*This is only strictly true when the number of data points used to obtain each parameter estimate is the same. Had there been different numbers of data points in each case, one could compute a more accurate estimate of the standard error of the difference by using a pooled variance estimate similar to that following Eq. (2.11) for linear regression, which is a weighted average, based on the degrees of freedom in each case:

$$s_{N_{max_N} - N_{max_H}} = \sqrt{\frac{MS_{resp}}{(n_N - 1)s_{P_N}^2} + \frac{MS_{resp}}{(n_H - 1)s_{P_H}^2}}$$

where

$$MS_{resp} = \frac{(n_N - p)s_{N_{max_N}}^2 + (n_H - p)s_{N_{max_H}}^2}{n_N + n_H - 2p}$$

s_{P_N} and s_{P_H} are the standard deviations of the pressure in the normal and hypertensive groups, respectively, and p is the number of parameters (in this example, 4).

the number of data points used to estimate the parameters in Eq. (10.19) for the normal and hypertensive dogs, respectively. We subtract 8 from the total number of data points because there are 4 parameters in each equation.

From the parameter estimates in Table 10-2, for N_{max} we compute

$$t = \frac{90.65 - 72.67}{\sqrt{1.74^2 + 1.36^2}} = \frac{17.98}{2.2084} = 8.142$$

which indicates a highly significant difference in N_{max} between normal and hypertensive dogs $(P < .001)$. None of the other parameter estimates were significantly different (Table 10-2). This analysis leads us to the conclusion that the only change between the normal and hypertensive dogs was a blunting of the maximum output of the nerves in the carotid sinus during hypertension, which could account for the reduction in the body's ability to reduce blood pressure.

Is the Model Parameterized in the Best Form?

Often, as we saw in the pharmacokinetics example above, there are several equivalent forms of an equation. These different forms contain different sets of parameters, but they are interchangeable if you know the relationships between the parameters. Such different parameterizations give you a choice of how you analyze the data, and often there will be one form whose parameters make more intuitive sense, or are more directly interpretable in terms of the process being investigated. We saw that the shape parameters C and D in Eq. (10.19) do not have a straightforward interpretation. There is an equivalent form of the logistic equation in which the parameters are not as abstract and are easier to interpret in terms of the physiology in question.

The logistic function given by Eq. (10.19) can be rewritten as

$$\hat{N} = N_{min} + \frac{N_{max} - N_{min}}{1 + e^{4S(P_{1/2} - P)/(N_{max} - N_{min})}} \tag{10.20}$$

Although this equation looks more complicated than the original equation, the interpretation of the parameters is simpler. N_{max} and N_{min} have the same meaning as before. However, instead of C and D, we have two new parameters, $P_{1/2}$ and S. $P_{1/2}$ is the pressure at which neural activity N is halfway between its minimum and maximum values, and S is the slope of the line tangent to the curve at $P_{1/2}$.[*] In terms of the physiological pressure control problem being investigated here, $P_{1/2}$ is the pressure control set point, and S is the sensitivity of the pressure controller to

---

[*]Comparing this equation with Eq. (10.19) reveals that $C = -4SP_{1/2}/(N_{max} - N_{min})$ and $D = 4S/(N_{max} - N_{min})$. For more details, see D. S. Riggs, *Control Theory and Physiological Feedback Mechanisms*, Huntington, NY, Krieger, 1976, p. 527.

TABLE 10-3 Parameter Estimates Describing Baroreceptor Control of Blood Pressure via Logistic Function in Normal and Hypertensive Dogs; Alternative Parameterization of the Model

Parameter	Parameter Estimate $\pm$ SE		t	P
	Normal Dogs	Hypertensive Dogs		
N_{max} %	90.6 ± 1.7	72.7 ± 1.4	8.14	$<.001$
N_{min} %	1.3 ± 5.0	3.3 ± 2.2	$-.35$	$>. 50$
$P_{1/2}$ mmHg	82.2 ± 3.9	97.0 ± 2.9	-3.04	$<.002$
S, mmHg^{-1}	$1.0 \pm .1$	$.8 \pm .1$	1.83	$\approx. 08$

changes in pressure. These parameters are more directly related to the question at hand, namely, how does the neural activity produced by the carotid sinus change in the presence of hypertension?

The computer output from fitting the data from normal dogs (in Table C-20, Appendix C) to this equation is in Fig. 10-15. Notice that the algorithm took about the same number of steps to converge as it did for the other parameterization. One of the additional benefits of this new parameterization is that, because it is easier to attach meaning to each of the parameters in the equation, it is easier to come up with good first guesses. The sums of squares and estimates of N_{max} and N_{min} are the same as before. The pressure set point $P_{1/2}$ is 82 mmHg for the normal dogs, and the sensitivity of the control system at the set point S is 1.0/mmHg. Now we have numbers, along with their associated standard errors, that relate directly to the question we originally posed: Is regulation of blood pressure reset in hypertension?

To answer this question, we repeated this same analysis for the data from hypertensive animals. The summary of this analysis, together with the analysis of the normal animals, is in Table 10-3. As before, we compare the parameter estimates with t tests; the results of these comparisons also appear in Table 10-3. As before, we find that there is no difference in the basal response N_{min} and that the maximum response N_{max} is significantly lower in the hypertensive dogs than the normal dogs. In contrast to the results we found with the original parameterization [Eq. (10.19)], we find that the shape of the response curve also changes during hypertension. Of most interest to our physiological question, the set point $P_{1/2}$ increases from 82 mmHg in the normal dogs to 97 mmHg in the hypertensive dogs ($P < .002$). There is also a suggestion that the sensitivity to neural activity to changes in blood pressure at the set point S falls from 1.0/mmHg in the normal dogs to .8/mmHg in the hypertensive dogs ($P \approx .08$). Thus, using this more appropriate parameterization

```
                    NON-LINEAR LEAST SQUARES ITERATIVE PHASE

              DEPENDENT VARIABLE: N        METHOD: MARQUARDT
```

ITERATION	NMIN	NMAX	PHALF	S	RESIDUAL SS
0	0	90.000000000	100.000000000	1.500000000	15891.327636758
1	5.777949130	91.369662539	88.790894296	0.618037804	9603.426909361
2	-1.461691	89.244343454	76.713211029	0.934590856	6406.931111845
3	3.523541068	90.572882132	84.118314263	0.999890394	6106.871658736
4	1.058773909	90.711419413	82.040396070	1.017419328	6058.463151998
5	1.391432312	90.649902066	82.289817939	1.031039639	6057.335379637
6	1.330937789	90.645582778	82.223996081	1.029662284	6057.305471484
7	1.342735603	90.644943298	82.233745865	1.029757147	6057.304979613
8	1.341008897	90.645082912	82.232413435	1.029710547	6057.304964555

NOTE: CONVERGENCE CRITERION MET.

```
         NON-LINEAR LEAST SQUARES SUMMARY STATISTICS    DEPENDENT VARIABLE N
```

SOURCE	DF	SUM OF SQUARES	MEAN SQUARE
REGRESSION	4	378497.45504	94624.36376
RESIDUAL	74	6057.30496	81.85547
UNCORRECTED TOTAL	78	384554.76000	
(CORRECTED TOTAL)	77	91037.95179	

PARAMETER	ESTIMATE	ASYMPTOTIC STD. ERROR	ASYMPTOTIC 95 % CONFIDENCE INTERVAL LOWER	UPPER
NMIN	1.34100890	5.0239210968	-8.669400019	11.351417812
NMAX	90.64508291	1.7411761730	87.175704084	94.114461740
PHALF	82.23241344	3.9093157942	74.442910192	90.021916679
S	1.02971055	0.1305218121	0.769639442	1.289781652

```
              ASYMPTOTIC CORRELATION MATRIX OF THE PARAMETERS
```

CORR	NMIN	NMAX	PHALF	S
NMIN	1.0000	-0.3007	0.7831	0.5599
NMAX	-0.3007	1.0000	0.0672	-0.6270
PHALF	0.7831	0.0672	1.0000	0.2504
S	0.5599	-0.6270	0.2504	1.0000

FIGURE 10-15 Results of fitting the better parameterized model given by eq. (10.20) to the baroreceptor response data of Fig. 10-12. Although the overall quality of fit given by the residual sum of squares is no different from that in the original model, the new parameters are more easily interpreted physiologically and can be estimated more precisely than in the original parameterization.

of the problem, we conclude that not only is the maximum output of the carotid sinus depressed in hypertension, but also that the set point of the response is shifted to higher blood pressures, and there is also a suggestion that the carotid sinus output is less sensitive to changes in blood pressure.

Why were we able to detect differences in the shape of the curve described in terms of $P_{1/2}$ and S but not C and D? $P_{1/2}$ and S are estimated at the "center" of the data, so can be estimated relatively precisely. The standard errors associated with these two parameters (given in Table 10-3) are only about 5 to 10 percent of the observed values. This situation contrasts with that for C and D (in Table 10-2), where the associated standard errors are 15 to 20 percent of the observed values. This increase in precision with the second parameterization is analogous to the improvements in the precision of estimates in multiple linear regression we obtained by centering the data in Chap. 4. This procedure also reduced the values of the correlations between the regression coefficients, which reflects the level of multicollinearity and thus redundant information in the different parameter estimates. This reduction in multicollinearity led to smaller standard errors of the parameter estimates, resulting in improved precision of the estimates, which then permitted us to detect changes in the shape, as well as maximum value, of the carotid sinus response.

SUMMARY

There are many instances when problems cannot be transformed into linear regression problems, which necessitates directly fitting a nonlinear equation to the data. In contrast to the situation that existed in linear regression, there is no general closed-form solution to the nonlinear regression problem. To estimate the parameters in the regression model, we must use iterative techniques in which the values of the parameters that minimize the sum of squared residuals are sought by making successive educated guesses at the parameters until it is not possible to find new values of the parameters that further reduce the residual sum of squares. There are many different algorithms for seeking the best parameter estimates in nonlinear regression; we discussed the steepest descent, Gauss-Newton, and Marquardt algorithms. All three of these methods require first guesses that are reasonably close to the final estimates.

Although each nonlinear regression problem is unique, once the method has converged, it is possible to approximate the nonlinear problem in the local region of the minimum residual sum of squares as a linear regression problem. It is then possible to use this approximation to obtain approximate estimates of the standard errors of the regression parameters and compute the regression diagnostics, such as leverage and Cook's distance as well as diagnostics for multicollinearity. These results can then be used to conduct *approximate* hypothesis tests and diagnostics for the regression.

We have concentrated on the case of a single continuous independent variable (time or pressure). The methods we have described can also be generalized directly to the case of several independent variables or dummy independent variables which encode the presence or absence of some condition.

Given the complexities of nonlinear regression, it is important to always directly compare the regression model containing the estimated parameters to the data. When there is a single independent variable, this process simply involves looking at a plot of the data with the regression equation superimposed on it. When there is more than one independent variable, you can use residual plots as we did in linear regression. Carefully checking the regression equation against the data is even more crucial in nonlinear regression problems than in linear regression.

PROBLEMS

10.1 In Prob. 4.4, you fit a curvilinear relationship between the change in rate of growth G in growth hormone H treated children and their native H levels at the time treatment started. Fit these data (Table D-15, Appendix D) to the equation, $\hat{G} = b_2 + b_0 e^{-b_1 H}$. Is there evidence that this is a better fit to the data than a linear regression? Evaluate the residuals and correct any problems, if necessary. Is the investigators' hypothesis that this should be a nonlinear relationship supported?

10.2 Surfactant is a substance secreted by lung cells lining the alveoli, the small sacs at the end of airways where gas exchange between air and blood occurs. Surfactant reduces surface tension and is essential for preventing collapse of alveoli. Surfactant is continually recycled by the cells lining the alveoli. This process includes the uptake of the phospholipids in surfactant and reusing them to synthesize new surfactant. This uptake of phospholipids has been studied in cell culture systems, where it has been shown that synthetic phospholipid vesicles are readily internalized. However, there may be composition or structural differences between synthetic phospholipid vesicles and natural surfactant. To see if there was such a difference, Fisher and his coworkers[*] studied the uptake of natural surfactant and synthetic phospholipid vesicles into cultured lung cells. They measured phosphatidylcholine transport P as a function of time t after adding phosphatidylcholine to the cell culture (the data are in Table D-35, Appendix D). Is the uptake of natural surfactant different from that of synthetic phospholipid? Use dummy variables.

[*]A. B. Fisher, A. Chander, and J. Reicherter, "Uptake and Degradation of Natural Surfactant by Isolated Rat Granular Pneumocytes," *Am. J. Physiol.* 253:C792–C796, 1987.

10.3   For the data analyzed in Prob. 10.2, two observations had Studentized residuals r_i greater than 2: $r_4 = 3.6$ and $r_7 = 2.43$. Calculate the corresponding Cook's distances.

10.4   When patients are given certain antibiotics, they sometimes get an inflammation of the colon that is usually caused by a bacterium, *Clostridium difficile*, which produces an endotoxin. The mechanism by which this endotoxin causes damage to the cells lining the colon is not clear. Therefore, Hecht and her coworkers[*] studied the effect of this endotoxin on the permeability of cultured intestinal cells. They measured permeability by two methods, the flux of mannitol across the layer of cells and the resistance of the layer of cells (the data are in Table D-36, Appendix D). *A.* Quantify the relationship between mannitol flux F and epithelial cell resistance R using a sum of exponentials. *B.* If a sum of exponentials proves unsatisfactory, propose an alternative model.

10.5   Do a complete analysis of the residuals and other regression diagnostics for the analysis of Prob. 10.4*B*.

10.6   In an example in this chapter, we presented data relating to the effect of carotid sinus baroreceptors on nerve discharge. These data related very specifically to the function of a single baroreceptor in terms of the signal it sends to the brain. There are many such baroreceptors feeding information to the brain, which uses the information to control heart rate and vascular resistance to keep blood pressure at the desired normal level. Such integrated responses can be studied using methods similar to those used in the earlier example. Wilkinson and his coworkers[†] studied the effect of changing blood pressure P using vasodilating and vasoconstricting drugs on the interval between heart beats I in dogs anesthetized with different levels of halothane. *A.* What is the relationship between I and P with .5% halothane anesthesia (the data are in Table D-37, Appendix D)? *B.* What is the relationship between I and F with 2% halothane anesthesia (the data are in Table D-38, Appendix D)? *C.* Does the baroreceptor sensitivity or set point change with the level of anesthesia?

10.7   Are there any influential observations in the data analyzed in Prob. 10.6*B*?

[*]G. Hecht, C. Pothoulakis, J. T. LaMont, and J. L. Madara, "*Clostridium difficile* Toxin A Perturbs Cytoskeletal Structure and Tight Junction Permeability of Cultured Human Intestinal Epithelial Monolayers," *J. Clin. Invest.* 82:1516–1524, 1988.

[†]P. L. Wilkinson, D. F. Stowe, S. A. Glantz, and J. V. Tyberg, "Heart Rate–Systemic Blood Pressure Relationship in Dogs during Halothane Anesthesia," *Acta Anaesthesiol. Scand.* 24:181–186, 1980.

10.8   The amount of protein in the diet modulates both total food intake and body weight gain. Many investigators have studied the relationship between dietary protein and food intake and weight gain, but most studies have ignored the effect of age. Toyomizu and coworkers[*] noted a prior study that quantitatively related growth and food intake to dietary protein including the effect of age. The equations used to quantify these effects contain several parameters, among them one relating to the food intake at maturity F and one relating to the body weight at maturity W. It was concluded that these two parameters were independent of the amount of dietary protein. However, Toyomizu and coworkers thought that the range of dietary proteins studied was too limited (14 to 36 percent of energy intake), and therefore they studied the growth and food intake of rats fed a much broader range of diets such that the fraction of protein in the diet P ranged from 5 to 80 percent of the gross energy available in the diet. Is W independent of dietary protein (the data are in Table D-39, Appendix D)? Toyomizu and coworkers used the equation $\hat{W} = (b_1 - b_2 P)(1 - e^{-b_3(P - b_4)})$.

10.9   Repeat the analysis of Prob. 10.8 for the dependent variable F, using starting values $b_1 = 5$, $b_2 = 1.6$, $b_3 = 300$, and $b_4 = .05$. Then repeat starting from $b_3 = 200$ and keeping the other starting values the same. Finally, repeat starting from $b_3 = 100$. Discuss the results of these three attempts. Constrast this result to the result of Prob. 10.8. Use Marquardt's method. (The data are in Table D-40, Appendix D.)

10.10 Repeat the analysis of Prob. 10.8 using the data in Table D-41, Appendix D. Compare this result with that of Prob. 10.9. Fully explain your answer.

[*]M. Toyomizu, K. Hayashi, K. Yamashita, and Y. Tomita, "Response Surface Analyses of the Effects of Dietary Protein on Feeding and Growth Patterns in Mice from Weaning to Maturity," *J. Nutr.* 118:86–92, 1988.

Regression with a Qualitative Dependent Variable

All the statistical methods we have developed so far have been for quantitative independent variables, measured on more-or-less continuous scales. The assumptions of linear regression—in particular that the mean value of the population at any combination of the independent variables be a linear function of the independent variables and that the variation about the plane of means be normally distributed—required that the dependent variable be measured on a continuous scale. In contrast, because we did not need to make any assumptions about the nature of the independent variables, we could incorporate qualitative information (such as whether or not a Martian was exposed to secondhand tobacco smoke) into the independent variables of the regression model. There are, however, many times when we would like to evaluate the effects of multiple independent variables on a *qualitative dependent variable*, such as the presence or absence of a disease. Because the methods that we have developed so far depend strongly on the continuous nature of the dependent variable, we will have to develop a new approach to deal with the problem of regression with a qualitative dependent variable.

This new approach is known as *logistic regression.** To develop logistic regression, we need to address five related issues:

- We need a dependent variable that represents the two possible qualitative outcomes in the observations. We will use a *dummy*

---

*We will develop logistic regression for the simplest (and most common) case of a dichotomous outcome, where the dependent variable can take on one of two values. For the generalization to polychotomous outcomes, see J. Fox, *Linear Statistical Models and Related Methods with Applications to Social Research*, New York, Wiley, 1984, pp. 311–332.

variable, which takes on values of 0 and 1, as the dependent variable.

- We need a continuous mathematical function to estimate the observed value of the dependent variable. We will use the *probability P of the observed value of the dependent variable being 1.*

- We need a way to relate the dependent variable to the independent variables. We will use the *logistic function.*

- We need a way to estimate the coefficients in the regression model because the ordinary least-squares criterion we have used so far is not relevant for a qualitative dependent variable. We will use *maximum likelihood estimation.*

- We need *statistical hypothesis tests* for the goodness of fit of the regression model and whether or not the individual coefficients in the model are significantly different from zero, as well as confidence intervals for the individual coefficients.

We will now develop logistic regression and illustrate how to apply it and interpret the results.* The logistic regression equation will permit us to estimate the probability that any individual has one outcome or the other.

OUR LAST VISIT TO MARS

Years after our first, ground-breaking study of the height and weight of Martians, we return one last time to receive an award for opening up friendly relations between Mars and earth and do one final study, this time at the Martians' request. In the intervening years, 120 Martians, selected at random from the population of Mars have visited earth to study. Of these 120 Martian students, 78 graduated and 42 dropped out without finishing their studies. The question is: Can we predict who will graduate based on how intelligent the student is?

---

*Logistic regression is closely related to discriminant analysis, which is another way to find the characteristics that are important for determining whether subjects end up in one group or another. One important factor in determining whether you should use discriminant analysis or logistic regression is that, unlike logistic regression, discriminant analysis assumes normally distributed independent variables. For a discussion of discriminant analysis and its relationship to logistic regression, see A. A. Afifi and V. Clark, *Computer-Aided Multivariate Analysis*, Belmont, CA, Lifetime Learning Publications, Chaps. 11 and 12. For a comparison of these two methods in the context of clinical and epidemiological studies, see L. A. Cupples, T. Herren, A. Schatzkin, and T. Colton, "Multiple Testing of Hypotheses in Comparing Two Groups," *Ann. Int. Med.* 100:122–129, 1984.

TABLE 11-1 Martian Graduation as a Function of Intelligence

Intelligence I, zorp	Number Graduating	Number Not Graduating	Proportion Graduating
1	0	2	0
2	0	2	0
3	0	5	0
4	0	3	0
5	1	6	.14
6	1	3	.25
7	1	2	.33
8	4	7	.36
9	5	4	.56
10	7	4	.64
11	7	2	.78
12	7	1	.88
13	11	1	.92
14	7	0	1
15	11	0	1
16	5	0	1
17	7	0	1
18	2	0	1
19	2	0	1

In this problem, the dependent variable is qualitative (graduated or not) and the independent variable (intelligence) is quantitative. Just as we did with qualitative independent variables, we define a dummy *dependent* variable G as

$$G = \begin{cases} 1 \text{ if student graduated} \\ 0 \text{ if student did not graduate} \end{cases}$$

Tables 11-1 and C-22 (Appendix C) present the data from this study, and Fig. 11-1A shows a plot of the data, with whether or not the student graduated G as the dependent variable and the student's intelligence I as the independent variable.

We need a mathematical function that varies continuously (so that it can be differentiated and otherwise manipulated mathematically) to describe the relationship between G and I. We will use the *logistic function*

$$P_G(I) = \frac{1}{1 + e^{-(\beta_0 + \beta_I I)}} \tag{11.1}$$

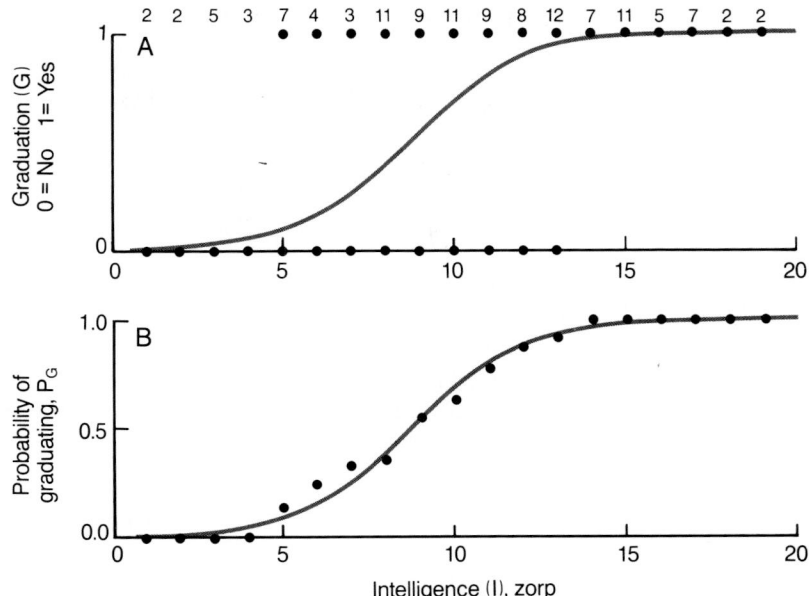

FIGURE 11-1 (*A*) There are only two potential outcomes in this study of whether or not Martians graduated from college as a function of intelligence. If the Martian graduated, $G = 1$, and if he or she did not graduate, $G = 0$. Examining this plot reveals that Martians of greater intelligence are more likely to graduate. The small numbers across the top are the total number of Martians at each level of intelligence. (*B*) This impression can be quantified because there are several Martians at each level of intelligence in the sample, so it becomes possible to compute the fraction of Martians who graduate at each level of intelligence, then fit these proportions with the logistic function to obtain a function for the probability of graduating P_G as a function of intelligence. When there are many observations at each level of the independent variable, one can simply use a nonlinear regression algorithm to fit the logistic equation to the observations. Unfortunately, this situation does not often occur, and one has only a few observations at each combination of the independent variables.

The logistic function has a direct interpretation: $P_G(I)$ is the *probability* that a Martian of intelligence I will graduate. Figure 11-1 shows this function. Note that the logistic function varies smoothly from 0 to 1 as I increases. The problem of logistic regression is to estimate the parameters β_0 and β_I to obtain the best estimate of the probability that a Martian of intelligence I will graduate.

From the observations we know whether or not each individual Martian graduated, so the associated *observed* value of the dependent variable is either 0 because he or she did not graduate or 1 because he or she did. There is no uncertainty associated with the individual observations; the uncertainty is associated with the effects of a given intelligence level. Logistic regression of the data in Fig. 11-1 reveals that our best estimates for the regression coefficients are $b_0 = -5.47$ and $b_I = .627/\text{zorp}$ (Fig. 11-2), so the estimated relationship between whether or not a Martian graduated and intelligence is

$$\hat{P}_G(I) = \frac{1}{1 + e^{-(-5.47 + .627/\text{zorp } I)}} \tag{11.2}$$

To better see how $\hat{P}_G(I)$ relates to the observed values of G, let us consider the observations at three different levels of intelligence.

First, consider Martians of low intelligence, say $I = 3$ zorp. None of the 5 individuals graduated; $G = 0$ for all of them. The observed proportion that graduated was $0/5 = 0$. Substituting the value of $I = 3$ zorp into Eq. (11.2) yields

$$\hat{P}_G(3 \text{ zorp}) = \frac{1}{1 + e^{-(-5.47 + .627/\text{zorp} \cdot 3 \text{ zorp})}} = .027$$

In other words, the estimated probability for a Martian of intelligence 3 zorp is only .027, which is virtually zero, in close agreement with the observations. Second, let us consider Martians of high intelligence, say 15 zorp. All of these 11 individuals graduated; $G = 1$ for all of them. The observed proportion that graduated was $11/11 = 1$. Substituting the value of $I = 15$ zorp into Eq. (11.2) yields

$$\hat{P}_G(15 \text{ zorp}) = \frac{1}{1 + e^{-(-5.47 + .627/\text{zorp} \cdot 15 \text{ zorp})}} = .981$$

In other words, the estimated probability for a Martian of intelligence 15 zorp is .981, which is virtually 1.

This situation contrasts with that for Martians of intermediate intelligence, say 7 zorp. Of the 3 Martians with intelligence 7 zorp, 1 graduated and 2 did not; $G = 1$ for 1 Martian and $G = 0$ for 2 Martians. The observed proportion of Martians that graduated was $1/3 = .333$. From Eq. (11.2), the predicted probability that a Martian of intelligence 7 zorp will graduate is

$$\hat{P}_G(7 \text{ zorp}) = \frac{1}{1 + e^{-(-5.47 + .627/\text{zorp} \cdot 7 \text{ zorp})}} = .25$$

In other words, the estimated probability of graduation for a Martian of intelligence 7 zorp is .25, a value intermediate between certain graduation (1) and certain failure (0). This predicted value compares with an observed proportion of Martians that graduated of .333 and reflects the fact that there is uncertainty in the outcome for individual Martians of intelligence 7 zorp.

Although the curve representing the logistic function on Fig. 11-1*A* does not pass through the points corresponding to the observations for Martians at intermediate levels of intelligence, it still provides a good "fit" to the data because the logistic function estimates the *probability of graduation* at each intelligence level. To better see the quality of the fit, we compute the proportion of Martians who graduated at each level of intelligence. Table 11-1 shows this computation, and Fig. 11-1*B* shows a plot of the observed proportion of Martians who graduated together with the logistic equation. Figure 11-1*B* shows that Eq. (11.1) provides a good description of the probability that a Martian will graduate as a function of intelligence.

Odds

In contrast to least-squares regression with a quantitative dependent variable, when we were trying to predict a continuous dependent variable, we are now trying to estimate the probability that a given individual will fall into one outcome class or the other. Thinking in terms of these probabilities makes it possible to interpret the coefficients in the logistic regression model, just as we were able to attach meaning to the coefficients in linear regression with quantitative dependent variables.

It is possible to simplify the interpretation of the coefficients in the logistic function if we change our focus from the probability of an event occurring to the *odds* of the event occurring.[*] Let P equal the probability of a given event occurring. The odds of that event occurring are defined as

$$\text{Odds} = \Omega = \frac{P}{1 - P} = \frac{\text{probability of event occurring}}{\text{probability of event not occurring}} \quad (11.3)$$

The probability of the event not occurring is $1 - P$ because there are only two possible outcomes, and the probabilities of all possible outcomes must add up to 1. For example, the probability of all Martians graduating, regardless of intelligence, is $78/120 = 65$, so the odds of graduating are

---

[*]The odds ratio is also an estimate of the *relative risk* in epidemiological studies of rare outcomes. For an introductory discussion of the relationship between the odds ratio and relative risk, see G. D. Friedman, *Primer of Epidemiology* (3rd ed.), New York, McGraw-Hill, 1987, pp. 106–107, 200–204. For a more detailed discussion of this issue and the use of logistic regression in epidemiological research, see N. E. Breslow and N. E. Day, *Statistical Methods in Cancer Research: Volume 1 — The Analysis of Case-Control Studies*, Lyon, International Agency for Research on Cancer, 1980, Chapter 6, "Unconditional Logistic Regression for Large Strata"; or D. G. Kleinbaum, L. L. Kupper, and H. Morgenstern, *Epidemiologic Research: Principles and Quantitative Methods*, New York, Van Nostrand Reinhold, 1982, Chaps. 20–24.

$$\text{Odds of graduating} = \frac{\text{probability of graduating}}{\text{probability of not graduating}} = \frac{.65}{1 - .65} = 1.86$$

We say that the odds are 1.86:1 that a Martian will graduate, or that the *odds ratio* is 1.86.

To see how using odds simplifies the interpretation of the coefficients in the logistic regression equation, substitute from Eq. (11.1) into Eq. (11.3) to obtain

$$\Omega_G = \frac{1/(1 + e^{-(\beta_0 + \beta_I I)})}{1 - 1/(1 + e^{-(\beta_0 + \beta_I I)})} = e^{(\beta_0 + \beta_I I)} = e^{\beta_0} e^{\beta_I I} \qquad (11.4)$$

Notice that the effects of the β_0 and β_I coefficients have now been separated, but instead of adding, they multiply. In terms of the estimates of β_0 and β_I,

$$\hat{\Omega}_G = e^{b_0} e^{b_I I}$$

Suppose a Martian has intelligence $I = 0$ zorp; his or her odds of graduating are

$$\hat{\Omega}_G(0 \text{ zorp}) = e^{-5.47} e^0 = .0042$$

which is not very high. e^{b_I} is the multiplicative change in the odds of graduating for each zorp increase in intelligence. Because $e^{b_I} = e^{.627} = 1.87$, we see that the odds of graduating increase, on the average, by a factor of 1.87 for each zorp of increase in intelligence. Thus, the predicted odds of a Martian of intelligence 1 zorp graduating is

$$\hat{\Omega}_G(1 \text{ zorp}) = e^{-5.47} e^{(.627/\text{zorp}) \cdot 1 \text{ zorp}} = .0042 \cdot 1.87 = .00785$$

and the predicted odds of a Martian of intelligence 6 zorp graduating are

$$\hat{\Omega}_G(6 \text{ zorp}) = e^{-5.47} e^{(.627/\text{zorp}) \cdot 6 \text{ zorp}} = e^{-5.47} (e^{.627})^6 = .0042 \cdot 1.87^6 = .18$$

The independent contributions of each factor to the odds become even more evident if we take the natural logarithm of both sides of Eq. (11.4) to obtain the log odds

$$\ln(\Omega_G) = \ln\left(\frac{P_G}{1 - P_G}\right) = \beta_0 + \beta_I I$$

$\ln(\Omega)$ is also known as the *logit* transformation, where

$$\text{logit } P = \ln\left(\frac{P}{1 - P}\right)$$

and people often write the logistic regression equation as

$$\text{logit } P_G = \beta_0 + \beta_I I$$

We now generalize this model to the problem of several independent variables.

The Multiple Logistic Equation

The logistic model we just developed directly generalizes to the case of k independent variables, $x_1, x_2, \ldots, x_k$:

$$P(x_1, x_2, \ldots, x_k) = \frac{1}{1 + e^{-(\beta_0 + \beta_1 x_1 + \beta_2 x_2 + \cdots + \beta_k x_k)}} \tag{11.5}$$

P is the probability of a *"successful" outcome* — the condition that corresponds to the dependent variable being equal to 1 as a function of the k independent variables $x_1, x_2, \ldots, x_k$. The independent variables can be quantitative or qualitative (using dummy variables). They can also represent transformed variables or interactions, just as we did with multiple linear regression.

The interpretation of the coefficients β_i in the multiple logistic equation is exactly the same as in the simple one-variable logistic equation we just discussed. As before, we can also write this equation in terms of the log odds

$$\ln (\Omega) = \ln \left(\frac{P}{1 - P} \right) = \beta_0 + \beta_1 x_1 + \beta_2 x_2 + \cdots + \beta_k x_k$$

or

$$\text{logit } P = \beta_0 + \beta_1 x_1 + \beta_2 x_2 + \cdots + \beta_k x_k \tag{11.6}$$

People often talk about the log odds or the change in log odds associated with each term in the logistic regression equation. To convert these log odds to changes in the actual odds, just raise e to that power. For example, suppose the two possible values of the dependent variable are the presence or absence of a disease, with the presence of the disease scored as 1. In this case, β_0 represents the log odds of disease risk for a person with a standard ($x_1 = x_2 = \cdots = x_k = 0$) set of independent variables, while β_i is the fraction by which the risk is increased (or decreased) for every unit change in x_i.

Now that we have a regression model, we need to estimate the regression coefficients β_i in Eq. (11.5), or, equivalently, Eq. (11.6).

ESTIMATING THE COEFFICIENTS IN A LOGISTIC REGRESSION

At first blush, the fact that the log odds are a linear function of the independent variables seems to suggest that we can transform the observed proportion of successful outcomes at each combination of the independent variables into an odds ratio, then use the methods of linear regression described in Chap. 3 to compute estimates of the parameters, $\beta_0, \beta_1, \ldots, \beta_k$. Occasionally, when we have many observations at each combination of the independent variables, we can compute an observed value of the odds of success for each such combination. It is possible to compute estimates for the parameters in the regression equation by minimizing the sum of squared deviations between the observed and predicted values of P for each combination of the independent variables. This procedure is equivalent to fitting the logistic function through the points in Fig. 11-1B.* Unfortunately, one rarely has enough replications at each combination of the independent variables to make this procedure feasible. More often, one is faced with the problem of observations collected under a wide variety of combinations of independent variables, with relatively few, if any, replications under any given (unique) combination of the independent variables. So it is difficult or impossible to obtain observed values for P at each unique combination of the independent variables. We therefore need to use some criterion other than minimizing the residual sum of squared deviations between the predicted proportion (or log odds) and the observed proportions (or log odds) at each combination of the independent variables in order to estimate the parameters in the logistic equation.

Maximum Likelihood Estimation

Rather than seeking estimates of the model parameters that minimize the residual sum of squares between the observed and predicted values of the dependent variable, we will use the procedure of *maximum likelihood estimation* to compute our estimates of the parameters in the regression model from the observations. *The maximum likelihood estimates of a set of parameters are those values most likely to give rise to the pattern of observations in the data.*

To understand the logic of maximum likelihood estimation, we need to change our focus when thinking about the logistic regression model given by Eq. (11.5). Each datum includes the observed values of the independent variables $X_{1i}, X_{2i}, \ldots, X_{ki}$ and the corresponding observed value of the dependent variable, Y_i. As already noted, the independent variables can represent continuous variables and qualitative variables, coded as dummy variables, as well as interactions between and transfor-

---

*Even in this case, however, you cannot do a straightforward linear regression of log odds on the independent variables because the equal variance assumption will be violated; it is necessary to do a weighted linear regression to obtain unbiased estimators of the parameters in the regression model. An alternative approach would be to use a nonlinear regression to estimate the parameters in Eq. (11.5) against the observed proportions (as they appear in Fig. 11-1B). For a detailed discussion of this issue, see J. Neter, W. Wasserman, and M. H. Kutner, *Applied Linear Regression Models*, Homewood, IL, Irwin, 1983, pp. 359–364.

mations of the native independent variables. The dependent variable Y_i takes on values of 0 or 1, depending on whether the outcome associated with the ith datum is a "failure" or a "success." We now consider this information as *fixed* for the problem at hand and turn our attention to the parameters in the logistic equation.

We can use Eq. (11.5) to compute the probability—or *likelihood*—of the dependent variable Y_i being 1 (a "success") given the observed values of the independent variables as a function of the parameters $\beta_0, \beta_1, \ldots, \beta_k$, as

$$L_{Y_i=1}(\beta_0, \beta_1, \ldots, \beta_k) = \frac{1}{1 + e^{-(\beta_0 + \beta_1 x_{1i} + \beta_2 x_{2i} + \cdots + \beta_k x_{ki})}}$$

Because the observation is given, *this likelihood is a function of the parameters β_j, not the observations*. Likewise, the probability of observing a failure (i.e., $Y_i = 0$) is

$$L_{Y_i=0}(\beta_0, \beta_1, \ldots, \beta_k) = 1 - L_{Y_i=1} \tag{11.7}$$

We take advantage of the facts that (1) Y_i must be either 1 or 0, (2) any expression raised to the first power is just the expression, and (3) any expression raised to the 0 power is 1 to write an equation for the likelihood (or probability) of obtaining a given observation:

$$L_i(\beta_0, \beta_1, \ldots, \beta_k) = L_{Y_i=1}^{Y_i} \cdot L_{Y_i=0}^{(1-Y_i)} \tag{11.8}$$

because when $Y_i = 1$, this equation reduces to

$$L_i(\beta_0, \beta_1, \ldots, \beta_k) = L_{Y_i=1}^{1} \cdot L_{Y_i=0}^{(1-1)} = L_{Y_i=1} \cdot 1 = L_{Y_i=1}$$

and, when $Y_i = 0$

$$L_i(\beta_0, \beta_1, \ldots, \beta_k) = L_{Y_i=1}^{0} \cdot L_{Y_i=0}^{(1-0)} = 1 \cdot L_{Y_i=0} = L_{Y_i=0}$$

Substitute from Eq. (11.7) into Eq. (11.8) to obtain

$$L_i(\beta_0, \beta_1, \ldots, \beta_k) = L_{Y_i=1}^{Y_i}(1 - L_{Y_i=1})^{(1-Y_i)} = \left(\frac{L_{Y_i=1}}{1 - L_{Y_i=1}}\right)^{Y_i}(1 - L_{Y_i=1})$$

This equation is the likelihood of obtaining the ith observation, $(X_{1i}, X_{2i}, \ldots, X_{ki}, Y_i)$ if the parameters of the regression model are $\beta_0, \beta_1, \ldots, \beta_k$.

Because the n observations are independent, the probability of observing all n data points is the product of the probabilities of observing each of the individual points. Hence, the likelihood of observing the n data points in the sample is the product of the likelihoods of observing the individual points:

$$L(\beta_0, \beta_1, \ldots, \beta_k) = \Pi \, L_i$$

where the Greek Π indicates multiplication over all n observations. Substitute from Eq. (11.5) to obtain

$$L(\beta_1, \beta_2, \ldots, \beta_k) = \Pi \left(e^{\beta_0 + \beta_1 x_{1i} + \cdots + \beta_k x_{ki}} \right)^{Y_i} \left(\frac{1}{1 + e^{-(\beta_0 + \beta_1 x_{1i} + \cdots + \beta_k x_{ki})}} \right)$$

The maximum likelihood estimates for the parameters, β_0, β_1, $\ldots$, β_k are the values that maximize $L(\beta_1, \beta_2, \ldots, \beta_k)$. In other words, *the maximum likelihood estimates for the parameters are those values that maximize the probability of observing the data at hand.*

Because the logarithm converts multiplication into addition and powers into products, the estimation problem can be simplified by taking the logarithm of the last equation to obtain the *log likelihood function:*

$$\ln L(\beta_0, \beta_1, \ldots, \beta_k) = \Sigma Y_i(\beta_0 + \beta_1 x_{1i} + \cdots + \beta_k x_{ki}) - \Sigma \ln (1 + e^{-(\beta_0 + \beta_1 x_{1i} + \cdots + \beta_k x_{ki})})$$

$$(11.9)$$

The same values of the parameters that maximize L will also maximize $\ln L$. The likelihood (or log likelihood) function is a nonlinear function of the model parameters, so it is necessary to use an iterative method similar to those discussed in Chap. 10 to determine the parameter estimates. The computer programs that are used to conduct logistic regressions have the necessary nonlinear parameter estimation algorithm built in, as well as the process of computing first guesses. Once the algorithm has converged, it can also generate approximate standard errors for the regression coefficients, which, may in turn, be used to conduct approximate tests of significance as we did for other nonlinear regressions in Chap. 10.

In sum, the maximum likelihood method involves finding the values of the population parameters that best match the observed sample, i.e., the hypothetical population parameters that are more likely than any other to generate the observed sample. Although this criterion may seem strange, it is widely used in applied statistics because it applies more generally than the least-squares criterion we have used through the rest of this book. When both criteria for parameter estimation—maximum likelihood and least squares—are applicable, they almost always lead to the same parameter estimates. For example, the sample mean is the maximum likelihood estimate for the population mean and the regression coefficients we computed in Chap. 3 are also the maximum likelihood estimates for the coefficients in a multiple regression.[*]

---

[*]This situation usually requires that the random component of the observations be normally distributed. For a general discussion of maximum likelihood estimators, see T. H. Wonnacott and R. J. Wonnacott, *Introductory Statistics for Business and Economics* (3rd ed); New York, Wiley, 1984, Chap. 18, "Maximum Likelihood Estimation (MLE)."

HYPOTHESIS TESTING IN LOGISTIC REGRESSION

Once we have computed the coefficients in a logistic equation, we are still left with the question of whether the regression model "explains" the data and whether or not individual coefficients in the regression equation are significantly different from zero. Although the specific test statistics we use to test these hypotheses are different from those we developed for least-squares regression, the general approach is completely analogous.

Testing the Logistic Equation

The best defense against error in logistic regression has nothing to do with statistical tests. It consists of dividing the data at random into two halves, analyzing one-half of the data, and then seeing if the coefficients obtained by analyzing the second half are similar.

In multiple linear regression, we tested the hypothesis that the regression equation did not provide any information about the dependent variable by examining the ratio of the variation associated with the regression equation compared to the residual variation about the regression plane,

$$F = \frac{MS_{\text{reg}}}{MS_{\text{res}}}$$

When the resulting value of F was large, we concluded that the regression equation "explained" some of the variance in the dependent variable. In essence, we asked whether the independent variables in the regression model contained any more information about the dependent variable than simply reporting its mean.

The *likelihood ratio test* serves a role analogous to this F test for testing the overall fit of a logistic regression. To construct the test statistic for the likelihood ratio test, we first compute the likelihood of obtaining the observations if the independent variables had no effect on whether or not the dependent variable was "success" or "failure" (i.e., whether Y was 1 or 0). In this case $\beta_1 = \beta_2 = \cdots = \beta_k = 0$, and the regression model given by Eq. (11.6) reduces to

$$\text{logit } P = \beta_0$$

and β_0 is the mean outcome, or proportion of successes. It is analogous to describing a set of continuous data in a linear regression problem with the mean of all values of the dependent variable using the linear regression equation containing only the constant term

$$y = \alpha_0 + \epsilon$$

In terms of the Martian example we discussed earlier in this chapter, this estimate, b_0, of β_0 is .619, so that the estimated probability of graduation is

$$\hat{P} = \frac{1}{1 + e^{-b_0}} = \frac{1}{1 + e^{-.619}} = .65$$

In other words, if we do not know anything about an *individual* Martian, our best estimate of the probability of that individual graduating is the mean value of G, .65, the proportion of all Martians in our sample who graduated (78/120).

The likelihood of obtaining the set of *all* the observations is simply the product of the individual likelihoods. Therefore,

$$L_0 = \Pi \, (e^{b_0})^{Y_i} \left(\frac{1}{1 + e^{b_0}} \right)$$

is the likelihood that our best predictor of the dependent variable is simply the constant term in the logistic regression equation if the probability of graduating does not depend on intelligence.

The effects of the other terms in the logistic function modify the predicted probability of a successful outcome, depending on the observed values of the independent variables. When we include the information in the independent variables, the likelihood L of obtaining the set of observations increases. The question is: Is the increase in likelihood we obtain by adding the independent variables to our regression model large enough to conclude that the independent variables significantly contribute to our ability to predict the dependent variable?

To answer this question, we compare the likelihood of obtaining the observations if the full model holds L with the likelihood of obtaining the observations if the independent variables had no effect on the outcome (the intercept-only model) L_0 by computing the test statistic

$$G_0^2 = 2 \ln \frac{L}{L_0} = 2(\ln L - \ln L_0)$$

If the pattern of observed outcomes is more likely to have occurred when the independent variables affect the outcome than when they do not, G_0^2 will be a large number. On the other hand, if the independent variables do not influence the outcome (i.e., the value of the dependent variable), then the likelihood of obtaining the observations given the information in the independent variables will be the same as that contained in the mean response b_0 and G_0^2 will be zero.

Because of sampling variation, G_0^2 will vary even if the null hypothesis of no effect of the independent variables is true, just as the values of F varied in multiple linear regression. We thus need to compare the observed value of the G_0^2 test statistics with an appropriate sampling distribution. It has been shown that the likelihood ratio test statistic approximately follows a χ^2 distribution with $k - 1$ degrees of freedom, with the quality of this approximation improving as the sample size grows.[*] A large value of the likelihood ratio test statistic (or correspond̓ing small P value) indicates that the independent variables significantly improve our ability to predict the outcome, the dependent variable.

Testing the Individual Coefficients

The likelihood ratio test just described is used to test whether the logistic regression equation significantly improves our ability to predict the outcome (dependent variable), compared with the proportion of successes, without taking into account any of the information contained in the independent variables. We would also like to test whether the coefficients associated with each independent variable are significantly different from zero. As with multiple linear regression, we have two approaches to this problem, one comparing the individual coefficients with their standard errors (analogous to the t tests in multiple regression) and another looking at the improvement in overall predictive ability of the regression equation with the addition of the independent variable in question (analogous to the incremental F tests in multiple regression).

The simplest test for the significance of an individual coefficient b_i is to *compare the observed value with its associated standard error*[†] s_{b_i} with the ratio

$$z = \frac{b_i}{s_{b_i}}$$

If the value of b_i is large compared to the uncertainty in its measurement, quantified with $s_{b_{i'}}$ we conclude that it is significantly different from zero and that the independent variable x_i contributes significantly to predicting the dependent variable. We compute the associated risk of a false positive—the P value—by comparing z with the standard normal

---

[*]Statisticians say that the likelihood ratio statistic is distributed *asymptotically* as a χ^2 distribution. Because this likelihood ratio test statistic G_0^2 asymptotically follows the χ^2 distribution, it is also sometimes called the *model chi square*.

[†]As with the other nonlinear regression methods described in Chap. 10, it is possible to compute approximate estimates of the standard errors associated with the individual regression coefficients by linearizing the logistic regression equation in the neighborhood of the best estimates of the regression coefficients. For the actual equations used to define these standard errors, see J. Fox, *Linear Statistical Models and Related Methods with Applications to Social Research*, New York, Wiley, 1984, pp. 308–310.

distribution (Table E-1, Appendix E, with an infinite number of degrees of freedom).

A second approach to testing whether or not the independent variable x_j improves the predictive ability of the regression equation is a form of the *likelihood ratio test* just described. Let L_{-j} be the likelihood associated with the regression model containing all the independent variables *except* x_j,

$$\text{logit } P = \beta_0 + \beta_1 x_1 + \cdots + \beta_{j-1} x_{j-1} + \beta_{j+1} x_{j+1} + \cdots + \beta_k x_k$$

and let L be the likelihood associated with the full regression equation. Just as when we tested the overall regression equation with a likelihood ratio test, we can test whether adding the independent variable x_j significantly improves the predictive ability of the equation by computing the test statistic

$$G_j^2 = 2 \ln \frac{L}{L_{-j}} = 2(\ln L - \ln L_{-j})$$

This test statistic is compared with the critical values of the χ^2 distribution with 1 degree of freedom (Table E-5, Appendix E). As with the overall test of significance described above, the χ^2 distribution becomes a more accurate description of the actual distribution of the likelihood ratio test as the sample size increases. Because G_j^2 approximately follows a χ^2 distribution, it is also sometimes called the *improvement chi square*.[*] This test is analogous to the incremental F test in linear regression.

In large samples both the z and likelihood ratio test statistics give comparable results. In both tests, a large value of the test statistic (or, equivalently, a small P value) means that the variable in question provides useful information for predicting the dependent variable.

The likelihood ratio test can be generalized further to test whether there is a significant improvement in the regression equation's predictive ability when several independent variables are added together. This situation most typically arises when one of the independent variables is a qualitative condition that has several levels (for example, different experimental conditions) encoded with dummy variables. Because there are several levels of the independent variable, it requires more than a single dummy variable to describe fully all the levels. In a logistic regression analysis, one often wants to treat all these dummy variables as a single variable and consider the effects of all of them on a single step. For example, suppose that, rather than just adding x_j, we added the m independent variables $x_j, x_{j+1}, \ldots, x_{j+m-1}$. Let L_{-j} be the likelihood asso-

---

[*]G_j^2 is sometimes also called the *approximate chi square* to remove or enter the variable in the regression equation.

ciated with the regression model excluding these m independent variables and L be the likelihood associated with the full regression equation including these m additional independent variables. We can then test whether this set of additional independent variables significantly improves our ability to predict the dependent variable using Eq. (11.6). The only difference is that the resulting test statistic is compared with the critical values of the χ^2 distribution with m degrees of freedom, rather than 1 degree of freedom.

Back to Mars

Figure 11-2 shows the computer output for the analysis of the data in Fig. 11-1*A* (and Table 11-1) using Eq. (11.1). As we noted when we first

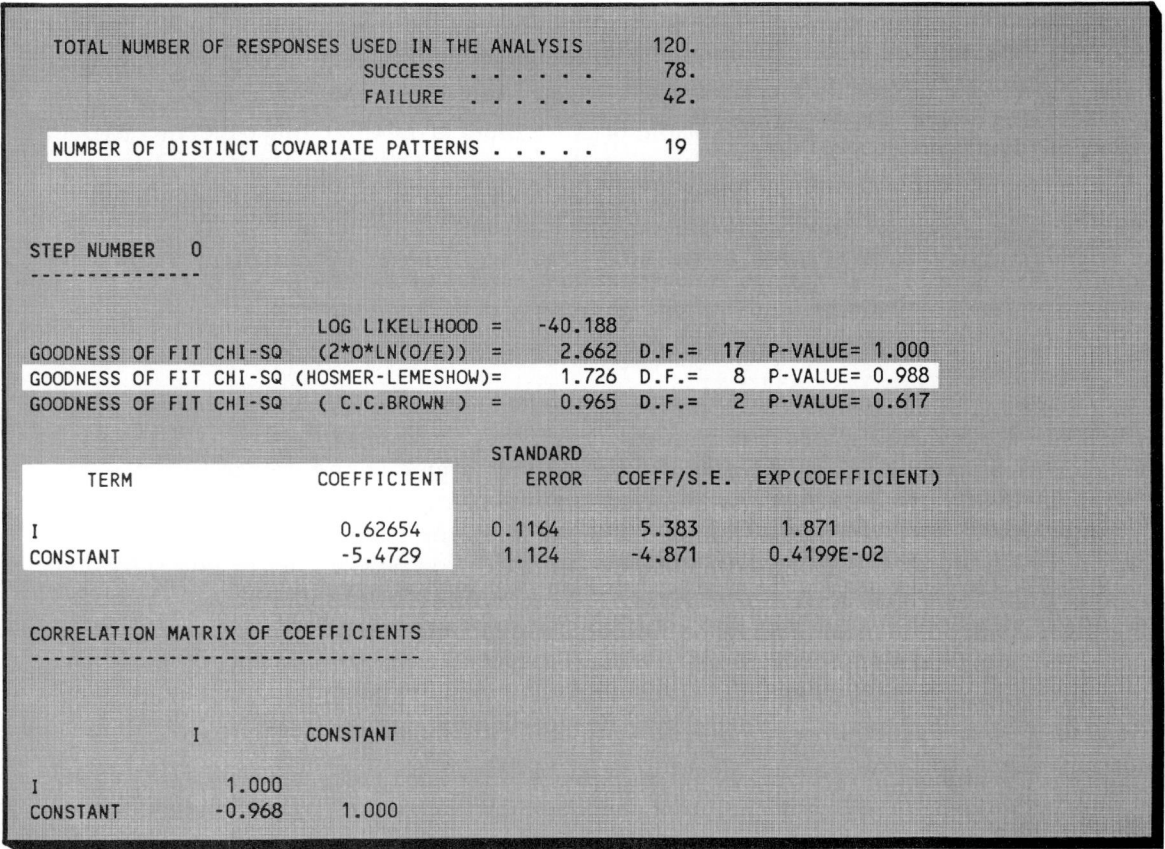

FIGURE 11-2 Results of logistic regression analysis of the probability of graduating as a function of intelligence for the data shown in Fig. 11-1.

introduced the Martian intelligence example, the constant in the logistic regression equation b_0 is -5.47 and the coefficient associated with intelligence b_I is $.6265/\text{zorp}$. The standard error associated with the independent variable, intelligence, is $.1164/\text{zorp}$, so we can test whether it is significantly different from zero by computing the ratio

$$z_{b_I} = \frac{b_I}{s_{b_I}} = \frac{.6265/\text{zorp}}{.1164/\text{zorp}} = 5.383$$

which is greater than 3.291, the critical value of the normal distribution for $\alpha = .001$. Hence, we can conclude that intelligence significantly affects the probability that a Martian will graduate from college $(P < .001)$.

We reach the same conclusion by considering the likelihood ratio test associated with adding the variable I to the logistic regression equation. The value of the likelihood ratio test statistic associated with the variable intelligence, G_I^2, is 75.01. This value exceeds 10.828, the critical value for the χ^2 distribution with 1 degree of freedom for $\alpha = .001$, so this test also leads to the conclusion that intelligence significantly affects whether or not a Martian graduates from college $(P < .001)$. As with simple linear regression, because there is only one independent variable, testing the significance of the effect of that variable is equivalent to testing the overall significance of the logistic regression equation.

IS THE LOGISTIC REGRESSION EQUATION AN APPROPRIATE DESCRIPTION OF THE DATA?

The methods developed so far permit us to test whether or not the logistic regression model permits one to predict the probability of a successful outcome better than simply using the overall probability of success, without taking into account the effects of the independent variables. We also developed methods to test whether or not individual independent variables contribute significantly to estimating the probability of a successful outcome. The question still remains whether the logistic model is a reasonable one with which to describe the data, particularly since it is often difficult to visualize the actual data set. We will present regression diagnostics for identifying influential data points analogous to those we developed in Chap. 4. We will also develop two similar approaches to testing the overall goodness of fit of the logistic equation using methods based on the χ^2 statistic. These statistics, plus plots of the residuals vs. the independent variables, can help determine whether or not the logistic model using the independent variables that are included is appropriate and how to modify the equation when it is not.

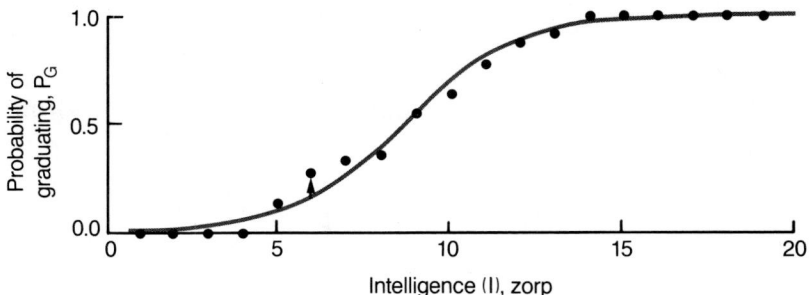

FIGURE 11-3 Because the logistic regression equation estimates the probability of a successful outcome P_G and the original dichotomous dependent variable G indicates whether or not a successful outcome occurred, it is possible to define a residual as the difference between P_i and $\hat{P}_i = P_G$, i.e., the difference between the observed proportion and the estimated probability of that outcome, indicated by the arrow on the figure. (Compare with Fig. 11-1B.)

Regression Diagnostics for Logistic Regression

As in linear models fit by least squares, it is desirable to assess the adequacy of a fitted linear logit model by examining the data for outliers and influential points. As we saw in Chap. 4, examining these diagnostics can point to specific shortcomings in the way the regression model is formulated and help guide us in developing a form of the regression model which better describes the data and leads to more accurate hypothesis tests. We will now extend many of the diagnostic methods developed in Chap. 4 to logistic regression.*

As with linear regression, there are three ways to define *residuals*. The *raw residual* is

$$e_i = P_i - \hat{P}_i$$

where P_i is the observed proportion of successful outcomes ($Y_i = 1$) and $\hat{P}_i$ is the probability of a success obtained from the logistic regression equation.† The raw residual is just the difference between what actually happened and the estimated probability of that occurrence. Figure 11-3

---

*For more formal derivations of many of these diagnostics, see J. Fox, *Linear Statistical Models and Related Methods with Applications to Social Research*, New York, Wiley, 1984, pp. 320–332; D. Pregibon, "Logistic Regression Diagnostics," *Ann. Stat.* 9:705–724, 1981; and J. M. Landwehr, D. Pregibon, and A. C. Shoemaker, "Some Graphical Procedures for Studying a Logistic Regression Fit," *1980 Proceedings of the Business and Economic Statistics Section, American Statistical Association*, pp. 15–20, 1980.

†An alternative definition is the difference between the actual outcome for an individual and the estimated probability of that outcome, i.e., $G - P_G$. Most computers produce residuals and residual diagnostics using either definition.

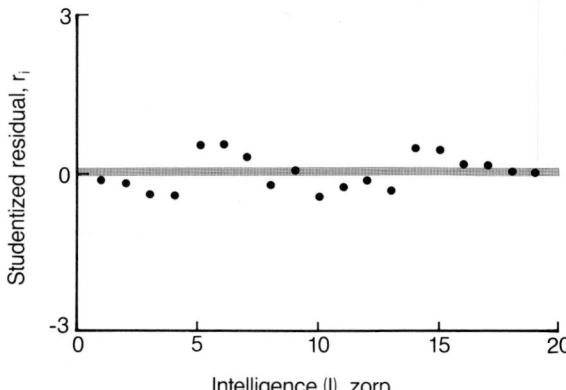

FIGURE 11-4 Residual plot for proportions corresponding to the analysis of Fig. 11-2 shows no problems with the regression model.

shows the same data on Martian intelligence and graduation as in Fig. 11-1*A*, with a typical raw residual indicated.

Because the standard deviation of a proportion is*

$$s_{\hat{p}_i} = \sqrt{\hat{P}_i^2 - (1 - \hat{P}_i)}$$

we define the *standardized residual* as

$$e_{s_i} = \frac{e_i}{s_{\hat{p}_i}} = \frac{P_i - \hat{P}_i}{\sqrt{\hat{P}_i(1 - \hat{P}_i)}}$$

Finally, one can also define *Studentized residuals* r_i and *Studentized deleted residuals* r_{-i} in ways analogous to the original definitions in Chap. 4. In all three cases, if the sample size is large, these residuals will approximately follow a normal distribution. Therefore, points with standardized or Studentized residuals exceeding about 2 should be examined for potential problems.

Figure 11-4 shows the studentized residuals plotted against intelligence for our study of Martian intelligence and graduation. It shows two data points with absolute values larger than 3, and a few with absolute values larger than 2. There is no suggestion of nonlinearity or unequal variances.

It is also possible to define quantities analogous to the leverage, and Cook's distance. The application of these influence diagnostics proceeds exactly as we used them in the multiple linear regression problem in Chap. 4. As we discussed in Chap. 4, they are best used to identify influential points for examination for possible data entry errors or problems in the way the regression model is formulated. Figure 11-5*A* and *B* shows

*For a derivation and justification of this result, see S. Glantz, *Primer of Biostatistics* (2nd ed.), New York, McGraw-Hill, 1987, pp. 109–110.

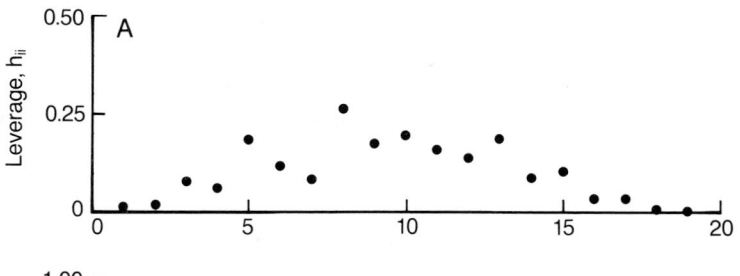

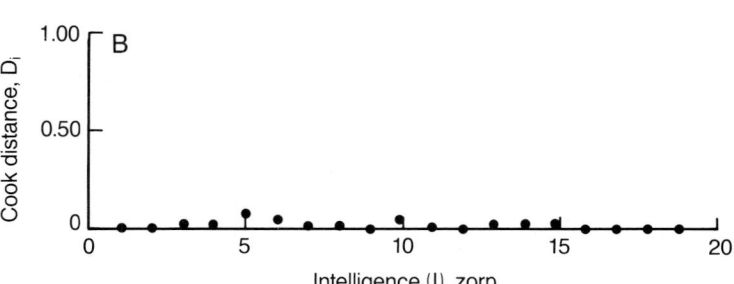

FIGURE 11-5 The leverage (*A*) and Cook's distance (*B*) corresponding to the analysis of Fig. 11-2 reveal no particularly influential points in the data on Martian graduation as a function of intelligence. There are no obvious problems with the logistic regression model as a description of these data.

the leverage and Cook's distance for the Martian data, plotted against subject number. The values of these statistics are unremarkable: None of the leverage values exceed the cutoff of twice the expected value, $2(p/n) = 2(2/120) = .033$, where p is the number of parameters; and none of the Cook's distances exceeds 1, and thus are all well below the threshold of 1 we suggested in Chap. 4.

As with nonlinear regression in general, the most commonly available diagnostic for *multicollinearity* is the correlation matrix of the parameter estimates in the logistic regression equation. Large values of these correlations (above about .9) indicate that multicollinearity may be complicating the parameter estimation process. When multicollinearity is present, one should use the same techniques for dealing with it that we presented in Chap. 5: centering the independent variables, collecting more data in a way that breaks up sample-based multicollinearities, or deleting redundant independent variables.

Goodness-of-Fit Testing

Because there is a single independent variable and several observations at each value of the independent variable in our study of Martian intelligence and graduation, we could estimate the proportion of successes at each value of the independent variable and plot the logistic regression equation vs. the observed proportions (in Fig. 11-1*B*), then look at this plot to see if it is reasonable. Unfortunately, it is not possible to make such plots when there are more than one or two independent variables.

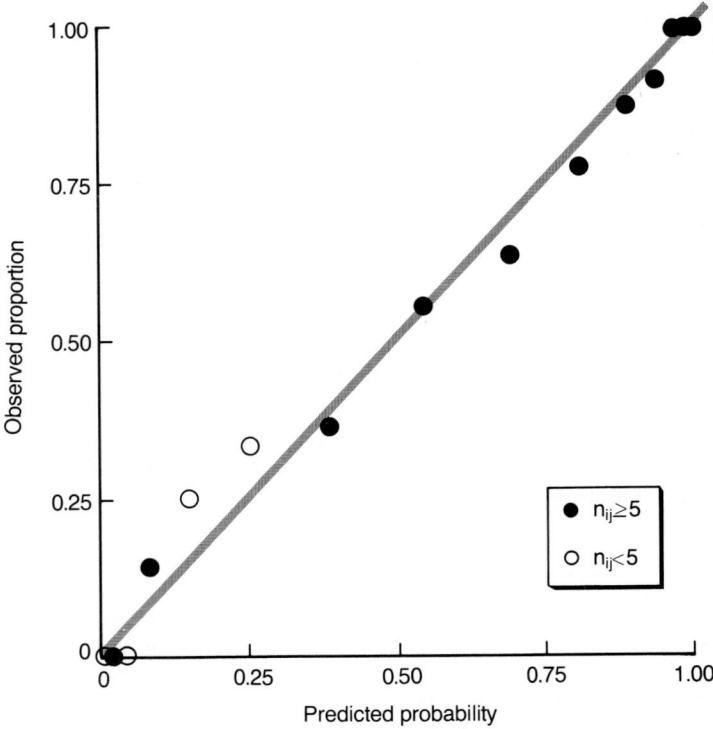

FIGURE 11-6 This plot of the observed portion of Martians graduated at any given intelligence as a function of the predicted probability, estimated from the logistic regression equation, reveals good agreement. Such plots of the observed vs. predicted proportions of successful outcomes can be useful aids in diagnosing whether or not a logistic regression equation is a good description of a set of data, so long as there are enough observations at each combination of the independent variables (known as *patterns*) to make it possible to compute a reliable observed proportion of successful outcomes for data exhibiting that pattern in the independent variables. Generally, one requires at least five observations in any given pattern to reasonably compute an estimate of the probability or the proportion of successful outcomes. When there are small data sets or a large number of patterns compared with the number of observations, plots such as this are not particularly meaningful because small changes in the data set can produce big changes in the appearance of this plot. The different symbols indicate the number of points used to estimate each of the proportions.

An alternative graphical approach to assess the goodness-of-fit qualitatively is to compute the observed proportion of successes within each pattern of the independent variables and plot that proportion versus the predicted probability of success for that combination of independent variables. If the logistic regression model is reasonable, the resulting plot will be a straight line with a slope of about 1. Figure 11-6 shows such a plot for our study of Martian intelligence and graduation; it reveals that the logistic equation is a reasonable description of the observations.

In addition to these graphical approaches, we would like more formal tests of goodness of fit. We will discuss two approaches for testing the goodness of fit: the *classic approach* and the *Hosmer-Lemeshow test*. Both approaches use the idea of comparing the observed number of individuals with each outcome with the number expected based on the logistic regression equation. These observed (O) and expected (E) numbers are combined to form a *goodness-of-fit chi square*:

$$\chi^2 = \sum \frac{(O - E)^2}{E} \tag{11.10}$$

Large values of this test statistic indicate that the observed and expected results are very different, so that the regression equation has a poor fit to the data (as with any statistical distribution, one must consider the degrees of freedom when defining "large"). In contrast, small values of this test statistic indicate that the observed and expected results are similar, so the regression equation provides a good fit to the observations. Alternatively, large P values (approaching 1) indicate a good fit and small P values (approaching 0) indicate a poor fit. Anytime the P value for the goodness-of-fit test even approaches statistical significance, you should be concerned that the logistic model, as formulated, is not a good description of the data. The difference between the classical and Hosmer-Lemeshow tests lies in how the observed and expected numbers in this equation are derived.

The *classic approach* begins with identifying different combinations of the independent variables, called *patterns*.* For example, if there are two dichotomous independent variables, sex and nicotine abuse, there are four distinct patterns: male nonsmoker, male smoker, female nonsmoker, and female smoker. For each pattern, we count the observed number of individuals, O, with success ($Y_i = 1$) and failure ($Y_i = 0$) outcomes and compare these numbers with the number expected (predicted by the regression equation) E, using the χ^2 statistic.

Suppose that there are N individuals in a particular pattern of whom n_1 are successes and n_0 are failures. Because the exposures to secondhand smoke are identical, the estimated probabilities $\hat{P}$ will apply equally to everyone in the group. $N\hat{P}$ therefore is the expected or predicted number of successes (cases with $Y_i = 1$), and $N(1 - \hat{P})$ is the expected number of failures (cases with $Y_i = 0$). Thus, the goodness of fit of the regression equation to the data is

---

*The patterns are analogous to the cells in a contingency table, which is also analyzed using a chi-square test statistic. For a discussion of contingency tables, see S. A. Glantz, *Primer of Biostatistics* (2nd ed.), New York, McGraw-Hill, 1987, pp. 119–132.

$$\chi^2 = \sum \left\{ \frac{(n_1 - N\hat{P})^2}{N\hat{P}} + \frac{[n_0 - N(1 - \hat{P})]^2}{N(1 - \hat{P})} \right\}$$

where the summation is over all patterns in the independent variables. An alternative formulation of this test statistic comes from writing it in the form of a likelihood ratio

$$\chi^2 = 2\sum \left[n_1 \ln \left(\frac{n_1}{N\hat{P}} \right) + n_0 \ln \left(\frac{n_0}{N(1 - \hat{P})} \right) \right] = 2 \, \Sigma \, O \ln \left(O/E \right)$$

When there are many data points so that the data are not "thin" (so that the predicted values of the successes and failures for many patterns exceed five), these two formulations will yield reasonably close numerical results. These χ^2 statistics can be converted to P values using the chi-square distribution with $n_p - k - 1$ degrees of freedom where n_p is the number of patterns and k is the number of independent variables in the regression equation. There are 19 patterns in the 120 data points we collected on Martian intelligence and graduation (Fig. 11-2), which is "thin" for using the classic goodness-of-fit test.

This classical approach works reasonably well when the number of distinct patterns is neither too small nor too large. If the number of patterns is too small, the test for goodness of fit is too coarse, whereas if the number of patterns is too large, the predicted number of successes for many of the patterns in the independent variables will be too small (less than 5), and so the distribution of the computed χ^2 is not reasonably described by the theoretical χ^2 distribution. This situation is particularly likely to occur when one or more of the independent variables are continuous or in clinical studies when the total sample size is relatively small (below 100 patients).

The *Hosmer-Lemeshow goodness-of-fit statistic* was developed to address the problem of having too few observations under each pattern in the independent variables using the classical approach. In this approach the probability of success $\hat{P}_i$ is calculated for every individual in the sample, and the resulting numbers are arranged in increasing order. The range of probability values is then divided into 10 groups (deciles).* For each decile, the observed numbers O of individuals with successful out-

---

*There do not have to be 10 groups; this number was first suggested by Lemeshow and Hosmer and seems to work well under most circumstances. When there are probability deciles for which there are no predicted observations for either outcome, either change the number of groupings to avoid empty cells or delete the empty deciles from the analysis and reduce the degrees of freedom accordingly. Another alternative is to divide the data into groups of $n/10$ members in each group after ranking the points in order of P_i. For more details, see S. Lemeshow and D. W. Hosmer, "A Review of Goodness-of-Fit Statistics for use in the Development of Logistic Regression Models," *Am. J. Epid.* 115:92–106, 1982.

comes $(Y_i = 1)$ are counted and the expected numbers E are calculated by adding the predicted probabilities $(\hat{P}_i)$ for all the individuals in each decile. Likewise, the number of individuals with failure outcomes $(Y_i = 0)$ is counted in each decile, and the expected numbers are computed by adding up the predicted probabilities of failure $(1 - \hat{P}_i)$ for each decile. As before, the goodness-of-fit statistic is calculated using Eq. (11.10), where the summation extends over both successes and failures in the ten deciles. As before, large values of χ^2 indicate a poor fit and small values indicate a good fit.

This value of χ^2 can be converted to a P value by comparing it to the theoretical χ^2 distribution with $g - 2$ degrees of freedom, where g is the number of groups, usually 10. Values of P near 1 indicate a good fit and values of P near 0 indicate a poor fit. As before, there need to be at least five predicted successes and failures in most deciles for the value of the χ^2 test statistic to be reasonably described by the theoretical χ^2 distribution.

The computer output in Fig. 11-2 shows both χ^2 values: the classic χ^2 is 2.662 with $n_p - k - 1 = 19 - 1 - 1 = 17$ degrees of freedom $(P = 1.000)$, and the Hosmer-Lemeshow goodness-of-fit chi square is 1.726 with $g - 2 = 10 - 2 = 8$ degrees of freedom, corresponding to a P value of .988. Hence, both the classic and Hosmer-Lemeshow goodness-of-fit tests fail to reveal any problems with the logistic regression model to describe how intelligence and graduation are related among our sample of Martians. Because the sample size, while large by the standards of many clinical investigations, still leads to the situation in which there are many cells (patterns) that have expected numbers of success less than 5 in the classic approach, the Hosmer-Lemeshow goodness-of-fit statistic is to be preferred over the classic goodness-of-fit test.

Finally, another measure of goodness of fit is the *deviance*, defined as

$$G^2 = -2 \ln L = -2 \Sigma [Y_i \ln \hat{P}_i + (1 - Y_i) \ln (1 - \hat{P}_i)]$$

The deviance is actually a likelihood ratio test that contrasts the fitted model with one that predicts each observation perfectly (and hence for which $L = 1$ and $\ln L = 0$). This statistic follows directly from Eq. (11.9) by setting $L_{Y_i=1} = \hat{P}_i$. G^2 follows an asymptotic χ^2 distribution with $n - k$ degrees of freedom. The deviance is also used to define another type of residual:

$$G_i = \pm\sqrt{-2[Y_i \ln \hat{P}_i + (1 - Y_i) \ln (1 - \hat{P}_i)]}$$

where the sign associated with G_i is the same as that assigned to the raw residual. Points associated with large values of G_i differ from what the logistic regression equation predicts.

Notice that all the methods for assessing goodness of fit in logistic regression require that the sample sizes be large enough to have at least five individuals in most groups (pattern or decile) for both outcomes in order to be able to convert the χ^2 statistic to a P value. As a practical matter, this requirement means that one cannot associate a P value with a goodness-of-fit statistic in logistic regression when the total sample size is below about 80 individuals, and *much larger sample sizes are desirable*. Although this restriction is rarely a problem in large epidemiological studies, it can be important when assessing the results of logistic regression when it is used to analyze the results of smaller clinical studies.* Anytime the appropriate goodness-of-fit statistic even approaches statistical significance, you should be concerned that the regression model is not appropriate.

ARE BONE CANCER PATIENTS RESPONDING TO CHEMOTHERAPY?

Patients with bone cancer are often treated with drugs to kill cancer cells (chemotherapy) before their cancers are surgically removed. In the past the characteristics of the cancer have been evaluated after it was removed in order to judge how well the chemotherapy was working. If the cancer is being killed by the drug, the tumor should be dying and remodeling. Thus, one would expect to find both evidence of tumor cell death and evidence that new normal tissue is growing where the tumor is dying. Although it has been assumed that these characteristics of cancer tissue death and new normal tissue formation are associated with the chemotherapy, it is not known how much of this change is, in fact, related to the chemotherapy versus how much is due to normal aging of the tumor.

To see if these characteristics were associated with presurgical chemotherapy, Misdorp and coworkers[†] studied the tissue characteristics

---

*Even if there is a large sample size, there can still be problems if there are many independent variables in the equation. For example, if there are 6 independent variables, each of which took on 1 of 2 values, there would be $2 \times 2 \times 2 \times 2 \times 2 \times 2 = 2^6 = 64$ cells. Thus, there would need to be a minimum of about $5 \cdot 64 = 320$ subjects to reasonably assess goodness of fit. If there were 8 independent variables, 6 of which were dichotomous and 3 of which took on 1 of 3 values, there would be $2 \times 2 \times 2 \times 2 \times 2 \times 2 \times 3 \times 3 = 576$ cells, which means we would need a minimum of about $5 \cdot 576 = 2880$ subjects. This computation assumes that data are distributed evenly over all the cells, which is often not the case.

†W. Misdorp, G. Hart, J. F. M. Delemarre, P. A. Voute, and J. W. van der Eijken, "An Analysis of Spontaneous and Chemotherapy-Associated Changes in Skeletal Osteosarcomas," *J. Pathol.* 156:119–128, 1988.

of the tumors removed from patients. These patients had been divided into two groups, one of which received chemotherapy for 3 to 5 weeks before surgical removal of the tumor and one of which did not. In particular, two tissue characteristics were found to be of most interest. The first characteristic was the amount of viable tumor and the second was the amount of fibroblasts (cells that form new tissue). We will use logistic regression to determine whether either of these variables is associated with those patients who received chemotherapy.

We have a dichotomous dependent variable, whether or not the patient received chemotherapy, which we code as

$$C = \begin{cases} 0 & \text{if no preoperative chemotherapy} \\ 1 & \text{if preoperative chemotherapy} \end{cases}$$

We have two independent variables, the amount of viable tumor tissue and the amount of fibroblasts. As is common in such clinical studies, these two variables were measured on ordinal scales. The tumor viability V was graded 1, 2, 3, or 4, with the lower score corresponding to the least amount of viable tissue (i.e., the most dead tissue in the tumor). The amount of fibroblasts F was given a score of 0 to 3, with 0 indicating no new tissue was present, 1 indicating a little new tissue was present, 2 a moderate amount of new tissue was present, and 3 indicating extensive new tissue was present.

The fact that the independent variables are measured on an ordinal scale complicates the formulation of this problem. The most straightforward approach would be to treat V and F as continuous variables and simply use the observed numbers in a logistic regression equation. In doing so, we implicitly assume that the odds of being in the chemotherapy group change by the same amount as the classification of the tumor moves from, say, 1 to 2, as it does if the classification changes from 3 to 4. Such assumptions may or may not be reasonable and need to be examined explicitly as part of the analysis. The other alternative is to treat ordinal independent variables as categorical variables and encode them with dummy variables. In doing so, one no longer has to assume that the change in odds of being in the "success" group changes uniformly as a patient moves from one clinical category to the next. The disadvantage of using dummy variables is that the ordinal information in the independent variables is lost. We will use the former approach in this problem.*

---

*Like all assumptions in a regression analysis, this one should be tested against the data. One can use the categorical approach as a check on how reasonable the assumptions behind the approach we are using here is by examining the regression coefficients for each dummy variable to see if they are about equal. If they are, the assumption is reasonable.

```
TOTAL NUMBER OF RESPONSES USED IN THE ANALYSIS        176.
                           SUCCESS  . . . . .           88.
                           FAILURE  . . . . . .         88.

  NUMBER OF DISTINCT COVARIATE PATTERNS . . . . .        12



STEP NUMBER   0
---------------


                     LOG LIKELIHOOD =  -104.609
GOODNESS OF FIT CHI-SQ  (2*O*LN(O/E))  =     14.801  D.F.=  8  P-VALUE= 0.063
GOODNESS OF FIT CHI-SQ (HOSMER-LEMESHOW)=    13.305  D.F.=  8  P-VALUE= 0.102
GOODNESS OF FIT CHI-SQ  ( C.C.BROWN )  =      8.466  D.F.=  2  P-VALUE= 0.015

                                   STANDARD
        TERM              COEFFICIENT    ERROR   COEFF/S.E.  EXP(COEFFICIENT)

V                          -0.51883     0.3013    -1.722        0.5952
F                           0.72299     0.4788     1.510        2.061
VF                         -0.36178E-01 0.2463    -0.1469       0.9645
CONSTANT                    0.29382     0.7009     0.4192       1.342


CORRELATION MATRIX OF COEFFICIENTS
----------------------------------


                 V        F        VF       CONSTANT

V             1.000
F             0.692    1.000
VF           -0.625   -0.920    1.000
CONSTANT     -0.937   -0.729    0.570      1.000
```

FIGURE 11-7 Results of logistic regression analysis for the data on bone cancer treatments as a function of tumor viability and fibroblast content. The P value associated with the Hosmer-Lemeshow goodness-of-fit χ^2 of .102 indicates that this model is not a particularly good fit to the data. In addition, the large values of the correlation between the F and VF coefficients indicates that multicollinearity may be a problem.

We begin with the logistic regression model

$$\hat{P}_C(V, F) = \frac{1}{1 + e^{-(b_0 + b_V V + b_F F + b_{VF} VF)}}$$

or, equivalently,

$$\text{logit } \hat{P}_C = b_0 + b_V V + b_F F + b_{VF} VF \qquad (11.11)$$

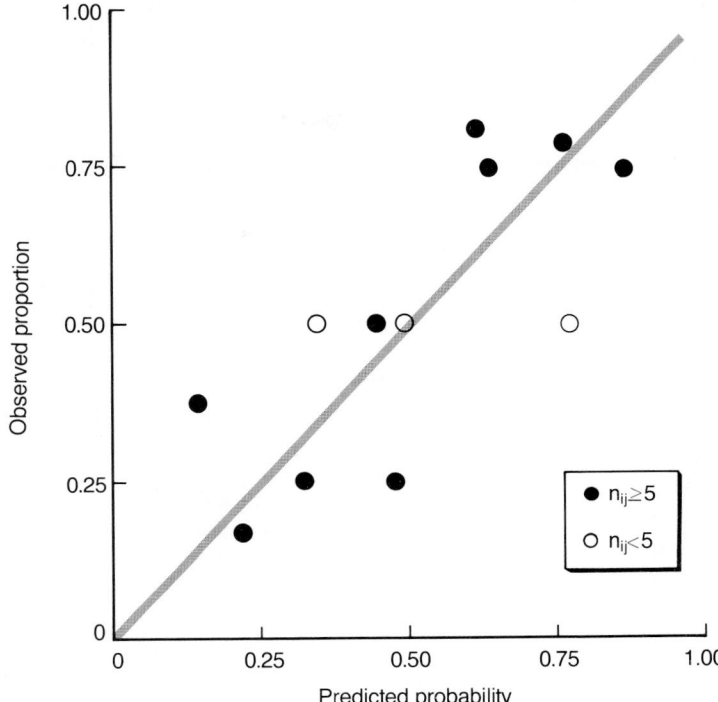

FIGURE 11-8 Plot of the observed proportions of patients with "successful" outcomes (i.e., those in the presurgical chemotherapy group) against the probability of success predicted by the logistic regression equation shows considerable scatter, which indicates that the regression model is not a particularly good description of these data.

where we have included both the effects of tumor viability V and fibroblasts F as well as allowed for an interaction between these two independent variables by including the term VF.

Figure 11-7 shows the results of fitting the data in Table C-23, Appendix C, using this equation. The results are not very encouraging. Both goodness-of-fit statistics show that the model is not a very good fit to the data; the classic goodness-of-fit χ^2 has a P value of .063 and the Hosmer-Lemeshow statistic has a P value of .102. Moreover, none of the regression coefficients is significantly different from zero. This lack of fit is also suggested by the wide scatter in the plot of observed vs. predicted probabilities (Fig. 11-8).

To help us decide if we can improve this model we consider the standardized residual plots in Fig. 11-9, which suggest possible nonlinear relationships between the standardized residuals and the two independent variables, particularly for V. These plots are of the residuals at each pattern (combination of V and F scores), not for individuals in the sample. As we have seen before, such nonlinear relationships can mask correlations between variables, so let us consider a model including quadratic terms in V and F:

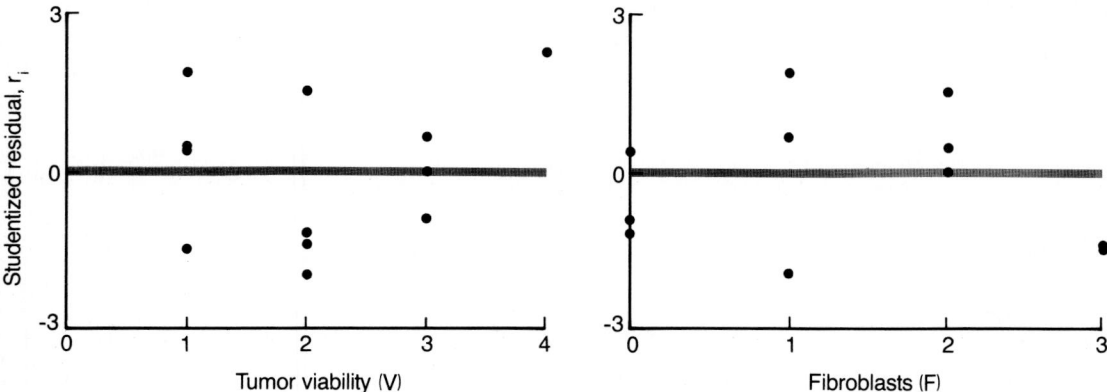

FIGURE 11-9 Plots of the Studentized residual, r_i vs. the two independent variables, tumor viability $V(A)$ and fibroblast content $F(B)$, suggest that there may be a nonlinear relationship between the independent variables and the probability of a successful outcome. This suggestion is particularly strong for tumor viability V. To see if these suspected nonlinearities will significantly improve the fit to the cancer data, we will add the quadratic terms to the logistic equation.

$$\text{logit } \hat{P}_C = b_0 + b_V V + b_F F + b_{V^2} V^2 + b_{F^2} F^2 + b_{VF} VF \qquad (11.12)$$

Figure 11-10 shows that this logistic regression equation produces a much better description of the data than did Eq. (11.12). The P value associated with the Hosmer-Lemeshow goodness-of-fit statistic has increased from .102 to .808. Moreover, two of the coefficients in the regression equation, $b_V = -3.26 \pm 1.30$ and $b_{V^2} = .600 \pm .256$ are significantly different from zero; the values of $z = b_i/s_{b_i}$ (in Fig. 11-10) exceed 1.96, the critical value for $P < .05$ for the standard normal distribution, and the chi square to remove statistics exceed 3.841, the critical value of $P < .05$ for the χ^2 distribution with 1 degree of freedom. As discussed earlier in this chapter, both these statistics provide approximate tests of the null hypothesis that the independent variable in question does not add information to predicting the outcome (whether or not the patient received chemotherapy). The coefficients associated with F, F^2, and VF interaction were not significantly different from zero.

Examining the residual plots in Fig. 11-11 shows that the standardized residuals no longer suggest a nonlinearity (compare with Fig. 11-9) and the plot of the observed proportions of patients who received chemotherapy vs. the predicted probability of receiving chemotherapy in Fig. 11-12 follows a reasonably straight line with less scatter than before (compare with Fig. 11-8), confirming the quantitative results of the goodness-of-fit tests. These results suggest that the tumor viability is the one variable associated with whether or not a patient received chemotherapy, albeit in a nonlinear way.

```
      TOTAL NUMBER OF RESPONSES USED IN THE ANALYSIS      176.
                                SUCCESS . . . . .          88.
                                FAILURE . . . . . .        88.

      NUMBER OF DISTINCT COVARIATE PATTERNS . . . . .      12



      STEP NUMBER    0
      ---------------


                         LOG LIKELIHOOD =  -100.543
      GOODNESS OF FIT CHI-SQ  (2*O*LN(O/E))  =     6.670  D.F.=  6  P-VALUE= 0.352
      GOODNESS OF FIT CHI-SQ (HOSMER-LEMESHOW)=    4.513  D.F.=  8  P-VALUE= 0.808
      GOODNESS OF FIT CHI-SQ  ( C.C.BROWN )  =     0.881  D.F.=  2  P-VALUE= 0.644

                                       STANDARD
             TERM         COEFFICIENT     ERROR    COEFF/S.E.   EXP(COEFFICIENT)

      V                     -3.2632       1.295     -2.520      0.3827E-01
      F                      1.0026       0.9454     1.061      2.725
      VF                     0.21926      0.3161     0.6936     1.245
      V2                     0.59981      0.2556     2.347      1.822
      F2                    -0.28759      0.2185    -1.316      0.7501
      CONSTANT               2.8364       1.537      1.845      17.05


      CORRELATION MATRIX OF COEFFICIENTS
      ------------------------------------


                    V          F         VF        V2        F2        CONSTANT

      V           1.000
      F           0.522      1.000
      VF         -0.694     -0.816      1.000
      V2         -0.974     -0.410      0.597     1.000
      F2         -0.188     -0.824      0.399     0.120     1.000
      CONSTANT   -0.953     -0.665      0.750     0.873     0.312      1.000
```

FIGURE 11-10 Results of a logistic regression output on the tumor bone cancer treatment data, including quadratic terms for both tumor viability and fibroblast content. Including the quadratic terms greatly improved the quality of the fit compared with the first model. The Hosmer-Lemeshow goodness-of-fit P value has increased to .808 (compared with .102 in Fig. 11-7). Unfortunately, adding the quadratic terms has introduced a serious structural multicollinearity as indicated by the correlation matrix of the estimated coefficients in the logistic regression equation.

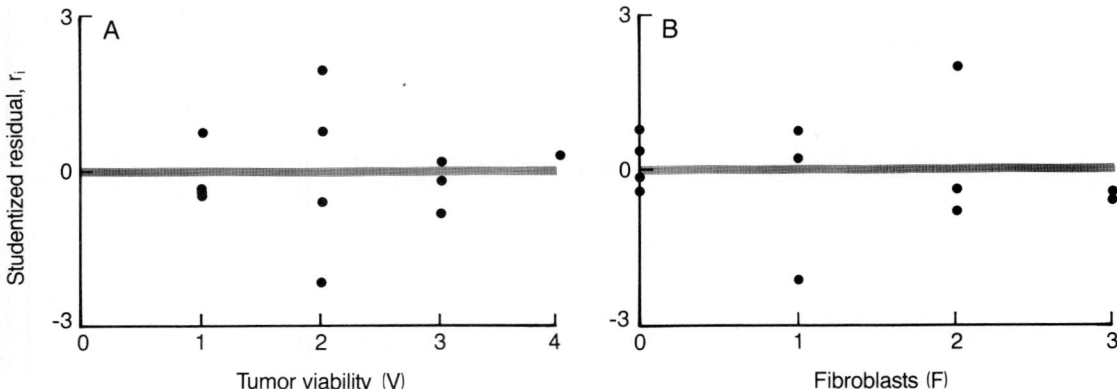

FIGURE 11-11 Plots of the residuals corresponding to the analysis in Fig. 11-10 reveal that the nonlinear patterns have been eliminated.

However, the high correlation between the estimate of b_V and b_{V^2} of $-.974$ (Fig. 11-10) is cause for concern because it may indicate a problem with multicollinearity. There is also a moderately high correlation, $-.82$, between the b_F and b_{F^2} coefficients and the b_F and b_{VF} coefficients, al-

FIGURE 11-12 The plot of the observed proportion of successful outcomes in the bone cancer treatment data vs. the predicted probability of success, corresponding to the analysis in Fig. 11-10, has been greatly improved by including the quadratic terms (compare with Fig. 11-8). Both this result and the larger P value associated with the Hosmer-Lemeshow goodness-of-fit χ^2 indicate that the logistic regression equation now reasonably describes the data. The remaining problem is the multicollinearity we have introduced by including the quadratic terms.

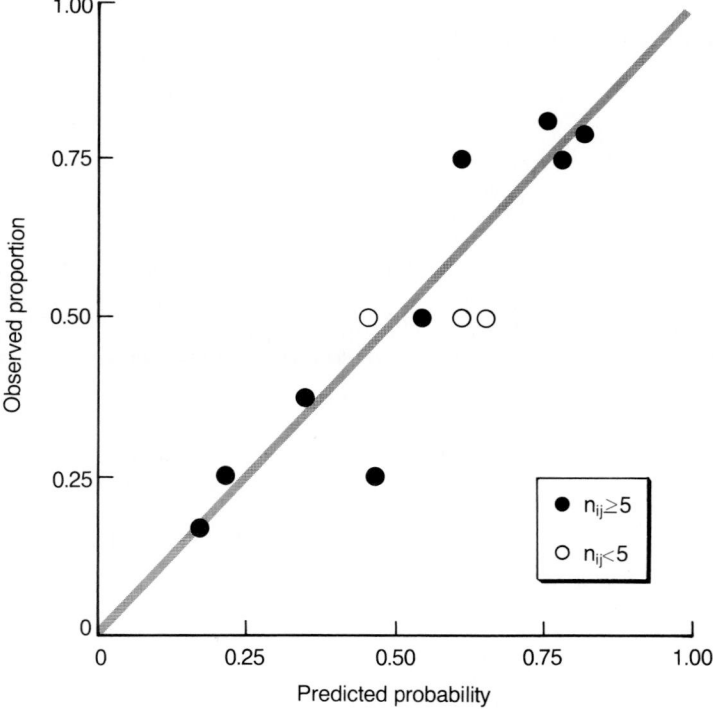

though this value alone would probably not be a cause for concern. These correlations are not surprising, because we have introduced a structural multicollinearity in the regression model by introducing the V^2, F^2, and VF terms. As we discussed in Chap. 5, the simplest way to resolve such structural multicollinearities is to center the native independent variables before computing the powers and interaction terms in the regression equation. To be sure that multicollinearity is not leading us astray in our analysis, let us repeat the analysis after centering the independent variables.

The mean value of viability observed in the data is 1.93 (some viable tumor tissue) and the mean value of fibroblast is 1.03 (little normal tissue). We use this information to center the native independent variables to obtain

$$V_c = V - \overline{V} = V - 1.93$$

and

$$F_c = F - \overline{F} = F - 1.03$$

The logistic regression equation, written in terms of the centered variables, becomes

$$\text{logit } \hat{P}_C = c_0 + c_V V_c + c_{V^2} V_c^2 + c_F F_c + c_{F^2} F_c^2 + c_{VF} V_c F_c \qquad (11.13)$$

Figure 11-13 shows the results of fitting this equation to these data. The correlation matrix of the regression coefficients shows that we have resolved any multicollinearity problems; all the correlations are well below the threshold for concern of about .9.

Because the mathematical function represented by Eq. (11.13) is the same as the mathematical function represented by Eq. (11.12), the goodness-of-fit statistics are the same as before, as are the residual patterns (not shown). As before, the viability and square of viability contribute significantly to producing the outcome (whether or not the patient received chemotherapy), indicating that there is a nonlinear effect of viability on predicting the outcome. However, one important thing is different now: the fibroblast score also significantly contributes to predicting the outcome. The higher the fibroblast score (i.e., the more remodeling to normal tissue), the more likely the patient was to have been receiving chemotherapy. The structural multicollinearity in the original model of Eq. (11.12) masked this effect.

Now that we have a reasonably good fit and have reduced the multicollinearity to an acceptable level, we can turn our attention to interpreting the results of the analysis. Retaining only the terms that are significantly different from zero, the relationship between the probability of receiving chemotherapy and the viability of the tumor and number of fibroblasts is

```
         TOTAL NUMBER OF RESPONSES USED IN THE ANALYSIS     176.
                                   SUCCESS  . . . . .        88.
                                   FAILURE  . . . . . .      88.

         NUMBER OF DISTINCT COVARIATE PATTERNS . . . . .     12


    STEP NUMBER    0
    ---------------

                           LOG LIKELIHOOD =  -100.543
    GOODNESS OF FIT CHI-SQ   (2*O*LN(O/E))  =     6.670  D.F.=  6  P-VALUE= 0.352
    GOODNESS OF FIT CHI-SQ (HOSMER-LEMESHOW)=     4.513  D.F.=  8  P-VALUE= 0.808
    GOODNESS OF FIT CHI-SQ  ( C.C.BROWN )   =     0.881  D.F.=  2  P-VALUE= 0.644

                                         STANDARD
         TERM              COEFFICIENT      ERROR    COEFF/S.E.   EXP(COEFFICIENT)

    Vc                    -0.71151        0.2564      -2.775        0.4909
    Fc                     0.81178        0.2071       3.919        2.252
    VcFc                   0.21926        0.3161       0.6936       1.245
    Vc2                    0.59981        0.2556       2.347        1.822
    Fc2                   -0.28759        0.2185      -1.316        0.7501
    CONSTANT              -0.33651E-01    0.2960      -0.1137       0.9669


    CORRELATION MATRIX OF COEFFICIENTS
    ----------------------------------


                 Vc        Fc       VcFc      Vc2       Fc2      CONSTANT

    Vc         1.000
    Fc         0.332     1.000
    VcFc       0.112     0.123     1.000
    Vc2       -0.282     0.160     0.597     1.000
    Fc2        0.037    -0.329     0.399     0.120     1.000
    CONSTANT   0.156     0.166    -0.289    -0.461    -0.669     1.000
```

FIGURE 11-13 Results of computing a logistic regression for the bone cancer chemotherapy data after centering the independent variables before computing the quadratic terms. Because centering does not affect the actual mathematical function given by the regression equation, the P value associated with the Hosmer-Lemeshow goodness-of-fit χ^2 remains unchanged at .808. The centering, as expected, has eliminated the structural multicollinearity problem; i.e., all of the values in the correlation matrix of the regression coefficients are now acceptably small. The pattern of residuals and observed vs. predicted proportions will be the same as when using the uncentered data (see Figs. 11-11 and 11-12).

$$\text{logit } \hat{P}_C = -.034 - .712V_c + .600V_c^2 + .812F_c$$

or, in terms of the odds of receiving chemotherapy,

$$\hat{\Omega}_C = e^{-.034} \, e^{-712V_c + .600V_c^2} \, e^{.812F_c}$$

We have combined the two terms of the quadratic function involving V_c to simplify the interpretation.

We begin with the constant term, $e^{b_0} = e^{-.034} = .967 \approx 1$. This term gives the odds that a patient will receive chemotherapy if his or her viability score and fibroblast score are average (i.e., if $V = \overline{V}$ so $V_c = 0$ and $F = \overline{F}$ so $F_c = 0$). Hence an "average" patient is equally likely to have received chemotherapy or not. This result is not surprising given the study design; it is comforting, however, that the model estimated this to be true.

Next, let us consider the effect of changes in the fibroblast score. $e^{b_F} = e^{.812} = 2.3$, which indicates that, at a given level of viability, each increase in fibroblast score of 1 unit increases the odds that a patient would have received chemotherapy by a factor of 2.3. Hence, a patient with moderate fibroblasts (a score of 2) has 2.3 times the odds of having received chemotherapy than an average patient (with a fibroblast score of $1.03 \approx 1$), and a patient with extensive fibroblasts (a score of 3) has about $2.3^2 = 5.3$ times the odds of having received chemotherapy than the average patient. Likewise, a patient with no fibroblasts (a score of 0) has about $2.3^{-1} = .4$ times the odds of an average patient of having received chemotherapy. Because there is no interaction between fibroblast score and tumor viability, these changes in the odds are independent of the level of tumor viability.

Finally, we consider the effect of the level of tumor viability. Table 11-2 shows the effects of different levels of tumor viability on the odds of receiving chemotherapy. Because there is a nonlinear effect of viability on the odds, the interpretation of this effect is more complex than the effects of fibroblasts. Table 11-2 shows that having a tumor viability score of 1 (little viable tissue) increases the chances of having received chemotherapy by a factor of 3.9, compared with the average patient (who

TABLE 11-2 Effect of Tumor Viability on Odds That a Bone Cancer Patient Received Chemotherapy

V	V_c	$-.712V_c + .812V_c^2$	$e^{-.712V_c + .812V_c^2}$
1	$-.93$	1.364	3.91
2	.07	$-.046$	.96
3	1.07	.168	1.18
4	2.07	2.005	7.40

has a tumor viability score of $1.93 \approx 2$). There is little change in the odds for patients who score 2 or 3, with the factor around 1. The odds of having received chemotherapy increase again, by a factor of 7.4, for patients with a viability score of 4 (indicating a large amount of viable tissue).

Do these results make sense? The analysis shows that when more fibroblasts forming new tissue are present, the patients were more likely to have received preoperative chemotherapy. This result makes sense because we would expect successful chemotherapy to kill the tumor, and that dead tumor tissue would be replaced with new tissue. Because of the nonlinearity in V, patients with both the largest and smallest amounts of viable tissue in the tumor are more likely to have received chemotherapy than patients with moderate amounts of viable tissue in the tumor (a score of 2 or 3). We expect successful chemotherapy to kill the tumor, so would expect that low viability would be associated with chemotherapy. Indeed, this is true. However, high viability is also associated with chemotherapy; perhaps because there is a group of patients that are not responding to chemotherapy, or perhaps there is good response to chemotherapy and the new tissue being formed as the tumor dies is counted as viable tissue.

STEPWISE LOGISTIC REGRESSION

One is often faced with the problem of screening a large number of potential independent variables to identify important determinants of a dichotomous dependent variable. We will now explore variable selection procedures that can be applied to the problem of logistic regression that are similar to the methods we developed in Chap. 6 for variable selection in multiple linear regression in which the dependent variable is continuous. These methods have the same benefits and problems as the analogous methods for multiple linear regression. In addition, given the requirement for large sample sizes, variable selection techniques in multiple logistic regression should be approached cautiously and only when there is an adequately large data set.

The test for whether a *single* independent variable improves the prediction forms the basis for a sequential variable selection technique for logistic regression. The principle is the same as that used in stepwise linear regression. Just as we used the incremental change in the F test statistic as a criterion for adding or deleting variables in a stepwise multiple linear regression procedure, we can use the improvement chi-square value as a guide for entering or deleting variables in a stepwise multiple logistic regression. The improvement chi square tests the hypothesis that the term entered (or removed) at that step significantly improves (or degrades) the predictive ability of the regression model in terms of maximizing the likelihood function. A large value of the improvement chi square (or small P value) indicates a significant change in the likelihood function associated with the variable in question.

Forward stepwise logistic regression begins with no variables in the equation and adds independent variables to the regression model in decreasing order of the associated improvement chi square until no further improvements are possible. *Backward stepwise logistic regression* begins with all variables in the equation and removes them in the order of their smallest improvement chi square until removing any of the variables in the equation would significantly degrade the predictive ability. As with sequential variable selection techniques for linear regression, forward stepwise logistic regression will tend to include fewer variables in the final equation than backward stepwise logistic regression will.

The results depend on the criteria selected to enter and remove variables from the equation—typically the P value associated with entering or removing the next variable—which you specify. As with stepwise linear regression, there are a variety of strategies one can follow in selecting these threshold values, P_{in} and P_{out}. P_{in} is the minimum P value associated with the improvement chi square, P_{enter}, for which a variable will be entered into the regression equation and P_{out} is the maximum P value associated with the reduction in chi square, P_{remove}, which will be consistent with deleting a variable from the regression equation. (These parameters play roles analogous to F_{in} and F_{out} in stepwise multiple linear regression.) Some investigators set $P_{in} = P_{out} = .05$ in an effort to develop an equation, all of whose independent variables meet the traditional $P < .05$ criterion for statistical significance. Other investigators set these threshold values higher, say around .10, to avoid missing potentially important variables. Sometimes people set P_{out} to be a larger value than P_{in}, so that once a variable enters the equation, it tends to stay, even if subsequently entered variables contain some of the same information in the variables already in the equation. The algorithm continues adding variables until there are no variables remaining out of the equation with $P_{enter} < P_{in}$ and no variables remaining in the equation with $P_{remove} > P_{out}$.

We developed several tools in Chap. 6 to help us decide what a good model was in the case of ordinary linear regression, such as the statistics $s_{y|x}$, R^2, and C_p. There is no analog to all possible subsets regression for logistic regression,[*] and we have only limited information about the quality of fit; we do not have an R^2 or C_p to help us with our search and the goodness-of-fit tests for logistic regression are not very powerful. Thus, we have only a partial tool kit with which to approach a model identification problem with logistic regression. This limitation makes

---

[*]Of course, you could manually repeat the logistic regression for all possible subsets, but this approach rapidly becomes impractical as the number of variables increases.

using available regression diagnostics, such as residual plots, leverage, Cook's distance, and indications of multicollinearity, particularly important when considering the results of a stepwise logistic regression.

The following example illustrates a strategy for finding a reasonable model. We will combine forward and backward stepwise logistic regression using the basic variables of the problem to isolate a set of candidate variables for the final model, then investigate possible interactions between these variables. When considering interactions we will first specify only those we think might be of interest a priori and then repeat the stepwise regression going both forward and backward. Then, based on those results, we will narrow down what we think the main effects are, force them into the model, and use stepwise procedures to help us select from all possible simple interactions.*

This strategy does not allow us to examine nearly as many possibilities as we could if we had something like all possible subsets regression. In addition, the result of following this strategy is not necessarily the best model—indeed, there is no generally accepted criterion for what the "best" logistic regression model is. However, we hope we will identify a good model that is interpretable. Moreover, this model will often be better than one obtained by simply dumping a set of observations into a stepwise logistic regression program, which is a recipe for obtaining questionable results.

Nuking the Heart

People develop chest pain (angina) and ultimately have heart attacks when the buildup of atherosclerotic plaque or clots blocks a coronary artery and restricts or stops the flow of blood to the heart muscle. Cardiologists have developed many diagnostic tests to measure the level of blockage by stenoses (narrowings) in the coronary arteries. The most definitive—and also the most invasive—of these tests is a cardiac catheterization, in which a tube is placed in the coronary artery in question and a radiopaque substance injected so that the artery (and any narrowings) can be seen with x-rays. Because cardiac catheterization is expensive and not without risks, various methods of noninvasively assessing cardiac function have been developed.

One of these noninvasive methods involves having a patient exercise while the blood flow in the heart muscle is imaged using the radioisotope thallium. Having the patient exercise dilates the coronary arteries to the

---

*For a detailed development of strategies for model building using stepwise logistic regression, with particular emphasis on how to treat interactions, see D. G. Kleinbaum, L. L. Kupper, and H. Morgenstern, *Epidemiologic Research: Principles and Quantitative Methods*, New York, Van Nostrand Reinhold, 1982, Chaps. 21–23.

maximal extent possible and the radioactive thallium is taken up into the heart muscle in proportion to the blood flow in different parts of the heart muscle. Thus, if there is a region that gets relatively too little blood under circumstances in which blood flow should be maximal in all regions (because of a narrowed or blocked coronary artery), there will be less thallium taken up and this lack of thallium can be detected in an image formed from its radioactive emissions. Exercise thallium testing is reasonably reliable when there is serious blockage of the coronary arteries, but there is some question as to what determines a positive thallium test in modest blockages, say below about 50 percent of the original artery diameter.

To determine what factors might be associated with a positive exercise thallium test in such patients, Brown and his coworkers[*] examined 100 people (55 men and 45 women) who had both exercise thallium tests and cardiac catheterizations that could be used to definitively describe the level of blockage in their coronary arteries. They sought to determine which variables were good predictors of the outcome of a positive thallium exercise test, with the ultimate goal of identifying the characteristics associated with positive tests in this group of patients with modest narrowing of their coronary arteries.

The dependent variable is

$$T = \begin{cases} 1 \text{ if a positive thallium scan} \\ 0 \text{ if a negative thallium scan} \end{cases}$$

They considered the following potential independent variables in their effort to predict who would have a positive thallium test:

S sex (0 = male, 1 = female)

D degree of stenosis (narrowing) of the coronary artery, as a percentage of original artery diameter, with 0 percent indicating no blockage and 100 percent indicating complete blockage

C chest pain during exercise test (0 = no, 1 = moderate, 2 = severe)

E ECG changes indicative of ischemia during the test (0 = no, 1 = yes)

R maximum heart rate during exercise (0 = heart rate did not rise to at least 85 percent of maximum predicted heart rate, 1 = heart rate exceeded 85 percent of predicted response.)

P use of the drug propranolol prior to the test (0 = no, 1 = yes)

[*]K. A. Brown, M. Osbakken, C. A. Boucher, H. W. Strauss, G. M. Pohost, and R. D. Okada, "Positive Exercise Thallium-201 Test Responses in Patients with Less than 50% Maximal Coronary Stenosis: Angiographic and Clinical Predictors," *Am. J. Cardiol.* 55:54–57, 1985.

```
        TOTAL NUMBER OF RESPONSES USED IN THE ANALYSIS      100.
                                SUCCESS  . . . . .            34.
                                FAILURE  . . . . . .          66.

        NUMBER OF DISTINCT COVARIATE PATTERNS . . . . .       45

  STEP NUMBER    0
  ---------------

                        LOG LIKELIHOOD =    -64.104
  GOODNESS OF FIT CHI-SQ   (2*O*LN(O/E))  =    79.718   D.F.=  44  P-VALUE= 0.001
  GOODNESS OF FIT CHI-SQ  ( C.C.BROWN )  =     0.000   D.F.=   0  P-VALUE= 1.000

                                           STANDARD
        TERM            COEFFICIENT         ERROR      COEFF/S.E.   EXP(COEFFICIENT)

  CONSTANT             -0.66329            0.2111       -3.142          0.5152


  STATISTICS TO ENTER OR REMOVE TERMS
  -----------------------------------
                     APPROX.              APPROX.
        TERM        CHI-SQ. D.F.         CHI-SQ. D.F.
                      ENTER               REMOVE        P-VALUE      LIKELIHOOD
                                                                       LOG

  D                   5.62    1                          0.0177      -61.2929
  P                   2.46    1                          0.1166      -62.8724
  R                   2.00    1                          0.1577      -63.1053
  E                   0.39    1                          0.5329      -63.9091
  S                   0.52    1                          0.4711      -63.8439
  C                   3.57    1                          0.0588      -62.3176
  CONSTANT                              10.42    1       0.0012      -69.3147
  CONSTANT                              IS IN          MAY NOT BE REMOVED.

  STEP NUMBER    1            D                        IS ENTERED
  ---------------

                        LOG LIKELIHOOD =    -61.293
  IMPROVEMENT CHI-SQUARE   ( 2*(LN(MLR) ) =     5.621   D.F.=   1  P-VALUE= 0.018
  GOODNESS OF FIT CHI-SQ   (2*O*LN(O/E))  =    74.096   D.F.=  43  P-VALUE= 0.002
  GOODNESS OF FIT CHI-SQ (HOSMER-LEMESHOW)=     5.677   D.F.=   4  P-VALUE= 0.225
  GOODNESS OF FIT CHI-SQ  ( C.C.BROWN )  =     5.428   D.F.=   2  P-VALUE= 0.066

                                           STANDARD
        TERM            COEFFICIENT         ERROR      COEFF/S.E.   EXP(COEFFICIENT)

  D                    0.32720E-01       0.1394E-01     2.348          1.033
  CONSTANT            -0.96882            0.2570        -3.770          0.3795


  CORRELATION MATRIX OF COEFFICIENTS
  ----------------------------------


                    D         CONSTANT

  D               1.000
  CONSTANT       -0.532       1.000


  STATISTICS TO ENTER OR REMOVE TERMS
  -----------------------------------
                     APPROX.              APPROX.
        TERM        CHI-SQ. D.F.         CHI-SQ. D.F.
                      ENTER               REMOVE        P-VALUE        LOG
                                                                    LIKELIHOOD

  D                                       5.62    1     0.0177      -64.1035
  P                   1.94    1                          0.1631      -60.3205
  R                   1.26    1                          0.2619      -60.6635
  E                   0.51    1                          0.4769      -61.0400
  S                   0.92    1                          0.3369      -60.8318
  C                   2.92    1                          0.0875      -59.8333
  CONSTANT                              15.88    1       0.0001      -69.2315
  CONSTANT                              IS IN          MAY NOT BE REMOVED.
```

A

```
STEP NUMBER   2              C                    IS ENTERED
---------------

                        LOG LIKELIHOOD =   -59.833
IMPROVEMENT CHI-SQUARE   ( 2*(LN(MLR) ) =     2.919  D.F.=  1  P-VALUE= 0.088
GOODNESS OF FIT CHI-SQ   (2*O*LN(O/E)) =    71.177  D.F.= 42  P-VALUE= 0.003
GOODNESS OF FIT CHI-SQ (HOSMER-LEMESHOW)=  15.199  D.F.=  8  P-VALUE= 0.055
GOODNESS OF FIT CHI-SQ   ( C.C.BROWN ) =     6.695  D.F.=  2  P-VALUE= 0.035

                                      STANDARD
          TERM          COEFFICIENT    ERROR    COEFF/S.E.  EXP(COEFFICIENT)


D                       0.31447E-01  0.1421E-01   2.213        1.032
C                       0.64486      0.3833       1.683        1.906
CONSTANT               -1.1158       0.2774      -4.022        0.3277


CORRELATION MATRIX OF COEFFICIENTS
----------------------------------


              D        C        CONSTANT

D           1.000
C          -0.001    1.000
CONSTANT   -0.497   -0.344     1.000


STATISTICS TO ENTER OR REMOVE TERMS
-----------------------------------
                  APPROX.          APPROX.
          TERM    CHI-SQ. D.F.     CHI-SQ. D.F.              LOG
                  ENTER            REMOVE        P-VALUE   LIKELIHOOD

D                                  4.97    1     0.0258    -62.3176
P                 1.94    1                      0.1632    -58.8613
R                 1.77    1                      0.1830    -58.9467
E                 0.48    1                      0.4891    -59.5941
S                 0.74    1                      0.3887    -59.4618
C                                  2.92    1     0.0875    -61.2929
CONSTANT                          18.64    1     0.0000    -69.1516
CONSTANT                          IS IN         MAY NOT BE REMOVED.


NO TERM PASSES THE REMOVE AND ENTER LIMITS (  0.1500  0.1000 ) .


SUMMARY OF STEPWISE RESULTS

STEP      TERM               LOG      IMPROVEMENT    GOODNESS OF FIT
NO.   ENTERED REMOVED   DF LIKELIHOOD CHI-SQUARE P-VAL CHI-SQUARE P-VAL
---   --------------- --- ---------- ---------- ----- ---------- ----
  0                            -64.104                    79.718 0.001
  1 D                  1       -61.293   5.621 0.018      74.096 0.002
  2 C                  1       -59.833   2.919 0.088      71.177 0.003
```

B

FIGURE 11-14 *A* and *B*. Results of a forward stepwise logistic regression of the degree of stenosis of the coronary artery *D*, which include whether or not a patient is taking the drug propranolol prior to the thallium exercise test *P*; the maximum heart rate χ^2 during exercise *R*; whether or not there were ECG changes indicating ischemia during the test *E*; the sex of the patient *S*; and whether or not the patient experienced chest pain during exercise *C*. The dependent variable is whether or not the results of the thallium scan were positive or negative, with a "successful" outcome established by a positive thallium test.

These variables were included as candidates because (1) they might indicate abnormally low blood flow, even though the stenoses were subcritical by angiographic criteria (C, R, and E); (2) there is the possibility that even subcritical stenoses are important limiters of flow (D); or (3) they might affect the heart's demands for oxygen (P). The full logistic regression model we will begin with is

$$\text{logit } \hat{P}_T = b_0 + b_S S + b_D D + b_C C + b_E E + b_R R + b_P P$$

The results of a forward stepwise regression to fit the data in Table C-24, Appendix C, to the logistic model are in Fig. 11-14. We will use reasonably liberal P_{in} and P_{out} values of .10 and .15, respectively, so as to not miss potentially important predictors in this first pass. The program took two steps, adding first the degree of stenosis D and then chest pain C, giving us the regression model

$$\text{logit } \hat{P}_t = b_0 + b_D D + b_C C \tag{11.14}$$

The algorithm stopped stepping because after adding these two variables, the next smallest P_{enter} was .16. If we wanted to be even more liberal, we could change our entry criteria and consider the effect of drug P ($P_{\text{enter}} = .16$) or even maximum heart rate response R ($P_{\text{enter}} = .18$ on the last step). For now, let us stow away this information about these additional variables and consider only the model with D and C included.

Because of the limited number of cases and the fact that some of the independent variables are measured on continuous scales, we will use the Hosmer-Lemeshow goodness-of-fit test to assess the overall model. After all, there are 45 distinct patterns in the independent variables. Taking 45 times the 2 possible outcomes, there will be 90 cells in the contingency table used to compute a classic goodness-of-fit test. With only 100 patients, one can hardly rely on the assumption that most of the cells will be predicted to have at least five patients in them! The Hosmer-Lemeshow statistic has a χ^2 of 15.199 with 8 degrees of freedom, leading to $P = .055$, which indicates a poor fit to the logistic model. This poor fit is obvious in the plot of the observed vs. predicted probabilities for each combination of these independent variables (Fig. 11-15).[*]

The plot of studentized residuals (Fig. 11-16) suggests that there is an interaction between the degree of stenosis and another variable (or variables), because the residuals are large for $D = 0$ (when interaction effects would be zero), then drops to a low value and increases as the level of stenosis increases. The other regression diagnostics are unremarkable (not shown), as is the correlation matrix of the parameter estimates.

[*]The usefulness of this plot also depends on sample size in much the same way that the goodness-of-fit χ^2 statistics do. If there are only 1 or 2 observations in each cell, we will not get anything even approaching a straight line in such a plot.

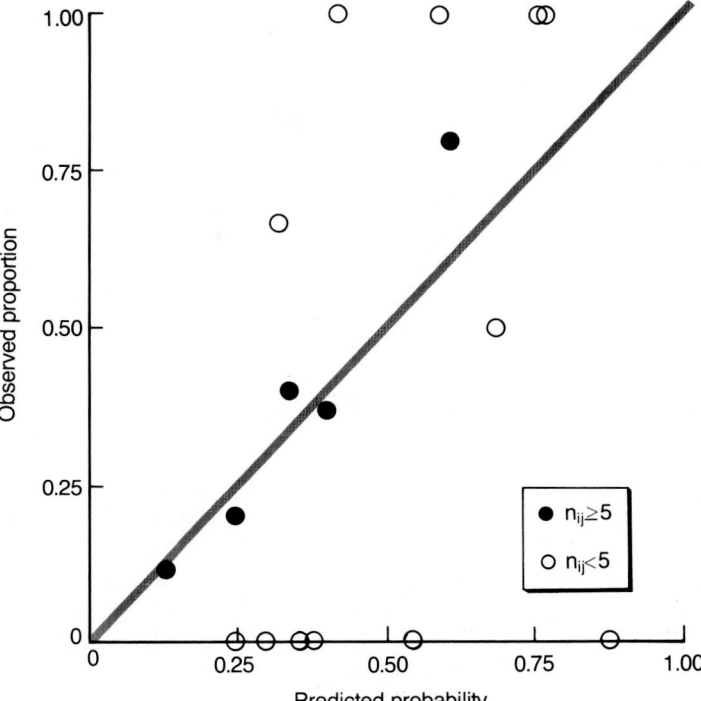

FIGURE 11-15 The plot of the observed proportions of patients with positive thallium scans vs. the predicted probability of a positive thallium scan for patients exhibiting different patterns of independent variables falls nowhere near a straight line, confirming the poor goodness of fit of the model selected in the analysis of Fig. 11-14 as indicated by the small P value of .055 associated with the Hosmer-Lemeshow goodness-of-fit χ^2.

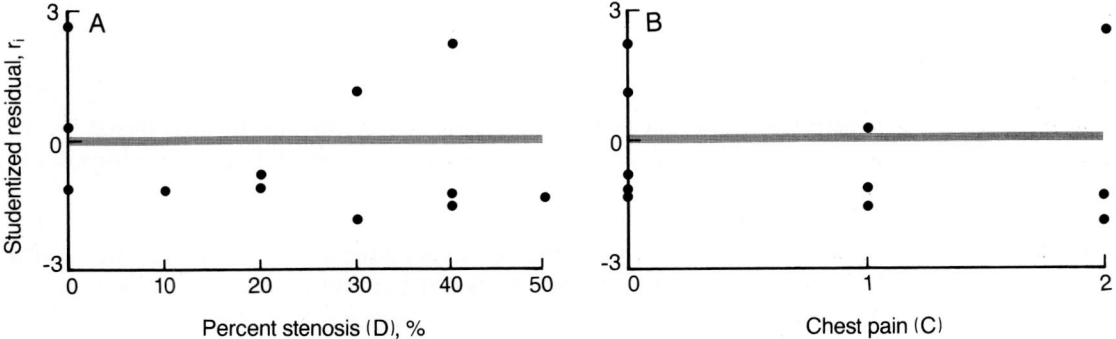

FIGURE 11-16 The patterns of residuals, particularly for the independent variable D representing the degree of stenosis, suggests that there may be an interaction effect between D and other independent variables which have not been considered in the model. We draw this conclusion because of the megaphone patterns in the residuals except at a value of 0 where they are broadly distributed.

The next step is to consider interactions. Brown and his coworkers decided to limit the number of interactions considered based on their knowledge of the clinical problem being investigated: Of the 15 possible first-order (two-way) interactions among these 6 variables, they only consider 4, *DP*, *DR*, *PR*, and *RE*.

The computer output from the forward stepwise logistic regression, now including these four interactions in the list of candidate independent variables, is in Fig. 11-17. Not surprisingly, this began much the same as before, with both D and C entering on the first two steps. Now, however, the algorithm continued stepping and added the interaction of degree of stenosis with propranolol (*DP*) ($P_{enter} = .09$), and then propranolol itself ($P_{enter} = .006$), resulting in the regression model

$$\text{logit } \hat{P}_T = c_0 + c_D D + c_C C + c_P P + c_{DP} DP \tag{11.15}$$

On step 2, just before the *DP* interaction was added in step 3, the P_{enter} for propranolol was only .16 (as we saw above). However, the addition of the interaction term caused the main effect of propranolol to enter on a subsequent step. Remember that the logistic regression stepwise algorithm adds the variable that produces the largest significant increase in the likelihood function *on that step*. Thus, the fact that the interaction term entered first is not disturbing, but it is comforting that both main effects associated with the interaction are in the model. It makes very little sense, under most circumstances, to consider interaction effects in the absence of the corresponding main effects.

By considering interactions, we have added more terms to the model, which means we have significantly increased the likelihood function. Thus, in terms of increased explanatory power, we have a better model. However, is it a good model? The answer is: Maybe. The Hosmer-Lemeshow goodness-of-fit statistic has a P value of .446, indicating a better fit than the original model. However, the plot of observed proportions of patients with each combination of independent variables vs. predicted probabilities in Fig. 11-18 shows that the fit is not good: There is still significant scatter in this plot. Thus, interpretations based on this model should be made with caution and we should continue looking for a better model.

Because forward and backward stepping can lead to different models, it is often useful to perform both procedures and compare the results. The computer output from a backward stepping run using the same list

FIGURE 11-17 Results of a forward stepwise logistic regression to predict the out- ▶ come of an exercise thallium study on the heart, this time including several interaction terms suggested by the investigators' knowledge of the clinical problem. Including the interaction terms has improved the fit. The Hosmer-Lemeshow goodness-of-fit now has an associated P value of .446. Some intermediate steps were omitted from this figure.

```
    TOTAL NUMBER OF RESPONSES USED IN THE ANALYSIS    100.
                            SUCCESS  . . . . .     34.
                            FAILURE  . . . . . .   66.

    NUMBER OF DISTINCT COVARIATE PATTERNS . . . . .    45

   STEP NUMBER   0
   ---------------
                        LOG LIKELIHOOD =    -64.104
   GOODNESS OF FIT CHI-SQ   (2*O*LN(O/E)) =     79.718  D.F.= 44  P-VALUE= 0.001
   GOODNESS OF FIT CHI-SQ   ( C.C.BROWN ) =      0.000  D.F.=  0  P-VALUE= 1.000

                                        STANDARD
          TERM           COEFFICIENT     ERROR    COEFF/S.E.  EXP(COEFFICIENT)

   CONSTANT               -0.66329      0.2111     -3.142       0.5152

   STATISTICS TO ENTER OR REMOVE TERMS
   -----------------------------------
                     APPROX.          APPROX.
          TERM       CHI-SQ. D.F.     CHI-SQ. D.F.                   LOG
                     ENTER            REMOVE          P-VALUE     LIKELIHOOD

   D                    5.62   1                      0.0177      -61.2929
   P                    2.46   1                      0.1166      -62.8724
   D*P                  0.23   1                      0.6325      -63.9892
   R                    2.00   1                      0.1577      -63.1053
   D*R                  0.31   1                      0.5786      -63.9493
   P*R                  1.03   1                      0.3090      -63.5861
   E                    0.39   1                      0.5329      -63.9091
   R*E                  1.51   1                      0.2186      -63.3467
   S                    0.52   1                      0.4711      -63.8439
   C                    3.57   1                      0.0588      -62.3176
   CONSTANT                           10.42   1       0.0012      -69.3147
   CONSTANT                           IS IN         MAY NOT BE REMOVED.
   STEP NUMBER   1            D                    IS ENTERED
   ---------------
                        LOG LIKELIHOOD =    -61.293
   IMPROVEMENT CHI-SQUARE   ( 2*(LN(MLR) ) =      5.621  D.F.=  1  P-VALUE= 0.018
   GOODNESS OF FIT CHI-SQ   (2*O*LN(O/E)) =     74.096  D.F.= 43  P-VALUE= 0.002
   GOODNESS OF FIT CHI-SQ (HOSMER-LEMESHOW)=      5.677  D.F.=  4  P-VALUE= 0.225
   GOODNESS OF FIT CHI-SQ   ( C.C.BROWN ) =      5.428  D.F.=  2  P-VALUE= 0.066

                                        STANDARD
          TERM           COEFFICIENT     ERROR    COEFF/S.E.  EXP(COEFFICIENT)

   D                    0.32720E-01  0.1394E-01    2.348        1.033
   CONSTANT            -0.96882      0.2570       -3.770        0.3795

   CORRELATION MATRIX OF COEFFICIENTS
   ----------------------------------
              D        CONSTANT

   D         1.000
   CONSTANT -0.532      1.000

   STATISTICS TO ENTER OR REMOVE TERMS
   -----------------------------------
                     APPROX.          APPROX.
          TERM       CHI-SQ. D.F.     CHI-SQ. D.F.                   LOG
                     ENTER            REMOVE          P-VALUE     LIKELIHOOD

   D                                  5.62   1       0.0177      -64.1035
   P                    1.94   1                      0.1631      -60.3205
   D*P                  2.46   1                      0.1166      -60.0615
   R                    1.26   1                      0.2619      -60.6635
   D*R                  1.02   1                      0.3114      -60.7805
   P*R                  0.87   1                      0.3510      -60.8580
   E                    0.51   1                      0.4769      -61.0400
   R*E                  1.82   1                      0.1777      -60.3844
   S                    0.92   1                      0.3369      -60.8318
   C                    2.92   1                      0.0875      -59.8333
   CONSTANT                           15.88   1       0.0001      -69.2315
   CONSTANT                           IS IN         MAY NOT BE REMOVED.
```

(continued on the next page)

```
STEP NUMBER   2              C                    IS ENTERED
----------------

                        LOG LIKELIHOOD =    -59.833
IMPROVEMENT CHI-SQUARE   ( 2*(LN(MLR) ) =     2.919  D.F.=   1  P-VALUE= 0.088
GOODNESS OF FIT CHI-SQ   (2*O*LN(O/E)) =    71.177  D.F.=  42  P-VALUE= 0.003
GOODNESS OF FIT CHI-SQ (HOSMER-LEMESHOW)=   15.199  D.F.=   8  P-VALUE= 0.055
GOODNESS OF FIT CHI-SQ   ( C.C.BROWN ) =     6.695  D.F.=   2  P-VALUE= 0.035

                                        STANDARD
      TERM              COEFFICIENT       ERROR     COEFF/S.E.  EXP(COEFFICIENT)


D                      0.31447E-01   0.1421E-01     2.213          1.032
C                      0.64486       0.3833         1.683          1.906
CONSTANT              -1.1158        0.2774        -4.022          0.3277


CORRELATION MATRIX OF COEFFICIENTS
----------------------------------

              D          C        CONSTANT


D            1.000
C           -0.001      1.000
CONSTANT    -0.497     -0.344      1.000


STATISTICS TO ENTER OR REMOVE TERMS
-----------------------------------
                 APPROX.        APPROX.
      TERM       CHI-SQ. D.F.   CHI-SQ. D.F.                 LOG
                 ENTER          REMOVE         P-VALUE    LIKELIHOOD


D                              4.97     1     0.0258      -62.3176
P                1.94    1                    0.1632      -58.8613
D*P              2.88    1                    0.0897      -58.3937
R                1.77    1                    0.1830      -58.9467
D*R              1.04    1                    0.3090      -59.3157
P*R              1.08    1                    0.2996      -59.2954
E                0.48    1                    0.4891      -59.5941
R*E              1.42    1                    0.2329      -59.1217
S                0.74    1                    0.3887      -59.4618
C                              2.92     1     0.0875      -61.2929
CONSTANT                      18.64     1     0.0000      -69.1516
CONSTANT                        IS IN          MAY NOT BE REMOVED.
STEP NUMBER   3              D*P                  IS ENTERED
----------------

                        LOG LIKELIHOOD =    -58.394
IMPROVEMENT CHI-SQUARE   ( 2*(LN(MLR) ) =     2.879  D.F.=   1  P-VALUE= 0.090
GOODNESS OF FIT CHI-SQ   (2*O*LN(O/E)) =    68.298  D.F.=  41  P-VALUE= 0.005
GOODNESS OF FIT CHI-SQ (HOSMER-LEMESHOW)=    6.839  D.F.=   8  P-VALUE= 0.554
GOODNESS OF FIT CHI-SQ   ( C.C.BROWN ) =    13.763  D.F.=   2  P-VALUE= 0.001

                                        STANDARD
      TERM              COEFFICIENT       ERROR     COEFF/S.E.  EXP(COEFFICIENT)

D                      0.54864E-01   0.2106E-01     2.605          1.056
D*P                   -0.42405E-01   0.2589E-01    -1.638          0.9585
C                      0.69240       0.3842         1.802          1.999
CONSTANT              -1.1336        0.2784        -4.071          0.3219


CORRELATION MATRIX OF COEFFICIENTS
----------------------------------


              D        D*P        C        CONSTANT

D            1.000
D*P         -0.711      1.000
C            0.054     -0.106     1.000
CONSTANT    -0.368      0.054    -0.343      1.000
```

FIGURE 11-17 (continued)

```
STATISTICS TO ENTER OR REMOVE TERMS
------------------------------------
                   APPROX.          APPROX.
        TERM       CHI-SQ. D.F.     CHI-SQ. D.F.                    LOG
                   ENTER            REMOVE         P-VALUE       LIKELIHOOD

D                                     7.78   1    0.0053         -62.2832
P                    7.43   1                      0.0064         -54.6775
D*P                                   2.88   1    0.0897         -59.8333
R                    1.65   1                      0.1991         -57.5691
D*R                  0.78   1                      0.3759         -58.0017
P*R                  2.52   1                      0.1122         -57.1321
E                    0.10   1                      0.7505         -58.3432
R*E                  0.81   1                      0.3688         -57.9899
S                    0.97   1                      0.3238         -57.9071
C                                     3.34   1    0.0678         -60.0615
CONSTANT                             19.14   1    0.0000         -67.9652
CONSTANT                                    IS IN       MAY NOT BE REMOVED.
STEP NUMBER   4            P                        IS ENTERED
----------------
                  LOG LIKELIHOOD =    -54.677
IMPROVEMENT CHI-SQUARE   ( 2*(LN(MLR) ) =     7.433  D.F.=   1  P-VALUE= 0.006
GOODNESS OF FIT CHI-SQ   (2*O*LN(O/E)) =     60.866  D.F.=  40  P-VALUE= 0.018
GOODNESS OF FIT CHI-SQ (HOSMER-LEMESHOW)=     7.871  D.F.=   8  P-VALUE= 0.446
GOODNESS OF FIT CHI-SQ   ( C.C.BROWN ) =     12.578  D.F.=   2  P-VALUE= 0.002

                                  STANDARD
        TERM         COEFFICIENT    ERROR     COEFF/S.E.   EXP(COEFFICIENT)

D                   0.77189E-01  0.2370E-01     3.256          1.080
P                     1.4740       0.5605        2.630          4.367
D*P                -0.85976E-01  0.3127E-01    -2.749          0.9176
C                    0.77109       0.4076        1.892          2.162
CONSTANT            -1.8896        0.4432       -4.264          0.1511
CORRELATION MATRIX OF COEFFICIENTS
----------------------------------

             D        P       D*P       C      CONSTANT

D          1.000
P          0.414    1.000
D*P       -0.767   -0.539    1.000
C          0.100    0.130   -0.165    1.000
CONSTANT  -0.538   -0.755    0.435   -0.310    1.000
STATISTICS TO ENTER OR REMOVE TERMS
------------------------------------
                   APPROX.          APPROX.
        TERM       CHI-SQ. D.F.     CHI-SQ. D.F.                    LOG
                   ENTER            REMOVE         P-VALUE       LIKELIHOOD

D                                    12.69   1    0.0004         -61.0238
P                                     7.43   1    0.0064         -58.3937
D*P                                   8.37   1    0.0038         -58.8613
R                    0.48   1                      0.4896         -54.4387
D*R                  0.68   1                      0.4084         -54.3358
P*R                  0.17   1                      0.6763         -54.5903
E                    0.46   1                      0.4970         -54.4468
R*E                  0.81   1                      0.3680         -54.2723
S                    0.15   1                      0.6942         -54.6002
C                                     3.77   1    0.0522         -56.5626
CONSTANT                             25.79   1    0.0000         -67.5703
CONSTANT                                    IS IN       MAY NOT BE REMOVED.
NO TERM PASSES THE REMOVE AND ENTER LIMITS ( 0.1500  0.1000 )
SUMMARY OF STEPWISE RESULTS

STEP    TERM                LOG        IMPROVEMENT     GOODNESS OF FIT
NO.   ENTERED REMOVED    DF LIKELIHOOD CHI-SQUARE P-VAL CHI-SQUARE P-VAL
---   ------- -------    --- ---------- ---------- ----- ---------- -----
  0                          -64.104                       79.718 0.001
  1  D                    1  -61.293     5.621 0.018        74.096 0.002
  2  C                    1  -59.833     2.919 0.088        71.177 0.003
  3  D*P                  1  -58.394     2.879 0.090        68.298 0.005
  4  P                    1  -54.677     7.433 0.006        60.866 0.018
```

FIGURE 11-7 (continued)

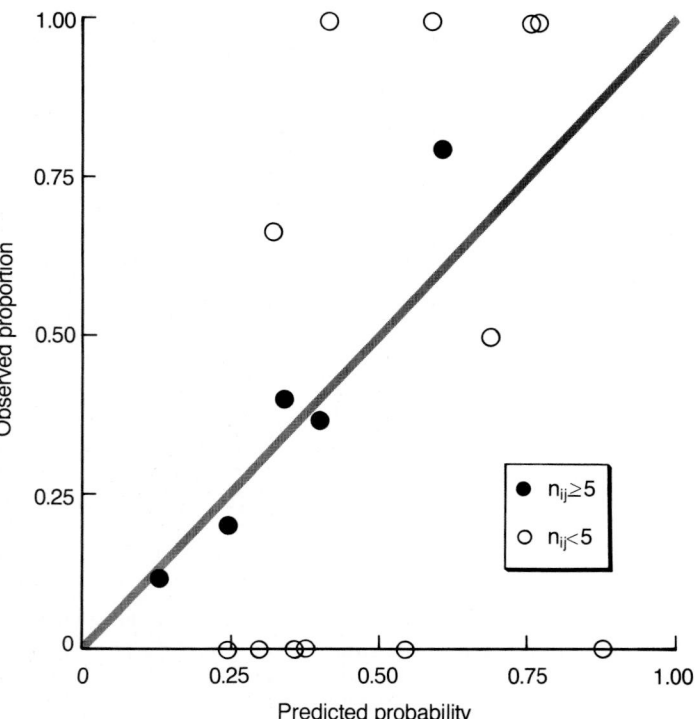

FIGURE 11-18 A plot of the predicted probability of a positive thallium scan vs. the observed proportion of patients with a positive scan for all patterns of observed independent variables corresponding to the analysis of Fig. 11-17. Although better than before, this plot still does not reveal a very good correspondence between the observations and the logistic equation. However, when assessing this plot it is important to note that there are many patterns that are exhibited in only one or two patients, which means that the observed proportions are not very reliable estimates of the actual proportions.

of potential predictor variables is in Fig. 11-19. We start with all variables in the model and remove them if they exceed our specified P_{remove} of .15, with the one decreasing the likelihood function the least removed first. The ischemic ECG changes are thrown out first, followed by the DR interaction, propranolol, sex, and finally, the RC interaction. Thus, we are left with the regression model

$$\text{logit } \hat{P}_T = d_0 + d_D D + d_C C + d_R R + DP + d_{PR} PR \qquad (11.16)$$

which is different from the equation we identified using forward stepping. Compared to Eq. (11.15), this model includes R, excludes P, and adds the PR interaction. Unlike Eq. (11.15) we now have two interaction effects involving propranolol (P), but propranolol itself is not included in the model. Thus, we have a model that is difficult, if not impossible, to interpret.

Even though the P value associated with the Hosmer-Lemeshow statistic is .800, indicating no significant lack of fit, examining a plot of the observed proportions vs. the predicted probabilities in Fig. 11-20 still shows a poor fit. In fact, it is not much better than we obtained with the last model, Eq. (11.15) (compare Figs. 11-18 and 11-20). This example

```
        TOTAL NUMBER OF RESPONSES USED IN THE ANALYSIS    100.
                               SUCCESS . . . . .  34.
                               FAILURE . . . . .  66.

        NUMBER OF DISTINCT COVARIATE PATTERNS . . . . .    45


    STEP NUMBER   0
    ---------------


                           LOG LIKELIHOOD =   -52.349
    GOODNESS OF FIT CHI-SQ  (2*O*LN(O/E))  =    56.208  D.F.= 34  P-VALUE= 0.010
    GOODNESS OF FIT CHI-SQ (HOSMER-LEMESHOW)=    5.389  D.F.=  8  P-VALUE= 0.715
    GOODNESS OF FIT CHI-SQ  ( C.C.BROWN )  =     0.169  D.F.=  2  P-VALUE= 0.919

                                      STANDARD
             TERM         COEFFICIENT   ERROR    COEFF/S.E.  EXP(COEFFICIENT)

    D                     0.74201E-01  0.2893E-01    2.564       1.077
    P                     0.38372      0.8054        0.4764      1.468
    D*P                  -0.73575E-01  0.3211E-01   -2.291       0.9291
    R                    -1.1234       0.8566       -1.311       0.3252
    D*R                  -0.83094E-02  0.3211E-01   -0.2587      0.9917
    P*R                   1.9948       1.124         1.775       7.351
    E                     0.32992E-02  0.9124        0.3616E-02  1.003
    R*E                  -1.1199       1.374        -0.8149      0.3263
    S                    -0.50544      0.5521       -0.9154      0.6032
    C                     0.88003      0.4387        2.006       2.411
    CONSTANT             -0.88197      0.7393       -1.193       0.4140


    CORRELATION MATRIX OF COEFFICIENTS
    -----------------------------------

                D       P      D*P      R      D*R     P*R      E     R*E      S       C    CONSTANT

    D         1.000
    P         0.272   1.000
    D*P      -0.666  -0.405   1.000
    R         0.249   0.494  -0.043   1.000
    D*R      -0.532  -0.089   0.056  -0.405   1.000
    P*R      -0.054  -0.670   0.054  -0.692   0.097   1.000
    E         0.049  -0.171  -0.019   0.106  -0.049   0.100   1.000
    R*E       0.016   0.215  -0.078  -0.049  -0.069  -0.293  -0.650   1.000
    S        -0.032   0.356  -0.040   0.204  -0.129  -0.342   0.064   0.152   1.000
    C        -0.060  -0.048  -0.096  -0.320   0.137   0.232  -0.115   0.087   0.011   1.000
    CONSTANT -0.393  -0.758   0.303  -0.637   0.293   0.531  -0.107  -0.066  -0.584  -0.013   1.000

    STATISTICS TO ENTER OR REMOVE TERMS
    -----------------------------------

                    APPROX.        APPROX.
         TERM      CHI-SQ. D.F.    CHI-SQ. D.F.                  LOG
                     ENTER          REMOVE        P-VALUE    LIKELIHOOD

    D                              7.89   1       0.0050     -56.2933
    P                              0.23   1       0.6327     -52.4629
    D*P                            5.61   1       0.0179     -55.1525
    R                              1.74   1       0.1869     -53.2197
    D*R                            0.07   1       0.7957     -52.3822
    P*R                            3.29   1       0.0698     -53.9925
    E                              0.00   1       1.0000     -52.3487
    R*E                            0.68   1       0.4112     -52.6863
    S                              0.85   1       0.3561     -52.7745
    C                              4.40   1       0.0358     -54.5509
    CONSTANT                       1.49   1       0.2226     -53.0923
    CONSTANT                       IS IN          MAY NOT BE REMOVED.
```

FIGURE 11-19 The results of a backward elimination using the same regression model and set of potential independent variables as in Fig. 11-17. This procedure selected a different set of variables for the "best" model than the forward selection analysis (in Fig. 11-17) did. (Figure 11-19 is continued on the next page.)

```
STEP NUMBER   5              R*E              IS REMOVED
----------------

                        LOG LIKELIHOOD =   -53.772
IMPROVEMENT CHI-SQUARE    ( 2*(LN(MLR) ) =    1.041  D.F.=  1  P-VALUE= 0.308
GOODNESS OF FIT CHI-SQ    (2*O*LN(O/E)) =   59.055  D.F.= 39  P-VALUE= 0.021
GOODNESS OF FIT CHI-SQ (HOSMER-LEMESHOW)=    4.590  D.F.=  8  P-VALUE= 0.800
GOODNESS OF FIT CHI-SQ    ( C.C.BROWN ) =    0.229  D.F.=  2  P-VALUE= 0.892

                                    STANDARD
        TERM          COEFFICIENT    ERROR    COEFF/S.E.  EXP(COEFFICIENT)

D                     0.61817E-01  0.2236E-01   2.764         1.064
D*P                  -0.66972E-01  0.2847E-01  -2.352         0.9352
R                     -1.5495      0.6445      -2.404         0.2124
P*R                    2.0710      0.7887       2.626         7.932
C                     0.98262      0.4284       2.294         2.671
CONSTANT             -0.84309      0.3549      -2.376         0.4304


CORRELATION MATRIX OF COEFFICIENTS
----------------------------------

              D         D*P        R         P*R        C        CONSTANT

D          1.000
D*P       -0.726     1.000
R         -0.137     0.242     1.000
P*R        0.224    -0.363    -0.672     1.000
C          0.060    -0.167    -0.280     0.296     1.000
CONSTANT  -0.305     0.044    -0.392    -0.047    -0.196     1.000

STATISTICS TO ENTER OR REMOVE TERMS
-----------------------------------

               APPROX.             APPROX.
            CHI-SQ. D.F.         CHI-SQ. D.F.               LOG
    TERM      ENTER               REMOVE       P-VALUE   LIKELIHOOD

D                                 8.95    1    0.0028    -58.2450
P             0.89    1                         0.3448    -53.3258
D*P                               6.15    1    0.0131    -56.8473
R                                 6.72    1    0.0095    -57.1321
D*R           0.20    1                         0.6516    -53.6702
P*R                               7.59    1    0.0059    -57.5691
E             0.30    1                         0.5870    -53.6246
R*E           1.04    1                         0.3077    -53.2518
S             1.03    1                         0.3109    -53.2588
C                                 5.76    1    0.0164    -56.6530
CONSTANT                          6.05    1    0.0139    -56.7987
CONSTANT                 IS IN            MAY NOT BE REMOVED.


NO TERM PASSES THE REMOVE AND ENTER LIMITS ( 0.1500  0.1000 ).


SUMMARY OF STEPWISE RESULTS

STEP       TERM               LOG     IMPROVEMENT     GOODNESS OF FIT
NO.   ENTERED REMOVED   DF  LIKELIHOOD CHI-SQUARE P-VAL CHI-SQUARE P-VAL
---   --------------    ---  --------- ---------- ----- ---------- -----
 0                           -52.349                      56.208 0.010
 1      E               1    -52.349    0.000 0.997       56.208 0.013
 2         D*R          1    -52.382    0.067 0.795       56.275 0.017
 3      P               1    -52.490    0.216 0.642       56.491 0.021
 4      S               1    -53.252    1.523 0.217       58.014 0.020
 5         R*E          1    -53.772    1.041 0.308       59.055 0.021
```

FIGURE 11-19 *(continued)*

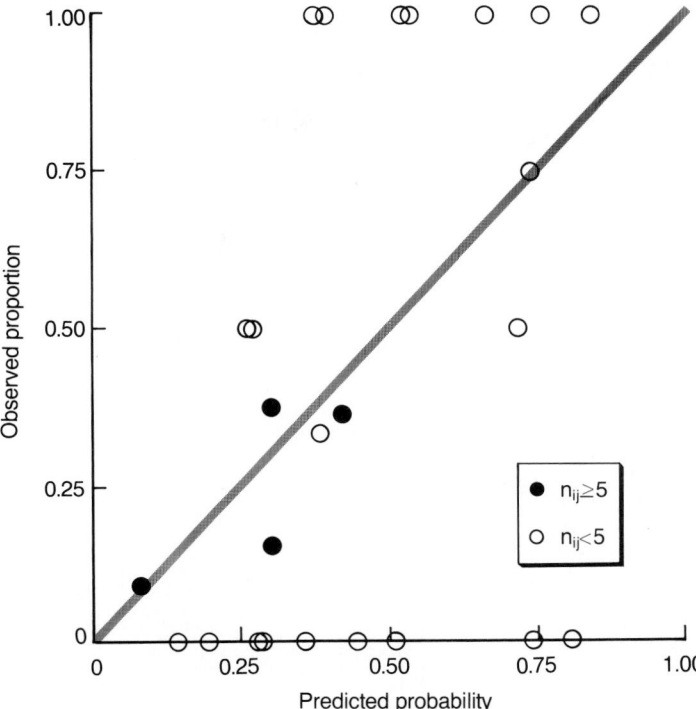

FIGURE 11-20 A plot of the observed proportions vs. the predicted probability of a positive thallium scan, corresponding to the analysis of Fig. 11-19, also reveals a poor fit between the logistic regression equation and the data, although, as in Fig. 11-18, many of the patterns in the data are exhibited by only a few patients, making reliable estimation of the proportions difficult.

illustrates the fact that just because the goodness-of-fit statistic has a large P value, it does not mean that the fit is a good one in the same sense that an R^2 near 1 means that an ordinary multiple linear regression has high predictive ability in individual cases.

Because we still seem to be having problems with finding a good model, let us continue by considering what we have learned from the three models identified so far [Eqs. (11.14), (11.15), and (11.16)]. Two variables, D and C, consistently enter and seem to form the kernel of the final model. Two other variables, P and R, either enter, depending on whether the method is forward or backward, or are close to entering (i.e., $P_{enter} < .2$). Furthermore, interactions involving these two variables enter. Based on these interactions, it seems that P might be more important than R, so a reasonable next step would be to consider the basic three-variable model

$$\text{logit } \hat{P}_T = f_0 + f_D D + f_C C + f_P P \tag{11.17}$$

and search for additional important interactions of these three variables. Therefore, we will force these three variables into the equation and then allow the algorithm to stepwise select from the three possible two-way interactions, DC, DP, and CP.

```
        TOTAL NUMBER OF RESPONSES USED IN THE ANALYSIS      100.
                            SUCCESS  . . . . .     34.
                            FAILURE  . . . . .     66.

        NUMBER OF DISTINCT COVARIATE PATTERNS . . . . .     18


    STEP NUMBER    0
    --------------

                        LOG LIKELIHOOD =    -58.861
    GOODNESS OF FIT CHI-SQ   (2*O*LN(O/E)) =    33.922  D.F.= 14  P-VALUE= 0.002
    GOODNESS OF FIT CHI-SQ (HOSMER-LEMESHOW)=   14.411  D.F.=  8  P-VALUE= 0.072
    GOODNESS OF FIT CHI-SQ  ( C.C.BROWN ) =     10.507  D.F.=  2  P-VALUE= 0.005

                                    STANDARD
          TERM            COEFFICIENT    ERROR    COEFF/S.E.  EXP(COEFFICIENT)

    D                   0.30386E-01  0.1441E-01    2.109          1.031
    C                     0.65536      0.3903       1.679          1.926
    P                     0.62092      0.4471       1.389          1.861
    CONSTANT             -1.4017       0.3578      -3.917          0.2462


    CORRELATION MATRIX OF COEFFICIENTS
    ----------------------------------


                  D         C         P        CONSTANT

    D           1.000
    C           0.001     1.000
    P          -0.021     0.046     1.000
    CONSTANT   -0.376    -0.296    -0.623      1.000


    STATISTICS TO ENTER OR REMOVE TERMS
    -----------------------------------
                    APPROX.            APPROX.
          TERM    CHI-SQ. D.F.       CHI-SQ. D.F.                    LOG
                    ENTER              REMOVE        P-VALUE      LIKELIHOOD

    D                                 4.50    1      0.0338       -61.1132
    D                                 IS IN            MAY NOT BE REMOVED.
    C                                 2.92    1      0.0876       -60.3205
    C                                 IS IN            MAY NOT BE REMOVED.
    P                                 1.94    1      0.1632       -59.8333
    P                                 IS IN            MAY NOT BE REMOVED.
    D*P         8.37    1                             0.0038       -54.6774
    D*C         9.01    1                             0.0027       -54.3545
    C*P         0.17    1                             0.6813       -58.7770
    CONSTANT                         18.59    1      0.0000       -68.1546
    CONSTANT                          IS IN            MAY NOT BE REMOVED.
```

FIGURE 11-21 Based on the previous model selection results, a final analysis was used that forced the independent variables D, P, and C into the model, then allowed a forward selection from a list of all possible pairwide interactions between these three independent variables. This final model produces a good description of the results as indicated by the P value of .928 associated with the Hosmer-Lemeshow goodness-of-fit χ^2.

The computer output from this stepwise selection procedure is in Fig. 11-21. All three interactions were selected, and so we have identified a model that is an extension of the second model identified above, Eq. (11.17).

```
STEP NUMBER   3              C*P                    IS ENTERED
---------------

                      LOG LIKELIHOOD =    -48.197
   IMPROVEMENT CHI-SQUARE   ( 2*(LN(MLR) ) =      4.252  D.F.=  1  P-VALUE= 0.039
   GOODNESS OF FIT CHI-SQ  (2*O*LN(O/E)) =     12.594  D.F.= 11  P-VALUE= 0.321
   GOODNESS OF FIT CHI-SQ (HOSMER-LEMESHOW)=    3.105  D.F.=  8  P-VALUE= 0.928
   GOODNESS OF FIT CHI-SQ   ( C.C.BROWN ) =      0.242  D.F.=  2  P-VALUE= 0.886

                                    STANDARD
           TERM          COEFFICIENT    ERROR    COEFF/S.E.  EXP(COEFFICIENT)

   D                       0.11365    0.3103E-01   3.663        1.120
   C                       1.5653     0.6966       2.247        4.784
   P                       1.5575     0.6584       2.365        4.747
   D*P                    -0.10709    0.3736E-01  -2.867        0.8984
   D*C                    -0.11946    0.4675E-01  -2.556        0.8874
   C*P                     2.9828     1.832        1.628       19.74
   CONSTANT               -2.2038     0.5307      -4.153        0.1104


CORRELATION MATRIX OF COEFFICIENTS
----------------------------------


              D        C        P        D*P      D*C      C*P      CONSTANT

   D         1.000
   C         0.257    1.000
   P         0.411    0.410    1.000
   D*P      -0.818   -0.204   -0.516    1.000
   D*C      -0.431   -0.354   -0.139    0.330    1.000
   C*P       0.314   -0.042   -0.022   -0.334   -0.821    1.000
   CONSTANT -0.526   -0.522   -0.799    0.431    0.211   -0.003    1.000


STATISTICS TO ENTER OR REMOVE TERMS
-----------------------------------
                   APPROX.        APPROX.
           TERM   CHI-SQ. D.F.   CHI-SQ. D.F.              LOG
                   ENTER          REMOVE       P-VALUE   LIKELIHOOD

   D                              20.32   1    0.0000    -58.3582
   D                              IS IN        MAY NOT BE REMOVED.
   C                               5.45   1    0.0196    -50.9223
   C                              IS IN        MAY NOT BE REMOVED.
   P                               6.17   1    0.0130    -51.2839
   P                              IS IN        MAY NOT BE REMOVED.
   D*P                             9.94   1    0.0016    -53.1660
   D*C                            12.78   1    0.0004    -54.5855
   C*P                             4.25   1    0.0392    -50.3234
   CONSTANT                       28.17   1    0.0000    -62.2809
   CONSTANT                       IS IN        MAY NOT BE REMOVED.


NO TERM PASSES THE REMOVE AND ENTER LIMITS (  0.1500  0.1000 ) .

SUMMARY OF STEPWISE RESULTS

STEP       TERM              LOG      IMPROVEMENT    GOODNESS OF FIT
NO.   ENTERED REMOVED   DF LIKELIHOOD CHI-SQUARE P-VAL CHI-SQUARE P-VAL
---  ------------------- --- ---------- ---------- ----- ---------- -----
  0                          -58.861                    33.922 0.002
  1    D*C            1    -54.355    9.014 0.003     24.908 0.024
  2    D*P            1    -50.323    8.062 0.005     16.846 0.155
  3    C*P            1    -48.197    4.252 0.039     12.594 0.321
```

FIGURE 11-21 *(continued)*

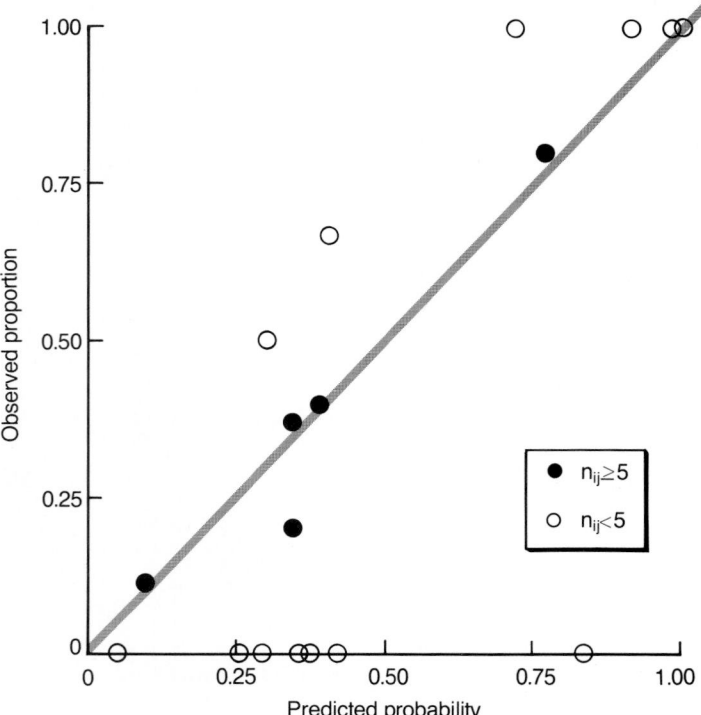

Figure 11-22 Although not great, the plot of observed proportions vs. the probability of a positive thallium test corresponding to the analysis in Fig. 11-21 is much closer to a straight line than in any of the previous models, indicating that the regression model presented in Fig. 11-21 is the best mdoel we have developed so far and is in fact a reasonable description of the data, particularly considering the small number of patients who exhibit any given pattern in the independent variables.

$$\text{logit } \hat{P}_T = f_0 + f_D D + f_C C + f_P P + f_{DP} DP + f_{DC} DC + f_{CP} CP \qquad (11.18)$$

The goodness of fit for this model is much better. The Hosmer-Lemeshow goodness-of-fit statistic now has a P value of .928, which indicates a good fit to the logistic function. More importantly, although it is not ideal, there has been a marked improvement in the plot of the observed vs. predicted probabilities (Fig. 11-22). Now, at least, we have something that fits well enough that we can begin to make interpretations.

All the coefficients f_i in Eq. (11.18) are significantly different from zero as judged by their associated P_{enter} values (all are smaller than .04), and 5 of the 6 f_i values are significant as judged by their associated z statistics (for f_{CP}, $P < .1$). Because the correlation matrix of these coefficients indicates no problem with multicollinearity and all the approximate standard errors are small, we have reasonable confidence in these parameter estimates and we can rewrite Eq. (11.18) as

$$\text{logit } \hat{P}_T = -2.2 + .114/\% \ D + 1.565C + 1.558P - .107/\% \ DP - .119/\% \ DC + 2.98CP$$

The intercept -2.2 estimates the log odds of having a positive thallium test when all values of the independent variables are 0, i.e., the patient has no abnormality as judged by these variables. The odds are $e^{-2.2} =$ .11 [(or in terms of the probability of having a positive test, $\hat{P}_T =$ $\hat{\Omega}_T/(1 + \hat{\Omega}_T) = .11/1.11 = .10$)]. Thus, "normal" patients are not likely to have a positive thallium, as we would expect. In general, the odds of having a positive thallium test increase as the degree of stenosis increases, degree of chest pain increases, and if the patient is on the drug propranolol. The interaction effects indicate that the odds of having a positive test further increases if the patient both has chest pain and is on propranolol, but that propranolol decreases the odds of having a positive test for any given degree of stenosis. The other interaction effect also indicates that having chest pain decreases the odds of having a positive test for any given degree of stenosis.

The value of $f_D = .114/\%$ quantifies the change in log odds for each one percent increase in the degree of stenosis. For example, if all other variables were 0, but the patient had a 40 percent stenosis, the odds of having a positive thallium would be

$$e^{-2.2}(e^{.114/\%})^{40\%} = 10.59$$

Thus, it would seem that even these mild, so-called subcritical, stenoses can have an impact on the results of the test.

We can use the other coefficients to estimate the odds (or, equivalently, the probability) that a patient with a given set of clinical scores will have a positive thallium test. For example, for a patient similar to the one above with a 40 percent stenosis, but who was taking propranolol, the odds of having a positive thallium test are

$$e^{f0}(e^{fd})^D(e^{fp})^P(e^{fDP})^{DP} = e^{-2.2}(e^{.114})^{40}(e^{1.558})^1(e^{-.107})^{40\cdot1} = .698$$

which is not much different from the "normal" patient who has all scores $= 0$, in which case the odds were $e^{-2.2} = .11$. Thus, the drug propranolol may mask otherwise positive tests, and so the patient should probably stop taking the drug before the test is administered.

The presence of the interaction effects greatly increases the goodness of fit to the logistic model, but makes the interpretation of the model more difficult. We have constrained the model to include all main effects for which interactions seemed important. It is possible, as in the results of the backward stepping in Fig. 11-19, to have interaction effects enter, but not the corresponding main effect. This is essentially a nonsense result, and thus we constrained the model. The tradeoff between the improved goodness of fit and the increased complexity of interpretation weighs heavily toward needing the interaction effects. However, just because an interaction effect enters does not mean it is particularly important. If the model fits well already, and you get only a small improvement with adding interactions, the increased complexity of interpretation is not worth it.

SUMMARY

Multiple logistic regression is a technique for analyzing problems in which there are one or more independent variables which determine an outcome that is measured with a dichotomous variable in which there are only two possible outcomes.* The independent variables can be continuous or qualitative (encoded as dummy variables), together with transformations of, and interactions between, the native independent variables.

To estimate the parameters in the logistic regression equation, we had to shift our focus from predicting the observed value of the dependent variable to predicting the probability of one of the possible outcomes of the dependent variable. We also had to develop a new method—the method of maximum likelihood—to compute these parameter estimates. The technique of multiple logistic regression, although very useful, is not as well developed as ordinary linear regression. As a result, the measures of goodness of fit are not as precise as in ordinary linear regression and the properties of the estimates of the coefficients in the equation and the associated hypothesis tests are approximate rather than exact. Furthermore, we do not have as complete a set of statistics, like R^2 and C_p, to help us in model identification and so we must rely more heavily on residuals and influence diagnostics.

These factors all combine to require much larger data sets than one needs to conduct an ordinary linear regression with the same number of independent variables. Interpreting the results of logistic regression also require careful examination of the available regression diagnostics and indicators of multicollinearity. Nevertheless, despite these limitations, logistic regression is an important tool for analyzing data from studies in which the outcome is a "success" or "failure," such as often occurs in clinical or epidemiological research.

PROBLEMS

11.1 *A.* Using the equation estimated for the example of Martian graduation, what are the odds that a Martian of 4 zorp will graduate? *B.* Of 12 zorp? *C.* From the answers to parts *A* and *B*, calculate the predicted probability that Martians with intelligences of 4 and 12 zorp will graduate.

---

*As noted earlier in this chapter, these techniques can be generalized for the case of more than two possible outcomes.

11.2 In the example of using stepwise (forward) regression to find the predictors of a positive exercise thallium test (nuking the heart), why does the degree of stenosis D enter the equation first (refer to Fig. 11-14)? Explain your answer.

11.3 Sudden infant death syndrome (SIDS) is the condition where infants die during their sleep without warning. Some people believe that SIDS is caused by cardiac arrest, perhaps much like sudden death in older adults who have coronary disease or abnormal cardiac rhythm. A new way to look at the control of cardiac rhythm is to analyze the frequencies at which important periodic fluctuations in heart rate occur. Spicer and coworkers[*] identified an important frequency component of the heart beat with a period of variation of about every 6 to 8 heart beats. They determined this component, called the 6 to 8 component of the heart rate power spectrum F_{68}, in many infants. They also recorded the heart rate R, birth weight W, gestation age at birth (weeks at term) G, and the age at the time of making the F_{68} determinations A. Of the many infants studied, 49 were later (after they were no longer at risk for SIDS) selected at random to serve as controls to be compared with data from 16 infants who died of SIDS. These data are in Table D-42, Appendix D. *A.* What are the important predictors of SIDS (S)? Is there evidence that the F_{68} component of the heart rate power spectrum is a predictor of SIDS? *B.* Evaluate the goodness of fit and the residuals.

11.4 Is there a problem with multicollinearity in the data analyzed in Prob. 11.3?

11.5 Adult respiratory distress syndrome (ARDS) is a complication in many critically ill patients. The usual diagnosis of ARDS is based on clinical findings of refractory respiratory failure and x-rays of the lungs showing fluid accumulation. Because it is known that the airways of the lungs are leaky toward plasma proteins in ARDS, accumulation of protein in the fluids in the lung might help diagnose ARDS, particularly because it is possible to noninvasively determine protein accumulation using nuclear imaging of the chest when plasma proteins have been radioactively labeled. Rocker and coworkers[†] used lung images obtained after labeling the plasma protein transferrin in patients meeting the clinical criteria for

---

[*]C. C. Spicer, C. J. Lawrence, and D. P. Southall, "Statistical Analysis of Heart Rates in Subsequent Victims of Sudden Infant Death Syndrome," *Stat. Med.* 6:159–166, 1987.

[†]G. M. Rocker, D. Pearson, M. Stephens, and D. J. Shale, "An Assessment of a Double-Isotope Method for the Detection of Transferrin Accumulation in the Lungs of Patients with Widespread Pulmonary Infiltrates," *Clin. Sci.* 75:47–52, 1988.

ARDS and patients who did not to calculate a lung protein accumulation index P that is larger as there is more protein in the lungs. They also recorded other characteristics of these patients, including their sex S, age A, x-ray lung fluid score X, and amount of oxygen in their blood O (the data are in Table D-43, Appendix D). *A.* Is there evidence that this noninvasive index of protein accumulation in the lungs is associated with ARDS? Include the other variables in the analysis to control for potential confounding. *B.* Interpret any significant regression coefficients.

11.6 Do a complete residual analysis on the logistic equation estimated in Prob. 11.5.

11.7 Is there any evidence that multicollinearity is a problem in the analysis done in Prob. 11.5? Explain your answer.

11.8 *A.* Using the results of Prob. 11.5, calculate the odds that a 65-year-old patient with low arterial oxygen, 6 kPa, and high protein index, 5.5, has ARDS. *B.* Make the same calculation for a 65-year-old patient with high oxygen, 20 kPa, and low protein index, .2.

A Brief Introduction to Matrices and Vectors

Modern regression has been developed using notation and concepts of matrix algebra. Indeed, many of the underlying theoretical results could not have been developed without matrix algebra. While we have developed most of the concepts in this book without resorting to matrix notation, some of the more advanced topics require presenting the equations in matrix notation. In addition, most of the results associated with multiple regression and related topics can be presented more compactly and simply using matrix notation. This appendix reviews the basic definitions and notation necessary to understand the use of matrices in this book. We will concentrate on notational, as opposed to computational or theoretical, issues because computers take care of the computations and there are many excellent books on the theory of linear algebra for those who would like to understand linear algebra in more detail.

DEFINITIONS

A *scalar* is an ordinary number, such as 17.

A *matrix* is a rectangular array of numbers with r rows and c columns. For example, let $\mathbf{X}$ be the 4×3 matrix

$$\mathbf{X} = \begin{bmatrix} 1 & 2 & 4 \\ 6 & 3 & 9 \\ 0 & -1 & 8 \\ 5 & 7 & 10 \end{bmatrix} = \begin{bmatrix} x_{11} & x_{12} & x_{13} \\ x_{21} & x_{22} & x_{23} \\ x_{31} & x_{32} & x_{33} \\ x_{41} & x_{42} & x_{43} \end{bmatrix} = [x_{ij}]$$

Each number in the matrix is called an *element*. Typically (but not always), the matrix is denoted with a boldface letter and the elements are denoted with the same letter in regular type. The subscripts on the symbol for the element denote its row and column. x_{ij} is the element in the ith row and jth column. For example, the element x_{23} is the element in the second row and the third column, 9.

A *vector* is a matrix with one column. For example, a 4×1 matrix, or 4 vector, is

$$\mathbf{y} = \begin{bmatrix} 17 \\ 23 \\ -9 \\ 38 \end{bmatrix} = \begin{bmatrix} y_1 \\ y_2 \\ y_3 \\ y_4 \end{bmatrix}$$

Because all vectors have a single column, we only need one subscript to define an element in a vector. For example, the first element in the vector $\mathbf{y}$ is $y_1 = 17$. There are no fundamental differences between matrices and vectors; a vector is just a matrix with one column.

A *square matrix* is one in which the number of rows equals the number of columns. A square matrix $\mathbf{S}$ is *symmetric* if $s_{ij} = s_{ji}$. A square matrix $\mathbf{D}$ is *diagonal* if all the elements off the diagonal are zeros, i.e., if $d_{ij} = 0$ unless $i = j$. For example,

$$\mathbf{S} = \begin{bmatrix} 1 & 2 & 3 \\ 2 & 4 & 5 \\ 3 & 5 & 6 \end{bmatrix}$$

is a 3×3 symmetric matrix, and

$$\mathbf{D} = \begin{bmatrix} 1 & 0 & 0 \\ 0 & 4 & 0 \\ 0 & 0 & 6 \end{bmatrix}$$

is a 3×3 diagonal matrix. A diagonal matrix in which all the diagonal elements are 1 is called the *identity matrix* and denoted with $\mathbf{I}$. For example, the 4×4 identity matrix is

$$\mathbf{I} = \begin{bmatrix} 1 & 0 & 0 & 0 \\ 0 & 1 & 0 & 0 \\ 0 & 0 & 1 & 0 \\ 0 & 0 & 0 & 1 \end{bmatrix}$$

The identity matrix plays a role similar to 1 in scalar arithmetic.

ADDING AND SUBTRACTING MATRICES

Two matrices can only be added and subtracted if they have the same number of rows and columns. You obtain the sum (or difference) by adding (or subtracting) the corresponding elements of each matrix. For example, if **A** and **B** are both 2×2 matrices, their sum, **C** = **A** + **B** will be the 2×2 matrix

$$\mathbf{C} = \mathbf{A} + \mathbf{B} = \begin{bmatrix} a_{11} & a_{12} \\ a_{21} & a_{22} \end{bmatrix} + \begin{bmatrix} b_{11} & b_{12} \\ b_{21} & b_{22} \end{bmatrix} = \begin{bmatrix} c_{11} & c_{12} \\ c_{21} & c_{22} \end{bmatrix}$$

in which

$$c_{ij} = a_{ij} + b_{ij}$$

For example,

$$\begin{bmatrix} 100 & 73 \\ 21 & -13 \end{bmatrix} + \begin{bmatrix} -13 & 2 \\ 98 & 0 \end{bmatrix} = \begin{bmatrix} 87 & 75 \\ 119 & -13 \end{bmatrix}$$

The usual rules for communitivity and associativity that apply to scalar addition and subtraction also apply to matrix addition and subtraction, i.e.,

$$\mathbf{A} + \mathbf{B} = \mathbf{B} + \mathbf{A}$$

and

$$(\mathbf{A} + \mathbf{B}) + \mathbf{C} = \mathbf{A} + (\mathbf{B} + \mathbf{C})$$

MATRIX MULTIPLICATION

To multiply (or divide) a matrix by a scalar, simply multiply each element by the scalar. For example, using the vector **y** defined above, 5**y** is just

$$5\mathbf{y} = 5 \begin{bmatrix} 17 \\ 23 \\ -9 \\ 38 \end{bmatrix} = \begin{bmatrix} 5y_1 \\ 5y_2 \\ 5y_3 \\ 5y_4 \end{bmatrix} = \begin{bmatrix} 85 \\ 115 \\ -45 \\ 190 \end{bmatrix}$$

Multiplication of matrices follows rules that are more complex than for addition and subtraction and scalar multiplication. These rules, which may seem strange at first, are defined to facilitate, among other things, writing simultaneous algebraic equations. Because of this motivation, we will begin with a set of three simultaneous algebraic equations in three unknowns, then define matrix multiplication from them. Suppose we have the three simultaneous equations in the unknowns b_1, b_2, and b_3:

$$y_1 = x_{11}b_1 + x_{12}b_2 + x_{13}b_3$$

$$y_2 = x_{21}b_1 + x_{22}b_2 + x_{23}b_3$$

$$y_3 = x_{31}b_1 + x_{32}b_3 + x_{33}b_3$$

in which the y_i's and x_{ij}'s are given. We define the vectors

$$\mathbf{y} = \begin{bmatrix} y_1 \\ y_2 \\ y_3 \end{bmatrix}$$

and

$$\mathbf{b} = \begin{bmatrix} b_1 \\ b_2 \\ b_3 \end{bmatrix}$$

and the square matrix

$$\mathbf{X} = \begin{bmatrix} x_{11} & x_{12} & x_{13} \\ x_{21} & x_{22} & x_{23} \\ x_{31} & x_{32} & x_{33} \end{bmatrix}$$

Finally, we define the multiplication of the matrices $\mathbf{X}$ and $\mathbf{b}$ to make the following statement equivalent to the simultaneous equations above:

$$\mathbf{y} = \mathbf{Xb}$$

or

$$\mathbf{y} = \begin{bmatrix} y_1 \\ y_2 \\ y_3 \end{bmatrix} = \begin{bmatrix} x_{11} & x_{12} & x_{13} \\ x_{21} & x_{22} & x_{23} \\ x_{31} & x_{32} & x_{33} \end{bmatrix} \begin{bmatrix} b_1 \\ b_2 \\ b_3 \end{bmatrix}$$

Specifically, for two matrices $\mathbf{A}$ and $\mathbf{B}$ to be multiplied together in the order $\mathbf{AB}$, the number of columns of $\mathbf{A}$ must equal the number of rows of $\mathbf{B}$. For example, if $\mathbf{A}$ is an $r_A \times c_A$ matrix and $\mathbf{B}$ is an $r_B \times c_B$ matrix, their product can only be formed if $c_A = r_B$. The resulting product matrix $\mathbf{C}$ will be an $r_A \times c_B$ matrix. Each element of the product matrix $\mathbf{C}$, c_{ij}, is computed according to

$$c_{ij} = \Sigma a_{ik} b_{kj}$$

where the summation is over k. In other words, c_{ij} is formed by taking the ith row of $\mathbf{A}$ and the jth column of $\mathbf{B}$, multiplying the first element of the specified row in $\mathbf{A}$ by the first element in the specified column in $\mathbf{B}$, multiplying second elements, and so on, and adding the products together. For example, if we multiply the 3×2 matrix $\mathbf{A}$ times a 2×2 matrix $\mathbf{B}$, we obtain

$$\begin{bmatrix} a_{11} & a_{12} \\ a_{21} & a_{22} \\ a_{31} & a_{32} \end{bmatrix} \begin{bmatrix} b_{11} & b_{12} \\ b_{21} & b_{22} \end{bmatrix} = \begin{bmatrix} a_{11}b_{11} + a_{12}b_{21} & a_{11}b_{12} + a_{12}b_{22} \\ a_{21}b_{11} + a_{22}b_{21} & a_{21}b_{12} + a_{22}b_{22} \\ a_{31}b_{11} + a_{32}b_{21} & a_{31}b_{12} + a_{32}b_{22} \end{bmatrix}$$

Here is a numerical example:

$$\begin{bmatrix} 1 & 4 \\ 2 & 5 \\ 3 & 6 \end{bmatrix} \begin{bmatrix} 7 & 8 \\ 9 & 10 \end{bmatrix} = \begin{bmatrix} 1 \cdot 7 + 4 \cdot 9 & 1 \cdot 8 + 4 \cdot 10 \\ 2 \cdot 7 + 5 \cdot 9 & 2 \cdot 8 + 5 \cdot 10 \\ 3 \cdot 7 + 6 \cdot 9 & 3 \cdot 8 + 6 \cdot 10 \end{bmatrix} = \begin{bmatrix} 43 & 48 \\ 59 & 66 \\ 75 & 84 \end{bmatrix}$$

Matrix multiplication is not, in general, communitive; $\mathbf{AB} \neq \mathbf{BA}$ in general. Matrix multiplication is associative; $(\mathbf{AB})\mathbf{C} = \mathbf{A}(\mathbf{BC})$.

INVERSE OF A MATRIX

The closest thing in matrix algebra to division is the inverse of a square matrix. Recall that we say that the inverse of a is $a^{-1} = 1/a$ because $a \cdot a^{-1} = 1$. In terms of matrices, suppose we have an $n \times n$ square matrix $\mathbf{A}$. If we can find another $n \times n$ square matrix $\mathbf{B}$ such that

$$\mathbf{AB} = \mathbf{I}$$

we say that $\mathbf{B}$ is the *inverse* of $\mathbf{A}$ and denote it $\mathbf{A}^{-1}$. Hence

$$\mathbf{AA}^{-1} = \mathbf{A}^{-1}\mathbf{A} = \mathbf{I}$$

where the identity matrix plays a role similar to 1 in scalar division. Only square matrices can have inverses, and not all square matrices have inverses. To have an inverse, the matrix $\mathbf{A}$ must be *nonsingular*, which is roughly comparable to requiring that the scalar 0 does not have an inverse.

The calculations of the inverse are complicated, but we have computers to take care of the arithmetic.

TRANSPOSE OF A MATRIX

The transpose of a matrix is the matrix obtained by interchanging the rows and columns of the matrix. We denote the transpose of the matrix $\mathbf{X}$ as $\mathbf{X}^T$. If the elements of $\mathbf{X}$ are x_{ij}, then the elements of $\mathbf{X}^T$ are x_{ji}. For example,

$$\begin{bmatrix} 1 & 4 \\ 2 & 5 \\ 3 & 6 \end{bmatrix}^T = \begin{bmatrix} 1 & 2 & 3 \\ 4 & 5 & 6 \end{bmatrix}$$

The transpose is useful for computing the sum of squared elements of a vector. For example, for the vector $\mathbf{y}$ above

$$\mathbf{y}^T\mathbf{y} = \begin{bmatrix} y_1 & y_2 & y_3 \end{bmatrix} \begin{bmatrix} y_1 \\ y_2 \\ y_3 \end{bmatrix} = y_1^2 + y_2^2 + y_3^2 = \Sigma y_i^2$$

EIGENVALUES AND EIGENVECTORS

All square matrices have special numbers and vectors associated with them called *eigenvalues* and *eigenvectors*. Let **A** be an $n \times n$ square matrix. There are n numbers λ_i, each associated with an n vector $\mathbf{p}_i$, which satisfy the equation

$$\mathbf{A}\mathbf{p}_i = \lambda_i \mathbf{p}$$

The eigenvalues and eigenvectors play a crucial role in principal components regression analysis and many aspects of matrix algebra. As with computing inverses, computing eigenvalues and the associated eigenvectors is quite involved, but we have computers to handle the arithmetic. For a general discussion of the many meanings of these two quantities, see any introductory linear algebra text.

Statistical Package Cookbook

This appendix provides copies of all the control files for Minitab, SAS, SPSS-X, and BMDP* that were used to generate the computer outputs in this book, together with some comments about the strengths and weaknesses of these programs. These files will permit you to reproduce the full computer printouts for the examples. In addition, these examples should provide a useful practical supplement to the documentation that accompanies these programs—which can range from overwhelming to frustratingly terse—to help you translate the advice we give you in this book into practical applications.

The appendix is organized according to the primary statistical technique being illustrated (e.g., regression, multicollinearity), with subsections giving the control files for working the different examples. We carry many of the examples through several chapters in the book, but all the control commands necessary to generate the entire example are contained in a single file, with comments noting which part of the example

---

*At least one of these programs (and often all) is available at virtually every academic or industrial computing site; they are also available for personal computers. The best source of documentation is your local computing facility. Should you wish to contact the software publisher directly, here are their addresses and phone numbers:

BMPD Statistical Software, Inc.
1440 Sepulveda Blvd.
Suite 316
Los Angeles, CA 90025
213–479–7799

SAS Institute, Inc.
Box 8000
Cary, NC 27511
919–467–8000

Minitab, Inc.
3801 Enterprise Drive
State College, PA 16801
814–238–3280

SPSS, Inc.
444 North Michigan Avenue
Chicago, IL 60611
312–329–3500

is being computed at each step. In most cases, we use all these programs for different examples. When we only use one or two of the programs for examples in the text, we also include sample control files to do these examples with the other programs, if they can do the necessary computations. As a result, this appendix should provide all the information you need to implement the computations we describe, no matter which of these programs you use.

These files, together with the input data files, are all included on a disk that can be ordered from the authors.* We used Minitab Release 6.1,† SAS Version 5, BMDP 1988 Release, and SPSS-X Release 3.1 on a DEC VAX running the VMS operating system. Some of the commands, particularly those relating to file handling, may be different for different operating systems. All these programs are available on a wide variety of mainframe computers as well as microcomputers. These programs have all been around for years, and the companies are very good about maintaining "downward compatibility" as new versions are released, so that old command files continue to work. Hence, there is little risk that these files will no longer work in the face of new releases. Some of the comments and helpful hints may change as the programs evolve, capabilities are added, and difficulties are fixed.

The file suffix identifies the computer package: ".BMD" indicates BMDP, ".MTB" indicates Minitab, ".SAS" indicates SAS, and ".SPS" indicates SPSS-X.

GENERAL COMMENTS ON SOFTWARE

The four computer packages we illustrate can be presumed to be accurate as far as the number crunching goes. If you have given all the correct commands, you understand what the program is doing and how to tell it what you want, and your data are correct, the numbers you get from the computer can be trusted. However, there is much more to using statistical software than letting the computer chew on some data.

Throughout the book we have cautioned that you should know what the software you use is doing before you interpret the numbers it gives you. This point cannot be overemphasized. As long as you have given legal commands and used data in a proper format, you can generate numbers. Most programs have certain protections built in to help you avoid errors, but the people who write these programs cannot anticipate all of

---

*To order the disk, send $19.95 per disk to Datadisk, SE 810 Meadowvale, Pullman, WA 99163. Specify disk format. Discounts on bulk orders.
†We used Minitab to develop most of the examples, then translated them into the other programs. We thank Minitab, Inc., for assisting us in this task. The consulting staffs at SAS, SPSS, and BMDP also provided helpful comments and assistance.

the silly things you might try to do. More important, even if you have not done something silly, you need to know how the program works to know what the numbers mean.

For example, if you use the SAS procedure GLM to analyze a mixed model analysis of variance (some factors fixed and others random) several things can go wrong, even if you have balanced data. To do a mixed model analysis, you need to give the program a command to tell it which of the factors is random. If, for example, you have factors A and B, which are fixed, and factor C, which is random, and you want to analyze the full model (including all interactions), you would give it a command something like

```
proc GLM;
     class A B C;
     model Y = A B C A*B A*C B*C A*B*C;
     random C A*C B*C A*B*C;
```

Any time an interaction involves a random factor, the interaction is usually considered to be random. Some programs, such as the Minitab ANOVA procedure, simply let you specify that factor C is random, and then the program takes care of the rest. However, you have to tell SAS GLM that the interactions are random as well. Having gotten this far, you now have to be careful about the hypothesis testing. If the model were fixed (none of the factors were random), the denominator of all F statistics would be MS_{res}, the residual, or error, mean square. However, fixed factors in mixed models have different error terms; for example, the denominator of the F statistic for factor A in the mixed model above is MS_{AC}, not MS_{res}. The problem occurs when using GLM as specified above because it computes F statistics as though all factors were fixed, even though you told it factor C was random. As a result, the F statistics in the computer printout will be wrong for factors A, B, and the A by B interaction. (The F test statistics for the random factor C, and any interactions involving C, have MS_{res} in the denominator of their F statistics, so they will be correct as reported by the computer.) You have to calculate the correct F statistics for factors A, B and the A by B interaction from the sums of squares and degrees of freedom reported in the computer output or give commands to tell SAS which sums of squares to use in calculating the F statistics for the fixed effects.

While all the computer packages do the basic things like multiple linear regression and analysis of variance of balanced fixed effects designs, not all packages do all things in the same way. The default conditions that the program uses in the absence of an alternative specification by the user for the different programs are different. These differences take the form of different default values for the entry and removal criteria for stepwise regression, convergence criteria for nonlinear regression, the model assumed for analysis of variance (such as that discussed above), and many other ways, all of which can affect the appropriateness and accuracy of the results you obtain with these programs. We have pointed out some of these differences in this appendix. As a general rule, you should know the defaults that your program is using and make sure you approve of them.

As already noted, different packages have different capabilities, and there are often several different ways to use any given package to solve a specific problem. For example, one can do multiple linear regressions with BMDP procedures 1R, 2R, and 9R and SAS procedures REG, STEP, and GLM. In other packages, many capabilities are included in a single procedure, such as SPSS REGRESSION, which does ordinary multiple linear regression, and stepwise regression using forward selection and backward elimination. As we noted above, all these packages do essentially the same thing, so they share a common set of words to describe what they do. Unfortunately, they do not always use these words to mean the same thing. (Indeed, there are sometimes conflicts in nomenclature between different procedures in the same package; the "Type II Sum of Squares" computed by the SAS REG and GLM procedures are different.) Because computers take key words quite literally, you need to be sure that you know what each word you use means to the program you are using. This bit of language ambiguity can lead to great confusion, particularly if you are using two or more of the different packages at the same time.

The final difference between these packages is how the commands are punctuated. They all use periods, semicolons, commas, and slashes differently. Most of the packages require that variables in lists be separated with commas or spaces; SAS requires spaces. SAS requires each statement to end with a semicolon; BMDP with a period. BMDP uses a slash to start a procedure request (called a paragraph); SAS and SPSS use a slash to define subcommands. Minitab uses a semicolon to define subcommands and a period to end subcommands. These differences can be a great source of confusion. Most of the time, incorrect punctuation simply stops the execution of the program, and the problem is obvious. Occasionally the program will run and produce answers but will not construct the analysis the way you wanted it to because of the punctuation errors. When you get strange results in complex analysis, check the punctuation.

A Note on User Interfaces

With the exception of Minitab, the software we illustrate was developed in a batch processing environment. Therefore, BMDP, SAS, and SPSS work from a list of command instructions submitted as an input file. All of these packages have developed interactive interfaces as batch processing has faded, but we still find them easier to use in their noninteractive mode, which is why we generated all the control files listed in this appendix. Minitab is of more recent vintage and was developed as an interactive program. Although it can run batch, it is most at home in an interactive environment. The "control files" listed for Minitab are really macros, which are lists of normally interactive commands that can be stored in a file for automatic execution in an interactive environment. The interface is the same, but the macro allows you to save time when repeating similar analyses many times.

REGRESSION

There are many ways to do linear regression with these four packages, most of which will be illustrated in the following command files, which are related to the examples presented in Chaps. 1 through 5. Because we used regression to do analysis of variance, many more examples of regression control files can be found in the remaining sections of this appendix.

Martian Height and Weight (Figs. 1-1 and 2-10; Data in Table 2-1)

```
NOTE   This is file MARSWH.MTB
name c1 'W'
name c2 'H'
READ c1 c2
 7.8 31
 8.3 32
 7.6 33
 9.1 34
 9.6 35
 9.8 35
11.8 40
12.1 41
14.7 42
13.0 46
end
REGRESS 'W' on 1 independent variable, 'H'
```

```
/COMMENT
      This is file MARSWH.BMD
      to be used with procedure 1R.
/PROBLEM
      title is 'Mars Weight vs. Height'.
/INPUT
      format is free.
      variables are 2.
```

```
/VARIABLE
     names are W, H.
/REGRESS
     depend is W.
     independ is H.
/END
  7.8 31
  8.3 32
  7.6 33
  9.1 34
  9.6 35
  9.8 35
 11.8 40
 12.1 41
 14.7 42
 13.0 46
```

```
*      This is file MARSWH.SAS;
DATA MARS;
input W H @@;
LABEL W = 'Weight'
       H = 'Height';
CARDS;
  7.8 31
  8.3 32
  7.6 33
  9.1 34
  9.6 35
  9.8 35
 11.8 40
 12.1 41
 14.7 42
 13.0 46
;
proc REG data = mars;
     model W = H;
```

```
*      This is in file MARSWH.SPS
TITLE 'MARS Weight vs Height'
DATA LIST FREE
   / W H
BEGIN DATA
  7.8 31
  8.3 32
  7.6 33
```

```
      9.1 34
      9.6 35
      9.8 35
     11.8 40
     12.1 41
     14.7 42
     13.0 46
     END DATA
     VAR LABELS H 'Height'
                W 'Weight'
     REGRESSION
         /variables H W
         /dependent = W
         /enter H
     FINISH
```

Martian Weight, Height, and Water Consumption
(Figs. 1-2, 3-1, 3-2, 3-6, and 4-16; Data in Table 1-1)

```
NOTE    This is file MARSWHW.MTB
READ 'marswhw.dat' c1-c3
name c1 'W'
name c2 'H'
name c3 'C'
REGRESS 'W' on 2 independent variables, 'H' and 'C', put Studentized &
residuals in c46 and predicted W in c44;
     resids c45.
name c44 'What'
name c45 'RawRes'
PLOT 'RawRes' vs. 'H'
PLOT 'RawRes' vs. 'C'
PLOT 'RawRes' vs. 'W'
PLOT 'RawRes' vs. 'What'
REGRESS 'W' on 2 independent variables, 'C' and 'H'
NOTE    The following is to produce residual plots for Chap. 4.
REGRESS 'W' on 1 variable 'H';
     resids c45.
name c45 'RawRes'
PLOT 'RawRes' versus 'C'
```

Heat Exchange in Gray Seals
(Figs. 2-11, 3-16, and 4-18; Data in Table C-1, Appendix C)

```
*       This is file SEALS.SAS;
title 'Seal Thermal Conductivity';
```

```
DATA SEALS;
      infile seals;
      input Cb Ta;
      T2 = Ta**2;
      LABEL Cb = 'Conductivity'
            Ta = 'Ambient Temperature'
            T2 = 'Ambient Temperature Squared';
proc REG data = seals;
      model Cb = Ta T2;
*     The following produces residual plots needed for;
*     the residual analysis in Chap. 4                    ;
proc REG data = seals;
      model Cb = Ta;
      output out = sealplot R = RawRes P = Chat;
proc PLOT data = sealplot;
      plot RawRes*Ta / vref = 0 vpos = 18 hpos = 60;
      plot RawRes*Chat / vref = 0 vpos = 18 hpos = 60;
```

Martian Secondhand Smoke Exposure
(Figs. 1-3, 3-7, and 3-8; Data in Table 3-1)

```
/COMMENT
      This is file MARSSMOK.BMD
      for procedure 1R.
/PROBLEM
      title is 'Martian Weight vs. Height with Smoke Exposure
      Dummy Variable'.
/INPUT
      file is 'marssmok.dat'.
      format is free.
      variables are 3.
/VARIABLE
      names are W, H, D.
/REGRESS
      depend is W.
      independ are H, D.
/END
```

Mechanisms of Toxic Shock
(Figs. 3-9 and 3-10; Data in Table C-2, Appendix C)

```
NOTE   This is file ENDOTOXN.MTB
READ 'endotoxn.dat' c1-c3
name c1 'P'
```

```
name c2 'A'
name c3 'E'
REGRESS 'P' on 2 independent variables 'A' and 'E'
```

Protein Synthesis in Newborns and Adults
(Figs. 3-11 and 3-12; Data in Table C-3, Appendix C)

```
*       This is file LEUCINE.SPS
TITLE 'Leucine Metabolism in Adults and Newborns'
DATA LIST FILE = leucine.dat LIST
     / logL logW A
VAR LABELS logL   'log Leucine'
           logW 'log Body Weight'
           A       'dummy variable for adult (=1)'
REGRESSION
     /variables logL logW A
     /dependent = logL
     /enter logW A
FINISH
```

Diabetes, Cholesterol, and the Treatment of High Blood Pressure
(Fig. 3-13; Data in Table C-4, Appendix C)

```
NOTE   This is file DIABETES.MTB
READ 'diabetes.dat' c1-c9
name c1 'H'
name c2 'B'
name c3 'D'
name c4 'S'
name c5 'A'
name c6 'W'
name c7 'T'
name c8 'C'
name c9 'G'
REGRESS 'H' on 8 variables 'B'-'G'
```

Baby Birds Breathing in Burrows
(Figs. 3-14 and 3-15; Data in Table C-5, Appendix C)

```
NOTE   This is file BABYBIRD.MTB
READ 'babybird.dat' c1-c3
name c1 'V'
name c2 'O'
name c3 'C'
REGRESS 'V' on 2 independent variables 'O' and 'C'
```

How Bacteria Adjust to Living in Salty Environments
(Figs. 3-17 through 3-20; Data in Table C-6, Appendix C)

```
*        This is file SALTBACT.SAS;
title 'Salty Bacteria';
DATA saltbact;
        infile saltbact;
        input W t S;
        I = t*S;
        LABEL W = 'Tritiated H20 production'
              t = 'time'
              S = 'NaCl concentration'
              I = 'interaction of Time and NaCl';
proc REG    data = saltbact;
        model W = S t I;
proc REG    data = saltbact;
        model W = S t;
```

The Response of Smooth Muscle to Stretching
(Figs. 3-21 and 3-22; Data in Table C-7, Appendix C)

```
*        This is file MUSCLE.SAS;
title 'Smooth muscle stretching';
DATA muscle;
        infile muscle;
        input deltaF deltaL F;
        deltaL2 = deltaL**2;
        deltaLxF = deltaL*F;
        LABEL deltaF = 'change in Force'
              deltaL = 'change in Length'
              deltaL2 = 'squared change in Length'
              deltaLxF = 'interaction of deltaL and Force';
proc REG   data = muscle;
        model deltaF = deltaL deltaL2 deltaLxF / noint;
```

Martian Intelligence
(Figs. 4-1 to 4-6, 4-10 to 4-15; Data in Table C-8, Appendix C)

```
NOTE   This is file MARSINT.MTB
NOTE   Edit the file name in READ for correct data file to input the
NOTE   other input files:
NOTE        marsint1.dat     (Data in Table C-8A, Appendix C)
```

```
NOTE        marsint2.dat     (Data in Table C-8B, Appendix C)
NOTE        marsint3.dat     (Data in Table C-8C, Appendix C)
NOTE        marsint4.dat     (Data in Table C-8D, Appendix C)
NOTE        marsint5.dat     (Data in Table C-8E, Appendix C)
READ 'marsint1.dat' c1 c2
name c1 'I'
name c2 'F'
NOTE   Set up column names for residual diagnostics.
NOTE   Raw residuals; from subcommand RESID.
name c45 'RawRes'
NOTE   Standardized residuals; to be computed from RawRes.
name c46 'StanRes'
NOTE   Studentized residuals; from extra column specified
NOTE   in REGRESS command. Minitab instructions refer to
NOTE   this command line option as "standardized residuals"
NOTE   even though they are what we call "Studentized
NOTE   residuals."
name c47 'StudRes'
NOTE   Studentized-deleted residuals; from subcommand TRESID.
name c48 'StDelRes'
NOTE   Leverage values (Hat matrix diagonal); from subcommand HI.
name c49 'Leverage'
NOTE   Cook's Distance; from subcommand COOKD.
name c50 'CookDist'
NOTE   Regression with appropriate command line modifier and
NOTE   subcommands to generate residual diagnostics.
REGRESS 'I' on 1 variable 'F' and put Studentized residuals in c47;
        resids c45;
        tresids c48;
        hi c49;
        cookd c50.
NOTE   Compute standardized residuals as RawRes/stan. dev. of
NOTE   RawRes
let c46 = 'RawRes'/STAN('RawRes').
NOTE   Display residual diagnostics.
print c45-c50
NOTE   Display distribution of residuals.
HISTOGRAM 'RawRes'
NOTE   Compute and display normal probability plot of residuals.
NSCORE 'RawRes' c44
PLOT 'RawRes' versus c44
NOTE   Display plot of residuals vs. independent variable F.
PLOT 'RawRes' versus 'F'
```

Water Movement across the Placenta
(Figs. 4-20 to 4-24; Data in Table C-9, Appendix C)

```
/COMMENT
        This is file PLACENTA.BMD
        for procedure 2R.
        This control file is to be used with BMD 2R, not 1R. The result
        is the same as far as the regression, but 1R does not give all of
        the residual diagnostics, so 2R must be used for extensive
        residual analysis. 2R is meant to be a stepwise routine, but by
        using FORCE = independent variable(s) in REGRESS paragraph
        and suppressing some intermediate results with NO ANOVA in
        the PRINT paragraph, you can get an ordinary regression, with a
        reasonably clear output. If you have three or more independent
        variables, you can also use BMDP 9R, all possible subsets
        regression, and set METHOD = NONE to do an ordinary
        regression. When you can use it, 9R is preferable to 2R for
        doing straight regression when you need fancy diagnostics.
/PROBLEM
        title is 'Placental water clearance data'.
/INPUT
        file is 'placenta.dat'.
        format is free.
        variables are 2.
/VARIABLE
        names are D, U.
/REGRESS
        title is 'Simple regression'.
        depend is D.
        independ is U.
        force = 1.
/PRINT
        no anova.
        diagnostics = dstresid,hatdiag,cook.
/PLOT
        size is 60,20.
        normal.
        yvar = D,residual,residual.
        xvar = U,U,predictd.
/END
/COMMENT
        We have to repeat the whole problem in order to repeat the plot
        and print paragraphs.
/INPUT
        file is 'placenta.dat'.
```

```
            format is free.
            variables are 2.
/VARIABLE
            added is 1.
            names are D, U, U2.
/TRANSFORM
            U2 = U**2.
/REGRESS
            title is 'Quadratic regression'.
            depend is D.
            independ are U, U2.
            force = 1.
/PRINT
            no anova.
            diagnostics = dstresid,hatdiag,cook.
/PLOT
            size is 60,20.
            normal.
            yvar = D,residual,residual.
            xvar = U,U,predictd.
/END
/COMMENT
            Because the transform paragraph can only be used once per
            problem, we have to start a new problem and input the data
            again to delete a case.
/INPUT
            file is 'placenta.dat'.
            format is free.
            variables are 2.
/VARIABLE
            names are D, U.
/TRANSFORM
            omit  =  23.
/REGRESS
            title is 'Simple regression repeated; deleting outlier'.
            depend is D.
            independ is U.
            force = 1.
/PRINT
            no anova.
            diagnostics = dstresid,hatdiag,cook.
/PLOT
            size is 60,20.
            normal.
```

```
        yvar = D,residual,residual.
        xvar = U,U,predictd.
/END
```

Cheaper Chicken Feed (Figs. 4-25 to 4-30; Data in Table C-10, Appendix C)

```
*        This is file CHICKEN.SPS
TITLE 'Methionine hydroxy analog intestinal uptake data'
DATA LIST FILE = chicken.dat LIST
     / R M
COMPUTE SQRTR = sqrt(R)
COMPUTE INVR = 1/R
COMPUTE logR = ln(R)
COMPUTE logM = ln(M)
VAR LABELS logR 'natural log of rate'
            logM 'natural log of Methionine'
            R    'Rate of uptake'
            M    'Methionine concentration'
            Invr 'reciprocal of rate'
            Sqrtr 'square root of rate'
REGRESSION
     /variables R M
     /dependent = R
     /enter M
     /residuals NORMPROB
     /casewise = all SDRESID LEVER COOK
     /scatterplot (*RES,M)
PLOT
     /hsize = 60
     /vsize = 20
     /plot = R Invr Sqrtr logR logR with M M M M logM (PAIR)
REGRESSION
     /variables logR logM
     /dependent = logR/ ENTER logM
     /residuals NORMPROB
     /casewise = all SDRESID LEVER COOK
     /scatterplot (*RES,logM)
FINISH
```

How the Body Protects Itself from Excess Copper and Zinc (Figs. 4-31 to 4-42; Data in Table C-11, Appendix C)

```
*        This is file COPZINC.SAS;
TITLE 'Metallothionein Regulation by Copper and Zinc';
```

```
DATA zinc;
      infile copzinc;
      input M Cu Zn;
      Zn2 = Zn**2;
      INVZn = 1/Zn;
      INVCu = 1/Cu;
      INVM = 1/M;
      logCu = log(Cu);
      logZn = log(Zn);
      logM = log(M);
LABEL M = 'Methallothionein mRNA'
      Cu = 'Dietary Copper level'
      Zn = 'Dietary Zinc level'
      Zn2 = 'Squared Zinc level'
      INVZn = 'Reciprocal of Zinc'
      INVCu = 'Reciprocal of Copper'
      INVM = 'Reciprocal of Metallothionein mRNA'
      logCu = 'natural log of Copper'
      logZn = 'natural log of Zinc'
      logM = 'natural log of Metallothionein mRNA';
proc REG   data = zinc;
      model M = Cu Zn;
      output P = Mhat R = RawRes STUDENT = StudRes
            RSTUDENT = StDelRes H = leverage COOKD = Cook;
proc PRINT;
      var StudRes StDelRes leverage Cook;
proc PLOT;
      plot RawRes*Cu/ Vref = 0 Vpos = 18 Hpos = 60;
      plot RawRes*Zn/ Vref = 0 Vpos = 18 Hpos = 60;
*     To get a normal probability plot of residuals, we have         ;
*     to use the RANK procedure with the NORMAL = BLOM               ;
*     option to generate normal probability scores for the residuals. ;
*     We then plot the residuals against their normals scores.        ;
*     (Note: This is identical to the procedure you would use in      ;
*     Minitab using the NSCORE command.)                             ;
proc RANK NORMAL = BLOM OUT = NORMPLT;
      var RawRes;
      ranks Nresid;
proc PLOT;
      plot RawRes*Nresid/ Vref = 0 Vpos = 18 Hpos = 60;
proc REG data = zinc;
      model M = Cu Zn Zn2;
      output P = Mhat R = RawRes STUDENT = StudRes
            RSTUDENT = StDelRes H = leverage COOKD = Cook;
```

```
proc PRINT;
      var StudRes StDelRes leverage Cook;
proc PLOT;
      plot RawRes*Cu/ Vref=0 Vpos=18 Hpos=60;
      plot RawRes*Zn/ Vref=0 Vpos=18 Hpos=60;
proc RANK NORMAL=BLOM OUT=NORMPLT;
      var RawRes;
      ranks Nresid;
proc PLOT;
      plot RawRes*Nresid/ Vref=0 Vpos=18 Hpos=60;
proc PLOT data=zinc;
      plot INVM*INVZn/ Vref=0 Vpos=18 Hpos=60;
      plot INVM*INVCu/ Vref=0 Vpos=18 Hpos=60;
      plot logM*logZn/ Vref=0 Vpos=18 Hpos=60;
      plot logM*logCu/ Vref=0 Vpos=18 Hpos=60;
      plot logM*INVZn/ Vref=0 Vpos=18 Hpos=60;
      plot logM*INVCu/ Vref=0 Vpos=18 Hpos=60;
proc REG data=zinc;
      model logM=INVCu INVZn;
      output P=Mhat R=RawRes STUDENT=StudRes
            RSTUDENT=StDelRes H=leverage COOKD=Cook;
proc PRINT;
      var StudRes StDelRes leverage Cook;
proc PLOT;
      plot RawRes*Cu/ Vref=0 Vpos=18 Hpos=60;
      plot RawRes*Zn/ Vref=0 Vpos=18 Hpos=60;
proc RANK NORMAL=BLOM OUT=NORMPLT;
      var RawRes;
      ranks Nresid;
proc PLOT;
      plot RawRes*Nresid/ Vref=0 Vpos=18 Hpos=60;
```

MULTICOLLINEARITY

Most regression programs have several multicollinearity diagnostics available to you. In addition, most allow some control over the tolerance to force the program to analyze increasingly multicollinear data. The essential thing to remember, however, is that forcing a program to analyze highly multicollinear data is asking for trouble.

Minitab has a two-level protection that issues a warning if the tolerance is less than .01 (R_i^2 greater than .99) and deletes the offending variable if the tolerance is less than .0001 (the default). BMDP and SPSS have a single level of protection that simply omits offending independent variables when the associated tolerance falls below .01 (BMD) or .0001

(SPSS). The action the routines take when tolerance is exceeded is also different: BMDP, Minitab, and SPSS will continue the regression analysis after omitting any independent variable with a tolerance greater than the specified value. SAS simply stops, with an explanation.

It is possible to reset the tolerance level in most of these programs by issuing the appropriate command. *You should never reset the tolerance to force the analysis of highly multicollinear data unless you know what you are doing and you tell people you did it when you report your results.* The way the different packages deal with tolerance is summarized below.

Martian Weight, Height, and Water Consumption, Showing Severe Multicollinearity (Figs. 5-1, 5-2, and 5-16; Data in Table 5-2)

```
NOTE   This is file MARS2WHW.MTB
NOTE   for analysis in Figs. 5-1 and 5-2.
READ 'mars2whw.dat' c1-c3
name c1 'W'
name c2 'H'
name c3 'C'
REGRESS 'W' on 2 independent variables 'H' and 'C';
        vif.
```

```
/COMMENT
        This is file MARS2WHW.BMD
        for procedure 4R.
        Because the output includes the eigenvalues and eigenvectors of
        the correlation matrix, it can be used for diagnostic purposes
        regardless of how you control which principal components enter
        the regression or whether you ultimately even use the principal
        components regression results.
/PROBLEM
        title is 'MARTIAN W vs. H and C; multicollinearity'.
/INPUT
        file is 'mars2whw.dat'.
        variables are 3.
        format is free.
/VARIABLE
        use = all.
        names are W, H, C.
/REGRESS
        title is 'PRINCIPAL COMPONENTS REGRESSION OF MARS
        W vs H and C'.
```

    depend is 1.
    independ are 2 to 3.
    eigen.
    limit = .01, .000000000001.
/COMMENT
    Program 4R allows you several ways to decide which
    components to include. The only way that makes sense, in
    terms of the development we have given in Chap. 5, is to
    exclude only the very smallest eigenvalues. Because we like to
    see sort of a "stepwise" result of including first the component
    with the largest eigenvalue, then adding the next largest, and so
    on through the smallest, we specify that the components to
    enter are selected by their eigenvalues (eigen. command in
    REGRESS paragraph above). To ensure all components are
    entered, you must set the eigenvalue entry limit below the value
    of the smallest eigenvalue. Thus, the limit command above
    specifies "limit = .01, .000000000001." The first number, .01, is
    the limit if you use the correlation method (so it does not apply
    here) and the second is the eigenvalue limit. Any component
    corresponding to an eigenvalue larger than the limit is included.
    The result is a list of several regression equations starting with
    the one including only the first component, then one obtained
    by adding each successive component until all components are
    in, corresponding to the ordinary least-squares solution.
/END

NOTE   This is file MARS2PCA.MTB
NOTE   If you simply want diagnostic information from principal
NOTE   components, Minitab has a command, PCA, that will do a
NOTE   a principal components analysis of the columns of data
NOTE   you specify.
READ 'mars2whw.dat' c1-c3
name c1 'W'
name c2 'H'
name c3 'C'
PCA c2 c3
NOTE   You can then save the principle components, inspect them,
NOTE   and decide which to include in your regression, then create
NOTE   the new variables manually and submit them to the
NOTE   regression procedure.

    Principal components regression can also be done with the SAS pro-
cedure RIDGEREG, which also does ridge regression. RIDGEREG is not
part of the main SAS package and is usually available only on IBM main-
frames. It is described in the *SUGI Supplemental Library User's Guide.*

The advantage of RIDGEREG, if you can use it, is that it prints out approximate standard errors for the principal component regression coefficients, which BMDP 4R does not.

Interaction between Heart Ventricles
(Figs. 5-6, 5-11 to 5-13; Data in Table C-12, Appendix C)

```
*       This is file VENTRICL.SAS;
*       Create data set from pulmonary artery constriction only      ;
DATA;
        infile vent1;
        input Aed Ped Dr;
*       set up to create centered variables by making copies         ;
*       of the variables to be centered                              ;
        cPed = Ped;
        cDr = Dr;
*       Use proc STANDARD with no STD specification to               ;
*       subtract means which overwrites cPed and cDr with centered   ;
*       values and retains other variables in unmodified form.       ;
proc STANDARD mean = 0;
        var cPed cDr;
DATA vent1;
*       Get last data set (from proc STANDARD);
        set;
        Ped2 = Ped**2;
        PedDr = Ped*Dr;
        cPed2 = cPed**2;
        cPedDr = cPed*cDr;
LABEL Aed = 'Left Ventricle end-diastolic area'
        Ped = 'Left Ventricle end-diastolic pressure'
        Dr = 'Right Ventricle end-diastolic dimension'
        Ped2 = 'Squared end-diastolic pressure'
        PedDr = 'Pressure by dimension interaction'
        cPed = 'Centered end-diastolic pressure'
        cPed2 = 'Squared centered end-diastolic pressure'
        cDr = 'Centered end-diastolic dimension'
        cPedDr = 'Centered pressure by dimension interaction';
*       Create data set from sequence of pulmonary artery and vena   ;
*       caval constrictions and releases (data in Table C-12B,       ;
*       Appendix C)                                                  ;
DATA temp1;
        infile vent2;
        input Aed Ped Dr;
        cPed = Ped;
        cDr = Dr;
```

```
            proc STANDARD mean = 0;
                  var cPed cDr;
            DATA vent2;
                  set;
                  Ped2 = Ped**2;
                  PedDr = Ped*Dr;
                  cPed2 = cPed**2;
                  cPedDr = cPed*cDr;
            LABEL Aed = 'Left Ventricle end-diastolic area'
                  Ped = 'Left Ventricle end-diastolic pressure'
                  Dr = 'Right Ventricle end-diastolic dimension'
                  Ped2 = 'Squared end-diastolic pressure'
                  PedDr = 'Pressure by dimension interaction'
                  cPed = 'Centered end-diastolic pressure'
                  cPed2 = 'Squared centered end-diastolic pressure'
                  cDr = 'Centered end-diastolic dimension'
                  cPedDr = 'Centered press. by dimension interaction';
      *     Regression of raw (i.e., uncentered) data from first data set:    ;
      *     Vent1 followed by centered data regression, including             ;
      *     collinearity diagnostics with /VIF (for VIF) and COLLINOINT       ;
      *     (for eigenvalues and condition indices from correlation matrix).  ;
            proc REG     data = vent1;
                  model Aed = Ped Ped2 Dr PedDr/ VIF COLLINOINT;
                  model Aed = cPed cPed2 cDr cPedDr/ VIF COLLINOINT;
      *     Regression of raw (i.e., uncentered) data from second data        ;
      *     set: Vent2 followed by centered data regression, including        ;
      *     collinearity diagnostics with /VIF (for VIF) and COLLINOINT       ;
      *     (for eigenvalues and condition indices from correlation matrix).  ;
            proc REG     data = vent2;
                  model Aed = Ped Ped2 Dr PedDr/ VIF COLLINOINT;
                  model Aed = cPed cPed2 cDr cPedDr/ VIF COLLINOINT;
```

Effect of Centering on Multicollinearity
(Figs. 5-7 to 5-10; Data in Table 5-4)

```
      *     This is file CENTER.SPS
      TITLE 'Simple example of effect of centering on regression'
      DATA LIST FILE = center.dat LIST
            / Y X
      *     Note:  The simplest way to center in SPSS is to compute the
      *            mean before the analysis and input it directly in the
      *            compute step.
      COMPUTE Xc = X - 5.5
```

```
COMPUTE X2 = X**2
COMPUTE Xc2 = Xc**2
VAR LABELS Y  = 'Y value'
              X  = 'X variable; raw'
              X2 = 'squared X'
              Xc = 'X variable; centered'
              Xc2 = 'squared centered X'
*        Regression with raw X value, X
REGRESSION
        /variables Y X X2
        /statistics = defaults tol
        /dependent = Y
        /enter X X2
*        Regression with centered X value, Xc
REGRESSION
        /variables Y Xc Xc2
        /statistics = defaults tol
        /dependent = Y
        /enter Xc Xc2
FINISH
```

VARIABLE SELECTION METHODS

There are many programs for variable selection.* We include examples of how to do these with all four computer packages, even though the results in the text were generated with Minitab. Often, the programs only have one main regression or stepwise command, and you select stepwise, forward selection, or backward elimination, using subcommands. This is how you implement forward and backward methods with the STEPWISE command in Minitab. A similar procedure is used to implement forward and backward methods in the stepwise algorithm of BMDP procedure 2R. Forward and backward methods are implemented as options in proc STEPWISE in SAS, and all three methods are options in the REGRESSION procedure in SPSS.

What Determines an Athlete's Time in a Triathlon
(Figs. 6-3 to 6-7; Data in Table C-13, Appendix C)

NOTE   This is file TRIATHLT.MTB
READ 'triathlt.dat' c1-c10

---

*Another variable selection method that we did not discuss in the SAS MAXR method, which is implemented as an option in procedure STEPWISE.

```
name c1 'T'
name c2 'A'
name c3 'W'
name c4 'E'
name c5 'Tr'
name c6 'Tb'
name c7 'Ts'
name c8 'Vr'
name c9 'Vb'
name c10 'Vs'
NOTE    All possible subsets regression, displaying best 4 regressions
BREGRESS 'T' on 'A' − 'Vs';
        best 4.
NOTE    Regression of "best" regression identified above
REGRESS 'T' on 5 independent variables 'A' 'E' 'Tr' 'Tb' 'Vr';
        vif.
NOTE    Regression of full model (i.e., all 9 independent variables)
REGRESS 'T' on 9 independent variables 'A' − 'Vs';
        vif.
NOTE    Stepwise regression
STEPWISE 'T' on c2-c10
NOTE    Forward selection
STEPWISE 'T' on c2-c10;
        fremove = 0.
NOTE    Backward elimination
STEPWISE 'T' on c2-c10;
        fenter = 10000;
        enter c2-c10.
```

The following are control files that show you how you might analyze the above triathlete data using BMDP, SAS, and SPSS.

```
/COMMENT
        This is file TRIATHLT.BMD
        for procedure 2R.
/PROBLEM
        title is 'Variable selection for triathlete data'.
/INPUT
        file is 'triathlt.data'.
        format is free.
        variables are 10.
/VARIABLES
        names are T, A, W, E, 'Tr', 'Tb', 'Ts', 'Vr', 'Vb', 'Vs'.
        use are all.
/COMMENT
        Stepwise regression.
```

```
/REGRESS
      depend = T.
      independ are 2 to 10.
      method = F.
      enter = 4.0,4.0.
      remove = 3.96,3.96.
/END
```

```
*       This is file TRIATHLT.SAS;
TITLE 'Variable selection methods for triathlete data';
DATA;
      INFILE triathlt;
      INPUT T A W E Tr Tb Ts Vr Vb Vs;
      LABEL T = 'time in minutes'
            A = 'age'
            W = 'weight'
            E = 'years experience'
            Tr = 'run training'
            Tb = 'bike training'
            Ts = 'swim training'
            Vr = 'run oxygen consumption'
            Vb = 'bike oxygen consumption'
            Vs = 'swim oxygen consumption';
proc STEPWISE;
      model T = A W E Tr Tb Ts Vr Vb Vs/STEPWISE SLENTRY = .05
            SLSTAY = .05;
      model T = A W E Tr Tb Ts Vr Vb Vs/FORWARD
            SLENTRY = .05;
      model T = A W E Tr Tb Ts Vr Vb Vs/BACKWARD SLSTAY = .05;
      model T = A W E Tr Tb Ts Vr Vb Vs/MAXR;
```

```
*       This is file TRIATHLT.SPS
TITLE 'Variable selections for triathlete data'
DATA LIST FILE = triathlt.dat LIST
      / T A W E Tr Tb Ts Vr Vb Vs
VAR LABELS   T = 'time in minutes'
             A = 'age'
             W = 'weight'
             E = 'years experience'
             Tr = 'run training'
             Tb = 'bike training'
             Ts = 'swim training'
             Vr = 'run oxygen consumption'
```

```
                    Vb = 'bike oxygen consumption'
                    Vs = 'swim oxygen consumption'
REGRESSION
        /variables = T A W E Tr Tb Ts Vr Vb Vs
        /criteria = FIN(4.0) FOUT(3.96) TOL(.0001)
        /dependent = T /stepwise
        /criteria = FIN(4.0)
        /dependent = T /forward
        /criteria = FOUT(3.96)
        /dependent = T /backward
FINISH
*       if Fout is not less than Fin, SPSS sets Fout = .9*Fin
```

ONE-WAY ANALYSIS OF VARIANCE

Because we use linear regression to approach most of our analyses of variance, many of these command files are simply additional examples of regression commands. However, in many of these files, we also include one or more ways of analyzing the same data using traditional analysis of variance commands.

Does Secondhand Tobacco Smoke Nauseate Martians? (Figs. 7-1 and 7-2; Data in Table 7-1)

```
NOTE   This is file MARSSICK.MTB
READ 'marssick.dat' c1 c2
name c1 'N'
name c2 'D'
print 'N' and 'D'
NOTE   Set up for later t test
unstack 'N' c11 c12;
        subscripts in 'D'.
name c11 'nosmoke'
name c12 'smoke'
NOTE   Traditional one-way analysis of variance
ONEWAY 'N' by groups described in 'D'
NOTE   Regression approach (with dummy variable) to do analysis of
NOTE   variance
REGRESS 'N' on 1 independent variable 'D'
NOTE   t test using pooled variance estimate
TWOSAMPLE t test of 'smoke' versus 'nosmoke';
        pooled.
```

Hormones and Depression (Figs. 7-3 and 7-4; Data in Table 7-5)

```
NOTE   This is file DEPRESS.MTB
READ 'depress.dat' c1-c4
name c1 'C'
name c2 'Dn'
name c3 'Dm'
name c4 'Group'
NOTE   Analysis using traditional one-way ANOVA
ONEWAY dependent variable in 'C' by groups in 'Group'
NOTE   Analysis using regression and 0, 1 coding of dummy variables
REGRESS 'C' on 2 independent variables 'Dn' and 'Dm'
NOTE   Generate 0, 1, −1 codes for dummy variables to see result of
NOTE   effects coding
Set data in c2
(1)16 (0)22 (−1)18
end
Set data in c3
(0)16 (1)22 (−1)18
end
NOTE   Repeat regression implementation of ANOVA
REGRESS 'C' on 2 independent variables 'Dn' and 'Dm'
```

Diet, Drugs, and Atherosclerosis
(Figs. 7-6 to 7-9; Data in Table C-14, Appendix C)

```
NOTE   This is file CHOLEST.MTB
READ 'cholest.dat' c1-c6
name c1 'P'
name c2 'G2'
name c3 'G3'
name c4 'G4'
name c5 'G5'
name c6 'Group'
NOTE   The following is the traditional one-way analysis of
NOTE   variance.
ONEWAY 'P' by the groups specified in 'Group'
NOTE   The following is the same analysis of variance using
NOTE   regression with dummy variables.
REGRESS 'P' on 4 independent variables 'G2'-'G5';
        resids c45.
name c45 'RawRes'
NOTE   Residuals are generated above to plot as a graphic check
NOTE   on the assumption of equality of variances.
PLOT 'RawRes' versus 'Group'
```

```
NOTE   This is file CHOLVAR.MTB
NOTE   Minitab macro to do modified Levene test for equality
NOTE   of variances in the groups of the rabbit plaque and
NOTE   cholesterol data.
READ 'cholest.dat' c1-c6
name c1 'P'
name c6 'Group'
NOTE   Break out each group into a separate column
unstack 'P' c11-c15;
       subscripts in 'Group'.
name c11 'Group1 Z'
name c12 'Group2 Z'
name c13 'Group3 Z'
name c14 'Group4 Z'
name c15 'Group5 Z'
NOTE   Center each group by the median.
NOTE   Then take the absolute values.
NOTE   Then set any value equal to zero to missing (i.e., *).
let c11=c11-median(c11)
let c12=c12-median(c12)
let c13=c13-median(c13)
let c14=c14-median(c14)
let c15=c15-median(c15)
absolute value of c11 and put result in c11
absolute c12 c12
absolute c13 c13
absolute c14 c14
absolute c15 c15
code (0) '*' c11-c15 c11-c15
NOTE   Print groups and then do ordinary one-way analysis of
NOTE   variance.
PRINT c11-c15
AOVONEWAY c11-c15
```

The following provide examples of how you might analyze the rabbit cholesterol data using BMDP, SAS, and SPSS.

```
/COMMENT
      This is file CHOLEST.BMD
      for procedure 1V.
/PROBLEM
      title is 'One-way ANOVA of rabbit cholesterol data'.
/INPUT
      file is 'cholest.dat'.
```

```
        format is free.
        variables are 6.
/VARIABLES
        names are P, G2, G3, G4, G5, GROUP.
        use are P, GROUP.
        grouping = GROUP.
/DESIGN
        dependent is P.
/END
```

```
*       This is file CHOLEST.SAS
title 'One-way ANOVA of rabbit cholesterol data';
DATA;
        INFILE cholest;
        INPUT P G2 G3 G4 G5 Group;
        LABEL P = 'percent aortic plaque'
                G2 = 'dummy variable = 1 if group 2'
                G3 = 'dummy variable = 1 if group 3'
                G4 = 'dummy variable = 1 if group 4'
                G5 = 'dummy variable = 1 if group 5'
                Group = 'Group number';
*       Traditional analysis of variance;
proc ANOVA;
        class Group;
        model P = Group;
        means group/bon snk;
```

```
*       This is file CHOLEST.SPS
TITLE 'One-way ANOVA of rabbit cholesterol data'
DATA LIST FILE = cholest.dat LIST
        / P G2 G3 G4 G5 Group
VAR LABELS P = 'percent aortic plaque'
                G2 = 'dummy variable = 1 if group 2'
                G3 = 'dummy variable = 1 if group 3'
                G4 = 'dummy variable = 1 if group 4'
                G5 = 'dummy variable = 1 if group 5'
                Group = 'Group number'
*       Analysis with the ONEWAY command
ONEWAY P by Group (1,5)
        /ranges = SNK
*       Analysis with the ANOVA command
ANOVA P by Group (1,5)
FINISH
```

TWO-WAY ANALYSIS OF VARIANCE

Personality Assessment and Faking High Gender Identification
(Figs. 8-1 to 8-3; Data in Tables 8-2, 8-5, and 8-7)

```
*       This is file GENDER.SAS;
TITLE 'Gender role playing effect on personality assessment';
*       Create data set;
DATA;
        infile gender;
        input S G R G1 R1;
        GR = G*R;
        GR1 = G1*R1;
        LABEL S = 'Personality Score'
                G = 'dummy variable  = 1 if female'
                R = 'dummy variable  = 1 if role play'
                G1 = 'dummy variable, 1 if male, −1 if female'
                R1 = 'dummy variable, 1 if honest, −1 if role'
                GR1 = 'interaction of G1 and R1'
                GR = 'interaction of G and R';
*       Analysis of variance using regression and dummy variables    ;
*       REFERENCE CELL CODES                                         ;
proc REG;
        model S = G R GR/SS1 VIF;
*       Traditional analysis of variance;
proc ANOVA;
        class G R;
        model S = G|R;
*       Analysis of variance using regression and dummy variables    ;
*       EFFECTS CODES                                                ;
proc REG;
        model S = G1 R1 GR1/SS1 VIF;
```

The Kidney, Sodium, and High Blood Pressure
(Figs. 8-4 to 8-8 and 8-10; Data in Table 8-12)

```
*       This is file KIDNEY.SAS;
title 'Two-way ANOVA--kidney ATPase data';
*       Create data set with no missing data;
DATA kidney;
        infile kidney;
        input A H N1 N2 Ngroup;
        I1 = H*N1;
```

```
        I2 = H*N2;
        LABEL A = 'Sodium-potassium ATPase'
              H = 'dummy variable  =  −1 if hypertensive'
              N1 = 'dummy variable  =  1 if DCT group'
              N2 = 'dummy variable  =  1 if CCD group'
              I1 = 'interaction of H and N1'
              I2 = 'interaction of H and N2'
              Ngroup = 'group labels for site factor';
*       Create data set with missing data (Data in Table 8-15);
DATA kidney1;
        infile kidnymis;
        input A H N1 N2 Ngroup;
        I1 = H*N1;
        I2 = H*N2;
        LABEL A = 'Sodium-potassium ATPase'
              H = 'dummy variable  =  −1 if hypertensive'
              N1 = 'dummy variable  =  1 if DCT group'
              N2 = 'dummy variable  =  1 if CCD group'
              I1 = 'interaction of H and N1'
              I2 = 'interaction of H and N2'
              Ngroup = 'group labels for site factor';
*       Create data set with missing cell (Data in Table 8-18);
DATA kidney2;
        infile kidmiscl;
        input A H N1 N2 Ngroup;
        I1 = H*N1;
        I2 = H*N2;
        LABEL A = 'Sodium-potassium ATPase'
              H = 'dummy variable  =  −1 if hypertensive'
              N1 = 'dummy variable  =  1 if DCT group'
              N2 = 'dummy variable  =  1 if CCD group'
              I1 = 'interaction of H and N1'
              I2 = 'interaction of H and N2'
              Ngroup = 'group labels for site factor';
*       Traditional analysis of variance;
proc ANOVA data = kidney;
        class H Ngroup;
        model A = H Ngroup H*Ngroup;
        means H Ngroup H*Ngroup/BON DUNCAN SNK;
*       Analysis of variance using regression and dummy variables;
proc REG  data = kidney;
        model A = H N1 N2 I1 I2/ SS1;
*       MISSING DATA: All cells have data                           ;
*       Analysis of variance using GLM procedure                   ;
```

```
        proc GLM data = kidney1;
              class H Ngroup;
              model A = H Ngroup H*Ngroup/solution;
              lsmeans H Ngroup H*Ngroup/stderr pdiff;
        *     MISSING DATA: All cells have data                                    ;
        *     Dummy variable implementation with regression, once with H ;
        *     last, once with N1 and N2 last, and once with interaction last  ;
        proc REG data = kidney1;
              model A = N1 N2 I1 I2 H/ SS1;
              model A = H I1 I2 N1 N2/ SS1;
              model A = H N1 N2 I1 I2/ SS1;
        *     MISSING DATA: one cell empty                                         ;
        *     Analysis of variance using GLM procedure                            ;
        proc GLM data = kidney2;
              class H Ngroup;
              model A = H Ngroup H*Ngroup/solution;
              lsmeans H Ngroup H*Ngroup/stderr pdiff;
        proc GLM data = kidney2;
              class H Ngroup;
              model A = H Ngroup /solution;
              lsmeans H Ngroup /stderr pdiff;
        *     MISSING DATA: one cell empty                                         ;
        *     Dummy variable implementation with regression, once with H ;
        *     last and once with N1 and N2 last                                    ;
        proc REG    data = kidney2;
              model A = N1 N2 H/ SS1;
              model A = H N1 N2/ SS1;
```

More on Missing Data

The SAS control file listed above contains all of the commands we used to analyze the missing data cases for the kidney data using SAS procedure GLM. We want the marginal sums of squares for each effect, and these are given directly by GLM if you ask for the TYPE III sums of squares. The other packages we used also give you the option of obtaining the marginal sums of squares. Some of the ways you can obtain these are*

SAS	GLM	Type III SS. OK for missing data with all cells filled and for empty cells if they are connected

---

*Other sources of information about what these various computer programs produce in the missing data case can be found in G. A. Milliken and D. E. Johnson, *Analysis of Messy Data. Volume 1: Designed Experiments*, New York, Van Nostrand Reinhold, 1984, pp. 156–158, 189–190; and S. R. Searle, *Linear Models for Unbalanced Data*, New York, Wiley, 1987, Chap. 12, "Comments on Computing Packages."

and no interactions. (SAS proc ANOVA will analyze missing data with filled cells, but it is not clear what the SS mean. It is recommended that you use GLM for missing data.)

BMDP	2V	Default SS. Will not handle empty cells.
Minitab	GLM*	Adjusted sum of squares ("Adj SS"). This sum of squares is computed as the default and used to calculate the reported F values.
SPSS	ANOVA	METHOD = UNIQUE. This method is also referred to as the regression approach in the SPSS documentation. Selecting this option limits some of the statistics you can request. MANOVA is a better general-purpose procedure when there are missing data.
	MANOVA	SSTYPE(UNIQUE) the default in release 3.1. Also OK for empty cells if connected and no interaction. If you include interaction in the empty cell situation, you can generate numbers, but MANOVA issues many warnings that you need to be careful in interpreting those numbers.

Control files for analyzing the kidney data with missing values for BMDP 2V and SPSS MANOVA are

/COMMENT
      This is file KIDNEY.BMD
      (for procedure 2V and KIDNYMIS.DAT.)
      2V handles unbalanced fixed effects models. The default sums of squares are what we have called the marginal sums of squares and what SAS GLM refers to as 'TYPE III' sums of squares. These are the appropriate sums of squares when there are missing data. This same program could analyze the full kidney data set as well. 2V cannot handle the empty cell case.
/PROBLEM
      title is 'Missing data in kidney example'.
/INPUT
      file is 'kidnymis.dat'.
      format is free.
      variables are 5.

*Minitab procedure GLM is available in Release 7.1.

```
/VARIABLES
      names are A, HYPER, N1, N2, NEPHRON.
      use are A, HYPER, NEPHRON.
/DESIGN
      depend = A.
      grouping = HYPER, NEPHRON.
/END


*      This is file KIDNEY.SPS
TITLE 'Two-way ANOVA with missing data'
DATA LIST FILE = kidnymis.dat LIST
      / A Hyper N1 N2 Nephron
VAR LABELS A = 'Na-K-ATPase activity'
            Hyper = 'normal or hypertensive rats'
            Nephron = 'nephron site'
*      MANOVA requires consecutive integer values for group codes,
*      so we need to recode the Hyper factor.
RECODE Hyper( - 1 = 0)
MANOVA A by Hyper (0,1) Nephron (1,3)
      /Design = Hyper Nephron Hyper by Nephron
*      The default sums of squares are SSTYPE(UNIQUE), which are
*      the marginal sums of squares need for missing data (what SAS
*      GLM calls the 'TYPE III' sums of squares).
*
*      Now, do a missing cell problem (no interaction).
DATA LIST FILE = kidmiscl.dat LIST
      / A Hyper N1 N2 Nephron
VAR LABELS A = 'Na-K-ATPase activity'
            Hyper = 'normal or hypertensive rats'
            Nephron = 'nephron site'
RECODE Hyper    ( - 1 = 0)
MANOVA A by Hyper (0,1) Nephron (1,3)
      /Error = W + R
      /Design = Hyper Nephron
*      The ERROR = W + R statement pools the residual and within-
*      cell errors to obtain the error term necessary to conduct tests
*      based on the marginal sums of squares. This command only
*      needs to be entered if there is one or more empty cell and
*      interaction left out of the model. If there are no empty
*      cells and all interactions are in the model, including the
*      ERROR = W + R command will not affect the residual sum
*      of squares.
FINISH
```

REPEATED MEASURES IN REGRESSION

Hypothetical Example Illustrating Repeated Measures in Regression (Figs. 9-2 to 9-4; Data in Table 9-1)

```
NOTE   This is file VOLFAKE.MTB
READ 'volfake.dat' c1-c5
name c1 'dog'
name c2 'x'
name c3 'y'
name c4 'D1'
name c5 'D2'
PLOT 'y' versus 'x'
LPLOT 'y' versus 'x' using labels in 'dog'
REGRESS c3 on 1 variable c2
REGRESS c3 on 3 variables c2 c4 c5
```

Measuring Heart Size with a Catheter (Figs. 9-1 and 9-5; Data in Table 9-2)

```
NOTE   This is file VOLUME.MTB
READ 'volume.dat' c1-c7
name c1 'Vp'
name c2 'Vt'
name c3 'D1'
name c4 'D2'
name c5 'D3'
name c6 'D4'
name c7 'Dog'
PLOT 'Vp' versus 'Vt'
LPLOT 'Vp' versus 'Vt' using labels in 'Dog'
NOTE   First do regression without dummy variables for subjects
REGRESS 'Vp' on 1 independent variable 'Vt'
NOTE   Now do regression with dummy variables for subjects
REGRESS 'Vp' on 5 independent variables 'D1'-'D4' 'Vt'
```

ONE-WAY REPEATED-MEASURES ANALYSIS OF VARIANCE

Hormones and Food (Figs. 9-7 and 9-8; Data in Table 9-4)

```
*      This is file EATING.SAS;
title 'One-way repeated measures -- eating and metabolism data';
DATA;
```

```
            infile eating;
            input R A1 A2 A3 A4 A5 Ee Es Group Subject;
            LABEL R = 'metabolic rate after eating'
                    Ee = 'dummy var. = 1 if bypass stomach'
                    Es = 'dummy var. = 1 if direct to stomach'
                    A1 = 'dummy variable = 1 if subject 1'
                    A2 = 'dummy variable = 1 if subject 2'
                    A3 = 'dummy variable = 1 if subject 3'
                    A4 = 'dummy variable = 1 if subject 4'
                    A5 = 'dummy variable = 1 if subject 5'
                    Group = 'Treatment group'
                    Subject = 'Subject number';
*           Do analysis of variance using regression with dummy variables ;
proc    REG;
            model R = A1-A5 Ee Es/ SS1;
*           Traditional ANOVA (with repeated                              ;
*           measures) using GLM with random subcommand to specify        ;
*           model: This will be the way we handle missing data.          ;
proc    GLM;
            class Group Subject;
            model R = Group Subject/solution;
            random Subject;
            lsmeans Group/stderr pdiff;
*           Use SORT and TRANSPOSE to convert data set                   ;
*           into form where each row contains the repeated               ;
*           observations from each subject, which is the                 ;
*           easiest way to get SAS to do traditional ANOVA.              ;
DATA;
            set;
            keep R Subject;
proc    SORT;
            by Subject;
proc    TRANSPOSE;
            Var R;
            by Subject;
DATA;
            set;
            keep COL1-COL3;
            rename COL1 = Enorm COL2 = Ee COL3 = Es;
*           Do traditional analysis of variance on transposed            ;
*           data set. Note that the model statement contains             ;
*           no "class" variables, because there is only one              ;
*           factor that, because it has repeated measures, is            ;
*           represented only by the 3 "dependent" variables,             ;
```

```
*       Enorm, Ee, Es, which identify the treatment groups.          ;
proc   ANOVA;
        model Enorm Ee Es = ;
        repeated TREATMT/nom;
*       Get data file with missing data for GLM ANOVA                 ;
DATA;
        infile eatmis1;
        input R A1 A2 A3 A4 A5 Ee Es Group Subject;
        LABEL R = 'metabolic rate after eating'
              Group = 'Treatment group'
              Subject = 'Subject number';
*       Traditional ANOVA (with repeated measures) using             ;
*       GLM with random subcommand to specify model                  ;
proc   GLM;
        class Group Subject;
        model R = Group Subject/solution;
        random Subject;
        lsmeans Group/stderr pdiff;

/COMMENT
        This is file EATING.BMD
        for use with procedure 2V.
        For simplicity, data are entered directly.
/PROBLEM
        title is 'Metabolic rate and eating data'.
/INPUT
        variables are 3.
        format is 'free'.
/VARIABLE
        names are 'Enorm', 'Ee', 'Es'.
/DESIGN
        dependent are 'Enorm', 'Ee', 'Es'.
        level is 3.
/END
   104  91 22
   106  94 14
   111 105 14
   114 106 15
   117 120 18
   139 111  8

NOTE   This is file EATING.MTB
NOTE   Data are entered directly for simplicity.
set c1
```

```
                104 106 111 114 117 139 91 94 105 106 120 111 22 14 14 15 18 8
                end
                set c2
                (1:3)6
                end
                set c3
                3(1:6)
                end
                name c1 'R'
                name c2 'EatGroup'
                name c3 'Subject'
                NOTE   We do a two-way anova with the eating methods as one factor
                NOTE   and the subjects as another factor, which we specify as
                NOTE   RANDOM, with the subcommand
                ANOVA 'R' = 'EatGroup' 'Subject';
                        random 'Subject'.


                *      This is file EATING.SPS
                *      For simplicity, data are entered directly
                TITLE    'Metabolic rate and feeding method data'
                DATA LIST Free
                        /Enorm Ee Es
                BEGIN DATA
                    104   91 22
                    106   94 14
                    111 105 14
                    114 106 15
                    117 120 18
                    139 111   8
                END DATA
                RELIABILITY
                        variables = Enorm, Ee, Es
                        /statistics = anova
```

TWO-WAY REPEATED-MEASURES ANALYSIS OF VARIANCE WITH REPEATED MEASURES ON ONE FACTOR

Is Alcoholism Associated with a History of Childhood Aggression? (Figs. 9-10 to 9-15; Data in Table 9-8)

```
                *      This is file DRINK.SAS;
                title 'drinking and agression data -- two factor with rpt meas
                        on one factor';
```

```
*       As set up, this file will analyze antisocial personality        ;
*       and drinking data for the AGGRESSION personality                ;
*       score (Table 9-8). To use with other scores (anger, depression, ;
*       well-being) edit infile appropriately and delete the            ;
*       last section for reading in and analyzing missing data.         ;
*       Additional data files are drinkang.dat (Table C-15,             ;
*       Appendix C) and drinkdep.dat (Table C-16, Appendix C).          ;
*       Create data set for AGGRESSION Score: no missing data           ;
DATA drinkagg;
       infile drinkagg;
       input S D A PASP1 PASP2 PASP3 PASP4 PASP5 PASP6 PASP7
           PnonASP1 PnonASP2 PnonASP3 PnonASP4 PnonASP5
           PnonASP6 PnonASP7 Group Subject;
       DA = D*A;
       LABEL S = 'Aggression score'
               D = 'Dummy variable = 1 if drinking'
               A = 'Dummy variable = 1 if ASP alcoholic'
               DA = 'Interaction of D and A'
               PASP1 = 'subject dummy for first ASP subject'
               PASP7 = 'subject dummy for seventh ASP subject'
               PnonASP1 = 'dummy for first non-ASP subject'
               PnonASP7 = 'dummy for seventh non-ASP subject';
*       Analysis of variance computed with regression and               ;
*       dummy variables                                                 ;
proc REG data = drinkagg;
       model S = PASP1-PASP7 PnonASP1-PnonASP7 A D DA/SS1;
*       Do analysis of variance with GLM procedure and the              ;
*       RANDOM subcommand to specify a three-way problem                ;
*       with the subjects factor (nested in ASP group, A) declared as   ;
*       random.                                                         ;
proc GLM;
       class A D Subject;
       model S = A|D Subject(A)/Solution;
       random Subject(A);
*       Modify data set to get it into form needed for                  ;
*       repeated measures using ANOVA or GLM and the                    ;
*       REPEATED subcommand.                                            ;
DATA;
       set;
       keep S A D Subject;
*       First sort the data set by the child aggression                 ;
*       factor A and Subjects: then transpose so that                   ;
*       each ROW of the data set contains the repeated                  ;
*       measurement over the drinking factor D.                         ;
```

```
proc SORT;
        by A Subject;
proc TRANSPOSE;
        Var S;
        by A Subject;
DATA anov;
        set;
        keep A COL1 COL2;
        rename COL1 = DO COL2 = D1;
*       Traditional ANOVA on transposed data set                        ;
proc   ANOVA data = anov;
        class A;
        model DO-D1 = A;
        repeated D/NOM;
*       Create data set for AGGRESSION Score: with MISSING data         ;
*       Delete this section when analyzing other scores                 ;
DATA drinkagm;
        infile drinkagm;
        input S D A PASP1 PASP2 PASP3 PASP4 PASP5 PASP6
            PASP7 PnonASP1 PnonASP2 PnonASP3 PnonASP4
            PnonASP5 PnonASP6 PnonASP7 Group Subject;
        DA = D*A;
        LABEL S = 'Aggression score'
                D = 'Dummy variable = 1 if drinking'
                A = 'Dummy variable = 1 if ASP alcoholic'
                DA = 'Interaction of D and A'
                PASP1 = 'subject dummy for first ASP subject'
                PASP7 = 'subject dummy for seventh ASP subject'
                PnonASP1 = 'dummy for first non-ASP subject'
                PnonASP7 = 'dummy for seventh non-ASP subject';
*       Analysis of variance on missing data                           ;
*       Ask for expected mean squares                                  ;
proc   GLM;
        class A D Subject;
        model S = A|D Subject(A)/Solution;
        random Subject(A);
        lsmeans A D A*D/stderr pdiff;
*       Analysis of variance using dummy variables; repeat with        ;
*       different order of effects to compute SS. In this case,        ;
*       we can get sums of squares, but cannot do complete            ;
*       analysis without expected mean squares from GLM above.        ;
proc   REG data = drinkagm;
        model S = PASP1-PASP7 PnonASP1-PnonASP7 A D DA/SS1;
        model S = PASP1-PASP7 PnonASP1-PnonASP7 D DA A/SS1;
        model S = PASP1-PASP7 PnonASP1-PnonASP7 A DA D/SS1;
```

```
/COMMENT
        This is file DRINKAGG.BMD
        for procedure 2V.
        This is how you would analyze the two-way repeated measures
        ANOVA, with repeated measures on only one of the factors
        illustrated with the personality (in this case AGGRESSION
        SCORE).
        For simplicity, data are entered directly.
/PROBLEM
        title is 'Agression score, drinking, and antisocial personality'.
/INPUT
        variables are 3.
        format is 'free'.
/VARIABLE
        names are SOBER, DRINK, 'Aggress'.
/DESIGN
        dependent are SOBER, DRINK.
        level is 2.
        grouping is 'Aggress'.
        name is 'PersScor'.
/END
  .81   .59 0
  .91  1.04 0
  .98  1.11 0
 1.08  1.13 0
 1.10  1.15 0
 1.16  1.16 0
 1.19  1.25 0
 1.44  1.70 0
  .72   .83 1
  .82   .99 1
  .89  1.17 1
 1.01  1.24 1
 1.10  1.33 1
 1.14  1.47 1
 1.24  1.59 1
 1.34  1.73 1


NOTE    This is file DRINKAGG.MTB
NOTE    This is how to do a two-way analysis of variance with
NOTE    repeated measures on only one of the factors using the
NOTE    ANOVA command with the RANDOM subcommand.
NOTE
NOTE    Remember we have factor A (no repeated measures), with
```

NOTE N subjects in each level of factor A. Then we have
NOTE   factor B (repeated measures). Both factors A and B are fixed.
NOTE
NOTE   As with all Minitab repeated-measures analyses of variance
NOTE   problems, we explicitly include the SUBJECTS as an additional,
NOTE   RANDOM, factor. In this case, it is slightly more complicated
NOTE   because the subjects are said to be nested within the
NOTE   nonrepeated factor A.
NOTE   The usual notation for these subjects within factor A effect
NOTE   is Subjects(A).
NOTE
NOTE   To illustrate this, we will use the antisocial personality and
NOTE   drinking example with the aggression score as the dependent
NOTE   variable. For simplicity, we will enter the data directly.
NOTE
set c1
0.81 0.91 0.98 1.08 1.10 1.16 1.19 1.44 0.72 0.82 0.89 1.01 1.10
1.14 1.24 1.34 0.59 1.04 1.11 1.13 1.15 1.16 1.25 1.70 0.83 0.99
1.17 1.24 1.33 1.47 1.59 1.73
end
set c2
(0,1,0,1)8
end
set c3
4(1:8)
end
set c4
(0,1)16
end
NOTE   There are 16 different subjects, 8 of them in one group (antisocial
NOTE   personality) and 8 in the other group (non-antisocial
NOTE   personality).
NOTE   Instead of numbering them from 1 to 16, we number them
NOTE   separately within each level of the antisocial personality factor
NOTE   (just as we coded the subject dummy variables to account for
NOTE   this nesting of subjects within this factor).
name c1 'AggrScor'
name c2 'ASPGroup'
name c3 'Subject'
name c4 'Drink'
NOTE   So, the factors are ASPgroup, which has two levels -- one group
NOTE   which has a diagnosis of ASP personality, and one group which
NOTE   does not have that diagnosis -- and Drink, which also has two
NOTE   levels, sober or drinking. The Drink factor has repeated

```
NOTE    measures; i.e., each subject is assessed while sober, then
NOTE    while drinking.
NOTE    The design of the analysis is then ASPGroup effect, Drink effect,
NOTE    Interaction of Drink and ASPGroup, Subjects(ASPGroup). The
NOTE    simplest way to specify this is a fully crossed model (i.e., all
NOTE    interactions are specified) using the general notation
NOTE    Y = A|Subj(A)|B.
NOTE    However, a fully crossed model has no degrees of freedom
NOTE    remaining, so we need to subtract out the highest
NOTE    interaction using a minus sign ( − ).
NOTE                Y = A|Subj(A)|B  −  B*Subj(A)
NOTE    Alternatively we could specify the model as
NOTE                Y = A B A*B Subj(A)
NOTE    In both cases, you have to specify that Subj(A) is RANDOM --
NOTE    it will not work if you specify only Subj as RANDOM.
NOTE    This generalizes to more complicated designs. For an example,
NOTE    see the Minitab Reference Manual.
NOTE    The ANOVA table in the output will contain an F test for the
NOTE    Subj(A) effect, but it is of no interest to us in this example.
ANOVA AggrScor = ASPGroup|Subject(ASPGroup)|Drink −          &
      Drink*Subject(ASPGroup);
      random Subject(ASPGroup).
```

TWO-WAY REPEATED-MEASURES ANALYSIS OF VARIANCE WITH REPEATED MEASURES ON BOTH FACTORS

Candy, Chewing Gum, and Tooth Decay (Figs 9-18 to 9-21; Data in Table 9-15)

```
*       This is file GUM.SAS;
title   'Two-way repeat measures ANOVA -- Gum data';
*       read in data set -- NO MISSING DATA (Table 9-15)
DATA;
        infile gum;
        input pH G T20 T30 S1 S2 S3 S4 S5 S6 S7 T Subj;
        GT20 = G*T20;
        GT30 = G*T30;
        GS1 = S1*G;
        GS2 = S2*G;
        GS3 = S3*G;
        GS4 = S4*G;
        GS5 = S5*G;
        GS6 = S6*G;
        GS7 = S7*G;
```

```
           T20S1 = S1*T20;
           T20S2 = S2*T20;
           T20S3 = S3*T20;
           T20S4 = S4*T20;
           T20S5 = S5*T20;
           T20S6 = S6*T20;
           T20S7 = S7*T20;
           T30S1 = S1*T30;
           T30S2 = S2*T30;
           T30S3 = S3*T30;
           T30S4 = S4*T30;
           T30S5 = S5*T30;
           T30S6 = S6*T30;
           T30S7 = S7*T30;
           LABEL pH = 'plaque pH'
                 G = 'dummy variable = -1 if bicarb gum'
                 T20 = 'dummy variable = 1 if 20 min'
                 T30 = 'dummy variable = 1 if 30 min'
                 GT20 = 'interaction of gum and T20'
                 GT30 = 'interaction of gum and T30'
                 S1 = 'dummy variable = 1 if subject 1'
                 S2 = 'dummy variable = 1 if subject 2'
                 S3 = 'dummy variable = 1 if subject 3'
                 S4 = 'dummy variable = 1 if subject 4'
                 S5 = 'dummy variable = 1 if subject 5'
                 S6 = 'dummy variable = 1 if subject 6'
                 S7 = 'dummy variable = 1 if subject 7'
                 GS1 = 'interaction of gum and subject1'
                 T20S1 = 'interaction of T20 and Subject1'
                 T30S7 = 'interaction of T30 and Subject7'
                 T = 'level of time factor'
                 Subj = 'Subject number';
*      Do analysis of variance using regression with dummy variables ;
proc REG;
      model pH = S1-S7 GS1-GS7 G T20 T30 GT20 GT30 T20S1-T20S7
      T30S1-T30S7 /ss1;
*      Do analysis of variance using proc GLM with RANDOM            ;
*      subcommand to specify the Subjects factor as random. Note     ;
*      that if there is an interaction involving a random command,   ;
*      the interaction is also a random effect and must be included in ;
*      the random list -- SAS does not take care of this automatically. ;
proc GLM;
      class G T Subj;
      model pH = G|T Subj*G Subj*T Subj/solution;
```

```
        random Subj Subj*G Subj*T;
        lsmeans G T G*T/stderr pdiff;
*       Use SORT and TRANSPOSE to convert data set                     ;
*       into form where each row contains the repeated                 ;
*       observations from each subject, which is the                   ;
*       easiest way to get SAS to do traditional ANOVA.                ;
DATA;
        set;
        keep pH Subj;
proc SORT;
        by Subj;
proc TRANSPOSE;
        Var pH;
        by Subj;
DATA;
        set;
        keep COL1-COL6;
        rename COL1 = NoGumT0 COL2 = NoGumT20
COL3 = NoGumT30 COL4 = GumT0 COL5 = GumT20
COL6 = GUMT30;
*       Do traditional analysis of variance on transposed             ;
*       data set. Notice that the model statement contains            ;
*       no "class" variables, because there is only one               ;
*       factor that, because it has repeated measures, is             ;
*       represented only by the 6 "dependent" variables,              ;
*       NoGumTn to GumTn, which identify the treatment groups.        ;
proc ANOVA;
        model NoGumT0 NoGumT20 NoGumT30 GumT0 GumT20
            GumT30 = ;
        repeated GUM2, TIME 3/nom;
*       REPEAT ANALYSIS for MISSING DATA (Table 9-17)                 ;
*       Read in data set -- MISSING DATA                             ;
DATA;
        infile gummis;
        input pH G T20 T30 S1 S2 S3 S4 S5 S6 S7 T Subj;
        LABEL pH = 'plaque pH'
                G = 'dummy variable = - 1 if bicarb gum'
                T = 'level of time factor'
                Subj = 'Subject number';
*       Do analysis of variance using proc GLM with RANDOM           ;
*       subcommand to specify the Subjects factor as random. Note    ;
*       that if there is an interaction involving a random command, the ;
*       interaction is also a random effect and must be included in the ;
*       random list — SAS does not take care of this automatically.  ;
```

```
        proc GLM;
              class G T Subj;
              model pH = G|T Subj*G Subj*T Subj/solution;
              random Subj*G Subj*T Subj;
              lsmeans G T G*T/stderr pdiff;



        /COMMENT
              This is file GUM.BMD
              for procedure 2V.
              For simplicity, data are entered directly.
        /PROBLEM
              title is 'Plaque pH and Gum Chewing data'.
        /INPUT
              variables are 6.
              format is 'free'.
        /VARIABLE
              names are NOGUMT0, NOGUMT20, NOGUMT30, GUMT0,
              GUMT20, GUMT30.
        /DESIGN
              dependents are NOGUMT0, NOGUMT20, NOGUMT30,
              GUMT0, GUMT20, GUMT30.
              levels are 2, 3.
              names are GUM, TIME.
        /END
        4.6 4.4 4.1 3.8 5.3 5.1
        4.2 4.7 4.7 4.1 5.8 5.1
        3.9 5.0 5.7 4.2 5.8 5.6
        4.6 5.2 5.6 4.2 6.9 5.6
        4.6 5.1 5.7 4.3 6.2 5.8
        4.7 5.3 5.3 4.4 6.3 5.9
        4.9 5.6 5.9 4.9 6.0 5.9
        5.0 6.0 5.9 4.5 7.2 6.2


        NOTE    This is file GUM.MTB.
        NOTE    This will illustrate a two-way analysis of variance, with
        NOTE     repeated measures on both factors, using Minitab and the
        NOTE    ANOVA command. As with all implementations of repeated
        NOTE    measures using Minitab, we specify an additional SUBJECTS
        NOTE    factor, which we specify as RANDOM with a subcommand.
        NOTE
        NOTE    We will use the plaque pH and gum data, which will be
        NOTE    entered directly for simplicity.
```

```
set c1
4.6  4.2  3.9  4.6  4.6  4.7  4.9  5.0  4.4  4.7  5.0  5.2  5.1  5.3  5.6
6.0  4.1  4.7  5.7  5.6  5.7  5.3  5.9  5.9  3.8  4.1  4.2  4.2  4.3  4.4
4.9  4.5  5.3  5.8  5.8  6.9  6.2  6.3  6.0  7.2  5.1  5.1  5.6  5.6  5.8
5.9  5.9  6.2
end
set c2
(0,1)24
end
set c3
(0,30,50,0,30,50)8
end
set c4
6(1:8)
end
name c1 'pH'
name c2 'Gum'
name c3 'Time'
name c4 'Subject'
NOTE   We specify the model by entering the desired factors and then
NOTE   use the RANDOM subcommand to specify the Subjects as a
NOTE   random effect.
ANOVA pH = GUM Time Subject Gum*Time Gum*Subject        &
       Time*Subject;
       random Subject.
```

This is one way to get SPSS MANOVA to do the repeated-measures gum problem.

```
*      This is file GUM.SPS
TITLE 'Two-way repeated-measures ANOVA of gum and mouth pH
       data'
DATA LIST FILE = gum.dat LIST
       / pH G T20 T30 S1 to S7 Time Subject
VAR LABELS pH = 'mouth pH'
       G  = 'dummy variable = −1 if chewing standard gum'
*      You have to specify the values of the levels of the grouping
*      factors and these must be consecutive integers. Because we
*      have G coded as (1, −1), we need to recode one of them to meet
*      this requirement.
RECODE G (−1 = 0)
MANOVA pH by G (0,1) Time (1,3) Subject (1,8)
       /design = G vs 1
                 G by Subject = 1
                 Time vs 2
```

<blockquote>
Time by Subject = 2
G by Time
Subject
</blockquote>

*   You need to tell MANOVA what to use for the denominators in
*   the *F* statistics, unless you want to calculate them by hand from
*   the output.
FINISH

How Similar Are the Mechanical Properties of the Whole Heart to Heart Muscle? (Figs. 9-22 to 9-26; Data in Table C-17, Appendix C)

```
NOTE    This if file TETANUS.MTB
NOTE    Read in data
READ    'tetanus.dat' c1-c4
name c1 'Timecon'
name c2 'Volume'
name c3 'Speed'
name c4 'subject'
NOTE    Do repeated-measures analysis of variance using general
NOTE    analysis of variance command, ANOVA, specifying subject by
NOTE    main effects interactions, and the subjects factor as
NOTE    RANDOM with a subcommand.
ANOVA c1 = c2 c3 c4 c2*c3 c2*c4 c3*c4;
        random c4.
NOTE    Break dependent variable into the four cells
unstack (c1,c3) (c11,c12) (c13,c14);
        subscripts in c2.
unstack (c11,c13) (c11,c13) (c12,c14);
        subscripts in c12.
name c11 'withfast'
name c12 'withslow'
name c13 'infufast'
name c14 'infuslow'
PRINT c11-c14
describe c11-c14
NOTE    Do Levene median procedure to check for equality of
NOTE    variances in the four cells.
NOTE
NOTE    Center to the median, take the absolute values, and,
NOTE    because each group has an odd sample size and n < 20,
NOTE    omit centered values of 0 (i.e., values corresponding
NOTE    to median of raw data in each cell).
let c21 = c11-median(c11)
let c22 = c12-median(c12)
```

```
let c23 = c13-median(c13)
let c24 = c14-median(c14)
name c21 'Zwithfas'
name c22 'Zwithslo'
name c23 'Zinfufas'
name c24 'Zinfuslo'
absolute c21 c21
absolute c22 c22
absolute c23 c23
absolute c24 c24
code (0) '*' c21-c24 c21-c24
PRINT c21-c24
AOVONEWAY c21-c24
NOTE   Try variance stabilization using log transform on 'Timecon'
NOTE   and repeating ANOVA.
let c5 = loge(c1)
name c5 'lnTcon'
ANOVA c5 = c2 c3 c4 c2*c3 c2*c4 c3*c4;
      random c4.
NOTE   Verify the effectiveness of the variance stabilization by
NOTE   separating data into the four cells and repeating the
NOTE   Levene test on the log transformed data.
unstack (c5,c3) (c31,c32) (c33,c34);
      subscripts in c2.
unstack (c31,c33) (c31,c33) (c32,c34);
      subscripts in c32.
name c31 'lnwitfas'
name c32 'lnwitslo'
name c33 'lninffas'
name c34 'lninfslo'
PRINT c31-c34
describe c31-c34
let c41 = c31-median(c31)
let c42 = c32-median(c32)
let c43 = c33-median(c33)
let c44 = c34-median(c34)
name c41 'Zlnwifas'
name c42 'Zlnwislo'
name c43 'Zlninfas'
name c44 'Zlninslo'
absolute c41 c41
absolute c42 c42
absolute c43 c43
absolute c44 c44
```

```
code (0) '*' c41-c44 c41-c44
PRINT c41-c44
AOVONEWAY c41-c44
```

NONLINEAR REGRESSION

As noted in Chap. 10, nonlinear regression programs usually do not report R^2. You can compute R^2, based on the reported sums of squares, as

$$R^2 = \frac{SS_{reg}}{SS_{tot}} = 1 - \frac{SS_{res}}{SS_{tot}}$$

If you use SAS PROC NLIN, you must compute

$$R^2 = 1 - \frac{SS_{res}}{\text{"corrected total" } SS}$$

The "uncorrected total" SS is simply Σy^2 and is not very useful. SS_{reg} is incorrect as reported because it is based on "uncorrected total" SS.

The SPSS NLR procedure reports an analysis of variance table identical to SAS PROC NLIN, but reports a correctly computed R^2. BMDP 3R only reports SS_{res}, and you must compute SS_{tot} yourself in order to compute R^2.

Martian Moods (Figs. 10-2 to 10-7; Data in Table C-18, Appendix C)

```
This is file MARSINVD.SAS;
options noovp;
title 'Martian Invasion Data';
DATA raw;
        infile marsinvd;
        input time I;
*       Grid Search;
proc NLIN;
        parameters Imax = 10 to 30 by 2 T = 10 to 30 by 2;
        model I = Imax*(cos(2*3.14159*time/T) + 1)/2;
        DER.Imax = (cos(2*3.14159*time/T) + 1)/2;
        DER.T = Imax*sin(2*3.14159*time/T)*(3.14159*time/T**2);
        output OUT = PlotGrid   PREDICTED = Ihat   RESIDUAL = resid;
proc PLOT data = PlotGrid;
        plot I*time = '+'   Ihat*time = 'P' / OVERLAY VPOS = 25;
        plot resid*time / VREF = 0 VPOS = 25;
```

```
*       Steepest descent;
*       Good first guess;
*       Global minimum;
proc NLIN METHOD = GRADIENT CONVERGENCE = .005
        data = raw;
        parameters Imax = 20    T = 20;
        model I = Imax*(cos(2*3.14159*time/T) + 1)/2;
        DER.Imax = (cos(2*3.14159*time/T) + 1)/2;
        DER.T = Imax*sin(2*3.14159*time/T)*(3.14159*time/T**2);
        output OUT = PlotG1    PREDICTED = Ihat    RESIDUAL = resid;
proc PLOT data = PlotG1;
        plot I*time = '+'   Ihat*time = 'P' / OVERLAY VPOS = 25;
        plot resid*time / VREF = 0 VPOS = 25;
*       Gauss-Newton;
proc NLIN METHOD = GAUSS CONVERGENCE = .00000001
        data = raw;
        parameters Imax = 20    T = 20;
        model I = Imax*(cos(2*3.14159*time/T) + 1)/2;
        DER.Imax = (cos(2*3.14159*time/T) + 1)/2;
        DER.T = Imax*sin(2*3.14159*time/T)*(3.14159*time/T**2);
        output OUT = PlotGaus    PREDICTED = Ihat    RESIDUAL = resid;
proc PLOT data = PlotGaus;
        plot I*time = '+'   Ihat*time = 'P' / OVERLAY VPOS = 25;
        plot resid*time / VREF = 0 VPOS = 25;
*       Marquardt;
proc NLIN METHOD = MARQUARDT CONVERGENCE = .00000001
        data = raw;
        parameters Imax = 20    T = 20;
        model I = Imax*(cos(2*3.14159*time/T) + 1)/2;
        DER.Imax = (cos(2*3.14159*time/T) + 1)/2;
        DER.T = Imax*sin(2*3.14159*time/T)*(3.14159*time/T**2);
        output OUT = PlotMarq    PREDICTED = Ihat    RESIDUAL = resid;
proc PLOT data = PlotMarq;
        plot I*time = '+'   Ihat*time = 'P' / overlay Vpos = 25;
        plot resid*time / Vref = 0 Vpos = 25;
```

Experimenting with Drugs
(Figs. 10-8 to 10-11; Data in Table C-19, Appendix C)

```
/COMMENT
        This is file CARNITIN.BMD
        for procedure 3R.
/PROBLEM
        title is 'Double exponential fit to Carnitine Data'.
```

```
/INPUT
        file is 'carnitin.dat'.
        variables are 2.
        format is 'free'.
/VARIABLE
        names are 'c(t)', t.
/REGRESS
        dependent is 'c(t)'.
        independent is t.
        parameters = 4.
        halvings = 10.
/FUNCTION
        f = A*exp( − t/tau1) + B*exp( − t/tau2).
        temp1 = exp( − t/tau1).
        temp2 = exp( − t/tau2).
        df1 = temp1.
        df2 = temp2.
        df3 = − t*A*temp1/tau1**2.
        df4 = − t*B*temp2/tau2**2.
/PARAMETERS
        names are A, B, tau1, tau2.
        initial is 600, 300, 100, 300.
/PLOT
        residual.
        variable = t.
        normal.
        size is 60, 20.
/END
```

Keeping Blood Pressure under Control
(Figs. 10-12 to 10-15; Data in Tables C-20 and C-21, Appendix C)

```
*       This is file PRESSURE.SAS;
options noovp;
title 'Baroreceptor control of blood pressure − nonlinear regression';
*       Normal data set (Data in Table C-20, Appendix C);
DATA normal;
        infile presnorm;
        input N P;
*       Hypertensive data set (Data in Table C-21, Appendix C)         ;
DATA hyper;
        infile preshypr;
        input N P;
*       Fit simple, MIN, MAX parameterization to normal data           ;
```

```
proc NLIN data = normal METHOD = MARQUARDT;
      parameters Nmax = 90 Nmin = 5 C = 4 D = - .1;
      model N = Nmax - (Nmax - Nmin)/(1 + exp( - (C + D*P)));
      X = exp( - (C + D*P));
      DER.Nmax = 1 - 1/(1 + X);
      DER.Nmin = 1/(1 + X);
      DER.C = - (Nmax - Nmin)*X/(1 + X)**2;
      DER.D = P*DER.C;
      output OUT = Plot1 PREDICTED = Nhat RESIDUAL = resid
            STUDENT = stdres H = hat;
proc PLOT data = Plot1;
      plot N*P = 'O' Nhat*P = 'P' / OVERLAY VPOS = 25;
      plot resid*P / VREF = 0 VPOS = 25;
      plot resid*Nhat / VREF = 0 VPOS = 25;
proc PRINT data = Plot1;
      var N Nhat resid stdres hat;
*     Fit simple, MIN, MAX parameterization to hypertensive data    ;
proc NLIN data = hyper METHOD = MARQUARDT;
      parameters Nmax = 70 Nmin = 5 C = 4 D = - .1;
      model N = Nmax - (Nmax - Nmin)/(1 + exp( - (C + D*P)));
      X = exp( - (C + D*P));
      DER.Nmax = 1 - 1/(1 + X);
      DER.Nmin = 1/(1 + X);
      DER.C = - (Nmax - Nmin)*X/(1 + X)**2;
      DER.D = P*DER.C;
      output OUT = Plot2 PREDICTED = Nhat RESIDUAL = resid
            STUDENT = stdres H = hat;
proc PLOT data = Plot2;
      plot N*P = 'O' Nhat*P = 'P' / OVERLAY VPOS = 25;
      plot resid*P / VREF = 0 VPOS = 25;
      plot resid*Nhat / VREF = 0 VPOS = 25;
proc PRINT data = Plot2;
      var N Nhat resid stdres hat;
*     Fit alternative parameterization to normal data               ;
proc NLIN data = normal METHOD = MARQUARDT;
      parameters Nmin = 0 Nmax = 90 Phalf = 100 s = 1.5;
      model N = Nmin + (Nmax - Nmin)/(1 + exp(4*s*(Phalf - P)/
            (Nmax - Nmin)));
      DeltaN = Nmax - Nmin;
      DeltaP = Phalf - P;
      X = exp(4*s*DeltaP/DeltaN);
      X1 = 1 + X;
      DER.Nmax = (X1 + X*4*s*DeltaP/DeltaN)/X1**2;
      DER.Nmin = 1 - (X1 + X*4*s*DeltaP/DeltaN)X1**2;
```

```
                    DER.Phalf = − X*4*s/X1**2;
                    DER.s = − X*4*DeltaP/X1**2;
                    output OUT = Plot3 PREDICTED = Nhat RESIDUAL = resid
                            STUDENT = stdres H = hat;
            proc PLOT data = Plot3;
                    plot N*P = 'O' Nhat*P = 'P' / OVERLAY VPOS = 25;
                    plot resid*P / VREF = 0 VPOS = 25;
                    plot resid*Nhat / VREF = 0 VPOS = 25;
            proc PRINT data = Plot3;
                    var N Nhat resid stdres hat;
    *           Fit alternative parameterization to hypertensive data                    ;
            proc NLIN data = hyper METHOD = MARQUARDT;
                    parameters Nmin = 0 Nmax = 70 Phalf = 100 s = 1.5;
                    model N = Nmin + (Nmax − Nmin)/(1 + exp(4*s*(Phalf − P)/
                            (Nmax − Nmin)));
                    DeltaN = Nmax − Nmin;
                    DeltaP = Phalf − P;
                    X = exp(4*s*DeltaP/DeltaN);
                    X1 = 1 + X;
                    DER.Nmax = (X1 + X*4*s*DeltaP/DeltaN)/X1**2;
                    DER.Nmin = 1 − (X1 + X*4*s*DeltaP/DeltaN)X1**2;
                    DER.Phalf = − X*4*s/X1**2;
                    DER.s = − X*4*DeltaP/X1**2;
                    output OUT = Plot4 PREDICTED = Nhat RESIDUAL = resid
                            STUDENT = stdres H = hat;
            proc PLOT data = Plot4;
                    plot N*P = 'O' Nhat*P = 'P' / OVERLAY VPOS = 25;
                    plot resid*P / VREF = 0 VPOS = 25;
                    plot resid*Nhat / VREF = 0 VPOS = 25;
            proc PRINT data = Plot4;
                    var N Nhat resid stdres hat;
```

LOGISTIC REGRESSION

Martian Graduations (Figs. 11-1 to 11-6; Data in Table C-22, Appendix C)

```
            /COMMENT
                    This is file MARSGRAD.BMD
                    for procedure LR.
            /PROBLEM
                    title is 'MARS' graduation data'.
            /INPUT
                    file is 'marsgrad.dat'.
```

```
        variables are 3.
        format is 'free'.
/VARIABLE
        names are 'G',I,SUBJ.
/REGRESS
        depend is 1.
        interval = 2.
        model = 2.
        constant=in.
        start=in.
        move=0.
        method = MLR.
/COMMENT
        In the REGRESS paragraph above, we put the variables in with
        the START command and did not let them move (MOVE=0).
        Thus, this is a straight logistic regression, even though the
        program was designed to do stepwise logistic regression.
/PRINT
        cases = 0.
        cells=both.
        plot.
/PLOT
        xvar=I,I,I,I.
        yvar=influenc,hatdiag,stdresid,deviance.
        size is 60,20.
/END
```

Are Bone Cancer Patients Responding to Chemotherapy?
(Figs. 11-7 to 11-13; Data in Table C-23, Appendix C)

```
/COMMENT
        This is file CANCER.BMD
        for procedure LR.
        This control file produces three logistic regressions: a simple
        two-variable regression with interaction, a quadratic regression
        with interaction, and a centered quadratic regression. Because
        procedure LR is set up to do stepwise regression, we get it to do
        a regular regression by putting all variables into the regression
        (START=IN) and not allowing them to move (MOVE=0).
/PROBLEM
        title is 'Bone cancer logistic regression — simple model'.
/INPUT
        file is 'cancer.dat'.
```

```
        variables are 4.
        format is 'free'.
/VARIABLE
        added are 1.
        use = all.
        names are C,V,F,SUBJ,VF.
/TRANSFORM
        VF = V*F.
/REGRESS
        depend is C.
        interval = V, F, VF.
        model = V, F, VF.
        constant = in.
        start = in,in,in.
        move = 0,0,0.
        rule = none.
        method = MLR.
/PRINT
        plot.
        cases = 0.
/PLOT
        xvar = V,F.
        yvar = stdresid,stdresid.
        size is 60,20.
/END
/PROBLEM
        title is 'Bone cancer logistic regression — quadratic model'.
/INPUT
        file is 'cancer.dat'.
        variables are 4.
        format is 'free'.
/VARIABLE
        added are 3.
        use = all.
        names are C,V,F,SUBJ,VF,V2,F2.
/TRANSFORM
        VF = V*F.
        V2 = V**2.
        F2 = F**2.
/REGRESS
        depend is C.
        interval = V, F, VF, V2, F2.
        model = V, F, VF, V2, F2.
        constant = in.
        start = in,in,in,in,in.
```

```
          move = 0,0,0,0,0.
          rule  = none.
          method  =  MLR.
/PRINT
          plot.
          cases  =  0.
/PLOT
          xvar = V,F.
          yvar = stdresid,stdresid.
          size is 60,20.
/END
/PROBLEM
          title is 'Bone cancer logistic regression — centered quadratic
                    model'.
/INPUT
          file is 'cancer.dat'.
          variables are 4.
          format is 'free'.
/VARIABLE
          added are 5.
          use  =  all.
          names are C,V,F,SUBJ,Vc,Fc,VcFc,Vc2,Fc2.
/TRANSFORM
          Vc = V − 1.9318.
          Fc = F − 1.0682.
          VcFc = Vc*Fc.
          Vc2 = Vc**2.
          Fc2 = Fc**2.
/REGRESS
          depend is C.
          interval  =  Vc, Fc, VcFc, Vc2, Fc2.
          model  =  Vc, Fc, VcFc, Vc2, Fc2.
          constant = in.
          start = in,in,in,in,in.
          move = 0,0,0,0,0.
          rule  = none.
          method  =  MLR.
/PRINT
          plot.
          cases  =  0.
/PLOT
          xvar = Vc,Fc.
          yvar = stdresid,stdresid.
             size is 60,20.
/END
```

Nuking the Heart (Figs. 11-14 to 11-22; Data in Table C-24, Appendix C)

```
/COMMENT
        This is file THALLIUM.BMD
        for procedure LR.
        This control language file contains multiple problems for
        the thallium imaging of the heart data.
/PROBLEM
        title is 'Heart thallium data — FORWARD without interactions'.
/INPUT
        file is 'thallium.dat'.
        variables are 8.
        format is 'free'.
/VARIABLE
        names are T,D,P,R,E,S,C,SUBJ.
/REGRESS
        depend is T.
        interval = T to C.
        model is D,P,R,E,S,C.
        constant = in.
        start = OUT,OUT,OUT,OUT,OUT,OUT.
        move = 5,5,5,5,5,5.
        rule = none.
        method = MLR.
        enter = .1, .1.
        remove = .15, .15.
        convergence = .000001.
/PRINT
        cases = 0.
        plot.
/PLOT
        xvar = D,C,D,D,D.
        yvar = stdresid,stdresid,influenc,hatdiag,deviance.
        size is 60, 20.
/END
/PROBLEM
        title is 'Heart thallium data — FORWARD with some
                interactions'.
/INPUT
        file is 'thallium.dat'.
        variables are 8.
        format is 'free'.
/VARIABLE
        names are T,D,P,R,E,S,C,SUBJ.
```

```
/REGRESS
      depend is T.
      interval  =  T to C.
      model is D*P,D*R,P*R,R*E,S,C.
      constant = in.
      start = OUT,OUT,OUT,OUT,OUT,OUT.
      move = 5,5,5,5,5,5.
      rule  =  none.
      method  =  MLR.
      enter  =  .1, .1.
      remove  =  .15, .15.
      convergence  =  .000001.
/PRINT
      cases  =  0.
      plot.
/PLOT
      xvar = D,D,D,D.
      yvar = stdresid,influenc,hatdiag,deviance.
      size is 60, 20.
/END
/PROBLEM
      title is 'Heart thallium data — BACKWARD, with some
              interaction'.
/COMMENT
      To do a backward selection, START with all variables IN.
/INPUT
      file is 'thallium.dat'.
      variables are 8.
      format is 'free'.
/VARIABLE
      names are T,D,P,R,E,S,C,SUBJ.
/REGRESS
      depend is T.
      interval  =  T to C.
      model is D*P,D*R,P*R,R*E,S,C.
      constant = in.
      start = in,in,in,in,in,in.
      move = 5,5,5,5,5,5.
      rule  =  none.
      method  =  MLR.
      enter  =  .1, .1.
      remove  =  .15, .15.
      convergence  =  .000001.
```

```
/PRINT
        cases = 0.
        plot.
/PLOT
        xvar = D,D,D,D.
        yvar = stdresid,influenc,hatdiag,deviance.
        size is 60, 20.
/END
/PROBLEM
        title is 'Heart thallium data — FORWARD, forcing, with
                interactions'.
/INPUT
        file is 'thallium.dat'.
        variables are 8.
        format is 'free'.
/VARIABLE
        names are T,D,P,R,E,S,C,SUBJ.
/COMMENT
        To force some effects, we START them IN and do not let
        them move (MOVE = 0 for those effects).
/REGRESS
        depend is T.
        interval = T to C.
        model is D,C,P,D*P,D*C,C*P.
        constant = in.
        start = in,in,in,OUT,OUT,OUT.
        move = 0,0,0,5,5,5.
        rule = none.
        method = MLR.
        enter = .1, .1.
        remove = .15, .15.
        convergence = .000001.
/PRINT
        cases = 0.
        plot.
/PLOT
        xvar = D,D,D,D.
        yvar = stdresid,influenc,hatdiag,deviance.
        size is 60, 20.
/END
```

All the logistic regression examples are worked with BMDP procedure LR because it is the most widely available stepwise logistic regression program. For straight logistic regressions, there is a SAS procedure called LOGIST, which is unsupported and therefore described in the *SUGI Supplemental Library User's Guide*.

APPENDIX

C

Data for Examples

TABLE C-1 Body Thermal Conductivity and Ambient Water Temperature for Study of Heat Exchange in Gray Seals

Thermal Conductivity C_b, W/(m²·°C)	Ambient Temperature T_a, °C	Thermal Conductivity C_b, W/(m²·°C)	Ambient Temperature T_a, °C
2.81	−40	1.18	−20
1.82	−40	1.63	−10
1.80	−40	1.26	−10
2.25	−30	0.93	−10
2.19	−30	2.22	10
2.02	−30	1.97	10
1.57	−30	1.74	10
1.40	−30	4.49	20
1.94	−20	4.33	20
1.40	−20	3.10	20

TABLE C-2 Prostacyclin Production, Arachidonic Acid, and Endotoxin Data for Study of Mechanisms of Toxic Shock

Prostacyclin P, ng/mL	Arachidonic Acid A, μM	Endotoxin E, ng/mL
19.2	10	0
10.8	10	0
33.6	10	0
11.9	10	10
15.9	10	10
33.3	10	10
81.1	10	50
36.7	10	50
58.0	10	50
60.8	10	100
50.6	10	100
69.4	10	100
30.8	25	0
27.6	25	0
13.2	25	0
38.8	25	10
37.0	25	10
38.3	25	10
65.2	25	50
66.4	25	50
63.2	25	50
49.9	25	100
89.5	25	100
60.5	25	100
102.9	50	0
57.1	50	0
76.7	50	0
70.5	50	10
66.4	50	10
76.3	50	10
83.1	50	50
61.7	50	50
101.5	50	50
86.2	50	100
115.9	50	100
102.1	50	100

TABLE C-3 Leucine Turnover, Body Weight, and Dummy Variable for Newborn ($A = 0$) or Adult ($A = 1$) for Study of Protein Metabolism

Log Leucine, log L	Log Body Weight, log W	A	Log Leucine, log L	Log Body Weight, log W	A
2.54	0.44	0	3.61	1.77	1
2.57	0.42	0	3.64	1.81	1
2.59	0.46	0	3.65	1.70	1
2.73	0.46	0	3.66	1.76	1
2.74	0.51	0	3.75	1.75	1
2.81	0.52	0	3.75	1.87	1
2.81	0.57	0	3.84	1.92	1
2.86	0.58	0	3.88	1.97	1
2.93	0.61	0	3.89	1.92	1
3.61	1.73	1	3.93	1.86	1

TABLE C-4 Plasma HDL-2, β-blocker Use, Alcohol Use, Smoking, Age, Weight, Plasma Triglycerides, Plasma C-peptide, and Blood Glucose Data for Study of Treatment of High Blood Pressure in Diabetic Patients

HDL-2 H, mmol/L	B	D	S	Age A, year	Weight W, kg/m^2	Triglyceride T, mmol/L	C-peptide C, mU/L	Glucose G, mmol/L
0.12	0	1	0	78	19.3	1.1	0.4	6.6
0.08	0	1	0	72	24.6	2.3	1.9	8.6
0.37	0	0	0	61	25.2	0.7	4.1	5.7
0.38	0	1	1	62	24.9	1.0	3.9	8.7
0.22	0	1	1	55	23.4	2.1	2.5	6.2
0.43	0	0	0	51	24.8	0.9	6.6	9.1
0.27	0	0	0	65	22.2	1.6	0.7	15.1
0.23	0	1	0	71	27.7	2.2	2.4	5.7
0.13	0	1	1	64	26.4	2.2	0.6	11.2
0.28	0	0	0	56	27.8	1.2	1.3	6.8
0.28	0	1	1	54	30.9	2.1	0.9	3.8
0.43	0	0	0	71	21.9	0.8	2.3	7.7
0.27	0	0	1	57	21.2	1.8	1.5	8.2
0.24	0	0	0	64	27.7	1.4	1.1	13.9
0.23	0	1	0	76	26.6	1.6	1.5	7.2
0.26	0	0	0	77	26.7	1.4	4.7	8.8
0.35	0	0	0	63	26.8	1.1	1.7	7.6
0.16	0	0	0	74	30.7	1.6	2.3	10.8
0.34	0	1	0	65	22.2	1.3	0.6	11.8
0.29	0	0	0	83	23.4	1.2	1.8	17.2
0.14	0	1	0	59	24.5	1.4	2.3	10.7

TABLE C-4 (*Continued*)   Plasma HDL-2, β-blocker Use, Alcohol Use, Smoking, Age, Weight, Plasma Triglycerides, Plasma C-peptide, and Blood Glucose Data for Study of Treatment of High Blood Pressure in Diabetic Patients

HDL-2 H, mmol/L	B	D	S	Age A, year	Weight W, kg/m^2	Triglyceride T, mmol/L	C-peptide C, mU/L	Glucose G, mmol/L
0.48	0	1	0	57	23.7	0.6	2.5	6.9
0.16	0	1	0	73	18.6	1.3	1.5	9.2
0.06	0	1	1	74	32.2	1.7	1.9	8.4
0.30	0	0	0	48	26.3	1.2	1.7	8.3
0.21	0	0	0	55	21.2	2.1	1.7	13.6
0.24	0	1	0	66	30.3	0.8	3.4	6.0
0.42	0	0	1	52	29.8	0.7	4.5	4.0
0.20	0	1	0	56	26.9	1.6	1.0	10.6
0.15	0	0	0	74	23.6	2.0	2.1	8.3
0.17	1	1	1	61	22.3	2.1	2.4	10.1
0.08	1	1	1	69	25.4	2.4	1.3	11.2
0.07	1	1	0	60	29.3	2.5	1.5	11.8
0.11	1	1	1	55	34.5	3.4	0.8	10.8
0.42	1	0	0	57	25.9	0.6	2.5	6.6
0.23	1	0	0	67	35.0	1.2	2.1	8.2
0.31	1	1	0	43	26.8	1.7	1.2	10.4
0.17	1	1	0	63	32.1	2.5	3.1	4.3
0.05	1	0	0	68	26.3	1.5	3.2	10.8
0.02	1	1	1	65	34.1	4.1	0.9	8.3
0.28	1	0	0	47	23.4	1.7	1.5	6.8
0.15	1	1	0	50	28.6	2.3	2.5	11.9
0.22	1	1	1	53	28.6	1.3	2.5	10.7
0.27	1	0	0	65	32.2	0.6	2.3	9.7
0.23	1	1	1	68	26.1	0.6	2.3	5.7
0.31	1	0	0	53	26.7	0.6	2.5	8.3
0.15	1	1	0	60	31.2	1.9	1.9	7.2
0.27	1	0	0	59	33.2	1.1	2.9	16.7
0.02	1	1	1	57	28.1	3.9	2.7	8.7
0.14	1	0	0	59	23.9	2.5	1.9	13.6
0.17	1	1	1	69	25.7	0.9	0.8	10.6
0.09	1	1	0	59	23.5	7.4	1.9	13.9
0.02	1	0	1	67	29.6	2.1	5.1	11.9
0.16	1	1	0	66	32.6	2.3	6.1	6.0
0.28	1	1	0	58	27.8	0.9	1.7	9.1
0.07	1	0	1	54	33.9	1.8	1.7	8.4
0.08	1	0	0	64	24.4	1.9	2.4	9.1
0.10	1	1	0	52	27.5	2.3	2.3	10.9
0.05	1	1	0	48	27.4	2.9	1.7	7.7
0.08	1	1	0	62	32.3	4.7	2.1	8.8
0.13	1	1	1	50	25.9	2.4	1.1	20.7
0.08	1	0	1	56	24.0	1.7	4.5	10.6

TABLE C-4 (*Continued*) Plasma HDL-2, β-blocker Use, Alcohol Use, Smoking, Age, Weight, Plasma Triglycerides, Plasma C-peptide, and Blood Glucose Data for Study of Treatment of High Blood Pressure in Diabetic Patients

HDL-2 H, mmol/L	B	D	S	Age A, year	Weight W, kg/m^2	Triglyceride T, mmol/L	C-peptide C, mU/L	Glucose G, mmol/L
0.02	1	1	0	70	24.5	4.5	0.8	15.1
0.15	1	0	0	58	35.7	1.5	3.2	17.0
0.17	1	1	0	62	29.6	1.1	3.4	8.4
0.23	1	0	0	62	27.4	1.0	4.7	8.6
0.17	1	1	0	68	30.3	1.5	3.9	9.2
0.24	1	1	0	53	29.5	0.7	2.4	9.2
0.02	1	0	1	68	29.7	1.5	1.8	11.2
0.03	1	1	0	62	25.7	2.4	4.1	10.8
0.20	1	1	1	58	29.8	0.6	1.1	7.6

TABLE C-5 Minute Ventilation, Inspired Oxygen, and Inspired Carbon Dioxide for Study of Respiratory Control in Nestling Bank Swallows

Minute Ventilation V, Percent Change	Inspired O$_2$ O, Volume %	Inspired CO$_2$ C, Volume %	Minute Ventilation V, Percent Change	Inspired O$_2$ O, Volume %	Inspired CO$_2$ C, Volume %
−49	19	0.0	−141	19	6.0
0	19	0.0	400	19	6.0
−98	19	0.0	253	19	6.0
148	19	0.0	203	19	6.0
49	19	0.0	146	19	9.0
49	19	0.0	244	19	9.0
−24	19	3.0	−51	19	9.0
25	19	3.0	687	19	9.0
−123	19	3.0	441	19	9.0
222	19	3.0	392	19	9.0
123	19	3.0	61	17	0.0
74	19	3.0	61	17	0.0
11	19	4.5	160	17	0.0
60	19	4.5	12	17	0.0
−88	19	4.5	−86	17	0.0
306	19	4.5	−37	17	0.0
158	19	4.5	−11	17	3.0
158	19	4.5	136	17	3.0
56	19	6.0	−110	17	3.0
105	19	6.0	235	17	3.0

TABLE C-5 (*Continued*)   Minute Ventilation, Inspired Oxygen, and Inspired Carbon Dioxide for Study of Respiratory Control in Nestling Bank Swallows

Minute Ventilation V, Percent Change	Inspired O_2 O, Volume %	Inspired CO_2 C, Volume %	Minute Ventilation V, Percent Change	Inspired O_2 O, Volume %	Inspired CO_2 C, Volume %
87	17	3.0	328	15	6.0
38	17	3.0	279	15	6.0
169	17	4.5	33	15	6.0
317	17	4.5	−164	15	6.0
22	17	4.5	8	15	9.0
71	17	4.5	205	15	9.0
169	17	4.5	549	15	9.0
−77	17	4.5	254	15	9.0
413	17	6.0	402	15	9.0
−128	17	6.0	352	15	9.0
266	17	6.0	183	13	0.0
118	17	6.0	−63	13	0.0
69	17	6.0	−14	13	0.0
216	17	6.0	84	13	0.0
−46	17	9.0	84	13	0.0
446	17	9.0	35	13	0.0
397	17	9.0	14	13	3.0
249	17	9.0	−85	13	3.0
692	17	9.0	63	13	3.0
151	17	9.0	112	13	3.0
73	15	0.0	161	13	3.0
−74	15	0.0	260	13	3.0
172	15	0.0	192	13	4.5
−25	15	0.0	−54	13	4.5
73	15	0.0	192	13	4.5
24	15	0.0	340	13	4.5
−98	15	3.0	45	13	4.5
148	15	3.0	94	13	4.5
99	15	3.0	439	13	6.0
247	15	3.0	95	13	6.0
50	15	3.0	−102	13	6.0
1	15	3.0	292	13	6.0
180	15	4.5	242	13	6.0
180	15	4.5	144	13	6.0
33	15	4.5	701	13	9.0
−66	15	4.5	406	13	9.0
82	15	4.5	258	13	9.0
328	15	4.5	160	13	9.0
574	15	6.0	455	13	9.0
131	15	6.0	−37	13	9.0

TABLE C-6 Tritiated Water Production, Time, and Sodium Chloride Concentration for Study of Hydrogenase Enzyme Activity in Halophilic Bacteria

Tritiated H_2O T, Counts $\times 10^5$	Time t, min	Sodium Chloride S, M/L	Tritiated H_2O T, Counts $\times 10^5$	Time t, min	Sodium Chloride S, M/L
0.28	30	3.4	1.44	90	1.7
0.41	60	3.4	1.97	120	1.7
0.59	90	3.4	0.47	30	1.0
0.73	120	3.4	1.15	60	1.0
0.47	30	2.3	1.83	90	1.0
0.80	60	2.3	2.63	120	1.0
1.09	90	2.3	0.80	30	0.5
1.46	120	2.3	1.50	60	0.5
0.47	30	1.7	2.06	90	0.5
0.98	60	1.7	2.55	120	0.5

TABLE C-7 Change in Force Resulting from a Change in Length at Different Steady-State Force Levels in Isolated Smooth Muscle

Force Change ΔF, dyne	Length Change ΔL, μm	Steady-State Force F, dyne	Force Change ΔF, dyne	Length Change ΔL, μm	Steady-State Force F, dyne
1.4	1.7	200	0.7	1.7	100
4.0	6.0	200	1.8	6.0	100
10.4	15.6	200	5.5	15.6	100
14.7	22.4	200	7.5	22.4	100
19.2	31.4	200	10.1	31.4	100
24.7	41.7	200	13.3	41.7	100
30.7	54.8	200	15.7	54.8	100
1.1	1.7	150	0.4	1.7	50
3.0	6.0	150	1.1	6.0	50
7.8	15.6	150	2.8	15.6	50
11.0	22.4	150	3.9	22.4	50
14.7	31.4	150	5.4	31.4	50
18.9	41.7	150	6.9	41.7	50
23.6	54.8	150	8.1	54.8	50

TABLE C-8 Martian Intelligence and Foot Size

Intelligence I, zorp	Foot Size F, cm	Intelligence I, zorp	Foot Size F, cm
A		**D**	
8.04	10	6.58	8
6.95	8	5.76	8
7.58	13	7.71	8
8.81	9	8.84	8
8.33	11	8.47	8
9.96	14	7.04	8
7.24	6	5.25	8
4.26	4	5.56	8
10.84	12	7.91	8
4.82	7	6.89	8
5.68	5	12.50	19
B		**E**	
9.14	10	8.04	10
8.14	8	6.95	8
8.74	13	13.00	7
8.77	9	8.81	9
9.26	11	8.33	11
8.10	14	9.96	14
6.13	6	7.24	6
3.10	4	4.26	4
9.13	12	10.84	12
7.26	7	4.82	7
4.74	5	5.68	5
C			
7.46	10		
6.77	8		
12.74	13		
7.11	9		
7.81	11		
8.84	14		
6.08	6		
5.39	4		
8.15	12		
6.42	7		
5.73	5		

TABLE C-9 Placental Water Clearance and Uterine Artery Blood Flow
for Study of Placental Exchange Mechanisms in Cows

Water Clearance (D_2O) D, L/min	Uterine Artery Flow U, L/min	Water Clearance (D_2O) D, L/min	Uterine Artery Flow U, L/min
0.19	0.27	2.13	3.27
0.10	0.19	1.50	2.27
0.27	0.38	1.85	2.72
0.66	0.98	2.03	2.90
0.81	1.20	1.73	2.81
0.65	0.98	1.47	2.26
0.76	1.08	2.38	3.55
0.69	1.02	1.86	2.37
0.75	1.18	2.89	5.10
0.86	1.24	3.00	3.92
0.59	0.90	3.48	10.04
1.99	2.96	2.29	4.38

TABLE C-10 Rate of Methionine Hydroxy Analog (MHA) Uptake in Isolated
Intestinal Segments and MHA Concentration for Study of Mechanisms
of Intestinal Transport of MHA

MHA Uptake Rate R, nmol/(g·min)	MHA Concentration M, mM/L	MHA Uptake Rate R, nmol/(g·min)	MHA Concentration M, mM/L
3550	100	1250	20
6250	100	975	20
5950	100	950	20
5550	100	875	20
6225	100	475	10
4000	100	450	10
4400	100	350	10
3850	100	375	10
3350	50	400	10
3050	50	425	10
2800	50	320	5
2650	50	325	5
2400	50	315	5
2325	50	320	5
2175	50	315	5
2925	50	310	5
1500	20	215	4
1275	20	220	4

TABLE C-10 (*Continued*)   Rate of Methionine Hydroxy Analog (MHA) Uptake in Isolated Intestinal Segments and MHA Concentration for Study of Mechanisms of Intestinal Transport of MHA

MHA Uptake Rate R, nmol/(g·min)	MHA Concentration M, mM/L	MHA Uptake Rate R, nmol/(g·min)	MHA Concentration M, mM/L
220	4	110	2
210	4	110	2
210	4	115	2
225	4	110	2
160	3	115	2
165	3	50	1
170	3	40	1
165	3	45	1
175	3	45	1
160	3	45	1
115	2	40	1

TABLE C-11 Metallothionein mRNA Production, Dietary Copper, and Dietary Zinc for Study of Metallothionein Regulation by Trace Elements

mRNA M, molecules/cell	Dietary Copper (Cu), mg/kg	Dietary Zinc (Zn), mg/kg	mRNA M, molecules/cell	Dietary Copper (Cu), mg/kg	Dietary Zinc (Zn), mg/kg
5.8	1	5	21.9	6	30
3.2	1	5	11.1	6	30
3.9	1	5	25.9	6	180
3.8	1	5	21.0	6	180
12.6	1	30	30.4	6	180
11.8	1	30	17.6	6	180
18.9	1	30	5.5	36	5
21.9	1	30	5.4	36	5
12.7	1	180	6.9	36	5
15.2	1	180	6.0	36	5
17.3	1	180	31.4	36	30
27.2	1	180	34.4	36	30
4.1	6	5	20.7	36	30
5.3	6	5	28.1	36	30
6.9	6	5	20.1	36	180
7.6	6	5	20.4	36	180
14.9	6	30	17.7	36	180
14.6	6	30	18.8	36	180

TABLE C-12 **Left Ventricular End-Diastolic Area, End-Diastolic Pressure, and Right Ventricular End-Diastolic Dimension during Vena Caval Constriction (*A*) and Sequential Vena Caval and Pulmonary Artery Constriction and Release (*B*)**

Left Ventricle		Right Ventricle
End-Diastolic Area A_{ed}, cm^2	End-Diastolic Pressure P_{ed}, mmHg	End-Diastolic Dimension D_R, mm
A		
29.273	17.635	27.579
29.245	17.966	27.526
29.261	17.808	27.514
29.302	18.344	27.593
29.292	17.839	27.549
29.303	17.461	27.539
29.295	17.855	27.546
29.328	17.508	27.511
29.295	17.635	27.477
29.320	18.454	27.548
29.277	18.234	27.756
29.132	17.461	28.047
28.868	17.083	28.655
28.630	16.799	29.187
28.447	16.247	29.557
28.302	16.247	29.873
28.170	16.342	30.187
28.023	15.664	30.423
27.908	14.781	30.575
27.852	15.427	30.790
27.725	15.049	30.919
27.680	15.585	31.085
27.593	15.601	31.200
27.520	15.175	31.350
27.435	14.608	31.426
27.405	15.349	31.599
27.363	14.813	31.660
27.336	15.065	31.807
27.278	15.285	31.826
27.224	14.355	31.874
27.216	14.513	31.965
27.204	14.434	32.010
27.202	14.939	32.125
27.170	14.844	32.120
27.165	14.418	32.120
27.139	14.277	32.097
27.146	14.671	32.025

TABLE C-12 (*Continued*)   Left Ventricular End-Diastolic Area, End-Diastolic Pressure, and Right Ventricular End-Diastolic Dimension during Vena Caval Constriction (*A*) and Sequential Vena Caval and Pulmonary Artery Constriction and Release (*B*)

Left Ventricle		Right Ventricle
End-Diastolic Area A_{ed}, cm^2	End-Diastolic Pressure P_{ed}, mmHg	End-Diastolic Dimension D_R, mm
27.132	14.560	32.175
27.168	14.954	32.172
27.176	15.112	32.176
27.161	14.166	32.191
27.153	14.387	32.177
27.183	14.608	32.139
27.208	14.718	32.199
27.212	14.623	32.165
27.276	15.191	32.074
27.684	14.781	31.227
28.070	15.664	30.452
28.375	16.547	29.851
28.605	16.689	29.535
28.767	17.366	29.198
28.908	17.477	28.926
29.014	17.603	28.655
29.145	17.808	28.531
29.188	18.407	28.355
29.304	18.990	28.259
29.339	19.132	28.066
29.406	19.179	27.961
29.442	18.959	27.877
29.505	18.817	27.812
29.509	19.179	27.699
29.547	19.511	27.667
29.558	20.094	27.654
29.607	20.173	27.649
29.608	19.984	27.602
29.621	19.826	27.558
29.634	19.668	27.579
29.633	19.274	27.570
29.638	19.747	27.569
B		
28.795	15.412	27.104
28.831	15.916	27.035
28.839	15.380	27.046
28.870	15.680	26.984

TABLE C-12 (*Continued*) Left Ventricular End-Diastolic Area, End-Diastolic Pressure, and Right Ventricular End-Diastolic Dimension during Vena Caval Constriction (*A*) and Sequential Vena Caval and Pulmonary Artery Constriction and Release (*B*)

Left Ventricle		Right Ventricle
End-Diastolic Area A_{ed}, cm^2	End-Diastolic Pressure P_{ed}, mmHg	End-Diastolic Dimension D_R, mm
28.873	15.806	26.929
28.967	15.821	27.016
28.968	16.105	27.001
28.984	16.121	27.005
29.052	16.783	26.642
29.087	16.547	26.714
28.998	15.900	27.025
28.803	15.821	27.651
28.483	15.601	28.383
28.267	15.002	28.882
27.989	14.954	29.376
27.852	14.418	29.749
27.605	14.355	30.144
27.501	13.914	30.511
27.290	13.961	30.914
27.128	13.677	31.115
26.957	13.268	31.520
26.805	13.804	31.678
26.665	13.394	32.069
26.510	12.590	32.216
26.439	12.653	32.508
26.302	12.495	32.743
26.172	13.047	33.039
26.161	12.653	33.007
26.293	12.779	32.418
26.565	12.684	31.512
26.770	12.322	30.677
26.926	12.085	29.928
26.953	11.612	29.388
26.907	11.313	28.936
26.851	11.502	28.703
26.734	11.045	28.490
26.638	10.619	28.376
26.506	10.288	28.212
26.409	10.272	28.146
26.287	10.430	28.187
26.139	9.610	27.999
26.017	9.878	27.927

Table C-12 (*Continued*)   Left Ventricular End-Diastolic Area, End-Diastolic
Pressure, and Right Ventricular End-Diastolic Dimension during Vena Caval
Constriction (*A*) and Sequential Vena Caval and Pulmonary Artery
Constriction and Release (*B*)

Left Ventricle		Right Ventricle
End-Diastolic Area A_{ed}, cm^2	End-Diastolic Pressure P_{ed}, mmHg	End-Diastolic Dimension D_R, mm
25.882	9.831	27.884
25.743	10.162	27.849
25.604	9.153	27.730
25.775	9.500	26.570
25.968	9.295	25.736
26.028	9.846	25.096
25.998	9.216	24.600
25.965	9.042	24.367
25.892	9.310	24.127
25.805	9.184	23.964
25.741	8.901	23.775
25.638	9.200	23.678
25.588	9.090	23.676
25.505	9.011	23.577
25.428	9.058	23.595
25.272	9.137	23.811
24.062	9.153	30.087
23.723	9.216	33.042
24.349	9.547	32.666
25.213	10.682	31.795
25.849	11.202	30.913
26.402	11.786	30.195
26.795	12.921	29.710
27.081	13.882	29.312
27.284	13.346	29.061
27.468	13.740	28.859
27.632	13.740	28.768
27.760	13.536	28.547
27.910	14.387	28.374
28.061	14.765	28.217
28.215	14.828	28.020
28.345	15.648	27.855
28.507	15.900	27.705
28.599	15.569	27.604
28.685	16.358	27.484
28.769	16.294	27.377
28.846	16.594	27.325

TABLE C-13 Half-Triathlon Performance Time, Age, Weight, Years Triathlon Experience, Amount of Training Running (R), Biking (B), and Swimming (S), and Maximum Oxygen Consumption while Running, Biking, and Swimming

Time t, min	Age A, years	Weight W, kg	Experience E, years	Training, km/wk T_R	T_B	T_S	Maximal O_2 Consumption, mL/(min·kg) V_R	V_B	V_S
322	36	73.0	3	48.4	306.5	9.7	55.3	58.1	54.9
283	32	60.3	1	69.4	330.6	14.5	70.3	67.7	62.0
286	33	61.2	2	80.6	443.5	14.5	69.4	66.3	59.0
335	44	67.1	2	32.3	241.9	4.8	60.1	57.4	44.2
329	32	70.7	2	61.3	222.6	8.1	60.2	59.3	56.0
410	37	78.0	1	64.5	322.6	12.9	55.6	54.7	52.4
362	38	72.1	1	24.2	145.2	2.4	61.0	60.8	50.2
345	43	72.6	3	48.4	282.3	9.7	59.6	58.0	54.6
460	43	65.3	0	40.3	177.4	2.4	49.6	47.4	43.6
327	37	76.2	1	40.3	241.9	17.7	57.2	49.8	48.4
301	46	75.3	5	88.7	314.5	9.7	60.1	58.0	50.6
334	36	65.3	2	48.4	153.2	1.6	66.3	61.5	52.6
365	43	69.8	3	24.2	80.6	3.2	62.7	55.4	53.4
324	36	73.8	3	46.0	302.6	7.7	53.4	60.5	53.2
285	31	59.9	1	69.5	334.0	16.2	66.4	68.9	64.5
287	33	60.7	2	82.1	436.2	10.8	65.0	65.6	61.1
336	44	67.9	2	33.8	251.4	9.6	67.9	57.7	42.9
332	33	69.4	2	61.6	215.3	4.4	61.8	60.4	55.1
408	36	77.9	1	66.3	314.1	8.6	59.1	60.0	50.5
363	38	71.7	1	25.8	144.2	1.9	56.8	58.4	47.4
349	44	72.7	3	48.7	275.1	6.1	56.2	52.7	51.5
463	43	65.7	0	37.3	179.4	3.4	48.6	49.7	46.0
328	36	78.5	1	39.0	235.9	14.7	56.8	51.1	50.8
300	47	75.8	5	85.7	315.1	10.0	59.0	58.8	48.5
336	36	63.6	2	47.8	152.4	1.2	62.1	62.3	53.7
368	41	68.2	3	25.1	85.0	5.4	63.4	57.5	50.9
322	35	72.5	3	49.2	310.2	11.6	54.1	59.5	57.0
282	31	60.7	1	70.0	335.0	16.7	67.5	65.6	66.1
285	32	62.1	2	81.9	442.3	13.9	66.8	61.0	61.6
336	42	66.6	2	38.8	235.8	1.7	58.7	57.4	44.5
332	32	69.6	2	63.7	225.6	9.6	64.0	53.6	57.6
407	37	79.5	1	67.1	336.2	19.7	54.0	57.5	49.7
364	40	72.2	1	25.8	148.4	4.0	60.7	60.6	52.8
344	44	73.0	2	50.7	288.7	12.9	60.4	60.9	56.3
464	43	66.0	0	41.3	175.4	1.4	47.5	46.0	41.3
325	37	76.4	1	40.8	242.9	18.2	55.6	50.4	50.0
301	47	74.3	6	87.2	317.9	11.4	65.1	58.4	49.0

TABLE C-13 (*Continued*)   **Half-Triathlon Performance Time, Age, Weight, Years Triathlon Experience, Amount of Training Running (R), Biking (B), and Swimming (S), and Maximum Oxygen Consumption while Running, Biking, and Swimming**

Time t, min	Age A, years	Weight W, kg	Experience E, years	Training, km/wk T_R	T_B	T_S	Maximal O_2 Consumption, mL/(min·kg) V_R	V_B	V_S
335	37	65.2	2	53.1	156.3	3.2	70.2	63.3	54.6
366	44	69.4	3	26.7	87.6	6.7	67.3	60.1	51.9
322	37	71.6	3	49.3	308.0	10.5	55.9	61.5	51.6
288	31	61.4	1	68.6	329.8	14.1	69.8	65.6	60.6
285	33	61.7	2	84.4	443.7	14.6	68.3	67.6	61.1
332	44	66.3	2	32.4	242.1	4.9	58.2	57.9	43.6
331	32	70.2	3	57.5	224.7	9.2	58.6	59.8	58.0
409	38	77.1	1	64.6	317.2	10.2	54.2	52.9	52.1
364	38	71.0	1	23.1	152.5	6.1	58.7	62.7	49.2
344	43	74.1	3	49.9	281.3	9.2	62.1	61.4	54.8
462	42	66.7	1	40.8	173.8	0.6	48.8	47.4	48.1
327	36	74.1	2	45.5	249.2	21.4	58.4	48.8	51.9
302	45	75.4	5	89.8	308.8	6.8	57.0	57.9	55.0
335	37	65.7	2	50.9	150.2	0.1	69.2	62.9	56.5
368	44	69.6	3	24.9	86.1	6.0	68.6	51.0	53.9
317	36	72.8	2	45.6	302.3	7.6	59.5	53.8	55.3
282	32	60.9	1	65.7	331.0	14.7	67.9	67.2	59.1
284	33	61.8	2	78.9	449.3	17.4	69.4	64.1	56.2
332	45	66.3	2	34.1	243.7	5.7	64.1	56.4	46.7
326	32	70.3	2	61.7	221.5	7.5	62.6	63.6	57.2
412	36	75.6	1	64.5	325.7	14.5	53.9	51.5	51.9
363	38	71.4	0	27.5	141.7	0.6	61.4	63.4	51.6
343	43	71.3	3	47.0	277.7	7.4	59.2	55.3	57.2
459	43	66.8	1	40.4	181.3	4.4	50.2	51.1	43.0
329	39	75.2	1	38.4	237.1	15.3	61.0	48.1	45.3
301	46	75.0	4	88.9	308.1	6.5	62.8	61.0	53.0
337	38	63.9	1	47.1	146.3	1.9	66.9	61.1	48.4
359	43	69.7	3	21.5	78.8	2.3	60.5	60.1	58.9

TABLE C-14 Percentage of Rabbit Aorta Covered by Atherosclerotic Plaque in 5 Treatment Groups and Associated Dummy Variables G_i Equal to 1 if Group i

Percent Plaque P	G_2	G_3	G_4	G_5	Group
22.0	0	0	0	0	1
14.9	0	0	0	0	1
16.8	0	0	0	0	1
68.0	0	0	0	0	1
78.0	0	0	0	0	1
42.0	0	0	0	0	1
39.4	0	0	0	0	1
28.0	0	0	0	0	1
85.7	1	0	0	0	2
87.0	1	0	0	0	2
74.6	1	0	0	0	2
86.7	1	0	0	0	2
70.6	1	0	0	0	2
46.4	0	1	0	0	3
100.0	0	1	0	0	3
89.6	0	1	0	0	3
95.4	0	1	0	0	3
85.0	0	1	0	0	3
54.8	0	1	0	0	3
66.4	0	0	1	0	4
63.9	0	0	1	0	4
46.5	0	0	1	0	4
10.8	0	0	1	0	4
21.0	0	0	1	0	4
60.0	0	0	1	0	4
45.7	0	0	0	1	5
67.1	0	0	0	1	5
89.6	0	0	0	1	5
100.0	0	0	0	1	5
51.7	0	0	0	1	5

TABLE C-15 Anger Score S (points) while Sober $(D = 1)$ or Drinking $(D = -1)$ in Alcoholic Subjects $(P_{ASP_i}$ and $P_{non\text{-}ASP_i})$ Diagnosed as Either Antisocial Personality $(A = -1)$ or Non-Antisocial Personality $(A = 1)$

S	D	A	$P_{non\text{-}ASP_i}$							P_{ASP_i}						
			1	2	3	4	5	6	7	1	2	3	4	5	6	7
0.83	1	1	1	0	0	0	0	0	0	0	0	0	0	0	0	0
1.09	1	1	0	1	0	0	0	0	0	0	0	0	0	0	0	0
1.12	1	1	0	0	1	0	0	0	0	0	0	0	0	0	0	0
1.17	1	1	0	0	0	1	0	0	0	0	0	0	0	0	0	0
1.26	1	1	0	0	0	0	1	0	0	0	0	0	0	0	0	0
1.34	1	1	0	0	0	0	0	1	0	0	0	0	0	0	0	0
1.36	1	1	0	0	0	0	0	0	1	0	0	0	0	0	0	0
1.45	1	1	-1	-1	-1	-1	-1	-1	-1	0	0	0	0	0	0	0
1.09	1	-1	0	0	0	0	0	0	0	1	0	0	0	0	0	0
1.21	1	-1	0	0	0	0	0	0	0	0	1	0	0	0	0	0
1.23	1	-1	0	0	0	0	0	0	0	0	0	1	0	0	0	0
1.23	1	-1	0	0	0	0	0	0	0	0	0	0	1	0	0	0
1.25	1	-1	0	0	0	0	0	0	0	0	0	0	0	1	0	0
1.34	1	-1	0	0	0	0	0	0	0	0	0	0	0	0	1	0
1.54	1	-1	0	0	0	0	0	0	0	0	0	0	0	0	0	1
1.60	1	-1	0	0	0	0	0	0	0	-1	-1	-1	-1	-1	-1	-1
1.25	-1	1	1	0	0	0	0	0	0	0	0	0	0	0	0	0
1.27	-1	1	0	1	0	0	0	0	0	0	0	0	0	0	0	0
1.39	-1	1	0	0	1	0	0	0	0	0	0	0	0	0	0	0
1.41	-1	1	0	0	0	1	0	0	0	0	0	0	0	0	0	0
1.53	-1	1	0	0	0	0	1	0	0	0	0	0	0	0	0	0
1.55	-1	1	0	0	0	0	0	1	0	0	0	0	0	0	0	0
1.63	-1	1	0	0	0	0	0	0	1	0	0	0	0	0	0	0
1.80	-1	1	-1	-1	-1	-1	-1	-1	-1	0	0	0	0	0	0	0
1.46	-1	-1	0	0	0	0	0	0	0	1	0	0	0	0	0	0
1.71	-1	-1	0	0	0	0	0	0	0	0	1	0	0	0	0	0
1.81	-1	-1	0	0	0	0	0	0	0	0	0	1	0	0	0	0
1.90	-1	-1	0	0	0	0	0	0	0	0	0	0	1	0	0	0
1.90	-1	-1	0	0	0	0	0	0	0	0	0	0	0	1	0	0
1.94	-1	-1	0	0	0	0	0	0	0	0	0	0	0	0	1	0
2.02	-1	-1	0	0	0	0	0	0	0	0	0	0	0	0	0	1
2.36	-1	-1	0	0	0	0	0	0	0	-1	-1	-1	-1	-1	-1	-1

TABLE C-16 Depression Score S (points) while Sober ($D = 1$) or Drinking ($D = -1$) in Alcoholic Subjects (P_{ASP_i} and $P_{\text{non-ASP}_i}$) Diagnosed as Either Antisocial Personality ($A = -1$) or Non-Antisocial Personality ($A = 1$)

S	D	A	$P_{\text{non-ASP}_i}$							P_{ASP_i}						
			1	2	3	4	5	6	7	1	2	3	4	5	6	7
1.21	1	1	1	0	0	0	0	0	0	0	0	0	0	0	0	0
2.06	1	1	0	1	0	0	0	0	0	0	0	0	0	0	0	0
0.88	1	1	0	0	1	0	0	0	0	0	0	0	0	0	0	0
1.23	1	1	0	0	0	1	0	0	0	0	0	0	0	0	0	0
1.86	1	1	0	0	0	0	1	0	0	0	0	0	0	0	0	0
0.61	1	1	0	0	0	0	0	1	0	0	0	0	0	0	0	0
0.68	1	1	0	0	0	0	0	0	1	0	0	0	0	0	0	0
2.13	1	1	-1	-1	-1	-1	-1	-1	-1	0	0	0	0	0	0	0
1.04	1	-1	0	0	0	0	0	0	0	1	0	0	0	0	0	0
1.20	1	-1	0	0	0	0	0	0	0	0	1	0	0	0	0	0
1.40	1	-1	0	0	0	0	0	0	0	0	0	1	0	0	0	0
1.88	1	-1	0	0	0	0	0	0	0	0	0	0	1	0	0	0
0.84	1	-1	0	0	0	0	0	0	0	0	0	0	0	1	0	0
1.52	1	-1	0	0	0	0	0	0	0	0	0	0	0	0	1	0
1.35	1	-1	0	0	0	0	0	0	0	0	0	0	0	0	0	1
1.13	1	-1	0	0	0	0	0	0	0	-1	-1	-1	-1	-1	-1	-1
1.56	-1	1	1	0	0	0	0	0	0	0	0	0	0	0	0	0
1.29	-1	1	0	1	0	0	0	0	0	0	0	0	0	0	0	0
1.89	-1	1	0	0	1	0	0	0	0	0	0	0	0	0	0	0
1.13	-1	1	0	0	0	1	0	0	0	0	0	0	0	0	0	0
1.82	-1	1	0	0	0	0	1	0	0	0	0	0	0	0	0	0
1.61	-1	1	0	0	0	0	0	1	0	0	0	0	0	0	0	0
1.23	-1	1	0	0	0	0	0	0	1	0	0	0	0	0	0	0
0.83	-1	1	-1	-1	-1	-1	-1	-1	-1	0	0	0	0	0	0	0
2.27	-1	-1	0	0	0	0	0	0	0	1	0	0	0	0	0	0
1.71	-1	-1	0	0	0	0	0	0	0	0	1	0	0	0	0	0
0.50	-1	-1	0	0	0	0	0	0	0	0	0	1	0	0	0	0
1.27	-1	-1	0	0	0	0	0	0	0	0	0	0	1	0	0	0
2.33	-1	-1	0	0	0	0	0	0	0	0	0	0	0	1	0	0
1.36	-1	-1	0	0	0	0	0	0	0	0	0	0	0	0	1	0
1.60	-1	-1	0	0	0	0	0	0	0	0	0	0	0	0	0	1
2.85	-1	-1	0	0	0	0	0	0	0	-1	-1	-1	-1	-1	-1	-1

TABLE C-17 Time Constants of Pressure Response to Sudden Left Ventricular Volume Infusion ($V = 1$) or Withdrawal ($V = 0$) during Fast ($S = 0$) and Slow ($S = 1$) Phases of Response

Time Constant T, ms	V	S	Subject	Time Constant T, ms	V	S	Subject
23	0	0	1	8	1	0	1
7	0	0	2	6	1	0	2
14	0	0	3	8	1	0	3
25	0	0	4	13	1	0	4
18	0	0	5	4	1	0	5
7	0	0	6	5	1	0	6
33	0	0	7	6	1	0	7
123	0	1	1	183	1	1	1
74	0	1	2	68	1	1	2
177	0	1	3	119	1	1	3
79	0	1	4	147	1	1	4
40	0	1	5	102	1	1	5
196	0	1	6	174	1	1	6
65	0	1	7	89	1	1	7

TABLE C-18 Interest of Martians in Invading Earth as a Function of Time

Time t, day	Interest in Invasion I, percent	Time t, day	Interest in Invasion I, percent
1	17.3	11	3.6
2	15.9	12	9.0
3	13.9	13	12.5
4	11.6	14	9.0
5	5.0	15	13.5
6	8.2	16	10.7
7	6.0	17	11.8
8	0.5	18	17.4
9	0.1	19	12.8
10	3.3	20	14.7

TABLE C-18 (*Continued*)   Interest of Martians in Invading Earth as a Function of Time

Time t, day	Interest in Invasion I, percent	Time t, day	Interest in Invasion I, percent
21	13.0	36	19.3
22	6.7	37	18.9
23	2.5	38	15.7
24	9.0	39	4.9
25	4.9	40	9.0
26	1.1	41	6.7
27	0.1	42	0.9
28	1.1	43	1.3
29	4.7	44	0.0
30	5.6	45	0.0
31	10.7	46	2.0
32	13.4	47	6.0
33	17.3	48	3.8
34	11.8	49	10.3
35	20.2	50	16.2

TABLE C-19 Blood Carnitine Concentration as a Function of Time after Intravenous Injection

Carnitine $c(t)$, μmol/L	Time t, min	Carnitine $c(t)$, μmol/L	Time t, min
1150	5	120	240
850	15	110	270
780	20	90	300
700	25	80	330
660	30	50	360
545	45	40	420
485	60	20	480
365	90	15	540
290	120	10	600
245	150	5	660
185	180	5	720
160	210		

TABLE C-20 Normalized Carotid Sinus Nerve Activity and Carotid Sinus
Blood Pressure for Study of Control of Blood Pressure in Normal Dogs

Nerve Activity N, %	Carotid Sinus Pressure P, mmHg	Nerve Activity N, %	Carotid Sinus Pressure P, mmHg
6.7	20	75.6	130
0.0	20	73.6	130
25.7	20	90.6	130
4.2	20	91.0	150
0.0	20	79.6	150
19.2	20	77.6	150
18.3	30	88.7	150
17.7	30	82.4	150
5.2	30	90.7	150
0.5	30	77.7	170
7.9	30	77.1	170
3.8	30	105.3	170
14.3	50	79.2	170
1.4	50	97.1	170
10.0	50	89.9	170
0.0	50	97.9	190
20.1	50	83.2	190
7.2	50	79.8	190
41.5	65	102.0	190
30.1	65	90.3	190
28.1	65	98.3	190
39.2	65	97.3	210
32.9	65	77.3	210
41.2	65	79.2	210
57.8	85	101.4	210
57.2	85	89.7	210
44.7	85	97.7	210
41.0	85	95.3	230
47.4	85	102.9	230
43.3	85	87.8	230
65.8	105	98.6	230
55.7	105	85.4	230
84.8	105	80.1	230
63.3	105	98.2	255
59.1	105	93.3	255
78.3	105	81.4	255
78.1	130	88.4	255
68.7	130	98.2	255
97.1	130	87.5	255

TABLE C-21 Normalized Carotid Sinus Nerve Activity and Carotid Sinus Blood Pressure for Study of Control of Blood Pressure in Renal Hypertensive Dogs

Nerve Activity N, %	Carotid Sinus Pressure P, mmHg	Nerve Activity N, %	Carotid Sinus Pressure P, mmHg
6.5	5	61.4	125
6.3	5	62.8	125
7.8	5	61.9	125
0.6	5	56.5	140
0.9	5	61.2	140
10.3	5	68.5	140
3.0	20	61.8	140
7.9	20	59.1	140
5.6	20	49.5	140
0.0	20	81.4	165
2.1	20	74.7	165
8.4	20	72.7	165
7.9	45	63.0	165
12.8	45	70.0	165
11.8	45	65.7	165
4.9	45	62.8	175
7.0	45	62.5	175
13.3	45	67.2	175
13.8	55	73.5	175
11.8	55	75.6	175
17.1	55	74.0	175
8.5	55	76.7	210
17.5	55	66.9	210
21.8	55	66.9	210
16.3	85	80.4	210
15.7	85	70.0	210
20.7	85	69.9	210
27.0	85	70.9	230
28.4	85	75.5	230
27.5	85	82.9	230
40.6	100	76.2	230
40.0	100	73.5	230
45.0	100	64.4	230
51.3	100	67.5	250
53.4	100	81.3	250
51.8	100	77.2	250
50.7	125	64.4	250
50.1	125	66.7	250
53.5	125	77.9	250

TABLE C-22 Martian Graduations ($G = 1$) as a Function of Intelligence

Graduation G	Intelligence I, zorp	Subject	Graduation G	Intelligence I, zorp	Subject
0	1	1	0	13	44
0	3	2	0	9	45
0	4	3	1	8	46
0	1	4	0	10	47
0	5	5	1	15	48
0	3	6	1	14	49
0	2	7	1	11	50
0	3	8	1	11	51
1	5	9	1	13	52
0	6	10	1	12	53
0	8	11	1	16	54
0	5	12	1	17	55
0	6	13	0	8	56
0	10	14	0	8	57
0	11	15	1	8	58
0	8	16	1	16	59
0	7	17	1	14	60
0	5	18	1	13	61
0	6	19	1	12	62
0	8	20	1	17	63
0	11	21	1	11	64
0	9	22	1	11	65
1	7	23	1	15	66
0	5	24	1	14	67
0	3	25	1	13	68
0	4	26	1	14	69
0	5	27	1	16	70
1	6	28	1	8	71
0	8	29	1	9	72
0	9	30	0	10	73
0	10	31	1	10	74
0	9	32	0	8	75
0	5	33	1	13	76
0	7	34	1	9	77
0	4	35	1	16	78
0	3	36	1	10	79
0	2	37	1	10	80
1	10	38	0	12	81
1	19	39	1	13	82
1	17	40	1	15	83
1	16	41	1	14	84
1	15	42	1	12	85
1	13	43	1	12	86

TABLE C-22 (*Continued*)   Martian Graduations ($G = 1$) as a Function of Intelligence

Graduation G	Intelligence I, zorp	Subject	Graduation G	Intelligence I, zorp	Subject
1	9	87	1	12	104
1	11	88	1	11	105
1	13	89	1	9	106
1	15	90	1	8	107
1	15	91	1	15	108
1	14	92	1	17	109
1	18	93	1	19	110
1	17	94	1	10	111
1	15	95	1	13	112
1	10	96	1	13	113
1	15	97	1	11	114
1	10	98	1	15	115
1	9	99	1	17	116
1	15	100	1	13	117
1	14	101	1	12	118
1	17	102	1	18	119
1	12	103	1	13	120

TABLE C-23 Treatment Group (Presurgical Chemotherapy, $C = 1$, vs. No Chemotherapy, $C = 0$) and Bone Tumor Characteristics Quantified as Viability and Number of New Fibroblasts

Group C	Viability V	Fibroblasts F	Subject
0	2	0	1
0	2	0	2
0	3	0	3
0	2	0	4
1	2	0	5
1	2	0	6
0	2	1	7
0	3	0	8
0	3	0	9
0	3	0	10
1	3	2	11
0	2	0	12
0	1	0	13
0	2	0	14
1	4	0	15

TABLE C-23 (*Continued*)   Treatment Group (Presurgical Chemotherapy, $C = 1$, vs. No Chemotherapy, $C = 0$) and Bone Tumor Characteristics Quantified as Viability and Number of New Fibroblasts

Group C	Viability V	Fibroblasts F	Subject
0	2	0	16
0	2	1	17
0	2	3	18
1	4	0	19
1	3	1	20
0	2	1	21
0	3	0	22
1	2	0	23
0	2	0	24
0	3	0	25
1	2	0	26
1	2	0	27
0	2	0	28
0	2	1	29
0	3	0	30
0	3	0	31
0	3	0	32
0	3	2	33
0	2	0	34
0	1	0	35
0	2	0	36
0	4	0	37
1	2	0	38
0	2	1	39
1	2	3	40
0	4	0	41
1	3	1	42
0	2	1	43
0	3	0	44
0	2	0	45
0	2	0	46
0	3	0	47
0	2	0	48
0	2	0	49
0	2	0	50
0	2	1	51
0	3	0	52
0	3	0	53
0	3	0	54
0	3	2	55
1	2	0	56

TABLE C-23 (*Continued*)   Treatment Group (Presurgical Chemotherapy,
C = 1, vs. No Chemotherapy, *C* = 0) and Bone Tumor Characteristics
Quantified as Viability and Number of New Fibroblasts

Group *C*	Viability *V*	Fibroblasts *F*	Subject
0	1	0	57
0	2	0	58
0	4	0	59
0	2	0	60
0	2	1	61
0	2	3	62
0	4	0	63
0	3	1	64
0	2	1	65
0	3	0	66
0	2	0	67
0	2	0	68
0	3	0	69
0	2	0	70
0	2	0	71
1	2	0	72
0	2	1	73
0	3	0	74
0	3	0	75
0	3	0	76
1	3	2	77
0	2	0	78
0	1	0	79
0	2	0	80
1	4	0	81
0	2	0	82
0	2	1	83
1	2	3	84
0	4	0	85
0	3	1	86
0	2	1	87
0	3	0	88
1	1	0	89
1	2	2	90
0	2	2	91
1	1	2	92
0	1	1	93
1	2	2	94
1	1	2	95
1	1	1	96
1	1	3	97

TABLE C-23 (*Continued*)   Treatment Group (Presurgical Chemotherapy, $C = 1$, vs. No Chemotherapy, $C = 0$) and Bone Tumor Characteristics Quantified as Viability and Number of New Fibroblasts

Group C	Viability V	Fibroblasts F	Subject
1	1	1	98
1	1	2	99
1	2	2	100
0	2	2	101
0	2	2	102
0	1	1	103
1	1	3	104
1	3	0	105
1	1	2	106
1	2	1	107
1	1	2	108
1	1	2	109
1	1	3	110
1	1	0	111
0	2	2	112
1	2	2	113
0	1	2	114
1	1	1	115
1	2	2	116
1	1	2	117
1	1	1	118
0	1	3	119
1	1	1	120
1	1	2	121
1	2	2	122
1	2	2	123
1	2	2	124
1	1	1	125
1	1	3	126
1	3	0	127
1	1	2	128
1	2	1	129
1	1	2	130
0	1	2	131
1	1	3	132
1	1	0	133
1	2	2	134
0	2	2	135
1	1	2	136
1	1	1	137

TABLE C-23 (*Continued*)   Treatment Group (Presurgical Chemotherapy,
$C = 1$, vs. No Chemotherapy, $C = 0$) and Bone Tumor Characteristics
Quantified as Viability and Number of New Fibroblasts

Group C	Viability V	Fibroblasts F	Subject
1	2	2	138
1	1	2	139
1	1	1	140
1	1	3	141
1	1	1	142
0	1	2	143
1	2	2	144
1	2	2	145
1	2	2	146
1	1	1	147
1	1	3	148
1	3	0	149
1	1	2	150
1	2	1	151
1	1	2	152
1	1	2	153
0	1	3	154
1	1	0	155
1	2	2	156
1	2	2	157
1	1	2	158
1	1	1	159
0	2	2	160
0	1	2	161
1	1	1	162
0	1	3	163
1	1	1	164
0	1	2	165
1	2	2	166
1	2	2	167
1	2	2	168
0	1	1	169
1	1	3	170
1	3	0	171
1	1	2	172
1	2	1	173
1	1	2	174
1	1	2	175
1	1	3	176

TABLE C-24 Positive ($T = 1$) or Negative ($T = 0$) Exercise Thallium Test and Patient Characteristics; Degree of Coronary Artery Stenosis, Propranolol Use ($P = 1$), Heart Rate Change during Exercise ($R = 0$ if < 85% Maximum Predicted Rate or $R = 1$ if > 85% Maximum Predicted Rate), Ischemia during Exercise ($E = 1$), Sex (Male = 1, Female = 1), and Chest Pain during Exercise (0 = None, 1 = Moderate, 2 = Severe)

Thallium Test T	Percent Stenosis D	Propranolol P	Heart Rate R	Ischemia E	Sex S	Chest Pain C	Subject
1	40	0	0	0	1	0	1
1	40	0	1	0	1	0	2
1	40	1	0	1	1	2	3
1	30	0	1	0	1	0	4
1	30	0	0	0	1	0	5
0	40	0	1	0	1	1	6
0	30	1	1	1	0	0	7
0	30	0	1	0	1	2	8
1	40	1	0	0	0	0	9
1	30	0	0	0	1	0	10
1	30	0	0	0	0	0	11
0	30	0	0	0	1	0	12
1	40	0	0	0	0	0	13
0	0	0	1	0	0	1	14
1	0	0	1	0	0	2	15
0	0	1	1	1	0	0	16
1	0	1	1	0	0	0	17
0	0	0	1	0	1	0	18
0	10	0	1	1	0	0	19
0	0	0	0	0	1	0	20
0	0	0	1	0	0	1	21
1	0	1	0	0	1	0	22
0	0	0	1	0	1	1	23
0	0	0	1	0	0	0	24
0	0	0	1	1	1	0	25
0	0	0	1	0	1	0	26
0	0	0	1	1	1	0	27
0	0	0	1	0	0	0	28
0	0	1	1	1	1	0	29
0	0	0	0	0	1	0	30
0	0	0	1	0	1	0	31
0	0	1	0	1	0	0	32
0	10	1	0	0	0	0	33
0	0	1	1	0	1	0	34
0	0	1	0	0	1	0	35
1	0	0	0	1	1	0	36
0	0	0	1	0	1	0	37
0	0	0	1	0	1	0	38

TABLE C-24 (*Continued*)   Positive (T = 1) or Negative (T = 0) Exercise Thallium Test and Patient Characteristics; Degree of Coronary Artery Stenosis, Propranolol Use (P = 1), Heart Rate Change during Exercise (R = 0 if < 85% Maximum Predicted Rate or R = 1 if > 85% Maximum Predicted Rate), Ischemia during Exercise (E = 1), Sex (Male = 1, Female = 1), and Chest Pain during Exercise (0 = None, 1 = Moderate, 2 = Severe)

Thallium Test T	Percent Stenosis D	Propranolol P	Heart Rate R	Ischemia E	Sex S	Chest Pain C	Subject
1	0	1	0	0	0	2	39
0	20	1	0	1	0	0	40
0	0	1	0	1	0	0	41
0	0	1	0	1	0	0	42
0	0	1	1	0	0	0	43
0	0	1	1	0	1	0	44
0	0	0	1	0	0	0	45
0	0	1	1	0	1	0	46
0	0	1	0	0	1	0	47
0	0	0	1	0	0	0	48
1	0	1	0	0	0	0	49
1	0	1	1	0	0	0	50
0	0	0	0	0	1	0	51
1	0	0	0	0	0	0	52
0	0	0	1	0	1	0	53
0	0	0	1	0	1	1	54
1	0	1	1	0	1	2	55
1	0	1	0	0	0	0	56
0	0	0	1	1	0	0	57
0	0	0	0	0	0	0	58
0	40	1	0	0	0	2	59
0	0	1	0	0	0	0	60
0	0	1	0	0	0	0	61
1	0	1	1	1	0	1	62
0	0	1	0	0	0	0	63
1	30	1	1	0	1	0	64
0	20	1	0	0	1	1	65
0	30	1	0	0	0	0	66
1	0	1	0	0	0	0	67
1	30	1	0	0	1	0	68
0	0	1	1	0	1	0	69
1	0	1	0	0	0	0	70
1	0	1	1	0	1	0	71
1	0	1	1	0	1	0	72
1	0	1	0	0	1	0	73
0	30	1	1	1	0	0	74
0	50	1	1	0	1	0	75
0	0	1	0	0	0	0	76

TABLE C-24 (*Continued*) Positive ($T = 1$) or Negative ($T = 0$) Exercise Thallium Test and Patient Characteristics; Degree of Coronary Artery Stenosis, Propranolol Use ($P = 1$), Heart Rate Change during Exercise ($R = 0$ if $< 85\%$ Maximum Predicted Rate or $R = 1$ if $> 85\%$ Maximum Predicted Rate), Ischemia during Exercise ($E = 1$), Sex (Male $= 1$, Female $= 1$), and Chest Pain during Exercise ($0 =$ None, $1 =$ Moderate, $2 =$ Severe)

Thallium Test T	Percent Stenosis D	Propranolol P	Heart Rate R	Ischemia E	Sex S	Chest Pain C	Subject
1	0	1	0	1	0	1	77
0	0	1	0	0	0	0	78
0	0	0	0	0	1	0	79
0	0	0	1	0	0	0	80
0	0	0	0	0	1	0	81
0	0	0	1	0	0	0	82
1	0	0	0	0	1	1	83
0	0	0	1	0	0	0	84
0	0	0	1	0	1	0	85
0	0	0	0	0	1	0	86
0	0	0	0	0	1	0	87
1	0	0	1	0	0	0	88
0	0	0	1	1	1	0	89
0	0	0	0	0	1	0	90
0	0	0	0	0	1	0	91
0	0	0	0	0	1	0	92
0	0	0	1	0	1	0	93
0	40	1	0	0	1	0	94
1	40	1	1	1	1	0	95
1	40	0	0	0	0	0	96
0	0	0	1	0	1	0	97
1	0	0	1	0	0	2	98
1	0	0	1	0	1	0	99
0	10	0	0	0	1	0	100

APPENDIX

D

Data for Problems

TABLE D-1 Level of Sedation and Plasma Cortisol during Valium Administration

Sedation Score S	Cortisol C, μg/dL
32	6.6
42	7.4
52	8.8
61	9.7
62	10.5
65	11.8
66	10.7

TABLE D-2 Breast Cancer Rate, Male Lung Cancer Rate, and Dietary Animal Fat Intake in Different Countries

Breast Cancer F, deaths/100,000	Male Lung Cancer L, deaths/100,000	Animal Fat A, g/day	Breast Cancer F, deaths/100,000	Male Lung Cancer L, deaths/100,000	Animal Fat A, g/day
25	76	119	15	66	103
26	71	100	13	14	30
24	44	120	12	44	76
25	48	135	10	27	38
25	49	143	10	37	38
23	47	120	9	32	41
23	47	81	9	41	39
22	68	99	10	17	39
22	52	100	9	30	40
19	48	92	8	16	62
19	24	95	9	43	40
19	48	120	8	15	38
18	53	80	7	11	20
17	37	100	5	7	18
17	43	40	5	9	22
16	49	88	4	19	18
17	20	107	1	5	10
16	68	64	1	2	15

TABLE D-3 Antibiotic Efficacy and Blood Levels during Different Dosing Intervals

Number of Colonies C, log CFU	Time above MIC M, % of 24 h	Dose Code*	Number of Colonies C, log CFU	Time above MIC M, % of 24 h	Dose Code*
1.30	0	0	−1.61	89	0
2.26	15	0	−1.80	89	0
2.25	18	0	−2.40	93	0
2.06	15	0	−3.31	99	0
2.06	18	0	−2.69	99	0
1.85	17	0	−2.54	99	0
1.23	21	0	−2.29	99	0
1.23	24	0	−2.02	99	0
0.89	21	0	−1.72	99	0
0.89	24	0	−1.52	99	0
0.40	29	0	−1.39	99	0
0.47	36	0	−1.06	99	0
0.07	29	0	2.47	10	1
0.24	36	0	2.33	13	1
−0.81	26	0	2.11	13	1
−1.19	32	0	2.00	11	1
−1.18	35	0	1.83	11	1
−0.97	35	0	1.92	18	1
−0.78	43	0	1.71	15	1
−1.03	44	0	1.53	18	1
−1.14	43	0	1.42	19	1
−1.36	43	0	1.23	19	1
−1.39	44	0	0.68	19	1
−0.92	53	0	0.75	22	1
−1.11	53	0	0.06	22	1
−1.19	50	0	−0.01	26	1
−1.39	50	0	−0.22	22	1
−1.55	53	0	−0.28	26	1
−1.73	53	0	−0.42	29	1
−1.83	59	0	−0.77	26	1
−1.91	59	0	−0.97	32	1
−2.05	63	0	−1.11	31	1
−1.04	63	0	−1.11	39	1
−0.92	65	0	−1.61	46	1
−1.58	65	0	−1.80	46	1
−1.61	71	0	−2.28	60	1
−1.91	71	0	−2.43	60	1
−2.13	79	0	−1.61	29	1
−2.21	79	0	−2.58	38	1
−2.73	76	0	−2.90	38	1
−3.61	76	0	−2.90	51	1
−0.89	88	0	−3.12	51	1

*Dose code: 0 = 1- to 4-h intervals, 1 = 6- to 12-h intervals.

TABLE D-4 Renal Vascular Resistance and Plasma Renin during Infusion of Solutions with Differing Concentrations of Ions

Change in Renal Vascular Resistance ΔR, mmHg/(mL·min·kg)	Change in Plasma Renin Δr, ng/(mL·h)	Group Code*	Change in Renal Vascular Resistance ΔR, mmHg/(mL·min·kg)	Change in Plasma Renin Δr, ng/(mL·h)	Group Code*
2.0	−28.1	1	1.7	0.3	2
5.2	−6.7	1	1.9	−6.0	2
3.9	−8.6	1	−2.1	−1.3	3
0.6	2.0	1	−3.1	12.6	3
4.4	−6.2	1	−1.4	2.5	3
−0.2	−4.4	1	−0.8	−0.5	3
−0.7	−16.6	1	−0.5	8.7	3
3.6	−13.1	1	−3.4	7.5	3
1.6	5.8	2	−1.9	15.0	3
0.8	1.6	2	−0.4	7.7	3
−0.1	−3.1	2	−2.4	12.8	4
4.9	−1.8	2	−1.6	−7.5	4
2.3	0.7	2	−4.0	22.4	4
0.1	−8.1	2	−5.1	17.3	4
1.2	0.9	2	−3.8	11.4	4

*Group code: 1 = NaCl plus arachidonic acid, 2 = NaCl only, 3 = dextrose, and 4 = sodium acetate.

TABLE D-5 Kidney Function and Dietary Calcium and Protein Intake in Patients with Total Parenteral Nutrition

Urinary Ca^{2+} U_{Ca}, mg/12 h	Dietary Ca^{2+} D_{Ca}, mg/12 h	Glomerular Filtration Rate G_{fr}, mL/min	Urinary Na^+ U_{Na}, meq/12 h	Dietary Protein D_P, g/day
220	554	63	54	80
182	303	43	99	48
166	287	45	14	53
162	519	58	55	70
137	249	53	37	82
136	142	70	37	36
128	184	85	110	56
113	150	49	78	25
100	383	56	79	101
90	391	51	1	76
75	90	91	134	28
71	0	58	29	0
60	279	39	1	41
60	249	41	17	37
43	208	50	3	41
42	182	46	35	50
37	0	42	54	0
29	125	69	1	42
24	0	36	20	0
22	170	30	7	54
15	0	66	1	0
3	0	25	0	0
1	82	16	2	27
36	0	66	83	0
31	125	40	6	42
21	0	45	16	0
12	0	23	3	0

TABLE D-6 Prediction of Muscle Fiber Composition with MR Spectroscopy

Muscle Fiber Type F, % fast twitch	T_1, ms	T_2, ms	Muscle Fiber Type F, % fast twitch	T_1, ms	T_2, ms
38	313	22.3	58	350	28.3
44	320	26.7	55	354	31.0
39	322	25.6	61	362	29.5
33	320	24.5	67	364	29.7
29	321	23.0	68	368	29.9
45	324	28.1	69	373	30.3
47	332	26.5	72	382	32.3
55	348	28.3	93	379	32.7

TABLE D-7 Coverage of Tobacco-Related News in Publications That Do, or Do Not, Accept Tobacco Advertising

Year	Advertising Revenue, % of Total Advertising Revenue	Smoking-Related Stories in the *Christian Science Monitor*	Average No. of Smoking Articles per Year per Publication
1959	1.25	6	0.13
1960	1.51	1	0.13
1961	1.53	5	0.05
1962	1.90	9	0.36
1963	2.14	22	0.36
1964	1.81	75	0.82
1965	2.20	23	0.28
1966	1.81	21	0.21
1967	1.86	29	0.46
1968	1.93	23	0.15
1969	2.32	57	0.56
1973	9.25	32	0.08
1974	10.86	18	0.08
1975	12.56	20	0.03
1976	11.56	19	0.05
1977	10.85	44	0.08
1978	9.64	30	0.18
1979	11.76	28	0.08
1980	11.44	10	0.13
1981	11.16	18	0.18
1982	11.18	19	0.13
1983	10.91	24	0.23
1959	0.00	6	0.36
1960	0.00	1	0.00
1961	0.00	5	0.27
1962	0.00	9	0.36
1963	0.00	22	0.27
1964	0.00	75	0.73
1965	0.00	23	0.27
1966	0.00	21	0.36
1967	0.00	29	0.27
1968	0.00	23	0.36
1969	0.00	57	0.18
1973	0.00	32	0.00
1974	0.00	18	0.00
1975	0.00	20	0.00
1976	0.00	19	0.36
1977	0.00	44	0.09
1978	0.00	30	0.27
1979	0.00	28	0.18
1980	0.00	10	0.55
1981	0.00	18	0.36
1982	0.00	19	0.09
1983	0.00	24	0.55

TABLE D-8 Blood Pressure and Plasma Norepinephrine during Different Levels of Exercise with and without the β-Blocker Xamoterol

Systolic Blood Pressure P, mmHg	Plasma Norepinephrine N, pg/mL	Exercise Group[*]	Drug Group[†]
134	335	1	0
146	226	1	0
132	339	1	0
101	160	1	0
96	132	1	0
144	214	1	0
167	238	2	0
136	533	2	0
156	598	2	0
108	307	2	0
109	357	2	0
152	180	2	0
136	572	3	0
147	382	3	0
126	528	3	0
182	539	3	0
117	326	3	0
192	911	3	0
148	1091	4	0
159	625	4	0
138	509	4	0
194	943	4	0
129	954	4	0
204	403	4	0
157	7	1	1
85	262	1	1
177	97	1	1
140	256	1	1
145	329	1	1
146	211	1	1
146	332	1	1
132	176	1	1
82	146	1	1
143	297	1	1
170	317	2	1
156	418	2	1
160	616	2	1

[*]Exercise code: 1 = no exercise, 2 = 33 percent of maximum capacity, 3 = 67 percent of maximum capacity, 4 = 100 percent of maximum capacity.
[†]Drug code: 0 = no drug, 1 = drug.

TABLE D-8 (*Continued*)   **Blood Pressure and Plasma Norepinephrine during Different Levels of Exercise with and without the β-Blocker Xamoterol**

Systolic Blood Pressure P, mmHg	Plasma Norepinephrine N, pg/mL	Exercise Group*	Drug Group†
106	19	2	1
191	740	2	1
152	490	2	1
142	693	2	1
153	637	2	1
114	293	2	1
115	46	2	1
180	476	3	1
108	1069	3	1
200	1098	3	1
163	504	3	1
168	761	3	1
169	736	3	1
169	628	3	1
155	1073	3	1
105	1222	3	1
166	88	3	1
175	832	4	1
197	1392	4	1
123	802	4	1
184	559	4	1
198	1946	4	1
196	905	4	1
144	1564	4	1
113	1493	4	1
109	61	4	1
174	1125	4	1

TABLE D-9 Leucine Accumulation Over Time at Different Cholesterol Content of Artificial Membranes

Leucine Uptake L, nmol/mg	Time t, min	Membrane Cholesterol C, fraction of total lipid	Leucine Uptake L, nmol/mg	Time t, min	Membrane Cholesterol C, fraction of total lipid
0.08	5	0.20	2.03	25	0.05
0.23	5	0.10	2.91	25	0.00
0.51	5	0.05	0.20	30	0.20
1.23	5	0.00	0.75	30	0.10
0.14	10	0.20	1.69	30	0.05
0.31	10	0.10	3.69	30	0.00
0.51	10	0.05	0.24	60	0.20
1.37	10	0.00	0.94	60	0.10
0.14	15	0.20	2.42	60	0.05
0.34	15	0.10	3.72	60	0.00
1.26	15	0.05	0.22	90	0.20
1.97	15	0.00	1.02	90	0.10
0.20	20	0.20	2.43	90	0.05
0.47	20	0.10	3.31	90	0.00
1.57	20	0.05	0.22	120	0.20
1.98	20	0.00	1.01	120	0.10
0.17	25	0.20	1.97	120	0.05
0.49	25	0.10	2.69	120	0.00

TABLE D-10 Comparison of RIA and Bioassay for Inhibin during the Estrus Cycle

RIA Level R, U/L	Bioassay Level B, U/L	Cycle Code*	RIA Level R, U/L	Bioassay Level B, U/L	Cycle Code*
295	260	0	745	1250	0
265	365	0	740	1295	0
230	485	0	740	1495	0
230	330	0	940	1950	0
195	335	0	840	2250	0
145	350	0	1010	2015	0
100	370	0	1250	2315	0
205	390	0	1385	2210	0
180	430	0	1430	2355	0
155	435	0	1755	2465	0
200	895	0	920	3215	0
330	825	0	2345	4915	0
380	820	0	100	300	1
430	755	0	245	230	1
470	1230	0	330	410	1
430	1270	0	255	555	1
425	1025	0	635	495	1
425	1110	0	805	355	1
335	1290	0	575	1020	1
685	895	0	640	1290	1
640	945	0	1000	1400	1
645	1025	0	1180	1025	1
580	1215	0	1795	635	1
970	1020	0	2370	960	1
1155	965	0	1965	2440	1
680	1255	0			

*Cycle code: 0 = early, 1 = midluteal.

TABLE D-11 Modulation of Pulmonary Stretch Receptor Nerve Discharge by the Central Nervous System

Normal Discharge S, spikes/s	Change in Discharge ΔS, spikes/s	Frequency F, Hz	Normal Discharge S, spikes/s	Change in Discharge ΔS, spikes/s	Frequency F, Hz
1.9	23.5	20	34.8	81.4	40
0.0	25.5	20	38.1	91.2	40
1.8	29.5	20	43.9	97.4	40
11.0	30.6	20	35.5	98.2	40
7.7	35.6	20	37.4	102.1	40
7.7	38.5	20	12.9	31.6	80
3.9	40.6	20	21.3	42.8	80
3.9	44.4	20	28.4	47.5	80
9.1	50.6	20	32.3	57.4	80
6.2	52.4	20	34.2	61.4	80
7.7	59.3	20	41.9	69.5	80
11.0	64.6	20	48.1	80.0	80
8.0	71.6	20	46.2	86.6	80
20.0	73.2	20	52.9	89.7	80
12.9	78.6	20	61.2	102.3	80
16.8	89.2	20	56.5	102.3	80
16.1	89.2	20	60.4	111.5	80
16.1	92.4	20	60.4	113.2	80
18.1	99.3	20	10.3	22.3	120
5.2	23.5	40	11.4	29.5	120
11.4	32.3	40	25.8	44.4	120
11.4	36.1	40	35.1	56.6	120
18.1	37.2	40	51.9	71.6	120
17.0	44.4	40	50.1	75.6	120
20.0	47.5	40	59.1	86.5	120
20.9	51.2	40	64.3	91.3	120
19.4	59.3	40	69.0	99.5	120
25.2	61.2	40	73.3	109.2	120
30.3	71.5	40	75.1	115.6	120
25.8	77.2	40			

TABLE D-12 Minute Ventilation, Inspired Oxygen, and Inspired Carbon Dioxide for Study of Respiratory Control in Adult Bank Swallows

Minute Ventilation V, Percent Change	Inspired O_2 O, Volume %	Inspired CO_2 C, Volume %	Minute Ventilation V, Percent Change	Inspired O_2 O, Volume %	Inspired CO_2 C, Volume %
−99	19	0.0	−263	17	3.0
−35	19	0.0	−3	17	4.5
207	19	0.0	−41	17	4.5
−130	19	0.0	140	17	4.5
−130	19	0.0	54	17	4.5
250	19	0.0	427	17	4.5
−34	19	3.0	70	17	4.5
−202	19	3.0	171	17	6.0
−310	19	3.0	347	17	6.0
−34	19	3.0	118	17	6.0
−11	19	3.0	5	17	6.0
186	19	3.0	152	17	6.0
256	19	4.5	387	17	6.0
−20	19	4.5	213	17	9.0
227	19	4.5	244	17	9.0
331	19	4.5	272	17	9.0
56	19	4.5	599	17	9.0
232	19	4.5	−81	17	9.0
220	19	6.0	78	17	9.0
15	19	6.0	−119	15	0.0
272	19	6.0	0	15	0.0
235	19	6.0	0	15	0.0
505	19	6.0	−112	15	0.0
65	19	6.0	−102	15	0.0
124	19	9.0	151	15	0.0
76	19	9.0	414	15	3.0
536	19	9.0	−42	15	3.0
346	19	9.0	114	15	3.0
313	19	9.0	58	15	3.0
5	19	9.0	200	15	3.0
118	17	0.0	419	15	3.0
−193	17	0.0	215	15	4.5
150	17	0.0	44	15	4.5
251	17	0.0	101	15	4.5
150	17	0.0	−45	15	4.5
−68	17	0.0	315	15	4.5
105	17	3.0	−1	15	4.5
155	17	3.0	195	15	6.0
158	17	3.0	−35	15	6.0
85	17	3.0	187	15	6.0
314	17	3.0	344	15	6.0

TABLE D-12 (*Continued*)   Minute Ventilation, Inspired Oxygen, and Inspired Carbon Dioxide for Study of Respiratory Control in Adult Bank Swallows

Minute Ventilation V, Percent Change	Inspired O_2 O, Volume %	Inspired CO_2 C, Volume %	Minute Ventilation V, Percent Change	Inspired O_2 O, Volume %	Inspired CO_2 C, Volume %
446	15	6.0	33	13	3.0
265	15	6.0	−173	13	4.5
485	15	9.0	−15	13	4.5
242	15	9.0	244	13	4.5
477	15	9.0	120	13	4.5
−15	15	9.0	430	13	4.5
492	15	9.0	377	13	4.5
113	15	9.0	572	13	6.0
41	13	0.0	60	13	6.0
144	13	0.0	176	13	6.0
−168	13	0.0	102	13	6.0
95	13	0.0	64	13	6.0
−11	13	0.0	468	13	6.0
−1	13	0.0	381	13	9.0
49	13	3.0	674	13	9.0
99	13	3.0	544	13	9.0
118	13	3.0	33	13	9.0
46	13	3.0	535	13	9.0
117	13	3.0	535	13	9.0

TABLE D-13   Growth Hormone and Insulin-like Growth Factor in Subjects with Acromegaly

Insulin-like Growth Factor/ Somatomedin-C I_{gf}, µg/L	Growth Hormone G_h, µg/L	Insulin-like Growth Factor/ Somatomedin-C I_{gf}, µg/L	Growth Hormone G_h, µg/L
61	12.4	511	18.5
297	10.9	590	18.5
327	14.2	646	17.4
359	4.4	660	17.4
373	6.6	692	11.4
426	7.8	715	9.6
433	18.6	753	17.7
462	11.5	706	24.9
491	15.0	496	54.8
502	9.2	745	67.9
526	10.3		

TABLE D-14 Urinary Ammonium and Anion Gap in Patients with Kidney Disease

Ammonium A, meq/L	Anion Gap U_{ag}, meq/L	Ammonium A, meq/L	Anion Gap U_{ag}, meq/L
105	−52	6	16
47	−30	3	22
44	−13	6	20
39	−16	11	27
31	−29	6	27
30	−23	6	29
14	−20	3	30
28	3	3	33
47	−41	3	35
33	−82	2	38
33	−15	4	35
33	−10	5	39
27	−21	9	38
26	−13	9	40
28	−15	11	35
11	2	12	32
11	6	12	35
6	4	14	35
9	10	11	38
17	11	11	40
21	13	14	47
16	20	14	56
12	22	10	56
3	10	10	61
3	13	7	66
1	18	2	66
3	16		

**TABLE D-15 Normalized Growth Rate and Growth Hormone
in Growth Hormone–Deficient Children**

Growth Rate G	Growth Hormone H, mU/L	Growth Rate G	Growth Hormone H, mU/L
3.4	2	5.1	59
6.2	5	4.7	76
6.6	4	3.2	73
8.4	8	2.6	70
8.8	2	0.7	87
10.0	2	2.3	103
2.5	16	2.6	103
3.7	13	3.4	100
7.6	20	3.2	105
7.6	26	2.9	114
8.0	32	3.6	118
2.9	42	2.9	125
2.9	43	1.7	118
1.6	51	0.4	152
2.8	51	2.8	165
3.9	47	2.4	170
2.8	62	1.7	179
4.0	64		

**TABLE D-16 Malate Dehydrogenase Activity (Reduction of Oxalacetate)
as a Function of pH in *C. aurantiacus***

Enzyme Activity A, % control	pH	Enzyme Activity A, % control	pH
17.4	4.30	49.4	6.21
41.1	4.99	45.2	6.21
44.5	5.00	42.5	6.21
48.4	5.42	54.8	6.39
50.9	5.73	48.4	6.39
55.0	5.73	45.0	6.39
54.8	5.96	58.2	6.60
64.8	6.84	56.2	6.60
68.5	6.85	51.3	6.60
43.0	5.80	60.1	6.79
48.4	5.96	60.6	7.00
48.4	6.02	74.6	7.00
43.0	6.02	79.2	7.00
40.6	6.02	72.4	7.18

TABLE D-16 (*Continued*) Malate Dehydrogenase Activity (Reduction of Oxalacetate) as a Function of pH in *C. aurantiacus*

Enzyme Activity A, % control	pH	Enzyme Activity A, % control	pH
75.3	7.18	81.9	7.51
79.7	7.42	75.8	7.48
80.2	7.38	76.3	7.63
84.1	7.38	81.9	7.66
84.4	7.60	81.2	7.76
85.6	7.63	85.3	7.76
89.5	7.60	84.6	7.94
91.4	7.80	91.4	7.94
96.8	7.80	94.1	8.08
100.2	7.80	96.6	8.08
70.4	7.00	91.9	8.17
67.0	7.00	97.3	8.17
68.5	7.27	86.1	8.29
75.8	7.27	55.0	8.49

TABLE D-17 Blood Alcohol Levels Determined with Breathalyzer and Computed from Self-Reports of Alcohol Consumption

Breathalyzer Level B, %	Self-Report Level S, %	Breathalyzer Level B, %	Self-Report Level S, %
0.00	0.01	0.07	0.08
0.01	0.01	0.08	0.09
0.01	0.02	0.08	0.09
0.01	0.02	0.08	0.10
0.02	0.04	0.10	0.09
0.04	0.04	0.10	0.11
0.05	0.03	0.11	0.12
0.05	0.05	0.19	0.12
0.05	0.06	0.13	0.17
0.06	0.06	0.12	0.18
0.08	0.05	0.11	0.20
0.08	0.07	0.10	0.22
0.05	0.08	0.11	0.24

TABLE D-18 Renal Excretion of cAMP (nmol/dL) in Three Groups of Patients

Normal	Myotonic Dystrophy	Nonmyotonic Dystrophy
0.61	0.43	0.48
0.61	0.68	0.60
0.77	0.69	0.69
0.77	0.75	0.69
0.77	0.75	0.78
0.85	0.75	0.78
0.96	0.82	0.91
1.01	0.95	0.98
1.01	1.11	1.01
1.11	1.23	1.07
1.11	1.30	1.07
1.28	1.30	1.27
	1.34	1.77
	1.44	
	1.44	
	1.50	
	1.50	
	1.68	
	2.00	
	2.51	
	2.98	
	3.31	
	3.37	
	3.70	
	4.65	

TABLE D-19 Biceps Circumference, cm, in Paraplegic Athletes and Normal Athletes of Different Body Types

Normal		Paraplegic
Ectomorphic	Mesomorphic	Paraplegic
27.9	38.4	36.2
25.6	36.5	34.5
33.8	41.9	37.4
32.1	37.4	39.7
30.2	35.8	43.1
31.7	35.3	33.5
32.5	43.3	40.1
26.9	36.6	36.6
28.8	35.8	41.2
32.2	33.8	34.4
30.6	34.9	35.0
28.7	38.6	39.9
33.2	39.5	30.8
31.9	36.3	40.5
30.1	42.6	35.6
30.7	41.6	39.6
28.3	35.5	29.0
31.8	41.1	39.3
29.9	36.6	37.5
28.2	37.0	39.0
29.0	42.5	40.8
28.5	39.5	37.0
28.2	30.7	
34.1	38.4	
29.6	36.2	
	35.7	
	43.6	
	33.1	
	34.9	
	37.5	

TABLE D-20 Salivary Proline-Rich Protein, mg/mL, with Different β-Receptor Blockers and Stimulators

Control	Isoproterenol	Dobutamine	Terbutaline	Metoprolol
39.7	185.6	173.7	37.6	73.9
52.6	141.8	97.0	52.9	60.8
34.7	174.0	123.6	50.2	74.1
43.3	158.4	166.7	34.4	59.1
48.0	177.1	115.9	45.6	75.2
31.1	178.1	182.6	52.9	27.2
	134.7	146.3	32.8	32.4
	148.5	166.3	57.3	39.2
	168.0	159.3	37.2	

TABLE D-21 DNA Specific Activity with Arachidonic Acid (AA), or Eicospentanoic Acid (EPA), with and without Tetradecanoylphorbol Acetate (TPA)

No TPA		TPA	
AA	EP	AA	EPA
269	208	314	252
256	223	336	275
274	217	332	195
225	180	373	198

TABLE D-22 cGMP Levels, pmol/g, in the Cerebellum and Pituitary with Different Stimulators of GABA: None (Control), Aminooxyacetic Acid (AOAA), Ethanolamine-*O*-Sulfate (EOS), and Muscimol (M)

Cerebellum				Pituitary			
Control	AOAA	EOS	M	Control	AOAA	EOS	M
11.2	3.8	7.4	4.5	1.7	1.1	0.7	1.0
11.4	4.6	0.5	8.6	1.9	1.1	0.7	0.7
7.9	5.3	9.3	8.7	1.5	0.7	0.4	1.0
12.9	10.7	0.9	1.1	1.4	1.0	1.0	1.0
10.7	7.0	6.3	7.7	1.5	1.0	0.7	1.2

TABLE D-23 cGMP Levels, pmol/g, in the Cerebellum and Pituitary with Different Stimulators of GABA: None (Control), Aminooxyacetic Acid (AOAA), Ethanolamine-O-Sulfate (EOS), and Muscimol (M) (Some Data Are Missing)

Cerebellum				Pituitary			
Control	AOAA	EOS	M	Control	AOAA	EOS	M
11.2	3.8	7.4	4.5	1.7	1.1	0.7	1.0
11.4	4.6	0.5	8.6	1.9		0.7	
7.9	5.3	9.3	8.7	1.5	0.7	0.4	
12.9	10.7	0.9	1.1	1.4	1.0	1.0	1.0
10.7	7.0	6.3	7.7	1.5	1.0	0.7	1.2

TABLE D-24 Number of Nursing Assessment Activities Performed by Masters and Nonmasters Nurses Who Have 7 Years of Experience or Less or More Than 7 Years of Experience

More Than 7 Years		7 Years or Less	
Nonmasters	Masters	Nonmasters	Masters
16	62	75	22
37	54	58	50
66	65	38	7
44	87	60	55
58	9	39	58
69	49	59	46
19	43	22	60
13	27	55	58
18	58	39	26
73	50	45	50
38	62	26	44
16	8	66	26
62	55	19	63
30	104	21	24
34	26	12	36
33	73	58	10
28	17	19	61
31	67	18	28
12	32	26	76
21	33	19	
43	56	36	
28	54	33	

TABLE D-24 (*Continued*)   **Number of Nursing Assessment Activities Performed by Masters and Nonmasters Nurses Who Have 7 Years of Experience or Less or More Than 7 Years of Experience**

More Than 7 Years		7 Years or Less	
Nonmasters	Masters	Nonmasters	Masters
25	71	2	
64	89	35	
9	72	62	
13	61	48	
36	46	44	
30	27	36	
34	50	28	
22	43	29	
10	67	17	
37	69	45	
67	108	66	
24	27	55	
29	68	20	
43		35	
35		26	
51		31	
39		55	
69		74	
89		43	
5		30	
17		22	
46		52	
6		69	
27		32	
37		4	
37		18	
61		52	
32		42	
41		50	
26		23	
32		55	
65		65	
2		18	
48		27	
39		29	
76		8	
46		74	
57		56	
62		51	

TABLE D-24 (*Continued*) **Number of Nursing Assessment Activities Performed by Masters and Nonmasters Nurses Who Have 7 Years of Experience or Less or More Than 7 Years of Experience**

More Than 7 Years		7 Years or Less	
Nonmasters	Masters	Nonmasters	Masters
71		32	
39		20	
51		28	
47		61	
35		42	
50		29	
		84	
		75	
		55	
		22	
		7	
		33	
		51	
		53	
		34	
		76	
		31	
		38	
		54	
		30	
		18	
		40	
		31	
		29	
		11	
		7	
		12	
		67	
		55	
		66	
		51	
		44	
		47	
		76	
		72	
		27	

TABLE D-25 Carcinogenicity (DNA Damage) in Four Tissues (Liver, Kidney, Lung, Skin) Caused by Three Carcinogens [Benzo(a)pyrine, Dibenzo(c,g)carbazole, and Acetylaminofluorene] Administered by Three Different Routes (Topical, Oral, and Subcutaneous)

DNA Damage	Carcinogen Code[*]	Route Code[†]	Tissue Code[‡]
0.7	1	1	1
0.5	1	1	1
0.6	1	1	1
0.5	1	1	1
0.4	1	1	1
0.4	1	1	1
0.4	1	1	1
0.5	1	1	1
1.1	1	1	2
1.2	1	1	2
0.7	1	1	2
1.3	1	1	2
1.3	1	1	2
1.2	1	1	2
1.3	1	1	2
1.3	1	1	2
1.1	1	1	3
1.2	1	1	3
0.7	1	1	3
1.1	1	1	3
1.3	1	1	3
1.2	1	1	3
1.3	1	1	3
1.3	1	1	3
100.9	1	1	4
132.4	1	1	4
106.0	1	1	4
126.8	1	1	4
107.6	1	1	4
83.8	1	1	4
107.0	1	1	4
107.8	1	1	4
1.0	1	2	1
1.4	1	2	1
0.9	1	2	1
1.2	1	2	1
1.2	1	2	1

[*]Carcinogen code: 1 = benzo(a)pyrine, 2 = dibenzo(c,g)carbazole, 3 = acetylaminofluorene.
[†]Route code: 1 = topical, 2 = oral, 3 = subcutaneous.
[‡]Tissue code: 1 = liver, 2 = kidney, 3 = lung, 4 = skin.

TABLE D-25 (*Continued*) Carcinogenicity (DNA Damage) in Four Tissues (Liver, Kidney, Lung, Skin) Caused by Three Carcinogens [Benzo(*a*)pyrine, Dibenzo(*c,g*)carbazole, and Acetylaminofluorene] Administered by Three Different Routes (Topical, Oral, and Subcutaneous)

DNA Damage	Carcinogen Code*	Route Code[†]	Tissue Code[‡]
0.9	1	2	1
1.1	1	2	1
1.4	1	2	1
1.1	1	2	2
0.7	1	2	2
0.8	1	2	2
1.2	1	2	2
1.2	1	2	2
0.9	1	2	2
1.2	1	2	2
1.2	1	2	2
2.3	1	2	3
2.0	1	2	3
1.7	1	2	3
1.5	1	2	3
2.0	1	2	3
1.4	1	2	3
1.9	1	2	3
2.2	1	2	3
4.8	1	2	4
4.1	1	2	4
3.3	1	2	4
4.0	1	2	4
3.5	1	2	4
5.0	1	2	4
5.2	1	2	4
3.7	1	2	4
0.1	1	3	1
0.2	1	3	1
0.2	1	3	1
0.3	1	3	1
0.2	1	3	1
0.3	1	3	1
0.3	1	3	1
0.2	1	3	1
0.3	1	3	2
0.6	1	3	2
0.4	1	3	2
0.5	1	3	2
0.6	1	3	2
0.4	1	3	2
0.5	1	3	2

TABLE D-25 (*Continued*)   Carcinogenicity (DNA Damage) in Four Tissues
(Liver, Kidney, Lung, Skin) Caused by Three Carcinogens [Benzo(*a*)pyrine,
Dibenzo(*c,g*)carbazole, and Acetylaminofluorene] Administered
by Three Different Routes (Topical, Oral, and Subcutaneous)

DNA Damage	Carcinogen Code*	Route Code[+]	Tissue Code[‡]
0.5	1	3	2
1.0	1	3	3
1.1	1	3	3
0.7	1	3	3
0.9	1	3	3
1.1	1	3	3
1.0	1	3	3
1.2	1	3	3
0.6	1	3	3
0.4	1	3	4
0.4	1	3	4
0.3	1	3	4
0.4	1	3	4
0.5	1	3	4
0.5	1	3	4
0.6	1	3	4
0.6	1	3	4
508.9	2	1	1
443.9	2	1	1
502.3	2	1	1
421.8	2	1	1
463.6	2	1	1
519.7	2	1	1
469.9	2	1	1
461.1	2	1	1
24.8	2	1	2
27.0	2	1	2
26.0	2	1	2
29.4	2	1	2
22.9	2	1	2
23.4	2	1	2
22.9	2	1	2
22.0	2	1	2
9.0	2	1	3
10.7	2	1	3
14.1	2	1	3
14.8	2	1	3
11.8	2	1	3
10.7	2	1	3
14.3	2	1	3
16.1	2	1	3
40.3	2	1	4

TABLE D-25 (*Continued*)   Carcinogenicity (DNA Damage) in Four Tissues (Liver, Kidney, Lung, Skin) Caused by Three Carcinogens [Benzo(*a*)pyrine, Dibenzo(*c,g*)carbazole, and Acetylaminofluorene] Administered by Three Different Routes (Topical, Oral, and Subcutaneous)

DNA Damage	Carcinogen Code*	Route Code[+]	Tissue Code[‡]
42.5	2	1	4
41.5	2	1	4
36.8	2	1	4
38.2	2	1	4
43.5	2	1	4
43.1	2	1	4
35.2	2	1	4
172.5	2	2	1
170.3	2	2	1
153.9	2	2	1
161.8	2	2	1
146.7	2	2	1
136.8	2	2	1
146.8	2	2	1
154.7	2	2	1
13.7	2	2	2
15.4	2	2	2
11.3	2	2	2
12.2	2	2	2
14.0	2	2	2
14.4	2	2	2
9.9	2	2	2
16.5	2	2	2
5.1	2	2	3
9.5	2	2	3
6.7	2	2	3
5.0	2	2	3
5.4	2	2	3
7.5	2	2	3
8.0	2	2	3
7.3	2	2	3
1.8	2	2	4
1.8	2	2	4
1.4	2	2	4
1.4	2	2	4
1.4	2	2	4
1.9	2	2	4
1.6	2	2	4
1.7	2	2	4
118.0	2	3	1
116.9	2	3	1
104.1	2	3	1

TABLE D-25 (*Continued*)   Carcinogenicity (DNA Damage) in Four Tissues (Liver, Kidney, Lung, Skin) Caused by Three Carcinogens [Benzo(*a*)pyrine, Dibenzo(*c,g*)carbazole, and Acetylaminofluorene] Administered by Three Different Routes (Topical, Oral, and Subcutaneous)

DNA Damage	Carcinogen Code*	Route Code[+]	Tissue Code[‡]
118.9	2	3	1
107.1	2	3	1
115.0	2	3	1
104.4	2	3	1
113.6	2	3	1
10.9	2	3	2
9.9	2	3	2
11.1	2	3	2
12.2	2	3	2
11.9	2	3	2
12.1	2	3	2
11.4	2	3	2
12.6	2	3	2
2.6	2	3	3
2.0	2	3	3
1.7	2	3	3
2.1	2	3	3
3.9	2	3	3
5.0	2	3	3
3.5	2	3	3
3.9	2	3	3
2.1	2	3	4
1.8	2	3	4
2.1	2	3	4
1.7	2	3	4
2.0	2	3	4
2.0	2	3	4
1.5	2	3	4
1.8	2	3	4
2.0	3	1	1
1.8	3	1	1
1.7	3	1	1
1.8	3	1	1
1.3	3	1	1
1.6	3	1	1
1.7	3	1	1
1.6	3	1	1
0.8	3	1	2
0.6	3	1	2
0.7	3	1	2
0.7	3	1	2
0.8	3	1	2

TABLE D-25 (*Continued*) Carcinogenicity (DNA Damage) in Four Tissues (Liver, Kidney, Lung, Skin) Caused by Three Carcinogens [Benzo(*a*)pyrine, Dibenzo(*c,g*)carbazole, and Acetylaminofluorene] Administered by Three Different Routes (Topical, Oral, and Subcutaneous)

DNA Damage	Carcinogen Code[*]	Route Code[†]	Tissue Code[‡]
0.8	3	1	2
0.5	3	1	2
0.8	3	1	2
0.3	3	1	3
0.5	3	1	3
0.3	3	1	3
0.4	3	1	3
0.4	3	1	3
0.5	3	1	3
0.6	3	1	3
0.4	3	1	3
3.1	3	1	4
2.4	3	1	4
2.0	3	1	4
2.0	3	1	4
2.6	3	1	4
2.2	3	1	4
2.7	3	1	4
2.2	3	1	4
4.7	3	2	1
5.2	3	2	1
5.5	3	2	1
3.9	3	2	1
5.3	3	2	1
4.1	3	2	1
4.4	3	2	1
4.2	3	2	1
1.7	3	2	2
1.6	3	2	2
1.6	3	2	2
1.6	3	2	2
1.8	3	2	2
1.7	3	2	2
1.8	3	2	2
1.8	3	2	2
0.5	3	2	3
0.6	3	2	3
0.4	3	2	3
0.4	3	2	3
0.5	3	2	3
0.6	3	2	3
0.4	3	2	3

TABLE D-25 (*Continued*)   **Carcinogenicity (DNA Damage) in Four Tissues (Liver, Kidney, Lung, Skin) Caused by Three Carcinogens [Benzo(*a*)pyrine, Dibenzo(*c,g*)carbazole, and Acetylaminofluorene] Administered by Three Different Routes (Topical, Oral, and Subcutaneous)**

DNA Damage	Carcinogen Code*	Route Code[†]	Tissue Code[‡]
0.6	3	2	3
0.9	3	2	4
0.7	3	2	4
0.7	3	2	4
0.6	3	2	4
0.6	3	2	4
0.6	3	2	4
0.8	3	2	4
0.8	3	2	4
4.4	3	3	1
4.1	3	3	1
4.8	3	3	1
3.3	3	3	1
4.4	3	3	1
2.8	3	3	1
3.1	3	3	1
4.1	3	3	1
0.6	3	3	2
0.9	3	3	2
0.8	3	3	2
1.0	3	3	2
0.8	3	3	2
0.5	3	3	2
1.0	3	3	2
1.1	3	3	2
0.8	3	3	3
0.9	3	3	3
0.9	3	3	3
1.1	3	3	3
0.8	3	3	3
1.0	3	3	3
1.0	3	3	3
1.0	3	3	3
0.4	3	3	4
0.5	3	3	4
0.3	3	3	4
0.5	3	3	4
0.5	3	3	4
0.6	3	3	4
0.5	3	3	4
0.6	3	3	4

TABLE D-26 Metabolic Rate (mL O$_2$/kg) in Dogs Late after Three Different Methods of Eating

Dog	Normal Eating	Stomach Bypassed	Eating Bypassed
1	93	25	101
2	134	19	107
3	127	26	104
4	115	28	110
5	142	38	112
6	123	35	106

TABLE D-27 Bacterial Counts at Four Times during the Day [Start Shift (SS), before Lunch (BL), after Lunch (AL), and End of Shift (ES)] in Two Different Groups of Operating Room Personnel: One Group Wears Same Scrubs All Day and the Other Puts on Fresh Scrubs Again after Lunch

Subject	Same Scrubs Group				Fresh Scrubs Group			
	SS	BL	AL	ES	SS	BL	AL	ES
1	23	22	26	30	13	23	6	17
2	20	6	27	31	13	15	7	52
3	16	11	32	26	14	24	8	15
4	19	20	26	37	17	23	10	47
5	24	13	18	14	10	16	8	37
6	30	11	28	34	19	23	4	54
7	20	13	24	27	21	21	12	22
8	12	9	31	21	6	22	9	16
9	19	7	21	27	14	30	9	9
10	22	18	25	30	11	10	10	45
11	19	11	19	28	8	17	7	4
12	19	14	29	23	17	28	11	69
13	16	18	27	24	13	26	8	42
14	24	19	20	22	18	26	5	54
15	19	17	26	30	13	22	10	33
16	38	15	35	27	10	29	6	42
17	28	14	27	27	14	25	9	49
18	20	16	33	25	8	24	9	10
19	14	25	18	25	10	19	8	22

TABLE D-28 Leucine Oxidation (Percent Contribution to Energy Needs) in Two Groups of Rat Muscles (Untrained and Trained) under Three Different Contraction Frequencies (0, 15, and 45 tetani/min)

Subject	Trained			Untrained		
	0	15	45	0	15	45
1	2.7	1.3	1.0	7.0	1.0	1.3
2	1.9	1.2	0.9	3.0	1.9	1.6
3	4.0	0.9	0.7	3.1	0.8	1.6
4	2.6	1.1	0.8	1.9	1.5	1.6
5	5.3	0.9	1.1	3.8	1.3	1.2
6	3.7	1.3	0.1	2.3	1.6	0.7
7	5.0	0.5	0.9	6.8	1.4	0.7
8	6.2	1.1	0.8	5.4	0.9	1.4
9	3.4	0.8	1.0	5.8	1.0	0.6
10	5.1	1.1	0.6	4.1	1.8	0.6
11	1.7	1.2	1.1	4.4	1.4	1.2
12	5.9	0.6	1.0	5.5	1.2	0.8
13	5.4	1.0	1.2	5.7	1.2	1.1

TABLE D-29 Study of Sociability before and after Drinking in Alcoholics Diagnosed as Either Antisocial Personality (ASP) or Non-ASP

Subject	Sober		Drinking	
	Non-ASP	ASP	Non-ASP	ASP
1	2.06	2.47	1.86	2.57
2	3.09	3.30	1.87	3.94
3	2.14	2.80	2.05	3.08
4	2.13	3.51	1.49	2.11
5	3.04	2.95	1.11	1.99
6	3.29	2.23	2.58	2.88
7	1.31	2.70	2.13	1.44
8	2.40	2.40	1.95	3.64

TABLE D-30 Study of Well-Being before and after Drinking in Alcoholics Diagnosed as Either Antisocial Personality (ASP) or Non-ASP

	Sober		Drinking	
Subject	Non-ASP	ASP	Non-ASP	ASP
1	2.52	2.60	2.12	2.15
2	2.90	2.09	2.48	2.26
3	2.17	2.38	2.60	2.20
4	2.65	3.24	2.13	2.02
5	1.93	2.63	2.29	2.81
6	3.05	2.59	2.74	2.59
7	3.16	2.90	2.53	2.31
8	2.84	2.69	2.59	2.70

TABLE D-31 Study of Well-Being before and after Drinking in Alcoholics Diagnosed as Either Antisocial Personality (ASP) or Non-ASP (Some Data Are Missing)

	Sober		Drinking	
Subject	Non-ASP	ASP	Non-ASP	ASP
1	2.06	2.47	1.86	2.57
2	3.09	3.30	1.87	3.94
3	2.14	2.80	2.05	
4	2.13	3.51	1.49	2.11
5	3.04	2.95		1.99
6	3.29	2.23	2.58	2.88
7	1.31	2.70	2.13	1.44
8	2.40	2.40	1.95	

TABLE D-32 Muscle Soreness in Subjects Three Times (6, 18, and 24 h) after Four Bouts of Exercise

Subject	Bout 1			Bout 2			Bout 3			Bout 4		
	6	18	24	6	18	24	6	18	24	6	18	24
1	2.3	1.4	3.0	0.8	1.7	1.2	0.8	1.2	1.0	0.8	1.3	0.9
2	2.5	2.8	3.1	1.3	1.7	1.6	1.2	1.2	1.5	1.1	1.4	1.7
3	2.9	2.9	3.1	1.8	2.1	2.5	1.2	1.6	2.4	1.2	1.6	2.3
4	3.3	3.1	3.2	2.0	2.4	2.7	1.3	1.9	2.5	1.3	1.8	2.4
5	3.3	3.3	4.2	2.2	2.5	3.3	1.5	2.0	2.5	1.3	2.1	2.4
6	3.9	3.8	5.8	2.4	3.3	3.4	1.7	2.8	3.2	1.3	2.3	3.3

TABLE D-33 Muscle Soreness in Subjects Three Times (6, 18, and 24 h) after Four Bouts of Exercise (Some Data Are Missing)

Subject	Bout 1			Bout 2			Bout 3			Bout 4		
	6	18	24	6	18	24	6	18	24	6	18	24
1	2.3	1.4	3.0	0.8	1.7	1.2	0.8	1.2	1.0	0.8	1.3	0.9
2	2.5	2.8		1.3	1.7	1.6	1.2	1.2		1.1	1.4	1.7
3	2.9	2.9			2.1	2.5	1.2	1.6	2.4	1.2		2.3
4	3.3	3.1		2.0	2.4	2.7	1.3	1.9	2.5	1.3		2.4
5	3.3	3.3	4.2	2.2	2.5		1.5	2.0	2.5	1.3	2.1	2.4
6	3.9	3.8	5.8	2.4	3.3	3.4	1.7	2.8		1.3	2.3	3.3

TABLE D-34 Two Methods of Measuring Blood Flow in the Wall of the Stomach in Four Dogs

Laser Doppler Flow L, volts	Hydrogen Clearance H, mL/(min·100 g)	Dog	Laser Doppler Flow L, volts	Hydrogen Clearance H, mL/(min·100 g)	Dog
2.3	17.7	1	3.4	47.3	3
2.3	17.7	1	3.4	50.0	3
2.6	27.2	1	3.6	46.2	3
3.9	35.9	1	3.6	46.8	3
3.8	39.7	1	4.3	65.2	3
3.9	41.9	1	4.5	62.0	3
4.4	46.8	1	4.8	56.3	3
3.9	49.5	1	4.9	57.6	3
3.8	51.1	1	5.2	62.8	3
5.4	64.2	1	5.6	58.2	3
5.4	68.5	1	7.7	68.5	3
2.4	12.2	2	7.9	69.6	3
2.4	16.3	2	8.0	65.2	3
2.3	39.1	2	8.9	72.9	3
4.9	43.5	2	9.0	68.5	3
4.9	51.7	2	10.2	74.8	3
4.1	52.7	2	10.3	76.7	3
4.1	56.8	2	2.3	31.3	4
4.3	55.7	2	2.5	35.1	4
2.3	59.3	2	2.7	35.9	4
3.3	63.1	2	2.9	29.9	4
3.9	69.3	2	3.2	30.7	4
4.5	70.7	2	4.3	34.8	4
5.5	76.1	2	4.6	32.9	4
5.0	82.9	2	5.5	38.6	4
6.0	70.7	2	6.4	41.9	4
6.9	63.9	2	7.6	37.5	4
11.5	103.0	2	8.5	44.9	4
3.0	49.5	3	10.4	40.8	4
3.1	45.9	3	11.0	47.6	4
3.3	47.8	3			

TABLE D-35 Phosphatidylcholine Transport by Pulmonary Cells Bathed with Synthetic Vesicles or Natural Surfactant

Phosphatidylcholine Choline Uptake P, % control	Time t, min	Group Code*
16	1	0
12	1	0
2	1	0
41	1	0
49	30	0
60	30	0
73	30	0
40	30	0
73	60	0
69	60	0
66	60	0
75	60	0
83	90	0
91	90	0
107	90	0
92	90	0
5	1	1
3	1	1
4	1	1
0	1	1
17	30	1
13	30	1
15	30	1
16	30	1
21	60	1
18	60	1
16	60	1
30	60	1
25	90	1
32	90	1
34	90	1
28	90	1
24	120	1
33	120	1
48	120	1
32	120	1

*Group code: 0 = natural surfactant, 1 = synthetic vesicles.

TABLE D-36 Mannitol Flux and Epithelial Resistance in Cultured Intestinal Cells Exposed to *C. difficile* Endotoxin

Mannitol Flux F, nmol/(h·cm²)	Resistance R, ohm/cm²	Mannitol Flux F, nmol/(h·cm²)	Resistance R, ohm/cm²
123.1	150	15.7	455
84.9	150	14.4	455
72.8	115	14.4	475
68.1	140	12.3	515
63.4	105	11.0	540
61.6	175	8.9	545
64.7	200	10.5	565
53.7	140	15.7	615
53.7	150	16.5	625
50.8	160	17.0	655
47.2	145	15.2	665
47.7	165	10.0	600
42.7	105	9.2	635
39.8	135	8.4	675
39.8	160	9.2	700
40.9	175	14.4	705
40.3	200	7.6	765
42.7	285	7.6	825
35.9	290	12.6	805
35.6	170	12.1	830
38.0	150	11.5	905
32.8	140	11.5	930
29.9	160	8.4	1010
27.0	195	10.0	1020
28.0	220	7.6	1030
31.4	235	6.6	1055
28.8	275	6.6	1080
25.4	245	8.9	1110
19.9	260	6.6	1145
11.5	250	7.3	1160
17.3	305	13.1	1200
21.5	300	7.1	1205
23.6	315	6.0	1205
18.3	375	6.6	1220
15.2	370	1.3	1235
13.6	390	10.5	1270
23.6	420	11.8	1290
24.4	425	1.8	1340
24.4	550	6.6	1395
19.1	520	4.7	1465

TABLE D-37 Heart Beat Interval and Systemic Arterial Pressure in Dogs Anesthetized with .5% Halothane

Heart Beat Interval I, ms	Systemic Arterial Pressure P, kPa	Heart Beat Interval I, ms	Systemic Arterial Pressure P, kPa
220	6.0	385	14.0
240	6.0	355	14.5
260	6.7	440	15.8
240	7.9	545	17.1
240	9.2	545	18.3
330	6.8	625	17.7
345	6.8	960	18.3
335	8.2	875	18.3
315	8.5	760	20.3
330	9.2	960	20.3
300	9.2	1060	21.0
300	10.0	1070	21.7
340	11.9	840	22.3
350	12.1	825	22.3
325	13.0	925	23.5
325	13.6	900	25.6
355	12.9	980	25.3
355	13.3	1000	25.3
440	13.2	1015	25.3

TABLE D-38 Heart Beat Interval and Systemic Arterial Pressure in Dogs Anesthetized with 2.0% Halothane

Heart Beat Interval I, ms	Systemic Arterial Pressure P, kPa	Heart Beat Interval I, ms	Systemic Arterial Pressure P, kPa
385	6.8	540	11.9
440	6.8	500	12.5
440	7.0	625	13.2
465	7.4	605	14.2
505	6.8	585	14.6
505	7.0	565	14.6
415	7.4	505	14.6
440	8.0	640	15.8
505	8.0	680	15.8
510	8.3	660	17.2
545	8.0	600	17.3
450	8.6	600	17.4
395	9.2	585	19.4
455	10.0	585	19.8
545	9.3	685	18.6
440	10.6	740	18.5
405	14.9	800	19.8
525	10.7	585	21.0
545	10.7	630	21.7
630	10.7	605	23.8
620	11.9	565	23.8
565	11.9	900	21.7
545	12.0		

TABLE D-39 Feeding Characteristic W and Dietary Protein as Fraction of Total Available Energy

W	Fraction of Energy in Protein	W	Fraction of Energy in Protein
13.63	0.05	38.75	0.40
20.52	0.05	40.80	0.40
10.59	0.05	39.24	0.40
18.87	0.05	40.63	0.50
36.23	0.10	40.73	0.50
32.62	0.10	37.39	0.50
28.95	0.10	40.11	0.50
39.98	0.15	38.23	0.50
43.17	0.15	42.88	0.50
42.51	0.15	38.54	0.60
42.43	0.15	38.58	0.60
32.41	0.15	37.82	0.60
38.44	0.20	43.30	0.60
40.32	0.20	36.58	0.60
36.75	0.20	36.78	0.60
36.45	0.20	36.86	0.70
38.50	0.20	37.15	0.70
36.94	0.20	36.11	0.70
42.32	0.30	34.63	0.70
37.79	0.30	41.24	0.70
38.63	0.30	46.98	0.80
40.39	0.30	39.79	0.80
38.87	0.30	32.46	0.80
40.74	0.40	41.51	0.80
42.62	0.40	40.23	0.80
39.05	0.40	39.63	0.80

TABLE D-40 Feeding Characteristic F and Dietary Protein as Fraction of Total Available Energy

F	Fraction of Energy in Protein	F	Fraction of Energy in Protein
2.42	0.05	4.16	0.40
2.71	0.05	4.47	0.40
2.25	0.05	4.64	0.40
3.26	0.05	4.10	0.50
5.76	0.10	3.88	0.50
5.12	0.10	4.84	0.50
4.15	0.10	3.99	0.50
5.45	0.15	4.63	0.50
4.37	0.15	5.13	0.50
6.30	0.15	4.41	0.60
5.22	0.15	4.15	0.60
4.83	0.15	4.26	0.60
4.19	0.20	3.40	0.60
4.87	0.20	4.54	0.60
4.38	0.20	3.26	0.60
4.86	0.20	3.65	0.70
4.18	0.20	4.22	0.70
3.99	0.20	2.76	0.70
3.77	0.30	4.57	0.70
4.59	0.30	3.96	0.70
5.23	0.30	3.91	0.80
4.91	0.30	5.15	0.80
4.69	0.30	3.96	0.80
4.82	0.40	3.33	0.80
4.00	0.40	4.56	0.80
5.03	0.40	3.45	0.80

TABLE D-41 **Feeding Characteristic *F* and Dietary Protein as Fraction of Total Available Energy (Alternative Data Set)**

F	Fraction of Energy in Protein	F	Fraction of Energy in Protein
2.42	0.05	3.77	0.30
2.71	0.05	4.59	0.30
2.25	0.05	5.23	0.30
3.26	0.05	4.91	0.30
4.43	0.06	4.69	0.30
4.91	0.06	4.82	0.40
4.07	0.06	4.00	0.40
5.00	0.06	5.03	0.40
5.59	0.07	4.16	0.40
5.10	0.07	4.47	0.40
5.12	0.07	4.64	0.40
6.13	0.07	4.10	0.50
4.53	0.08	3.88	0.50
5.92	0.08	4.84	0.50
3.95	0.08	3.99	0.50
4.81	0.08	4.63	0.50
4.76	0.09	5.13	0.50
4.80	0.09	4.41	0.60
4.96	0.09	4.15	0.60
5.15	0.09	4.26	0.60
5.76	0.10	3.40	0.60
5.12	0.10	4.54	0.60
4.15	0.10	3.26	0.60
5.45	0.15	3.65	0.70
4.37	0.15	4.22	0.70
6.30	0.15	2.76	0.70
5.22	0.15	4.57	0.70
4.83	0.15	3.96	0.70
4.19	0.20	3.91	0.80
4.87	0.20	5.15	0.80
4.38	0.20	3.96	0.80
4.86	0.20	3.33	0.80
4.18	0.20	4.56	0.80
3.99	0.20	3.45	0.80

TABLE D-42 Occurrence of SIDS and Characteristics of Infants

Power Spectrum Factor F_{68}	Age at F_{68} Measurement A, months	Heart Rate R, beats/min	Birth Weight W, g	Gestational Age G, weeks	SIDS S $(1 = $ SIDS$)$
0.291	2	115.6	3060	39	0
0.277	3	108.2	3570	40	0
0.390	3	114.2	3950	41	0
0.339	3	118.8	3480	40	0
0.248	3	76.9	3370	39	0
0.342	3	132.6	3260	40	0
0.310	2	107.7	4420	42	0
0.220	3	118.2	3560	40	0
0.233	2	126.6	3290	38	0
0.309	2	138.0	3010	40	0
0.355	2	127.0	3180	40	0
0.309	2	127.7	3950	40	0
0.250	2	106.8	3400	40	0
0.368	3	142.1	2410	38	0
0.223	2	91.5	2890	42	0
0.364	5	151.1	4030	40	0
0.335	4	127.1	3770	42	0
0.356	9	134.3	2680	40	0
0.374	5	114.9	3370	41	0
0.152	3	118.1	3370	40	0
0.356	4	122.0	3270	40	0
0.394	15	167.0	3520	41	0
0.250	4	107.9	3340	41	0
0.422	5	134.6	3940	41	0
0.409	7	137.7	3350	40	0
0.241	6	112.8	3350	39	0
0.312	6	131.3	3000	40	0
0.196	6	132.7	3960	40	0
0.266	7	148.1	3490	40	0
0.310	7	118.9	2640	39	0
0.351	11	133.7	3630	40	0
0.420	11	141.0	2680	38	0
0.366	10	134.1	3580	40	0
0.503	17	135.5	3800	39	0
0.272	15	148.6	3350	40	0
0.291	18	147.9	3030	40	0
0.308	22	162.0	3940	42	0
0.235	34	146.8	4080	40	0
0.287	36	131.7	3520	40	0
0.456	17	149.0	3630	40	0
0.284	19	114.1	3290	40	0

TABLE D-42 (*Continued*)   Occurrence of SIDS and Characteristics of Infants

Power Spectrum Factor F_{68}	Age at F_{68} Measurement A, months	Heart Rate R, beats/min	Birth Weight W, g	Gestational Age G, weeks	SIDS S (1 = SIDS)
0.239	24	129.2	3180	40	0
0.191	24	144.2	3580	40	0
0.334	16	148.1	3060	40	0
0.321	20	108.2	3000	37	0
0.450	18	131.1	4310	40	0
0.244	40	129.7	3975	40	0
0.173	40	142.0	3000	40	0
0.304	41	145.5	3940	41	0
0.409	2	139.7	3740	40	1
0.626	3	121.3	3005	38	1
0.383	2	131.4	4790	40	1
0.432	11	152.8	1890	38	1
0.347	5	125.6	2920	40	1
0.493	8	139.5	2810	39	1
0.521	6	117.2	3490	38	1
0.343	5	131.5	3030	37	1
0.359	5	137.3	2000	41	1
0.349	24	140.9	3770	40	1
0.279	30	139.5	2350	40	1
0.409	16	128.4	2780	39	1
0.388	19	154.2	2980	40	1
0.372	16	140.7	2120	38	1
0.314	2	105.5	2700	39	1
0.405	40	121.7	3060	41	1

TABLE D-43 Risk Factors Associated with ARDS in Critically Ill Patients

ARDS (1 = ARDS)	Sex S (1 = female)	Age A, year	Arterial Oxygen O, kPa	Protein Accumulation Index P	X-ray Score X
1	0	53	8.2	0.11	6
0	0	66	5.8	0.39	6
0	1	71	9.1	0.33	6
1	1	67	8.5	0.87	6
0	1	63	9.4	0.74	6
0	0	39	8.0	−0.32	6
0	0	45	14.8	0.57	6
1	0	44	8.5	2.75	6

TABLE D-43 (*Continued*) **Risk Factors Associated with ARDS in Critically Ill Patients**

ARDS (1 = ARDS)	Sex S (1 = female)	Age A, year	Arterial Oxygen O, kPa	Protein Accumulation Index P	X-ray Score X
0	1	64	11.7	0.55	6
0	1	65	14.8	0.35	5
0	0	60	9.8	0.74	5
0	1	64	25.5	0.35	6
1	1	23	12.6	2.30	6
0	0	63	11.0	0.41	6
1	1	36	9.8	1.34	6
0	0	67	6.2	2.05	6
0	0	72	10.6	0.88	6
0	1	76	23.8	0.68	5
1	1	16	7.2	5.74	6
1	0	73	6.8	4.94	6
1	1	68	9.0	2.81	6
1	0	63	7.0	1.86	6
0	0	73	14.3	0.48	4
1	0	19	10.7	2.33	6
1	0	61	8.7	0.45	6
1	0	75	6.6	1.24	6
0	0	41	7.7	−0.69	4
1	1	55	10.1	3.17	6
0	0	66	19.2	0.51	5
0	0	43	15.7	−0.22	6
0	0	64	19.2	0.17	6
1	0	50	9.2	1.31	4
0	1	59	10.0	1.34	6
1	0	65	10.7	0.75	4
0	1	72	10.1	0.96	6
1	1	42	6.1	2.63	5
1	1	62	11.1	2.52	6
1	0	74	7.2	2.33	6
1	0	66	8.0	0.14	6
1	0	75	5.9	0.76	6
0	0	55	17.5	1.07	6
0	0	79	10.2	0.79	5
1	0	26	7.5	0.72	6
1	0	52	7.8	0.41	6
1	0	19	10.7	2.33	6
0	0	72	10.6	0.88	6
1	0	50	9.2	1.31	4
0	0	55	17.5	1.07	6

TABLE D-43 (*Continued*)   Risk Factors Associated with ARDS in Critically Ill Patients

ARDS (1 = ARDS)	Sex S (1 = female)	Age A, year	Arterial Oxygen O, kPa	Protein Accumulation Index P	X-ray Score X
1	0	63	7.0	1.86	6
1	0	74	7.2	2.33	6
1	0	66	8.0	0.14	6
1	1	23	12.6	2.30	6
1	1	55	10.1	3.17	6
0	1	63	9.4	0.74	6
0	0	41	7.7	−0.69	4
0	1	65	14.8	0.35	5
0	1	59	10.0	1.34	6
1	0	65	10.7	0.75	4
0	0	63	11.0	0.41	6
1	0	26	7.5	0.72	6
1	0	44	8.5	2.75	6
0	0	73	14.3	0.48	4
0	1	71	9.1	0.33	6
0	1	64	11.7	0.55	6
1	0	75	5.9	0.76	6
1	1	36	9.8	1.34	6
1	0	53	8.2	0.11	6
0	0	66	5.8	0.39	6
1	0	52	7.8	0.41	6
1	0	75	6.6	1.24	6
0	0	64	19.2	0.17	6
0	1	76	23.8	0.68	5
0	0	39	8.0	−0.32	6
0	1	64	25.5	0.35	6
0	1	72	10.1	0.96	6
0	0	67	6.2	2.05	6
1	1	62	11.1	2.52	6
1	1	16	7.2	5.74	6
1	1	68	9.0	2.81	6
0	0	66	19.2	0.51	5
0	0	60	9.8	0.74	5
0	0	43	15.7	−0.22	6
0	0	45	14.8	0.57	6
0	0	79	10.2	0.79	5
1	1	42	6.1	2.63	5
1	0	61	8.7	0.45	6
1	1	67	8.5	0.87	6
1	0	73	6.8	4.94	6

APPENDIX

E

Statistical Tables

TABLE E-1 Critical Values of t (Two-Tailed)

ν	\multicolumn{9}{c}{Probability of Greater Value P}								
	0.50	0.20	0.10	0.05	0.02	0.01	0.005	0.002	0.001
1	1.000	3.078	6.314	12.706	31.821	63.657	127.321	318.309	636.619
2	0.816	1.886	2.920	4.303	6.965	9.925	14.089	22.327	31.599
3	0.765	1.638	2.353	3.182	4.541	5.841	7.453	10.215	12.924
4	0.741	1.533	2.132	2.776	3.747	4.604	5.598	7.173	8.610
5	0.727	1.476	2.015	2.571	3.365	4.032	4.773	5.893	6.869
6	0.718	1.440	1.943	2.447	3.143	3.707	4.317	5.208	5.959
7	0.711	1.415	1.895	2.365	2.998	3.499	4.029	4.785	5.408
8	0.706	1.397	1.860	2.306	2.896	3.355	3.833	4.501	5.041
9	0.703	1.383	1.833	2.262	2.821	3.250	3.690	4.297	4.781
10	0.700	1.372	1.812	2.228	2.764	3.169	3.581	4.144	4.587
11	0.697	1.363	1.796	2.201	2.718	3.106	3.497	4.025	4.437
12	0.695	1.356	1.782	2.179	2.681	3.055	3.428	3.930	4.318
13	0.694	1.350	1.771	2.160	2.650	3.012	3.372	3.852	4.221
14	0.692	1.345	1.761	2.145	2.624	2.977	3.326	3.787	4.140
15	0.691	1.341	1.753	2.131	2.602	2.947	3.286	3.733	4.073
16	0.690	1.337	1.746	2.120	2.583	2.921	3.252	3.686	4.015
17	0.689	1.333	1.740	2.110	2.567	2.898	3.222	3.646	3.965
18	0.688	1.330	1.734	2.101	2.552	2.878	3.197	3.610	3.922
19	0.688	1.328	1.729	2.093	2.539	2.861	3.174	3.579	3.883
20	0.687	1.325	1.725	2.086	2.528	2.845	3.153	3.552	3.850

TABLE E-1 (*Continued*) **Critical Values of *t* (Two-Tailed)**

					Probability of Greater Value, P				
ν	0.50	0.20	0.10	0.05	0.02	0.01	0.005	0.002	0.001
21	0.686	1.323	1.721	2.080	2.518	2.831	3.135	3.527	3.819
22	0.686	1.321	1.717	2.074	2.508	2.819	3.119	3.505	3.792
23	0.685	1.319	1.714	2.069	2.500	2.807	3.104	3.485	3.768
24	0.685	1.318	1.711	2.064	2.492	2.797	3.091	3.467	3.745
25	0.684	1.316	1.708	2.060	2.485	2.787	3.078	3.450	3.725
26	0.684	1.315	1.706	2.056	2.479	2.779	3.067	3.435	3.707
27	0.684	1.314	1.703	2.052	2.473	2.771	3.057	3.421	3.690
28	0.683	1.313	1.701	2.048	2.467	2.763	3.047	3.408	3.674
29	0.683	1.311	1.699	2.045	2.462	2.756	3.038	3.396	3.659
30	0.683	1.310	1.697	2.042	2.457	2.750	3.030	3.385	3.646
31	0.682	1.309	1.696	2.040	2.453	2.744	3.022	3.375	3.633
32	0.682	1.309	1.694	2.037	2.449	2.738	3.015	3.365	3.622
33	0.682	1.308	1.692	2.035	2.445	2.733	3.008	3.356	3.611
34	0.682	1.307	1.691	2.032	2.441	2.728	3.002	3.348	3.601
35	0.682	1.306	1.690	2.030	2.438	2.724	2.996	3.340	3.591
36	0.681	1.306	1.688	2.028	2.434	2.719	2.990	3.333	3.582
37	0.681	1.305	1.687	2.026	2.431	2.715	2.985	3.326	3.574
38	0.681	1.304	1.686	2.024	2.429	2.712	2.980	3.319	3.566
39	0.681	1.304	1.685	2.023	2.426	2.708	2.976	3.313	3.558
40	0.681	1.303	1.684	2.021	2.423	2.704	2.971	3.307	3.551
42	0.680	1.302	1.682	2.018	2.418	2.698	2.963	3.296	3.538
44	0.680	1.301	1.680	2.015	2.414	2.692	2.956	3.286	3.526
46	0.680	1.300	1.679	2.013	2.410	2.687	2.949	3.277	3.515
48	0.680	1.299	1.677	2.011	2.407	2.682	2.943	3.269	3.505
50	0.679	1.299	1.676	2.009	2.403	2.678	2.937	3.261	3.496
52	0.679	1.298	1.675	2.007	2.400	2.674	2.932	3.255	3.488
54	0.679	1.297	1.674	2.005	2.397	2.670	2.927	3.248	3.480
56	0.679	1.297	1.673	2.003	2.395	2.667	2.923	3.242	3.473
58	0.679	1.296	1.672	2.002	2.392	2.663	2.918	3.237	3.466
60	0.679	1.296	1.671	2.000	2.390	2.660	2.915	3.232	3.460
62	0.678	1.295	1.670	1.999	2.388	2.657	2.911	3.227	3.454
64	0.678	1.295	1.669	1.998	2.386	2.655	2.908	3.223	3.449
66	0.678	1.295	1.668	1.997	2.384	2.652	2.904	3.218	3.444
68	0.678	1.294	1.668	1.995	2.382	2.650	2.902	3.214	3.439
70	0.678	1.294	1.667	1.994	2.381	2.648	2.899	3.211	3.435

TABLE E-1 (*Continued*)   **Critical Values of *t* (Two-Tailed)**

	Probability of Greater Value *P*								
ν	0.50	0.20	0.10	0.05	0.02	0.01	0.005	0.002	0.001
72	0.678	1.293	1.666	1.993	2.379	2.646	2.896	3.207	3.431
74	0.678	1.293	1.666	1.993	2.378	2.644	2.894	3.204	3.427
76	0.678	1.293	1.665	1.992	2.376	2.642	2.891	3.201	3.423
78	0.678	1.292	1.665	1.991	2.375	2.640	2.889	3.198	3.420
80	0.678	1.292	1.664	1.990	2.374	2.639	2.887	3.195	3.416
90	0.677	1.291	1.662	1.987	2.368	2.632	2.878	3.183	3.402
100	0.677	1.290	1.660	1.984	2.364	2.626	2.871	3.174	3.390
120	0.677	1.289	1.658	1.980	2.358	2.617	2.860	3.160	3.373
140	0.676	1.288	1.656	1.977	2.353	2.611	2.852	3.149	3.361
160	0.676	1.287	1.654	1.975	2.350	2.607	2.846	3.142	3.352
180	0.676	1.286	1.653	1.973	2.347	2.603	2.842	3.136	3.345
200	0.676	1.286	1.653	1.972	2.345	2.601	2.839	3.131	3.340
∞	0.675	1.282	1.645	1.960	2.326	2.576	2.807	3.090	3.291

Adapted from J. H. Zar, *Biostatistical Analysis*, 2nd ed., 1984, pp. 484–485. Reprinted by permission of Prentice-Hall, Inc., Englewood Cliffs, N.J.

TABLE E-2 Critical Values of F Corresponding to $P < .05$ (Lightface) and $P < .01$ (Boldface)

ν_n

ν_d	1	2	3	4	5	6	7	8	9	10	11	12	14	16	20	24	30	40	50	75	100	200	500	∞
1	161	200	216	225	230	234	237	239	241	242	243	244	245	246	248	249	250	251	252	253	253	254	254	254
	4052	**4999**	**5403**	**5625**	**5764**	**5859**	**5928**	**5981**	**6022**	**6056**	**6082**	**6106**	**6142**	**6169**	**6208**	**6234**	**6261**	**6286**	**6302**	**6323**	**6334**	**6352**	**6361**	**6366**
2	18.51	19.00	19.16	19.25	19.30	19.33	19.36	19.37	19.38	19.39	19.40	19.41	19.42	19.43	19.44	19.45	19.46	19.47	19.47	19.48	19.49	19.49	19.50	19.50
	98.49	**99.00**	**99.17**	**99.25**	**99.30**	**99.33**	**99.36**	**99.37**	**99.39**	**99.40**	**99.41**	**99.42**	**99.43**	**99.44**	**99.45**	**99.46**	**99.47**	**99.48**	**99.48**	**99.49**	**99.49**	**99.49**	**99.50**	**99.50**
3	10.13	9.55	9.28	9.12	9.01	8.94	8.88	8.84	8.81	8.78	8.76	8.74	8.71	8.69	8.66	8.64	8.62	8.60	8.58	8.57	8.56	8.54	8.54	8.53
	34.12	**30.82**	**29.46**	**28.71**	**28.24**	**27.91**	**27.67**	**27.49**	**27.34**	**27.23**	**27.13**	**27.05**	**26.92**	**26.83**	**26.69**	**26.60**	**26.50**	**26.41**	**26.35**	**26.27**	**26.23**	**26.18**	**26.14**	**26.12**
4	7.71	6.94	6.59	6.39	6.26	6.16	6.09	6.04	6.00	5.96	5.93	5.91	5.87	5.84	5.80	5.77	5.74	5.71	5.70	5.68	5.66	5.65	5.64	5.63
	21.20	**18.00**	**16.69**	**15.98**	**15.52**	**15.21**	**14.98**	**14.80**	**14.66**	**14.54**	**14.45**	**14.37**	**14.24**	**14.15**	**14.02**	**13.93**	**13.83**	**13.74**	**13.69**	**13.61**	**13.57**	**13.52**	**13.48**	**13.46**
5	6.61	5.79	5.41	5.19	5.05	4.95	4.88	4.82	4.78	4.74	4.70	4.68	4.64	4.60	4.56	4.53	4.50	4.46	4.44	4.42	4.40	4.38	4.37	4.36
	16.26	**13.27**	**12.06**	**11.39**	**10.97**	**10.67**	**10.45**	**10.29**	**10.15**	**10.05**	**9.96**	**9.89**	**9.77**	**9.68**	**9.55**	**9.47**	**9.38**	**9.29**	**9.24**	**9.17**	**9.13**	**9.07**	**9.04**	**9.02**
6	5.99	5.14	4.76	4.53	4.39	4.28	4.21	4.15	4.10	4.06	4.03	4.00	3.96	3.92	3.87	3.84	3.81	3.77	3.75	3.72	3.71	3.69	3.68	3.67
	13.74	**10.92**	**9.78**	**9.15**	**8.75**	**8.47**	**8.26**	**8.10**	**7.98**	**7.87**	**7.79**	**7.72**	**7.60**	**7.52**	**7.39**	**7.31**	**7.23**	**7.14**	**7.09**	**7.02**	**6.99**	**6.94**	**6.90**	**6.88**
7	5.59	4.74	4.35	4.12	3.97	3.87	3.79	3.73	3.68	3.63	3.60	3.57	3.52	3.49	3.44	3.41	3.38	3.34	3.32	3.29	3.28	3.25	3.24	3.23
	12.25	**9.55**	**8.45**	**7.85**	**7.46**	**7.19**	**7.00**	**6.84**	**6.71**	**6.62**	**6.54**	**6.47**	**6.35**	**6.27**	**6.15**	**6.07**	**5.98**	**5.90**	**5.85**	**5.78**	**5.75**	**5.70**	**5.67**	**5.65**
8	5.32	4.46	4.07	3.84	3.69	3.58	3.50	3.44	3.39	3.34	3.31	3.28	3.23	3.20	3.15	3.12	3.08	3.05	3.03	3.00	2.98	2.96	2.94	2.93
	11.26	**8.65**	**7.59**	**7.01**	**6.63**	**6.37**	**6.19**	**6.03**	**5.91**	**5.82**	**5.74**	**5.67**	**5.56**	**5.48**	**5.36**	**5.28**	**5.20**	**5.11**	**5.06**	**5.00**	**4.96**	**4.91**	**4.88**	**4.86**
9	5.12	4.26	3.86	3.63	3.48	3.37	3.29	3.23	3.18	3.13	3.10	3.07	3.02	2.98	2.93	2.90	2.86	2.82	2.80	2.77	2.76	2.73	2.72	2.71
	10.56	**8.02**	**6.99**	**6.42**	**6.06**	**5.80**	**5.62**	**5.47**	**5.35**	**5.26**	**5.18**	**5.11**	**5.00**	**4.92**	**4.80**	**4.73**	**4.64**	**4.56**	**4.51**	**4.45**	**4.41**	**4.36**	**4.33**	**4.31**
10	4.96	4.10	3.71	3.48	3.33	3.22	3.14	3.07	3.02	2.97	2.94	2.91	2.86	2.82	2.77	2.74	2.70	2.67	2.64	2.61	2.59	2.56	2.55	2.54
	10.04	**7.56**	**6.55**	**5.99**	**5.64**	**5.39**	**5.21**	**5.06**	**4.95**	**4.85**	**4.78**	**4.71**	**4.60**	**4.52**	**4.41**	**4.33**	**4.25**	**4.17**	**4.12**	**4.05**	**4.01**	**3.96**	**3.93**	**3.91**
11	4.84	3.98	3.59	3.36	3.20	3.09	3.01	2.95	2.90	2.86	2.82	2.79	2.74	2.70	2.65	2.61	2.57	2.53	2.50	2.47	2.45	2.42	2.41	2.40
	9.65	**7.20**	**6.22**	**5.67**	**5.32**	**5.07**	**4.88**	**4.74**	**4.63**	**4.54**	**4.46**	**4.40**	**4.29**	**4.21**	**4.10**	**4.02**	**3.94**	**3.86**	**3.80**	**3.74**	**3.70**	**3.66**	**3.62**	**3.60**
12	4.75	3.88	3.49	3.26	3.11	3.00	2.92	2.85	2.80	2.76	2.72	2.69	2.64	2.60	2.54	2.50	2.46	2.42	2.40	2.36	2.35	2.32	2.31	2.30
	9.33	**6.93**	**5.95**	**5.41**	**5.06**	**4.82**	**4.65**	**4.50**	**4.39**	**4.30**	**4.22**	**4.16**	**4.05**	**3.98**	**3.86**	**3.78**	**3.70**	**3.61**	**3.56**	**3.49**	**3.46**	**3.41**	**3.38**	**3.36**
13	4.67	3.80	3.41	3.18	3.02	2.92	2.84	2.77	2.72	2.67	2.63	2.60	2.55	2.51	2.46	2.42	2.38	2.34	2.32	2.28	2.26	2.24	2.22	2.21
	9.07	**6.70**	**5.74**	**5.20**	**4.86**	**4.62**	**4.44**	**4.30**	**4.19**	**4.10**	**4.02**	**3.96**	**3.85**	**3.78**	**3.67**	**3.59**	**3.51**	**3.42**	**3.37**	**3.30**	**3.27**	**3.21**	**3.18**	**3.16**
14	4.60	3.74	3.34	3.11	2.96	2.85	2.77	2.70	2.65	2.60	2.56	2.53	2.48	2.44	2.39	2.35	2.31	2.27	2.24	2.21	2.19	2.16	2.14	2.13
	8.86	**6.51**	**5.56**	**5.03**	**4.69**	**4.46**	**4.28**	**4.14**	**4.03**	**3.94**	**3.86**	**3.80**	**3.70**	**3.62**	**3.51**	**3.43**	**3.34**	**3.26**	**3.21**	**3.14**	**3.11**	**3.06**	**3.02**	**3.00**
15	4.54	3.68	3.29	3.06	2.90	2.79	2.70	2.64	2.59	2.55	2.51	2.48	2.43	2.39	2.33	2.29	2.25	2.21	2.18	2.15	2.12	2.10	2.08	2.07
	8.68	**6.36**	**5.42**	**4.89**	**4.56**	**4.32**	**4.14**	**4.00**	**3.89**	**3.80**	**3.73**	**3.67**	**3.56**	**3.48**	**3.36**	**3.29**	**3.20**	**3.12**	**3.07**	**3.00**	**2.97**	**2.92**	**2.89**	**2.87**
16	4.49	3.63	3.24	3.01	2.85	2.74	2.66	2.59	2.54	2.49	2.45	2.42	2.37	2.33	2.28	2.24	2.20	2.16	2.13	2.09	2.07	2.04	2.02	2.01
	8.53	**6.23**	**5.29**	**4.77**	**4.44**	**4.20**	**4.03**	**3.89**	**3.78**	**3.69**	**3.61**	**3.55**	**3.45**	**3.37**	**3.25**	**3.18**	**3.10**	**3.01**	**2.96**	**2.98**	**2.86**	**2.80**	**2.77**	**2.75**

713

TABLE E-2 (*Continued*) Critical Values of *F* Corresponding to *P* < .05 (Lightface) and *P* < .01 (Boldface)

v_d	1	2	3	4	5	6	7	8	9	10	11	12	14	16	20	24	30	40	50	75	100	200	500	∞
17	4.45 **8.40**	3.59 **6.11**	3.20 **5.18**	2.96 **4.67**	2.81 **4.34**	2.70 **4.10**	2.62 **3.93**	2.55 **3.79**	2.50 **3.68**	2.45 **3.59**	2.41 **3.52**	2.38 **3.45**	2.33 **3.35**	2.29 **3.27**	2.23 **3.16**	2.19 **3.08**	2.15 **3.00**	2.11 **2.92**	2.08 **2.86**	2.04 **2.79**	2.02 **2.76**	1.99 **2.70**	1.97 **2.67**	1.96 **2.65**
18	4.41 **8.28**	3.55 **6.01**	3.16 **5.09**	2.93 **4.58**	2.77 **4.25**	2.66 **4.01**	2.58 **3.85**	2.51 **3.71**	2.46 **3.60**	2.41 **3.51**	2.37 **3.44**	2.34 **3.37**	2.29 **3.27**	2.25 **3.19**	2.19 **3.07**	2.15 **3.00**	2.11 **2.91**	2.07 **2.83**	2.04 **2.78**	2.00 **2.71**	1.98 **2.68**	1.95 **2.62**	1.93 **2.59**	1.92 **2.57**
19	4.38 **8.18**	3.52 **5.93**	3.13 **5.01**	2.90 **4.50**	2.74 **4.17**	2.63 **3.94**	2.55 **3.77**	2.48 **3.63**	2.43 **3.52**	2.38 **3.43**	2.34 **3.36**	2.31 **3.30**	2.26 **3.19**	2.21 **3.12**	2.15 **3.00**	2.11 **2.92**	2.07 **2.84**	2.02 **2.76**	2.00 **2.70**	1.96 **2.63**	1.94 **2.60**	1.91 **2.54**	1.90 **2.51**	1.88 **2.49**
20	4.35 **8.10**	3.49 **5.85**	3.10 **4.94**	2.87 **4.43**	2.71 **4.10**	2.60 **3.87**	2.52 **3.71**	2.45 **3.56**	2.40 **3.45**	2.35 **3.37**	2.31 **3.30**	2.28 **3.23**	2.23 **3.13**	2.18 **3.05**	2.12 **2.94**	2.08 **2.86**	2.04 **2.77**	1.99 **2.69**	1.96 **2.63**	1.92 **2.56**	1.90 **2.53**	1.87 **2.47**	1.85 **2.44**	1.84 **2.42**
21	4.32 **8.02**	3.47 **5.78**	3.07 **4.87**	2.84 **4.37**	2.68 **4.04**	2.57 **3.81**	2.49 **3.65**	2.42 **3.51**	2.37 **3.40**	2.32 **3.31**	2.28 **3.24**	2.25 **3.17**	2.20 **3.07**	2.15 **2.99**	2.09 **2.88**	2.05 **2.80**	2.00 **2.72**	1.96 **2.63**	1.93 **2.58**	1.89 **2.51**	1.87 **2.47**	1.84 **2.42**	1.82 **2.38**	1.81 **2.36**
22	4.30 **7.94**	3.44 **5.72**	3.05 **4.82**	2.82 **4.31**	2.66 **3.99**	2.55 **3.76**	2.47 **3.59**	2.40 **3.45**	2.35 **3.35**	2.30 **3.26**	2.26 **3.18**	2.23 **3.12**	2.18 **3.02**	2.13 **2.94**	2.07 **2.83**	2.03 **2.75**	1.98 **2.67**	1.93 **2.58**	1.91 **2.53**	1.87 **2.46**	1.84 **2.42**	1.81 **2.37**	1.80 **2.33**	1.78 **2.31**
23	4.28 **7.88**	3.42 **5.66**	3.03 **4.76**	2.80 **4.26**	2.64 **3.94**	2.53 **3.71**	2.45 **3.54**	2.38 **3.41**	2.32 **3.30**	2.28 **3.21**	2.24 **3.14**	2.20 **3.07**	2.14 **2.97**	2.10 **2.89**	2.04 **2.78**	2.00 **2.70**	1.96 **2.62**	1.91 **2.53**	1.88 **2.48**	1.84 **2.41**	1.82 **2.37**	1.79 **2.32**	1.77 **2.28**	1.76 **2.26**
24	4.26 **7.82**	3.40 **5.61**	3.01 **4.72**	2.78 **4.22**	2.62 **3.90**	2.51 **3.67**	2.43 **3.50**	2.36 **3.36**	2.30 **3.25**	2.26 **3.17**	2.22 **3.09**	2.18 **3.03**	2.13 **2.93**	2.09 **2.85**	2.02 **2.74**	1.98 **2.66**	1.94 **2.58**	1.89 **2.49**	1.86 **2.44**	1.82 **2.36**	1.80 **2.33**	1.76 **2.27**	1.74 **2.23**	1.73 **2.21**
25	4.24 **7.77**	3.38 **5.57**	2.99 **4.68**	2.76 **4.18**	2.60 **3.86**	2.49 **3.63**	2.41 **3.46**	2.34 **3.32**	2.28 **3.21**	2.24 **3.13**	2.20 **3.05**	2.16 **2.99**	2.11 **2.89**	2.06 **2.81**	2.00 **2.70**	1.96 **2.62**	1.92 **2.54**	1.87 **2.45**	1.84 **2.40**	1.80 **2.32**	1.77 **2.29**	1.74 **2.23**	1.72 **2.19**	1.71 **2.17**
26	4.22 **7.72**	3.37 **5.53**	2.98 **4.64**	2.74 **4.14**	2.59 **3.82**	2.47 **3.59**	2.39 **3.42**	2.32 **3.29**	2.27 **3.17**	2.22 **3.09**	2.18 **3.02**	2.15 **2.96**	2.10 **2.86**	2.05 **2.77**	1.99 **2.66**	1.95 **2.58**	1.90 **2.50**	1.85 **2.41**	1.82 **2.36**	1.78 **2.28**	1.76 **2.25**	1.72 **2.19**	1.70 **2.15**	1.69 **2.13**
27	4.21 **7.68**	3.35 **5.49**	2.96 **4.60**	2.73 **4.11**	2.57 **3.79**	2.46 **3.56**	2.37 **3.39**	2.30 **3.26**	2.25 **3.14**	2.20 **3.06**	2.16 **2.98**	2.13 **2.93**	2.08 **2.83**	2.03 **2.74**	1.97 **2.63**	1.93 **2.55**	1.88 **2.47**	1.84 **2.38**	1.80 **2.33**	1.76 **2.25**	1.74 **2.21**	1.71 **2.16**	1.68 **2.12**	1.67 **2.10**
28	4.20 **7.64**	3.34 **5.45**	2.95 **4.57**	2.71 **4.07**	2.56 **3.76**	2.44 **3.53**	2.36 **3.36**	2.29 **3.23**	2.24 **3.11**	2.19 **3.03**	2.15 **2.95**	2.12 **2.90**	2.06 **2.80**	2.02 **2.71**	1.96 **2.60**	1.91 **2.52**	1.87 **2.44**	1.81 **2.35**	1.78 **2.30**	1.75 **2.22**	1.72 **2.18**	1.69 **2.13**	1.67 **2.09**	1.65 **2.06**
29	4.18 **7.60**	3.33 **5.42**	2.93 **4.54**	2.70 **4.04**	2.54 **3.73**	2.43 **3.50**	2.35 **3.33**	2.28 **3.20**	2.22 **3.08**	2.18 **3.00**	2.14 **2.92**	2.10 **2.87**	2.05 **2.77**	2.00 **2.68**	1.94 **2.57**	1.90 **2.49**	1.85 **2.41**	1.80 **2.32**	1.77 **2.27**	1.73 **2.19**	1.71 **2.15**	1.68 **2.10**	1.65 **2.06**	1.64 **2.03**
30	4.17 **7.56**	3.32 **5.39**	2.92 **4.51**	2.69 **4.02**	2.53 **3.70**	2.42 **3.47**	2.34 **3.30**	2.27 **3.17**	2.21 **3.06**	2.16 **2.98**	2.12 **2.90**	2.09 **2.84**	2.04 **2.74**	1.99 **2.66**	1.93 **2.55**	1.89 **2.47**	1.84 **2.38**	1.79 **2.29**	1.76 **2.24**	1.72 **2.16**	1.69 **2.13**	1.66 **2.07**	1.64 **2.03**	1.62 **2.01**
32	4.15 **7.50**	3.30 **5.34**	2.90 **4.46**	2.67 **3.97**	2.51 **3.66**	2.40 **3.42**	2.32 **3.25**	2.25 **3.12**	2.19 **3.01**	2.14 **2.94**	2.10 **2.86**	2.07 **2.80**	2.02 **2.70**	1.97 **2.62**	1.91 **2.51**	1.86 **2.42**	1.82 **2.34**	1.76 **2.25**	1.74 **2.20**	1.69 **2.12**	1.67 **2.08**	1.64 **2.02**	1.61 **1.98**	1.59 **1.96**
34	4.13 **7.44**	3.28 **5.29**	2.88 **4.42**	2.65 **3.93**	2.49 **3.61**	2.38 **3.38**	2.30 **3.21**	2.23 **3.08**	2.17 **2.97**	2.12 **2.89**	2.08 **2.82**	2.05 **2.76**	2.00 **2.66**	1.95 **2.58**	1.89 **2.47**	1.84 **2.38**	1.80 **2.30**	1.74 **2.21**	1.71 **2.15**	1.67 **2.08**	1.64 **2.04**	1.61 **1.98**	1.59 **1.94**	1.57 **1.91**

v_n

TABLE E-2 (*Continued*) Critical Values of F Corresponding to $P < .05$ (Lightface) and $P < .01$ (Boldface)

ν_n

ν_d	1	2	3	4	5	6	7	8	9	10	11	12	14	16	20	24	30	40	50	75	100	200	500	∞
36	4.11 **7.39**	3.26 **5.25**	2.86 **4.38**	2.63 **3.89**	2.48 **3.58**	2.36 **3.35**	2.28 **3.18**	2.21 **3.04**	2.15 **2.94**	2.10 **2.86**	2.06 **2.78**	2.03 **2.72**	1.98 **2.62**	1.93 **2.54**	1.87 **2.43**	1.82 **2.35**	1.78 **2.26**	1.72 **2.17**	1.69 **2.12**	1.65 **2.04**	1.62 **2.00**	1.59 **1.94**	1.56 **1.90**	1.55 **1.87**
38	4.10 **7.35**	3.25 **5.21**	2.85 **4.34**	2.62 **3.86**	2.46 **3.54**	2.35 **3.32**	2.26 **3.15**	2.19 **3.02**	2.14 **2.91**	2.09 **2.82**	2.05 **2.75**	2.02 **2.69**	1.96 **2.59**	1.92 **2.51**	1.85 **2.40**	1.80 **2.32**	1.76 **2.22**	1.71 **2.14**	1.67 **2.08**	1.63 **2.00**	1.60 **1.97**	1.57 **1.90**	1.54 **1.86**	1.53 **1.84**
40	4.08 **7.31**	3.23 **5.18**	2.84 **4.31**	2.61 **3.83**	2.45 **3.51**	2.34 **3.29**	2.25 **3.12**	2.18 **2.99**	2.12 **2.88**	2.07 **2.80**	2.04 **2.73**	2.00 **2.66**	1.95 **2.56**	1.90 **2.49**	1.84 **2.37**	1.79 **2.29**	1.74 **2.20**	1.69 **2.11**	1.66 **2.05**	1.61 **1.97**	1.59 **1.94**	1.55 **1.88**	1.53 **1.84**	1.51 **1.81**
42	4.07 **7.27**	3.22 **5.15**	2.83 **4.29**	2.59 **3.80**	2.44 **3.49**	2.32 **3.26**	2.24 **3.10**	2.17 **2.96**	2.11 **2.86**	2.06 **2.77**	2.02 **2.70**	1.99 **2.64**	1.94 **2.54**	1.89 **2.46**	1.82 **2.35**	1.78 **2.26**	1.73 **2.17**	1.68 **2.08**	1.64 **2.02**	1.60 **1.94**	1.57 **1.91**	1.54 **1.85**	1.51 **1.80**	1.49 **1.78**
44	4.06 **7.24**	3.21 **5.12**	2.82 **4.26**	2.58 **3.78**	2.43 **3.46**	2.31 **3.24**	2.23 **3.07**	2.16 **2.94**	2.10 **2.84**	2.05 **2.75**	2.01 **2.68**	1.98 **2.62**	1.92 **2.52**	1.88 **2.44**	1.81 **2.32**	1.76 **2.24**	1.72 **2.15**	1.66 **2.06**	1.63 **2.00**	1.58 **1.92**	1.56 **1.88**	1.52 **1.82**	1.50 **1.78**	1.48 **1.75**
46	4.05 **7.21**	3.20 **5.10**	2.81 **4.24**	2.57 **3.76**	2.42 **3.44**	2.30 **3.22**	2.22 **3.05**	2.14 **2.92**	2.09 **2.82**	2.04 **2.73**	2.00 **2.66**	1.97 **2.60**	1.91 **2.50**	1.87 **2.42**	1.80 **2.30**	1.75 **2.22**	1.71 **2.13**	1.65 **2.04**	1.62 **1.98**	1.57 **1.90**	1.54 **1.86**	1.51 **1.80**	1.48 **1.76**	1.46 **1.72**
48	4.04 **7.19**	3.19 **5.08**	2.80 **4.22**	2.56 **3.74**	2.41 **3.42**	2.30 **3.20**	2.21 **3.04**	2.14 **2.90**	2.08 **2.80**	2.03 **2.71**	1.99 **2.64**	1.96 **2.58**	1.90 **2.48**	1.86 **2.40**	1.79 **2.28**	1.74 **2.20**	1.70 **2.11**	1.64 **2.02**	1.61 **1.96**	1.56 **1.88**	1.53 **1.84**	1.50 **1.78**	1.47 **1.73**	1.45 **1.70**
50	4.03 **7.17**	3.18 **5.06**	2.79 **4.20**	2.56 **3.72**	2.40 **3.41**	2.29 **3.18**	2.20 **3.02**	2.13 **2.88**	2.07 **2.78**	2.02 **2.70**	1.98 **2.62**	1.95 **2.56**	1.90 **2.46**	1.85 **2.39**	1.78 **2.26**	1.74 **2.18**	1.69 **2.10**	1.63 **2.00**	1.60 **1.94**	1.55 **1.86**	1.52 **1.82**	1.48 **1.76**	1.46 **1.71**	1.44 **1.68**
60	4.00 **7.08**	3.15 **4.98**	2.76 **4.13**	2.52 **3.65**	2.37 **3.34**	2.25 **3.12**	2.17 **2.95**	2.10 **2.82**	2.04 **2.72**	1.99 **2.63**	1.95 **2.56**	1.92 **2.50**	1.86 **2.40**	1.81 **2.32**	1.75 **2.20**	1.70 **2.12**	1.65 **2.03**	1.59 **1.93**	1.56 **1.87**	1.50 **1.79**	1.48 **1.74**	1.44 **1.68**	1.41 **1.63**	1.39 **1.60**
70	3.98 **7.01**	3.13 **4.92**	2.74 **4.08**	2.50 **3.60**	2.35 **3.29**	2.23 **3.07**	2.14 **2.91**	2.07 **2.77**	2.01 **2.67**	1.97 **2.59**	1.93 **2.51**	1.89 **2.45**	1.84 **2.35**	1.79 **2.28**	1.72 **2.15**	1.67 **2.07**	1.62 **1.98**	1.56 **1.88**	1.53 **1.82**	1.47 **1.74**	1.45 **1.69**	1.40 **1.62**	1.37 **1.56**	1.35 **1.53**
80	3.96 **6.96**	3.11 **4.88**	2.72 **4.04**	2.48 **3.56**	2.33 **3.25**	2.21 **3.04**	2.12 **2.87**	2.05 **2.74**	1.99 **2.64**	1.95 **2.55**	1.91 **2.48**	1.88 **2.41**	1.82 **2.32**	1.77 **2.24**	1.70 **2.11**	1.65 **2.03**	1.60 **1.94**	1.54 **1.84**	1.51 **1.78**	1.45 **1.70**	1.42 **1.65**	1.38 **1.57**	1.35 **1.52**	1.32 **1.49**
100	3.94 **6.90**	3.09 **4.82**	2.70 **3.98**	2.46 **3.51**	2.30 **3.20**	2.19 **2.99**	2.10 **2.82**	2.03 **2.69**	1.97 **2.59**	1.92 **2.51**	1.88 **2.43**	1.85 **2.36**	1.79 **2.26**	1.75 **2.19**	1.68 **2.06**	1.63 **1.98**	1.57 **1.89**	1.51 **1.79**	1.48 **1.73**	1.42 **1.64**	1.39 **1.59**	1.34 **1.51**	1.30 **1.46**	1.28 **1.43**
120	3.92 **6.85**	3.07 **4.79**	2.68 **3.95**	2.45 **3.48**	2.29 **3.17**	2.18 **2.96**	2.09 **2.79**	2.02 **2.66**	1.96 **2.56**	1.91 **2.47**	1.87 **2.40**	1.84 **2.34**	1.78 **2.23**	1.73 **2.15**	1.66 **2.03**	1.61 **1.95**	1.56 **1.86**	1.50 **1.76**	1.46 **1.70**	1.39 **1.61**	1.37 **1.56**	1.32 **1.48**	1.28 **1.42**	1.25 **1.38**
∞	3.84 **6.63**	2.99 **4.60**	2.60 **3.78**	2.37 **3.32**	2.21 **3.02**	2.09 **2.80**	2.01 **2.64**	1.94 **2.51**	1.88 **2.41**	1.83 **2.32**	1.79 **2.24**	1.75 **2.18**	1.69 **2.07**	1.64 **1.99**	1.57 **1.87**	1.52 **1.79**	1.46 **1.69**	1.40 **1.59**	1.35 **1.52**	1.28 **1.41**	1.24 **1.36**	1.17 **1.25**	1.11 **1.15**	1.00 **1.00**

Note: ν_n = degrees of freedom for numerator; ν_d = degrees of freedom for denominator.
Reprinted by permission from *Statistical Methods*, 7th ed., by G. W. Snedecor and W. G. Cochran. © 1980 by Iowa State University Press, Ames, I.A. 50010.

TABLE E-3 Critical Values of Dunnett's q'

ν	$p = 2$	3	4	5	6	7	8	9	10	11	12	13	16	21
						$\alpha_T = 0.05$								
5	2.57	3.03	3.29	3.48	3.62	3.73	3.82	3.90	3.97	4.03	4.09	4.14	4.26	4.42
6	2.45	2.86	3.10	3.26	3.39	3.49	3.57	3.64	3.71	3.76	3.81	3.86	3.97	4.11
7	2.36	2.75	2.97	3.12	3.24	3.33	3.41	3.47	3.53	3.58	3.63	3.67	3.78	3.91
8	2.31	2.67	2.88	3.02	3.13	3.22	3.29	3.35	3.41	3.46	3.50	3.54	3.64	3.76
9	2.26	2.61	2.81	2.95	3.05	3.14	3.20	3.26	3.32	3.36	3.40	3.44	3.53	3.65
10	2.23	2.57	2.76	2.89	2.99	3.07	3.14	3.19	3.24	3.29	3.33	3.36	3.45	3.57
11	2.20	2.53	2.72	2.84	2.94	3.02	3.08	3.14	3.19	3.23	3.27	3.30	3.39	3.50
12	2.18	2.50	2.68	2.81	2.90	2.98	3.04	3.09	3.14	3.18	3.22	3.25	3.34	3.45
13	2.16	2.48	2.65	2.78	2.87	2.94	3.00	3.06	3.10	3.14	3.18	3.21	3.29	3.40
14	2.14	2.46	2.63	2.75	2.84	2.91	2.97	3.02	3.07	3.11	3.14	3.18	3.26	3.36
15	2.13	2.44	2.61	2.73	2.82	2.89	2.95	3.00	3.04	3.08	3.12	3.15	3.23	3.33
16	2.12	2.42	2.59	2.71	2.80	2.87	2.92	2.97	3.02	3.06	3.09	3.12	3.20	3.30
17	2.11	2.41	2.58	2.69	2.78	2.85	2.90	2.95	3.00	3.03	3.07	3.10	3.18	3.27
18	2.10	2.40	2.56	2.68	2.76	2.83	2.89	2.94	2.98	3.01	3.05	3.08	3.16	3.25
19	2.09	2.39	2.55	2.66	2.75	2.81	2.87	2.92	2.96	3.00	3.03	3.06	3.14	3.23
20	2.09	2.38	2.54	2.65	2.73	2.80	2.86	2.90	2.95	2.98	3.02	3.05	3.12	3.22
24	2.06	2.35	2.51	2.61	2.70	2.76	2.81	2.86	2.90	2.94	2.97	3.00	3.07	3.16
30	2.04	2.32	2.47	2.58	2.66	2.72	2.77	2.82	2.86	2.89	2.92	2.95	3.02	3.11
40	2.02	2.29	2.44	2.54	2.62	2.68	2.73	2.77	2.81	2.85	2.87	2.90	2.97	3.06
60	2.00	2.27	2.41	2.51	2.58	2.64	2.69	2.73	2.77	2.80	2.83	2.86	2.92	3.00
120	1.98	2.24	2.38	2.47	2.55	2.60	2.65	2.69	2.73	2.76	2.79	2.81	2.87	2.95
∞	1.96	2.21	2.35	2.44	2.51	2.57	2.61	2.65	2.69	2.72	2.74	2.77	2.83	2.91
						$\alpha_T = 0.01$								
5	4.03	4.63	4.98	5.22	5.41	5.56	5.69	5.80	5.89	5.98	6.05	6.12	6.30	6.52
6	3.71	4.21	4.51	4.71	4.87	5.00	5.10	5.20	5.28	5.35	5.41	5.47	5.62	5.81
7	3.50	3.95	4.21	4.39	4.53	4.64	4.74	4.82	4.89	4.95	5.01	5.06	5.19	5.36
8	3.36	3.77	4.00	4.17	4.29	4.40	4.48	4.56	4.62	4.68	4.73	4.78	4.90	5.05
9	3.25	3.63	3.85	4.01	4.12	4.22	4.30	4.37	4.43	4.48	4.53	4.57	4.68	4.82
10	3.17	3.53	3.74	3.88	3.99	4.08	4.16	4.22	4.28	4.33	4.37	4.42	4.52	4.65
11	3.11	3.45	3.65	3.79	3.89	3.98	4.05	4.11	4.16	4.21	4.25	4.29	4.30	4.52
12	3.05	3.39	3.58	3.71	3.81	3.89	3.96	4.02	4.07	4.12	4.16	4.19	4.29	4.41
13	3.01	3.33	3.52	3.65	3.74	3.82	3.89	3.94	3.99	4.04	4.08	4.11	4.20	4.32
14	2.98	3.29	3.47	3.59	3.69	3.76	3.83	3.88	3.93	3.97	4.01	4.05	4.13	4.24
15	2.95	3.25	3.43	3.55	3.64	3.71	3.78	3.83	3.88	3.92	3.95	3.99	4.07	4.18
16	2.92	3.22	3.39	3.51	3.60	3.67	3.73	3.78	3.83	3.87	3.91	3.94	4.02	4.13
17	2.90	3.19	3.36	3.47	3.56	3.63	3.69	3.74	3.79	3.83	3.86	3.90	3.98	4.08
18	2.88	3.17	3.33	3.44	3.53	3.60	3.66	3.71	3.75	3.79	3.83	3.86	3.94	4.04
19	2.86	3.15	3.31	3.42	3.50	3.57	3.63	3.68	3.72	3.76	3.79	3.83	3.90	4.00
20	2.85	3.13	3.29	3.40	3.48	3.55	3.60	3.65	3.69	3.73	3.77	3.80	3.87	3.97
24	2.80	3.07	3.22	3.32	3.40	3.47	3.52	3.57	3.61	3.64	3.68	3.70	3.78	3.87
30	2.75	3.01	3.15	3.25	3.33	3.39	3.44	3.49	3.52	3.56	3.59	3.62	3.69	3.78
40	2.70	2.95	3.09	3.19	3.26	3.32	3.37	3.41	3.44	3.48	3.51	3.53	3.60	3.68
60	2.66	2.90	3.03	3.12	3.19	3.25	3.29	3.33	3.37	3.40	3.42	3.45	3.51	3.59
120	2.62	2.85	2.97	3.06	3.12	3.18	3.22	3.26	3.29	3.32	3.35	3.37	3.43	3.51
∞	2.58	2.79	2.92	3.00	3.06	3.11	3.15	3.19	3.22	3.25	3.27	3.29	3.35	2.42

Reproduced from C. W. Dunnett, "New Tables for Multiple Comparisons with a Control," *Biometrics* 20:482–491, 1964. With permission from The Biometric Society.

TABLE E-4 Critical Values of Student-Newman-Keuls q

v_d	$p = 2$	3	4	5	6	7	8	9	10
				$\alpha_T = 0.05$					
1	17.97	26.98	32.82	37.08	40.41	43.12	45.40	47.36	49.07
2	6.085	8.331	9.798	10.88	11.74	12.44	13.03	13.54	13.99
3	4.501	5.910	6.825	7.502	8.037	8.478	8.853	9.177	9.462
4	3.927	5.040	5.757	6.287	6.707	7.053	7.347	7.602	7.826
5	3.635	4.602	5.218	5.673	6.033	6.330	6.582	6.802	6.995
6	3.461	4.339	4.896	5.305	5.628	5.895	6.122	6.319	6.493
7	3.344	4.165	4.681	5.060	5.359	5.606	5.815	5.998	6.158
8	3.261	4.041	4.529	4.886	5.167	5.399	5.597	5.767	5.918
9	3.199	3.949	4.415	4.756	5.024	5.244	5.432	5.595	5.739
10	3.151	3.877	4.327	4.654	4.912	5.124	5.305	5.461	5.599
11	3.113	3.820	4.256	4.574	4.823	5.028	5.202	5.353	5.487
12	3.082	3.773	4.199	4.508	4.751	4.950	5.119	5.265	5.395
13	3.055	3.735	4.151	4.453	4.690	4.885	5.049	5.192	5.318
14	3.033	3.702	4.111	4.407	4.639	4.829	4.990	5.131	5.254
15	3.014	3.674	4.076	4.367	4.595	4.782	4.940	5.077	5.198
16	2.998	3.649	4.046	4.333	4.557	4.741	4.897	5.031	5.150
17	2.984	3.628	4.020	4.303	4.524	4.705	4.858	4.991	5.108
18	2.971	3.609	3.997	4.277	4.495	4.673	4.824	4.956	5.071
19	2.960	3.593	3.977	4.253	4.469	4.645	4.794	4.924	5.038
20	2.950	3.578	3.958	4.232	4.445	4.620	4.768	4.896	5.008
24	2.919	3.532	3.901	4.166	4.373	4.541	4.684	4.807	4.915
30	2.888	3.486	3.845	4.102	4.302	4.464	4.602	4.720	4.824
40	2.858	3.442	3.791	4.039	4.232	4.389	4.521	4.635	4.735
60	2.829	3.399	3.737	3.977	4.163	4.314	4.441	4.550	4.646
120	2.800	3.356	3.685	3.917	4.096	4.241	4.363	4.468	4.560
∞	2.772	3.314	3.633	3.858	4.030	4.170	4.286	4.387	4.474

(continued on next page)

TABLE E-4 (*Continued*)   **Critical Values of Student-Newman-Keuls** *q*

ν_d	$p = 2$	3	4	5	6	7	8	9	10
				$\alpha_T = 0.01$					
1	90.03	135.0	164.3	185.6	202.2	215.8	227.2	237.0	245.6
2	14.04	19.02	22.29	24.72	26.63	28.20	29.53	30.68	31.69
3	8.261	10.62	12.17	13.33	14.24	15.00	15.64	16.20	16.69
4	6.512	8.120	9.173	9.958	10.58	11.10	11.55	11.93	12.27
5	5.702	6.976	7.804	8.421	8.913	9.321	9.669	9.972	10.24
6	5.243	6.331	7.033	7.556	7.973	8.318	8.613	8.869	9.097
7	4.949	5.919	6.543	7.005	7.373	7.679	7.939	8.166	8.368
8	4.746	5.635	6.204	6.625	6.960	7.237	7.474	7.681	7.863
9	4.596	5.428	5.957	6.348	6.658	6.915	7.134	7.325	7.495
10	4.482	5.270	5.769	6.136	6.428	6.669	6.875	7.055	7.213
11	4.392	5.146	5.621	5.970	6.247	6.476	6.672	6.842	6.992
12	4.320	5.046	5.502	5.836	6.101	6.321	6.507	6.670	6.814
13	4.260	4.964	5.404	5.727	5.981	6.192	6.372	6.528	6.667
14	4.210	4.895	5.322	5.634	5.881	6.085	6.258	6.409	6.543
15	4.168	4.836	5.252	5.556	5.796	5.994	6.162	6.309	6.439
16	4.131	4.786	5.192	5.489	5.722	5.915	6.079	6.222	6.349
17	4.099	4.742	5.140	5.430	5.659	5.847	6.007	6.147	6.270
18	4.071	4.703	5.094	5.379	5.603	5.788	5.944	6.081	6.201
19	4.046	4.670	5.054	5.334	5.554	5.735	5.889	6.022	6.141
20	4.024	4.639	5.018	5.294	5.510	5.688	5.839	5.970	6.087
24	3.956	4.546	4.907	5.168	5.374	5.542	5.685	5.809	5.919
30	3.889	4.455	4.799	5.048	5.242	5.401	5.536	5.653	5.756
40	3.825	4.367	4.696	4.931	5.114	5.265	5.392	5.502	5.559
60	3.762	4.282	4.595	4.818	4.991	5.133	5.253	5.356	5.447
120	3.702	4.200	4.497	4.709	4.872	5.005	5.118	5.214	5.299
∞	3.643	4.120	4.403	4.603	4.757	4.882	4.987	5.078	5.157

Source: H. L. Harter, *Order Statistics and Their Use in Testing and Estimation,* Vol. I: *Tests Based on Range and Studentized Range of Samples from a Normal Population,* U.S. Government Printing Office, Washington, D.C., 1970.

TABLE E-5 Critical Values of χ^2

ν	Probability of Greater Value P							
	.50	.25	.10	.05	.025	.01	.005	.001
1	.455	1.323	2.706	3.841	5.024	6.635	7.879	10.828
2	1.386	2.773	4.605	5.991	7.378	9.210	10.597	13.816
3	2.366	4.108	6.251	7.815	9.348	11.345	12.838	16.266
4	3.357	5.385	7.779	9.488	11.143	13.277	14.860	18.467
5	4.351	6.626	9.236	11.070	12.833	15.086	16.750	20.515
6	5.348	7.841	10.645	12.592	14.449	16.812	18.548	22.458
7	6.346	9.037	12.017	14.067	16.013	18.475	20.278	24.322
8	7.344	10.219	13.362	15.507	17.535	20.090	21.955	26.124
9	8.343	11.389	14.684	16.919	19.023	21.666	23.589	27.877
10	9.342	12.549	15.987	18.307	20.483	23.209	25.188	29.588
11	10.341	13.701	17.275	19.675	21.920	24.725	26.757	31.264
12	11.340	14.845	18.549	21.026	23.337	26.217	28.300	32.909
13	12.340	15.984	19.812	22.362	24.736	27.688	29.819	34.528
14	13.339	17.117	21.064	23.685	26.119	29.141	31.319	36.123
15	14.339	18.245	22.307	24.996	27.488	30.578	32.801	37.697
16	15.338	19.369	23.542	26.296	28.845	32.000	34.267	39.252
17	16.338	20.489	24.769	27.587	30.191	33.409	35.718	40.790
18	17.338	21.605	25.989	28.869	31.526	34.805	37.156	42.312
19	18.338	22.718	27.204	30.144	32.852	36.191	38.582	43.820
20	19.337	23.828	28.412	31.410	34.170	37.566	39.997	45.315
21	20.337	24.935	29.615	32.671	35.479	38.932	41.401	46.797
22	21.337	26.039	30.813	33.924	36.781	40.289	42.796	48.268
23	22.337	27.141	32.007	35.172	38.076	41.638	44.181	49.728
24	23.337	28.241	33.196	36.415	39.364	42.980	45.559	51.179
25	24.337	29.339	34.382	37.652	40.646	44.314	46.928	52.620
26	25.336	30.435	35.563	38.885	41.923	45.642	48.290	54.052
27	26.336	31.528	36.741	40.113	43.195	46.963	49.645	55.476
28	27.336	32.020	37.916	41.337	44.461	48.278	50.993	56.892
29	28.336	33.711	39.087	42.557	45.722	49.588	52.336	58.301
30	29.336	34.800	40.256	43.773	46.979	50.892	53.672	59.703
31	30.336	35.887	41.422	44.985	48.232	52.191	55.003	61.098
32	31.336	36.973	42.585	46.194	49.480	53.486	56.328	62.487
33	32.336	38.058	43.745	47.400	50.725	54.776	57.648	63.870
34	33.336	39.141	44.903	48.602	51.966	56.061	58.964	65.247
35	34.336	40.223	46.059	49.802	53.203	57.342	60.275	66.619

TABLE E-5 (*Continued*)   Critical Values of χ^2

				Probability of Greater Value P				
ν	.50	.25	.10	.05	.025	.01	.005	.001
36	35.336	41.304	47.212	50.998	54.437	58.619	61.581	67.985
37	36.336	42.383	48.363	52.192	55.668	59.893	62.883	69.346
38	37.335	43.462	49.513	53.384	56.896	61.162	64.181	70.703
39	38.335	44.539	50.660	54.572	58.120	62.428	65.476	72.055
40	39.335	45.616	51.805	55.758	59.342	63.691	66.766	73.402
41	40.335	46.692	52.949	56.942	60.561	64.950	68.053	74.745
42	41.335	47.766	54.090	58.124	61.777	66.206	69.336	76.084
43	42.335	48.840	55.230	59.304	62.990	67.459	70.616	77.419
44	43.335	49.913	56.369	60.481	64.201	68.710	71.893	78.750
45	44.335	50.985	57.505	61.656	65.410	69.957	73.166	80.077
46	45.335	52.056	58.641	62.830	66.617	71.201	74.437	81.400
47	46.335	53.127	59.774	64.001	67.821	72.443	75.704	82.720
48	47.335	54.196	60.907	65.171	69.023	73.683	76.969	84.037
49	48.335	55.265	62.038	66.339	70.222	74.919	78.231	85.351
50	49.335	56.334	63.167	67.505	71.420	76.154	79.490	86.661

Adapted from J. H. Zar, *Biostatistical Analysis*, 2nd ed., 1984, pp. 479–482. Reprinted by permission of Prentice-Hall, Inc., Englewood Cliffs, N.J.

Solutions to Problems

CHAPTER 2

2.1 Slope: .444 g/cm; standard error of slope: .0643 g/cm; intercept: -6.01 g; standard error of intercept: 2.39 g (not available in BMDP); correlation coefficient: .925; standard error of the estimate: .964 g; F: 47.7.

2.2 Yes. A simple regression yields $r = .96$ ($t = 7.8$ with 5 degrees of freedom; $P < .001$) and $s_{y|x} = 3.9$.

2.3 Yes. Regress female breast cancer rate B on male lung cancer rate L to obtain $\hat{B} = 4.00 + .282\ L$, $P = .02$.

2.4 The regression equation is $\hat{C} = 1.36$ log CFU $-.04$ log CFU/% M, with $R^2 = .649$, $s_{y|x} = .87$ log CFU, and $F = 96.1$ with 1 and 52 degrees of freedom. Both the slope ($t = -9.8$; $P < .001$; 95 percent confidence interval, $-.049$ to $-.032$) and intercept ($t = 5.21$; $P < .001$; 95 percent confidence interval, .84 to 1.88) are highly significant. This linear model provides a significant reduction in the observed variability of the dependent variable.

2.5 A. Based on the mean data, the regression equation is $\Delta\hat{R} = .167$ mmHg/(mL·min·kg) $- .276$ (mmHg·h)/(mL·min·ng) Δr, with $r = .957$, $s_{y|x} = .97$ mmHg/(mL·min·kg), and a t statistic for the slope of -4.63 with 2 degrees of freedom ($P = .044$). Thus, the change in vascular resistance seems to be significantly related to the change in renin production by the kidney. **B.** Using the raw data points, we find the similar regression equation $\Delta\hat{R} = .183$ mmHg/(mL·min·kg) $- .162$ (mmHg·h)/(mL·min·ng) Δr, with $r = .632$, $s_{y|x} = 2.15$ mmHg/(mL·min·kg), and the t statistic for the slope is -4.32 with 28 degrees of freedom ($P < .001$). Thus, we still conclude, based on the t statistic, that the change in vascular resistance is significantly related to the change in renin production by the kidney. **C.** Whereas the regression equations are similar, the equation computed using the group mean values had a much higher correlation than the one based on the raw data (.957 vs. .632). Likewise, the estimate of the variability about the regression plane $s_{y|x}$ is much smaller when computed using the mean values [.97 vs. 2.15 mmHg/(mL·min·kg)]. The regression equation computed from the group means seems to provide a much better prediction of the data than the one computed from the raw data. The difference in the results of the two regression analyses lies in the difference between the mean values and the raw data. By computing the means of each group, then analyzing these four data points with regression, we have effectively thrown away most of the variability in the observations. Thus, $s_{y|x}$ is artificially reduced and is no longer an unbiased estimate of the population variability around the plane of means. As a result, r is inflated. To understand why r and $s_{y|x}$ change but not the regression equation, you need to consider the assumptions behind the linear regression model. The first assumption is that the mean value of the dependent variable at any given value of the independent variable changes linearly with the independent variable. The second assumption is that there is a constant random variation around the line of means. By "binning" the data and computing means *before* completing the regression computations, you greatly reduce the variation in the points used to compute the regression equation. (In fact, if there were no sampling variation, the residual variation in the "binned" sample means about the line of means would be zero!) Because you are estimating the same line of means in both cases, the regression equations are similar. However, because you threw out most of the variation about the line of means by averaging the data in the first part of the problem, you underestimated the residual variation. In this example the overall conclusion about the relationship between renal vascular resistance and renin production did not change because the increase in degrees of freedom associated with the raw data compensated for the increase in $s_{y|x}$ in the hypothesis tests about the slope of the relationship. This fortuitous situation does not often occur, and the kind of binning of the data illustrated here often leads to conclusions of a statistically significant relationship when the data do not justify such a conclusion. **D.** The analysis based on the mean values is incorrect and gives an artificially rosy picture of the goodness of the regression fit to the data. The analysis based on all the data points is correct. The use of regression to analyze mean values of many points, rather than to analyze all the points themselves, is one of the most common abuses of simple linear regression.

2.6 Linear regressions between U_{Ca} and the four other variables yield these correlations:

	R^2	r	P
D_{Ca}	.58	.76	<.001
D_p	.40	.63	<.001
U_{Na}	.24	.49	.009
G_{fr}	.17	.41	.034

Based on these four simple correlation analyses, all four variables have significant correlations with urinary calcium. Based on the relative magnitudes of the correlations and the associated P values, D_{Ca} and D_p are more strongly related to U_{Ca}. However, it is extremely difficult to draw firm conclusions about the relative importance of these four variables, because we have not taken into account the multivariate structure of the data analysis. To obtain a clearer view of which of these four independent variables are most strongly influencing U_{Ca}, we need to do a multiple regression analysis (see Probs. 3.8 and 6.4).

2.7 A. The regression equation is $\hat{F} = -172.5\% + .657\%/\text{ms } T_1$, with $r = .931$, $s_{y|x} = 6.4\%$, and $F = 90.7$ with 1 and 14 degrees of freedom ($P < .001$). Both regression coefficients are highly significant. The correlation is high and $s_{y|x}$ is relatively small (11 percent of the mean percent of fast-twitch fibers in the sample). Thus, MR spectroscopy has promise for identifying the muscle fiber composition of athletes so that they can concentrate on sports for which they are physiologically well suited. **B.** The slope is positive, indicating that muscle bundles with a higher percentage of fast-twitch fibers have longer relaxation times. The intercept does not make physiological sense; there cannot be a negative percentage of muscle fibers that are fast-twitch. The problem is that the range of measured relaxation times is from about 300 to 400 ms, far away from the origin. Thus, estimates of the regression intercept represent a large extrapolation beyond the range of the data. The fact that the intercept cannot be negative indicates that the actual relationship between the fraction of fast-twitch fibers and T_1 must be nonlinear over the entire range of fractions of fast-twitch fibers. The good correlation between F and T_1 indicates that a linear model is a good description of the data over the range of observations and is useful for predicting muscle fiber composition over the range of observed fiber types.

CHAPTER 3

3.1 $\hat{W} = b_0 + b_H H + b_D D + b_{HD} HD$. To see that the coefficient b_{HD} relates to the change in slope and the coefficient b_D relates to the change in intercept, first solve the equation with $D = 0$ and then solve the equation with $D = 1$, collecting all terms.

3.2 The fact that the dummy variable only takes on two values (0 and 1) does not alter the ability to treat it like any other numeric variable. The slope of the regression plane in the dummy variable dimension indicates the mean difference between the two conditions encoded by the dummy variable. If this slope is significant, the two conditions are significantly different. Numerically the slope is the change in intercept between the two regression lines; that is, the magnitude of the parallel shift in the Y value between the lines when moving the X value 1 unit (the change in value of the dummy variable from 0 to 1).

3.3 The regression equation is $\hat{W} = -15.9$ g $+ .759$ g/cm $H - .252$ g/(cups·day) C. The coefficients in this equation differ from those in Eq. (3.1) because the values of water consumption are different. The point is that the independent variables in multiple regression are free to take on any values on continuous or discrete measurement scales. We originally had only three discrete values of water consumption to make it easier to graphically illustrate the transition from simple regression to multiple regression, and from two to three dimensions in Fig. 3-1. You can just as easily draw a three-dimensional picture like Fig. 3-1C using the water consumption data in this problem as we did using the original discrete values of water consumption. It would not have been possible to use these new data to draw Fig. 3-1B.

3.4 .31 g, labeled "ROOT MSE" in the output.

3.5 The regression equation is $\hat{A} = 5.47 - 0.0027Y - 0.0126P + 0.00535C$, where A = average number of articles per publication per year, Y = year, P = average percentage of advertising revenues derived from tobacco, C = number of articles related to smoking in the *Christian Science Monitor* that year. The effect of time Y is not statistically significant ($P = .462$), while the effect of cigarette advertising P and other coverage C are significantly different from zero ($P = .044$ and $.001$, respectively). The effect of cigarette advertising is large, particularly after 1971, when cigarette advertising was banned from radio and television. The percentage of cigarette advertising increased to about 10 percent in those publications that accepted it, indicating a reduction of about .12 stories per year related to smoking, after taking into account changes in the "newsworthiness" of the story as indexed by C. This change of .12 compares with an average coverage of only about .25 stories per publication per year.

3.6 $R^2 = .268$ and $F = 21.44$ with 2 numerator and 117 denominator degrees of freedom $(P < .001)$. The large F value (and corresponding small P value) indicate that the regression equation provides a significantly better prediction of the observations than simply the mean value of the dependent variable $\overline{V}$. However, it would be incorrect to conclude that this was a particularly "good" fit because the R^2 is only .268, indicating that the independent variables account for only 27 percent of the total variation observed in V. Thus, although the regression equation does provide some description of the data, there is great uncertainty about the value of the dependent variable for any combination of the independent variables, as reflected in the large value of $s_{y|x} = 157$ percent.

3.7 **A.** We solve this problem by fitting the data collected with and without the xamoterol separately, then compare these two regression lines. (The problem could also be solved with a single regression using a dummy variable to encode the xamoterol groups, to account for an intercept shift, and the dummy by N interaction, to account for a slope change.) Based on the mean data, the regression equation without xamoterol is $\hat{P} = 111$ mmHg $+ .069$ (mmHg·mL)/pg N, with $r = .999$ and $s_{y|x} = 2.1$ mmHg. The regression equation with xamoterol is $\hat{P} = 131$ mmHg $+ .031$ (mmHg·mL)/pg N, with $r = .961$ and $s_{y|x} = 4$ mmHg. A t test for difference in intercept $(t = 3.94$ with 4 degrees of freedom; $P = .017)$ and slope $(t = -4.26$ with 4 degrees of freedom; $P = .013)$ are both significant, indicating that the regression lines are different. They cross at a value of norepinephrine of 525 pg/mL. (To see this, equate the two regression equations and solve for norepinephrine.) Thus, at values of norepinephrine lower than 525 pg/mL, blood pressure is higher at any given level of norepinephrine when xamoterol is present, indicating that xamoterol has a stimulating effect. Conversely, at epinephrine levels higher than 525 pg/mL, blood pressure is lower at any given level of norepinephrine when xamoterol is present, indicating that xamoterol has a blocking effect. **B.** Based on all the data points, not the mean values, the regression equation without xamoterol is $\hat{P} = 122$ mmHg $+ .046$ (mmHg·mL)/pg N with $r = .432$ and $s_{y|x} = 27$ mmHg. The regression equation with xamoterol is $\hat{P} = 137$ mmHg $+ .021$ (mmHg·mL)/pg N with $r = .326$ and $s_{y|x} = 30$ mmHg. t tests for difference in intercept $(t = 1.13$ with 60 degrees of freedom; $P = .262)$ and slope $(t = -1.12$ with 60 degrees of freedom; $P = .267)$ are both not significant, indicating that the regression lines are not different. Thus, based on all the data, we conclude that xamoterol had no effect on the relationship between blood pressure and norepinephrine. This conclusion is opposite the one we reached based on the mean data points. The difference is that by doing the regression on the mean data points, we threw away most of the variability and thus greatly inflated the correlation coefficient and deflated $s_{y|x}$. Because the t test for difference in slope and intercept depends on $s_{y|x}$, the artificially low value of $s_{y|x}$ led us to calculate large t values and conclude the regression lines were different. **C.** The first analysis is incorrect. Regression should be done using all the raw data points so that $s_{y|x}$ is an unbiased estimate of the true underlying population variability. Only when t tests for differences in slopes and intercepts are based on correct estimates of variability can the hypothesis tests be meaningful. Regression using mean values, rather than all the data points, is one of the most common abuses of statistics in the biomedical literature and, as in this case, leads to incorrect conclusions, particularly if regression slopes and intercepts are compared using t tests. (See the discussion of Prob. 2.5.)

3.8   **A.** The multiple regression equation relating U_{Ca} to D_{Ca}, D_p, U_{Na}, and G_{fr} is $\hat{U}_{Ca} = -5.73$ mg/12 h $+ .345\ D_{ca} - .511$ (mg·day)/(12 h·g) $D_p + .498$ mg/meq $U_{Na} + .427$ (mg·min)/(12 h·mL) G_{fr}. $R^2 = .735$ ($R = .86$), $F = 15.27$ with 4 and 22 degrees of freedom ($P < .001$), and $s_{y|x} = 34.2$ mg/12 h. Only two of the variables have statistically significant effects, D_{ca} ($P = .001$) and U_{Na} ($P = .035$). Of the four independent variables, only U_{Na} and D_{Ca} are significantly related to U_{Ca}. **B.** This conclusion is different from what we obtained in Prob. 2.6, where the consideration of four separate univariate correlation analyses led us to conclude that all four independent variables were related to U_{Ca}, with D_{Ca} and D_p most strongly related to U_{Ca}. **C.** The present multivariate analysis differs from the collection of univariate analyses because it simultaneously considers the effects of the four variables. Thus, when taking into account the information provided by D_{Ca}, which had the strongest univariate correlation, D_p is no longer correlated, but U_{Na} is. Not only is the multivariate approach preferable, but it could also be argued that of these two approaches, it is the only one that is correct.

3.9   This problem is much like the salty bacteria example presented in this chapter. The rate of leucine transport is the slope of the relationship between leucine accumulation L and time t. The first step in solving this problem is to examine a plot of the data. This plot reveals that there are four relationships between L and t, one for each of the cholesterol levels C. However, unlike the salty bacteria example, these relationships are not linear. They rise initially, level off, then begin to fall as t increases. The initial slopes appear to be steeper at lower cholesterol contents, but the amount of curvature also seems to be more pronounced at lower cholesterol contents. Thus, both the "slope" and the curvature depend on the cholesterol level, so there is a more complex interaction here than we have treated before. The simplest treatment of these data is to model the basic function (at any constant level of C) as a quadratic, $\hat{L} = a_t t + a_{t^2} t^2$, which has no intercept because, by the design of the experiment, there should be no accumulation of leucine at time $t = 0$. Let both a_t and a_{t^2} depend on C, which yields the model $\hat{L} = b_t t + b_{t^2} t^2 + b_{tC} tC + b_{t^2 C} t^2 C$, where b_{tC} quantifies the interaction effect of cholesterol to change the initial slope (transport rate) of the curves, and $b_{t^2 C}$ quantifies the effect of cholesterol to change the amount of curvature. In terms of the physiological question, b_{tC} is of most interest because the initial slope relates most closely to the transport rate. If the investigators' hypothesis is correct, b_{tC} should be significant and negative, indicating a decreased transport rate as cholesterol content goes up (fluidity goes down). Multiple regression using the above equation yields $\hat{L} = .11$ nmol/(mg·min) $t - .001$ nmol/(mg·min^2) $t^2 - .57$ min^{-1} $tC + .004$ min^{-2} $t^2 C$, with $R^2 = .933$, $s_{y|x} = .45$ nmol/mg, and all parameter estimates are statistically significant ($P < .001$). Of most importance, b_{tC} is negative, $-.57$ min^{-1}, which is what we would expect if the hypothesis being investigated were true. Thus, the transport of leucine slows as membrane fluidity decreases, probably because the proteins that serve as "carriers" to move leucine across the membrane are less mobile as membrane fluidity decreases. The significant coefficient of $t^2 C$, .004 min^{-2}, indicates that the curvature decreases as C increases. (Although this effect can be clearly seen in a plot of the data, the curvature is a by-product of the experimental design and unrelated to the experimental hypothesis.)

3.10 A. To solve this problem, write one multiple regression equation with dummy variables to encode the effect of the two different groups on the slope and intercept between the lines describing the two phases, $\hat{R} = b_0 + b_B B + b_P P + b_{BP} BP$, where P is a dummy variable equal to 0 if early phase and 1 if midluteal phase. From the parameters estimated for this equation, we find that the intercept of the early phase line is $b_0 = 71.2$ U/L, whereas the intercept of the midluteal phase line is $b_0 + b_P = 71.2 + 266.6 = 337.8$ U/L. Similarly, the slope for the early phase line is $b_B = .46$, whereas the slope for the midluteal phase is $b_B + b_{BP} = .46 + .21 = .67$. (You should get the same result, within roundoff error, if you use two separate simple regressions.) The t statistic associated with b_P, which tests for a difference in intercept, is 1.31 with 47 degrees of freedom ($P = .198$), and the t test associated with b_{BP}, which tests for a difference in slopes, is 1.16 ($P = .25$). Thus, neither the intercept nor the slope of these lines is significantly different from one another, and within the sampling variability present in these data, the two assays are similarly related in both phases of the cycle. **B.** No. Having concluded the lines are not different in the early and midluteal phases, we use simple regression to estimate a common slope and intercept and compute a t statistic for the difference in the slope from one. The resulting regression equation is $\hat{R} = 210 + .44B$, with the standard error of the slope $= .066$. We test whether this slope is significantly different from 1 by computing $t = (1 - .44)/.066 = 8.48$ with $51 - 2 = 49$ degrees of freedom, so $P < .001$, and we conclude that the pooled common slope is different from 1, contrary to what one would expect if these two assays were measuring the same thing. The fact that the slope is less than 1 means that the new RIA method underestimates inhibin compared to the standard bioassay method.

3.11 Yes. The regression equation (assuming a linear model) is $\hat{C} = 1.35 \log CFU - .04 \log CFU/\% M + 1.63 \log CFU S - .07 \log CFU/\% MS$, where S is a dummy variable equaling 0 if 1- to 4-h dosing and equaling 1 if 6- to 12-h dosing, with $R^2 = .741$ and all regression coefficients highly significant ($P < .001$). The coefficient of S indicates that the intercept of the regression equation for the 6- to 12-h dosing group is higher than that of the 1- to 4-h group by 1.63 log CFU. Likewise, the coefficient of the MS interaction indicates that the slope of the 6- to 12-h group is steeper by $-.07 \log CFU/\%$ than the slope of the 1- to 4-h group, which has a slope of $-.04 \log CFU/\%$. Thus, both the slope and the intercept of the lines describing these two groups of data are different.

3.12 The regression equation is $\hat{F} = -149\% + .446\%/\text{ms } T_1 + 1.8\%/\text{ms } T_2$, with $R^2 = .885$, $s_{y|x} = 6.15\%$, and $F = 49.98$ with 2 and 13 degrees of freedom ($P < .001$). The regression coefficient associated with T_1 is significantly different from zero ($t = 2.79$; $P = .015$), but the coefficient associated with T_2 is not ($t = 1.45$; $P = .17$). Thus, adding T_2 to the predictive equation does not significantly add to the ability to predict muscle fiber composition using MR spectroscopy.

3.13 **A.** A plot of ΔS vs. S reveals a family of four lines with increasing slope as F increases. Thus, we fit an interaction model to these data and obtain the regression equation $\Delta \hat{S} = -3.34$ spikes/s $+ .194S - .006$ spikes/(s·Hz) $F + .005/\text{Hz } SF$, with $R^2 = .947$. **B.** The intercept is not significantly different from zero ($t = -1.1$ with 55 degrees of freedom; $P = .28$), which means that external stimulation changes stretch receptor discharge only when a basal level of discharge is present. Similarly, the coefficient associated with F is not significantly different from zero ($t = -.14$ with 55 degrees of freedom; $P = .89$). Thus, the intercept is zero (when S and F are zero, ΔS is zero). Because the coefficient associated with F is also zero, the intercept does not change as F changes. Thus, the four lines have a common intercept at the origin. The coefficient associated with S is positive, indicating that the change in stretch receptor output due to nerve signals coming from the brain ΔS depends on the basal level of nerve activity when there are no brain signals S. The significant coefficient for the SF term indicates that there is an interaction such that the sensitivity of ΔS to S increases by .005 for every 1-Hz increase in F. For example, for a frequency of stimulation F of 120 Hz, the sensitivity of ΔS to S is $.194 + .005(120) = .794$, whereas it is only $.194 + .005(20) = .294$ when F is 20 Hz. Thus, the change in nerve discharge from stretch receptors in the lung ΔS depends on the basal level of discharge of the stretch receptors S. Furthermore, the strength of this dependence increases as the frequency of nerve input into the lung increases.

3.14 The regression equation is $\hat{V} = 198\% - 12.3O + 33.4C$, with $R^2 = .27$ ($F = 17.3$ with 2 and 117 degrees of freedom; $P < .001$). These numbers are slightly different from what we obtained for the nestling swallows, but the overall conclusions are the same. The sensitivity of V to oxygen has a t statistic $= -1.74$ with 117 degrees of freedom ($P = .08$). Given the great scatter in this data, we would not be willing to take this as evidence that oxygen *does not* influence ventilation, but rather only that we do not have sufficient evidence to conclude that oxygen does influence ventilation in adult bank swallows, especially because oxygen is known to influence ventilation. The sensitivity to carbon dioxide is much greater and is highly significantly different from zero ($t = 6.34$; $P < .001$).

3.15 The regression equation is $\hat{V} = 198\% - 12.3O + 33.4C - 112\% B + 7BO - 2.3BC$, with $R^2 = .269$ and $s_{y|x} = 166\%$. The intercept of the adult plane is 112 percent above the intercept of the baby plane ($t = -.71$ with 234 degrees of freedom; $P = .48$), the adults are more sensitive to changes in oxygen (the baby's slope is 7 units less negative than the adult's slope, $t = .73$; $P = .46$), and the adults are slightly more sensitive to changes in carbon dioxide by 2.3 units ($t = -.32$; $P = .75$). None of these changes is statistically significantly different from zero. There is so much scatter in the data that we cannot estimate the parameters with much precision (all standard errors for the estimates of changes between adults and babies are greater than the estimated changes), and so we conclude, based on the analysis in this problem, that the ventilatory control sensitivity to oxygen and carbon dioxide is not different between babies and adults.

3.16 -26.3 to .7. Because this interval includes zero, we cannot conclude that this parameter is different from zero with $P < .05$. (The P value for a test of significance of this parameter is .08.) However, looking at the confidence interval in a less strict sense gives us the opportunity to apply some judgment. Clearly, it just barely includes zero. Given what we know about respiratory physiology, and given the tremendous scatter of data in this experiment, we would be inclined to argue that this is a significant effect if we were writing a paper based on these data. At the very least, we would not conclude that we had demonstrated *no* effect.

3.17 **A.** Yes. Do a multiple regression of breast cancer rate B on animal fat A and male lung cancer rate L to obtain $\hat{B} = 1.45 + .132/(g \cdot day) A + .104L$. $F = 100.92$ with 2 numerator and 33 denominator degrees of freedom ($P < .001$). For b_A, $t = 7.75$ with 33 degrees of freedom ($P < .001$), and for b_L, $t = 3.15$ with 33 degrees of freedom ($P = .004$). Hence, the effect of male lung cancer rates on female breast cancer rates is not the result of a confounding of these two independent variables. **B.** No. The regression equation including interaction is $\hat{B} = 0.89 + 0.143/(g \cdot day) A + .125L - 0.000323/(g \cdot day) AL$. b_A and b_L remain significantly different from zero ($P < .001$ and $P = .047$, respectively), but the A by L interaction effect b_{AL} is not significantly different from zero ($t = -0.42$ with 32 degrees of freedom; $P = 0.675$).

CHAPTER 4

4.1 There are two influential points. Point no. 1 has a leverage value of .28, which is greater than the cutoff for concern of twice the average value of $h = (k + 1)/n = 2[(1 + 1)/30] = .133$. Point no. 7 has a leverage, .123, near this value. These two points can easily be located in a plot of the raw data or of the residuals against the dependent variable. Because both points have relatively small Studentized residuals r_{-i}, they have Cook's distances below our threshold of concern. There is a slight suggestion of nonlinearity, and a quadratic fit increases the R^2 from .40 to .495 and reduces $s_{y|x}$ from 2.15 to 2.01 mmHg/(mL·min·kg). However, most of the evidence for a nonlinearity comes from the two points already identified above as influential. Thus, caution should be used when trying to draw physiological interpretations based on a nonlinear fit. In this situation, we would stick to linear regression.

4.2 **A.** $\hat{I}_{gf} = 441$ µg/L $+ 4G_h$, with a low correlation of .35, which is not significantly different from zero ($P = .12$). Thus, there is no evidence of a link between these hormones. Even when these investigators correlated I_{gf} to the 24-h average G_f level, there was a poor correlation. Thus, like those before them, these investigators found a poor linear correlation between these two hormones. How can one reconcile this finding with the underlying physiology? One could postulate that acromegaly changes the relationship between these hormones, but there is no good evidence for that. Another possibility is that the analysis assumes an incorrect model. A plot of the data shows no evidence of nonlinearity, yet there are some troubling aspects to the use of a linear model. In this case, one would expect I_{gf} to be zero if G_h is zero. However, the regression intercept for these data is an I_{gf} value of 441 µg/L. Hence, while there is no evidence the linear model has been incorrectly applied to these data, the large positive intercept makes no physiological sense. A final possibility to explain the lack of correlation is that the observations were made over the wrong (or too limited) range. In this experiment, the data are all at high G_h levels—by definition because this is a disease of overproduction of G_h. As in many biochemical processes, there is a nonlinear and saturable relationship between I_{gf} and G_h such that I_{gf} is zero when G_h is zero, increases as G_h increases at low to intermediate G_h values, but levels off (saturates) at high G_h values. The data collected in subjects with acromegaly show no correlation because their values of I_{gf} and G_h are in the saturated end of the nonlinear relationship. When these 21 subjects were treated to lower their G_h (data not shown), the expected nonlinear relationship became evident as data were collected at lower G_h values. Thus, the linear model was correct over the limited high range of G_h values, but the data were collected over too limited a range. **B.** There are two obvious influential points (nos. 20 and 21 in the data table). Both of these points have leverages, .33 and .58, respectively, that exceed $2(k + 1)/n = 4/21 = .19$. Point no. 20 has a modest Cook's distance of .36 (not large enough to be of great concern), whereas point no. 21 has a very small Cook's distance. Both points are influential but do not seriously affect the physiological conclusion (deleting both points only raises the correlation to .43 with $P = .07$). **C.** Because Cook's distance depends on both the leverage and the standardized residual, even though point no. 21 has high leverage, its Cook's distance is small because of the small residual. Remember, leverage only indicates the *potential* for influence.

4.3   **A.** A plot of A vs. U_{ag} and the corresponding regression equation, $\hat{A} = 22.5$ meq/L $- .41U_{ag}$, both suggest that higher U_{ag} is associated with lower A. **B.** $r = .71$, $F = 51.7$ with 1 and 51 degrees of freedom $(P < .001)$, and $s_{y|x} = 12.5$ meq/L, compared to an average A of 16.6 meq/L. Thus, we conclude A decreases with U_{ag}, but the amount of variability around the regression line is high enough that precise prediction of ammonium from U_{ag} in a given patient is not possible. **C.** There are two obviously influential data points. Point no. 1 has an r_{-i} of -7.53, indicating that it is a large outlier. This outlier can also be easily identified in a normal probability plot of the residuals or a plot of residuals against U_{ag}. This point also has a leverage value, .11, which exceeds the expected value of twice the expected value of h, $2(k + 1)/n = 4/53 = .08$. Together, the large residual and leverage cause this point to have a large Cook's distance of 1.65. Point no. 10 has a large leverage of .21, a relatively large r_{-i} of -2.1, and a Cook's distance of .55, which is an order of magnitude larger than the other points, indicating modest influence. Plots of residuals against U_{ag} and A against U_{ag} show evidence of a nonlinearity. **D.** The regression equation for a quadratic model is $\hat{A} = 19.9$ meq/L $- .42U_{ag} + .0025$ meq/L U_{ag}^2. A quadratic fit offers little improvement in fit: the coefficient of the quadratic term is marginally insignificant $(t = 1.93; P = .059)$, $s_{y|x}$ decreases from 12.5 to 12.2 meq/L, and r increases from .71 to .73. Both problem points are still big problems: point no. 10 now has a Cook's distance of 15.3 because it has a far larger r_{-i}, -7.47, than in the linear model. Point no. 1 still has a large r_{-i} of 7.11, large leverage of .15, and Cook's distance of 1.53. A residual plot now shows no evidence of a systematic trend. Thus, a quadratic model is better but shows little improvement over the linear model because of the two large outliers.

4.4   A plot of the data suggests a slight nonlinearity that is somewhat masked by the scatter in the data. A plot of the residuals against H from the linear regression of these data $(\hat{G} = 6.1 - .03$ L/mU H with $R^2 = .429; P < .001)$ also indicates a systematic trend and suggests that there are also unequal variances. Otherwise, the residuals are unremarkable: two have slightly larger than expected leverage values and one has a Studentized value r_{-i} of 2.41. The trend looks like an exponential decay $G = b_0 e^{-b_1 H}$, which can be fit with linear regression after a $\ln(G)$ transformation of the data. The fit of this equation is actually slightly worse, $R^2 = .398; P < .001$. Examination of the data reveals why: This exponential function assumes decay to $G = 0$ as $H \to \infty$, whereas the data suggest an asymptotic approach to a value of $G > 0$ as $H \to \infty$. Thus, a better exponential model would be $G = b_2 + b_0 e^{-b_1 H}$, where b_2 is a level of G which is asymptotically approached as H goes to infinity. Unfortunately, this model is not linearizable via a simple transformation, so we must use nonlinear regression (see Prob. 10.1).

4.5 A plot of the data shows a nearly linear increase up to a pH of about 8 and then an apparent fall off. Thus, the optimum pH appears to be about 8. However, there is really only one point that suggests that a maximum has been reached. As we have cautioned before, one needs to be very careful in making interpretations based on one or two points. This point could represent one point on a steep drop off or it could be an outlier as a result of some experimental problem. In terms of influence, a simple linear regression shows this point to be associated with an r_{-i} of -6.3 and a Cook's distance of .85. However, in terms of thinking about whether there is a maximum in this function, the Cook's distance from a linear regression greatly understates the influence of the point in one's judgment of whether or not there is a peak in the function. Although it is not unreasonable in terms of the dependence of enzyme function on pH to have a very steep fall in activity above some pH value, additional data should be obtained to better define the steep descent and ensure that this single point is not just an outlier.

4.6 **A.** Yes. Using the same regression model as in Prob. 3.10, we find $\hat{R} = 71.2 \text{ U/L} + .46 \text{ B} + 152 \text{ U/L } P + .35PB$. There is a statistically significant difference in slopes between the two phases ($P = .046$); the slope during the early phase is .46 (the same as before) and is increased to .81 ($b_B + b_{BP} = .46 + .35$) in the midluteal phase. This leads us to a different conclusion than with the correct data. Two things about the transposition led to this change in conclusion. First, the transposition caused us to obtain a higher estimate of the slope in the midluteal phase, thus the change in slopes (the numerator of the t statistic for b_{BP}) is larger. Second, it caused a reduction in $s_{y|x}$ and, thus, in the standard error of the slope (which affects the denominator of the t statistic). Hence, we calculate a higher t value, which exceeds the critical value of t for a significance level of .05. **B.** The two points involved in the transposition have Cook's distances that would not ordinarily be of concern, .32 and .19. One has a high leverage value, but that only indicates a potential for influence, and the Cook's distances indicate neither one is actually of much influence. It is therefore unlikely that you would detect this transposition error by examining the regression diagnostics. Sometimes transposition errors can explain why a point appears as an outlier, but regression diagnostics cannot be used as a substitute for careful checking of the numbers to guard against data entry errors.

4.7 Two points, numbers 37 and 38, have leverage values exceeding the cutoff of $2(k + 1)/n = 2(2)/38 = .105$. Point 37 has a Cook's distance of .69, which, although not large enough to cause particular concern, is 1 to 2 orders of magnitude larger than the other values of Cook's distance. This point has only a marginally elevated leverage of .14. Point 38 has a very small Cook's distance, .0002, in spite of having a much larger leverage of .43. The difference between these two points is that, although point 38 has the largest leverage (the plot shows it well removed from the rest of the data along the bioassay axis), it lies along the trend established by the remaining points and thus has a very small Studentized residual, $r_{-38} = .024$. The combination of high leverage and low residual leads to no actual influence as judged by Cook's distance. In contrast, point 37 is not so far removed from the others along the bioassay axis, so it has a lower leverage and thus a lower potential for influence, but it lies well off the trend established by the other points and so has a large Studentized residual, $r_{-37} = -3.26$. The combination of lower potential influence but high residual leads to greater actual influence. Hence, in order for the potential influence quantified by leverage to be translated into actual influence, the point must also be an outlier, far removed from the trend established by the rest of the data.

4.8 A linear regression of Breathalyzer alcohol level B against that estimated from self-reporting of consumption S yields the regression equation, $\hat{B} = .026 + .522S$, with $s_{y|x} = .029$ and $R^2 = .588$. The slope, .522, is significantly different from the expected slope of 1 ($t = 5.36$; $P < .001$; 95 percent confidence interval, .34 to .71). Thus, as a group, social drinkers tend to overestimate their alcohol consumption. However, examination of a plot of the data, as well as residual diagnostics from the linear regression, indicate a potential problem. Above actual consumptions of about .10 (often the legal limit for driving), a relatively good correspondence between self-reported alcohol consumption and the Breathalyzer test begins to break down. Point 21 is an outlier with $r_{-21} = 4.9$. (This outlier is readily apparent in any residual plot, including a normal probability plot of the residuals.) This point and points 25 and 26 have relatively high Cook's distance values (though not large enough to be of great concern all by themselves). Points 25 and 26, as well as point 24, have high leverages (exceeding the threshold value of .15). All of these data points are from subjects with blood alcohol levels at or above .10. The pattern of the residuals suggests a nonlinear relationship, with the curve flattening at higher perceived blood alcohol levels. For some reason, the ability to recall drinking breaks down in those subjects with higher alcohol consumption. Perhaps their recall is impaired or, perhaps, they are having fun by inflating their estimates. Regardless of the reason, self-recognition would seem to work at a blood alcohol level where it is least needed, those with blood alcohols less than the legal limit, whereas those individuals impaired enough to be legally drunk cannot accurately determine their consumption.

4.9   First, order the residuals from most negative to most positive. Assign rank 1 to the most negative and rank $n = 26$ to the most positive. From the ranks, compute the cumulative frequencies, $(\text{rank} - .5)/n$: $(1 - .5)/26 = .019$, $(2 - .5)/26 = .058, \ldots$, $(26 - .5)/26 = .981$. Plot each cumulative frequency against the corresponding residual on normal probability paper.

4.10   The regression equation is $\hat{B} = .003 + .86S$, with $s_{y|x} = .015$ and $R^2 = .76$. The intercept of this line is not significantly different from zero ($t = .34$ with 15 degrees of freedom; $P = .74$), and the slope, .86, is not significantly different from 1 ($t = 1.10$; $P = .29$; 95 percent confidence interval, .60 to 1.13). Thus, this lower range of data indicates reasonably accurate self-reporting. In this situation, there are many possible reasons why the accuracy of self-reporting will break down at higher alcohol consumption levels: subjects are more impaired and cannot recall accurately or may be more likely to lie about consumption. Thus, one is probably justified in removing these subjects for part of the analysis.

4.11   The regression equation is $\hat{B} = b_0 + b_S S + b_D(S - .11)D$, where D is a dummy variable $= 1$ if $S > .11$ and 0 if $S \leq .11$, where .11 is the "break point" in the relationship as determined by visual inspection of a plot of the data. Fitting this equation to the data of Table D-17, Appendix D, we find $\hat{B} = -.0072 + 1.12S - 1.14(S - .11)D$ with $R^2 = .784$, $s_{y|x} = .02$, and $F = 41.6$. The intercept b_0 and slope b_S (which describe the line when $S \leq .11$, i.e., when $D = 0$) are not significantly different from 0 ($t = -.71$ with 23 degrees of freedom; $P = .48$) and 1 ($t = .82$; $P = .42$), respectively, which indicates that self-reporting accurately reflects blood alcohol as determined by a Breathalyzer. The equation that describes the range of self-reported alcohol above .11 is given by $\hat{B} = (b_0 - .11b_D) + (b_S + b_D)S = .118 - .02S$. As expected from looking at the data, this line segment is virtually flat, the slope is only $-.02$. This procedure of fitting two different segments with one equation is known as *splining*.

4.12   The regression diagnostics show nothing unusual. The leverage value threshold for concern is $2(1 + 1)/54 = .074$. Only data point

1 has a residual with a leverage this large, .089. This point has a small r_{-1} of $-.066$, and so its Cook's distance is very low, .0002. Only one point has a Studentized residual greater than 2, $r_{-41} = -2.3$. None of these are striking, and so, from the viewpoint of these regression diagnostics, there are no problems with these data in regard to influential points and/or outliers.

4.13 The linear model has an F value of 96.1 $(P < .001)$, which indicates a relatively good fit. However, there is still a lot of "unexplained variance"; the R^2 is only .65, indicating that $1 - .65$, or 35 percent, of the variance in the dependent variable is unaccounted for by this regression model. A plot of the raw data suggests a curvilinear relationship. This is confirmed by a plot of the residuals against M. A quadratic relationship improves the R^2 to .78 $(r = .88)$ and reduces $s_{y|x}$ from .87 to .69 log CFU, giving the equation $\hat{C} = 3.13$ log CFU $-.12$ log CFU/% $M + .0007$ log CFU/%^{2} M^2. The regression coefficient associated with the square of M is highly significant $(t = 5.61; P < .001)$, indicating that this quadratic model provides a significant improvement in fit over the linear model. A plot of the residuals against the dependent variable now shows no systematic trends, which also indicates a better fit to these data.

4.14 **A.** When we used a simple linear regression, we found no obviously influential points. However, when we fit a quadratic equation to these data, point 1 becomes highly influential (its Cook's distance increases from .0002 to 1.43). Effectively it has about the same leverage as before (although it is numerically different because this is a different regression model), .3 compared to a maximum expected value of .11. However, it has a much larger value of r_{-1}, -3.5 (compared to $-.066$ for the linear model). **B.** When we used the linear model (which, as shown in Prob. 4.13 is not appropriate), this point was very near the line that best fit all of the data, giving it a small Studentized residual. When the quadratic model is used, this point is far removed from the curve that best describes the data which gives this point a large Studentized residual that, when coupled with the relatively high leverage, gives the point great influence.

4.15 The regression diagnostics such as leverage and Cook's distance are unremarkable. However, there is the suggestion of a systematic trend in both the raw data and a plot of the residuals against cortisol. A quadratic fit to these data increases R^2 to .987 (compared to .925 for the linear fit) and reduces $s_{y|x}$ from 3.9 to 1.8. The residual pattern no longer shows sytematic trends. Thus, these data are best described by a nonlinear relationship.

CHAPTER 5

5.1 **A.** No. VIF = 1.9 for both independent variables, which is well below our threshold for concern of 10. **B.** Including the interaction, the variance inflation factors for A, L, and AL are 6.0, 6.0, and 16.4, respectively, indicating that including the interaction may introduce a structural multicollinearity. Centering the data yields variance inflation factors of 2.0, 2.0, and 1.0, indicating that the multicollinearity was structural and is resolved. The interaction term fails to reach statistical significance, even after the centering ($t = -.42$ with 32 degrees of freedom, $P = .675$).

5.2 The variance inflation factor for T_1 is calculated from the correlation between T_1 and T_2, .91, as follows: VIF $= 1/(1 - r^2) = 1/(1 - .91^2) = 1/(1 - .828) = 5.8$. Because there are only two independent variables, the calculation for T_2 is identical.

5.3 No. Although the correlation between T_1 and T_2 is relatively high, .91, the variance inflation factors of 5.8 are below 10, our threshold for concern about serious multicollinearity. Likewise, the condition index of the correlation matrix of independent variables is only 21 (the eigenvalues are 1.9098 and .0902), below our threshold for concern of 100.

5.4 Because the range of values over which T_a was observed is already nearly centered on zero, T_a has a low correlation with T_a^2, so structural multicollinearity is not a problem in these data.

5.5 The variance inflation factors range from 1.6 to 4.6, well below our threshold for concern. Similarly, the eigenvalues of the correlation matrix indicate no serious problem: the smallest eigenvalue is .116 and the condition index is 18.4, well below our threshold for concern. Thus, multicollinearity is not a serious problem in this data set.

5.6 **A.** Structural multicollinearity is virtually absent in these data: the correlation between U_{ag} and U_{ag}^2 is only .15, and the variance inflation factors associated with these two variables are only 1.0. **B.** The correlation between U_{ag} and U_{ag}^2 is only .15 because the data are effectively already centered; the values of U_{ag} range from about -80 to $+70$ and are relatively evenly centered around zero. Thus, centering has little effect on these data. When data are measured on a scale that already centers them, structural multicollinearity will not be a problem.

5.7 **A.** Refer to the answer to Prob. 3.11 for the regression equation. The values of Studentized residuals, leverage, and Cook's distances are unremarkable. However, a plot of the residuals vs. M shows a definite systematic trend (as does the raw data), suggesting a nonlinearity. **B.** The simplest approach to account for this nonlinearity is to use a quadratic equation, as in Prob. 4.13. Because the raw data suggest that the two different relationships also have two different curvilinearities, the cross-product terms MS (where S is the dummy variable used to distinguish the dosing groups) and M^2S should be investigated. This regression equation is $\hat{C} = 3.42 \log \text{CFU} - .14 \log \text{CFU/\%} \ M + .00099 \log \text{CFU/\%}^2 \ M^2 + .38 \log \text{CFU} \ S - .033 \log \text{CFU/\%} \ MS - .0000013 \log \text{CFU/\%}^2 \ M^2S$, with $R^2 = .83$ and $F = 76.33$ ($P < .001$). However, only two of the regression coefficients are significant, $b_M = -.14 \log \text{CFU/\%}$ ($P < .001$) and $b_{MS} = -.033 \log \text{CFU/\%}$ ($P = .003$). Thus, there appears to be no significant curvilinearity (or change in curvilinearity) between the two groups: the two groups are described by lines with a common intercept (b_S is not significant; $P = .34$) but different slopes (b_M estimates the slope for the 1- to 4-h group and $b_M + b_{MS}$ estimates the slope of the 6- to 12-h group). Compared to the linear fit analyzed in Prob. 3.11, R^2 is higher (.83 vs .74), indicating that the nonlinear model is a better fit than the linear model. The residuals for the linear model suggested the need for the nonlinear model. Looking at the residuals for the nonlinear model suggests no systematic trends, also indicating the superiority of the linear model. **C.** The fact that the nonlinear model seems superior, yet neither regression coefficient relating to the nonlinearity is significant, $b_{M^2} = .00099$ ($P = .17$) and $b_{M^2S} = -.0000013$ ($P = .77$), suggests a problem with multicollinearity. The correlation matrix of the independent variables, where several pairwise correlations between these independent variables exceed .8, variance inflation factors — M has a VIF of 169, M^2 has a VIF of 912, and M^2S has a VIF of 341 — and the condition number of the correlation matrix of 494 (the largest eigenvalue is 3.0616 and the smallest is .0062) all verify a serious multicollinearity. Because this is largely a structural multicollinearity problem, the best solution is to center M and S before taking the squares and cross-products. If we repeat the regression analysis on the centered data, only one VIF remains above 10 (for M^2S, which has a VIF of 15) and all pairwise correlations are below our threshold of concern. The smallest eigenvalue of the correlation matrix is now .033, and the condition index is only 82. Thus, the multicollinearity has been greatly reduced. All regression coefficients are significant when they are estimated using the centered data.

5.8   **A.** 6. **B.** The condition index, $\lambda_1/\lambda_2 = 2116$. **C.** The condition index exceeds our threshold for concern of about 100; there is harmful multicollinearity.

5.9 The five independent variables are M, S, M^2, MS, and M^2S. For the native variables and their cross-products, the five corresponding eigenvalues are 3.0616, 1.6299, .2825, .0197, and .0062. The last two are quite small, and the condition index is 494, which exceeds the threshold of concern of 100. In contrast, for the centered variables and their cross-products, the five corresponding eigenvalues are 2.7262, 1.5572, .4169, .2666, and .0331. Only the last of these is particularly small, but it has a condition index of only 82, less than our threshold for concern. Thus, centering has greatly reduced the structural multicollinearity in these data.

5.10 **A.** The condition index is the largest eigenvalue divided by the smallest, $3.0616/.0062 = 494$. **B.** You are clearly justified in deleting the last principal component, which has a condition index of 494 and a magnitude less than .01. As a rule of thumb, components associated with eigenvalues greater than .01 should not be deleted (although in this case the condition number for the next largest eigenvalue is 155, just above our threshold for concern). The R^2 for the ordinary least-squares regression (including all components) is .85. R^2 only decreases to .84 when the principal component associated with the smallest eigenvalue (.0062) is deleted. Deleting the components associated with the two smallest eigenvalues (.0197 and .0062) greatly decreases R^2 to .67, an unacceptably large reduction. **C.** Given the ad hoc nature of principal components regression and the lack of standard errors to go with the parameter estimates if you use BMDP 4R (if you use SAS PROC RIDGEREG, you can obtain standard errors), it should not be used if a better alternative is available. In this case, the multicollinearity is structural, and so you should mitigate the multicollinearity with the more straightforward centering procedure (see the answers to Probs. 5.7 and 5.9).

5.11 The variance inflation factors for C and C^2 are 153.5, corresponding to a correlation of .997 between C and C^2. Thus, multicollinearity is a problem. Because this is purely a structural multicollinearity problem, center C to obtain $C^* = C - \overline{C}$, compute C^{*2}, and reanalyze. Doing so reduces the variance inflation factors to 1.1 (corresponding to a correlation of $-.31$ between C^* and C^{*2}. All regression parameters are still highly significant, and we make the same interpretations as before. In this case, the multicollinearity inflated the standard error for b_C (4.8 for uncentered data, .4 for centered data), but not to so great an extent as to cause the associated t statistic to lead us to an incorrect conclusion.

5.12 Plot C vs. C^2 and see that they are nearly linearly related $(r = .997)$. After centering C to obtain C^*, the plot of C^* vs. C^{*2} shows the parabola we expect and is associated with a reduction of r to $-.31$.

CHAPTER 6

6.1 .989

6.2 79.6

6.3 Yes. Consider the model $\hat{B} = b_0 + b_A A + b_L L + b_{AL} AL$ where B = female breast cancer rate, A = animal fat intake, and L = male lung cancer rate. Both forward and backward stepwise methods end with a model containing A and L. All possible subsets regression yields

			Variables		
k	R^2	C_p	A	L	AL
1	81.7	9.8	●		
1	77.2	20.3			●
1	60.4	58.7		●	
2	85.9	2.2	●	●	
2	84.2	6.3	●		●
2	77.2	22.2		●	●
3	86.0	4.0	●	●	●

Only the model containing both A and L has $C_p \approx p = k + 1$, indicating that leaving out one of these variables produces biased results. (Remember that, by construction of C_p, $C_p = p = k + 1$ for the full model containing all the variables, i.e., the last model in the list.)

6.4 All variable selection methods (stepwise, forward selection, backward elimination, and all possible subsets regression) choose the same regression model: $\hat{U}_{Ca} = 7.13$ mg/12 h + $.265D_{Ca}$ + .601 mg/meq U_{Na}. These are the same two independent variables that were significantly related to U_{Ca} in Prob. 3.8. This result is not surprising because there is no serious multicollinearity among the independent variables (Prob. 5.5).

6.5 There are 14 candidate variables: the 4 measured variables, 6 possible two-way cross-products, and 4 squared variables. (Some people would also include cross-products with squared terms.) Forward selection and backward elimination select different models. This result is probably due to multicollinearity, which is a severe problem in this situation: the variance inflation factors range from 97 to 1651, well above our threshold for concern, and there are several very small eigenvalues (the smallest is .0002, which has a condition index of

40,975, well above our threshold of 100). Centering helps the multicollinearity some, but there are so many squares and cross-products that even after centering, multicollinearity is still a serious problem: the VIFs range from 5 to 215, an order of magnitude improvement after centering, but are still far too large, and the smallest eigenvalue of the correlation matrix is .0022, again an order of magnitude improvement, but still with a condition index of 2080. After centering, both forward selection and backward elimination yield the same model, which is more comforting, but not enough to overcome concern with the remaining multicollinearity. For both the raw and centered data, all possible subsets regression is not very informative because all the cross-products and squared terms make C_p behave oddly; it is almost always $<< k + 1$ and is often negative. These difficulties illustrate that mindless use of variable selection methods is a poor substitute for more thoughtful variable selection based on the measured variables, use of residual plots suggesting the need to include linearizing transformations or interactions, and your knowledge of the substance of the problem being investigated. This exercise, in which a bunch of variables is thrown into a variable selection procedure, is a popular abuse of variable selection procedures that you should avoid.

CHAPTER 7

7.1 Yes. A one-way analysis of variance of the three groups yields $F = 5.56$ with 2 and 47 degrees of freedom ($P = .007$). All possible pairwise comparisons with the Student-Newman-Keuls test yield the following: myotonic dystrophy vs. normal ($q = 3.86$; $p = 3$; $P < .05$); myotonic dystrophy vs. nonmyotonic dystrophy ($q = 3.84$; $p = 2$; $P < .05$); normal vs. nonmyotonic dystrophy ($q = .11$; $p = 2$; $P > .05$). Thus, the mean cAMP in the myotonic dystrophy group is higher than the mean cAMP in either nonmyotonic dystrophy or normals, consistent with hyperparathyroidism accompanying the myotonic dystrophy. To compute the answer with regression and dummy variables, define two dummy variables, $M = 1$ if myotonic dystrophy and 0 otherwise, and $N = 1$ if nonmyotonic dystrophy and 0 otherwise. The regression equation would be $c\hat{A}MP = .91 + .78M + .02N$.

7.2 Using the Levene median test, we find that there is evidence of unequal variances, specifically the variance of the myotonic dystrophy group is much larger than that of the other two groups (the F statistic for the Levene test is 4.83 with 2 and 46 degrees of freedom; $P = .013$). Attempts to stabilize the variance are not fruitful: we tried logarithmic and square root transformations of the dependent variable, which yielded P values for the Levene median test of .023 and .011, respectively. A closer examination of the data reveals that the problem here is not really unequal variances but violation of the normality assumption, which is discussed in Prob. 7.3.

7.3   We did not include any formal tests for normality in this chapter. However, there are enough data points that you can look at a histogram of the data to get a rough idea of whether or not the data are normally distributed. In particular, the data from the myotonic dystrophy group are skewed to higher values and do not look normally distributed. Another clue that these are not normally distributed is that the mean and median of this group are quite different. There are no particular transformations that can be done to correct this problem. However, one can use a nonparametric statistic, the Kruskal-Wallace test, to reanalyze these data. The Kruskal-Wallace test is discussed in several introductory statistics books.* Reanalyzing these data with the Kruskal-Wallace test yields $P \approx .02$, and so we reach the same conclusion as before, but using a correct method.

7.4   Yes. $F = 50.5$ with 2 and 74 degrees of freedom; $P < .001$. A Student-Newman-Keuls test shows a significant difference between mesomorphs and ectomorphs ($q = 13.04$; $p = 3$; $P < .01$) and paraplegics and ectomorphs ($q = 11.47$; $p = 2$; $P < .01$), but not between mesomorphs and paraplegics ($q = .64$; $p = 2$; $P > .05$). Thus, paraplegics are like normal mesomorphs in terms of their biceps circumference.

7.5   A Levene median test yields $F = 1.7$ with 2 and 74 degrees of freedom and $P = .19$. Thus, there is no evidence of unequal variances and we need not transform and reanalyze the data.

7.6 $F = 81.5$ with 4 and 36 degrees of freedom ($P < .001$), indicating that at least one of the drugs affects saliva protein content. Multiple comparisons could be done using either Dunnett's test or all possible comparisons with a Student-Newman-Keuls test. In this situation, a Dunnett's test will tell us what we need to know: which drugs caused a difference from the control group? We first compare control to isoproterenol and calculate $q' = 12.03$ with $p = 5$, and 36 degrees of freedom ($P < .01$). We next compare control vs. dobutamine ($q' = 10.54$; $p = 4$; 36 degrees of freedom; $P < .01$), followed by control vs. metoprolol ($q' = 1.32$; $p = 3$; 36 degrees of freedom; $P > .05$). Because this last test is not significant, we do

---

*For example, see S. A. Glantz, *Primer of Biostatistics* (2nd ed.), New York, McGraw-Hill, 1987, pp. 310–317.

not test for the smaller difference of control vs. terbutaline. Thus, both isoproterenol, which affects both β_1- and β_2-receptors, and dobutamine, which affects only β_1-receptors, increase the amount of proline-rich proteins in saliva. In contrast, terbutaline, which affects only the β_2-receptor, does not change this protein content. Thus, the β_1-receptor, but not the β_2-receptor, regulates the proline-rich protein content of saliva.

7.7 A Levene median test yields $F = 3.17$ with 4 and 33 degrees of freedom, which is associated with $P = .026$. Thus, there is evidence that we have unequal variances in these data.* Inspection of the data reveals that the standard deviations tend to increase as the mean increases, the classic situation in which a logarithmic transform usually works. Accordingly, we take the natural logarithms of the data and repeat the Levene median test, which yields $F = 1.9$ $(P = .14)$. Thus, we have stabilized the variances. We repeat the analysis of variance on the transformed data and compute $F = 62.8$ $(P < .001)$, which yields the same conclusion as before. We recompute the Dunnett's test based on the logarithmic transformed data and reach the same conclusions as before.

7.8 Here are the results of the Student-Newman-Keuls tests:

Comparison	Difference in Means	p	q	$P < .05$?
Isoproterenol				
vs. Control	121.34	5	17.01	Yes
vs. Terbutaline	118.37	4	18.56	Yes
vs. Metoprolol	107.67	3	16.37	Yes
vs. Dobutamine	14.98	2	2.35	No
Dobutamine				
vs. Control	106.36	4	14.91	Yes
vs. Terbutaline	103.39	3	16.21	Yes
vs. Metoprolol	92.69	2	14.10	Yes
Metoprolol				
vs. Control	13.67	3	1.87	No
vs. Terbutaline	10.70			Do not test
Terbutaline				
vs. Control	2.97			Do not test

*Some statisticians would argue that one should use a .01 level of significance for testing for homogeneity of variance, which would mean that we could accept the results of the analysis in this problem. We still prefer a more traditional level of .05. See G. E. Milliken and D. A. Johnson, *Analysis of Messy Data: Volume 1; Designed Experiments*, New York, Van Nostrand Reinhold, 1984, pp. 23–28.

These tests indicate that the data fall into two distinct groups: isoproterenol and dobutamine exert similar effects to increase salivary protein; and metoprolol, terbutaline, and control (saline injection) are similar (i.e., metoprolol and terbutaline do not change salivary protein).

7.9 Each of the 10 tests is done at a level of .05/10 = .005 with 36 degrees of freedom. Thus, the critical value of t is 2.99.

7.10 The critical value of t for hypothesis testing (α = .005 and 36 degrees of freedom) is 2.99. See the answer for Prob. 7.8 for values of Student-Newman-Keuls q statistics, which are equal to $\sqrt{2}\, t$. The conclusions are the same as reached using the Student-Newman-Keuls test in Prob. 7.8.

CHAPTER 8

8.1 Yes. This is a two-way design with the type of fat, arachidonic acid (AA) or eicosapentanoic acid (EPA), as one factor and the presence or absence of TPA as the other factor. If you do this with multiple regression, you code a dummy variable $A = 1$ if AA and -1 if EPA, and a dummy variable $T = 1$ if no TPA and -1 if TPA. Also include a cross-product AT to account for the interaction. Using either regression or traditional methods, $F_T = 14.7$ with 1 and 12 degrees of freedom ($P = .002$), $F_A = 32.7$ with 1 and 12 degrees of freedom ($P < .001$), and $F_{AT} = 4.7$ with 1 and 12 degrees of freedom ($P = .051$). Thus, the two different fats differ in their effects on DNA synthesis; TPA causes the expected increase in DNA synthesis, on the average. Most important to the biological question, the significant interaction between tumor promotion and type of fat means that the two types of fat differ in their carcinogenicity. Thus, diets high in n-3 fatty acids are less likely to cause or facilitate cancer than diets high in n-6 fatty acids.

8.2 After centering the data of each cell to the median of that cell and taking the absolute values of the centered data, we do an ordinary one-way analysis of variance and calculate $F = 1.72$ with 3 and 12 degrees of freedom ($P = .22$). Thus, there is no evidence we have violated the assumption of equal variances.

8.3 Yes. This is a two-way analysis of variance with the site (pituitary vs. cerebellum) as one main effect and the drug (control, AOAA, EOS, or M) as the other main effect. If this is done with multiple regression, encode the site effect with a dummy variable, $S = 1$ if cerebellum or -1 if pituitary, and encode the drug effect with three dummy variables: $C = 1$ if control, -1 if M, and 0 otherwise; $A = 1$ if AOAA, -1 if M, and 0 otherwise; and $E = 1$ if EOS, -1

if M, and 0 otherwise. Also include three cross-products of dummy variables to account for the interaction: SC, SA, and SE. Whether you use regression or traditional methods, $F_S = 75.8$ with 1 and 32 degrees of freedom $(P < .001)$, $F_{Drug} = 4.75$ with 3 and 32 degrees of freedom $(P < .01)$, and $F_{SDrug} = 2.65$ with 3 and 32 degrees of freedom $(P = .07)$. The significant drug effect indicates that at least one of the three agonists affects cGMP levels. Comparison of the three drugs to control, averaged across both sites, with Dunnett's test shows all three drugs affect cGMP: for control vs. EOS, $q' = 3.53$ with 32 degrees of freedom and $p = 4$ $(P < .01)$; for control vs. M, $q' = 2.74$ with 32 degrees of freedom and $p = 3$ $(P < .05)$; for control vs. AOAA, $q' = 2.66$ with 32 degrees of freedom and $p = 2$ $(P < .05)$.

8.4 **A.** If you use regression, define the dummy variables as given in the answer to Prob. 8.3. Compute the regression three times, once with the interaction dummy variables entered last, once with the three drug effect dummy variables entered last, and once with the site dummy variable entered last. Sum the sequential sums of squares associated with each effect when it is entered last and divide by the number of dummy variables summed to obtain the mean square for that effect, for example, $MS_{Drug} = (35.802 + 9.439 + 20.457)/3 = 21.9$ with 3 degrees of freedom. (You could also do this computation with an analysis of variance program that gives you the correct incremental sums of squares.) $F_S = 61.39$ with 1 and 29 degrees of freedom (3 fewer than in Prob. 8.4 because we are missing 3 observations) $(P < .001)$, $F_{Drug} = 4.24$ with 3 and 29 degrees of freedom $(P = .01)$, and $F_{SDrug} = 2.38$ with 1 and 29 degrees of freedom $(P = .09)$. **B.** The conclusions from the analysis of variance remain the same as in Prob. 8.3. Repeating the Dunnett's test, we find: control vs. EOS, $q' = 3.36$ with 29 degrees of freedom and $p = 4$ $(P < .01)$; control vs. AOAA, $q' = 2.20$ with 29 degrees of freedom and $p = 3$ $(P < .05)$; and control vs. M, $q' = 1.84$ with 29 degrees of freedom and $p = 2$ $(P > .05)$. One of these conclusions is different than that reached in Prob. 8.3: we now conclude that, on the average, M did not significantly change cGMP levels.

8.5 **A.** In order to analyze this as a two-way design, you must assume no interaction. Given the lack of strong evidence for interaction in Probs. 8.3 and 8.4, this assumption seems reasonable. To do this with regression, define dummy variables as in Prob. 8.3, excluding interaction, and repeat the analysis twice, once with the three drug dummy variables last and once with the site dummy variable last. $F_S = 47.29$ with 1 and 30 degrees of freedom $(P < .001)$ and $F_{Drug} = 3.73$ with 3 and 30 degrees of freedom $(P = .02)$. **B.** These conclusions are the same as before, and we are assuming no inter-action, which is consistent with the previous conclusion.

8.6   To do the computations with regression, the equation would be $\hat{A} = b_0 + b_E E + b_M M + b_{EM} EM$, where A is the number of assessment activities, E is a dummy variable defined as -1 if > 7 years experience and 1 if ≤ 7 years, and M is a dummy variable $= 1$ if master's and -1 if not master's. Because the cells have unequal sample sizes, we must use the marginal sums of squares to compute the F statistics using three regression analyses with each of the main effects, E and M, and the interaction, EM, entered last, in turn. (You could also use an analysis of variance procedure that will report the appropriate sum of squares for unbalanced data.) $F_{EM} = 4.13$ with 1 and 214 degrees of freedom ($P = .04$). Therefore we conclude there is a significant interaction effect. The confounding effect of experience is not significant ($F_E = 2.37$ with 1 and 214 degrees of freedom; $P = .13$), whereas the level of education is significant ($F_M = 7.12$ with 1 and 214 degrees of freedom; $P < .008$). Thus, on the average, those with a master's-level education report significantly more nursing assessment activities than those without. On the average, there is no difference in the number of assessment activities with experience, but the interaction effect indicates that experience adds to the difference due to education: that is, the difference between levels of education is much greater for master's-level education than for nonmaster's. To see this most readily, graph the data in a form similar to that shown in Fig. 8-5.

8.7   Using the Levene median test to compare the four cells, we compute an F of $.46$ ($P = .77$). Therefore, we do not reject the null hypothesis of equal variances and conclude that we do not have a problem with unequal variances. Remember to delete the value corresponding to the median in the group with 19 subjects.

8.8   If you use regression with dummy variables to do this analysis of variance, following the atherosclerotic rabbits example, you need to define 4 dummy variables; for example, $D_1 = 1$ if normal CCD group, 0 otherwise; $D_2 = 1$ if normal OMCD, 0 otherwise; $D_3 = 1$ if hypertensive DCT, 0 otherwise; and $D_4 = 1$ if hypertensive CCD, 0 otherwise. The regression equation is $\hat{A} = 21 + 4.0D_1 - 11.0D_2 + 25.4D_3 - 8.65D_4$. $F = 18.1$ with 4 and 13 degrees of freedom ($P < .001$), so there is at least one statistically significant difference among the 5 cell means. (You should also get this F value if you do the computations with a traditional analysis of variance.) Pairwise comparisons using the Student-Newman-Keuls statistic reveal that the hypertensive and normal OMCD sites are not different, and neither is different from the hypertensive CCD site. The three sites from normal animals are different from each other. Thus, there is a site effect, and hypertension has an effect at the CCD site, but not the OMCD site. When we used two-way analysis of variance with the missing cell, we found, as with this one-way analysis, a significant site effect. In contrast to the finding from this one-way analysis, we did not find a significant hypertension effect with the two-way analysis.

8.9 **A.** Yes. This is a three-way analysis of variance with 3 carcinogens, 3 routes of administration, and 4 tissues (a $3 \times 3 \times 4$ analysis). The complete analysis of variance model includes the three main effects, chemical, route, and tissue; three two-way interactions, chemical by route, chemical by tissue, and route by tissue, and one three-way interaction, chemical by route by tissue. Evidence that the different carcinogens have different effects comes from the main effect for chemical, $F_C = 139,405/44 = 3168$ with 2 and 252 degrees of freedom ($P \ll .001$). Thus, we conclude that these three carcinogens have different abilities to cause DNA damage. **B.** Yes. All interaction terms are highly significant. To answer this question, we focus on the interactions involving the chemical factor. The chemical by route interaction, $F_{CR} = 28,894/44 = 652$ with 4 and 252 degrees of freedom ($P < .001$), and the chemical by tissue interaction, $F_{CT} = 117,375/44 = 2668$ with 6 and 252 degrees of freedom ($P \ll .001$), are both highly significant. Thus, the different chemicals do not have uniform effects in all tissues, and the amount of DNA damage caused by the chemicals depends on the route of administration. Finally, there is evidence that the ability of the chemicals to cause DNA damage is not only different in different tissues (the chemical by tissue interaction), but this difference in turn depends on the route. This evidence comes from the significant three-way interaction, $F_{CTR} = 27,902/44 = 630$ with 12 and 252 degrees of freedom ($P < .001$). Thus, these three carcinogens have different tissue specificities, and the tissue specificity depends on the route of entry into the body.

8.10 **A.** In Prob. 8.9, the sum of squares associated with the three-way interaction term was 334,823 (with 12 degrees of freedom), and SS_{wit} was 11,162 (with 252 degrees of freedom), yielding $MS_{wit} = 44$. When the three-way interaction term is left out, its sum of squares and the associated mean square are lumped into SS_{wit} and MS_{wit}. Thus, MS_{wit} increases to 1311 with 264 degrees of freedom, an increase of 2 orders of magnitude. This means that all F statistics will be smaller because the denominator MS_{wit} is much larger. For example, the F statistic for the main effect of chemical F_C decreases from 3168 (with 2 and 252 degrees of freedom) to 106 (with 2 and 264 degrees of freedom; $P < .001$). Thus, in this case, our statistical conclusion regarding the main effect of chemical did not change.

Likewise, all other F statistics are markedly reduced, but all are large enough to exceed the critical value for $\alpha = .001$, and none of our statistical conclusions change from what we found in Prob. 8.9. **B.** Leaving out the three-way interaction term removes an important piece of evidence when we ask the question: Do the different chemicals have different mechanisms of action? The three-way interaction of chemical by route by tissue means that the different chemicals have different effects on the tissues and that this difference in tissue specificity depends on the route of entry into the body. Although it is often tempting to leave out higher-order interactions because they are hard to interpret, sometimes the interpretation is important for answering a question and the interaction should not be removed.

CHAPTER 9

9.1 **A.** If you compute the one-way repeated-measures analysis of variance using regression and dummy variables, define the dummy variables the same way we did in the example of early metabolic rate changes. The regression equation is $\hat{R} = 122.33 - 12.83A_1 + .83A_2 - .17A_3 - 1.5A_4 + 11.5A_5 - 93.83E_E - 15.67E_S$. $F = 175$ with 2 and 10 degrees of freedom $(P < .001)$. (You should get the same result with a traditional computation of this analysis of variance.) A Dunnett's test for comparison of the eating-bypassed and stomach-bypassed group means vs. the control group mean shows that late metabolic rate is significantly different from normal in the group in which the dogs ate but food bypassed the stomach, but not in the group in which the dogs did not eat but had food placed directly into the stomach. **B.** This is the opposite of the conclusion we reached when we analyzed the early metabolic response in the example earlier in the chapter. **C.** When we put the two analyses together, we conclude that the act of eating is what is important for the early phase of digestion (when eating is bypassed, early metabolic rate greatly decreases), but that the distension of the stomach by food is what controls the late phase of digestion (when the animals eat but have the stomach bypassed so that it does not fill, late metabolic rate greatly decreases).

9.2 This is a two-way analysis of variance with repeated measures on both factors. In the two-way design, one factor is the same as before, the type of eating, and has three levels: normal eating, stomach bypassed, or eating bypassed. The second factor is early or late phase of digestion. We include interaction in the model. The regression approach to the computations would require defining several dummy variables; for example, $L = 1$ if late data, -1 if early data;

$E_n = 1$ if normal eating, 0 if stomach bypassed, -1 if eating bypassed; $E_E = 1$ if stomach bypassed, 0 if normal eating, -1 if eating bypassed; and $A_i = 1$ if animal i, -1 if animal 6, 0 otherwise ($i = 1$ to 5). Following the procedure outlined for the chewing gum data, we would form the interactions between L and E_N and E_E, and between L and the A_i and between E_N and E_E and the A_i. From the sequential sums of squares associated with these dummy variables in the regression output, we calculate $MS_L = 513.8$, $MS_{\text{Eat}} = 12,211.9$, $MS_{\text{interaction}} = 21,042.5$, $MS_{\text{subj} \times L} = 49.8$, $MS_{\text{subj} \times \text{Eat}} = 97.5$, and $MS_{\text{res}} = 80.4$. From these mean squares, we calculate $F_L = 10.3$ with 1 and 5 degrees of freedom ($P = .024$), $F_{\text{Eat}} = 125$ with 2 and 10 degrees of freedom ($P < .001$), and $F_{\text{interaction}} = 262$ with 2 and 10 degrees of freedom ($P < .001$). Because there is a significant interaction effect, our usual practice would be to proceed to all possible comparisons with the Student-Newman-Keuls test. However, we did not present the methods for making such comparisons. In this case, the F statistics, coupled with a look at a plot of the data, show clearly what is happening in terms of the original physiological question: the early digestive phase is initiated by the act of eating, whereas the late digestive phase is initiated by distension of the stomach.

9.3 This is a two-factor design with repeated measures on one factor (time) but not on the other (scrub suit group). Thus, the subjects are nested in the scrub suit group factor. To do this with regression, define 18 dummy variables to encode the 19 subjects in the fresh scrubs group, $S_{F_i} = 1$ if subject i, -1 if subject 19, and 0 otherwise. Similarly, encode 18 dummy variables S_{S_i} for the 19 subjects in the same old scrubs group. There are two groups, so define 1 dummy variable G as 1 if fresh scrubs and -1 if same old scrubs. To encode the four times, define three dummy variables, $T_i = 1$ if time i, -1 if last time, and 0 otherwise. Finally, compute the interaction effects GT_1, GT_2, and GT_3. (You could also do this with a repeated-measures, or mixed-model, analysis of variance procedure.) The interaction effect indicates that the change in bacterial counts over time is different in the two groups ($F_{GT} = 24.2$ with 3 and 108 degrees of freedom; $P < .001$). The main effect of time is also highly significant ($F_T = 25.9$ with 3 and 108 degrees of freedom; $P < .001$), indicating a significant trend in bacterial counts with time (a graph of the cell means shows a trend toward increasing bacterial counts as the shift progresses). The difference between scrub suit groups is marginally insignificant ($F_G = 3.86$ with 1 and 36 degrees of freedom; $P = .057$). Thus, there is a tendency for one group (the group that did not put on fresh scrubs) to have higher bacterial counts than the other.

9.4 No. This is a two-way design with repeated measures on one factor (stimulation at 0, 15, or 45 tetani per minute) yielding F statistics as follows: for interaction $F = .07$ with 2 and 48 degrees of freedom ($P = .931$); for the stimulation effect $F = 86.9$ with 2 and 48 degrees of freedom ($P < .001$); for the training effect $F = 3.12$ with 1 and 24 degrees of freedom ($P = .09$). Thus neither the test for interaction nor a main effect of training is statistically significant, and we therefore conclude that there is insufficient evidence to support the hypothesis that training affects the contribution of leucine to total energy needs during exercise.

9.5 **A.** This problem can be done with regression using dummy variables (as defined in the example in this chapter) or with traditional methods of repeated-measures or mixed-model analysis of variance. $F_A = MS_A/MS_{\text{subj}(A)} = 2.82625/.3674 = 7.69$ with 1 and 14 degrees of freedom ($P = .02$), $F_D = MS_D/MS_{\text{res}} = .8224/.4097 = 2.01$ with 1 and 14 degrees of freedom ($P = .18$), and $F_{DA} = MS_{DA}/MS_{\text{res}} = .4301/.4097 = 1.05$ with 1 and 14 degrees of freedom ($P = .32$). Thus, sociability is affected by antisocial personality but not by drinking, and the effect of antisocial personality is not affected by drinking (there is no interaction). **B.** The analysis of well-being score is computationally identical to the analysis of the sociability score in part A above. $F_A = .08$ with 1 and 14 degrees of freedom ($P = .78$), $F_D = 4.75$ with 1 and 14 degrees of freedom ($P = .047$), and $F_{DA} = .04$ with 1 and 14 degrees of freedom ($P = .85$). Thus, feelings of well-being are unaffected by antisocial personality diagnosis but are affected by drinking. There is no interaction, indicating that the effect of drinking on feelings of well-being is the same in both personality groups.

9.6 **A.** For this analysis, you need to obtain both the marginal sums of squares and the expected mean squares. The computations for F_D and F_{AD} are unaffected by the missing data, except you have to use the marginal sums of squares: $F_D = 3.24$ with 1 and 11 degrees of freedom ($P = .10$) and $F_{AD} = .1$ with 1 and 11 degrees of freedom ($P = .75$). Because the missing data affect the coefficients associated with $\sigma^2_{\text{subj}(A)}$ in the expected mean squares for A and subject (A), we must compute approximate F_A and denominator degrees of freedom according to Eqs. (9.8) and (9.9): $F_A = 3.63$ with 1 and 16.3 (round down to 16) degrees of freedom ($P = .075$). (This compares to an uncorrected F of 3.56 with 1 and 14 degrees of freedom; $P = .08$. Thus, the correction had little effect on our results.) **B.** We reach generally the same conclusions as before. The only difference is that the P value associated with antisocial personality is now .08, marginally nonsignificant. Thus, we no longer have strong evidence that the sociability score is different in the two personality groups.

9.7 Yes. A two-way analysis of variance with repeated measures on both factors yields the following: for the bout effect, $F = 94.7$ with 3 and 15 degrees of freedom ($P < .001$); for the effect of time after each bout, $F = 19.5$ with 2 and 10 degrees of freedom ($P < .001$); and for the interaction of time after each bout with bout group, $F = 2.49$ with 6 and 30 degrees of freedom ($P = .045$). Thus, based on the analysis of variance, we can conclude that there is a difference in muscle soreness between the bout groups (inspection of the data shows that soreness is uniformly higher after bout 1 than it is after the other three bouts), confirming the first hypothesis. There is a trend in soreness with time, and the significant interaction means that the time course of soreness is different after different bouts. Because the hypothesis was that the usual increase in soreness with time following an exercise bout is reduced in subsequent bouts (after bout 1), we elected to do eight Bonferroni t tests to compare 6 vs. 18 h and 6 vs. 24 h in each of the four exercise bouts. (There are a large number of possible comparisons; however, we should be selective about which ones we really need to make.) We compute t and make each test against the critical values of t for $\alpha = .05/8 = .0063$ with 10 degrees of freedom, 3.443. For the denominator of the t test, use the mean square from the denominator of the F statistic for the time effect $MS_{subj \times time}$, .21039. Of these eight t tests, only one was statistically significant, that associated with the hypothesis test for the difference in soreness between time 24 h and 6 h following bout 4. Given the highly significant F statistic for a time effect and the significant interaction effect, this seems surprising. It is not really: the conservatism (lack of power) of so many Bonferroni t tests in relation to the power of the analysis of variance almost guarantees such a result. In any case, very little can be said about the trends in soreness, except that there is some evidence to suggest the trend in soreness is different after one bout than after any other.

9.8 **A.** As in all two-way designs, missing data require that the F statistics be calculated from the marginal sums of squares. In addition, because this is a repeated-measures, or mixed-model, analysis of variance, the missing data also affect the coefficients of the σ^2_{bout}, $\sigma^2_{bout \times subj}$, σ^2_{time}, and $\sigma^2_{time \times subj}$ terms in the expected mean squares from which F_{bout} and F_{time} are calculated. Thus, both of these F statistics and denominator degrees of freedom must be approximated using Eqs. (9.17) to (9.20): $F_{bout} = 110$ with 3 and 18.45 (round down to 18) degrees of freedom ($P < .001$), and $F_{time} = 20.8$ with 2 and 10.39 (round down to 10) degrees of freedom ($P < .001$). $F_{bout \times time}$ does not involve random factors and thus can be calculated directly from the marginal sums of squares: $F_{bout \times time} = 7.13$

with 6 and 21 degrees of freedom ($P < .001$). **B.** Our overall conclusions are the same as reached in Prob. 9.7.

9.9   Use multiple regression with dummy variables to account for between-dog differences in slope and intercept. Define three dummy variables for the four dogs, $D_i = 1$ if dog i, -1 if dog 4, and 0 otherwise. The estimated regression equation is $\hat{L} = -4.5$ V + .2 V/(mL·min·100 g) H + 5.7 V D_1 + 4.8 V D_2 − 3 V D_3 − .15 V/(mL·min·100 g) HD_1 − .13 V/(mL·min·100 g) HD_2 + .02 V/(mL·min·100 g) HD_3, with $R^2 = .76$ and $s_{y|x} = 1.3$ V. All regression coefficients are significant except for the D_3 by H interaction ($t = .56$ with 53 degrees of freedom; $P = .58$). The sum of squares associated with the between-dogs variability is 29.7 + 30.4 + 1.5 + 24.3 + 66.4 + .5 = 152.8, so $MS_{\text{bet dogs}} = 152.8/6 = 25.5$. We can compute an F statistic to test for significant between-dogs variability using $MS_{\text{bet dogs}}/MS_{\text{res}} = 25.5/1.59 = 16.02$ with 6 and 53 degrees of freedom ($P < .001$). Thus, the t statistics for the regression coefficients related to the dummy variables and the F statistic for significant between-dogs variability indicate that the relationship between these two ways of measuring gastric blood flow is different from dog to dog. Therefore, a calibration factor for the laser Doppler flow probe will have to be calculated for each subject.

9.10 $R^2 = .33$ (compared to .76 with the dummy variables) and $s_{y|x} = 2$ V (compared to 1.2 with the dummy variables). The regression equation is $\hat{L} = 1$ V + .0787 V/(mL·min·100 g) H, with $F = 29.2$ with 1 and 59 degrees of freedom ($P < .001$). Leaving out the dog dummies greatly increases the residual sum of squares and so R^2 decreases.

CHAPTER 10

10.1   The equation is $\hat{G} = 1.94 + 5.44e^{-.021H}$ with $R^2 = .52$, a considerable improvement over the .40 for the exponential model we fit in Prob. 4.4 that decayed to zero rather than to a value b_2 greater than zero. Examination of leverage values and Cook's distances reveals no striking influential points, but a plot of the residuals vs. H reveals unequal variances. We can try to stabilize the variance with one of the transformations in Table 4-7, $\ln(G)$ or $\sqrt{G}$. In this case, because we are using nonlinear regression, we do not desire the transformation to change the nature of the function, so we take either the log or square root of both sides of the equation (the specification of derivatives must take this into account). The $\ln(G)$ transformation stabilizes the variance, but the R^2 actually decreases to .42 because there are some small values of G (below 1) that are heavily weighted by the $\ln(G)$ transform and appear as outliers with Studentized residuals r_i of -2.6 (point 23) and -3.1 (point 32). The $\sqrt{G}$ transformation fares better: $\hat{G} = 1.83 + 5.27e^{-.021H}$, with $R^2 = .50$, no suggestion of unequal variances, and no outliers (all $r_i < 2$). Because it stabilizes the variance without creating outliers, this transformation yields our best unbiased estimates of the parameters, which are all significantly different from zero. The t value for b_1, which reflects that the nonlinear change in the rate of decay is 2.03 yields $P \approx .051$ (the 95 percent confidence interval is $-.0001$ to .0417). In this case, it was worth the extra effort to use nonlinear regression to fit the most appropriate model, which provides evidence that a nonlinear model of the relationship between G and H describes these data better than a linear model, thus confirming the investigators' hypothesis.

10.2 The basic function is $\hat{P} = a(1 - e^{-bt})$, where a is the level of P to which the function asymptotically approaches as t gets large and b is the rate of growth of the curve toward the asymptote. Define a dummy variable D to be 1 if synthetic vesicles and 0 if natural surfactant. Let both a and b be affected by D and rewrite the equation as $\hat{P} = (b_0 + b_1D)(1 - e^{-(b_2 + b_3D)t})$. If b_1 is significantly different from zero, there is a significant displacement of the two curves, and if b_3 is significant, the two curves have different shapes (amounts of curvature). This equation is treated no differently than any other when using nonlinear regression routines, and you must specify the partial derivatives with respect to each parameter and initial guesses for each parameter estimate. The regression equation is $\hat{P} = (100\% - 56.5\% \; D)(1 - e^{-(.025\%/min - .012\%/minD)t})$, with $R^2 = .853$. All parameter estimates except b_3 are significantly different from zero. (The 95 percent confidence interval for b_3 is $-.0383$ to .0136. Because this interval includes zero, we cannot conclude that there is a shape change.) Thus, natural surfactant is taken up at a different rate than synthetic vesicles, but by a similar kinetic process.

10.3 The corresponding leverage values are $h_{44} = .00159$ and $h_{77} = .193$. From r_i and h_{ii}, calculate Cook's distances, $D_4 = .007$ and $D_7 = .37$, from the relationship given in Chap. 4, in which k is the regression degrees of freedom (the number of parameters $- 1$).

10.4 **A.** A plot of the data suggests that a sum of at least two exponentials is needed, $\hat{F} = b_0e^{-R/R_1} + b_1e^{-R/R_2}$. Regardless of the initial guess, the analysis has a difficult time converging. Once it has converged, there are two principal problems: (1) the standard errors of the parameter estimates are huge, and (2) the parameter estimates are highly correlated. These problems are related and reflect an ill conditioning analogous to multicollinearity. For example, starting from initial guesses of $b_0 = 100$, $b_1 = 20$, $R_1 = 125$, and $R_2 = 500$ using SAS NLIN with Marquardt's method, the solution is never reached: on the first iteration, λ is increased 20 times without a decrease in the residual sums of squares, and so the program prints a solution without ever taking a real step. The 95 percent confidence intervals for the parameter estimates are very wide and include zero: b_0, -43 to 244; b_1, -190 to 230; R_1, -182 to 432; and R_2, -1712 to 2512. Other initial guesses will lead to similar problems. A sum of three exponentials has even worse problems. **B.** The difficulties in finding a good nonlinear solution in part A reflect a model misspecification. In fact, you do not even need nonlinear regression to solve this problem. Careful examination of the data show that the function is better represented as a hyperbola than as a sum of exponentials. If a $1/F$, $1/R$ transformation is used, there is still a nonlinearity. However, $1/F$ vs. R shows a straight line (this corresponds to the function $y = 1/(b_0 + b_1 x)$. The resulting regression equation is $1/\hat{F} = .011$ (h·cm²)/nmol $ + .00011$ cm⁴/(h·nmol·ohm) R, with $R^2 = .786$, $s_{y/x} = .022$ nmol/(h·cm²), and $F = 278$ with 2 and 76 degrees of freedom $(P < .001)$. Thus, a simple hyperbola, which can be fit without resorting to nonlinear regression, is a better fit to these data.

10.5 Examination of the residuals reveals no glaring problems. A few observations have Studentized residuals r_{-i} greater than 2, and two of these are greater than 3 (r_{-71} and r_{-76}). The last four observations have leverage values greater than the threshold for concern of $2(2/78) = .051$, but none of these potential problems gets translated into a Cook's distance that warrants concern. The normal probability plot of the residuals is largely unremarkable, although the few points with larger r_{-i} are easily detected as small deviations at one end of an otherwise tight line. A plot of the residuals against the independent variable or predicted dependent variable shows uniform bands around zero. Overall, there is nothing to be concerned about.

10.6 **A.** Using the second, more useful parameterization, given in the example of keeping blood pressure under control, we fit the equation:

$$\hat{I} = I_{max} + \frac{I_{max} - I_{min}}{1 + e^{[4S(P_{1/2} - P)/(I_{max} - I_{min})]}}$$

and estimate $I_{min} = 302 \pm 23$ ms ($t = 13.1$ with 34 degrees of freedom; $P < .001$), $I_{max} = 962 \pm 33$ ms ($t = 29.2$; $P < .001$), $P_{1/2} = 17.3 \pm .4$ kPa ($t = 43.3$; $P < .001$), and $S = 125 \pm 34$ ms/kPa ($t = 3.68$; $P < .001$). Thus, all parameter estimates are significantly different from zero. The pressure control set point is $P_{1/2} = 17.3$ kPa, and the baroreceptor sensitivity is $S = 125$ ms/kPa. **B.** Using a similar function for the data collected under 2 percent halothane, we estimate $I_{min} = 412 \pm 144$ ms ($t = 2.86$ with 41 degrees of freedom; $P = .007$), $I_{max} = 678 \pm 62$ ms ($t = 10.9$; $P < .001$), $P_{1/2} = 12.2 \pm 3.5$ kPa ($t = 3.49$; $P = .001$), and $S = 19.8 \pm 16$ ms/kPa ($t = 1.25$; $P = .218$). All parameters except S are significant. Thus, baroreceptor activity is essentially abolished at a 2 percent level of halothane anesthesia. **C.** Yes. The set point changes significantly from 17.3 to 12.2 kPa ($t = 125.6$ with 75 degrees of freedom; $P < .001$). A similar t statistic reveals that S from the 2.0 percent halothane data is significantly different from S from the .5 percent halothane data. The overall conclusion is that increasing halothane depresses baroreceptor function to the point that they are almost inoperative.

10.7 No. Two observations have leverage values greater than the threshold of $2(4/45) = .18$: points 43 and 44 have a leverage of .236. Because both of these points have small r_i ($-.97$ and -1.56, respectively), the Cook's distances are not large enough to be of concern (.10 and .25, respectively). (If the program you use does not report Cook's distance, it can be calculated from the relationship between D_i, r_i, and h_{ii} given in Chap. 4.) Two other observations have moderately large Studentized residuals, $r_{17} = -2.6$ and $r_{45} = 3.3$. Neither has an overly large Cook's distance (.20 and .52, respectively).

10.8 No. Using nonlinear regression, we compute the equation $\hat{W} = (40.1 - 1.5P)(1 - e^{-26.8(P-.03)})$, with $R^2 = .83$. All parameter estimates are significantly different from zero except b_2, which has a 95 percent confidence interval of -3.75 to 6.75. Overall, this is a good fit which had little trouble converging. A plot of the data shows a rapid rise over the range of P from .05 to .15 and then a slow approach to an asymptote as P increases from .15 to .8. Thus, the reason previous investigators concluded that W was independent of P was that the range of P studied (.14 to .36) was in the flat region of the relationship.

10.9 One cannot answer the question with regard to F because the nonlinear regression algorithm cannot find a satisfactory solution, regardless of the starting values. The convergence patterns differ markedly as the starting value of b_3 is changed, but the final results are always unacceptable. The standard errors of the parameter estimates are sometimes huge, resulting in broad confidence intervals. In some cases the standard errors are zero, resulting in a confidence interval with no width. The parameter estimates are highly correlated (when their correlation can be estimated at all) indicating a severe ill-conditioning problem analogous to multicollinearity. When starting from a first guess of $b_3 = 300$, b_3 is estimated to be 1,142,198 with a standard error of zero (using Marquardt's method). When starting from a first guess of $b_3 = 200$, b_3 is estimated to be 1944, a decrease of 3 orders of magnitude, also with a standard error of zero. When starting from a first guess of $b_3 = 100$, b_3 is estimated to be 262 with a standard error of 1,525,554, resulting in a 95 percent confidence interval of $-3,067,064$ to 3,067,587. The other parameter estimates are relatively stable and uninteresting compared to b_3. The trouble is that b_3 relates to the rapid growth phase of the relationship between F and P. In these data, this phase is so rapid, compared to the spacing of the data points along the P axis, that any number of an order of magnitude of 10^2 or larger will do, so the parameter cannot be estimated with any degree of reliability at all. Because this parameter exerts its effect in the range of between .05 and .10, over which there are no data collected, there are essentially no data in the region needed for parameter estimation. You must have sufficient data in the regions where the parameter exerts its effects on the function in order to estimate the parameter. In contrast, the rate of growth of the function in Prob. 10.8 was slower, and there was sufficient data over the region of its predominant effect so that parameter estimation was not a problem. (See Prob. 10.10 for a follow-up to this problem.)

10.10 Now, the problem readily converges to a much more believable answer. The correlation matrix of the parameter estimates now shows only one large correlation, .95 between b_3 and b_4, and all the standard errors of the parameter estimates are small compared to the analysis of Prob. 10.9. None of the 95 percent confidence intervals include zero, so we conclude that all parameter estimates are significantly different from zero $(P < .05)$. The difference between this data set and the one used in Prob. 10.9 is that this set includes values of F at P levels of .06, .07, .08., and .09, thus filling in the region over which b_3 exerts its predominant effect on the function. Now that we have sufficient data in this region, the algorithm has a good chance of reliably estimating the parameter b_3, which is now estimated to be 202 ± 97.

CHAPTER 11

11.1 A. .052 **B.** 7.8 **C.** Solve Eq. (11.3) for P and compute probabilities of .05 and .89, respectively.

11.2 It enters first because it has the largest χ^2 for improvement, 5.62. Thus, of the candidate independent variables examined when only the intercept is in the equation, it is associated with the largest increase in the likelihood function (most improves the predictive ability of the equation).

11.3 A. Using stepwise logistic regression,[*] both birth weight W and the ECG component F_{68} are found to be the important predictors of SIDS, and both are highly significant $(P < .001)$. The equation is logit $\hat{S} = -1.15 + 16.2F_{68} - .0018W$. Thus, there is evidence that the ECG power spectrum is predictive of SIDS. Specifically, a high amplitude of the power spectrum at a periodicity of 6 to 8 beats per minute is associated with an increased probability of SIDS. The negative effect of birth weight simply accounts for the potential confounding of a known predictor of SIDS, low birth weight. The odds of dying of SIDS are given by the equation $\hat{\Omega}_S = e^{-1.15}e^{16.2F_{68}}e^{-.0018W}$. The intercept gives an estimate of the odds of dying of SIDS in the overall sample $e^{-1.15} = .32$ (this compares reasonably well with the observed proportion of SIDS cases in this sample, $16/65 = .25 = e^{-1.4}$). The odds of dying of SIDS increase by a factor of $e^{16.2} = 10,850$ for a one unit change in F_{68} (F_{68} ranges between about .2 and .6), whereas the odds of dying of SIDS increase by a factor of $e^{-.0018} = .998$ for every 1-g decrease in birth weight. For example, the odds of dying of SIDS for a high-birth-weight (say 4000-g) infant who has a low 68 factor of .2 are $e^{-1.15}e^{16.2(.2)}e^{-.0018(4000)} = .006$. On the other hand, the odds of dying of SIDS for a low-birth-weight (say 2000-g) infant who has a high 68 factor of .6 is $e^{-1.15}e^{16.2(.6)}e^{-.0018(2000)} = 144$. **B.** Because there are only 65 subjects, each with a unique combination of values of the independent variables, plots of the observed proportion vs. predicted probabilities are of little use in assessing goodness of fit. The Hosmer-Lemeshow goodness-of-fit χ^2 is associated with a P value of .35, which does not identify any serious problems. Another way to assess the goodness of fit (which we did not talk about) is to ask for a classification table,[†] which assesses how well the equation classifies the subjects into SIDS or non-SIDS. In this example, the table is

---

[*]Spicer and coworkers actually used discriminant analysis to answer this question. They reached the same general conclusions we reached using logistic regression.

[†]In BMDP LR, you specify the COST option in the /PRINT paragraph.

		Actual Outcome	
		SIDS	No SIDS
Predicted outcome	SIDS	7	3
	No SIDS	9	46

which indicates that the logistic equation we estimated correctly classified most of those who did not die (3 of the 49 were predicted to die of SIDS). The model did not do so well in correctly classifying those who died of SIDS: 9 of the 16 SIDS deaths were misclassified as nondeath. Thus, this logistic regression equation is not a very specific predictor of those infants who are at high risk of SIDS. The residual diagnostics indicate one outlier with a Studentized residual r_i of almost 7. This same point has no unusual leverage but has relatively high influence as judged by the BMDP LR "influence" statistic of .76, which is analogous to Cook's distance for ordinary regression.

11.4 No. The correlation matrix of the parameter estimates shows no high correlations.

11.5 **A.** Yes. Stepwise logistic regression yields the equation logit ARDS = 6.65 + 1.55P − .5O − .06 age. t statistics calculated from the asymptotic standard errors of the coefficients indicate that all coefficients are highly significant. The Hosmer-Lemeshow goodness-of-fit χ^2 is associated with a P value of .302, indicating no significant lack of fit of these data to the logistic model. **B.** The coefficient of P is positive, indicating that high lung protein is associated with a high probability of having ARDS. Both age and oxygen have negative effects. Thus, higher oxygen is associated with a lower probability of ARDS (which makes sense because ARDS is defined clinically in terms of low oxygenation). Similarly, higher age is associated with a lower probability of ARDS. This is somewhat surprising, physiologically, and may simply reflect the fact that in this sample ARDS patients tended to be younger. These coefficients can be interpreted in terms of the multiplicative change in odds relative to the intercept: for example, the odds of being in the ARDS group increase by a factor of $e^{1.55} = 4.73$ for each 1 unit increase in the protein accumulation index.

11.6 There is one highly influential combination of predictor variables associated with two ARDS patients who are 67 years old. The Studentized residual r_i for this combination is −6.4. Although the leverage is only moderately high (.08), this large residual has a high influence value (analogous to Cook's distance) of 2.87. A few other combinations have leverages exceeding our threshold for concern of $2(4/88) = .09$, but none of these combinations actually have large influences.

11.7 No. The correlation matrix of the parameter estimates shows no pairwise correlation with an absolute value higher than .76, well below our threshold for concern.

11.8 A. Rewrite the logit equation to obtain $\hat{\Omega}_{\text{ARDS}} = e^{6.65}e^{1.55P}e^{-.5O}e^{-.06A}$, and calculate $\hat{\Omega}_{\text{ARDS}} = 250{,}549$. **B.** .001.

Index

Page references followed by an *italic n.* indicate footnotes.

Adjusted R^2 (*see* R^2_{adj})

Adult respiratory distress syndrome (ARDS), 567

Adults, protein synthesis in, compared with newborns, 74–78

Alcohol:
- and antisocial personality disorder, 461–462
- and childhood aggression, 410–420
  - example, 610–615,
- and driving, 179

All possible subsets regression:
- description, 239
- example, 257–261
- motivation, 255–256

Amino acids, 157

Analysis of covariance, 72*n.*

Analysis of variance:
- fixed effects in, 422
- mixed model in, 422
- one-way repeated measures (*see* One-way repeated-measures analysis of variance)
- one-way or single factor in, 272
  - (*See also* One-way analysis of variance)
- purpose, 272
- random effects in, 422
- as regression problem, 283–284
- repeated measures (*see* Repeated-measures analysis of variance)
- two-way (*see* Two-way analysis of variance)

Anesthesia, 510

Antibiotics, 47, 108

Association:
- versus causality, 36
- identified by regression analysis, 13

Assumptions:
- analysis of variance, 452
  - testing in, 307–309
- check before accepting results of an analysis, 458

Assumptions (*Cont.*):
- multiple regression, 110–111
- repeated-measures analysis of variance, 400–405
- test with residuals, 111

Atherosclerosis, 303–307, 309–310

Auxiliary regression to identify multicollinearities, 193–194

Baby birds:
- breathing in burrows, 86–88
  - example, 583
- compared with adults, 109
- overall goodness of fit, 105

Backward elimination in multiple linear regression, 264–265

Bacteria:
- contamination of surgical patients, 460–461
- in salty environment, 94–100
  - example, 584
- thermophilic, 178–179
- treatment to reduce, 47, 108

Baroreceptor response, described with logistic equation, 502

Benzodiazepine tranquilizers, 46

Best regression model:
- criteria for selecting:
  - C_p, 252–253
  - coefficient of determination, 246
  - comparison of alternative criteria, 254–255
  - independent validation of model with new data, 248–249
  - predicted residual error sum of squares (PRESS), 250
  - R^2, 246
  - R^2_{adj}, 247
  - standard error of the estimate, 247–248
- no clear definition, 245–246
- techniques for defining, 239

Beta blockers and blood pressure, 83–86, 106–107

Between groups variance in one-way analysis of variance, 281

Bias:
due to model misspecification, 240–242, 250–252
evaluation of, 107–108
example, 242–244
in restricted regression procedures, 233n.

Bivariate linear regression (see Simple linear regression)

Blood pressure, 336–341, 500–508, 624–626

BMDP:
logistic regression, 626–632
missing data, 604–605
multicollinearity, 591–592
nonlinear regression, 623–624
one-way analysis of variance, 600–601
one-way repeated-measures analysis of variance, 609
principal components regression, 591–592
regression, 39–42, 578, 579–580, 582, 586–588
two-way repeated-measures analysis of variance:
with repeated measures on both factors, 618
with repeated measures on one factor, 613
variable selection techniques, 596–597

Body protection from excess copper and zinc, 164–176
example, 588–590

Bone cancer, 536–546, 627–629

Bone growth, 177

Bonferroni t test:
compared to Dunnett's test, 297
following one-way analysis of variance, 294–295
following two-way analysis of variance, 344–345
with missing data, 362

Breast cancer:
and dietary fat, 33–36
and involuntary smoking, 46–47, 236, 270

Breathalyzer, 179–180

Breathing, control of, 108–109

C_p:
all possible subsets regression, 256n.
assumptions, 253

C_p (Cont.):
compared with other measures of quality of regression fit, 255
definition, 252–253
practical limitations, 254

Cancer:
bone, 536–546, 627–629
breast (see Breast cancer)
chemotherapy, 627–630

Candy, chewing gum and tooth decay, 435–444, 450–452
example, 615–620

Carcinogenic compounds in secondhand smoke, 380

Catch, the, 231–234

Catheter used to measure heart size, 382–392, 607

Causality:
versus association, 36
identified by regression analysis, 13

Centering:
to define standardized variables, 221–227
to eliminate structural multicollinearity, 198–207
need to maintain intercept term, 207n.
no effect on final regression equation, 204
in sequential variable selection, 271

Central limit theorem, 24n.

Chi square:
improvement, in logistic regression, 526
table of critical values, 719–720
(See also Goodness of fit)

Chicken feed, cheaper, 157–163
example, 588

Cholesterol:
and diabetes, 83
effect on cell membranes, 107

Clusters of data points, effect on correlation coefficient, 75–78

Coefficient of determination:
compared with other criteria for quality of regression fit, 254–255
and correlation coefficient, 38
as criterion for selecting regression model, 246
definition, 38
and explained variance, 38
in multiple regression, 69
in nonlinear regression, 492n., 622

Colon, inflammation of, 510

Compartmental models, 495n.
Completely randomized designs:
  compared with repeated-measures design, 398–399
  definition, 272
Computer:
  to determine expected means squares in the presence of missing data, 425n.
  differences between statistical packages, 576–579
  incorrect value of F in the presence of missing data in analysis of variance, 429n.
  produce misleading results in unbalanced design, 351
Computer printouts, meaningless, 9
Condition index, 225
Condition number, 225
Confidence interval:
  for experimental effect computed using dummy variables, 85
  joint confidence region, 58n.
  for regression coefficients, 25, 58
  in nonlinear regression, 493n.
  in simple linear regression, 26, 27
  for slope, 27
Confounding variable:
  controlling for, 109
  definition, 36
Connected data, 366–367
Consequences of having two pumps in one heart, 226
Control group, comparisons against, 297
Convergence criterion in nonlinear regression, 482n., 483–484
Cook's distance:
  definition, 140
  example of use, 154
  influence of data point on regression coefficients, 136
  large value, 141
  in logistic regression, 531
  motivation, 137–140
  in nonlinear regression, 493
  relationship to Studentized residual and leverage, 140
Copper and zinc, excess, body protection and, 164–176
  example, 588–590
Correctly specified regression model:
  definition, 111
  importance, 241

Correlation:
  of independent variables of multiple regression and multicollinearity, 59
  of orthogonal independent variables, 228
  of parameter estimates: in multiple regression, 82–83
    in nonlinear regression, 494
  and statistical independence, 59
  (See also Correlation coefficient; Correlation matrix)
Correlation coefficient:
  and coefficient of determination, 38
  and correlation matrix, 222
  definition, 36
  effect of clusters of data points on, 75–78
  among independent variables, as indicator of multicollinearity, 191
  interpretation of values, 37
  in multiple regression, 68–69
  Pearson product moment correlation, 36
  of regression coefficients, 194
  relationship with slope of regression line, 38–39
  small value does not necessarily mean good fit to data, 116
  Spearman rank, 37n.
  as strength of association of two variables, 10
  and sums of squares, 38
Correlation matrix:
  and correlation coefficient, 222
  principal components of, 222–223
  and variance inflation factor, 222n.
Criteria for selecting best regression model (see Best regression model, criteria for selecting)

Data entry errors, 142
Deep-fat fried egg yolks, 2
Deleting predictor variables to eliminate multicollinearity, 207–208
Dependent variable:
  definition, 13
  dummy, 512
  in logistic regression, 512
Depression, 295–297
Design matrix, definition, 81
Deviance in logistic regression, 535
DFBETAS, 141n.
DFFITS, 141n.
Diabetes, cholesterol, and treatment of high blood pressure, 83–86
  example, 583

Diagonal matrix, 570

Diet, drugs and atherosclerosis, 303–307, 309–310
  example, 599–601

Dietary fat and breast cancer, 33–36, 378

Drinking:
  and antisocial personality, 428–430
  and driving, 179

Drugs, experimenting with, 494–500, 623–624

Dummy variables:
  to compare regression lines, 28n.
  dependent variable in logistic regression, 512
  effects coding, 328
  encoding:
    experimental conditions in multiple regression, 84
    exposure to secondhand tobacco smoke, 6, 69–71
    subject:
      in regression, 381–390
      in repeated-measures analysis of variance, 394
  for formulation of analysis of variance as regression problem, 273, 284–286, 394–395, 413–414, 438–440
  in logistic regression, 537, 549
  for one-way repeated-measures analysis of variance, 391–400
  reference cell coding, 328
  for repeated observations in the same individuals, 381
  for shift in regression line, 75–78
  for two-way analysis of variance:
    interaction, 321
    main effects, 320–321
  for two-way repeated-measures analysis of variance:
    with repeated measures on both factors, 438–440
    with repeated measures on one factor, 413–414

Dunnett's test:
  versus Bonferroni t test, 297
  following one-way analysis of variance, 297–299
  following two-way analysis of variance, 344–345
  with missing data, 362
  versus Student-Newman-Keuls test, 301
  table of critical values, 716–717

Eating and metabolism, 460

Effects coding:
  to avoid multicollinearity among dummy variables, 333–334
  compared with reference cell coding, 332–337
  definition, 328–329
  example, 330, 337–339
  interpretation in linear regression, 331, 387
  necessary in two-way analysis of variance, 335, 411–413
  necessary in unbalanced designs, 357–358

Egg yolks, deep-fat fried, 2

Eigenvalue:
  of correlation matrix, 223
  of a matrix, 574
  and principal components, 216, 223, 224, 229
  as variation about principal component axis, 224

Eigenvector:
  of a matrix, 574
  in principal components analysis, 216, 223

Element of a matrix, 569

Endotoxin and toxic shock, 72–74

Equal variance:
  in analysis of variance, 452
  in multiple regression, 111
  tested with Levene median test, 456–457

Error surface:
  global minimum, 477
  in nonlinear regression, 472–473
  relative minimum, 477

Error:
  in analysis of variance, 281
  in entering data, 142
  in regression, 14, 54
  effect of transformation, 149–150

Escherichia coli, 47

Expected mean square:
  identified in computer printouts, 425n.
  and missing data, 425–427, 448–449
  in one-way analysis of variance, 282
  in two-way repeated-measures analysis of variance:
    with repeated measures on both factors, 446–448
    with repeated measures on one factor, 421–423

Expected value:
  definition, 251
  of mean, 251

Expected value (*Cont.*):
  of mean square in analysis of variance, 282, 421–427, 446–449
  unbiased estimator, 23, 240–241
Experimenting with drugs, 623–624
Explained variance and coefficient of determination, 38, 69
Exponential function:
  contrasted with quadratic, 466
  in pharmacokinetics, 495
Extraneous variables in regression models, 241, 244

F:
  based on expected mean squares, 281–283, 421–427, 446–449
  Greenhouse-Geisser correction in repeated-measures analysis of variance, 403
  Huynh-Feldt correction in repeated-measures analysis of variance, 403
  incremental sum of squares in multiple regression, 66, 80
  missing data in two-way analysis of variance:
    with repeated measures on both factors, 449
    with repeated measures on one factor, 426–427
  multicollinearity effect on goodness of fit in regression, 189
  one-way analysis of variance, 279–281
  overall goodness of fit:
    in linear regression, 65, 79–80
    in nonlinear regression, 492
  problems in computing in unbalanced designs, 351
  proper construction of test from computer printouts, 429n., 577
  simple linear regression, 32
  single-factor repeated-measures analysis of variance, 397–398
  and t test in simple linear regression, 33
  table of critical values, 713–715
  two-way analysis of variance, 317–318, 326–327, 332, 339–341
  two-way repeated-measures analysis of variance:
    with repeated measures on both factors, 434–435
    with repeated measures on one factor, 408–409
F_{enter}, 263

F_{in}, 263–265, 266n.
F_{out}, 265, 266n.
F_{remove}, 265
Factor in analysis of variance, 272
Family:
  definition, 292, 294, 302–303
  in two-way analysis of variance, 345
First guess for nonlinear regression, 483
Fixed effects in analysis of variance, 422
Food intake and weight, 511
Frank-Starling mechanism, 382–383
Forward selection:
  definition, 262–264
  in stepwise regression, 265–266
  in stepwise logistic regression, 546–548

Gauss-Newton method:
  mathematical development, 488–490
  in nonlinear regression, 478–482
  relationship to Marquardt's method, 482
  standard errors for hypothesis test in nonlinear regression, 491–492
Gender identification, faking, and personality assessment, 319–332
  example, 602
Geometrically connected data, 366–367
Global minimum in nonlinear regression, 477
Goodness of fit:
  effect of multicollinearity, 189
  in logistic regression, 513, 531–536
  for multiple regression plane, 63
  tested with F, 79–80
  tested with Hosmer-Lemeshow statistic in logistic regression, 533–534
Gradient method (*see* Steepest descent, method of)
Greenhouse-Geisser correction, 403
Grey seals, heat exchange in, 42–45, 89–92
  example, 581–582
Grid search:
  description, 470–471
  as first step in nonlinear regression, 478n.
Growth hormone:
  and bone deformity, 177
  deficiency, 178, 509

Halobacteroides acetoethylicus, 95
Hat matrix, 131n.
Heart:
  getting more data to eliminate multicollinearity, 211–216

Heart (*Cont.*):
    importance of left ventricular volume, 383
    two pumps in, consequences of, 194–199,
        208–209
    and heart muscle properties, 453–459
Heart nuking, 548–566
    example, 630–632
Heart size measured with a catheter, 382–392
    example, 607
Heart ventricles, interaction, between,
        194–195, 208–216, 226–227
    example, 593–594
Heat exchange in grey seals, 42–45, 89–92
    example, 581–582
Homogeneity of variance in one-way analy-
        sis of variance, 308–309
Hormones:
    and depression, 286–292, 295–297, 599
    and food, 392–398, 607–610
Hosmer-Lemeshow statistic, 533–535
Huynh-Feldt condition, 403
Hypertension, 83–86, 500–508
Hypothesis test:
    concerning coefficients in multiple
        regression, 58
    effect of using wrong sum of squares,
        360–361
    importance of statistical independence,
        181
    incremental sum of squares in multiple
        regression, 66
    individual coefficients in logistic regres-
        sion, 525–527
    nonlinear regression, 504–505
    one-way analysis of variance, 280–281
    one-way repeated-measures analysis of
        variance, 397–398
    overall regression fit in multiple regres-
        sion, 63–65
    two-way analysis of variance, 317, 332,
        339–341
        with missing data, 357–358
    two-way repeated-measures analysis of
        variance:
        with repeated measures on both factors,
            434–435, 447
        with repeated measures on one factor,
            408–409, 426–427

Identity matrix, definition, 570
Incremental R^2 in sequential variable selec-
    tion, 267

Incremental sum of squares:
    in backward elimination, 264
    definition, 65
    in forward selection, 263
    in general multiple linear regression, 80
    hypothesis test concerning, 66
    and t tests of individual regression coeffi-
        cients, 67–68
Independence, assumption in multiple
        regression, 111
Independent validation of regression model
        with new data, 248–249
Independent variable:
    definition, 13
    deleted, to reduce multicollinearity,
        207–208
    orthogonal, 228
    qualitative, 512–513
    scales of, 217
    selection procedures, 239
    standardized, 217–221
Indicator variable (*see* Dummy variables)
Inflammation of the colon, 510
Influential observation:
    data entry errors, 142
    what to do with it, 142
Intelligence:
    effect of foot size on, 112–119
    of Martians, 513–517
Interaction:
    effect on sums of squares associated with
        dummy variables in analysis of var-
        iance, 334–335
    graphical interpretation:
        in multiple regression, 96
        in two-way analysis of variance, 324,
            343
    between heart ventricles, 593–594
    hypothesis test, 341
    ignored, when empty cells in two-way
        analysis of variance, 366, 371
    interpretation:
        of main effects in the face of, 344
        in multiple regression, 95
        in two-way analysis of variance,
            341–344
    in logistic regression, 564–565
    in multiple regression, 94
    in nonlinear regression, 101–103
    as problem with regression model, 145
    in two-way analysis of variance, 317, 321
Intercept in simple linear regression, 20
    standard error of, 25

Internally Studentized residual:
  relationship to externally Studentized residual, 134n.
  (*See also* Studentized residual)
Inverse matrix, 573
Involuntary smoking (*see* Secondhand smoke)
Iterative method, definition, 473

Joint confidence region, 58n.

Kidneys and high blood pressure, 47–48, 336–341, 353–357, 371–377, 602–606
  example, 624–626
Kolmogorov-Smirnov test, 130n.

λ:
  parameter in Marquardt's method, 482
  (*See also* Eigenvalue)
Least squares estimation:
  compared to maximum likelihood, 522
  in simple linear regression, 20
Left ventricle, importance of volume, 383
Leucine, 75
Levene median test, 309–311, 456–457
Leverage:
  and Cook's distance, 140
  definition, 130–132
  hat matrix diagonal, 131n.
  interpretation, 132
  large value, 132
  in linear regression, 154
  in logistic regression, 531
  in nonlinear regression, 493
  point, 118, 152
  and PRESS statistic, 250n.
  and raw residual, 250n.
  and Studentized residual, 140
Likelihood function, 521
Likelihood ratio test in logistic regression, 526
Line of means:
  in simple linear regression, 11–13
  variation about, 13
Linear regression:
  assumptions, 13–16
  contrasted with nonlinear regression, 464
  derivation of slope and intercept to minimize residual sum of squares, 20n.

Linear regression (*Cont.*):
  line of means, 11–13
  population, 11
  population parameters, 18
  repeated observations in the same individuals, 381–382, 384–387
Logarithmic transformation:
  example, 75–76
  to stabilize variances in analysis of variance, 458
Logistic equation:
  alternative formulation, 505
  to describe baroreceptor response, 502
  interpretation, 515–516
  multiple, 519
Logistic regression:
  BMDP, 626–632
  Cook's distance, 531
  dependent variable, 512
  deviance in, 535
  dummy variables, 537, 549
  goodness of fit testing, 531–536, 534–535
  Hosmer-Lemeshow statistic, 534–535
  hypothesis test, 513
  interpretation of coefficients, 564–565
  large sample size, need for, 536
  leverage, 531
  likelihood ratio test in, 526
  multicollinearity, 531, 543
  parameter estimation with maximum likelihood, 513, 520–522
  patterns, 533
  residual, 529
  standardized residual, 530
  stepwise, 546–548
  Studentized deleted residual, 530
  Studentized residual, 530
  testing individual coefficients, 525–527
  testing overall goodness of fit, 523–525
Logit, relationship to odds, 518

Main effects:
  interpretation in presence of interactions, 344
  in two-way analysis of variance, 321
Mallows' C_p (*see* C_p)
Marginal sum of squares:
  and t tests in multiple regression, 68
  called type III in SAS GLM, 359
  in unbalanced designs, 352
Marquardt's method:
  and Gauss-Newton method, 482

Marquardt's method (*Cont.*):
  mathematical development, 490–491
  for nonlinear regression, 482–483
  and steepest descent, 482
Martian:
  height, weight and water consumption of,
      51–53, 184–189, 225–226, 231,
      242–244
  intelligence of, 112–119, 136–139,
      513–517
  interest in invading Earth, 468–470
  principal components to diagnose multi-
      collinearity, 225–226
  secondhand smoke, exposure to, 6–7,
      69–72, 274–276, 277–278
Matrix:
  addition, 570–571
  correlation, 222
  definition, 569
  diagonal, 570
  eigenvalues, 574
  eigenvectors, 574
  element of, 569
  hat, 131*n*.
  identity, 570
  inverse, 573
  multiplication of, 571–573
  for nonlinear regression, 486
  representation of multiple linear regres-
      sion equations, 80–82
  square, 570
  subtraction, 570–571
  symmetric, 570
  transpose, 573
Maximum likelihood estimation:
  compared to least squares, 522
  in logistic regression, 513, 520–522
MAXR, 266*n*.
Mean, expected value, 251
Mean square error, 32
Mean square residual, 32
Meaningless computer printouts, 9
Measuring heart size with a catheter,
    382–392, 607
Mechanical properties of whole heart versus
      heart muscle, 453–459, 620–622
Mechanisms of toxic shock, 582–583
Metabolism:
  changes in protein synthesis with age,
      74–78
  following eating, 392–398
Metallothionein:
  affected by dietary copper and zinc, 164

Methionine, in chicken feed, 157–158
  and messenger RNA, 166
Method of steepest descent (*see* Steepest de-
    scent, method of)
MHA (methionine hydroxy analog) in
    chicken feed, 158
Minimum variance estimate, 241
Minitab:
  missing data, 604–605
  multicollinearity, 591. 592
  one-way analysis of variance,
      598–600
  one-way repeated-measures analysis
      of variance, 609–610
  regression, 579, 581–585
  repeated-measures regression, 607
  simple linear regression, 39–42
  two-way repeated-measures analysis of
      variance:
    with repeated measures on both factors,
        618–619, 620–622
    with repeated measures one factor,
        613–615
Misleading conclusions due to nonlinear re-
    lationship in data, 44
Missing data:
  BMDP, 605
  effect on expected mean squares in two-
      way analysis of variance:
    with repeated measures on both factors,
        446–449
    with repeated measures on one factor,
        421–428
  effect on multiple comparisons after two-
      way repeated-measures analysis of
      variance:
    with repeated measures on both factors,
        452
    with repeated measures on one factor,
        430–431
  empty cells in two-way analysis of vari-
      ance, 365–366
  Minitab, 605
  SAS, 604
  SPSS, 605
Misspecified model (*see* Model misspecifica-
    tion)
Mixed model, in analysis of variance,
    422
Model misspecification:
  bias, 250–252
  definition, 111, 117
  effects of, 111, 240–244

Model misspecification (*Cont.*):
  identified from residual plot, 119–121
  and multicollinearity, 184
Model overspecification and multicollinearity, 184
Model parameterization, alternative forms, 505–506
Model underspecification, bias due to, 250–252
MR spectroscopy, 49, 108
MS_{res}, 32
Multicollinearity:
  auxiliary regressions, 193–194
  BMDP, 591–593
  cause of, 181
  classic signs of, 189
  computer control files, 591–595
  detected by correlations among independent variables, 191, 194
  detected by variance inflation factors, 189
  effect of centering, 594–595
  example, 182
  imprecise parameter estimates, 104
  in logistic regression, 531, 543
  Minitab, 591
  in model selected by all possible subsets regression, 261
  no effect on value of regression equation for prediction, 181–182
  in nonlinear regression problems, 494
  in polynomial regression, 88*n*.
  potential to avoid, with sequential variable selection techniques, 262
  principal components analysis to identify, 216
  qualitative suggestions of, 190
  resolved, by deleting predictor variables, 207–208
  SAS, 593–594
  SPSS, 594–595
  structural, 182
    (*See also* Structural multicollinearity)
  when unbalanced data, 351–352
  variables involved, identified with auxiliary regression, 193–194
    (*See also* Sample-based multicollinearity; Structural multicollinearity)
Multiple comparisons:
  ambiguities in, 300
  Bonferroni *t* test, 294–295
  Dunnett's test, 297–299
  family of tests, 292, 302–303
  indication of significant differences in plots, 346*n*., 348

Multiple comparisons (*Cont.*):
  rationale, 292–294
  in single-factor repeated-measures analysis of variance, 399–400
  Student-Newman-Keuls test, 300–302
  in two-way analysis of variance, 344–345
    with empty cells, 368–370, 375–376
  in two-way analysis of variance with missing data, 362–364
  in two-way repeated-measures analysis of variance:
    with repeated measures on both factors, 445, 452
    with repeated measures on one factor, 420–421, 430–431
Multiple correlation coefficient, definition, 68–69
Multiple logistic equation, 519
Multiple regression:
  advantages in formulation of two-way analysis of variance, 318–319
  assumptions of, 54, 110–111
  backward elimination in, 264–265
  coefficient vector, 81
  computer control files, 579–590
  confidence levels for regression coefficients, 58
  definition of residuals, 119
  different sets of data can give rise to same regression equation, 113–117
  equation, written in terms of principal components, 229
  equations in matrix form, 80–82
  for formulation:
    of analysis of variance, 273, 283–284
    of one-way analysis of variance, 283–386
    one-way repeated-measures analysis of variance, 391–398
    two-way analysis of variance, 320–325, 353
    two-way repeated-measures analysis of variance with repeated measures on both factors, 438–440
      with repeated measures on one factor, 411–414
  forward selection, 262–264
  general formulation, 79–80
  hypothesis test concerning coefficients, 58
  incremental sum of squares, 65, 80
  interactions between independent variables, 94
  leverage point, 118

Multiple regression (*Cont.*):
   model misspecification, 117
   multicollinearity, 59–63, 181–182
   overall goodness of fit test, 79–80, 189
   parameter vector, 80
   plane of means, 54
   prediction, effect of multicollinearity,
      181–182
   regression plane, 51, 52, 53
   relationship to simple linear regression,
      50, 54
   residual sum of squares, 55
   standard error:
      of estimate, 79
      of regression coefficient, 57–58, 79
   in standardized variables, 221
   stepwise regression, 265–266
   t test for coefficients, 79, 189
   variability about regression plane, 57
   violating assumptions, 110
Multivariate methods:
   analyzed with computers, 39
   based on multiple linear regression, 8
   benefits over univariate methods, 8
   rationale for, 1–2
Myotonic dystrophy, 313

Neural transmitters, 378–379
Newborn, protein synthesis compared with
      adults, 74–78
Newsworthiness and tobacco advertising,
      105
Nonlinear least squares (*see* Nonlinear
      regression)
Nonlinear regression:
   BMDP, 623–624
   computer control files, 622–626
   confidence intervals, 493n.
   contrasted with linear regression, 464
   convergence criterion in, 482n.,
      483–484
   Cook's distance, 493
   correlation of parameter estimates, 494
   difficulties in parameter estimation, 468
   error surface, 472–473
   example, 43, 102, 466, 469, 500–503
   first guess, 483
   Gauss-Newton method, 478–482
   global minimum in, 477
   goodness of fit, 492
   grid search, 470–471
   hypothesis testing, 491–493, 504–505

Nonlinear regression (*Cont.*):
   interpretation of computer outputs,
      496–497
   intrinsically nonlinear, 93
   leverage, 493, 497–498
   linear regression problem, 88–89, 92–93
   Marquardt's method, 482–483
   in matrix notation, 486
   multicollinearity, 494
   normal probability plot, 498–499
   as problem, with regression model, 145
   reasons for using, 465–468
   regression diagnostic, 493–494
   relative minimum, 477
   residual plots, 498, 477n.
   SAS, 622–626
   standardized residuals, 493
   steepest descent, method of, 473–478
   Studentized residuals, 493
   treated as linear regression problem,
      88–89
   variance inflation factors in, 494
Nonlinear relationship:
   determined by a single point, 152
   identified, using residuals, 43, 45n.
   leads to misleading results in simple
      linear regression, 44
Nonorthogonality in unbalanced two-way
      analysis of variance design, 352
Normal probability plot:
   construction, 126–128
   definition, 125
   interpretation, 125, 128–129
   in nonlinear regression, 498–499
   paper, 126
   testing normality of residuals with t test,
      130
Normality:
   assumption in multiple regression, 111
   Kolmogorov-Smirnov test, 130n.
   Shapiro-Wilk test, 130n.
   tested with standardized residuals,
      125–129
Nuking the heart, 548–566, 630–632
Nurse practitioners, 379
Nutrition:
   and calcium, 270
   effects of dietary zinc and copper,
      164–176

Observational studies and sample-based
      multicollinearity, 184

Odds:
    definition, 517
    relationship to logit, 518
One-way analysis of variance:
    assumptions, 312
    basic approach, 279–281
    BMDP, 600–601, 605–606
    computer control files, 598–601
    to deal with empty cells in two-way anal-
        ysis of variance, 365
    definition, 272
    expected mean square, 282
    Minitab, 598–600
    as regression problem, 284–286
    SAS, 601, 602–604
    SPSS, 601, 606
    testing assumptions, 307–309
    as univariate statistical procedure, 8
    within groups variance in, 280
    within subjects variation in, 391
One-way repeated-measures analysis of
        variance:
    BMDP, 609, 613
    computer control files, 607–610
    description, 390–391
    hypothesis test, 397–398
    interpretation, 393–397
    Minitab, 609–610, 613–615
    multiple comparisons testing, 399–400
    SAS, 607–609
    SPSS, 610
    treatments encoded with dummy vari-
        ables, 394–395
Order of entry (see Incremental sum of
        squares)
Orthogonal variables:
    definition, 228
    dummy variables with effects coding, 333
Outlier, identified:
    with Cook's distance, 141
    with residual plot, 120–121
    with standardized residual, 124
    with Studentized residual, 135
Overspecified model, 241

P_{in}, 547
P_{out}, 547
Paraplegic, 313–314
Parenteral nutrition, 48–49
Partitioning the variance:
    in one-way repeated-measures analysis of
        variance, 390–391

Partitioning the variance (*Cont.*):
    in two-way repeated-measures analysis of
        variance:
        with repeated measures on one factor,
            405–406
        with repeated measures on both factors,
            432–433
Patterns, 533
Pearson product moment correlation
        coefficient, (see Correlation
        coefficient)
Personality assessment and faking gender
        identification, 319–332, 602
Pharmacokinetics:
    compartmental models, 495n.
    and nonlinear regression, 465, 494–500
Physical training and muscle metabolism,
        461
Placenta, water movement across, 151–157
    example, 586–588
Plane of means in multiple linear regression,
        51
Polynomial regression:
    centering, to eliminate structural multi-
        collinearity, 183, 200–207
    example, 88–90, 102, 156, 167
    structural multicollinearity, 182
Population parameters:
    for logistic regression, 519
    for multiple linear regression, 54
    for simple linear regression, 11
Prediction, effect of multicollinearity,
        181–182
Predictor variable (see Independent variable)
PRESS:
    compared with other criteria for quality of
        regression fit, 255
    to evaluate quality of regression model,
        249–250
    relationship to raw residual and leverage,
        250
    residual, 141n., 250
Principal components:
    condition index, 225
    condition number, 225
    coordinate system, 218–219, 224
    definition, 223
    to diagnose multicollinearity, 216,
        225–226
    eigenvalue, 216, 224, 229
    eigenvector, 216
    to justify effects coding in analysis of vari-
        ance, 333

Principal components regression:
  to compensate for multicollinearity,
      228–230
  done by computer, 230
  example, 231
  limitations, 231–234
  restricted regression procedure, 234
Problems with regression models, 145–146
  (*See also* Interaction; Model misspecifica-
      tion; Transformation)
Prostacyclin and toxic shock, 72–74
Protein synthesis in newborns and adults,
      74–78
  example, 583

q, table of critical values, 717–718
q', table of critical values, 716

R (*see* Multiple correlation coefficient)
R^2 (*see* Coefficient of determination)
R^2_{adj}:
  compared with other criteria for quality of
      regression fit, 254–255
  definition, 247
Radioimmunoassay to measure inhibin,
      179
Random effects in analysis of variance,
      422
Raw data, 170
  value of plots, 46
Redundant information:
  as cause of multicollinearity, 181
  indicated by correlations between inde-
      pendent variables in multiple
      regression, 191
  quantified with variance inflation factor,
      192
Reference cell coding:
  compared with effects coding, 332–337
  definition, 328
  interpretation in linear regression, 387n.
Regression:
  BMDP, 579–580, 586–588
  Minitab, 579, 582–585
  SAS, 580, 584, 588–590
  SPSS, 580–581, 583, 588
  sum of squares, in simple linear regres-
      sion, 29
Regression coefficients:
  correlation as multicollinearity diagnostic,
      194

Regression coefficients (*Cont.*):
  hypothesis testing:
      in linear regression, 58
      in logistic regression, 525–527
      in nonlinear regression, 492–493
  standard error of (*see* Standard error of
      regression coefficient)
  standardized, 221
  strange values, sign of multicollinearity,
      189
  unbiased estimates, 240–241
Regression diagnostics:
  leverage, 132
  for linear regression, 112
  for logistic regression, 529–531
  for nonlinear regression, 493–494
  value of different diagnostic statistics, 141
  (*See also* Cook's distance; Leverage; Nor-
      mal probability plot; Standardized
      residual; Studentized residual)
Regression model:
  best, 245–246
      no clear definition of, 245–246
  correctly specified, 111, 241
  extraneous variables, 241, 244
  misspecification, example, 170–176
  overspecified, 241
  underspecified, 241
Regression sum of squares:
  and correlation coefficient, 38
  for multiple regression, 63
Relative minimum:
  in nonlinear regression, 477
  ways to check for, 485
Renin, 48
Reparameterization:
  in nonlinear regression, 467
  in pharmacokinetics model, 495
Repeated-measures analysis of variance:
  assumptions, 400–405
  compared with completely randomized
      designs, 398–399
  computer control files, 607
  definition, 381
  Greenhouse-Geisser correction, 403
  Huynh-Feldt correction, 403
  observations in the same individual, ex-
      perimental design, 381
  one-way, 391–400
      (*See also* One-way repeated-measures
      analysis of variance)
  to reduce number of experimental sub-
      jects, 399

Repeated-measures analysis of variance (*Cont.*):
  two-way:
    with repeated measures on both factors, 431–435
    with repeated measures on one factor, 404–410
Repeated-measures regression, 381–390
Residual:
  to construct normal probability plots, 126–128
  definition, 111, 119
  effect of transformation on, 149–150
  in logistic regression, 529
  model misspecification, 119–121
  in nonlinear regression, 477n., 498–499
  normality tested, 130
  outlier, 120–121, 124
  PRESS, 250
  standardized, 123–125
  to test regression assumptions, 111
  variance in analysis of variance, 281
Residual diagnostic as complement to variable selection technique, 240
Residual mean square, definition, in simple linear regression, 32
Residual sum of squares:
  and correlation coefficient, 38
  to define error surface in nonlinear regression, 472–473
  for multiple regression, 55
  standard error of the estimate, 22
  in multiple regression, 57
Response of muscle to stretching, 584
Response variable (*see* Dependent variable)
Restricted regression:
  bias in, 233n.
  principal components, 234
  procedures, 232n., 233n.
  ridge regression, 233n.
RNA, and metallothionein, 166

Saliva, 314
Sample:
  need for large sample size in logistic regression, 536
  in simple linear regression, 15
Sample-based multicollinearity:
  collecting additional data to resolve, 210–211
  definition, 182
  example, 194–199
  and model overspecification, 184

Sample-based multicollinearity (*Cont.*):
  need for new data, 183
  in observational studies, 184
  origins, 183
  resolution, by getting more data under different experimental conditions, 184
SAS:
  analysis of variance, 577
  missing data, 604–605
  nonlinear regression, 622–626
  one-way analysis of variance, 601
  one-way repeated-measures analysis of variance, 607–609
  regression, 578, 580–582, 584, 588–590
  simple linear regression, 39–42
  two-way analysis of variance, 602, 602–604
  two-way repeated-measures analysis of variance with repeated measures on both factors, 615–618
  type III sum of squares, 359
  variable selection techniques, 597
Scalar, definition, 569
Scientific reasoning, compared to statistical reasoning, 36
Screening data, 239
Seals, 42–45, 89–92
Secondhand smoke:
  and breast cancer, 46–47, 236, 270
  carcinogenic compounds in, 380
  and lung cancer, 109
  and Martian height, 6–7
  and Martian weight, 69–72, 104
  and nausea in Martians, 274–278, 598
Sequential sum of squares in two-way analysis of variance, 334
Sequential variable selection:
  advantages and pitfalls, 270
  backward elimination, 264–265
  definition, 262
  forward selection, 262–264
  interpretation of results, 266–267
  stepwise regression, 265–266
Shapiro-Wilk test, 130n.
Shift in regression line, encoded with dummy variables, 7, 70
Simple linear regression:
  comparing slopes, 27–28
  computation of intercept, 20
  F test of overall regression, 32
  hypothesis testing concerning slope, 26–27

Simple linear regression (*Cont.*):
  interpolation and extrapolation problems, 75–78
  misleading results due to nonlinearity, 44
  partitioning of total sum of squares, 31
  regression sum of squares, 29
  relationship between F and t, 33
  relationship to multiple regression, 50, 54
  residual mean square, 32
  residuals, 18
  sample, 15
  slope, 20
  standard error of the estimate, 22
  standard errors of regression coefficients, 22–25
  univariate statistical procedure, 8
  value of plotting data, 46
Single-factor analysis of variance (*see* One-way analysis of variance)
Single-factor repeated-measures analysis of variance (*see* One-way repeated-measures analysis of variance.)
Slope:
  comparing different regression lines, 27–28
  confidence intervals, 27
  hypothesis testing regarding, 26–27
  relationship to correlation coefficient, 38–39
  simple linear regression, 20
  standard error in simple linear regression, 25
Smooth muscle, response to stretching, 100–103
  example, 584
SNK-test (*see* Student-Newman-Keuls test)
Sodium, effects on blood pressure via kidneys, 336–341
Software packages to do multivariate statistical analysis, 39
  (*See also specific programs*)
Spearman rank correlation coefficient, 37n.
SPSS:
  analysis of variance with missing data, 606
  missing data, 604–605
  multicollinearity, 594–595
  one-way analysis of variance, 601
  one-way repeated-measures analysis of variance, 610
  regression, 580–581, 583, 588
  simple linear regression, 39

SPSS (*Cont.*):
  two-way repeated-measures analysis of variance with repeated measures on both factors, 619–620
  variable selection techniques, 597–598
Square matrix definition, 570
SS_{reg} (*see* Regression sum of squares)
SS_{res} (*see* Residual sum of squares)
Standard deviation, relationship to total sum of squares, 31n.
Standard error:
  of the estimate:
    approximate estimates in nonlinear regression, 491–492
    as criterion for selecting best regression model, 247–248, 254–255
    in multiple regression, 57, 79
    in simple linear regression, 22
    small value does not necessarily mean good fit to data, 116
  of regression coefficient:
    for general multiple linear regression, 79
    interpretation, 181
    in multiple regression, 57–58
    reduced by centering independent variables, 205–206
    relationship to variance inflation factor, 191–192
    and variance covariance matrix of parameter estimates, 82–83
Standardized deviates, 217–220
Standardized independent variables defined in terms of principal components, 229
Standardized regression coefficients, 221
Standardized residual:
  definition, 123
  to identify outliers, 124
  in logistic regression, 530
  in nonlinear regression, 493
  to test for normality, 125–129
Statistical independence:
  and correlation, 59
  necessary for unambiguous hypothesis testing in multiple regression, 181
Statistical reasoning, compared to scientific reasoning, 36
Steepest descent, method of:
  definition, 473
  to estimate parameters in nonlinear regression, 473–478
  mathematical developments, 487–488
  necessity to adjust step size, 475

Steepest descent, method of (*Cont.*):
  relationship to Marquardt's method, 482
  step size in, 475
Stepwise regression:
  definition, 265–266
  description, 239
  logistic regression, 546–548, 552–565
Stomach and intestines, 462–463
Structural multicollinearity:
  controlled with centering, 183, 200–207
  definition, 182
  examples, 194–199
  in logistic regression, 543
  origin, 182
Student-Newman-Keuls test:
  compared with Dunnett's test, 301
  compared with t test, 301
  following one-way analysis of variance, 300–302
  following two-way analysis of variance, 344–345, 349
  with missing data, 362
  table of critical values, 717–718
Studentized deleted residual:
  example of use, 154
  in logistic regression, 530
Studentized residual:
  definition, 134
  internally versus externally, 135
  in logistic regression, 530
  in nonlinear regression, 493
  relationship to leverage and Cook's distance, 140
Sudden infant death syndrome, 567
Sum of squares:
  computed, when missing data, 604–605
  partitioning:
    in simple linear regressions, 31
    in single factor repeated-measures analysis of variance, 390–391
    in two-way repeated-measures analysis of variance:
      with repeated measures on both factors, 432–433
      with repeated measures on one factor, 405–406
  SAS type III, 359
  sequential, in two-way analysis of variance, 334
  in simple linear regression, 18–19
  (*See also* Regression sum of squares; Residual sum of squares; Total sum of squares)

Surfactant, 509
Symmetric matrix, definition, 570

t test:
  Bonferroni, 294–295
  to compare two groups, 273–274
  compared with Student-Newman-Keuls test, 301
  conducted as regression problem, 275, 277–279
  for coefficients in general multiple linear regression, 79
  for individual regression coefficients, effect of multicollinearity, 189
  relationship with F test in simple linear regression, 33
  relationship to incremental sums of squares in multiple regression, 67–68
  relationship of tests of individual coefficients in multiple regression with overall goodness of fit, 63
  table of critical values, 710–712
  to test coefficients in multiple regression, 58
  to test for normality of residuals, 130
  to test whether coefficients are significantly different from zero in nonlinear regression, 493
  univariate statistical procedures, 8
Test of normality (*see* Normality)
Testing assumptions of one-way analysis of variance, 307–309
Thermophilic bacteria, 179
Time constant, 465–466
Tobacco:
  effects of advertising on publication of tobacco related news stories, 105
  (*See also* Secondhand smoke)
Tolerance, 192–193*n.*
Total sum of squares:
  and correlation coefficient, 38
  definition, 29–31
  relationship to standard deviation, 31*n.*
  relationship to variance, 31*n.*
Toxic shock, 72–74
  example, 582–583
Transformation:
  commonly used, 148–150
  to control structural multicollinearity, 183
  effect on residuals, 149–150

Transformation (*Cont.*):
    example to stabilize variances, 175–176
    to introduce nonlinearity into regression
        model, 92–93, 147–150
    to stabilize variance, 147; 160–163
Transpose of a matrix, 573
Triathlon:
    determinants of athlete's time, 256–261
    evaluated with sequential variable selec-
        tion, 267–269
    example, 595–598
Two variable linear regression (*see* Simple
    linear regression)
Two-way analysis of variance:
    advantage, compared to one-way analysis
        of variance, 317–318
    advantages of regression formulation,
        318–319
    analyzed with multiple regression,
        357–358
    balanced design, 350
    benefit of regression approach in unbal-
        anced data, 351
    computer control files, 602–606
    definition, 272, 316
    effect of empty cells, 365–366
    effect of using wrong sum of squares,
        360–361
    example, 319–328, 336–341
    geometrically connected data, 366–367
    graphical interpretation in terms of regres-
        sion, 324–325
    hypothesis test, 326–327, 332, 339–341
    interaction, 317, 321, 341–344
    main effects, 321
    with missing data:
      and empty cells, 365–366
      but no empty cells, 352–353
    multiple comparisons, 344–345
      with empty cells, 368–370
      with missing data, 362–364
    SAS, 602
    with repeated measures:
      on both factors, 431–435
      on one factor, 404–408
Two-way repeated-measures analysis of vari-
    ance:
    with repeated measures on both factors:
      basic design, 445
      BMDP, 618
      computer control files, 615–622
      hypothesis test, 434–435
      Minitab, 618–622

Two-way repeated measures analysis of vari-
    ance (*Cont.*):
      missing data, 446–449
      partitioning of sum of squares, 432–433
      SAS, 615–618
      SPSS, 619–620
    with repeated measures on one factor:
      basic design, 404–405
      BMDP, 613
      computer control files, 610–615
      hypothesis test, 408, 409
      Minitab, 613–615
      missing data, 421–428
      multiple comparisons, 420–421
      partitioning sum of squares, 405–406
      SAS, 610–612
Type III sum of squares, 359

Unbalanced data:
    advantage of regression approach to analy-
        sis of variance, 351
    analysis with multiple regression,
        353–357
    in two-way analysis of variance, 350–351
    in two-way repeated-measures analysis of
        variance:
      with repeated measures on both factors,
        446–449
      with repeated measures on one factor,
        421–428
    pitfalls in analysis, 351
Unbiased estimator, definition, 23, 240–241
Underspecified regression model, 241
Unequal sample sizes (*see* Unbalanced data)
Univariate statistical methods:
    analyzing data piecemeal, 8
    limitations of, 1
Urinary calcium and nutrition, 270
Urine, 178

Valium, 46
Variable selection techniques:
    BMDP, 596–597
    caveats, 239–240
    computer control files, 595–598
    Minitab, 595–596
    SAS, 597
    SPSS, 597–598
    stepwise logistic regression, 546–548
    (*See also* All possible subsets regression;
        Stepwise regression)

Variance:
    explained by regression equation, 64
    homogeneity of, 308–309
    partitioning (*see* Partitioning the variance)
    in relationship to total sum of squares,
        31*n*.
    transformation to stabilize, 147
Variance covariance matrix of parameter es-
        timates in multiple regression, 82
Variance inflation factor(s):
    alternate strategies, 160–163
    to check results of all possible subsets
        regression for multicollinearity, 261
    definition, 192
    to detect multicollinearity, 189
    in nonlinear regression, 494
    relationship to correlation matrix, 222*n*.
    relationship to tolerance, 192*n*.–193*n*.

Variance stabilizing transformation:
    example, 147, 150
    in repeated-measures analysis of variance,
        458
Vector, definition, 570

Water:
    canal, 3–6
    movement across the placenta, 586–588
Within groups variance in one-way analysis
        of variance, 280
Within subjects variation in one-way analy-
        sis of variance, 391

z for hypothesis test in logistic regression,
        525–526